HEISENBERG'S QUANTUM MECHANICS

HEISENBERG'S QUANTUM MECHANICS

MOHSEN RAZAVY
University of Alberta, Canada

World Scientific
NEW JERSEY • LONDON • SINGAPORE • BEIJING • SHANGHAI • HONG KONG • TAIPEI • CHENNAI

Published by

World Scientific Publishing Co. Pte. Ltd.
5 Toh Tuck Link, Singapore 596224
USA office: 27 Warren Street, Suite 401-402, Hackensack, NJ 07601
UK office: 57 Shelton Street, Covent Garden, London WC2H 9HE

British Library Cataloguing-in-Publication Data
A catalogue record for this book is available from the British Library.

HEISENBERG'S QUANTUM MECHANICS

ISBN-13 978-981-4304-10-8
ISBN-10 981-4304-10-7
ISBN-13 978-981-4304-11-5 (pbk)
ISBN-10 981-4304-11-5 (pbk)

Printed in Singapore by World Scientific Printers.

Dedicated to my great teachers

A.H. Zarrinkoob, M. Bazargan and J.S. Levinger

Preface

There is an abundance of excellent texts and lecture notes on quantum theory and applied quantum mechanics available to the students and researchers. The motivation for writing this book is to present matrix mechanics as it was first discovered by Heisenberg, Born and Jordan, and by Pauli and bring it up to date by adding the contributions by a number of prominent physicists in the intervening years. The idea of writing a book on matrix mechanics is not new. In 1965 H.S. Green wrote a monograph with the title "*Matrix Mechanics*" (Nordhoff, Netherlands) where from the works of the pioneers in the field he collected and presented a self-contained theory with applications to simple systems.

In most text books on quantum theory, a chapter or two are devoted to the Heisenberg's matrix approach, but due to the simplicity of the Schrödinger wave mechanics or the elegance of the Feynman path integral technique, these two methods have often been used to study quantum mechanics of systems with finite degrees of freedom.

The present book surveys matrix and operator formulations of quantum mechanics and attempts to answer the following basic questions: (a) — why and where the Heisenberg form of quantum mechanics is more useful than other formulations and (b) — how the formalism can be applied to specific problems? To seek answer to these questions I studied what I could find in the original literature and collected those that I thought are novel and interesting. My first inclination was to expand on Green's book and write only about the matrix mechanics. But this plan would have severely limited the scope and coverage of the book. Therefore I decided to include and use the wave equation approach where it was deemed necessary. Even in these cases I tried to choose the approach which, in my judgement, seemed to be closer to the concepts of matrix mechanics. For instance in discussing quantum scattering theory I followed the determinantal approach and the LSZ reduction formalism.

In Chapter 1 a brief survey of analytical dynamics of point particles is presented which is essential for the formulation of quantum mechanics, and an understanding of the classical-quantum mechanical correspondence. In this part of the book particular attention is given to the question of symmetry and conservation laws.

In Chapter 2 a short historical review of the discovery of matrix mechanics is given and the original Heisenberg's and Born's ideas leading to the formulation of quantum theory and the discovery of the fundamental commutation relations are discussed.

Chapter 3 is concerned with the mathematics of quantum mechanics, namely linear vector spaces, operators, eigenvalues and eigenfunctions. Here an entire section is devoted to the ways of constructing Hermitian operators, together with a discussion of the inconsistencies of various rules of association of classical functions and quantal operators.

In Chapter 4 the postulates of quantum mechanics and their implications are studied. A detailed review of the uncertainty principle for position-momentum, time-energy and angular momentum-angle and some applications of this principle is given. This is followed by an outline of the correspondence principle. The question of determination of the state of the system from the measurement of probabilities in coordinate and momentum space is also included in this chapter.

In Chapter 5 connections between the equation of motion, the Hamiltonian operator and the commutation relations are examined, and Wigner's argument about the nonuniqueness of the canonical commutation relations is discussed. In this chapter quantum first integrals of motion are derived and it is shown that unlike their classical counterparts, these, with the exception of the energy operator, are not useful for the quantal description of the motion.

In Chapter 6 the symmetries and conservation laws for quantum mechanical systems are considered. Also topics related to the Galilean invariance, mass superselection rule and the time invariance are studied. In addition a brief discussion of classical and quantum integrability and degeneracy is presented.

Chapter 7 deals with the application of Heisenberg's equations of motion in determining bound state energies of one-dimensional systems. Here Klein's method and its generalization are considered. In addition the motion of a particle between fixed walls is studied in detail.

Chapter 8 is concerned with the factorization method for exactly solvable potentials and this is followed by a brief discussion of the supersymmetry and of shape invariance.

The two-body problem is the subject of discussion in Chapter 9, where the properties of the orbital and spin angular momentum operators and determination of their eigenfunctions are presented. Then the solution to the problem of hydrogen atom is found following the original formulation of Pauli using Runge–Lenz vector.

In Chapter 10 methods of integrating Heisenberg's equations of motion are presented. Among them the discrete-time formulation pioneered by Bender and collaborators, the iterative solution for polynomial potentials advanced by Znojil and also the direct numerical method of integration of the equations of motion are mentioned.

The perturbation theory is studied in Chapter 11 and in Chapter 12 other methods of approximation, mostly derived from Heisenberg's equations of motion are considered. These include the semi-classical approximation and variational method.

Chapter 13 is concerned with the problem of quantization of equations of motion with higher derivatives, this part follows closely the work of Pais and Uhlenbeck.

Potential scattering is the next topic which is considered in Chapter 14. Here the Schrödinger equation is used to define concepts such as cross section and the scattering amplitude, but then the deteminantal method of Schwinger is followed to develop the connection between the potential and the scattering amplitude. After this, the time-dependent scattering theory, the scattering matrix and the Lippmann–Schwinger equation are studied. Other topics reviewed in this chapter are the impact parameter representation of the scattering amplitude, the Born approximation and transition probabilities.

In Chapter 15 another feature of the wave nature of matter which is quantum diffraction is considered.

The motion of a charged particle in electromagnetic field is taken up in Chapter 16 with a discussion of the Aharonov–Bohm effect and the Berry phase.

Quantum many-body problem is reviewed in Chapter 17. Here systems with many-fermion and with many-boson are reviewed and a brief review of the theory of superfluidity is given.

Chapter 18 is about the quantum theory of free electromagnetic field with a discussion of coherent state of radiation and of Casimir force.

Chapter 19, contains the theory of interaction of radiation with matter.

Finally in the last chapter, Chapter 20, a brief discussion of Bell's inequalities and its relation to the conceptual foundation of quantum theory is given.

In preparing this book, no serious attempt has been made to cite all of the important original sources and various attempts in the formulation and applications of the Heisenberg quantum mechanics.

I am grateful to my wife for her patience and understanding while I was writing this book, and to my daughter, Maryam, for her help in preparing the manuscript.

Edmonton, Canada, 2010

Contents

Chapter 1

A Brief Survey of Analytical Dynamics

1.1 The Lagrangian and the Hamilton Principle

We can formulate the laws of motion of a mechanical system with N degrees of freedom in terms of Hamilton's principle. This principle states that for every motion there is a well-defined function of the N coordinates q_i and N velocities $\dot{q}_i$ which is called the Lagrangian, L, such that the integral

$$S = \int_{t_1}^{t_2} L\left(q_i, \dot{q}_i, t\right) dt, \tag{1.1}$$

takes the least possible value (or extremum) when the system occupies positions $q_i(t_1)$ and $q_i(t_2)$ at the times t_1 and t_2 [1],[2].

The set of N independent quantities $\{q_i\}$ which completely defines the position of the system of N degrees of freedom are called generalized coordinates and their time derivatives are called generalized velocities.

The requirement that S be a minimum (or extremum) implies that L must satisfy the Euler–Lagrange equation

$$\frac{\partial L}{\partial q_i} - \frac{d}{dt}\left(\frac{\partial L}{\partial \dot{q}_i}\right) = 0, \quad i = 1, \cdots N. \tag{1.2}$$

The mathematical form of these equations remain invariant under a point transformation. Let us consider a non-singular transformation of the coordinates from the set of N $\{q_i\}$ s to another set of N $\{Q_i\}$ s given by the equations

$$Q_i = Q_i\left(q_1, \cdots, q_N\right), \quad i = 1, \cdots N, \tag{1.3}$$

and its inverse transform given by the N equations

$$q_j = q_j\left(Q_1, \cdots, Q_N\right), \quad j = 1, \cdots N. \tag{1.4}$$

Now let $\mathcal{F}\left(q_1, \cdots, q_N, \dot{q}_1, \cdots, \dot{q_N}\right)$ be a twice differentiable function of $2N$ variables $q_1, \cdots, q_N, \dot{q}_1, \cdots, \dot{q_N}$. We note that this function can be written as a function of Q_j s and $\dot{Q}_j$ s if we replace q_i s and $\dot{q}_i$ s by Q_j s and $\dot{Q}_j$ s using Eq. (1.4). Now by direct differentiation we find that

$$\begin{aligned}
&\left(\frac{\partial}{\partial q_i} - \frac{d}{dt}\frac{\partial}{\partial \dot{q}_i}\right)\mathcal{F}\left(q_i(Q_j), \dot{q}_i(Q_j, \dot{Q}_j)\right) \\
&= \sum_{j=1}^{N}\left(\frac{\partial Q_j}{\partial q_i}\right)\left(\frac{\partial}{\partial Q_j} - \frac{d}{dt}\frac{\partial}{\partial \dot{Q}_j}\right)\mathcal{F}\left(q_i(Q_j), \dot{q}_i(Q_j, \dot{Q}_j)\right), \quad i = 1, \cdots N.
\end{aligned} \tag{1.5}$$

Thus if $L(Q_1, \cdots \dot{Q}_N)$ has a vanishing Euler–Lagrange derivative i.e.

$$\left[\frac{\partial}{\partial Q_j} - \frac{d}{dt}\left(\frac{\partial}{\partial \dot{Q}_j}\right)\right] L = 0, \tag{1.6}$$

then Eq. (1.5) implies that

$$\frac{\partial L}{\partial q_i} - \frac{d}{dt}\left(\frac{\partial L}{\partial \dot{q}_i}\right) = 0, \quad i = 1, \cdots N. \tag{1.7}$$

This result shows that we can express the motion of the system either in terms of the generalized coordinates q_i and generalized velocities $\dot{q}_i$ or in terms of Q_j and $\dot{Q}_j$.

For simple conservative systems for which potential functions of the type $V(q_1, \cdots, q_N, t)$ can be found, the Lagrangian L has a simple form:

$$L = T(q_1, \cdots, q_n;\ \dot{q}_1, \cdots, \dot{q}_N) - V(q_1, \cdots, q_N, t), \tag{1.8}$$

where T is the kinetic energy of the particles in the system under consideration and V is their potential energy. However given the force law acting on the i-th particle of the system as $F_i(q_1, \cdots, q_N;\ \dot{q}_1, \cdots, \dot{q}_N)$, in general, a unique Lagrangian cannot be found. For instance we observe that the Euler–Lagrange derivative of any total time derivative of a function $\mathcal{F}$ of q_i, $\dot{q}_i$ i.e. $\frac{d}{dt}\mathcal{F}(q_i, \dot{q}_i, t)$ is identically zero;

$$\left[\frac{\partial}{\partial q_i} - \frac{d}{dt}\left(\frac{\partial}{\partial \dot{q}_i}\right)\right]\frac{d\mathcal{F}(q_i, \dot{q}_i, t)}{dt} \equiv 0, \quad i = 1, \cdots N. \tag{1.9}$$

Therefore we can always add a total time derivative $\frac{d\mathcal{F}}{dt}$ to the Lagrangian without affecting the resulting equations of motion.

The inverse problem of classical mechanics is that of determination of the Lagrangian (or Hamiltonian) when the force law $F_j(q_i, \dot{q}_i, t)$ is known. The necessary and sufficient conditions for the existence of the Lagrangian has been studied in detail by Helmholtz [3]–[6]. In general, for a given set of F_j s, L satisfies a linear partial differential equation. To obtain this equation we start with the Euler–Lagrange equation (1.2), find the total time derivative of $\frac{\partial L}{\partial \dot{q}_i}$ and then replace $\ddot{q}_i$ by $\frac{F_i}{m_i}$. In this way we obtain

$$\frac{\partial^2 L}{\partial \dot{q}_i \partial t} + \sum_j \left(\frac{F_j}{m_j}\right)\left(\frac{\partial^2 L}{\partial \dot{q}_i \partial \dot{q}_j}\right) + \sum_j \left(\frac{\partial^2 L}{\partial \dot{q}_i \partial q_j}\right)\dot{q}_i \dot{q}_j - \frac{\partial L}{\partial q_i} = 0, \quad i = 1, \cdots N. \tag{1.10}$$

This set of equations yield the Lagrangian function. But as was stated earlier L is not unique even for conservative systems. The advantage of the Lagrangian formulation is that it contains information about the symmetries of the motion which, in general, cannot be obtained from the equations of motion alone.

For instance let us consider the Lagrangian for the motion of a free particle. In a reference frame in which space is homogeneous and isotropic and time is homogeneous, i.e. an inertial frame, a free particle which is at rest at a given instant of time, always remains at rest. Because of the homogeneity of space and time, the Lagrangian L cannot depend either on the position of the particle $\mathbf{r}$ nor on time t. Thus it can only be a function of velocity $\dot{\mathbf{r}}$. Now if the velocity of the particle is $\dot{\mathbf{r}}$ relative to a frame $\mathcal{S}$, then in another frame $\mathcal{S}'$ which is moving with a small velocity $\mathbf{v}$ with respect to $\mathcal{S}$ the velocity is $\dot{\mathbf{r}}'$, and the Lagrangian is

$$L'\left(\dot{\mathbf{r}}'^2\right) = L\left[(\dot{\mathbf{r}} + \mathbf{v})^2\right] \approx L\left[\dot{\mathbf{r}}^2\right] + 2\dot{\mathbf{r}} \cdot \mathbf{v}\frac{\partial L}{\partial \dot{\mathbf{r}}^2}, \tag{1.11}$$

where we have ignored higher order terms in $\mathbf{v}$. Since the equation of motion should have the same form in every frame, therefore the difference between $L'\left(\dot{\mathbf{r}}'^2\right)$ and $L\left(\dot{\mathbf{r}}^2\right)$ must be a total time derivative (Galilean invariance). For a constant $\mathbf{v}$ this implies that $\frac{\partial L}{\partial \dot{\mathbf{r}}^2}$ must be a constant and we choose this constant to be $\frac{m}{2}$. Thus we arrive at a unique Lagrangian for the motion of a free particle.

$$L = \frac{1}{2} m \dot{\mathbf{r}}^2. \tag{1.12}$$

As a second example let us consider a system consisting of two particles each of mass m interacting with each other with a potential $V(|\mathbf{r}_1 - \mathbf{r}_2|)$, where $\mathbf{r}_1$ and $\mathbf{r}_2$ denote the positions of the two particles. The standard Lagrangian according to Eq. (1.8) is

$$L_1 = \frac{1}{2} m \left(\dot{\mathbf{r}}_1^2 + \dot{\mathbf{r}}_2^2\right) - V(|\mathbf{r}_1 - \mathbf{r}_2|), \tag{1.13}$$

and this generates the equations of motion

$$m\frac{d^2\mathbf{r}_i}{d\,t^2} = -\nabla_i V(|\mathbf{r}_1 - \mathbf{r}_2|), \qquad i = 1, 2. \tag{1.14}$$

A Lagrangian equivalent to L_1, is given by [8]

$$L_2 = m\left(\dot{\mathbf{r}}_1 \cdot \dot{\mathbf{r}}_2\right) - V(|\mathbf{r}_1 - \mathbf{r}_2|), \tag{1.15}$$

and this L_2 also yields the equations of motion (1.14). However the symmetries of the two Lagrangians L_1 and L_2 are different. The Lagrangian L_1 is invariant under the rotation of the six-dimensional space $\mathbf{r}_1$ and $\mathbf{r}_2$, whereas L_2 is not.

The requirement of the invariance under the full Galilean group which includes the conservation of energy, the angular momentum and the motion of the center of mass, restricts the possible forms of the Lagrangian (apart from a total time derivative) but still leaves certain arbitrariness. Here we want to investigate this point and see whether by imposition of the Galilean invariance we can determine a unique form for the Lagrangian or not.

Consider a system of N pairwise interacting particles with the equations of motion

$$m_j \frac{d^2\mathbf{r}_j}{d\,t^2} = -\nabla_j V, \quad j = 1, 2, \cdots, N, \tag{1.16}$$

where V depends on the relative coordinates of the particles $\mathbf{r}_j - \mathbf{r}_k$ and hence

$$\sum_j \nabla_j V = 0. \tag{1.17}$$

This means that the forces are acting only between the particles of the system. Thus from (1.16) we have the law of conservation of the total linear momentum;

$$\frac{d}{dt}\sum_j m_j \mathbf{v}_j = 0, \tag{1.18}$$

where $\mathbf{v}_j = \frac{d}{dt}\mathbf{r}_j$. The conservation law (1.18) also follows from the Lagrangian

$$L = \sum_j \frac{1}{2} m_j \mathbf{v}_j^2 - V. \tag{1.19}$$

Now under the Galilean transformation $\mathbf{v}_j \to \mathbf{v}_j + \mathbf{v}$, and L will change to L' where $L' - L$ is a total time derivative. Therefore the resulting equations of motion, (1.16), will remain unchanged. Since $\sum_j m_j \mathbf{v}_j$ is constant, Eq. (1.18), we can add any function of $\sum_j m_j \mathbf{v}_j$ to the Lagrangian without affecting the equations of motion. If we denote this new Lagrangian which is found by the addition of the constant term $F\left(\sum_j m_j \mathbf{v}_j\right)$ to L by $L[F]$, then we observe that if in $L[F]$ we replace $\mathbf{v}_i$ by $\mathbf{v}_i + \mathbf{v}$, then $L'[F] - L[F]$ will not be a total time derivative unless F is of the form

$$F = \frac{1}{2\mu}\sum_i \left(m_i \mathbf{v}_i\right)^2, \tag{1.20}$$

where μ is a constant with the dimension of mass. From this result it follows that the general form of $L[F]$, can be rejected on the ground that it is not invariant under the full Galilean group [9],[10]. With the addition of the term $\frac{1}{2\mu}\sum_i (m_i \mathbf{v}_i)^2$, the resulting Lagrangian is now Galilean invariant but is dependent on the parameter μ is [9],[10];

$$L(\mu) = \sum_j \frac{1}{2} m_j \mathbf{v}_j^2 + \frac{1}{2\mu} \sum_i (m_i \mathbf{v}_i)^2 - V. \tag{1.21}$$

What is interesting about this Lagrangian is that for any two values of the parameter μ, say μ' and μ'', $L(\mu') - L(\mu'')$ is not a total time derivative, and in this sense the two Lagrangians are inequivalent.

The equations of motion derived from (1.21) are given by

$$\frac{d}{dt}\left(m_k \mathbf{v}_k + \frac{m_k}{\mu} \sum_j m_j \mathbf{v}_j\right) = -\nabla_k V, \quad k = 1, 2, \cdots, N. \tag{1.22}$$

Noting that $\frac{d}{dt}\sum_j m_j \mathbf{v}_j$ is zero, Eqs. (1.22) are the same as the equations of motion (1.16). However the relation between the canonical momentum $\mathbf{p}_k$ and the velocity $\mathbf{v}_k$ is now more complicated:

$$\mathbf{p}_k = \frac{\partial L}{\partial \mathbf{v}_k} = m_k \mathbf{v}_k + \frac{m_k}{\mu} \sum_j m_j \mathbf{v}_j, \tag{1.23}$$

or solving for $\mathbf{v}_k$ in terms of $\mathbf{p}_k$ we have

$$\mathbf{v}_k = \frac{\mathbf{p}_k}{m_k} - \frac{\sum_j \mathbf{p}_j}{\mu + \sum_j m_j}. \tag{1.24}$$

Velocity-Dependent Forces — If the velocity-dependent forces are such that the system can be described by a Lagrangian of the form

$$L = \frac{1}{2} m \mathbf{v}^2 - V(\mathbf{r}, \mathbf{v}, t), \tag{1.25}$$

then the generalized force F_i can be written as

$$F_i = -\frac{\partial V}{\partial x_i} + \frac{d}{dt}\left(\frac{\partial V}{\partial v_i}\right), \tag{1.26}$$

a result which follows from the Euler–Lagrange equation. Now according to Helmholtz for the existence of the Lagrangian such a generalized force can be at most a linear function of acceleration, and it must satisfy the Helmholtz identities [3]:

$$\frac{\partial F_i}{\partial \dot{v}_j} = \frac{\partial F_j}{\partial \dot{v}_i}, \tag{1.27}$$

$$\frac{\partial F_i}{\partial v_j} + \frac{\partial F_j}{\partial v_i} = \frac{d}{dt}\left(\frac{\partial F_i}{\partial \dot{v}_j} + \frac{\partial F_j}{\partial \dot{v}_i}\right), \tag{1.28}$$

and

$$\frac{\partial F_i}{\partial x_j} - \frac{\partial F_j}{\partial x_i} = \frac{1}{2}\frac{d}{dt}\left(\frac{\partial F_i}{\partial v_j} - \frac{\partial F_j}{\partial v_i}\right), \tag{1.29}$$

If we assume as in (1.25) that the generalized force is independent of acceleration then from (1.28) it follows that

$$\frac{\partial^2 F_i}{\partial v_j \partial v_k} = 0. \tag{1.30}$$

By integrating this equation once we get

$$\frac{\partial F_i}{\partial v_j} = \frac{e}{c}\sum_k \varepsilon_{ijk} B_k(\mathbf{r}, t), \tag{1.31}$$

where c and e are constants, and $\mathbf{B}(\mathbf{r}, t)$ is a vector function. The symbol ε_{ijk} denotes the totally antisymmetric tensor (Levi–Civita symbol) defined by

$$\varepsilon_{123} = \varepsilon_{231} = \varepsilon_{312} = 1, \tag{1.32}$$

$$\varepsilon_{321} = \varepsilon_{213} = \varepsilon_{132} = -1. \tag{1.33}$$

Now we integrate (1.31) a second time and we find that the generalized force is velocity-dependent and of Lorentz form

$$\mathbf{F}(\mathbf{r}, \mathbf{v}, t) = e\mathbf{E}(\mathbf{r}, t) + \frac{e}{c}\mathbf{v} \wedge \mathbf{B}(\mathbf{r}, t), \tag{1.34}$$

where $\mathbf{E}(\mathbf{r}, t)$ is a vector function of $\mathbf{r}$ and t. The two vector functions $\mathbf{E}(\mathbf{r}, t)$ and $\mathbf{B}(\mathbf{r}, t)$ are not independent of each other. From the three Helmholtz conditions, Eq. (1.27) is trivially satisfied in this case since $\mathbf{F}$ is independent of $\dot{\mathbf{v}}$. But in order to satisfy (1.29) we find that $\mathbf{E}$ and $\mathbf{B}$ cannot be chosen arbitrarily. To get the connection between these two vectors we substitute (1.34) in (1.29) and we find

$$\left(\nabla \wedge \mathbf{E} + \frac{1}{c}\frac{\partial \mathbf{B}}{\partial t}\right)_k = v_k(\nabla \cdot \mathbf{B}). \tag{1.35}$$

Since the velocities and coordinates are independent, each side of (1.35) must be equal to zero, and thus we find two "Maxwell" type equations [7]

$$\nabla \wedge \mathbf{E} = -\frac{1}{c}\frac{\partial \mathbf{B}}{\partial t}, \tag{1.36}$$

and

$$\nabla . \mathbf{B} = 0. \tag{1.37}$$

1.2 Noether's Theorem

The symmetries and the conservation laws associated with a given Lagrangian can be found by applying Noether's theorem to the Lagrangian of a system of particles. Let us consider the change in the Lagrangian under an infinitesimal transformation of the generalized coordinates

$$q_i \to q_i + \delta q_i, \quad i = 1, \cdots, N, \tag{1.38}$$

which to the order of δq_i is

$$\begin{aligned} & L(q_i + \delta q_i, \dot{q}_i + \delta \dot{q}_i) - L(q_i, \dot{q}_i) \\ = & \sum_j \delta q_j \frac{\partial L(q_i, \dot{q}_i)}{\partial q_j} + \sum_j \frac{d\delta q_j}{dt} \frac{\partial L(q_i, \dot{q}_i)}{\partial \dot{q}_j} \\ = & \sum_j \left[\frac{\partial L(q_i, \dot{q}_i)}{\partial q_j} - \frac{d}{dt} \frac{\partial L(q_i, \dot{q}_i)}{\partial \dot{q}_j} \right] \delta q_j + \frac{d}{dt} \sum_j \delta q_j \frac{\partial L(q_i, \dot{q}_i)}{\partial \dot{q}_j}. \end{aligned} \tag{1.39}$$

If the Lagrangian remains unchanged under this transformation, i.e.

$$L(q_i + \delta q_i, \dot{q}_i + \delta \dot{q}_i) - L(q_i, \dot{q}_i) = 0, \tag{1.40}$$

then the quantity

$$\mathcal{N} = \sum_j \delta q_j \frac{\partial L(q_i, \dot{q}_i)}{\partial \dot{q}_i}, \tag{1.41}$$

which is called the Noether charge is conserved. Here we assumed that L satisfies the Euler–Lagrange equation, and therefore each term in the square bracket in (1.39) is zero.

When $L\left(q_i, \dot{q}_i, t\right)$ does not depend on a particular coordinate, say q_k, then clearly the momentum conjugate to this coordinate, $p_k = \frac{\partial L}{\partial \dot{q}}$, is conserved. This particular coordinate is called ignorable or cyclic coordinate.

As an example let us consider the infinitesimal transformation

$$\begin{cases} \mathbf{r}_1 + \boldsymbol{\epsilon} \\ \mathbf{r}_2 + \boldsymbol{\epsilon} \end{cases}, \tag{1.42}$$

that is an infinitesimal space translation with $\boldsymbol{\epsilon}$ an arbitrary small vector. Using the Lagrangian (1.13) we find that

$$\mathcal{N} = \boldsymbol{\epsilon} \cdot m\left(\dot{\mathbf{r}}_1 + \dot{\mathbf{r}}_2\right), \tag{1.43}$$

is the conserved quantity. Since $\boldsymbol{\epsilon}$ is an arbitrary small vector, therefore the total momentum $\mathbf{P} = m\left(\dot{\mathbf{r}}_1 + \dot{\mathbf{r}}_2\right)$ remains constant. Now both of the Lagrangians L_1 and L_2, Eqs. (1.13) and (1.15) have the same Noether charge given by (1.43), but as we mentioned earlier they have different properties under space rotation.

1.3 The Hamiltonian Formulation

In classical mechanics the Hamiltonian is defined as a function of the canonical coordinates and momenta which satisfies Hamilton's principle

$$\delta \int_{t_1}^{t_2} \left\{ \sum_{i=1}^{N} \dot{q}_i p_i - H(p_i, q_i) \right\} dt = 0, \tag{1.44}$$

i.e. the integral in (1.44) between any two arbitrary times t_1 and t_2 for the actual path must be minimum (or extremum) [2].

This requirement gives us $2N$ Hamilton's canonical equations [1],[2]

$$\dot{q}_i = \frac{\partial H(p_i, q_i)}{\partial p_i}, \quad i = 1, 2, \cdots N, \tag{1.45}$$

and

$$\dot{p}_i = -\frac{\partial H(p_i, q_i)}{\partial q_i} \quad i = 1, 2, \cdots N. \tag{1.46}$$

These first order differential equations for $q_i(t)$ and $p_i(t)$ determine the position and momentum of the i-th degree of freedom as a function of time if we assume that the initial conditions

$$q_i(t = 0) = q_0(0), \quad \text{and} \quad p_i(t = 0) = p_0(0), \tag{1.47}$$

are known.

By eliminating p_i s between the two sets of equations (1.45) and (1.46) we find the Newton equations of motion

$$m_i \ddot{q}_i = F_i \left(q_1, q_2 \cdots q_N, \frac{\partial H}{\partial p_1} \cdots \frac{\partial H}{\partial p_N} \right), \quad i = 1, 2, \cdots N. \tag{1.48}$$

If we only require that the Hamiltonian generate the correct equations of motion in coordinate space, viz Eq. (1.48), then by differentiating (1.45) with respect to time and substituting for $\dot{q}_i$, $\dot{p}_i$ and $\ddot{q}_i$ using Eqs. (1.45), (1.46) and (1.48) we find

$$\begin{aligned} \ddot{q}_i &= \sum_{k=1}^{N} \left(\frac{\partial H}{\partial p_k} \frac{\partial^2 H}{\partial p_i \partial q_k} - \frac{\partial H}{\partial q_k} \frac{\partial^2 H}{\partial p_i \partial p_k} \right) \\ &= \frac{1}{m_i} F_i \left(q_1, \cdots q_N, \frac{\partial H}{\partial p_i} \cdots \frac{\partial H}{\partial p_N} \right), \quad i = 1, 2, \cdots N. \end{aligned} \tag{1.49}$$

The solutions of this set of nonlinear partial differential equations yield the desired Hamiltonians (H is not unique as will be shown later) and each of these Hamiltonians generates the motion in coordinate space (1.48). In this formulation, the canonical momenta p_i s are dummy variables and are related to the coordinates and generalized velocities via Eqs. (1.45).

q-equivalent Hamiltonians — The complete solution of (1.49) even for a simple one-dimensional motion with the force law $F(q,\dot{q})$ is not known. However if we assume that F is derivable from a velocity-independent potential function, i.e.

$$F(q) = -\left(\frac{\partial V(q)}{\partial q}\right), \tag{1.50}$$

then we can solve Eq. (1.49) in the following way:
We write Eq. (1.49) as [11]

$$-\frac{1}{m}\frac{\partial V}{\partial q} = \frac{\partial^2 H}{\partial q \partial p}\frac{\partial H}{\partial p} - \frac{\partial^2 H}{\partial p^2}\frac{\partial H}{\partial q}, \tag{1.51}$$

or rearranging the terms we have

$$\frac{\partial}{\partial q}\left\{\frac{m}{2}\left(\frac{\partial H}{\partial p}\right)^2 + V(q)\right\} = m\frac{\partial^2 H}{\partial p^2}\frac{\partial H}{\partial q}. \tag{1.52}$$

We note that the quantity inside the curly bracket in (1.52) is the Hamiltonian H_0 which is also the total energy of the particle. Thus if $G(H_0)$ is an arbitrary function of H_0, we have

$$\frac{\partial G}{\partial p} = \frac{dG}{dH_0}m\dot{q}\frac{\partial \dot{q}}{\partial p} = \frac{dG}{dH_0}m\frac{\partial^2 H}{\partial p^2}\frac{\partial H}{\partial p}. \tag{1.53}$$

By substituting for $m\frac{\partial^2 H}{\partial p^2}$ from Eqs. (1.52) in (1.53) it follows that

$$\frac{\partial G}{\partial p}\frac{\partial H}{\partial q} = \frac{\partial H}{\partial p}\left(\frac{dG}{dH_0}\frac{\partial H_0}{\partial q}\right) = \frac{\partial H}{\partial p}\frac{\partial G}{\partial q}, \tag{1.54}$$

and therefore

$$H = G\left[\frac{m}{2}\left(\frac{\partial H}{\partial p}\right)^2 + V(q)\right]. \tag{1.55}$$

The canonical momentum, p, in this case is not equal to the mechanical momentum unless $G(H_0) = H_0$. To find the relation between p and $m\dot{q}$, we write the Lagrangian L as

$$L = \dot{q}p - H(q,p), \tag{1.56}$$

and then use the definition of p to find it in terms of $\dot{q}$

$$p = \frac{\partial L}{\partial \dot{q}} = p + \dot{q}\frac{\partial p}{\partial \dot{q}} - \frac{dG}{dH_0}m\dot{q}. \tag{1.57}$$

Thus we have

$$\frac{1}{m}\frac{\partial p}{\partial \dot{q}} = \frac{dG}{dH_0}, \tag{1.58}$$

or

$$p = \int \frac{dG}{dH_0} m d\dot{q} + g(q), \tag{1.59}$$

where $g(q)$ is an arbitrary function of q. For systems with two or more degrees of freedom or when F_i is not derivable from a potential this method does not work (see also [12],[13].)

We can also try to determine the Hamiltonian $H(p_i, q_i)$ such that the canonical equations (1.45) and (1.46) yield the motion of the system in phase space. In this case, p_i s are not dummy variables but are directly related to the generalized velocities, i.e. instead of (1.46) we find the solutions of (1.45) and

$$\dot{p}_i = -\frac{\partial H(p_i, q_i)}{\partial q_i} = F_i \left(q_1, q_2 \cdots q_N, \frac{\partial H}{\partial p_1} \cdots \frac{\partial H}{\partial p_N} \right) \quad i = 1, 2, \cdots N. \tag{1.60}$$

Even this set of equations of motion will not give us a unique Hamiltonian. Thus we can have a set of q-equivalent (or coordinate equivalent) or a smaller set of pq-equivalent for a given set of forces F_i. For the classical description of motion any member of either set is acceptable.

Let us consider two simple examples:

(1) - Assuming that $V(q)$ is positive for all values of q and taking $G(H_0) = H_0^{\frac{1}{2}}$, by solving (1.55) we get [11]

$$H(p, q) = A\,(V(q))^{\frac{1}{2}} \cosh\left(\sqrt{\frac{2}{m}} \lambda p \right). \tag{1.61}$$

(2) - If we choose H to be the inverse positive square root of energy, $G(H_0) = H_0^{-\frac{1}{2}}$ we find [11]

$$H(p, q) = \left[\frac{1 - \left(\frac{2}{m}\right) p^2 (V(q))^2}{V(q)} \right]^{\frac{1}{2}}. \tag{1.62}$$

***pq*-equivalent Hamiltonians** — Now let us examine pq-equivalent Hamiltonians. For a one-dimensional case the equations of motion in coordinate space is

$$m \frac{d^2 q}{d\,t^2} + \frac{\partial V(q)}{\partial q} = 0, \tag{1.63}$$

and the equations for the mechanical momentum $m\dot{q} = p$ can be written as

$$m \frac{d^2 p}{d\,t^2} + \frac{\partial^2 V(q)}{\partial q^2} p = 0. \tag{1.64}$$

The solution of these equations subject to the initial conditions give us $q(t)$ and $p(t)$.

The case of the harmonic oscillator $V(q) = \frac{1}{2}m\omega^2 q^2$ is particularly simple since the two equations uncouple, and we have

$$m\frac{dq}{dt}d\left(\frac{dq}{dt}\right) = -m\omega^2 q dq, \tag{1.65}$$

and

$$m\frac{dp}{dt}d\left(\frac{dp}{dt}\right) = -m\omega^2 p dp. \tag{1.66}$$

These equations can be integrated to yield

$$\frac{1}{2}\left[\left(\frac{\partial H}{\partial p}\right)^2 + \omega^2 q^2\right] = C(H), \tag{1.67}$$

$$\frac{1}{2}\left[\left(\frac{\partial H}{\partial q}\right)^2 + \omega^2 p^2\right] = D(H), \tag{1.68}$$

where C and D are arbitrary functions of H. Rather than trying to find the most general solution of (1.67) and (1.68), let us consider a class of solutions which we can find in the following way [14]. Let

$$H = \Gamma(H_0) = \Gamma\left(\frac{p^2}{2m} + \frac{1}{2}m\Omega^2 q^2\right), \tag{1.69}$$

then by differentiating H with respect to p and q and substituting in $C(H)$ and $D(H)$ we find that if

$$\Omega = \frac{\omega}{\frac{\partial \Gamma}{\partial H_0}}, \tag{1.70}$$

then

$$C(H) = \frac{1}{m}\left(\frac{\omega}{\Omega}\right)^2 H^{-1}(H_0), \tag{1.71}$$

and

$$D(H) = m\omega^2 H^{-1}(H_0), \tag{1.72}$$

where Ω is an energy-dependent frequency. Depending on our choice of Γ we have an infinite set of pq-equivalent Hamiltonians. For instance we can choose the Hamiltonian to be

$$H = \Gamma(H_0) = \frac{H_0^j}{\varepsilon^{j-1}}, \tag{1.73}$$

where ε is a constant with the dimension of energy.

Galilean Invariant Hamiltonians — The same type of ambiguity which we found for the Lagrangian $L(\mu)$, Eq. (1.21), also appears in the classical Hamiltonian formulation. Thus for a system of interacting particles when $L(\mu)$ and $\mathbf{p}_k$ are given by (1.21) and (1.23), the corresponding Hamiltonian obtained from the definition $H(\mu) = \sum_k \mathbf{p}_k \cdot \mathbf{v}_k - L(\mu)$ is

$$H(\mu) = \sum_k \frac{\mathbf{p}_k^2}{2m_k} - \frac{\left(\sum_j \mathbf{p}_j\right)^2}{2\mu + 2\sum_j m_j} + V. \tag{1.74}$$

1.4 Canonical Transformation

A set of transformations

$$\begin{cases} p_j \to p_j(P_i, Q_i) \\ q_j \to q_j(P_i, Q_i) \end{cases} \quad j = 1, 2, \cdots N, \tag{1.75}$$

is called canonical if the form of Hamilton's equations of motion, Eqs. (1.45) and (1.46), are preserved. Thus the canonical transformation is, in a way, an extension of the point transformation of N space coordinates $q_i \to q_i(Q_j)$ to the transformation of $2N$ coordinates of $(p_i(P_j, Q_j), q_i(P_j, Q_j))$ space, called the phase space and is given by (1.75). Under the point transformation the Euler–Lagrange equations are left unchanged, whereas under the canonical transformation the form of the canonical equations (1.45) and (1.46) remain the same. Thus if $H(p_i, q_i)$ is transformed to $K(P_j, Q_j)$, then Eqs. (1.45) and (1.46) are replaced by

$$\dot{Q}_i = \frac{\partial K(P_j, Q_j)}{\partial P_i}, \quad \dot{P}_i = -\frac{\partial K(P_j, Q_j)}{\partial Q_i}, \quad i = 1, 2, \cdots N, \tag{1.76}$$

respectively.

We can derive the connection between H and K by noting that according to the Hamilton principle Eq. (1.44)

$$\delta \int L dt = \delta \int \left(\sum_j p_j dq_j - H(p_i, q_i, t) dt \right) = 0, \tag{1.77}$$

and when the same principle is applied to $K(P_j, Q_j)$ we find

$$\delta \int L' dt = \delta \int \left(\sum_j P_j dQ_j - K(P_i, Q_i, t) dt \right) = 0. \tag{1.78}$$

As we have seen earlier the difference between L and L' must be a total time derivative $\frac{d\mathcal{F}_1}{dt}$. Thus

$$\sum_i p_i dq_i - H dt = \sum_i P_i dQ_i - K dt + d\mathcal{F}_1, \tag{1.79}$$

where $\mathcal{F}_1$ is called the generating function for the canonical transformation. Rewriting (1.79) as

$$d\mathcal{F}_1(q_i, Q_i, t) = \sum_i (p_i dq_i - P_i dQ_i) + (K - H) dt, \tag{1.80}$$

we have

$$p_i = \frac{\partial \mathcal{F}_1}{\partial q_i}, \tag{1.81}$$

$$P_i = -\frac{\partial \mathcal{F}_1}{\partial Q_i}, \tag{1.82}$$

and

$$K = H + \frac{\partial \mathcal{F}_1}{\partial t}. \tag{1.83}$$

A different but more useful case is when the generating function depends on old coordinates q_i and the new momenta P_i. By writing (1.80) as

$$d\mathcal{F}_2 = d\left(\mathcal{F}_1(q_i, Q_i, t) + \sum_i P_i Q_i\right) = \sum_i p_i dq_i + (K - H)dt, \tag{1.84}$$

we obtain

$$p_i = \frac{\partial \mathcal{F}_2}{\partial q_i}, \quad Q_i = \frac{\partial \mathcal{F}_2}{\partial P_i}, \quad K = H + \frac{\partial \mathcal{F}_2}{\partial t}. \tag{1.85}$$

An interesting example of the generating function of this type, $\mathcal{F}_2$, is for the motion of the particle in a time-dependent harmonic oscillator potential where the Hamiltonian is

$$H(p,q,t) = \frac{1}{2m}\left[p^2 + \Omega^2(t)q^2\right]. \tag{1.86}$$

For this problem the generating function $\mathcal{F}_2$ is

$$\begin{aligned} \mathcal{F}_2 &= \frac{m}{2}\frac{\dot{\rho}(t)}{\rho(t)}q^2 \pm \frac{q}{\rho(t)}\left(2P - \frac{q^2}{\rho^2(t)}\right)^{\frac{1}{2}} \\ &\pm P\sin^{-1}\left[\frac{q}{\rho(t)\sqrt{2P}}\right] + \left(n + \frac{1}{2}\right)\pi P, \end{aligned} \tag{1.87}$$

where n is an integer and where

$$-\frac{\pi}{2} \le \sin^{-1}\left[\frac{q}{\rho(t)\sqrt{2P}}\right] \le \frac{\pi}{2}. \tag{1.88}$$

In Eq. (1.87) the $\pm$ signs are taken according to whether $\left(p - m\frac{\dot{\rho}(t)}{\rho(t)}q\right)$ is positive or negative. The function $\rho(t)$ appearing in Eqs. (1.87) and (1.88) is a solution of the nonlinear differential equation

$$m^2\ddot{\rho}(t) + \Omega^2(t)\rho(t) - \frac{1}{\rho^3(t)} = 0, \tag{1.89}$$

with the boundary conditions $\rho(0) = 1$ and $\dot{\rho}(0) = 0$. In this problem $\mathcal{F}_2$ depends explicitly on time, and therefore the new Hamiltonian $K(P,Q,t)$ is given by

$$K(P,Q,t) = H(p,q,t) + \frac{\partial \mathcal{F}_2}{\partial t} = \frac{1}{m}\left(\frac{P}{\rho^2(t)}\right). \tag{1.90}$$

We note that

$$Q = \frac{\partial \mathcal{F}_2}{\partial q} = -\tan^{-1}\frac{p\rho^2(t)}{q} - m\rho(t)\dot{\rho}(t), \tag{1.91}$$

and

$$P = \frac{1}{2}\left[\frac{q^2}{\rho^2(t)} + (\rho(t)p - mq\dot{\rho}(t))^2\right]. \tag{1.92}$$

The variable Q does not appear in $K(P, Q, t)$, therefore Q is a cyclic coordinate, and the momentum conjugate to it, P, is a constant of motion [15].

Let us also note that for the special generating function $\mathcal{F}_3 = \sum_i g_i(q_j, t)P_i$, the canonical transformation reduces to the point transformation (1.3).

1.5 Action-Angle Variables

A very important property of the canonical transformation is that it leaves the volume in phase space unchanged. We want to show this important property of the transformation for a system with one degree of freedom. To prove this result which is referred to as Liouville theorem, let us consider the canonical transformation $Q = Q(p, q)$, $P = P(p, q)$. By taking the time derivative of Q we have

$$\dot{Q} = \frac{\partial Q}{\partial p}\dot{p} + \frac{\partial Q}{\partial q}\dot{q} = -\frac{\partial Q}{\partial p}\frac{\partial H}{\partial q} + \frac{\partial Q}{\partial q}\frac{\partial H}{\partial p}. \tag{1.93}$$

Since the transformation is canonical and is independent of time, therefore the new Hamiltonian is just the old Hamiltonian written in terms of Q and P, $H(p, q) \to H(P, Q)$. The form of the Hamilton canonical equations are the same whether expressed in terms of (p, q) or (P, Q), Eq. (1.76). Thus

$$\dot{Q} = \frac{\partial H}{\partial P} = \frac{\partial H}{\partial q}\frac{\partial q}{\partial P} + \frac{\partial H}{\partial p}\frac{\partial p}{\partial P}. \tag{1.94}$$

By comparing (1.93) and (1.94) we find that

$$\frac{\partial q}{\partial P} = -\frac{\partial Q}{\partial p}, \quad \text{and} \quad \frac{\partial p}{\partial P} = \frac{\partial Q}{\partial q}. \tag{1.95}$$

Similarly we calculate $\dot{P}$ by differentiating $P(p, q)$ with respect to time and comparing the result with $\dot{P} = -\frac{\partial H}{\partial Q}$. In this way we get another pair of equations:

$$\frac{\partial P}{\partial p} = \frac{\partial q}{\partial Q}, \quad \text{and} \quad \frac{\partial P}{\partial q} = -\frac{\partial p}{\partial Q}. \tag{1.96}$$

Now let us consider the transformed phase space area

$$\int P dQ = \int J dq dp, \tag{1.97}$$

where J is the Jacobian of the transformation

$$J = \frac{D(P,Q)}{D(p,q)} = \begin{bmatrix} \frac{\partial P}{\partial p} & \frac{\partial P}{\partial q} \\ \frac{\partial Q}{\partial p} & \frac{\partial Q}{\partial q} \end{bmatrix}. \tag{1.98}$$

This Jacobian J can also be written as

$$J = \frac{\frac{D(P,Q)}{D(P,q)}}{\frac{D(p,q)}{D(P,q)}} = \frac{\left(\frac{\partial Q}{\partial q}\right)}{\left(\frac{\partial p}{\partial P}\right)} = 1, \tag{1.99}$$

where in getting the last step we have used (1.95). Therefore Eq. (1.97) with $J = 1$ shows that under canonical transformation the area (or volume) in phase space is preserved.

Definition of the Action-Angle Variables — We define the action-angle variables which we denote by $\{\theta,\ I\}$ in the following way: We first calculate the area under the curve $p(q, E)$, where E is the energy of the particle

$$\mathcal{A}(E) = \oint p(q,\ E)dq = 2\int_{q_1}^{q_2} [2m(E - V(q))]^{\frac{1}{2}}\, dq. \tag{1.100}$$

Then we find the area under the curve $I(E)$;

$$\mathcal{A}'(E) = \oint I d\theta = \int_0^{2\pi} I d\theta = 2\pi I. \tag{1.101}$$

Since the transformation $\{p,\ q\}$ to $\{\theta, I\}$ is canonical therefore $\mathcal{A}(E) = \mathcal{A}'(E)$ and from Eqs. (1.100) and (1.101) we have

$$I = \frac{1}{2\pi}\oint p(E,q)dq. \tag{1.102}$$

Equation (1.102) defines the action variable. The conjugate of action variable is θ which varies between $-\pi$ and π. This relation can be generalized to a system with N degrees of freedom where for the k-th degree of freedom we have

$$I_k = \frac{1}{2\pi}\oint p_k(E, q_1, \cdots, q_N)dq_k. \tag{1.103}$$

This action is conjugate to the angle variable θ_k, where θ_k changes between $-\pi < \theta_k \le \pi$. Thus all angle variables are periodic with period 2π.

For a system of two degrees of freedom, the phase space is four-dimensional and the toroid is a two-dimensional surface lying in a three-dimensional energy shell. We can generalize this concept to a system of N degrees of freedom in the following way:

For the bounded motion of an integrable system we can transform the phase space coordinates $(\mathbf{p},\ \mathbf{q}) = (p_1,\ p_2, \cdots p_N;\ q_1,\ q_2, \cdots q_N)$ to $(\mathbf{I},\ \boldsymbol{\theta}) = (I_1,\ I_2, \cdots I_N;\ \theta_1,\ \theta_2, \cdots \theta_N)$, where $\boldsymbol{\theta}$ is an N-dimensional angle variable,

$-\pi \leq \theta_k \leq \pi$. By this transformation the Hamiltonian becomes a function of $\mathbf{I}$, $H = H(\mathbf{I})$, i.e. all θ_k s become cyclic coordinates. Thus we have

$$\dot{\theta}_k = \frac{\partial H}{\partial I_k} = \frac{\partial E(\mathbf{I})}{\partial I_k} = \omega_k(\mathbf{I}), \tag{1.104}$$

and

$$\dot{I}_k = -\frac{\partial H}{\partial \theta_k} = 0. \tag{1.105}$$

Since $\mathbf{I}$ is a constant vector in N-dimensions, Eq. (1.104) can be integrated to yield

$$\theta_k = \omega_k t + \delta_k. \tag{1.106}$$

In this relation ω_k is a characteristic angular frequency of the motion and δ_k is a phase shift. For a completely separable system ω_k depends on I_k alone. Since the motion is periodic in θ_k, therefore $\theta_k(\mathbf{q}, \mathbf{I})$ s are all multi-valued functions of the coordinates $\mathbf{q}$. Now any single valued function $A(\mathbf{p},\ \mathbf{q})$, when expressed in terms of $\mathbf{I}$ and $\boldsymbol{\theta}$ is a periodic function of the angle variables with each variable having a period 2π. Hence we can expand the function $A(\mathbf{I},\ \boldsymbol{\theta})$ of the dynamical variables, I_i, θ_i in terms of the Fourier series

$$A(\mathbf{I},\ \boldsymbol{\theta}) = \sum_{j_1=-\infty}^{\infty} \cdots \sum_{j_N=-\infty}^{\infty} A_{j_1 j_2 \cdots j_N} \exp\left[i\left(j_1\theta_1 + j_2\theta_2 + \cdots + j_n\theta_N\right)\right]. \tag{1.107}$$

By substituting from (1.106) in (1.107) we have

$$A(\mathbf{I},\ \boldsymbol{\theta}) = \sum_{j_1=-\infty}^{\infty} \cdots \sum_{j_N=-\infty}^{\infty} A_{j_1 j_2 \cdots j_N} \exp\left[it\left(j_1\frac{\partial E}{\partial I_1} + j_2\frac{\partial E}{\partial I_2} + \cdots + j_n\frac{\partial E}{\partial I_N}\right)\right]. \tag{1.108}$$

We note that each term in this sum is a periodic function of time with the frequency

$$\sum_{l=1}^{N} j_l \frac{\partial E}{\partial I_l}, \tag{1.109}$$

but these frequencies are not generally commensurable, therefore the sum is not a periodic function of time. In particular we can choose A to be either p_k or q_k which shows that p_k and q_k are also non-periodic functions of time.

For an integrable system of N degrees of freedom, the motion of a phase point $(p_1,\ p_2 \cdots p_N; q_1,\ q_2 \cdots q_N)$ in $2N$-dimensional phase space will be confined to an invariant toroid of N dimensions. That is this invariant toroid occupy the whole phase space of bounded integrable motions.

If the system of N degrees of freedom is not integrable, subject to certain conditions we can still express $\mathbf{p}$ and $\mathbf{q}$ in terms of θ_k s. The parametric representation of $\mathbf{p}$ and $\mathbf{q}$ as functions of N-dimensional angle variable $\boldsymbol{\theta}$

$$P = P_{\mathcal{T}}(\boldsymbol{\theta}) = \left(\mathbf{p}_{\mathcal{T}}(\boldsymbol{\theta}),\ \mathbf{q}_{\mathcal{T}}(\boldsymbol{\theta})\right), \tag{1.110}$$

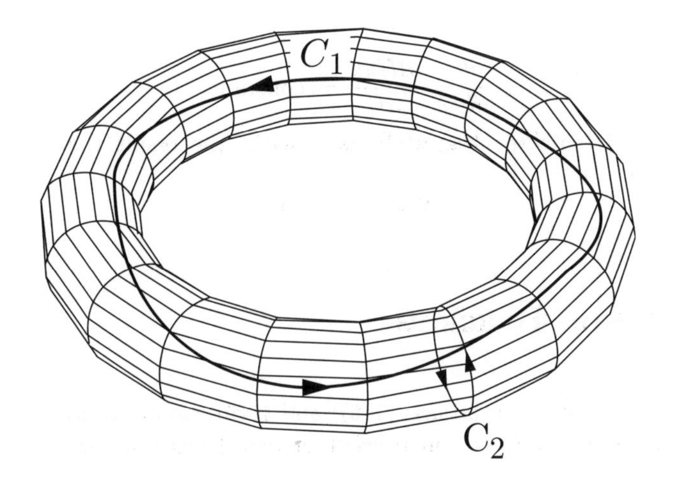

Figure 1.1: The phase space point $(\theta_1, \theta_2, I_1, I_2)$ for a system with two degrees of freedom is periodic in each θ_1 and θ_2 and this point moves in a region which is a toroid. The two curves C_1 and C_2 define the action integrals I_1 and I_2.

defines an N-dimensional toroid, $\mathcal{T}$, in the $2N$-dimensional phase space [16]–[18]. The action in this case is

$$I_k = \oint_{\mathcal{C}_k} \mathbf{p} \cdot d\mathbf{q}, \tag{1.111}$$

where $\mathbf{p}$ and $\mathbf{q}$ are the N-dimensional vectors and $\mathcal{C}_k$ s are N independent closed curves on toroid [17]. In the case of a conservative system with two degrees of freedom such a toroid is shown in Fig. 1.1.

For the bounded motion of a system the action-angle variables makes the description of the motion very simple [16]–[18]. Let us consider a motion of one degree of freedom with the Hamiltonian

$$H(p, q) = \frac{p^2}{2m} + V(q) = E, \tag{1.112}$$

where E, the total energy, is a constant. If we solve (1.112) for $p(q, E)$ we find the two-valued function

$$p(q, E) = \pm\sqrt{2m(E - V(q))}. \tag{1.113}$$

Since this canonical variable is multivalued, we try to find another set of conjugate variables (θ, I) in such a way that H becomes independent of θ, and I and becomes a constant of motion. Suppose that we have found these variables, then we have the canonical equations of motion

$$\frac{dI}{dt} = 0 = -\frac{\partial H(I)}{\partial \theta}, \tag{1.114}$$

and

$$\frac{d\theta}{dt} = \frac{\partial H(I)}{\partial I} = \text{constant}. \tag{1.115}$$

Equation (1.115) can be integrated easily with the result that θ is a linear function of time

$$\theta = \omega(I)t + \delta. \tag{1.116}$$

1.6 Poisson Brackets

The Hamilton canonical equations (1.45) and (1.46) can be written in a concise form if we introduce the Poisson bracket of two dynamical variables $u(p_i, q_i)$ and $v(p_i, q_i)$ by

$$\{u(p_i, q_i), v(p_i, q_i)\}_{p_i,q_i} = \sum_{i=1}^{N} \left(\frac{\partial u}{\partial q_i}\frac{\partial v}{\partial p_i} - \frac{\partial u}{\partial p_i}\frac{\partial v}{\partial q_i}\right). \tag{1.117}$$

From this definition it readily follows that

$$\{q_i, p_j\} = \delta_{ij}, \quad \{q_i, q_j\} = \{p_i, p_j\} = 0. \tag{1.118}$$

Also by direct differentiation we find

$$\{u, v\}_{p_i,q_i} = \sum_{j=1}^{N} \{u, v\}_{P_j,Q_j} \{Q_j, P_j\}_{p_i,q_i}. \tag{1.119}$$

If the transformation

$$Q_j \to Q_j(p_i, q_i, t), \quad \text{and} \quad P_j \to P_j(p_i, q_i, t), \tag{1.120}$$

is canonical then

$$\sum_i (p_i dq_i - P_i dQ_i), \tag{1.121}$$

is a complete differential and we have

$$\{Q_j, P_j\}_{p_i,q_i} = \delta_{ij}. \tag{1.122}$$

Thus (1.119) reduces to

$$\{u, v\}_{p_i,q_i} = \{u, v\}_{P_i,Q_i}. \tag{1.123}$$

This result shows that the Poisson bracket $\{u, v\}_{p_i,q_i}$ remains unchanged by a canonical transformation of one set of canonical variables to another. To simplify the notation we suppress the subscripts p_i, q_i and denote the Poisson bracket by $\{u, v\}$.

Some of the other properties of Poisson bracket are as follows:
(1) - Skew symmetry

$$\{u, v\} = -\{v, u\}. \tag{1.124}$$

(2) - Linearity

$$\{\alpha u + \beta v, w\} = \alpha\{u, w\} + \beta\{v, w\}, \tag{1.125}$$

where α and β are constants.
(3) - Leibniz property

$$\{uv, w\} = u\{v, w\} + v\{u, w\}, \tag{1.126}$$

where u, v and w are functions of p_i and q_i.
(4) - Jacobi identity

$$\{\{u, v\}, w\} + \{\{v, w\}, u\} + \{\{w, u\}, v\} = 0. \tag{1.127}$$

The first three results follow directly from the definition of the Poisson bracket and the proof of the last relation is straightforward but lengthy (see Ref. [1]).

Now let Φ be some function of the p_i s and q_i s and time, then its total time derivative is given by

$$\frac{d\Phi}{dt} = \frac{\partial \Phi}{\partial t} + \sum_j \left(\frac{\partial \Phi}{\partial q_j}\dot{q}_j + \frac{\partial \Phi}{\partial p_j}\dot{p}_j\right). \tag{1.128}$$

Substituting for $\dot{q}_j$ and $\dot{p}_j$ from canonical equations (1.45) and (1.46) we obtain

$$\frac{d\Phi}{dt} = \frac{\partial \Phi}{\partial t} + \{H, \Phi\}. \tag{1.129}$$

If Φ does not explicitly depend on time and remains a constant of motion, then

$$\{H, \Phi\} = 0, \tag{1.130}$$

and Φ is called an integral of motion.

Written in terms of the Poisson brackets the equation of motion (1.49) has a simple form

$$\frac{1}{m_i}F_i = \{H, \{H, q_i\}\}. \tag{1.131}$$

Poisson Brackets for Galilean Invariant Hamiltonian — Again it is interesting to examine the Poisson bracket found from the Lagrangian (1.21) and its Hamiltonian counterpart (1.74). According to Eq. (1.23) the relation between $\mathbf{p}_k$ and $\mathbf{v}_k$ involves the velocity of the other particles. For simplicity let us consider the case of two interacting particles in one dimension, where from (1.24) we have

$$v_1 = \frac{p_1}{m_1} - \frac{p_1 + p_2}{\mu + m_1 + m_2}, \quad \text{and} \quad v_2 = \frac{p_2}{m_2} - \frac{p_1 + p_2}{\mu + m_1 + m_2}. \tag{1.132}$$

From these relations we can calculate the Poisson brackets

$$\{x_1,\, m_1 v_1\} = \left(1 + \frac{m_2}{\mu}\right)\left(\frac{\mu}{\mu + m_1 + m_2}\right), \tag{1.133}$$

$$\{x_1,\, m_2 v_2\} = \left(-\frac{m_2}{\mu}\right)\left(\frac{\mu}{\mu + m_1 + m_2}\right), \tag{1.134}$$

and similar relations for $\{x_2,\, m_2 v_2\}$ and $\{x_2,\, m_1 v_1\}$. Thus for an arbitrary value of μ all the coordinates x_i fail to have a vanishing Poisson bracket with all the canonical momenta $m_i x_i$. Therefore, in quantum theory, for the Hamiltonian $H(\mu)$, none of these quantities can be sharply measured. As (1.133) shows the Poisson bracket of the coordinate x_1 and the mechanical momentum $m_1 v_1$ is zero if we choose $\mu = -m_2$. This result indicates that if the Lagrangian such as (1.21) is used as a classical basis for the quantum mechanical formulation then it is possible to violate the uncertainty principle $\Delta x \Delta(m_1 v_1)$ [10].

1.7 Time Development of Dynamical Variables and Poisson Brackets

Let us consider a dynamical quantity $u(p_i, q_i)$ and let us assume that its initial value $u(p_i(0), q_i(0))$ is known. We want to find the time evolution of this quantity. Suppressing the canonical coordinates and momenta we write $u(\Delta t)$ as a Taylor series

$$u(\Delta t) = u(0) + \frac{\Delta t}{1!}\left(\frac{du}{dt}\right)_{t=0} + \frac{(\Delta t)^2}{2!}\left(\frac{d^2 u}{d\,t^2}\right)_{t=0} + \cdots. \tag{1.135}$$

We express the derivatives of u in terms of the Poisson brackets

$$\left(\frac{du}{dt}\right)_{t=0} = \{u, H\}_{t=0}, \quad \left(\frac{d^2 u}{d\,t^2}\right)_{t=0} = \{\{u, H\}, H\}_{t=0}, \cdots. \tag{1.136}$$

Substituting these derivatives in (1.135) we obtain

$$u(\Delta t) = u(0) + \frac{\Delta t}{1!}\{u, H\}_{t=0} + \frac{(\Delta t)^2}{2!}\{\{u, H\}, H\}_{t=0} + \cdots. \tag{1.137}$$

This last relation shows that $u(\Delta t)$ can be expressed in terms of the initial values $p_i(0)$ and $q_i(0)$. For instance if we want to solve the one-dimensional problem of a harmonic oscillator with the Hamiltonian

$$H = \frac{1}{2m}p^2 + \frac{1}{2}m\omega^2 q^2, \tag{1.138}$$

and determine the coordinate $q(t)$ in terms of $q(0)$ and $p(0)$ we first find the Poisson brackets

$$\{q, H\}_{t=0} = \frac{p(0)}{m}, \tag{1.139}$$

$$\{\{q, H\}, H\}_{t=0} = -\omega^2 q(0), \tag{1.140}$$

$$\{\{\{q, H\}, H\}, H\} = -\frac{\omega^2}{m} p(0), \tag{1.141}$$

and so on. Then from the expansion

$$q(t) = q(0) + \frac{\Delta t}{1!}\{q, H\}_{t=0} + \frac{(\Delta t)^2}{2!}\{\{q, H\}, H\}_{t=0} + \cdots, \tag{1.142}$$

we get

$$\begin{aligned} q(t) &= q(0)\left[1 - \frac{\omega^2(\Delta t)^2}{2!} + \frac{\omega^4(\Delta t)^4}{4!} - \cdots\right] \\ &+ \frac{p(0)}{m}\left[\Delta t - \frac{\omega^2(\Delta t)^3}{3!} + \cdots\right] \\ &= q(0)\cos(\omega\Delta t) + \frac{p(0)}{m\omega}\sin(\omega\Delta t), \end{aligned} \tag{1.143}$$

a solution which is valid for all values of (Δt). We will see later that the quantum version of (1.137) can be used to solve the Heisenberg equation of motion.

1.8 Infinitesimal Canonical Transformation

If the generator of the canonical transformation $\mathcal{F}_2(p_i, q_i, t)$ is of the form

$$\mathcal{F}_2(P_i, q_i, t) = \sum_i P_i q_i + \varepsilon G(P_i, q_i, t), \tag{1.144}$$

where ε is a very small positive number, then according to (1.85) we have

$$Q_i = \frac{\partial \mathcal{F}_2}{\partial P_i} = q_i + \varepsilon\frac{\partial G}{\partial P_i} \to q_i + \varepsilon\frac{\partial G}{\partial p_i}, \tag{1.145}$$

and

$$p_i = \frac{\partial \mathcal{F}_2}{\partial q_i} = P_i + \varepsilon\frac{\partial G}{\partial q_i}. \tag{1.146}$$

i.e. the canonical coordinates and momenta are changed by an amount proportional to ε. In this case G is called the generator of the infinitesimal transformation. We can also write (1.145) and (1.146) as

$$\delta q_i = Q_i - q_i = \varepsilon\frac{\partial G}{\partial p_i}, \tag{1.147}$$

and

$$\delta p_i = P_i - p_i = -\varepsilon \frac{\partial G}{\partial q_i}. \tag{1.148}$$

The infinitesimal transformation generated by G takes a simple form if we write it in terms of the Poisson bracket. Let us consider an arbitrary dynamical variable $\Lambda(p_i, q_i, t)$ and determine how it will be changed by an infinitesimal canonical transformation. We note that after such a transformation $\Lambda(p_i, q_i, t)$ changes to

$$\Lambda(p_i + \delta p_i, q_i + \delta q_i, t) = \Lambda(p_i, q_i, t) + \sum_{i=1}^{N} \left(\frac{\partial \Lambda}{\partial q_i} \delta q_i + \frac{\partial \Lambda}{\partial p_i} \delta p_i \right)$$

$$= \Lambda(p_i, q_i, t) + \varepsilon \sum_{i=1}^{N} \left(\frac{\partial \Lambda}{\partial q_i} \frac{\partial G}{\partial p_i} - \frac{\partial \Lambda}{\partial p_i} \frac{\partial G}{\partial q_i} \right) = \Lambda(p_i, q_i, t) + \varepsilon \{\Lambda, G\}, \tag{1.149}$$

where we have used Eqs. (1.147) and (1.148). Denoting the difference $\Lambda(p_i + \delta p_i, q_i + \delta q_i, t) - \Lambda(p_i, q_i, t)$ by $\delta\Lambda(p_i, q_i, t)$, this last relation can be written as

$$\delta\Lambda(p_i, q_i, t) = \varepsilon\{\Lambda, G\}. \tag{1.150}$$

In particular if we choose $\Lambda(p_i, q_i, t)$ to be the Hamiltonian of the system, H, then

$$\delta H = \varepsilon\{H, G\} = \varepsilon \left(\frac{dG}{dt} - \frac{\partial G}{\partial t} \right). \tag{1.151}$$

Here Eq. (1.129) has been used to write H in terms of the time derivatives of G. This relation shows that if G does not depend on time explicitly, and if H is invariant under the infinitesimal transformation, then $\frac{dG}{dt} = 0$ and G is a constant of motion (i.e. a conserved quantity).

Among the constants of motion for a system of interacting particles the following constants are of particular interest:

(a) - The infinitesimal generator for space translation of two interacting particles, when the potential between them is $V(|\mathbf{r}_1 - \mathbf{r}_2|)$ can be written as

$$\boldsymbol{\epsilon} \cdot \mathbf{G} = \boldsymbol{\epsilon} \cdot (\mathbf{p}_1 + \mathbf{p}_2). \tag{1.152}$$

(b) - The generator for the rotation about the axis $\mathbf{a}$ by an angle $\delta\theta$ is

$$\delta\theta G = \delta\theta \mathbf{a} \cdot (\mathbf{r} \times \mathbf{p}). \tag{1.153}$$

(c) - For the Galilean transformation of a system of n particles we have

$$\begin{cases} \mathbf{r}_i \to \mathbf{r}_i + \mathbf{v}t \\ \mathbf{p}_i \to \mathbf{p}_i + m_i\mathbf{v} \end{cases}, \qquad i = 1, \cdots n, \tag{1.154}$$

and the generator for transformation by infinitesimal velocity $\delta\mathbf{v}$ is given by

$$\delta\mathbf{v}\cdot\mathbf{N} = \delta\mathbf{v}\cdot\sum_{i=1}^{n}(\mathbf{p}_i t - m_i\mathbf{r}_i)\,. \tag{1.155}$$

In the case of the Galilean transformation given by Eqs. (1.154) and (1.155), the vector $\mathbf{N}$ for a system of n interacting particles isolated from the rest of universe is a constant of motion

$$\frac{d\mathbf{N}}{dt} = \sum_{i=1}^{n}(\mathbf{p}_i - m_i\dot{\mathbf{r}}_i) + \sum_{i}^{n}\frac{d\mathbf{p}_i}{dt} = 0. \tag{1.156}$$

Let us note that while under this transformation the equations of motion remain invariant, the Hamiltonian H will change. For example for a single particle, $H = \frac{p^2}{2m}$, changes to H' where

$$H' = \frac{1}{2m}(\mathbf{p} + m\delta\mathbf{v})^2\,. \tag{1.157}$$

Combining these transformations we obtain the most general form of infinitesimal transformation of space-time displacements for a system of n particles. Such a transformation is given by

$$\epsilon G = \delta\boldsymbol{\epsilon}\cdot\mathbf{P} + \delta\boldsymbol{\omega}\cdot\mathbf{J} + \delta\mathbf{v}\cdot\mathbf{N} - \delta t H, \tag{1.158}$$

where $\mathbf{P} = \sum_{i=1}^{n}\mathbf{p}_i$ is the total momentum of the system, $\mathbf{J}$ is its total angular momentum, H is the total Hamiltonian, and $\mathbf{N}$ which is called the boost, describes the Galilean transformation.

1.9 Action Principle with Variable End Points

Earlier we discussed the standard approach for obtaining the equations of motion from the action principle. However it is possible to perform a more general variation, where in addition to $q_j(t)$ s we vary the time and the end points t_1 and t_2. The transformation of $q_j(t)$ and t are as follows:

$$q_j(t) \to q_j'(t) = q_j(t) + \delta q_j, \quad t \to t' = t + \delta t. \tag{1.159}$$

These variations generate a variation δS of the action

$$\delta S = \int_{t_1+\delta t_1}^{t_2+\delta t_2} L\left(q_j + \delta q_j,\ \dot{q}_j + \delta\dot{q}_j,\ t + \delta t\right)dt - \int_{t_1}^{t_2} L\left(q_j,\ \dot{q}_j, t\right)dt. \tag{1.160}$$

Using the usual technique of the calculus of variation we find δS to be

$$\delta S = \int_{t_1}^{t_2} \sum_{j=1}^{N} \left(\frac{\partial L}{\partial q_j} - \frac{d}{dt}\frac{\partial L}{\partial \dot{q}_j} \right) \delta q_j dt + \left[\sum_{j=1}^{N} p_j \tilde{\delta} q_j - H\delta t \right]_{t_1}^{t_2}, \tag{1.161}$$

where in (1.161) H and $\tilde{\delta} q_j$ are defined by

$$H = \sum_{j=1}^{N} p_j \dot{q}_j - L, \tag{1.162}$$

and

$$\begin{aligned} \tilde{\delta} q_j &= q_j'(t+\delta t) - q_j(t) = q_j'(t) - q_j(t) + \dot{q}_j \delta t \\ &= \delta q_j + \dot{q}_j \delta t. \end{aligned} \tag{1.163}$$

The first term on the right hand-side of (1.161) is zero since L satisfies the Euler–Lagrange equation (1.2), therefore

$$\delta S = \varepsilon (G(t_2) - G(t_1)), \tag{1.164}$$

where

$$\varepsilon G(t) = \sum_{j=1}^{N} p_j \tilde{\delta} q_j - H\delta t. \tag{1.165}$$

This quantity $\varepsilon G(t)$ is the generator of a canonical transformation, i.e. for any dynamical variable $\Lambda(p_k, q_k)$ the Poisson bracket Eq. (1.150) is satisfied by Λ and $\varepsilon G(t)$.
Let us consider the following two cases:

(1) - If

$$q_j \to q_j + \delta q_j, \quad p_j \to p_j, \quad \text{and} \quad t \to t, \tag{1.166}$$

then

$$\varepsilon G(t) = \sum_{j=1}^{N} p_j \tilde{\delta} q_j, \tag{1.167}$$

and

$$\{q_j,\, G(t)\} = \delta q_j, \quad \text{and} \quad \{p_j,\, G(t)\} = 0, \tag{1.168}$$

as is expected.

(2) - If $\tilde{\delta} q_j = \tilde{\delta} p_j = 0$ and $t \to t + \delta t$ then

$$\{q_j,\, \varepsilon G(t)\} = \{q_j,\, -H\delta t\} = -\frac{\partial H}{\partial p_j}\delta t = -\dot{q}_j \delta t, \tag{1.169}$$

and

$$\{p_j,\, \varepsilon G(t)\} = \{p_j,\, -H\delta t\} = \frac{\partial H}{\partial q_j}\, \delta t = -\dot{p}_j \delta t. \tag{1.170}$$

Since $\tilde{\delta} q_j = \tilde{\delta} p_j = 0$,

$$\tilde{\delta} q_j = \delta q_j + \dot{q}_j \delta t = 0, \quad \tilde{\delta} p_j = \delta p_j + \dot{p}_j \delta t = 0, \tag{1.171}$$

we have

$$\delta q_j = \{q_j,\ \varepsilon G(t)\}, \tag{1.172}$$

and

$$\delta p_j = \{p_j,\ \varepsilon G(t)\}. \tag{1.173}$$

These results show that $\varepsilon G(t)$ is indeed the generator of infinitesimal transformation [19],[20].

Quantum Version of the Action Principle for Cases Where $\{q_j,\ \dot{q}_k\}$ Depends on q_k s — It is important to note that this formulation of the Schwinger action principle when applied to quantum mechanical problems yields the correct result in the cases where the commutator

$$[q_j,\ \dot{q}_k], \tag{1.174}$$

is a *c*-number, i.e. $\left(\frac{i\hbar}{m}\delta_{jk}\right)$. As we have seen before, for a general curvilinear coordinate, the Hamiltonian is a complicated quadratic function of p_j, and thus in quantum theory the corresponding commutator, (see Eq. (1.174)), will depend on q_k and is a *q*-number. For these and also the problems related to dissipative forces depending on velocity, the action integral corresponding to the classical Hamilton's principle should be modified.

The classical Hamiltonian function which we defined by the variation of the action integral, Eq. (1.44), is now replaced by the quantum action integral

$$\mathcal{S}\left(q^i, \dot{q}^i, \pi_i, t_1, t_2\right) = \int_{t_1}^{t_2} \left\{ \sum_k \frac{1}{2}\left[\pi_k,\ \dot{q}^k\right]_+ + \mathcal{D}\left(q^i, \pi_i\right) - H(q_i, \pi_i, t) \right\} dt, \tag{1.175}$$

where $[\ ,\ ]_+$ denotes the anticommutator

$$\left[\pi_k,\ \dot{q}^k\right]_+ = \pi_k \dot{q}^k + \dot{q}^k \pi_k, \tag{1.176}$$

and all of the operators are assumed to be Hermitian. In Eq. (1.175) the Hermitian operator $\mathcal{D}$ is introduced such that the *q*-number variation $\delta(\mathcal{D} - H)$ of $(\mathcal{D} - H)$ differs from

$$\frac{1}{2}\left[\frac{\partial H}{\partial q^k},\ \delta q^k\right]_+ + \frac{1}{2}\left[\frac{\partial H}{\partial \pi_k},\ \delta \pi_k\right]_+ + \frac{\partial H}{\partial t}\delta t, \tag{1.177}$$

by a total time derivative of an operator $\mathcal{G}$.

This operator, $\mathcal{D}\left(q^i, \pi_i\right)$, is uniquely determined by the variation δH of the Hamiltonian. Thus if t is not varied, then the variation δ_t of $\mathcal{D}$ is such that the relation

$$\begin{aligned} \delta_t \mathcal{D} - \frac{d\mathcal{G}}{dt} &= \delta_t H - \frac{1}{2}\sum_k \left[\frac{\partial H}{\partial q^k},\ \delta q^k\right]_+ \\ &- \frac{1}{2}\sum_k \left[\frac{\partial H}{\partial \pi_k},\ \delta \pi_k\right]_+, \end{aligned} \tag{1.178}$$

holds. Furthermore just as in classical dynamics we assume that δ and $\frac{d}{dt}$ commute with each other;

$$\frac{d}{dt}\delta q^k = \delta\left(\frac{dq^k}{dt}\right), \quad \frac{d}{dt}(\delta t) = \delta\left(\frac{dt}{dt}\right) = 0. \tag{1.179}$$

Under these conditions the variation of $\mathcal{S}$ in the action integral (1.175) will depend on the end points t_1 and t_2;

$$\delta\mathcal{S} = J(t_2) - J(t_1). \tag{1.180}$$

Now let us consider the δ variation of $\mathcal{S}$ [21],[22].

$$\begin{aligned}
\delta\mathcal{S} &= \int_{t_1}^{t_2} d\left\{\frac{1}{2}\sum_k \left[\pi_k,\ \delta q^k\right]_+ + (\mathcal{D} - H)\delta t\right\} \\
&+ \int_{t_1}^{t_2} \frac{1}{2}\sum_k \left[dq^k - \frac{\partial H}{\partial \pi_k}dt,\ \delta\pi_k\right]_+ - \int_{t_1}^{t_2} \frac{1}{2}\sum_k \left[d\pi_k + \frac{\partial H}{\partial q^k}dt,\ \delta q^k\right]_+ \\
&+ \int_{t_1}^{t_2} \left(dH - \frac{\partial H}{\partial t}dt\right)\delta t.
\end{aligned} \tag{1.181}$$

By assuming that the variations δq^k, $\delta\pi_k$ and δt are all independent, from (1.181) it follows that

$$\frac{d}{dt}q^k = \frac{\partial H}{\partial \pi_k}, \tag{1.182}$$

$$\frac{d}{dt}\pi_k = -\frac{\partial H}{\partial q^k}, \tag{1.183}$$

and

$$\frac{d}{dt}H = -\frac{\partial H}{\partial t}, \tag{1.184}$$

$$J = \frac{1}{2}\sum_k \left[\pi_k,\ \delta q^k\right]_+ + (\mathcal{D} - H)\delta t. \tag{1.185}$$

Thus we have obtained Heisenberg's equations of motion from the action integral without assuming that $\left[q^j,\ \dot{q}^k\right]$ is a c-number. We note that here q^k and π_j are canonically conjugate variables, i.e.

$$\left[q^j,\ \pi_k\right] = i\hbar\delta^j_k, \tag{1.186}$$

while the other commutators of q^j, q^k and π_j, π^k vanish.

Equations (1.182)–(1.184) are Heisenberg's equations of motion, however unlike the c-number variation discussed earlier, (1.184) cannot be derived from (1.182) and (1.183). Moreover the canonical commutation relation that we found by the Schwinger variational principle cannot be obtained directly from these equations.

Examples — Let us consider two forms of Hamiltonian for the motion of a particle given in the curvilinear coordinates (see Eq. (3.216)). The first one is

$$H\left(q^j, \pi_j\right) = \frac{1}{2m}\sum_{i,j}\frac{1}{\sqrt{g}}\left(\pi_i\sqrt{g}\,g^{ij}\pi_j\right) + V\left(q^j\right), \tag{1.187}$$

where g and g^{ij} are functions of q^j s. The second form is that of the Hamiltonian

$$H\left(q^j, \pi_j\right) = \sum_{i,j}\frac{1}{2m}\left(\pi_i\, g^{ij}\pi_j\right) + V\left(q^j\right), \tag{1.188}$$

which has been used to describe the velocity-dependent forces between two nucleons [12]. For these Hamiltonians the velocity operator is defined by

$$\dot{q}^k = \frac{\partial H}{\partial \pi_k} = \frac{1}{2}\left[\pi_j,\, g^{jk}\right]_+. \tag{1.189}$$

Using (1.189) we can determine the commutator of q^j and $\dot{q}^k$ operator which is defined by

$$\left[q^j,\, \dot{q}^k\right] = i\hbar g^{jk}, \tag{1.190}$$

and is not a c-number.

1.10 Symmetry and Degeneracy in Classical Dynamics

When $G(q_i, p_i)$ is a first integral of motion which does not depend explicitly on time then from (1.151) it follows that $\{H, G\} = 0$. Now suppose that we have found a number of independent first integrals of motion $G_1(q_i, p_i), \cdots, G_n(q_i, p_i)$ then each of these G_i s can be used to generate a group of infinitesimal canonical transformation (1.149) and (1.150), i.e.

$$\delta q_k = \varepsilon\{q_k, G_i\}, \quad \delta p_i = \varepsilon\{p_k, G_i\}. \tag{1.191}$$

These continuous transformations will make a group provided that the first integrals, G_k, satisfy the conditions

$$\{G_k, G_m\} = \sum_i C^i_{km} G_i, \tag{1.192}$$

where C^i_{km} are constants which may also depend on the total energy. In what follows we consider three specific examples with different symmetry groups.

Isotropic Two-Dimensional Harmonic Oscillator — As a first example consider the simple case of an isotropic harmonic oscillator given by the Hamiltonian

$$H = \frac{1}{2m}\left(p_1^2 + p_2^2\right) + \frac{1}{2}m\omega^2\left(q_1^2 + q_2^2\right). \tag{1.193}$$

There are three first integrals of motion G_1, G_2 and G_3;

$$G_1 = \frac{1}{2}\left(q_1 p_2 - p_1 q_2\right), \tag{1.194}$$

$$G_2 = \frac{1}{4}\left[\frac{1}{m\omega}\left(p_1^2 - p_2^2\right) + m\omega\left(q_1^2 - q_2^2\right)\right], \tag{1.195}$$

and

$$G_3 = \frac{1}{2}\left(\frac{1}{m\omega} p_1 p_2 + m\omega q_1 q_2\right). \tag{1.196}$$

The first equation G_1 is the angular momentum and the second one is the energy difference between the two oscillators which is known as correlation [23], [24].

The set of functions G_1, G_2 and G_3 is closed under the Poisson bracket operation, i.e,

$$\{G_i, G_j\} = G_k, \tag{1.197}$$

where the indices form a cyclic permutation of 1, 2 and 3. In the case of the isotropic harmonic oscillator (1.193) we have a conserved second rank tensor with the components

$$T_{ij} = \frac{p_i p_j}{2m} + \frac{1}{2} m\omega^2 x_i x. \tag{1.198}$$

By differentiating T_{ij} with respect to t and then substituting for p_i and $\dot{p}_i$ from the equations of motion we find that this derivative is zero and T_{ij} is a constant. Here a second rank tensor rather than a vector is conserved since there are two axes of symmetry, the semi-major as well as the semi-minor axes. On the other hand in the Kepler problem which we will consider next, the conserved quantity is a vector and there is one axis of symmetry.

Two-Dimensional Kepler Problem — For the Kepler problem in two dimensions with the Hamiltonian

$$H = \frac{1}{2m}\left(p_1^2 + p_2^2\right) - \frac{Ze^2}{\left(q_1^2 + q_2^2\right)^{\frac{1}{2}}}, \tag{1.199}$$

we have the following first integrals

$$L_3 = q_1 p_2 - q_2 p_1, \tag{1.200}$$

$$R_1 = -\frac{L_3 p_2}{mZe^2} + \frac{q_1}{\left(q_1^2 + q_2^2\right)^{\frac{1}{2}}}, \tag{1.201}$$

and

$$R_2 = \frac{L_3 p_1}{mZe^2} + \frac{q_2}{\left(q_1^2 + q_2^2\right)^{\frac{1}{2}}}. \tag{1.202}$$

The vector $\mathbf{R}$ with components R_1 and R_2 is the two-dimensional form of the Runge–Lenz vector (see the next section) [2]. By calculating the Poisson brackets for L_3, R_1 and R_2 we find

$$\{L_3, R_1\} = R_2, \tag{1.203}$$

$$\{R_2, L_3\} = R_1, \tag{1.204}$$

and

$$\{R_1, R_2\} = \frac{1}{mZ^2e^4}(-2H)L_3. \tag{1.205}$$

We can transform these to the standard form of (1.192) by choosing

$$G_1 = \frac{\sqrt{m}Ze^2}{\sqrt{2|E|}}R_1, \tag{1.206}$$

$$G_2 = \frac{\sqrt{m}Ze^2}{\sqrt{2|E|}}R_2, \tag{1.207}$$

and

$$G_3 = L_3, \tag{1.208}$$

noting that E which is the total energy of the particle is negative.

The Runge–Lenz vector $\mathbf{R}$, with the components given by (1.201) and (1.202) can be written as a vector equation

$$\mathbf{R} = \frac{1}{m}(\mathbf{p} \wedge \mathbf{L}) - \frac{Ze^2\boldsymbol{\rho}}{\rho}, \tag{1.209}$$

where $\boldsymbol{\rho}$ is the position vector with components q_1 and q_2. This vector is directed from the focus of the orbit to its perihelion (or the point of closest approach). By multiplying (1.209) by $\boldsymbol{\rho}$ we find

$$\boldsymbol{\rho} \cdot \mathbf{R} = \rho R \cos\phi = \frac{1}{m}\boldsymbol{\rho} \cdot (\mathbf{p} \wedge \mathbf{L}) - Ze^2\rho. \tag{1.210}$$

Using the vector identity

$$(\mathbf{p} \wedge \mathbf{L}) \cdot \boldsymbol{\rho} = \mathbf{L} \cdot (\boldsymbol{\rho} \wedge \mathbf{p}) = \mathbf{L}^2, \tag{1.211}$$

from Eq. (1.210) we obtain the equation of orbit

$$\rho(\phi) = \frac{\mathbf{L}^2}{Ze^2 + R\cos\phi} = \frac{d}{1 + \varepsilon\ \cos\phi}, \tag{1.212}$$

where $d = \frac{\mathbf{L}^2}{Ze^2}$ is the semi-latus rectum of the ellipse and $\varepsilon = \frac{R}{mZe^2}$ is its eccentricity. Note that R vanishes for a circle ($\varepsilon = 0$) since there is no unique major axis.

Two-Dimensional Anisotropic Harmonic Oscillator — The problem of a special two-dimensional anisotropic oscillator provides an example where the symmetries and the group structure of the classical motion is not carried over to the quantized motion.

Consider the Hamiltonian of this system for the general case which is given by

$$H = \left(\frac{p_1^2}{2m_1} + \frac{1}{2}m_1\omega_1^2 q_1^2\right) + \left(\frac{p_2^2}{2m_2} + \frac{1}{2}m_2\omega_2^2 q_2^2\right). \tag{1.213}$$

If the ratio of the two frequencies (ω_1/ω_2) is rational then the classical orbit closes on itself. Let us assume that there are integers m and n such that

$$n\omega_1 = m\omega_2 = \tau, \tag{1.214}$$

For this case we introduce two constants $\alpha_1 = m_1\omega_1$ and $\alpha_2 = m_2\omega_2$ and two variables b_i and b_i^*

$$b_i = \frac{p_i - i\alpha_i q_i}{\sqrt{2\alpha_i}}, \quad b_i^* = \frac{p_i + i\alpha_i q_i}{\sqrt{2\alpha_i}}, \quad i = 1, 2, \tag{1.215}$$

and write the two first integrals as

$$F_1 = \frac{1}{2}\left[b_1^n (b_2^m)^* + (b_1^n)^* b_2^m\right], \tag{1.216}$$

and

$$F_2 = \frac{-i}{2}\left[b_1^n (b_2^m)^* - (b_1^n)^* b_2^m\right]. \tag{1.217}$$

The three G_i s which form the group (1.192) are given by [25]

$$G_1 = \frac{\sqrt{\omega_1\omega_2}}{\tau} F_1 \left(b_1 b_1^*\right)^{-\frac{1}{2}(n-1)} \left(b_2 b_2^*\right)^{-\frac{1}{2}(m-1)}, \tag{1.218}$$

$$G_2 = \frac{\sqrt{\omega_1\omega_2}}{\tau} F_2 \left(b_1 b_1^*\right)^{-\frac{1}{2}(n-1)} \left(b_2 b_2^*\right)^{-\frac{1}{2}(m-1)}, \tag{1.219}$$

and

$$G_3 = \frac{1}{2\tau}\left(b_2 b_2^* - b_1 b_1^*\right). \tag{1.220}$$

We note that here G_1 and G_2 are not always algebraic and for some m and n values they are multivalued functions. When any one of the G_i s is not algebraic, then the existence of its quantum counterpart is in doubt [25].

1.11 Closed Orbits and Accidental Degeneracy

For the three-dimensional Kepler problem in addition to the angular momentum vector, there is the Runge–Lenz vector, **R**, which is conserved [24]. This vector, **R**, is in the direction of the semi-major axis of the elliptic orbit, and since it is a constant vector the orbit is closed and there is no precession of the axis. The vector **R**, the angular momentum vector **L** and the associated elliptic orbit of the motion are shown in Fig. 1.2. More about this vector will be mentioned when we discuss Pauli's method of obtaining the spectra of the hydrogen atom by the matrix method in Sec. 9.8. First, let us consider the types of attractive central potentials that lead to closed orbits. According to Bertrand's theorem the

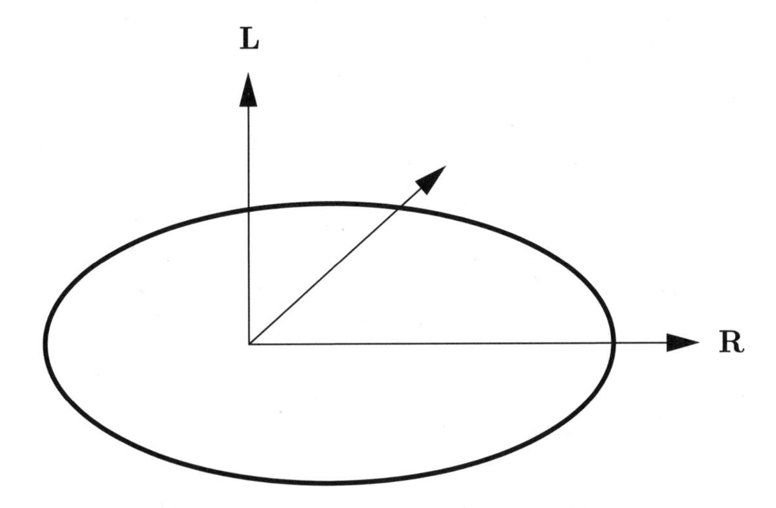

Figure 1.2: The Runge–Lenz vector, **R** and the angular momentum vector **L** for an elliptic orbit of the classical problem with $\frac{-k}{r}$ potential (Kepler problem).

only central forces giving closed orbits for all bound particles are the Coulomb potential (or Kepler problem) and the three-dimensional harmonic oscillator [26]–[28].

For noncentral forces we can have accidental degeneracy for a wide class of potentials [29],[30]. For instance consider the two-dimensional motion of a particle in the xy-plane when the Hamiltonian is given by

$$H = \frac{1}{2m}\left(p_x^2 + p_y^2\right) + \frac{1}{2}m\omega^2 x^2\left[\frac{\theta(x)}{(1+\lambda_1)^2} + \frac{\theta(-x)}{(1-\lambda_1)^2}\right] + \frac{1}{2}m\omega^2 y^2\left[\frac{\theta(y)}{(1+\lambda_2)^2} + \frac{\theta(-y)}{(1-\lambda_2)^2}\right], \quad (1.221)$$

where $\theta(x)$ is the step function

$$\theta(x) = \begin{cases} 0 & \text{for } x < 0 \\ 1 & \text{for } x > 0 \end{cases}, \quad (1.222)$$

and λ_1 and λ_2 satisfy the inequalities

$$-1 < \lambda_1 < 1, \quad -1 < \lambda_2 < 1. \quad (1.223)$$

We can write (1.221) in terms of the action-angle variables defined by [2]

$$I_x = \frac{1}{2\pi}\oint \left[2m(\alpha - V_1(x))\right]^{\frac{1}{2}} dx, \quad (1.224)$$

and

$$I_y = \frac{1}{2\pi}\oint \left[2m(E - \alpha - V_2(y))\right]^{\frac{1}{2}} dy, \quad (1.225)$$

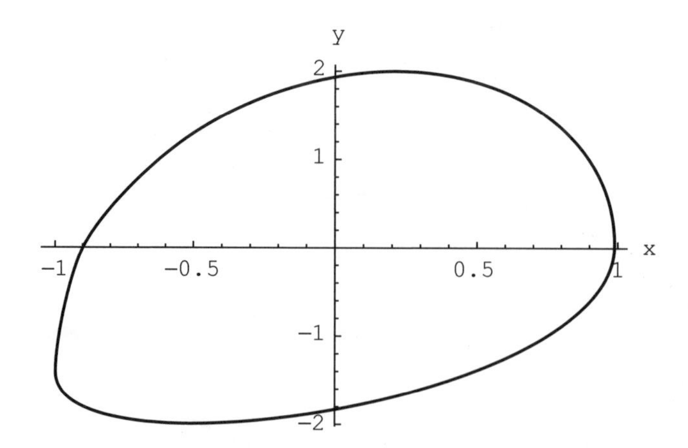

Figure 1.3: The closed orbit found from the Hamiltonian (1.221) for $m = 1$, $\omega = 1$, $\lambda_1 = 0.4$ and $\lambda_2 = 0.2$. Apart from the energy there are no apparent symmetries or conserved quantities for this motion.

where E is the total energy and α is the separation constant. The potentials $V_1(x)$ and $V_2(x)$ are given by the third and fourth terms in (1.221) respectively. By carrying out the integrations in (1.224) and (1.225) and replacing E by H we find a simple expression for the Hamiltonian

$$H = \omega(I_x + I_y). \tag{1.226}$$

This clearly shows the degeneracy of motion since the two periods given by

$$\nu_x = \frac{1}{2\pi}\frac{\partial H}{\partial I_x} = \frac{\omega}{2\pi}, \quad \nu_y = \frac{1}{2\pi}\frac{\partial H}{\partial J_x} = \frac{\omega}{2\pi}, \tag{1.227}$$

are equal. A closed orbit for the noncentral force derived from the Hamiltonian (1.221) is displayed in Fig. 1.3.
The example that we have just considered is for a potential which is separable in Cartesian coordinates, but one can construct similar cases in other coordinates [29].

As we will discuss later these classically degenerate systems may or may not remain degenerate in quantum mechanics. In other words closed orbits in classical motion do not necessarily imply accidental degeneracies in their quantum spectra.

1.12 Time-Dependent Exact Invariants

For some time-dependent potentials, Noether's theorem may not be an easy way to find conserved quantities. This is the case for example, when we want to determine a first integral of motion which explicitly depends on time. However we find the conserved quantity by the application of a method devised by Lewis and Leach [31].

Let us consider the one-dimensional Hamiltonian for a particle of unit mass

$$H = \frac{1}{2}p^2 + V(q,t), \tag{1.228}$$

and denote the invariant first integral by $I(p,q,t)$. Since $I(p,q,t)$ is invariant, we have

$$\frac{dI}{dt} = \frac{\partial I}{\partial t} + \{I, H\} = \frac{\partial I}{\partial t} + \left(\frac{\partial I}{\partial q}\frac{\partial H}{\partial p} - \frac{\partial I}{\partial p}\frac{\partial H}{\partial q}\right) = 0. \tag{1.229}$$

For the general solution of (1.229) when H is given by (1.228) we assume that $I(p,q,t)$ can be expanded as a power series in p. However this procedure, unless the power series terminates, does not give us a simple expression for $I(p,q,t)$. Therefore let us limit our investigation to the possibility of obtaining a function $I(p,q,t)$ which is at most quadratic in p;

$$I(p,q,t) = p^2 f_2(q,t) + p f_1(q,t) + f_0(q,t). \tag{1.230}$$

By substituting (1.228) and (1.230) in (1.229) and equating the coefficients of p^3, p^2, p and p^0 equal to zero we find

$$\frac{\partial f_2(q,t)}{\partial q} = 0, \tag{1.231}$$

$$\frac{\partial f_2(q,t)}{\partial t} + \frac{\partial f_1(q,t)}{\partial q} = 0, \tag{1.232}$$

$$\frac{\partial f_0(q,t)}{\partial q} + \frac{\partial f_1(q,t)}{\partial t} - 2f_2(q,t)\frac{\partial V(q,t)}{\partial q} = 0, \tag{1.233}$$

and

$$\frac{\partial f_0(q,t)}{\partial t} - f_1(q,t)\frac{\partial V(q,t)}{\partial q} = 0. \tag{1.234}$$

Integrations of (1.231) and (1.232) are trivial:

$$f_2(q,t) = 2a(t), \quad f_1(q,t) = b(t) - 2\dot{a}(t)q, \tag{1.235}$$

where $a(t)$ and $b(t)$ are arbitrary functions of time and where the dot denote differentiation with respect to time. If we substitute for $f_1(q,t)$ and $f_2(q,t)$ in (1.233) we get

$$\frac{\partial f_0(q,t)}{\partial q} - 4a(t)\frac{\partial V(q,t)}{\partial q} + \dot{b}(t) - 2\ddot{a}(t)q = 0. \tag{1.236}$$

This equation can easily be integrated with the result that

$$V(q,t) = \frac{1}{4a(t)}\left[f_0(q,t) - \ddot{a}(t)q^2 + \dot{b}(t)q\right] + g(t), \tag{1.237}$$

where again $g(t)$ is an arbitrary function of t. If we substitute (1.235) and (1.237) in (1.233) we find

$$(2\dot{a}(t)q - b(t))\left(\frac{\partial f_0(q,t)}{\partial q} - 2\ddot{a}(t)q + \dot{b}(t)\right) + 4a(t)\frac{\partial f_0(q,t)}{\partial t} = 0. \quad (1.238)$$

This is a partial differential equation of first order for $f_0(q,t)$.

Using the method of characteristics we solve (1.238) to get the general solution of this partial differential equation [32]

$$f_0(q,t) = \mathcal{G}\left[\frac{q}{\sqrt{a(t)}} + \frac{1}{4}\int^t \frac{b(t')dt'}{a^{\frac{3}{2}}(t')}\right] + \frac{1}{2a(t)}\left(\dot{a}(t)q - \frac{1}{2}b(t)\right)^2. \quad (1.239)$$

In thisexpression $\mathcal{G}$ is an arbitrary function of its argument. It is simpler to rewrite (1.239) in terms of $\rho_1(t)$, $\alpha(t)$ and $\tilde{\mathcal{G}}(z)$ where these functions of time are defined by:

$$\rho_1(t) = 2\sqrt{a(t)}, \quad (1.240)$$

$$\alpha(t) = -\frac{\rho_1(t)}{8}\int^t \frac{b(t')dt'}{a^{\frac{3}{2}}(t')}, \quad (1.241)$$

and

$$\tilde{\mathcal{G}}(z) = \mathcal{G}(2z). \quad (1.242)$$

Equation (1.237) shows that the general form of the potential consists of two parts: the fixed part which is a quadratic function of q and an arbitrary part

$$V(q,t) = -F(t)q + \frac{1}{2}\Omega^2(t)q^2 + \frac{1}{\rho_1^2(t)}\mathcal{U}\left(\frac{q-\alpha(t)}{\rho_1(t)}\right), \quad (1.243)$$

where $\mathcal{U}$ is an arbitrary function of its argument.

Now comparing (1.243) with (1.237) we find that the arbitrary functions $F(t), \Omega^2(t)$, $\rho_1(t)$ and $\alpha(t)$ are all related by the following relations:

$$\ddot{\rho}_1(t) + \Omega^2(t)\rho_1(t) - \frac{K}{\rho_1^3(t)} = 0, \quad (1.244)$$

$$\ddot{\alpha}(t) + \Omega^2(t)\alpha(t) = F(t). \quad (1.245)$$

with K a constant. In terms of the functions $\alpha(t)$ and $\rho_1(t)$ we can write the invariant function $I(p,q,t)$ as

$$\begin{aligned} I(p,q,t) &= \frac{1}{2}\left[\rho_1(t)\left(p - \dot{\alpha}(t)\right) - \dot{\rho}_1(t)(q-\alpha(t))\right]^2 \\ &+ \frac{1}{2}K\left(\frac{q-\alpha(t)}{\rho_1(t)}\right)^2 + \mathcal{U}\left(\frac{q-\alpha(t)}{\rho_1(t)}\right). \end{aligned} \quad (1.246)$$

For instance if the Hamiltonian is given by

$$H = \frac{1}{2}p^2 + \frac{1}{2}\Omega^2(t)q^2, \quad (1.247)$$

and if we also choose $\mathcal{G}$ to be

$$\tilde{\mathcal{G}} = \frac{1}{2}x^2, \tag{1.248}$$

then $\rho_1(t)$ satisfies the differential equation

$$\ddot{\rho_1}(t) + \Omega^2(t)\rho_1(t) - \frac{1}{\rho_1^3} = 0. \tag{1.249}$$

A scale transformation given by

$$\rho_1(t) = \frac{1}{\sqrt{\Omega_0}}\rho(t). \tag{1.250}$$

where Ω_0 is a constant, changes (1.249) to the form

$$\ddot{\rho}(t) + \Omega^2(t)\rho(t) - \frac{\Omega_0^2}{\rho^3} = 0. \tag{1.251}$$

This is the form that later we will use in the corresponding quantum mechanical problem.

Bibliography

[1] L.D. Landau and E.M. Lifshitz, *Mechanics*, (Pergamon Press, 1960), p. 136.

[2] H. Goldstein, C. Poole and J. Safco, *Classical Mechanics*, Third Edition (Addison-Wesley, San Francisco, 2002).

[3] H. Helmholtz, Über die physikaliche Bedeutung Prinzips der kleinsten Wirkung, J. Reine und Angew. Math. 100, 137 (1887).

[4] R.M. Santilli, *Foundations of Theoretical Mechanics*, (Springer, New York, 1987).

[5] P. Havas, The range of application of the Lagrange formalism, Nuovo Cimento. 5, 363 (1957).

[6] P. Havas, The connection between conservation laws and the invariance groups: Folklore, fiction and fact, Acta Physica Austriaca, 38, 145 (1973).

[7] R.J. Hughes, On Feynman's proof of the Maxwell equations, Am. J. Phys. 60, 301 (1992).

[8] Y. Takahashi, Lecture notes, University of Alberta, (unpublished) (1991).

[9] E.H. Kerner, How certain is the uncertainty principle?, Contemp. Phys. 36, 329 (1995).

[10] E.H. Kerner, An essential ambiguity of quantum theory, Found. Phys. Lett. 7, 241 (1994).

[11] F.J. Kennedy, Jr. and E.H. Kerner, Note on the inequivalence of classical and quantum Hamiltonians, Am. J. Phys. 33, 463 (1965).

[12] D.G. Currie and E.J. Saletan, q-equivalent particle Hamiltonians, I: The classical one-dimensional case, J. Math. Phys. 7, 967 (1966).

[13] G. Gelman and E.J. Saletan, q-equivalent particle Hamiltonians, II: The two-dimensional classical oscillator, Nuovo Cimento. 18B, 53 (1973).

[14] M. Razavy, Some counterexamples in quantum mechanics, Iranian J. Sci. Tech. 23, 329 (2001).

[15] H.R. Lewis, Jr., Classical and quantum systems with time-dependent harmonic-oscillator-type Hamiltonians, Phys. Rev. Lett. 18, 510 (1967).

[16] For a simple account of the action-angle variables see I. Percival and D. Richards, *Introduction to Dynamics*, (Cambridge University Press, Cambridge, 1982), Chapter 7.

[17] I.C. Percival, Variational principles for the invariant toroids of classical dynamics, J. Phys. A. 7, 794 (1974).

[18] I.C. Percival and N. Pomphrey, Vibrational quantization of polyatomic molecules, Mol. Phys. 31, 97 (1976).

[19] J. Schwinger, The theory of quantized fields. I, Phys. Rev. 82, 914 (1951).

[20] See P. Roman, *Advanced Quantum Mechanics*, (Addison-Wesley, Reading 1965), Chapter 1.

[21] M.Z. Shaharir, The modified Hamilton-Schwinger action principle, J. Phys. A7, 553 (1974).

[22] H.A. Cohen and M.Z. Shaharir, The action principle in quantum mechanics, Int. J. Theo. Phys. 11, 289 (1974).

[23] V.A. Dulock and H.V. McIntosh, On the degeneracy of the two-Dimensional harmonic oscillator, Am. J. Phys. 33, 109 (1965).

[24] H.V. McIntosh, Symmetry and Degeneracy in *Group Theory and its applications*, edited by E.M. Loebl, (Academic Press, New York, 1971), Vol. II, p. 28.

[25] J.M. Jauch and E.L. Hill, On the problem of degeneracy in quantum mechanics, Phys. Rev. 57, 641 (1940).

[26] L.S. Brown, Forces giving no orbit precession, Am. J. Phys. 46, 930 (1978).

[27] D.M. Fradkin, Existence of the dynamic symmetries O_4 and SU_3 for all classical central potential problems, Prog. Theor. Phys. 37, 798 (1967).

[28] D.F. Greenberg, Accidental degeneracy, Am. J. Phys. 34, 1101 (1966).

[29] M. Razavy, Quantization of periodic motions with nonprecessing orbits and the accidental degeneracy of the spectral lines, Hadronic J. 7, 493 (1984).

[30] J.F. Marko and M. Razavy, Dynamical symmetries of a class of action equivalent Hamiltonians, Lett. Nuovo Cimento 40, 533 (1984).

[31] H.R. Lewis and P.G.L. Leach, A direct approach to finding exact invariants for one-dimensional time-dependent classical Hamiltonian, J. Math. Phys. 23, 2371 (1982).

[32] See for instance, C.R. Chester, *Techniques in Partial Differential Equations*, (McGraw-Hill, New York, 1971), Chapter 8.

Chapter 2

Discovery of Matrix Mechanics

Historians of science unanimously agree that the new quantum mechanics was born on July 29, 1925 with the publication of a paper by Heisenberg under the title of "Über quantentheoretische Umdeutung kinematischer und mechanischer Beziehungen" [1]–[10]. The novel ideas contained in this paper paved the way for a complete departure from the classical description of atomic physics, and advanced a new formulation of the laws of micro-physics. First and foremost Heisenberg thought of replacing the classical dynamics of the Bohr atom by a formulation based exclusively on relations between quantities that are actually observables. Thus he abandoned the idea of determining the coordinates and momenta of the electron as functions of time, but at the same time he retained the mathematical form of the second law of motion. Now if Bohr's orbits are not observables, the spectral lines and their intensities are. In the classical Larmor formula for radiation from an accelerating electron the energy radiated by the electron, in cgs units is [11]

$$P = \left(\frac{2e^2}{3c^3}\right) \ddot{x}^2, \tag{2.1}$$

where e is the charge of the electron, c the speed of light and $\ddot{x}$ is the acceleration. For a harmonically bound electron, the average power for the α-th harmonic with amplitude $x_\alpha(n)$ is

$$P = \frac{4e^2}{3c^3} \left[\alpha\, \omega(n)\right]^4 |x_\alpha(n)|^2. \tag{2.2}$$

For such a bound electron, the position $x(n,t)$, where the stationary state is labeled by n, can be written as a Fourier series

$$x(n,t) = \sum_{\alpha=-\infty}^{\infty} a_\alpha e^{i\alpha\omega(n)t}. \tag{2.3}$$

Heisenberg observed that according to the correspondence principle, the α-th component of the classical motion corresponds to the quantum-mechanical transition from the state n to the state $n-\alpha$ [7],[8]. Guided by this principle, Heisenberg replaced the classical component

$$a_\alpha e^{i\omega(n)t}, \tag{2.4}$$

by

$$a(n, n-\alpha) e^{i\omega(n,\, n-\alpha)t}. \tag{2.5}$$

In order to account for the transition from one stationary state n to another $(n-\alpha)$, he replaced (2.3) by

$$x \to a(n, n-\alpha) e^{i\omega(n,\, n-\alpha)t}, \quad \text{or} \quad x_{m,n} \to a(n,m) e^{i\omega(n,\, m)t}, \tag{2.6}$$

and rather than summing over transition components as was done in (2.3), Heisenberg represented the position by a set of transition components, x_{mn}. Then he argued that the quantum mechanical analogue of (2.2) can be found by replacing $x_\alpha(n)$ by x_{mn} and $\alpha\omega(n)$ by $\omega(n,m)$. In revising classical dynamics, the next problem that he encountered was the question of multiplication of transition amplitudes x_{mn}. Such an operation was needed in the calculation of the potential energy of the oscillator, $\frac{1}{2}\omega^2 x^2$, or in the solution of the problem of anharmonic oscillator. Here Heisenberg assumed that the frequencies combine according to the Ritz combination principle.

There were two additional assumptions advanced by Heisenberg regarding the dynamics of the motion. First he proposed that the equations of motion should have the same mathematical form as the classical Newton's second law, with x_{mn} substituting for x in the equation $m\ddot{x} = f(x)$.
The second was the modification of the old Bohr–Sommerfeld quantization rule

$$\oint p dq = \oint m\dot{x}^2 dt = 2\pi\hbar n. \tag{2.7}$$

This equation expressed in terms of the Fourier series for $x(n,t)$, Eq. (2.3) takes the form

$$n\hbar = m \sum_\alpha |a_\alpha(n)|^2 \alpha^2 \omega(n). \tag{2.8}$$

But how this rule can be expressed when there is a transition from the state n to $n-\alpha$. Here Heisenberg made the bold assumption that what matters is the difference between $\oint pdq$ evaluated for the state n and the neighboring state $n-1$, viz,

$$\left[\oint pdq\right]_n - \left[\oint pdq\right]_{n-1}. \tag{2.9}$$

For Heisenberg the presence of the integer n in (2.8) seemed to be an arbitrary condition, and he concluded that and Eq. (2.7) which fixes the Bohr orbit must be replaced by a new condition and that the new condition must be about the transition between states.

By differentiating (2.8) with respect to n Heisenberg found

$$\hbar = m \sum_{\alpha} \alpha \frac{d}{dn} \left[|a_{\alpha}(n)|^2 \alpha\omega(n) \right] . \tag{2.10}$$

This differential relation was changed to a difference relation

$$\hbar = 2m \sum_{\alpha=0}^{\infty} \left\{ |a(n+\alpha)|^2 \omega(n+\alpha, n) - |a(n-\alpha)|^2 \omega(n-\alpha, n) \right\} . \tag{2.11}$$

This relation supplements the Planck–Bohr frequency condition, i.e. it relates the amplitudes of different lines within an atomic spectrum.

Returning to the problem of radiation form an electron we note that the quantum mechanical analogue of (2.2), according to Heisenberg, can be found by replacing $x_{\alpha}(n)$ by x_{nm} and $\alpha\omega(n)$ by $\omega(n,m)$. Heisenberg also assumed that the quantum-mechanical power can be written as the product of the transition probability per unit time $A(n,m)$ where

$$A(n,m) = \frac{4e^2}{3\hbar c^3} \left[\omega(n,m) \right]^3 |x_{nm}|^2, \tag{2.12}$$

times the emitted energy given by Planck–Bohr relation $E_n - E_m = \hbar\omega(n,m)$,

$$P(n,m) = A(n,m)\hbar\omega(n,m) = \frac{4e^2}{3c^3} \left[\omega(n,m) \right]^4 |x_{nm}|^2. \tag{2.13}$$

This equation shows how x_{nm} can be related to the observables $P(n,m)$ and $\omega(n,m)$ [1]–[7].

We note that in this formulation for the harmonic motion the set of numbers x_{nm} replace the Fourier coefficients of the classical position $x_{\alpha}(n)$. But Heisenberg asked how this can be generalized to the case of the anharmonic oscillator problem, e.g. [1]

$$\ddot{x} + \omega_0^2 x + \lambda x^2 = 0, \tag{2.14}$$

i.e. how to express $\left(x^2\right)$ in terms of the observables x_{nm} and $\omega(n,m)$. Here he assumed that the frequencies combine according to the Ritz combination principle

$$\omega(n,j) + \omega(j,n) = \omega(n,m). \tag{2.15}$$

Thus if one writes

$$x_{nm}(t) = x_{nm} e^{i\omega_{nm} t}. \tag{2.16}$$

then (2.15) can be satisfied provided that

$$\left(x^2\right)_{nm}(t) = \sum_j x_{nj}(t) x_{jm}(t). \tag{2.17}$$

By reading the manuscript of Heisenberg's paper Born recognized that (2.17) is the law of multiplication of matrices. Once this discovery was made Born and his assistant P. Jordan started to investigate how the basic laws of matrix mechanics should be formulated.

Born considered a dynamical system with one degree of freedom and then rather than writing the second law of motion as in (2.14) he assumed a Hamiltonian formulation $\hat{H}(\hat{p}, \hat{q})$ when both $\hat{p}$ and $\hat{q}$ are time-dependent matrices (matrices are shown by $\hat{}$)

$$\hat{p} = \left\{p_{nm} e^{i\omega(n,m)t}\right\}, \tag{2.18}$$

$$\hat{q} = \left\{q_{nm} e^{i\omega(n,m)t}\right\}, \tag{2.19}$$

where $\omega(n, m)$ denotes the frequency of transition between two stationary states with energies E_n and E_m, so that

$$\hbar\omega(n, m) = E_n - E_m. \tag{2.20}$$

From this relation it follows that $\omega(n, n) = 0$, and thus the diagonal elements of $\hat{p}$ and $\hat{q}$ defined by (2.18) and (2.19) are independent of time. We know that in the Fourier expansion of any dynamical function of time for a periodic motion the constant term is equal to the time average value of that function. Now from the Bohr correspondence principle (see Sec. 12.8) we may conclude that for any dynamical variable f, the diagonal element f_{nn} of the matrix representing f can be interpreted as the average value of f when the system is in the state n. The off-diagonal elements, f_{mn}, on the other hand, account for the transition between the states m and n. The converse of the result that the diagonal elements are constant in time is that if the total time derivative of a matrix is zero, then that matrix must be diagonal. Note that if $\frac{d\hat{H}}{dt} = 0$, it does not follow that $\hat{H}$ is diagonal unless the spectrum of $\hat{H}$ is non-degenerate, i.e. $\omega(n, m) \neq 0$ for $m \neq n$. For a conservative system the Hamiltonian is a constant of motion $\frac{d\hat{H}}{dt} = 0$, therefore we can represent H as a diagonal matrix, with elements $H(n, n)$. This $H(n, n)$ is the energy of the system when it is in the state n, i.e. E_n. Thus Eq. (2.20) can be written as

$$\omega(n, m) = \frac{1}{\hbar}[H(n, n) - H(m, m)]. \tag{2.21}$$

Following Heisenberg, Born also assumed that the Hamilton canonical equations of motion Sec. 1.3 preserve their form, and that $\hat{H}$, $\hat{p}$ and $\hat{q}$ satisfy the equations

$$\frac{d\hat{q}}{dt} = \frac{\partial \hat{H}}{\partial \hat{p}}, \tag{2.22}$$

and

$$\frac{d\hat{p}}{dt} = -\frac{\partial \hat{H}}{\partial \hat{q}}, \tag{2.23}$$

corresponding to the classical equations (1.45) and (1.46).

Substituting for $\hat{q}$ from (2.19) we have

$$\left(\frac{\partial \hat{q}}{\partial t}\right)_{nm} = \frac{i}{\hbar}\left[(E_n - E_m)q(n,m)e^{i\omega(n,m)t}\right] = \frac{i}{\hbar}\left(\hat{H}\hat{q} - \hat{q}\hat{H}\right)_{nm}, \tag{2.24}$$

or

$$\hat{H}\hat{q} - \hat{q}\hat{H} = -i\hbar\frac{\partial \hat{H}}{\partial \hat{p}}. \tag{2.25}$$

Thus the commutator

$$\left(\hat{f}\hat{q} - \hat{q}\hat{f}\right) = \left[\hat{f},\, \hat{q}\right] = -i\hbar\frac{\partial \hat{f}}{\partial \hat{p}}, \tag{2.26}$$

is valid for $\hat{f} = \hat{H}$ and also for $\hat{f} = \hat{q}$. But if (2.25) is valid for two matrices $\hat{f}_1$ and $\hat{f}_2$, then it is also valid for $\hat{f} = \hat{f}_1 + \hat{f}_2$ and $\hat{g} = \hat{f}_1\hat{f}_2$. Now if we solve $\hat{H}$ for $\hat{p}$, $\hat{p} = \hat{p}\left(\hat{q}, \hat{H}\right)$, then choose $\hat{p}$ to be $\hat{f}$ in (2.26) we find

$$(\hat{p}\hat{q} - \hat{q}\hat{p}) = -i\hbar, \tag{2.27}$$

which is the fundamental commutation relation in quantum mechanics [12]-[13].

The next important contribution to the matrix mechanics came with the complete solution of the hydrogen atom problem with the help of the operator version of the Runge–Lenz vector. Shortly after the discovery of the fundamental commutation relation, Pauli succeeded in obtaining the exact solution to the operator equation for the motion of a charged particle in a Coulomb field (Sec. 9.9) and found the Balmer formula for the discrete spectrum of this system [14].

According to van der Waarden "Pauli's paper convinced physicists that Quantum Mechanics is correct" [15]. Soon after the publication of Pauli's paper Schrödinger in a series of papers introduced his famous wave equation and applied the wave mechanics to a number of problems including the harmonic oscillator and the hydrogen atom. While these early attempts, whether in its matrix formulation or as a wave equation were powerful methods for producing answers, the underlying physical principle was still missing.

The final clue to this puzzle came in a paper of Born entitled " Zur Quantenmechanik der Stoβvorgänge" [16]. In this work where Born formulated the quantum theory of scattering, in a footnote he made the suggestion that the square of the absolute value of the wave function, $|\psi(\mathbf{r})|^2$ measures in some way the probability of finding the particle at the point $\mathbf{r}$.

As later Born observed "We free forces of their classical duty of determining directly the motion of the particles and allow them instead to determine the probability of states" [5].

Very few books have been written on matrix mechanics [17]–[21] and among these the charming little book by Green [18] is exceptional in its broad coverage of the subject.

2.1 Equivalence of Wave and Matrix Mechanics

Soon after the discovery of the wave equation, Schrödinger and independently, Eckart proved that the two formulations are equivalent [22]–[24]. Later the Dirac transformation theory provided a covering theory from which one or the other formulation can be derived. The exact meaning of this equivalence has been examined carefully, for instance by Hanson [26]. In this section we will outline the original Schrödinger's proof in some detail.

Consider a "well-ordered" operator $\mathcal{F}(q_k, p_k)$, Secs. 3.5–3.8, which depends on q_k and p_k of the general form

$$\mathcal{F}(q_k, p_k) = f(q_k)p_r p_s p_t g(q_k)p(r')h(q_k)p_{r''}p_{s''}\cdots. \tag{2.28}$$

To this $\mathcal{F}$ we assign the following operator

$$\mathcal{F} = f(q_k)(-i\hbar)^3\frac{\partial^3}{\partial q_r\partial q_s\partial q_t}g(q_k)(-i\hbar)\frac{\partial}{\partial q_{r'}}h(q_k)(-i\hbar)^2\frac{\partial^2}{\partial q_{r''}\partial q_{s''}}\cdots, \tag{2.29}$$

with the differential operators acting on all the factors to the right. In particular the action of $\mathcal{F}$ on a function $u(q_k)$ yields another function of q_k s, which we denote by $(\mathcal{F}, u)$. If $\mathcal{G}(q_k, p_k)$ is another well-ordered operator, in general, $(\mathcal{F}\mathcal{G}, u)$ will not be equal to $(\mathcal{G}\mathcal{F}, u)$. Now let us choose a complete set of orthonormal functions in the whole q space, $q = (q_1, q_2\cdots, q_k, \cdots)$:

$$u_\alpha(q)\sqrt{\rho(q)}, \quad u_\beta(q)\sqrt{\rho(q)}, \cdots, \quad u_\mu(q)\sqrt{\rho(q)}, \cdots \tag{2.30}$$

where the integration is over all q space and where $\rho(q)$ is a density function such that [25]

$$\int u_\mu(q)u_\nu(q)\rho(q)dq = \begin{cases} 0 & \mu \neq \nu \\ 1 & \mu = \nu \end{cases}. \tag{2.31}$$

Using this set of functions we define a matrix $\mathcal{F}_{\mu\nu}$, from the operator $\mathcal{F}$ and the complete set of orthonormal functions $u_\mu(q)$, with the elements

$$\mathcal{F}_{\mu\nu} = \int \rho(q)u_\mu(q)(\mathcal{F}, u_\nu(q))dq. \tag{2.32}$$

We also define another operator $\bar{F}$ which Schrödinger calls "gewẅallze" or "rolled over operator" by

$$\bar{F} = (-1)^\tau\cdots(-i\hbar)^2\frac{\partial^2}{\partial q_{r''}\partial q_{s''}}h(q_k)(-i\hbar)\frac{\partial}{\partial q_{r'}}g(q_k)(-i\hbar)^3\frac{\partial^3}{\partial q_r\partial q_s\partial q_t}g(q_k)f(q_k), \tag{2.33}$$

where τ is the number of derivatives in $\mathcal{F}$. Assuming that $u_\alpha(q)\cdots u_\mu(q)\cdots$ and their derivatives vanish at the boundaries, by means of partial integration we can show that $\mathcal{F}$ can also be written as

$$\mathcal{F}_{\mu\nu} = \int u_\mu(q)(\bar{F}, \rho(q)u_\nu(q))dq. \tag{2.34}$$

From this equation it follows that

$$\sum_{\beta} \mathcal{F}_{\mu\beta}\mathcal{G}_{\beta\nu} = \sum_{\beta} \int u_{\beta}(q)(\bar{F}, \rho(q)u_{\mu}(q))dq \times \int \rho(q_1)u_{\beta}(q_1)(\bar{G}, \rho(q_1)u_{\nu}(q_1))dq_1. \tag{2.35}$$

By carrying out the summation over β, and using the completeness of $u_{\alpha}(q)$ s (2.35) transforms into

$$\sum_{\beta} \mathcal{F}_{\mu\beta}\mathcal{G}_{\beta\nu} = \int (\bar{F}, \rho(q)u_{\mu}(q))(\bar{G}, u_{\nu}(q))dq. \tag{2.36}$$

This result shows that the quantities $\mathcal{F}_{\mu\nu}$ and $\mathcal{G}_{\mu\nu}$ defined by (2.34) satisfy the rule of product of matrices.

Now let us consider the matrices $(q_j)_{\mu\nu}$ and $(p_j)_{\mu\nu}$, where $p_j = -i\hbar\frac{\partial}{\partial q_j}$;

$$(q_j)_{\mu\nu} = \int (q_j, \rho(q)u_{\mu}(q))u_{\nu}(q)dq, \tag{2.37}$$

and

$$(p_j)_{\mu\nu} = -i\hbar \int (q_j, \rho(q)u_{\mu}(q))\frac{\partial}{\partial q_k}u_{\nu}(q)dq. \tag{2.38}$$

From these equations we obtain the canonical commutation relation

$$(p_j q_k - q_k p_j)_{\mu\nu} = -i\hbar\delta_{jk} \int \rho(q)u_{\mu}(q)u_{\nu}(q)dq = -i\hbar\delta_{\mu\nu}\delta_{kj}. \tag{2.39}$$

This is the commutation relation (2.27), here derived using wave mechanical formalism. We can also derive Eq. (2.25) from the wave equation:

$$Hu_{\mu}(q) = E_{\mu}u_{\mu}(q), \tag{2.40}$$

where $u_{\mu}(q)$ is the eigenfunction of the Hamiltonian with the eigenvalue E_{μ}. Using the orthogonality of $\{u_{\mu}(q)\}$ s we have

$$H_{\mu\nu} = E_{\nu} \int \rho(q)u_{\mu}(q)u_{\nu}(q)dq = E_{\nu}\delta_{\mu\nu}. \tag{2.41}$$

Since $H_{\mu\nu}$ is a diagonal matrix, we have

$$(q_l H)_{\mu\nu} = \sum_{\beta}(q_l)_{\mu\beta}H_{\beta\nu} = E_{\nu}(q_l)_{\mu\nu}, \tag{2.42}$$

$$(Hq_l)_{\mu\nu} = \sum_{\beta} H_{\mu\beta}(q_l)_{\beta\nu} = E_{\mu}(q_l)_{\mu\nu}. \tag{2.43}$$

These are equivalent to Eqs. (2.24) and (2.25). Similarly we can derive

$$i\hbar\left(\frac{dp_l}{dt}\right)_{\alpha\beta} = [p,\ H]_{\alpha\beta}, \tag{2.44}$$

from the wave equation (2.40).

Now H in (2.40) is a suitably symmetrized Hamiltonian, and according to Schrödinger we need to show that this H is the same wave operator which gives us the wave equation. To this end we will consider a Hamiltonian with the general form of

$$H = T(q_k, p_k) + V(q_k), \tag{2.45}$$

where T is a quadratic function of p_k s. Schrödinger now derives his wave equation from the variational principle

$$\delta J_1 = \delta\int\left\{\hbar^2 T\left(q_k, \frac{\partial\psi(q)}{\partial q_k}\right) + \psi^2(q)V(q_k)\right\}\frac{1}{\sqrt{\Delta_p}}dq = 0, \tag{2.46}$$

subject to the subsidiary condition

$$\delta J_2 = \int\frac{\psi^2(q)}{\sqrt{\Delta_p}}dq = 1, \tag{2.47}$$

where Δ_p is the discriminant of the quadratic form T. Multiplying (2.47) with the Lagrange's multiplier $(-E)$, adding it to (2.46), and carrying out the variation he obtains

$$\int\left\{-\frac{\hbar^2}{2}\sum_l\frac{\partial}{\partial q_l}\left[\frac{1}{\sqrt{\Delta_p}}T_{pl}\left(q_l, \frac{\partial\psi(q)}{\partial q_l}\right)\right] + \frac{1}{\sqrt{\Delta_p}}\left(V(q_l) - E\right)\psi(q)\right\}$$
$$\times \quad \delta\psi(q)dq = 0. \tag{2.48}$$

Thus from the Euler–Lagrange equation for this problem he finds the following partial differential equation which is the same as the Schrödinger equation;

$$-\frac{\hbar^2}{2\sqrt{\Delta_p}}\sum_l\frac{\partial}{\partial q_l}\left[\frac{1}{\sqrt{\Delta_p}}T_{pl}\left(q_l, \frac{\partial\psi(q)}{\partial q_l}\right)\right] - (V(q_l) - E)\psi(q) = 0. \tag{2.49}$$

Bibliography

[1] W. Heisenberg, Über quantentheoretische Umdeutung kinematischer und mechanischer Beziehungen, Z. Physik 33, 879 (1925). The English translation of this paper can be found in B.L. van der Waerden's book *Sources of Quantum Mechanics*, (Dover Publications, New York, 1967) p. 261. See also *W. Heisenberg, Collected Works*, edited by W. Blum, H.-P. Dürr and H. Rechenberg, Series A1, (Springer, Berlin, 1985).

[2] An excellent account of the Heisenberg reasoning leading to the discovery of matrix mechanics can be found in D.C. Cassidy's book, *Uncertainty The Life and Science of Werner Heisenberg*, (W.H. Freeman, New York, 1992).

[3] E.T. Whittaker, *A History of the Theories of Aether & Electricity*, Vol. II, (Harper Torchbooks, 1960).

[4] J. Mehra and H. Rechenberg, *The Historical Development of Quantum Theory, Vol. 3, The Foundation of Matrix Mechanics and its Modifications, 1925-1926*, (Springer, New York, 2001).

[5] A. Pais, *Inward Bound: Of Matter and Forces in the Physical World*, (Oxford University Press, 1986).

[6] W.A. Fedak and J.J. Prentis, The 1925 Born and Jordan paper "On quantum mechanics", Am. J. Phys. 77, 128 (2009).

[7] W.A. Fedak and J.J. Prentis, Quantum jumps and classical harmonics, Am. J. Phys. 70, 332 (2002).

[8] N. Bohr, On the quantum theory of line spectra, reprinted in B.L. van der Waerden, *Sources of Quantum Mechanics*, (North-Holland, Amsterdam, 1967) paper 3.

[9] I.J.R. Aitchison, D.A. MacManus and T.M. Snyder, Understanding Heisenberg's "magical" paper of July 1925: A new look at the calculational details, Am. J. Phys. 72, 1370 (2004).

[10] J. Bernstein, Max Born and quantum theory, Am. J. Phys. 73, 999 (2005).

[11] See for instance, J. Schwinger, L.L. DeRoad, Jr. K.A. Milton and W-Y Tsai, *Classical Electrodynamics* (Perseus Books, Reading, 1998), p. 355.

[12] M. Born and P. Jordan, Zur Quantenmechanik, Z. Physik 34, 858 (1925). The English translation of this paper can be found in B.L. van der Waerden, *Sources of Quantum Mechanics*, (North-Holland, Amsterdam, 1967), paper 13. For a detailed analysis of the content of this paper, see W.A. Fedak and J.J. Prentis, Am. J. Phys. 77, 128 (2009).

[13] M. Born, W. Heisenberg and P. Jordan, Zur Quantenmechanik II , Z. Physik 35, 557 (1926). English translation in van der Waerden's book, paper 15.

[14] W. Pauli, Über das Wasserstoffspectrum vom Standpunkt der neuen Quantenmechanik, Z. Physik 36, 336 (1926). English translation in van der Waerden's book, paper 16.

[15] B.L. van der Waerden, *Sources of Quantum Mechanics*, (North-Holland, Amsterdam, 1967).

[16] M. Born, Zur Quantenmechanik der Stoßvorgänge, Z. Phys. 37, 863 (1926).

[17] S-I. Tomonaga, *Quantum Mechanics*, Vol. I, (North-Holland, Amsterdam, 1962).

[18] H.S. Green, *Matrix Mechanics*, (P. Noordhoff, Netherlands, 1965).

[19] T.F. Jordan, *Quantum Mechanics in Simple Matrix Form*, (John Wiley & Sons, New York, 1986).

[20] H. Umezawa and G. Vitiello, *Quantum Mechanics*, (Bibliopolis, 1985).

[21] M. Taketani and M. Nagasaki, *The Formation and Logic of Quantum Mechanics*, Vol. III, (World Scientific, 2001).

[22] E. Schrödinger Über das Verhältnis der Heisenberg-Born-Jordaschen Quantenmechanik zu der meinen, Ann. d. Physik, 387, 734 (1926).

[23] C. Eckart, Operator calculus and the solution of the equations of quantum dynamics, Phys. Rev. 28, 711 (1926).

[24] C. Eckart, Note on the correspondence principle in the new quantum theory, Proc. Natl. Acad. Sci. USA, 12, 684 (1926).

[25] For a comprehensive treatment of the properties of orthogonal functions see G. Sansone, *Orthogonal Functions*, (Interscience Publishers, New York, 1959).

[26] N.R. Hanson, Are wave mechanics and matrix mechanics equivalent theories? Czech. J. Phys. B11, 693 (1961).

Chapter 3

Mathematical Preliminaries

In this chapter we want to recall some of the important mathematical definitions and results related to linear vector space and matrix calculus that will be needed for the formulation of the matrix mechanics [1]–[4].

3.1 Vectors and Vector Spaces

We define a vector space as a space whose elements known as vectors satisfy the following set of axioms:
Let $\mathcal{F}$ be a field of complex numbers, or a field of scalars, then the vector space over the field $\mathcal{F}$ is the set of vectors $\{f\}$ where the members of the set $\{f\}$ must satisfy the following axioms of addition and multiplication.
For addition we have

(1) - Closure: if f and g are vectors, then $f+g$ is also a vector.

(2) - Associativity of addition:

$$f+(g+h)=(f+g)+h. \tag{3.1}$$

(3) - Commutativity:

$$f+g=g+f. \tag{3.2}$$

(4) - Existence of the null vector 0:

$$f+0=f. \tag{3.3}$$

(5) - Inverse element $(-f)$:

$$f+(-f)=0. \tag{3.4}$$

For multiplication by scalars b and c we have

(1) - Closure: if f is a vector belonging to the vector space, then cf is also a vector belonging to the same vector space.

(2) - Distributivity for scalar multiplication:

$$c(f+g) = cf + cg. \tag{3.5}$$

(3) - Distributivity for scalar multiplication over vector addition:

$$(b+c)f = bf + cf. \tag{3.6}$$

(4) - Associativity of scalar multiplication:

$$c(bf) = (cb)f. \tag{3.7}$$

(5) - Existence of an identity scalar element 1:

$$1(f) = f. \tag{3.8}$$

We will denote an element f of the vector space by the symbol $|f\rangle$ which we call a ket. We also associate with every ket $|f\rangle$ a new object which we call bra and we denote it by $\langle f|$. The bra which is associated with the ket $a|f\rangle$ is $a^*\langle f|$, and the one associated with $b|f\rangle + c|g\rangle$ is $b^*\langle f| + c^*\langle g|$. Using these notations we can write the scalar product of $\langle f|$ and $|g\rangle$ as $\langle f|g\rangle$.

Having defined the vector space, let us consider N vectors, $f^1, f^2 \cdots, f^N$ belonging to this space.

If the equation

$$c_1 f^{(1)} + c_2 f^{(2)} + \cdots + c_N f^{(N)} = 0, \tag{3.9}$$

is satisfied if and only if $c_1 = c_2 = \cdots = c_N = 0$, then $f^{(1)}, f^{(2)} \cdots f^{(N)}$ form a set of linearly independent vectors.

The vectors that we have defined so far can have finite or infinite components. In quantum theory we are dealing with both finite- and infinite-dimensional vector spaces. That is the number of independent vectors spanning the space can be finite or infinite. In addition we are dealing with vectors in Hilbert space, i.e. we assume that any pair of vectors f and g have a scalar product $\langle f|g\rangle$ and a norm which is defined by

$$||f|| = \sqrt{\langle f|f\rangle}. \tag{3.10}$$

The scalar product $\langle f|g\rangle$ can be a real or a complex number, and has the following properties:

(1) - $\quad \langle f|f\rangle \geq 0$ and $\langle f|f\rangle = 0$ if and only if $f = 0$. (3.11)

(2) - $$\langle f|g^{(1)} + g^{(2)}\rangle = \langle f|g^{(1)}\rangle + \langle f|g^{(2)}\rangle. \tag{3.12}$$

(3) -
$$\langle f|g\rangle = \langle g|f\rangle^*, \tag{3.13}$$

where $\langle g|f\rangle^*$ denotes the complex conjugate of $\langle g|f\rangle$. If c is a complex number then $\langle cf\rangle = c^*\rangle f|g\langle$ and $\langle f|cg\rangle = c\langle f|g\rangle$.

(4) - Schwarz inequality

$$|\langle f|g\rangle|^2 \leq \langle f|f\rangle\langle g|g\rangle. \tag{3.14}$$

If vectors f and g are represented by finite or infinite column matrices

$$\begin{bmatrix} f_1 \\ f_2 \\ \vdots \\ f_n \\ \vdots \end{bmatrix}, \qquad \begin{bmatrix} g_1 \\ g_2 \\ \vdots \\ g_n \\ \vdots \end{bmatrix}, \tag{3.15}$$

where f_i is the i-th component of the vector f, then

$$\langle f|g\rangle = \left(f^t\right)^* g = f^\dagger g = \sum_i f_i^* g_i. \tag{3.16}$$

We use $\langle f|$ to denote the complex conjugate of the row vector f and $|g\rangle$ as a column vector g. The sum in (3.16) can run over a finite or an infinite number of components. The superscript t on f indicates the transpose of the matrix, and the superscript $\dagger$ denotes the complex conjugate of the transpose.

A set $\{f^{(k)}\}$ in a vector space is complete provided that any vector in the same space can be written as a linear combination of the vectors from the set $\{f^{(k)}\}$. Now suppose that we have a set of independent vectors $\{f^{(k)}\}$ which is complete, then we can always construct a set of vectors $\{e^{(k)}\}$ so that they are orthogonal to each other, i.e. $\langle e^{(j)}|e^{(k)}\rangle = 0$, when $j \neq k$. The construction of this set is as follows:

We normalize $f^{(1)}$ and call it $e^{(1)}$

$$e^{(1)} = \frac{f^{(1)}}{||f^{(1)}||}, \quad \text{thus} \quad \left\langle e^{(1)}|e^{(1)}\right\rangle = 1. \tag{3.17}$$

Then we subtract from $f^{(2)}$ the component $\langle e^{(1)}|f^{(2)}\rangle e^{(1)}$ and normalize the result to get $e^{(2)}$;

$$e^{(2)} = \frac{f^{(2)} - \langle e^{(1)}|f^{(2)}\rangle e^{(1)}}{||f^{(2)} - \langle e^{(1)}|f^{(2)}\rangle e^{(1)}||}. \tag{3.18}$$

This vector $e^{(2)}$ is clearly normalized and is also orthogonal to $e^{(1)}$. We can proceed in this way and find $e^{(n)}$ from the relation

$$e^{(n)} = \frac{\left[f^{(n)} - \sum_{k=1}^{n-1} \langle e^{(k)}|f^{(n)}\rangle e^{(k)}\right]}{\left|\left|\left[f^{(n)} - \sum_{k=1}^{n-1} \langle e^{(k)}|f^{(n)}\rangle e^{(k)}\right]\right|\right|}. \tag{3.19}$$

Again it can easily be shown that

$$\left\langle e^{(n)}|e^{(j)}\right\rangle = 0, \quad \text{for } j = 1, \cdots, n-1, \quad \left\langle e^{(n)}|e^{(n)}\right\rangle = 1. \tag{3.20}$$

A function $f(x)$ of a real variable x defined over a finite interval $a \le x \le b$ or over an infinite interval $(-\infty \le x \le \infty$ or $0 \le x \le \infty)$ is also a vector provided $cf(x)$ and $f(x) + g(x)$ are also defined over the same interval. In addition the scalar product is defined by

$$\langle f|g\rangle = \int_a^b f^*(x)g(x)dx. \tag{3.21}$$

The set of vectors $f^{(1)}(x), f^{(2)}(x), \cdots f^{(N)}(x) \cdots$ are called independent if any relation of the form $\sum_i c_i f^{(i)}(x) = 0$ between them implies that all the c_i s are zero. The norm (or the length) of the vector $f(x)$ is defined by

$$||f|| = \left[\int_a^b f^*(x)f(x)dx\right]^{\frac{1}{2}}, \tag{3.22}$$

or in the matrix form with components $f_1, f_2, \cdots$ the norm is given by

$$||f|| = [\langle f|f\rangle]^{\frac{1}{2}} = \left[\sum_i f_i^* f_i\right]^{\frac{1}{2}}. \tag{3.23}$$

If the scalar product of two non-zero vectors f and g is zero

$$\langle f|g\rangle = 0, \tag{3.24}$$

we say that g is orthogonal to f.

In quantum theory we work with vectors in Hilbert space and this Hilbert space is of crucial importance in the mathematical formulation of quantum mechanics. Hilbert space is the space of square integrable functions $f(x)$, $g(x)\cdots$, where each member is a complex valued function defined on the real interval (a, b), i.e. has a finite norm, $\| f \| < \infty$. This space will be denoted by $\mathcal{L}_2$. The scalar product of two vectors $f(x)$ and $g(x)$ in $\mathcal{L}_2$ is defined by Eq. (3.21) and satisfies the properties (1)–(4) of scalar products.

We define the distance between the two functions $f(x)$ and $g(x)$ by

$$d(f,\ g) = \| f(x) - g(x) \|, \tag{3.25}$$

and this makes $\mathcal{L}_2$ a metric space. The space $\mathcal{L}_2$ is complete, i.e. a space where every Cauchy series has a limit. That is if we consider a series of vectors $|f^{(n)}(x)\rangle$ of $\mathcal{L}_2$ such that $\| f^{(n)} - f^{(j)} \| \to 0$ as $n,\ j \to \infty$, then there exists a vector $|f(x)\rangle$ of $\mathcal{L}_2$ such that $\| f^{(n)} - f \| \to 0$ for $n \to \infty$. The dimension of a Hilbert space can be finite and equal to N, or it can be denumerably or nondenumerably infinite.

For the Hilbert space $\mathcal{L}_2$ one can find a denumerable infinite set of base functions (or vectors) $e^{(1)}(x)$ which are orthogonal

$$\left\langle e^{(k)}|e^{(j)}\right\rangle = \int_a^b e^{(k)\,*}(x)e^{(j)(x)}dx = \delta_{kj}. \tag{3.26}$$

We can expand any element of the Hilbert space $\mathcal{L}_2$ in terms of the set $e^{(k)}(x)$ as

$$f(x) = \sum_{k=1}^{\infty} c_k e^{(k)}(x). \tag{3.27}$$

This expansion must be understood in the following sense. Consider the partial sum

$$f^{(n)}(x) = \sum_{k=1}^{n} c_k e^{(k)}(x), \tag{3.28}$$

then the expansion (3.27) holds provided

$$\| f(x) - f^{(n)}(x) \| < \epsilon, \qquad \text{for} \qquad n > N(\epsilon). \tag{3.29}$$

We call the set $\{e^{(n)}\}$ a complete set if

$$\sum_{k=1}^{\infty} c_k e^{(k)}(x) \neq 0, \qquad \text{for all } x, \tag{3.30}$$

unless all c_k s are zero. The completeness of the set $\{e^{(k)}(x)\}$ can be expressed by the relation

$$\sum_{k=1}^{\infty} e^{(k)\,*}(x)e^{(k)}\left(x'\right) = \delta\left(x - x'\right), \tag{3.31}$$

where $\delta\left(x - x'\right)$ is the Dirac delta function (see below). In some problems the Hilbert spaces needed do not have a denumerably infinite base, e.g. when the range of x is infinite $-\infty \leq x \leq \infty$ or $0 \leq x \leq \infty$. When this is the case we need a nondenumerably infinite set of functions to expand a given function belonging to $\mathcal{L}_2$. Denoting the base function by $e(k, x)$, where k is continuous and varies from $-\infty$ to ∞, we can express the completeness relation by

$$\int_{-\infty}^{\infty} e^*(k, x)e\left(k, x'\right)dk = \delta\left(k - k'\right). \tag{3.32}$$

A function $f(x)$ belonging to $\mathcal{L}_2$ can then be expanded as

$$f(x) = \int_{-\infty}^{\infty} c(k)e(k, x)dk, \tag{3.33}$$

where

$$c(k) = \langle e(k, x)|f(x)\rangle = \int_{-\infty}^{\infty} e^*(k, x)f(x)dx. \tag{3.34}$$

In many problems we encounter Hilbert spaces where the base is neither entirely countable nor entirely continuous. If this is the case then the completeness condition reads

$$\sum_{j=1}^{N} e^{(j)\,*}(x)e^{(j)}(x') + \int_a^\infty e^*(k,x)e(k,x')\,dk = \delta(k-k'), \tag{3.35}$$

where N, the number of discrete elements $e^{(j)}(x)$ may be finite or infinite.

Dirac's δ-function — The Dirac δ-function that we introduced in (3.32) is not a proper function, rather it is a distribution [5]. However for practical purposes we may consider it a function. This δ-function :has the following properties:

$$\delta(x) = 0, \qquad x \neq 0, \tag{3.36}$$

$$\int_{-\infty}^{\infty} \delta(x)\,dx = 1, \tag{3.37}$$

$$\int_{-\infty}^{\infty} f(x)\delta(x)\,dx = f(0), \tag{3.38}$$

if $f(x)$ is a continuous function around $x = 0$,

$$\delta(ax) = \frac{1}{|a|}\delta(x), \tag{3.39}$$

$$x\delta'(x) = -\delta(x), \tag{3.40}$$

and finally

$$\delta[f(x)] = \sum_{j=1}^{N} \frac{1}{|f'(x_j)|}\delta(x-x_j), \tag{3.41}$$

where x_j s are the zeros of $f(x)$.

Matrix Representation of an Operator — We assume that we have a complete set of unit vectors $e^{(j)}$, then any vector f in this space can be expanded as

$$f = \sum_j f_j e^{(j)}. \tag{3.42}$$

The operator A acting on f will give us a vector g

$$g = Af. \tag{3.43}$$

This vector can also be expanded in terms of the basis set or unit vectors $e^{(k)}$, viz,

$$g = \sum_k g_k e^{(k)}. \tag{3.44}$$

Substituting (3.42) and (3.43) in (3.44) and multiplying the result by $e^{(n)}$ we obtain

$$g_n = \sum_k \left\langle e^{(n)}|A|e^{(k)}\right\rangle f_k. \tag{3.45}$$

This relation shows that the operator A can be represented by its matrix element A_{nk};

$$A_{nk} = \left\langle e^{(n)}|A|e^{(k)}\right\rangle. \tag{3.46}$$

3.2 Special Types of Operators

Among the operators which we encounter in quantum mechanics the followings are of special interest:

(1) - **Projection (or Idempotent) Operator** — A projection operator is a self-adjoint operator P with the property that $P^2 = P$. The eigenvalues of such an operator are either zero or one. If $|n\rangle$ is the normalized eigenfunction of a Hermitian operator A with discrete eigenvalues λ_n, then

$$P_n|\alpha\rangle = |n\rangle\langle n|\alpha\rangle, \tag{3.47}$$

is a projection operator since we have

$$P_n^2|\alpha\rangle = P_n|n\rangle\langle n|\alpha\rangle = |n\rangle\langle n|\alpha\rangle = P_n|\alpha\rangle. \tag{3.48}$$

(2) - **Hermitian Operator** — An operator A is said to be Hermitian if $A = A^\dagger$, or expressed in another way for arbitrary vectors f and g in Hilbert space we have the relation

$$\langle Af|g\rangle = \langle f|Ag\rangle. \tag{3.49}$$

If f and g are basis vectors for the vector space then

$$\langle e_k|A|e_j\rangle = \langle e_j|A|e_k\rangle^*. \tag{3.50}$$

or

$$A_{kj} = (A_{jk})^*. \tag{3.51}$$

That is the two elements symmetric with respect to the diagonal elements of the matrix are complex conjugates of each other.

An important property of the Hermitian operators is that their eigenvalues are all real numbers. Let $|n\rangle$ be the eigenfunction of λ_n;

$$A|n\rangle = \lambda_n|n\rangle, \tag{3.52}$$

then

$$\langle n|A|n\rangle = \lambda_n\langle n|n\rangle = \lambda_n. \tag{3.53}$$

Also from the Hermiticity property of A, i.e. $A = A^\dagger$ we get

$$\langle n|A^\dagger = \langle n|A = \lambda_n^*\langle n|. \tag{3.54}$$

Therefore

$$\langle n|A^\dagger|n\rangle = \langle n|A|n\rangle = \lambda_n^*\langle n|n\rangle = \lambda_n^*. \tag{3.55}$$

Comparing (3.53) and (3.55) we find that $\lambda = \lambda^*$, and thus λ is real.

(3) - **Unitary Operator** — We define a unitary operator U by the property that its inverse, U^{-1}, is equal to its adjoint $U^\dagger$ and this implies that

$$U^\dagger U = UU^\dagger = 1. \tag{3.56}$$

Now let us consider two arbitrary vectors $|f\rangle$ and $|g\rangle$ which under the action of U transform to vectors $|\tilde{f}\rangle$ and $|\tilde{g}\rangle$;

$$|\tilde{f}\rangle = U|f\rangle, \tag{3.57}$$

$$|\tilde{g}\rangle = U|g\rangle. \tag{3.58}$$

The scalar product $\langle\tilde{f}|\tilde{g}\rangle$ is given by

$$\langle\tilde{f}|\tilde{g}\rangle = \langle f|U^\dagger U|g\rangle = \langle f|g\rangle. \tag{3.59}$$

This scalar product is preserved under unitary transformation. In particular if $\tilde{g} = \tilde{f}$ then

$$\langle\tilde{f}|\tilde{f}\rangle = \langle f|f\rangle, \tag{3.60}$$

i.e. the norm of a vector is also preserved.

If an operator A is Hermtian, then the operator e^{iA} is unitary. This follows from the fact that $A^\dagger = A$. Thus if $U = e^{iA}$, then

$$U^\dagger = e^{-iA^\dagger} = e^{-iA}, \tag{3.61}$$

and

$$U^\dagger U = e^{-iA^\dagger}e^{iA} = e^{-iA}e^{iA} = 1, \tag{3.62}$$

where 1 denotes a unit matrix or operator. Next we consider the action of a unitary operator on a basis set $\{e_k\}$ which we assume to be orthonormal. If the unitary operator acts on $|e_k\rangle$ we obtain a new vector $|\tilde{e}_k\rangle$,

$$|\tilde{e}_k\rangle = U|e_k\rangle, \tag{3.63}$$

and for this new vector we find the scalar product $\langle\tilde{e}_j|\tilde{e}_k\rangle$ to be

$$\langle\tilde{e}_j|\tilde{e}_k\rangle = \langle e_j|U^\dagger U|e_k\rangle = \langle e_j|e_k\rangle = \delta_{jk}. \tag{3.64}$$

Thus the new set $\{\tilde{e}_j\}$ form an orthonormal set of basis vectors.

We can find the matrix representation of the operator U by defining

$$U_{jk} = \langle e_j|U|e_k\rangle, \tag{3.65}$$

and this matrix U_{jk} is a unitary matrix since

$$\langle e_j|U^\dagger U|e_k\rangle = \sum_n \langle e_j|U^\dagger|e_n\rangle\langle e_n|U|e_k\rangle = \sum_n \delta_{jn}\delta_{nk} = \delta_{jk}, \tag{3.66}$$

or written in the matrix form Eq. (3.66) becomes

$$\sum_n U^\dagger_{jn} U_{nk} = \delta_{jk}. \tag{3.67}$$

Just as infinitesimal canonical transformation Sec. 1.8 are important in classical mechanics, infinitesimal unitary transformation play a very important role in quantum mechanics. We write such a transformation as

$$U = 1 + i\varepsilon G, \tag{3.68}$$

where ε is a real infinitesimal number, G is a Hermitian operator and 1 represents the unit matrix. The Hermitian conjugate of U and the inverse of U, up to the order ε, have simple forms:

$$U^\dagger = 1 - i\varepsilon G, \tag{3.69}$$

$$U^{-1} = U^\dagger = 1 - i\varepsilon G, \tag{3.70}$$

and to this order we have

$$UU^\dagger = U^\dagger U = 1. \tag{3.71}$$

Let A be an arbitrary operator which under the infinitesimal unitary transformation changes to A', then A is related to A' by

$$A \to A' = (1 + i\varepsilon G)A(1 - i\varepsilon G) = A + i\varepsilon[G,\ A], \tag{3.72}$$

where $[G,\ A]$ is the commutator of G and A, and in (3.72) only terms of order ε have been retained. Denoting the change from the action of this unitary transformation by δA we have

$$\delta A = A' - A = i\varepsilon[G,\ A]. \tag{3.73}$$

This expression is the quantum analogue of the classical expression (1.150).

Baker–Campbell–Hausdorff Formula — If the commutator $[A, B]$ is a c-number, $[A, B] = \lambda$, then

$$e^{A+B} = e^A e^B e^{-\frac{\lambda}{2}}. \tag{3.74}$$

To find this result let us consider a function $f(\xi)$ defined by

$$f(\xi) = e^{\xi A} e^{\xi B}. \tag{3.75}$$

By differentiating (3.75) with respect to ξ while preserving the order of operators we have

$$\frac{1}{f(\xi)}\frac{df(\xi)}{d\xi} = (A + B + [A, B]\xi) = (A + B + \lambda\xi). \tag{3.76}$$

Now we integrate the differential equation (3.76) subject to the condition $f(0) = 1$, and then set $\xi = 1$ to obtain (3.74).

This is a special case of the general Baker–Campbell–Hausdorff formula. This formula can be written as

$$e^A B e^{-A} = B + [A, B] + \frac{1}{2}[A, [A, B]\,] + \cdots = \sum_{n=0}^{\infty} \frac{1}{n!}[A, [A, [A, \cdots]\,]\,]B, \quad (3.77)$$

or in the form of

$$e^A e^B = e^C, \quad (3.78)$$

where

$$C = A + B + \frac{1}{2}[A, B] + \frac{1}{12}[\,[A, B], B] + \frac{1}{12}[\,[B, A], A] + \cdots. \quad (3.79)$$

Note that both (3.77) and (3.79) reduce to (3.74) provided that

$$[A, [A, B]\,] = [B, [A, B]\,] = 0. \quad (3.80)$$

3.3 Vector Calculus for the Operators

A number of vector identities in classical mechanics and electrodynamics will assume different forms when the components of these vectors are linear operators. Let $\mathbf{A}$ and $\mathbf{B}$ be two non-commutative three-dimensional vectors with components (A_x, A_y, A_z) and (B_x, B_y, B_z) respectively, and let us write the commutator $[\mathbf{A},\ \mathbf{B}]$ as a dyadic [6],[7]

$$[\mathbf{A},\ \mathbf{B}] = [A_x,\ B_x]\mathbf{ii} + [A_x,\ B_y]\mathbf{ij} + \cdots. \quad (3.81)$$

By writing for $\mathbf{AB} - \mathbf{BA}$ as well as $\mathbf{BA} - \mathbf{AB}$ using Eq. (3.81) we find that

$$[\mathbf{A},\ \mathbf{B}] = \mathbf{AB} - (\mathbf{BA})^t = \mathbf{AB} - \mathbf{BA} - (\mathbf{B} \wedge \mathbf{A}) \wedge \hat{\mathbf{I}}, \quad (3.82)$$

where $\hat{\mathbf{I}}$ is the unit dyadic

$$\hat{\mathbf{I}} = \mathbf{ii} + \mathbf{jj} + \mathbf{kk}, \quad (3.83)$$

and the superscript t denotes the transposed dyadics. From (3.82) we find that in general $[\mathbf{B},\ \mathbf{A}] \neq -\,[\mathbf{A},\ \mathbf{B}]$, but

$$[\mathbf{B},\ \mathbf{A}] = -\,[\mathbf{A},\ \mathbf{B}]^t. \quad (3.84)$$

Also the commutator of $\mathbf{A}$ with a scalar operator Q is

$$[Q,\ \mathbf{A}] = -[\mathbf{A},\ Q] = [Q,\ A_x]\mathbf{i} + [Q,\ A_y]\mathbf{j} + [Q,\ A_z]\mathbf{k}. \quad (3.85)$$

Now from (3.82) we find the formulas [7]

$$\mathbf{A} \cdot (\mathbf{B} \wedge \mathbf{C}) = (\mathbf{A} \wedge \mathbf{B}) \cdot \mathbf{C}, \quad (3.86)$$

$$\mathbf{A}\wedge(\mathbf{B}\wedge\mathbf{C}) = \mathbf{B}(\mathbf{A}\cdot\mathbf{C})-(\mathbf{A}\cdot\mathbf{B})\mathbf{C}-[\mathbf{B},\ \mathbf{A}]\cdot\mathbf{C}$$
$$= (\mathbf{A}\cdot\mathbf{C})\mathbf{B}-(\mathbf{A}\cdot\mathbf{B})\mathbf{C}-\mathbf{A}\cdot[\mathbf{C},\ \mathbf{B}], \tag{3.87}$$

$$(\mathbf{A}\wedge\mathbf{B})\wedge\mathbf{C} = \mathbf{B}(\mathbf{A}\cdot\mathbf{C})-(\mathbf{A}\cdot\mathbf{B})\mathbf{C}-[\mathbf{B},\ \mathbf{A}]\cdot\mathbf{C}$$
$$= \mathbf{A}(\mathbf{C}\cdot\mathbf{B})-\mathbf{A}(\mathbf{B}\cdot\mathbf{C})-\mathbf{A}\cdot[\mathbf{C},\ \mathbf{B}], \tag{3.88}$$

$$[(\mathbf{A}\cdot\mathbf{B}),\ \mathbf{C}] = \mathbf{A}\cdot[\mathbf{B},\ \mathbf{C}]-[\mathbf{C},\ \mathbf{A}]\cdot\mathbf{B}, \tag{3.89}$$

$$[(\mathbf{A}\wedge\mathbf{B}),\ \mathbf{C}] = \mathbf{A}\wedge[\mathbf{B},\ \mathbf{C}]-([\mathbf{C},\ \mathbf{A}]\wedge\mathbf{B})^t, \tag{3.90}$$

and

$$[(\mathbf{A}\wedge\mathbf{B}),\ Q] = \mathbf{A}\wedge[\mathbf{B},\ Q]+[\mathbf{A},\ Q]\wedge\mathbf{B}. \tag{3.91}$$

As an example of the application of these formulas let us determine the commutation relations of the angular momentum with the position and momentum operators. We start with the basic commutators which we write as

$$[\mathbf{r},\ \mathbf{r}]=0,\quad [\mathbf{p},\ \mathbf{p}]=0,\quad [\mathbf{r},\ \mathbf{p}]=-[\mathbf{p},\ \mathbf{r}]=i\hbar\hat{\mathbf{I}}. \tag{3.92}$$

To find the commutator $[\mathbf{L},\ \mathbf{r}]$ or $[\mathbf{L},\ \mathbf{p}]$ we use (3.90) to get

$$[\mathbf{L},\ \mathbf{r}] = [\mathbf{r}\wedge\mathbf{p},\ \mathbf{r}] = -i\hbar\mathbf{r}\wedge\hat{\mathbf{I}} = [\mathbf{r},\ \mathbf{L}]. \tag{3.93}$$

and

$$[\mathbf{L},\ \mathbf{p}] = [\mathbf{r}\wedge\mathbf{p},\ \mathbf{p}] = \left(i\hbar\hat{\mathbf{I}}\wedge\mathbf{p}\right)^t = -i\hbar\mathbf{p}\wedge\hat{\mathbf{I}} = [\mathbf{p},\ \mathbf{L}]. \tag{3.94}$$

These follow from the fact that

$$\mathbf{A}\wedge\hat{\mathbf{I}} = \hat{\mathbf{I}}\wedge\mathbf{A} = -\left(\hat{\mathbf{I}}\wedge\mathbf{A}\right)^t. \tag{3.95}$$

Now if we set $\mathbf{A}=\mathbf{B}=\mathbf{r}$ and $\mathbf{C}=\mathbf{L}$ from (3.89) we find

$$\left[\mathbf{r}^2,\ \mathbf{L}\right] = \mathbf{r}\cdot[\mathbf{r},\ \mathbf{L}]-[\mathbf{L},\ \mathbf{r}]\cdot\mathbf{r} = -i\hbar\left[\mathbf{r}\cdot\left(\hat{\mathbf{I}}\wedge\mathbf{r}\right)-\left(\mathbf{r}\wedge\hat{\mathbf{I}}\right)\cdot\mathbf{r}\right]=0. \tag{3.96}$$

In a similar way we get

$$\left[\mathbf{p}^2,\ \mathbf{L}\right]=0. \tag{3.97}$$

3.4 Construction of Hermitian and Self-Adjoint Operators

For simple conservative systems where the classical Hamiltonian has the general form

$$H=\sum_{j=1}^{n}\frac{\mathbf{p}_j^2}{2m_j}+V(\mathbf{r}_1\cdots\mathbf{r}_n), \tag{3.98}$$

a Hermitian quantum mechanical operator can be constructed by the simple rule of replacing $\mathbf{p}_j$ by $(-i\hbar\nabla_\mathbf{j})$. However there are instances where this simple rule is inadequate. That is either the resulting operator is not Hermitian or by different ordering of factors we may get many Hermitian operators. Among the interesting physical problems where we encounter such a difficulty are the formulation of velocity-dependent dissipative systems [8], the motion of a particle with position-dependent mass (or effective mass) [9]–[13], or in the theory of nuclear forces [11],[12]. For these and similar problems we want to discuss the rules for the construction of Hermitian operators and then investigate the self-adjointness of the resulting operators [13].

The basic dynamical relation which will be used in our formulation are the Cartesian coordinates and momenta which satisfy the commutation relation

$$p_j q_k - q_k p_j = -i\hbar\delta_{jk}. \tag{3.99}$$

Now let us consider any classical function of the form $p_c^n F(q_c)$, where the subscript c refers to the fact that p_c and q_c are c-numbers and not operators. Classically the order in which we write p_c and q_c does not matter. However in quantum mechanics the order that the operators enter in any dynamical observable is important.

We call an operator to have the normal form if all powers of q precede all powers of p. Making use of the commutation relation (3.99) we can write

$$\begin{aligned} p^n F(q) &= \sum_{j=0}^{n} \frac{n!}{j!(n-j)!}(-i\hbar)^j \frac{d^j F}{dq^j} p^{n-j} \\ &= \exp\left(-i\hbar\frac{\partial^2}{\partial q \partial p}\right) F(q) p^n, \end{aligned} \tag{3.100}$$

where using (3.99) we have found (3.100) by induction [14].

In obtaining a quantum operator corresponding to a classical quantity which is a function of p_c and q_c we want to satisfy the following requirements:

(a) - The resulting operator be Hermitian. If this operator can be written as a power series in p and q, then this requirement implies that

$$\sum_{j,k} \alpha_{jk} q^j p^k = \sum_{j,k} \alpha_{jk}^* p^k q^j. \tag{3.101}$$

(b) - As $\hbar \to 0$, all coordinates and momenta commute and quantum mechanical operators must reduce to the corresponding classical functions.

(c) - If possible we want to have a unique Hermitian operator for a given classical function of p and q.

There are a number of different rules proposed for the construction of Hermitian operators, and we will discuss some of the well-known rules below. But let us emphasize that as Margenau and Cohen have shown no rule of ordering can be consistently used to derive quantum operators from their classical functions [15].

Regarding the ambiguity of quantization the following argument has been advanced by van Hove [16]. If we compare the group of canonical transformations of classical mechanics and the corresponding group of unitary transformation of quantum mechanics, we observe that the absence of isomorphism between these two groups is the reason for ambiguities appearing in every rule of quantizing a classical system.

3.5 Symmetrization Rule

This is the simplest way of constructing the quantum mechanical operator from a given classical function. Thus for a classical function $q_c^s p_c^r$ we write

$$\mathcal{O}_S(q_c^s p_c^r) = \frac{1}{2}(p^r q^s + q^s p^r), \tag{3.102}$$

where on the right-hand side q and p are operators. To bring $\mathcal{O}\left(q_c^s p_c^r\right)$ to the normal form we use Eq. (3.100) and find that

$$\mathcal{O}_S(q_c^s p_c^r) = \frac{1}{2}\left[1 + \exp\left(-i\hbar\frac{\partial^2}{\partial q \partial p}\right)\right] q^s p^r, \tag{3.103}$$

This relation shows the similarity of this rule to the Weyl's rule which will be considered next. It also shows that the symmetrization rule has the same defects as the Weyl's rule [13].

3.6 Weyl's Rule

Let us denote the classical function by $T(p_c, q_c)$ and its Fourier transform by $g(\tau, \theta)$, i.e.

$$g(\tau, \theta) = \frac{1}{2\pi}\int T(p_c, q_c)\exp[-i(\tau p_c + \theta q_c)]\, dp_c dq_c. \tag{3.104}$$

Since the operators p and q satisfy the commutation relation, from Eq. (3.74) we obtain

$$\exp[i(\tau p + \theta q)] = \exp\left(\frac{i}{2}\hbar\tau\theta\right) e^{i\theta q} e^{i\tau p}. \tag{3.105}$$

Now according to Weyl the ordered quantum operator $O_W(T)$ is given by [17]

$$\mathcal{O}_W(T) = \frac{1}{2\pi}\int g(\tau, \theta) e^{\frac{i}{2}\hbar\tau\theta} e^{i\theta q} e^{i\tau p} d\tau d\theta. \tag{3.106}$$

McCoy observed that the Weyl rule can also be expressed as a differential operator [18]. In the classical function $T(p_c, q_c)$ we arrange the factors in such a way that q_c always precedes p_c, then the Weyl rule can be written as

$$\mathcal{O}_W(T) = \exp\left(-\frac{i\hbar}{2}\frac{\partial^2}{\partial q \partial p}\right) T(p, q), \tag{3.107}$$

where the exponential operator acts only on $T(p, q)$ and that we preserve the orders of p and q in T.

To generalize this rule to more than one degree of freedom we replace

$$\exp\left(-\frac{i\hbar}{2}\frac{\partial^2}{\partial q \partial p}\right), \tag{3.108}$$

by

$$\exp\left(-\frac{i\hbar}{2}\sum_k \frac{\partial^2}{\partial q_k \partial p_k}\right). \tag{3.109}$$

From Eqs. (3.100) and (3.107) it follows that if T is a real function then $O_W(T)$ will be Hermitian and that in the limit of $\hbar \to 0$ we recover the classical function.

Similar to the argument of van Hove that we cited earlier, Uhlhorn has shown that if the classical dynamics is regarded as a theory of canonical transformation in phase space, and quantum theory as a theory of unitary transformation in state-vector space, A_2, then these two groups of transformations are not isomorphic, but they contain subgroups which are isomorphic. However the classical point transformations can also be represented by a unitary transformation in the vector space A_1 (set of complex valued phase functions). Now according to the Weyl quantization rule the vector spaces A_1 and A_2 are isomorphic and therefore can be regarded as different representations of the same vector space A [19].

As a simple example of Weyl's rule let us consider $T(p_c, q_c)$ to be a single term polynomial

$$T_{r,s}(p_c, q_c) = p_c^r q_c^s, \tag{3.110}$$

with r and s positive integers. The Weyl–McCoy rule gives us

$$T_{r,s}(p, q) = \frac{1}{2^r}\sum_{j=0}^{r}\begin{pmatrix} r \\ j \end{pmatrix} p^{r-j} q^s p^j = \frac{1}{2^s}\sum_{j=0}^{s}\begin{pmatrix} s \\ j \end{pmatrix} q^{s-j} p^r q^j. \tag{3.111}$$

We can also write $T_{r,s}(p, q)$ in its normal form

$$T_{r,s}(p, q) = \frac{1}{2^r}\sum_{j=0}^{r}\begin{pmatrix} r \\ j \end{pmatrix} p^{r-j} q^s p^j = \sum_{k=0}^{s}\left(\frac{-i\hbar}{2}\right)^k k!\begin{pmatrix} r \\ k \end{pmatrix}\begin{pmatrix} s \\ k \end{pmatrix} q^{s-k} p^{r-k}. \tag{3.112}$$

This result is different from the result of ordering suggested by Born and Jordan which is [14];

$$T_{r,s}(p,q) = \frac{1}{r+1}\sum_{j=0}^{r} p^{r-j}q^s p^j. \tag{3.113}$$

While the Weyl–McCoy rule of association gives us a unique operator, it is not free of problems and sometimes we get strange and even unacceptable results [13].

For instance consider the square of any component of angular momentum **L**, say L_k (see Sec. 9.1). By applying Weyl–McCoy rule (3.107) to L_k^2 we have

$$\mathcal{O}_W(L_k^2) = [O_W(L_k)]^2 + \frac{1}{2}\hbar^2, \tag{3.114}$$

i.e. the Weyl ordered square of angular momentum is not the same as the square of the ordered form of angular momentum (for the operator form of L_k see Sec. 9.1.

Even stranger result is obtained if we choose the Hamiltonian to be

$$H = \frac{p_c^2}{2m} + \frac{1}{4}\lambda q_c^4. \tag{3.115}$$

By applying the operation (3.107) to this Hamiltonian and its square we find [13]

$$\mathcal{O}_W\left(H^2\right) = [O(H)]^2 + \frac{3\lambda\hbar^2}{4m}q^2. \tag{3.116}$$

Thus the square of energy is not a constant of motion and does not commute with the Hamiltonian. Since this is a serious defect of this type of ordering, one can look into the possibility of generalization of Weyl's rule in the form of

$$\mathcal{O}_W(T) = \left[\sum_k \beta_k \frac{\partial^{2k}}{\partial q^k \partial p^k}\right] T(p,q), \tag{3.117}$$

so that for the simple Hamiltonian

$$H = \frac{p_c^2}{2m} + V(q_c), \tag{3.118}$$

we get the relation

$$\mathcal{O}_W\left(H^n\right) = \left[\mathcal{O}_W(H)\right]^n, \tag{3.119}$$

for all n. This equation can be satisfied for $n = 1$ and $n = 2$ provided that

$$\beta_0 = 1, \quad \beta_1 = -\frac{1}{2}i\hbar, \quad \text{and} \quad \beta_2 = -\frac{1}{4}\hbar^2, \tag{3.120}$$

but for $n = 3$, no matter what we choose for β_3 and β_4 it is impossible to satisfy (3.119) [13].

3.7 Dirac's Rule

If we define the Poisson bracket of the two dynamical variables $u(p_c, q_c)$ and $v(p_c, q_c)$ by (1.117), then according to Dirac the commutator of $u(p, q)$ and $v(p, q)$ as quantum mechanical operators is given by [20]

$$[u(p,q), v(p,q)] = i\hbar \mathcal{O}_D\left(\{u, v\}_{p_c, q_c}\right), \tag{3.121}$$

where the operators in (3.121) must be self-adjoint. This rule which is widely used in quantum mechanics does not satisfy the requirement of uniqueness [13].

The non-uniqueness of ordering can be observed for the simple Hamiltonian for the one-dimensional motion

$$H = \frac{p_c^2}{2m}\exp(-3\gamma q_c), \tag{3.122}$$

where using the Dirac rule we have

$$\mathcal{O}_{D1}\left(\frac{1}{2m}p^2 e^{-3\gamma q}\right) = \frac{1}{18im\gamma\hbar}\left[p^3,\ e^{-3\gamma q}\right]$$
$$= \frac{1}{2m}e^{-3\gamma q}\left(p^2 + 3i\gamma p - 3\hbar^2\gamma^2\right) = H_1, \tag{3.123}$$

and

$$\mathcal{O}_{D2}\left(\frac{1}{2m}p^2 e^{-3\gamma q}\right) = \frac{1}{6im\gamma\hbar}\left[e^{-\gamma q}pe^{-\gamma q},\ pe^{-\gamma q}p\right]$$
$$= \frac{1}{2m}e^{-3\gamma q}\left(p^2 + 3i\gamma p - 2\hbar^2\gamma^2\right) = H_2. \tag{3.124}$$

Thus

$$H_2 = H_1 + \frac{1}{2m}\hbar^2\gamma^2\exp(-3\gamma q), \tag{3.125}$$

and from this relation it follows that

$$[H_2, H_1] = \frac{3\hbar^3\gamma^3}{2m^2}e^{-6\gamma q}(-3\gamma\hbar + 2ip), \tag{3.126}$$

i.e. either H_1 or H_2 may be regarded as a constant of motion but not both, and this is not correct. We note that the operators $p^3, e^{-3\gamma q}, e^{-\gamma q}pe^{-\gamma q}$ and $pe^{-\gamma q}p$ are all Hermitian operators [21].

From the Dirac's rule of association and the properties of the Poisson brackets discussed in Sec. 1.6 we can deduce the following rules for the commutators:

Properties of the Commutators — The commutators of a set of operators A, B and C satisfy the following relations which correspond to the properties of the Poisson brackets discussed in Sec. 1.6 [20]:

(1) - Skew symmetry

$$[A,\ B] = -[B,\ A], \quad [A,\ A] = 0. \tag{3.127}$$

(2) - Linearity

$$[\alpha A + \beta B,\ C] = \alpha[A,\ C] + \beta[B,\ C], \tag{3.128}$$

where α and β are constants.

(3) - Leibniz property

$$[AB,\ C] = A[B,\ C] + B[A,\ C]. \tag{3.129}$$

(4) - Jacobi identity

$$[A,\ [B,\ C]\,] + [B,\ [C,\ A]\,] + [C,\ [A,\ B]\,] = 0. \tag{3.130}$$

Commutators and Poisson Brackets — The parallelism between the commutators and the Poisson brackets as given in the Dirac rule of association (3.121) needs further investigation. In fact we have to define what we mean by the Poisson bracket on the right-hand side and how we should take the derivatives. For a function $u(\xi)$ depending on the non-commutating set of operators $\xi = (\xi_1, \xi_2 \cdots \xi_{2N})$ we can define a more general derivative, called Fréchet derivative, involving a differentiable but otherwise arbitrary function $v(\xi)$ in the following way [22],[23]:

$$\frac{\partial u}{\partial \xi_i}\{v\} \equiv \lim_{\varepsilon \to 0} \frac{1}{\varepsilon}\left[u(\xi_1 \cdots \xi_i + \varepsilon v, \cdots \xi_{2N}) - u(\xi)\right], \tag{3.131}$$

From this definition we find that

$$\frac{\partial (u_1 u_2)}{\partial \xi_i}\{v\} = u_1 \frac{\partial u_2}{\partial \xi_i}\{v\} + \frac{\partial u_1}{\partial \xi_i}\{v\}\, u_2, \tag{3.132}$$

and

$$\frac{\partial (a u_1 + b u_2)}{\partial \xi_i}\{v\} = a\frac{\partial u_1}{\partial \xi_i}\{v\} + b\frac{\partial u_2}{\partial \xi_i}\{v\}, \tag{3.133}$$

where a and b are constants. The essential result of this rule of taking derivative is that the commutator $[u,\ v]$ can be expressed as

$$[u,\ v] = \sum_{i,j} \frac{\partial u}{\partial \xi_j}\left\{\frac{\partial v}{\partial \xi_i}[\xi_i,\ \xi_j]\right\}. \tag{3.134}$$

To prove this result by induction we start with the relation

$$\sum_i \frac{\partial u}{\partial \xi_i}\{[\xi_i,\ v]\} = [u,\ v], \tag{3.135}$$

and note that this equality is true for $u =$ a constant, and for $u = \xi_i$. Next let us assume that (3.135) is correct for $u(\xi)$, where $u(\xi)$ is of the general form of the product of various factors of ξ_i. We consider the Fréchet derivative,

$$\begin{aligned} & \sum_j \frac{\partial(\xi_j u)}{\partial \xi_i}\{\xi_j,\, v\} \\ = & \lim_{\varepsilon\to 0}\left(\frac{1}{\varepsilon}[\xi_j + \varepsilon[\xi_j,\, v] - \xi_j]u(\xi)\right) + \sum_i \xi_j \frac{\partial u(\xi)}{\partial \xi_i}\{[\xi_i,\, v]\} \\ = & ([\xi_j,\, v]u + \xi_j[u,\, v]) = [\xi_j u,\, v]. \end{aligned} \tag{3.136}$$

Thus if (3.135) is true for u then (3.136) shows that the same relation is also true for $\xi_j u$. Already we have noticed that $u = 1$ and $u = \xi_i$ satisfy (3.135), therefore for any $u(\xi)$ which can be written as a product of factors of ξ_i this relation is satisfied.

By replacing u by v and v by ξ_i in (3.135) we obtain

$$\sum_j \frac{\partial v}{\partial \xi_j}\{[\xi_j,\, \xi_i]\} = [v,\, \xi_i]. \tag{3.137}$$

Now if we substitute this result in the right-hand side of (3.134) we have

$$[u,\, v] = \sum_{i,j} \frac{\partial u}{\partial \xi_i}\left\{\frac{\partial v}{\partial \xi_j}\{[\xi_j,\, \xi_i]\}\right\}. \tag{3.138}$$

This identity establishes a relation between the Poisson brackets and the commutators. In the special case when the set ξ_i s can be written as N mutually commuting q_i s and N mutually commuting p_i s, then (3.136) with $[q_j,\, p_k] = i\hbar\delta_{jk}$ becomes

$$[u,\, v] = i\hbar \sum_j \left[\frac{\partial u}{\partial q_j}\left\{\frac{\partial v}{\partial p_j}\right\} - \frac{\partial u}{\partial p_j}\left\{\frac{\partial v}{\partial q_j}\right\}\right], \tag{3.139}$$

where $\frac{\partial v}{\partial p_j}$ and $\frac{\partial v}{\partial q_j}$ are the usual partial derivatives. This form of the Poisson bracket is appropriate for non-commuting operators and reduces to the standard form when the variables commute. We observe that for the general case of non-commuting operators we can expand the right-hand side of (3.139)

$$[u,\, v] = i\hbar\{u,\, v\} + \mathcal{O}\left(\hbar^2\right), \tag{3.140}$$

i.e. in general we have terms proportional to $\hbar^2$, $\hbar^3$ etc. for the commutator.

An alternative way of writing the commutator $[u,\, v]$ is by the direct use of the definition of the Fréchet derivative. Thus for a system with one degree of freedom we have

$$[u,\, v] = i\hbar \lim_{\varepsilon\to 0}\frac{1}{\varepsilon}\left[u\left(q + \varepsilon\frac{\partial v}{\partial p}, p\right) - u\left(q, p + \varepsilon\frac{\partial v}{\partial q}\right)\right], \tag{3.141}$$

where it is assumed that the orders of the factors of p and q in both u and v are known and are preserved in the right-hand side of (3.141). A simple example of the application of (3.141) is provided by considering the case in which $u(p,q) = pq$ and $v = v(p,q)$, then [23]

$$[u,\ v] = i\hbar\left(p\frac{\partial v(p,q)}{\partial p} - \frac{\partial v(p,q)}{\partial q}q\right) = -[v(p,q),\ pq]. \tag{3.142}$$

3.8 Von Neumann's Rules

In the Heisenberg formulation of quantum mechanics the classical functions are replaced by matrices, therefore one can argue that the following rules should be used to find quantum mechanical operators [24]:

(a) - If $\mathcal{O}_V(A_c) = A$ then $\mathcal{O}_V[f(A_c)] = f(A)$.

(b) - If $\mathcal{O}_V(B_c) = B$ and $\mathcal{O}_V(C_c) = C$, then $\mathcal{O}_V(B_c + C_c) = B + C$, B and C may or may not commute with each other.

Von Neumann's rules do not give us a unique operator for a given classical function of p_c and q_c. For instance consider the function $G(p_c, q_c) = p_c^2 q_c^2$, then

$$\mathcal{O}_V\left(p_c^2 q_c^2\right) = q^2p^2 - 2i\hbar qp - \frac{1}{4}\hbar^2, \tag{3.143}$$

and

$$\mathcal{O}_V\left(p_c^2 q_c^2\right) = q^2p^2 - 2i\hbar qp - \hbar^2. \tag{3.144}$$

In addition to these basic rules there are other variants essentially with similar defects [13]. But all these rules are linear, i.e. they satisfy condition (b) of Von Neumann. The lack of uniqueness is of course a major defect of Von Neumann's and Dirac's rules, but if we insist on linearity and uniqueness then we must abandon Von Neumann's first rule.

3.9 Self-Adjoint Operators

As we have seen earlier, with physically measurable quantities such as momentum, angular momentum and Hamiltonian we can associate Hermitian operators. The time evolution of the state or observables of the system are given by the unitary operator $U(t) = \exp(iHt)$. We require that $U(t)$ be a unitary operator so that the probability is conserved. However the Hermiticity of H does not guarantee a unique solution for the dynamics of the problem. Here we want to study this question and in particular show that only when H is self-adjoint operator in a given domain, then the dynamics is uniquely defined. A Hermitian H, in general, may lead to several motions or to none. Thus the condition of self-adjointness is more restrictive (and thus stronger) than the condition of Hermiticity. That is, there are additional conditions that must be satisfied for a Hermitian operator to be self-adjoint [25],[26].

Given a Hermitian operator there are three possible situations that we can have:

(a) - The operator is by itself self-adjoint and no other condition is needed. Thus $\hat{O}$ is self-adjoint if

$$\left\langle g,\, \hat{O}f\right\rangle = \left\langle \hat{O}^{\dagger}g, f\right\rangle. \tag{3.145}$$

For instance the momentum operator $p = -i\hbar\frac{d}{dx}$ defined on the Hilbert space $(-\infty,\, +\infty)$ is self-adjoint since [27],[28]

$$\int_{-\infty}^{+\infty} g^*(x)(-i\hbar)\frac{df(x)}{dx}dx = \int_{-\infty}^{+\infty}\left(-i\hbar\frac{dg(x)}{dx}\right)^* f(x)dx. \tag{3.146}$$

(b) - Some extra conditions for self-adjointless are required, and depending on the form of these additional conditions we can have different dynamics.

(c) - There are some cases where it is impossible to make Hermitian operators self-adjoint, and therefore no acceptable quantum mechanical solution can be found for these problems [26].

To begin our discussion we need to define the concepts of self-adjointness and self-adjoint extension of an operator.

Let us consider two linear operators $\hat{O}$ and $\hat{P}$ defined on two dense subspaces $\mathcal{D}\left(\hat{O}\right)$, $\mathcal{D}\left(\hat{P}\right)$ of the Hilbert space $\mathcal{L}_2$. Now if $\mathcal{D}\left(\hat{P}\right) \supset \mathcal{D}\left(\hat{O}\right)$ and $\hat{P}f = \hat{O}f$ for all $f \in \mathcal{D}\left(\hat{O}\right)$, then $\hat{P} \supset \hat{O}$ and $\hat{P}$ is called an extension of $\hat{O}$. In addition if $\hat{O}^* \supset \hat{O}$ we call $\hat{O}$ Hermitian, and when $\hat{O} = \hat{O}^*$ then we call $\hat{O}$ self-adjoint. Furthermore if $\hat{O}^{**} = \hat{O}^*$, $\hat{O}$ is called essential self-adjoint.

In the case where the operator $\hat{O}$ is Hermitian but is not self-adjoint, its departure from self-adjointness is measured by two numbers which we call deficiency indices (n_+, n_-). These are defined as the number of independent solutions of

$$\hat{O}^* f_\pm = \pm i f_\pm, \tag{3.147}$$

respectively. Let us suppose that both n_+ and n_- have nonzero values, then the operator $\hat{P}$ which is the extension of $\hat{O}$ is defined by

$$\hat{P}\left[f + \alpha(f_+ + f_-)\right] = \hat{O}f + i\alpha(f_+ - f_-), \tag{3.148}$$

with α being an arbitrary complex number. The deficiency indices of $\hat{P}$ are then $(n_+ - 1, n_- - 1)$.

As the first example we consider the problem of self-adjointness of the momentum operator $\hat{O} = -i\frac{d}{dx}$ when it acts on a domain Ω. Equation (3.147) for this operator has the solution

$$f_\pm(x) \sim \exp\left(\mp x\right). \tag{3.149}$$

(1) - For Ω being the entire x-axis, neither f_+ nor f_- is square integrable. Therefore $n_+ = n_- = 0$ and $\hat{O}$ is self-adjoint on this domain.

(2) - When $0 \leq x \leq \infty$, $f_- = \exp x$ is not square integrable but f_+ is. Thus $n_+ = 1$ and $n_- = 0$ and $\hat{O}$ is not self-adjoint and has no self-adjoint extension.

When the particle moves in a box of unit length $0 \leq x \leq 1$ the condition (3.145) becomes

$$\begin{aligned} &\int_0^1 \left[g^*(x)\left(-i\frac{\partial f(x)}{\partial x}\right) - f(x)\left(-i\frac{\partial g(x)}{\partial x}\right)^*\right] dx \\ &= -i\left[g^*(1)f(1) - g^*(0)f(0)\right] = 0, \end{aligned} \tag{3.150}$$

This relation is satisfied if

$$f(1) = e^{2\pi i\beta} f(0), \quad 0 \leq \beta < 1, \tag{3.151}$$

for all functions in the domain $0 \leq x \leq 1$.

In the case of motion of a free particle in the segment $0 \leq x \leq 1$ with periodic boundary condition $f(0) = f(1)$ we can choose $f(x)$ to be the eigenfunctions of the Hamiltonian operator,

$$H = -\frac{\hbar^2}{2m}\frac{d^2}{d\,x^2}, \tag{3.152}$$

then $f(x)$ will be a set of twice differentiable functions given by

$$f_n(x) = \frac{1}{\sqrt{2}}\exp(2i\pi n x), \quad n = 0, \pm 1, \pm 2 \cdots. \tag{3.153}$$

On the other hand for antiperiodic boundary condition, $f(1) = -f(0)$, we find

$$f_n(x) = \frac{1}{\sqrt{2}}\exp(i\pi(2n+1)x), \quad n = 0, \pm 1, \pm 2 \cdots. \tag{3.154}$$

Now let us examine the self-adjointness of the Hamiltonian operator (3.152) when the particle can move in a box of unit length $0 \leq x \leq 1$. The condition of the self-adjointness of H implies that

$$-\frac{\hbar^2}{2m}\int_0^1 \psi^*(x)\frac{d^2\phi(x)}{d\,x^2}dx = -\frac{\hbar^2}{2m}\int_0^1\left(\frac{d^2\psi(x)}{d\,x^2}\right)^*\phi(x)dx, \tag{3.155}$$

or by integrating by parts this condition becomes

$$\left[\psi^*(x)\frac{d\phi(x)}{dx} - \frac{d\psi^*(x)}{dx}\phi(x)\right]_0^1 = 0. \tag{3.156}$$

Equation (3.156) holds for an arbitrary $\psi(x)$ and $\phi(x)$ provided

$$\begin{bmatrix} -\psi'(1) \\ \psi'(0) \end{bmatrix} = M\begin{bmatrix} \psi(1) \\ \psi(0) \end{bmatrix}, \tag{3.157}$$

where the minus sign in front of $\psi'(1)$ is introduced for convenience. In (3.157) M must be a 2×2 Hermitian matrix in order that (3.156) be satisfied. The matrix M depends on four independent real quantities, and in general, we can express it as the matrix

$$M = \begin{bmatrix} \rho + \gamma & -\rho e^{i\theta} \\ -\rho e^{-i\theta} & \rho + \beta \end{bmatrix}, \tag{3.158}$$

where ρ, β, θ and γ are all real and thus M is Hermitian i.e. $M_{ij} = M_{ji}^*$. Now let us write the probability current $j(x)$ associated with the wave function $\psi(x)$ as

$$j(x) = \frac{\hbar}{2im}\left(\psi^*(x)\frac{d\psi(x)}{dx} - \psi(x)\frac{d\psi^*(x)}{dx}\right). \tag{3.159}$$

From this relation and (3.158) it follows that

$$j(0) = j(1) = \frac{\hbar\rho}{2im}\left[\psi^*(1)\psi(0)e^{i\theta} - \psi(1)\psi^*(0)e^{-i\theta}\right]. \tag{3.160}$$

This expression shows that the probability is conserved since $j(0) = j(1)$. However the current at the position of the walls is zero if and only if $\rho = 0$, and when this happens then we have $-\psi'(1) = \gamma\psi(1)$ and $-\psi'(0) = \beta\psi(0)$. Therefore there is no possibility of current flowing from one wall to the other. Let us consider the other limit when $\rho \to \infty$, and θ, γ and β all go to zero, then from (3.157) and (3.158) we find $\psi(1) = \psi(0)$ and $\psi'(1) = \psi'(0)$. This boundary condition means that we have no reflection at the walls, and if a wave packet collides with one of the walls it will instantly appear at the other. For the general case when ρ is not zero or infinity, we have partial reflection and partial transmission of the wave packet.

Next let us study the question of the self-adjoint extension of the motion of a particle in a box. Setting the units so that $\hbar^2 = 2m = 1$, we can find the deficiency indices that we defined earlier from the solution of the differential equation

$$-\frac{d^2}{d\,x^2}f_\pm(x) = \pm i f_\pm(x). \tag{3.161}$$

The solution of this equation is

$$f_\pm = \exp(\alpha x), \tag{3.162}$$

where α is given by

$$\alpha = \pm\sqrt{\pm i}. \tag{3.163}$$

Thus we have four square integrable solutions and the deficiency indices are $(n_+ = 2, n_- = 2)$ corresponding to the reflected and transmitted waves for the two directions. If we let one side of the box go to infinity $(0 \le x \le \infty)$ then the solutions with $\alpha = \sqrt{i}$ and $\alpha = -\sqrt{-i}$ will no longer be square integrable and we have $(n_+ = 1, n_- = 1)$, viz, a one parameter family of dynamics depending on the reflection law at the wall at $x = 0$. If we also move the other wall to $-\infty$, i.e. $-\infty \leq x \leq \infty$, then there will be no square integrable solution of (3.161). This means that in this case $(n_+ = 0, n_- = 0)$ and there are no reflections from the boundaries [25], [29], [30].

An important point which is useful in this connection is the correspondence between the classical solution of the equation of motion and the exitance of a self-adjoint Hamiltonian or a Hamiltonian with self-adjoint extension for the corresponding quantum problem. This connection can be used to recognize those quantum mechanical Hamiltonians that are troublesome. Let us consider some specific cases where we can solve the classical motion exactly, or find the solution in the form of a quadrature. Here again we set $\hbar = 1$.

(1) - As a first case consider the problem of a particle of unit mass moving in a quartic potential $V(q) = \frac{1}{4}q^4$ and thus the Hamiltonian has the simple form

$$H = \frac{1}{2}p^2 + \frac{1}{4}q^4 = E > 0. \tag{3.164}$$

Since $p = \dot{q}$, we have

$$\dot{q}^2 = 2E - \frac{1}{2}q^4, \tag{3.165}$$

and by solving the differential equation (3.165) we obtain

$$t + c = \pm \int^{q(t)} \frac{dq}{\sqrt{2E - \frac{1}{2}q^4}}. \tag{3.166}$$

For this problem we have a well-defined and unique solution for all times and the motion is periodic with an energy-dependent period $T(E)$ [26]. Now by solving the equation

$$-\frac{1}{2}\frac{d^2}{d\,q^2}f_\pm + \frac{1}{4}q^4 f_\pm = \pm i f_\pm, \tag{3.167}$$

we find that the deficiency indices are $(n_+ = 0,\ n_- = 0)$, and therefore the Hamiltonian is self-adjoint.

Among other physically important cases where the Hamiltonians are self-adjoint are the problems of the harmonic oscillator

$$H = -\frac{1}{2m}\frac{d^2}{d\,x^2} + \frac{1}{2}m\omega^2 x^2, \tag{3.168}$$

and that of the hydrogen atom

$$H = -\frac{1}{2mr^2}\frac{\partial}{\partial r}\left(r^2\frac{\partial}{\partial r}\right) - \frac{Ze^2}{r} + \frac{\ell(\ell+1)}{2mr^2}, \tag{3.169}$$

where the negative energy eigenvalues of H form a discrete spectrum.

Another important example is the motion of a charged particle in a uniform electric field $\mathcal{E}$, where

$$H = -\frac{1}{2m}\frac{d^2}{d\,x^2} + e\mathcal{E}x. \tag{3.170}$$

For this motion the deficiency indices are ($n_+ = 0,\ n_- = 0$) and therefore H is self-adjoint.

(2) - If instead of the attractive potential in (3.164) we have a repulsive quartic potential $V(q) = -\frac{1}{4}q^4$, then assuming that the energy of the particle is E, it takes a finite time for the particle to move from the classical turning point $q = q_0$ where its energy is E to infinity

$$t = \int_{q_0}^{\infty} \frac{dq}{2\sqrt{E - V(q)}} \sim \int_{q_0}^{\infty} \frac{dq}{q^2} < \infty. \tag{3.171}$$

When reflected from infinity the particle reaches q_0 in a finite time, therefore the reflection law at infinity is important in determining the dynamics. If we solve the corresponding equation for $f_{\pm}(q)$ for this potential we find the deficiency indices to be ($n_+ = 2,\ n_- = 2$). The spectrum of the self-adjoint extension of this motion is discrete [26].

We can use the same method to consider the spectra of the operator $\frac{1}{2}\left(p^2 - q^2\right)$. This operator arises in the theory of self-acceleration of charged particles [31]. van Kampen studied the spectral decomposition of this operator in detail and concluded that the spectrum is continuous and extends from $-\infty$ to $+\infty$. Just as in the previous example we find the deficiency indices for this operator to be ($n_+ = 0,\ n_- = 0$) and thus H is self-adjoint.

(3) - For the classical Hamiltonian $H = \frac{2p}{q}$, when $-\infty < x \le 0$ or $0 \le x < \infty$ the classical equation of motion is $\dot{q} = \frac{2}{q}$ which can be solved for q with the result

$$q^2 = 4t + c. \tag{3.172}$$

This solution shows that for $t < -\frac{c}{4}$, $q(t)$ is imaginary and therefore there is no classical solution for all times.

In quantum mechanical formulation of the problem we note that if we write a Hermitian Hamiltonian of the form

$$H = -i\left(\frac{1}{q}\frac{d}{dq} + \frac{d}{dq}\frac{1}{q}\right), \tag{3.173}$$

then the deficiency indices are found from the solutions of

$$-\frac{2i}{q}\frac{df_{\pm}(q)}{dq} + \frac{i}{q^2}f_{\pm}(q) = \pm i f_{\pm}(q). \tag{3.174}$$

By solving the differential equation (3.174) we find

$$f_{\pm}(q) = \sqrt{|q|}\exp\left(\mp\frac{q^2}{4}\right). \tag{3.175}$$

Since in the domains of $-\infty < x \le 0$ or $0 \le x < \infty$, only $f_+(q)$ is square integrable therefore the deficiency indices are ($n_+ = 1, n_- = 0$) hence this Hamiltonian is not self-adjoint nor has a self-adjoint extension.

3.10 Momentum Operator in a Curvilinear Coordinates

For a number of applications we need to find the mathematical form of the momentum operator and the kinetic energy in curvilinear coordinates. This problem can be formulated in different ways, for example, for the kinetic energy we can directly transform the Laplacian operator from Cartesian to the desired coordinate system. But here we try to find the linear and angular momentum operators in arbitrary curvilinear coordinate systems [32]–[34].

Position and Momentum Operators in Cartesian Coordinates — For the description of a physical system it is convenient to introduce a particular coordinate system in Hilbert space $\mathcal{L}_2$, that is to choose a representation. The representation in which $\mathbf{r}$, the position operator, is diagonal is called the position representation, but we can equally use a representation in which the linear momentum $\mathbf{p}$ is diagonal. If we choose the position representation then the state vector $|\psi\rangle$ is specified by its components $\langle\mathbf{r}|\psi\rangle$ along the ket $|\mathbf{r}\rangle$. The eigenvalue in this case is

$$\hat{\mathbf{r}}|\mathbf{r}\rangle = \mathbf{r}|\mathbf{r}\rangle, \tag{3.176}$$

where $\hat{\mathbf{r}}$ is the position operator with $\mathbf{r}$ being its eigenvalue. Here the spectrum of the operator $\hat{\mathbf{r}}$ consists of points in three-dimensional space.

The completeness relation for the eigenvectors can be written as

$$\int |\mathbf{r}\rangle\langle\mathbf{r}|\, d^3r = 1. \tag{3.177}$$

The form of the momentum operator $\mathbf{p}$ in this, i.e. position representation is obtained from the commutation relation $[p_j x_k - x_k p_j] = -i\hbar\delta_{jk}$, Eq. (3.99). Thus by calculating the matrix element of this commutator we find

$$\langle\mathbf{r}'\,|[x_k,\, p_j]|\,\mathbf{r}\rangle = i\hbar\delta_{kj}\delta\left(\mathbf{r}-\mathbf{r}'\right) = \left(x'_k - x_k\right)\langle\mathbf{r}'\,|\mathbf{p}|\,\mathbf{r}\rangle. \tag{3.178}$$

Recalling that $x\frac{d}{dx}\delta(x) = -\delta(x)$, we get

$$\langle\mathbf{r}'\,|p_j|\,\mathbf{r}\rangle = -i\hbar\frac{\partial}{\partial x_j}\langle\mathbf{r}'|\mathbf{r}\rangle. \tag{3.179}$$

From this result we conclude that the momentum operator in the position representation is given by

$$\mathbf{p} = -i\hbar\nabla. \tag{3.180}$$

The Momentum Operator — We note that simple expression (3.180) for the momentum operator is valid only in rectangular coordinates (see Eqs. (3.220) and (3.221) below).

In a curvilinear coordinates $\left(q^1, q^2, q^3\right)$ with the base vectors $(\mathbf{e}_1, \mathbf{e}_2, \mathbf{e}_3)$ we can write a small displacement as

$$d\mathbf{r} = \sum_{n=1}^{3} \mathbf{e}_n dq^n, \tag{3.181}$$

and in this system the components of the momentum operator (3.180) can be expressed as [32]

$$p_n = -\frac{i\hbar}{2}\left[\mathbf{e}_n \cdot \nabla + \nabla \cdot (\mathbf{e}_n \cdots)\right]$$
$$= -i\hbar\left[\mathbf{e}_n \cdot \nabla + \frac{1}{2}\nabla \cdot \mathbf{e}_n\right], \tag{3.182}$$

where $\nabla \cdot (\mathbf{e}_n \cdots)$ means that ∇ operator acts on every term on its right side. Note that since in any point transformation the new momenta depend linearly on the old ones, therefore there is no ambiguity in constructing the Hermitian operators, and the method of symmetrization can be used- which is what we have done in getting Eq. (3.182).

Introducing g_{ij} as the metric tensor and g as the determinant of g_{ij},

$$g = \text{Det } g_{ij} = \begin{vmatrix} g_{11} & g_{12} & g_{13} \\ g_{21} & g_{22} & g_{23} \\ g_{31} & g_{32} & g_{33} \end{vmatrix}, \tag{3.183}$$

we can express the divergent of a vector $\mathbf{V}$ as

$$\nabla \cdot \mathbf{V} = \frac{1}{J}\sum_{k=1}^{3}\frac{\partial\left(V^k J\right)}{\partial q^k}, \qquad J = \sqrt{g}, \tag{3.184}$$

where J is the Jacobian for the volume element in the curvilinear coordinate system and the vector $\mathbf{V}$ is expressed in terms of its components by

$$\mathbf{V} = \sum_{k=1}^{3} V^k \mathbf{e}_k. \tag{3.185}$$

If we choose $\mathbf{V} = \mathbf{e}_n$, then $V^k = \delta^k_n$, and the divergence of $\mathbf{e}_n$ becomes

$$\nabla \cdot \mathbf{e}_n = \frac{\partial \ln J}{\partial q^n} = \sum_{j=1}\Gamma^j_{nj}. \tag{3.186}$$

In this relation $\sum_{j=1}\Gamma^j_{nj}$ is the contracted Christoffel symbol, where the Christoffel symbol Γ^k_{nj} is defined by [35]

$$\Gamma^k_{nj} = \sum_l \frac{1}{2}g^{kl}\left(\frac{\partial g_{nl}}{\partial x^j} + \frac{\partial g_{lj}}{\partial x^n} - \frac{\partial g_{nj}}{\partial x^l}\right). \tag{3.187}$$

Now we have $\mathbf{e}_n \cdot \nabla = \frac{\partial}{\partial q^n}$, and therefore

$$p_n = -i\hbar\left(\frac{\partial}{\partial q^n} + \frac{1}{\sqrt{J}}\frac{\partial}{\partial q^n}\sqrt{J}\right) = -i\hbar\frac{1}{\sqrt{J}}\frac{\partial(\sqrt{J}\cdots)}{\partial q^n}. \tag{3.188}$$

The self-adjointness condition for p_n implies that

$$\begin{aligned}&\langle\psi_1|p_n|\psi_2\rangle - \langle\psi_2|p_n|\psi_1\rangle^* \\ = &-i\hbar\int dq^1\int dq^2\int\frac{\partial}{\partial q^n}\left(\psi_1^* J\psi_2\right)dq^3 = 0,\end{aligned} \tag{3.189}$$

where the vectors $|\psi_1\rangle$ and $|\psi_2\rangle$ are assumed to be in the Hermitian domain of p_n. Next we define the set of $\{\mathbf{e}^n\}$ s which is the set of reciprocal base vectors by

$$\mathbf{e}^n = \sum_{j=1}^{3} g^{nj}\mathbf{e}_j, \quad \mathbf{e}_n = \sum_{j=1}^{3} g_{nj}\mathbf{e}^n, \quad \mathbf{e}^n\cdot\mathbf{e}_j = \delta^n_j. \tag{3.190}$$

From the components p_n, Eq. (3.182), we obtain $\mathbf{p}$;

$$\mathbf{p} = -i\hbar\nabla = -\frac{i\hbar}{2}\sum_{n=1}^{3}\left[\mathbf{e}^n(\mathbf{e}_n\cdot\nabla) + \nabla\cdot(\mathbf{e}_n\mathbf{e}^n\cdots)\right]. \tag{3.191}$$

To simplify this expression we take the divergence of $\sum_{n=1}^{3}\mathbf{e}_n\mathbf{e}^n\cdots$

$$\nabla\cdot\left(\sum_{n=1}^{3}\mathbf{e}_n\mathbf{e}^n\cdots\right) = \sum_{n=1}^{3}\mathbf{e}^n\nabla\cdot(\mathbf{e}_n\cdots) + \sum_{n=1}^{3}\mathbf{e}_n\cdot\nabla\left(\mathbf{e}^n\right). \tag{3.192}$$

By substituting (3.192) in (3.191) we find $\mathbf{p}$

$$\mathbf{p} = \sum_{n=1}^{3}\mathbf{e}^n p_n - \frac{i\hbar}{2}\sum_{n=1}^{3}\mathbf{e}_n\cdot\nabla\left(\mathbf{e}^n\right). \tag{3.193}$$

If we combine this relation with (3.187) and note that

$$\nabla\cdot\left(\sum_{n=1}^{3}\mathbf{e}_n\mathbf{e}^n\right) = 0, \tag{3.194}$$

we obtain

$$\mathbf{p} = \sum_{n=1}^{3}\mathbf{e}^n\pi_n, \tag{3.195}$$

where from (3.188) we have an expression for π_n

$$\pi_n = p_n + i\hbar\frac{1}{\sqrt{J}}\frac{\partial\sqrt{J}}{\partial q^n} = -i\hbar\frac{\partial}{\partial q^n}, \tag{3.196}$$

Comparing (3.196) with (3.188) we get a simple form for π_n;

$$\pi_n = \sqrt{J}\,p_n\left(\frac{1}{\sqrt{J}}\right). \tag{3.197}$$

In a similar way we can substitute the expression

$$\sum_{n=1}^{3} \mathbf{e}^n(\mathbf{e}_n \cdot \nabla) = \sum_{n=1}^{3} \left[\mathbf{e}_n \cdot \nabla\left(\mathbf{e}^n \cdots\right) - \mathbf{e}_n \cdot \nabla\left(\mathbf{e}^n\right)\right], \tag{3.198}$$

in (3.191) and write $\mathbf{p}$ as

$$\mathbf{p} = \sum_{n=1}^{3} \left[p_n \mathbf{e}^n + \frac{i\hbar}{2}\mathbf{e}_n \cdot \nabla\left(\mathbf{e}^n\right)\right] = \sum_{n=1}^{3} \pi_n^\dagger \mathbf{e}^n. \tag{3.199}$$

where $\pi_n^\dagger$ which is the Hermitian adjoint of π_n is given by

$$\pi_n^\dagger = p_n - i\hbar \frac{1}{\sqrt{J}} \frac{\partial \sqrt{J}}{\partial q^n}. \tag{3.200}$$

Again using (3.188) we can express $\pi_n^\dagger$ as

$$\begin{aligned} \pi_n^\dagger &= -i\hbar\left[\frac{\partial}{\partial q^n} + \frac{1}{J}\frac{\partial J}{\partial q^n}\right] \\ &= -i\hbar \frac{1}{J}\frac{\partial}{\partial q^n}(J \cdots) = \frac{1}{\sqrt{J}}\, p_n \sqrt{J}, \end{aligned} \tag{3.201}$$

and this result is in agreement with (3.197).

Thus the momentum operator can be written in two different ways

$$\mathbf{p} = \sum_{n=1}^{3} \mathbf{e}_n \pi^n, \tag{3.202}$$

or

$$\mathbf{p} = \sum_{n=1}^{3} \pi^{n\,\dagger} \mathbf{e}_n, \tag{3.203}$$

where the contravariant components of $\mathbf{p}$ are:

$$\pi^n = \sum_{j=1}^{3} g^{nj}\pi_j = \sqrt{J} p^n \left(\frac{1}{\sqrt{J}}\right), \tag{3.204}$$

$$\pi^{n\dagger} = \sum_{j=1}^{n} \pi_j g^{jn} = \frac{1}{\sqrt{J}} p^{n\,\dagger}\sqrt{J}, \tag{3.205}$$

and

$$p^n = \sum_{j=1}^{3} g^{nj} p_j, \qquad p^{n\,\dagger} = \sum_{j=1}^{3} p_j\, g^{jn}. \tag{3.206}$$

The Kinetic Energy — Having found $\mathbf{p}$, let us write the kinetic energy of a particle of mass m in any curvilinear coordinate system

$$\begin{aligned} T = \frac{\mathbf{p}^2}{2m} &= -\frac{\hbar^2}{2m}\nabla^2 = \sum_{n=1}^{3}\pi_n^\dagger \pi^n = \sum_{n=1}^{3}\pi^{n\,\dagger}\pi_n \\ &= \frac{\hbar^2}{2m}\sum_{n=1}^{3}\frac{1}{\sqrt{J}}p_n J\, p^n\left(\frac{1}{\sqrt{J}}\right) \\ &= \frac{\hbar^2}{2m}\sum_{n=1}^{3}\frac{1}{\sqrt{J}}p^{n\,\dagger} J\, p_n\left(\frac{1}{\sqrt{J}}\right). \end{aligned} \tag{3.207}$$

We can simplify the above expression for T when we deal with an orthogonal curvilinear coordinate system. For such a system g_{ij} s are given by [35]

$$g_{11} = h_1^2, \quad g_{22} = h_2^2, \quad g_{33} = h_3^2, \quad g_{ij} = 0 \quad \text{for} \quad i \neq j. \tag{3.208}$$

Also

$$g^{11} = \frac{1}{h_1^2}, \quad g^{22} = \frac{1}{h_2^2}, \quad g^{33} = \frac{1}{h_3^2}, \quad g^{ij} = 0 \quad \text{for} \quad i \neq j, \tag{3.209}$$

where the quantity h_n is the scale factor for the coordinate q_n. Thus we have

$$J = h_1 h_2 h_3, \quad p^n = \frac{1}{h_n^2}p_n, \quad p^{n\,\dagger} = p_n\frac{1}{h_n^2}. \tag{3.210}$$

and

$$x_n = h_n^2 x^n, \qquad \mathbf{e}_n = h_n^2\mathbf{e}^n. \tag{3.211}$$

By substituting for g_{ij} s the expressions for linear momentum and kinetic energy reduce to

$$\mathbf{p} = \sum_{n=1}^{3}\mathbf{e}^n\sqrt{J}\,p_n\left(\frac{1}{\sqrt{J}}\right) = \sum_{n=1}^{3}\mathbf{e}_n\left(\frac{1}{\hbar_n^2}\right)\sqrt{J}\,p_n\left(\frac{1}{\sqrt{J}}\right). \tag{3.212}$$

and

$$T = \frac{\mathbf{p}^2}{2m} = \frac{1}{2m}\sum_{n=1}^{3}\left(\frac{1}{\sqrt{J}}\right)p_n\left(\frac{J}{h_n^2}\right)p_n\left(\frac{1}{\sqrt{J}}\right). \tag{3.213}$$

In view Eq. (3.188) we can write the kinetic energy as

$$T = -\frac{\hbar^2}{2m}\sum_{n=1}^{3}\frac{1}{J}\frac{\partial}{\partial q^n}\left(\frac{J}{h_n^2}\frac{\partial}{\partial q^n}\right). \tag{3.214}$$

For a general curvilinear coordinates with the line element

$$ds^2 = \sum_{i,j=1}^{3} g_{ij}dx^i dx^j, \tag{3.215}$$

the kinetic energy can be expressed as

$$T = \frac{1}{2m}\sum_{i,j=1}^{3}\frac{1}{\sqrt{g}}\left(\pi_i\sqrt{g}\,g^{ij}\,\pi_j\right), \tag{3.216}$$

and this operator is self-adjoint with respect to integration over the volume $\sqrt{g}\,dq^1dq^2dq^3$;

$$\langle\psi_1|T|\psi_2\rangle = \langle\psi_2|T|\psi_1\rangle^*, \tag{3.217}$$

exactly as in Eq. (3.189).

Canonical Momentum and Kinetic Energy in Spherical Polar Coordinates — As an example let us consider expressions for $\mathbf{p}$ and $\frac{\mathbf{p}^2}{2m}$ is spherical polar coordinates. For this system

$$\begin{cases} q^1 = r, \quad q^2 = \theta, \quad q^3 = \phi, \\ x^1 = r, \qquad x^2 = x^3 = 0 \end{cases}. \tag{3.218}$$

In addition we have

$$h_1 = 1, \quad h_2 = r, \quad h_3 = r\sin\theta, \quad J = r^2\sin\theta. \tag{3.219}$$

Using these in (3.188) we obtain

$$p_r = -i\hbar\left(\frac{\partial}{\partial r}+\frac{1}{r}\right), \tag{3.220}$$

$$p_\theta = -i\hbar\left(\frac{\partial}{\partial\theta}+\frac{1}{2}\cot\theta\right), \tag{3.221}$$

and

$$p_\phi = -i\hbar\frac{\partial}{\partial\phi}. \tag{3.222}$$

Also from (3.212) we find an alternative expression for $\mathbf{p}$

$$\begin{aligned}\mathbf{p} &= \mathbf{e}^r r\,p_r\left(\frac{1}{r}\right) + \mathbf{e}^\theta\,\sqrt{\sin\theta}\,p_\theta\left(\frac{1}{\sqrt{\sin\theta}}\right) + \mathbf{e}^\phi p_\phi \\ &= \frac{1}{r}\,p_r\,r\,\mathbf{e}_r + \frac{1}{r^2}\left[\frac{1}{\sqrt{\sin\theta}}\,p_\theta\sqrt{\sin\theta}\,\mathbf{e}_\theta + \frac{1}{\sin^2\theta}p_\phi\mathbf{e}_\phi\right].\end{aligned} \tag{3.223}$$

The kinetic energy operator T can be found from (3.223);

$$\begin{aligned}T &= \frac{\mathbf{p}^2}{2m} = \frac{1}{2m}\left\{\frac{1}{r}\,p_r\left(r^2p_r\,\frac{1}{r}\right)\right. \\ &+ \frac{1}{2m\,r^2}\left[\frac{1}{\sin\theta}\,p_\theta\left(\sin\theta\,p_\theta\,\frac{1}{\sqrt{\sin\theta}}\right) + \frac{1}{\sin^2\theta}p_\phi^2\right]\end{aligned} \tag{3.224}$$

Angular Momentum in Curvilinear Coordinates — We can obtain the components of the angular momentum $\mathbf{L} = \mathbf{r} \wedge \mathbf{p}$ in the same way. Thus if we start with

$$\mathbf{L} = \mathbf{r} \wedge \mathbf{p} = \sum_{n,j=1}^{3} x_n \mathbf{e}^n \wedge \mathbf{e}^j \pi_j, \tag{3.225}$$

and substitute for π^j from (3.204) we find [32]

$$\mathbf{L} = \frac{1}{\sqrt{j}} \left[\left(p^{3\dagger} x^2 - p^{2\dagger} x^3\right) \mathbf{e}^1 + \left(p^{1\dagger} x^3 - p^{3\dagger} x^1\right) \mathbf{e}^2 + \left(p^{2\dagger} x^1 - p^{1\dagger} x^2\right) \mathbf{e}^3 \right] J^{\frac{3}{2}}. \tag{3.226}$$

In particular in spherical polar coordinates we have

$$\mathbf{L} = \left[-\frac{p_\phi}{\sin^2\theta} \mathbf{e}^\theta + \left(p_\theta - i\frac{\hbar}{2} \cot\theta \right) \mathbf{e}^\phi \right] r \sin\theta, \tag{3.227}$$

where $J = h_1 h_2 h_3 = r^2 \sin\theta$. From this relation we can calculate $\mathbf{L}^2$;

$$\mathbf{L}^2 = \left(p_\theta - i\frac{\hbar}{2} \cot\theta \right) \left(p_\theta + i\frac{\hbar}{2} \cot\theta \right) + \frac{1}{\sin^2\theta} p_\phi^2. \tag{3.228}$$

If we substitute (3.228) in the expression for the kinetic energy T we find (See also Eq. (9.26))

$$T = \frac{\mathbf{p}^2}{2m} = \frac{1}{2m} \left(p_r^2 + \frac{\mathbf{L}^2}{r^2} \right). \tag{3.229}$$

The expression found for the kinetic energy in curvilinear coordinates, Eq. (3.216) is important in the formulation of the quantum theory of collective phenomena, such as the surface oscillations of a nucleus [36].

3.11 Summation Over Normal Modes

In a number of problems, for example the quantization of free electromagnetic field, it is convenient to formulate the problem in a large cubic box of volume L^3, and then take the limit of L going to infinity. Here by imposing the boundary conditions such as the condition of perfect reflectivity at the walls of the box we obtain a set of normal modes

$$k_x = \frac{2n_x\pi}{L}, \quad k_y = \frac{2n_y\pi}{L}, \quad \text{and} \quad k_z = \frac{2n_z\pi}{L}, \tag{3.230}$$

where n_x, n_y and n_z are integers. The vector $\mathbf{n} = (n_x, n_y, n_z)$ represents a normal mode in the lattice space. Let us consider a small volume located between n_x and $n_x + \Delta n_x$, n_y and $n_y + \Delta n_y$ and n_z and $n_z + \Delta n_z$. The number of lattice points which is the number of normal modes in this volume is $\Delta n_x \Delta n_y \Delta n_z$. From (3.230) we find

$$L\Delta k_x = 2\pi\Delta n_x, \quad L\Delta k_y = 2\pi\Delta n_y, \quad \text{and} \quad L\Delta k_z = 2\pi\Delta n_z. \tag{3.231}$$

Thus

$$\Delta n_x \Delta n_y \Delta n_z = \frac{L^3}{(2\pi)^3}\Delta k_x \Delta k_y \Delta k_z. \tag{3.232}$$

In the limit of $L \to \infty$, $\Delta k_x, \Delta k_y$ and Δk_z all go to zero and the summation over all normal modes can be replaced by integration

$$\sum_{\mathbf{k}} \to \frac{L^3}{(2\pi)^3}\int d^3k. \tag{3.233}$$

Bibliography

[1] J.D. Jackson, *Mathematics for Quantum Mechanics; an introductory survey of operators, eigenvalues and linear vector spaces*, (W.A. Benjamin, 1962).

[2] F.W. Byron and R.W. Fuller, *Mathematics of Classical and Quantum Physics*, (New York, Dover 1992).

[3] G. Fano, *Mathematical Methods of Quantum Mechanics*, (McGraw-Hill, New York, 1971).

[4] A.Z. Capri, *Nonrelativistic Quantum Mechanics*, Third Edition, (World Scientific, 2002).

[5] B. Friedman, *Principles and Techniques of Applied Mathematics*, (John Wiley & Sons, New York, 1957), p. 135.

[6] See for example, P.M. Morse and H. Feshbach, *Methods of Theoretical Physics*, Part I (McGraw-Hill, New York, 1953), p. 54.

[7] G.H. Shortley and G.E. Kimball, Analysis of non-commuting vectors with application to quantum mechanics and vector calculus, Proc. Natl. Acad. Sci. USA, 20, 82 (1934).

[8] See for instance, M. Razavy, *Classical and Quantum Disatlsipative Systems*, (Imperial College Press, 2006).

[9] J.-M. Levy-Leblond, Elementary quantum models with position-dependent mass, Eur. J. Phys. 13, 215 (1992).

[10] A. deSouza Dutra, Ordring ambiguity versus representation, J. Phys. A 39, 203 (2006).

[11] A.E.S. Green, T. Sawada and D.S. Saxson, *The Nuclear Independent Model*, (Academic Press, 1968).

[12] M. Razavy, G. Field and J.S. Levinger, Analytical solutions for velocity-dependent nuclear potentials, Phys. Rev. 125, 269 (1962).

[13] J.R. Shewell, On the formation of quantum-mechanical operators, Am. J. Phys. 27, 16 (1959).

[14] M. Born and P. Jordan: Zur Quantenmechanic. Z. Phys. 34, 858 (1925). Engish translation can be found in B.L. van der Waerden, *Sources of Quantum Mechanics*, (North-Holland, Amsterdam, 1967).

[15] H. Margenau and L. Cohen, Probabilities in quantum mechanics, in *Studies in Foundations, Methodology and Philosophy of Science*, edited by M. Bunge, (Springer Verlag, Berlin, 1967), Vol. 2, p. 71.

[16] L. van Hove, On the problem of the relation between the unitary transformations of quantum mechanics and the canonical transformations of classical mechanics, Mem. Acad. R. Belg. 37, 610 (1951).

[17] H. Weyl, *The Theory of Groups and Quantum Mechanics*, (Dover Publications, New York, 1950), p. 275.

[18] N.H. McCoy, On the function in quantum mechanics which corresponds to a given function in classical mechanics, Proc. Natl. Acad. Sci. USA, 18, 674 (1932).

[19] U. Uhlhorn, On the connection between transformations in classical mechanics and in quantum mechanics, Ark. Fysik, 11, 87 (1956).

[20] P.A.M. Dirac, *The Principles of Quantum Mechanics*, Third Edition, (Oxford University Press, 1947), Chapters IV and V.

[21] M. Razavy, Wave equation for dissipative force quadratic in velocity, Phys. Rev. A36, 482 (1987).

[22] F.J. Bloore, Identity of commutator and Poisson bracket, J. Phys. A6, L7 (1973).

[23] D. Park, *Classical Dynamics and Its Quantum Analogues*, Second Edition, (Springer-Verlag, Berlin 1990).

[24] J. Von Neumann, *Mathematical Foundations of Quantum Mechanics*, (Princeton University Press, Princeton, 1955), p. 313.

[25] A.S. Wightman, in *Cargese Lectures in Theoretical Physics* edited by M. Levy, (Gordon and Birch, New York, 1964), p. 262.

[26] C.J. Zhu and J.R. Klauder, Classical symptoms of quantum illnesses, Am. J. Phys. 61, 605 (1993).

[27] A.Z. Capri, Self-adjointness and spontaneously broken symmetry, Am. J. Phys. 45, 823 (1977).

[28] M. Carreau, E. Farhi and S. Gutmann, Functional integral for a free particle in a box, Phys. Rev. D42, 1194 (1990).

[29] C.J. Zhu and J.R. Klauder, The self-adjointness of Hermitian Hamiltonians, Found. Phys. 23, 617 (1993).

[30] G. Bonneau, J. Faraut and G. Vatent, Self-adjoint extensions of operators and teaching of quantum mechanics, Am. J. Phys. 69, 322 (2001).

[31] N.G. van Kampen, The spectral decomposition of the operator $(p^2 - q^2)$, Physica, 24, 545 (1958).

[32] B. Leaf, Momentum operators for curvilinear coordinate system, Am. J. Phys. 47, 811 (1979).

[33] H. Essén, Quantization and independent coordinates, Am. J. Phys. 46, 983 (1978).

[34] Yu.I. Ponomarev, Quantum momentum in curvilinear coordinates, Soviet J. Phys. 21, 129 (1978).

[35] See for example, H. Lass, *Vector and Tensor Analysis*, (McGraw-Hill, New York, 1950), p. 289.

[36] See I.M. Eisenberg and W. Greiner, *Nuclear Theory I, Nuclear Models*, Third Edition, (North-Holland, Amsterdam, 1987).

Chapter 4

Postulates of Quantum Theory

In this chapter we want to study the postulates of quantum mechanics. These postulates enable us to discuss certain aspects of the motion which arise naturally in classical dynamics [1]–[3]:

(1) - In classical mechanics the state of a system at a given time t_0 is given by a set of $2N$ dynamical variables q_j s and p_j s, $j = 1, 2 \cdots N$, where N is the number of degrees of freedom of the system. In quantum theory we need to know how to describe the state of the system mathematically.

(2) - Classically if all the q_j s and p_j s at the time t are given, then all of the properties of the system can be determined. Moreover we can predict the result of any measurement preformed at t with certainty. The question arises on how we can predict the result of measurements in quantum mechanics and if all of the observables of the motion can be determined with certainty.

(3) - The Hamilton canonical equations of motion

$$\dot{q}_j = \frac{\partial H}{\partial p_j}, \quad \text{and} \quad \dot{p}_j = -\frac{\partial H}{\partial q_j}, \tag{4.1}$$

enable us to find q_j s and p_j s at any given time once the initial conditions $q_j(t_0)$ s and $p_j(t_0)$ s are known. We want to find the corresponding quantum-mechanical equations which give us the state of the system at a later time if the initial state of the system is given.

Before stating the postulates of quantum mechanics let us first consider different ways that we can represent the state of a system. From elementary quantum mechanics we know about the Schrödinger equation, especially in position space, and we also know about the similarities between the solutions of the Schrödinger equation, $\psi_n(x)$, and the vectors in linear vector space. This connection between $\psi_n(x)$, the infinite dimensional complex vectors $\{c_n\}$ defined by (3.28) and the bracket notation of Dirac motivates us to view all three as simply different representation of the same quantum state vector. Thus we use the notations

$$\text{“ket” vector} \longleftrightarrow |\psi_n\rangle \longleftrightarrow \psi_n(x) \longleftrightarrow c_n, \tag{4.2}$$

$$\text{“bra” vector} \longleftrightarrow \langle\psi_n| \longleftrightarrow \psi_n^*(x) \longleftrightarrow c_n^*, \tag{4.3}$$

interchangeably throughout the book.

Now let us briefly state the postulates of quantum mechanics [1]–[3].

(1) - The state of a physical system at $t = t_0$ is defined in terms of a ket, or a row vector $|\psi_0\rangle$ belonging to the vector space of states. The space of states is a vector space, therefore it follows that the superposition of two sates is again a state of the system. The space of states also contains the concept of the scalar product. As we observed in Chapter 3 This scalar product associates a complex number to any two states $|\psi\rangle$ and $|\phi\rangle$

$$\langle\phi|\psi\rangle = \langle\psi|\phi\rangle^*. \tag{4.4}$$

(2) - Every measurable physical quantity is described by a self-adjoint operator A acting on the state space.

(3) - The only possible result of the measurement of an observable A is one of the eigenvalues a_n of A. A very important consequence of this postulate is that not all physical observables can be measured simultaneously.

(4) - When a measurement of a physical quantity is made on a given state $|\psi\rangle$, the probability of obtaining an eigenvalue a_n, denoted by $\mathcal{P}(a_n)$ is given by

$$\mathcal{P}(a_n) = |\langle\phi_n|\psi\rangle|^2, \tag{4.5}$$

where $|\phi_n\rangle$ is an eigenstate of A

$$A|\phi_n\rangle = a_n|\phi_n\rangle, \tag{4.6}$$

and where both $|\psi\rangle$ and $|\phi_n\rangle$ are normalized vectors,

$$\langle\phi_n|\phi_n\rangle = \langle\psi|\psi\rangle = 1. \tag{4.7}$$

If there are degenerate states $|\phi_n^j\rangle$, $j = 1, \cdots l$ for a given a_n then $\mathcal{P}(a_n)$ is given by

$$\mathcal{P}(a_n) = \sum_{j=1}^{l} \left|\langle\phi_n^j|\psi\rangle\right|^2. \tag{4.8}$$

When the spectrum of A is continuous then (4.5) is replaced by

$$d\mathcal{P}_k = |\langle\phi_k|\psi\rangle|^2 dk, \tag{4.9}$$

$|\phi_k\rangle$ being the eigenvector corresponding to the continuous eigenvector k of the operator A.

(5) - If by measuring the physical quantity represented by the operator A on the system in state $|\psi\rangle$ we find a_n, the state of the system just after the measurement is the normalized eigenstate $|\phi_n\rangle$.

(6) - The time evolution of any physical quantity described by the operator $F(p, q, t)$ is described by the Heisenberg equation

$$i\hbar\frac{dF}{dt} = i\hbar\frac{\partial F}{\partial t} + [F,\ H]. \tag{4.10}$$

If $F(p(t), q(t))$ and $H(p(t), q(t))$ are not explicitly time-dependent, we can express the time development of $F(t)$ in terms of the unitary transformation $e^{\frac{iHt}{\hbar}}$ as

$$F\left(t'\right) = \exp\left[\frac{iH\left(t'-t\right)}{\hbar}\right] F(t) \exp\left[\frac{-iH\left(t'-t\right)}{\hbar}\right], \tag{4.11}$$

where $F\left(t'\right) = F\left(p\left(t'\right), q\left(t'\right)\right)$ and $F(t) = F(p(t), q(t))$. We can show the equivalence of (4.10) and (4.11) by setting $t' = t + dt$ and expanding both sides of Eq. (4.11).

Heisenberg and Schrödinger Pictures — In the Schrödinger picture the states are represented by time-dependent bra and ket vectors but the operators representing the observables are independent of time.

In the Heisenberg picture of quantum mechanics the state vector $|\psi\rangle$ does not change with time, but the observables satisfy the Heisenberg equation (4.10). Now if we have an observable given by a self-adjoint operator F, and $\frac{\partial F}{\partial t} = 0$, then the expectation value of F for a given state $|\psi(t)\rangle$ is given by

$$\langle F\rangle_t = \langle\psi(t)|F|\psi(t)\rangle. \tag{4.12}$$

If the Hamiltonian does not depend explicitly on time, the time-dependent state $|\psi(t)\rangle$ satisfies the Schrödinger equation

$$|\psi(t)\rangle = \exp\left(-\frac{iHt}{\hbar}\right)|\psi(0)\rangle. \tag{4.13}$$

From Eqs. (4.12) and (4.13) we find that

$$\langle F\rangle_t = \langle\psi(0)|e^{\frac{iHt}{\hbar}} F e^{\frac{-iHt}{\hbar}}|\psi(0)\rangle, \tag{4.14}$$

therefore

$$F(t) = e^{\frac{iHt}{\hbar}} F e^{\frac{-iHt}{\hbar}}. \tag{4.15}$$

We observe that at $t = 0$, the states and the operators are the same in both pictures, and both pictures give the same results for time-dependent expectation values.

Between these two pictures, Heisenberg's is closer to the classical description of motion. It has also another advantage in relation to the Schrödinger picture when we are dealing with quantum field theory, viz, the spatial and temporal dependence of the field operators are treated on the same footing.

Mean Value of an Observable — As the fourth postulate implies, in quantum theory we can only predict probabilities, i.e. we can only determine the outcome of ensemble measurements. In other words we have to perform a very large number of experiments on identical noninteracting systems where we have prepared all these systems to be in the same state. To relate the result of N identical experiments to the prediction of the theory we introduce the idea of the mean value of the observable A in the state $|\psi\rangle$, which we denote by $\langle\psi|A|\psi\rangle$ [2]. This quantity is defined as the average of the results obtained when N measurements of this observable is performed on systems all in the state $|\psi\rangle$. Here again the ket or the vector $|\psi\rangle$ is assumed to be normalized. If we repeat this experiment N times, assuming that the system each time is in the state $|\psi\rangle$ and if the eigenvalues a_n is found $\mathcal{N}$ times, then the probability is given by

$$\mathcal{P}(a_n) = \frac{\mathcal{N}(a_n)}{N} \quad \text{as} \quad N \to \infty. \tag{4.16}$$

Since each member of the ensemble must exhibit one of the eigenvalues a_n, therefore

$$\sum_n \mathcal{N}(a_n) = N, \tag{4.17}$$

or

$$\sum_n \mathcal{P}(a_n) = 1. \tag{4.18}$$

The mean value of an observable A is the sum of values measured divided by the number of experiments N. If $\mathcal{N}$ experiments have produced the same result a_n, then the mean value $\langle|A|\rangle$ is given by

$$\langle\psi|A|\psi\rangle = \frac{1}{N}\sum_n a_n \mathcal{N}(a_n). \tag{4.19}$$

As N becomes very large then according to (4.16) and (4.19) we have

$$\langle\psi|A|\psi\rangle = \sum_n a_n \mathcal{P}(a_n). \tag{4.20}$$

For a system where A has a continuous spectrum we replace the summation in (4.20) by integration

$$\langle\psi|A|\psi\rangle = \int k \, d\mathcal{P}(k). \tag{4.21}$$

The Density Matrix — When a system is isolated and can be described by a Hamiltonian we can, in principle, assign a wave function to it, and then the time dependence of this wave function can be found from the time-dependent Schrödinger equation. Now suppose we have a system A which is interacting with another system B, and that the total system $A + B$ is closed, i.e. there is no interaction between the whole system and the rest of the universe. Let us denote the total wave function describing the system $A + B$ by $\Psi(Q, x)$, where

x denotes the set of coordinates of the system A and Q the coordinates of B. This wave function cannot be written as a product of x and of Q, otherwise we are dealing with two noninteracting systems. Now for any physical quantity pertaining to A, which is represented by the operator $\mathcal{F}$, and this operator acts on the coordinate x, we define the mean value by [4]

$$\bar{\mathcal{F}} = \int\int \Psi^*(x,Q)\mathcal{F}\Psi(Q,x)dQdx, \tag{4.22}$$

where dx and dQ denote integrations over the whole x-space and the whole Q-space. We also define the density matrix $\rho(x,\, x')$ by

$$\rho(x,\, x') = \int \Psi^*(Q,x')\,\Psi(Q,x)dQ. \tag{4.23}$$

From the definition of $\rho(x',\, x)$ if follows that the density matrix is Hermitian since its matrix elements satisfy the condition

$$\rho^*(x,\, x') = \rho(x',\, x). \tag{4.24}$$

The description in terms of the density matrix is the most general form that we can describe the properties of a system. This general form reduces to the particular form of description in terms of the wave function when $\rho(x',\, x)$ can be written as a simple product of two wave functions

$$\rho(x',\, x) = \Psi^*(x')\,\Psi(x). \tag{4.25}$$

States which can be described by wave functions are called "pure" states, but the the states which can be described only by a density matrix are called "mixed states". Thus the probability density as a function of the coordinates x for a mixed state can be found from the diagonal elements of the density matrix

$$\rho(x,\, x) = \int |\Psi(Q,x)|^2 dQ. \tag{4.26}$$

It must be emphasized that in the case of a pure state $\Psi(x)$ is an eigenfunction of some operator, however for mixed states there is no operator whose eigenvalues are related to the wave function. When the wave function is properly normalized, by setting $x = x'$, we have

$$\int \rho(x,x,)\,dx = 1. \tag{4.27}$$

For a pure state, the time dependence of the density matrix can be found from the time dependent Schrödinger equation. Thus we write

$$\rho(x',\, x,\, t) = \Psi^*(x',t)\,\Psi(x,t). \tag{4.28}$$

Now by differentiating (4.28) with respect to time and noting that $i\hbar\frac{\partial\Psi}{\partial t} = H\Psi$, we find that

$$\begin{aligned} i\hbar\frac{\partial\rho\left(x', x, t\right)}{\partial t} &= i\hbar\Psi^*\left(x', t\right)\frac{\partial\Psi(x,t)}{\partial t} + i\hbar\Psi\left(x, t\right)\frac{\partial\Psi^*\left(x', t\right)}{\partial t} \\ &= \Psi^*\left(x', t\right)H\Psi(x,t) - \Psi\left(x, t\right)H'^{\,*}\Psi\left(x'^{\,*}, t\right) \\ &= \left(H - H'^{\,*}\right)\rho\left(x', x, t\right). \end{aligned} \tag{4.29}$$

In this relation H which is the Hamiltonian of the system acts on the function of x, whereas H' which is the same Hamiltonian as H but it acts on x'.

Let $\psi_n(x)$ denote the n-th time independent eigenfunction of H corresponding to the eigenvalue E_n. We expand $\rho\left(x', x\right)$ in terms of this set of eigenfunctions:

$$\rho\left(x', x, t\right) = \sum_n\sum_j c_{nj}\psi_n^*\left(x'\right)\psi_n(x)e^{\frac{i}{\hbar}(E_n - E_j)t}. \tag{4.30}$$

This expansion is valid whether the state is pure or is mixed. However when the state is pure then c_{nj} s can be written as a product of two factors $c_{nj} = c_j^* c_n$. To show this we write the wave function for the pure state as

$$\Psi(x,t) = \sum_j c_j\psi_n(x)e^{-\frac{i}{\hbar}E_j t}, \tag{4.31}$$

and by substituting this expansion in (4.28) and comparing the result with (4.30) we find that

$$c_{jn} = c_n^* c_j. \tag{4.32}$$

If $\Psi(x,t)$ in (4.31) is normalized we have

$$\int |\Psi(x,t)|^2 dx = \sum_j |c_j|^2 = 1. \tag{4.33}$$

Using this result we calculate $\left(c^2\right)_{jn}$;

$$\begin{aligned} \left(c^2\right)_{jn} &= \sum_k c_{jk}c_{kn} = \sum_k c_k^* c_j c_n^* c_k \\ &= c_j c_n^* = c_{jn} \end{aligned} \tag{4.34}$$

that is the square of the density matrix is equal to the density matrix itself, a result which is valid only for pure states.

As we can see from Eq. (4.26) the diagonal elements $\rho(x,x)$ of the density matrix are positive, and these give us the probability distribution for coordinates. Now if we write a quadratic form constructed from the coefficients c_{nj}, i.e.

$$\sum_n\sum_j c_{nj}z_n^* z_j, \tag{4.35}$$

where z_n s are arbitrary complex quantities this quadratic form must be positive. The well-known results of the theory of quadratic forms imposes the following condition on the c_{nj} [3]–[5]:

$$c_{nn} \geq 0, \tag{4.36}$$

and

$$c_{nn}c_{jj} \geq |c_{nj}|^2. \tag{4.37}$$

In Chapter 20 we will see an application of the density matrix in the quantum theory of measurement.

Time Derivative in Heisenberg's Equations — Next we want to discuss the time derivative of an observable which we hope is itself an observable. For a free particle of mass m and momentum $\mathbf{p}$, we know the Hamiltonian is $\frac{\mathbf{p}^2}{2m}$. The velocity of this particle can be measured by a time-of-flight experiment [12]. If the position of the particle at t_1 is $\mathbf{r}_1$ and at t_2 is $\mathbf{r}_2$, then the velocity between the two observations is

$$\mathbf{v} = \frac{d\mathbf{r}}{dt} = \frac{\mathbf{r}_2 - \mathbf{r}_1}{t_2 - t_1}. \tag{4.38}$$

Can this definition be generalized and be applied to other quantum mechanical operators representing observables? This question was studied in some detail by Fulling [6]. He examined the result that one finds from the measurement at two different times and compared it with the definition of time derivative according to Heisenberg.

Let $A(t)$ be an operator corresponding to an observable of the system, then using the definition of time derivative we have

$$\frac{dA(t)}{dt} \approx \frac{A(t+\Delta t) - A(t)}{\Delta t}, \tag{4.39}$$

where Δt is sufficiently small time interval. When $\frac{dA(t)}{dt}$ represents a conserved quantity, then $A(t)$ is proportional to t and (4.39) becomes exact as in (4.38).

Now consider a simple system where the Hamiltonian is given by

$$2\omega S_y, \tag{4.40}$$

where S_y is the y component of the spin of a particle, Eq. (9.115). In this case the time derivative of S_z from the Heisenberg equation, (4.10), is

$$\frac{dS_z(t)}{dt} = i[H,\, S_z] = 2i\omega[S_y,\, S_z] = -2\omega S_x. \tag{4.41}$$

According to this relation $\frac{dS_z(t)}{dt}$ is a bounded observable taking one of the eigenvalues of the operator $-2\omega S_x$, i.e. $\pm\omega$. If at $t = 0$ the particle is in the state with spin up, $|\uparrow\rangle$, (see Eq. (9.114)), then at a later time the state of the particle is given by

$$e^{-2i\omega S_y t}|\uparrow\rangle = \cos\omega t|\uparrow\rangle + \sin\omega t|\downarrow\rangle, \tag{4.42}$$

a result which can easily be found from the matrix representation of S_y, Eq. (9.115), and the spin up and spin down states (9.116).

Now suppose we measure S_z at a very small positive time Δt, then the probability of finding S_z in spin down state $|\downarrow\rangle$ is proportional to $\sin^2 \omega\Delta t$. In this case the measured change in S_z per unit time is

$$\frac{dS_z(t)}{dt} \approx \frac{S_z(\Delta t) - S_z(0)}{\Delta t} = -\frac{1}{\Delta t}. \tag{4.43}$$

Since we can make Δt as small as we please, the right-hand side of (4.43) becomes arbitrarily large, and not equal to $\pm\omega$. The other possible result of measurement is zero. Thus one reaches the important conclusion that "two measurements of an observable in rapid succession do not constitute a measurement of the time derivative according to Heisenberg's equations of motion" [6].

Compatibility of Simultaneous Measurement — An important theorem regarding the compatibility of the measurement of two observables can be stated in the following way:

If A and B are two operators with the same set of eigenvectors then A and B commute. To show this let us denote the common eigenvectors of A and B by $|a_n, b_m\rangle$. The set $|a_n, b_m\rangle$ form a complete set and we can expand any ket $|\psi\rangle$ in terms of $|a_n, b_m\rangle$. Now consider the result of AB operating on an arbitrary ket $|\psi\rangle$

$$\begin{aligned} AB|\psi\rangle &= A\left(\sum_{n,m} B|a_n, b_m\rangle\langle a_n, b_m|\psi\rangle\right) \\ &= A\sum_{n,m} b_m|a_n, b_m\rangle\langle a_n, b_m|\psi\rangle \\ &= \sum_n\sum_m a_n b_m|a_n, b_m\rangle\langle a_n, b_m|\psi\rangle, \end{aligned} \tag{4.44}$$

and

$$\begin{aligned} BA|\psi\rangle &= B\left(\sum_{n,m} A|a_n, b_m\rangle\langle a_n, b_m|\psi\rangle\right) \\ &= B\sum_{n,m} a_n|a_n, b_m\rangle\langle a_n, b_m|\psi\rangle \\ &= \sum_n\sum_m b_m a_n|a_n, b_m\rangle\langle a_n, b_m|\psi\rangle, \end{aligned} \tag{4.45}$$

therefore

$$(AB - BA)|\psi\rangle = 0, \quad \text{or} \quad [A,\ B] = 0. \tag{4.46}$$

The converse of this result is also true, that is if two operators commute then they possess the same eigenvector. Let $|n\rangle$ be an eigenvector of A so that

$A|n\rangle = a_n|n\rangle$, then $AB|n\rangle = A(B|n\rangle)$ and $B(A|n\rangle) = B(a_n|n\rangle)$. Therefore if $AB = BA$ we have

$$A(B|n\rangle) = a_n(B|n\rangle), \tag{4.47}$$

or $B|n\rangle$ is an eigenvector of A. This can also be true if the ket $B|n\rangle$ is proportional to $|n\rangle$ or

$$B|n\rangle = b_n|n\rangle. \tag{4.48}$$

Here we have implicitly assumed that there is no degeneracy, but if there are degenerate states the theorem remains valid.

4.1 The Uncertainty Principle

In his seminal paper of 1927, Heisenberg rejected the idea of strict observability of the trajectory of an electron (position as well as momentum) as it was assumed in Bohr's model. Instead he considered the observable trajectory (e.g. in the Wilson cloud chamber) as a discrete sequence of imprecisely defined positions. Soon after the publication of his paper, most of the physicists regarded the uncertainty relations as an integral part of the foundation of quantum theory [7].

In classical mechanics, according to Laplace, if one knows the exact position and velocity of the particle at a given time, then one can predict the position and the velocity of the particle at any future time provided that all of the forces acting on the particle is known precisely [8]. This is one of the simplest and widely-accepted principles of causality in mechanics. Now the uncertainty principle invalidates this principle and replaces it with the following observation: In quantum theory we can calculate only a range of possibilities for the position and velocity of the electron for later times, one of which will result from the motion of a given electron. However, as was observed by Heisenberg, the predictions of quantum mechanics are statistical in nature [9].

Let us consider the limitations placed on the precision of simultaneous measurement of a pair of conjugate quantities such as position and momentum. We denote the mean value of an observable corresponding to the self-adjoint operator A when the system is in the state $|k\rangle$ by $\langle k|A|k\rangle$. The dispersion of the result about $\langle k|A|k\rangle$ is found from the root-mean-square deviation, ΔA, defined by

$$\Delta A = \left[\left\langle k\left|(A - \langle k|A|k\rangle)^2\right|k\right\rangle\right]^{\frac{1}{2}} = \left[\langle k\left|A^2\right|k\rangle - \langle k|A|k\rangle^2\right]^{\frac{1}{2}}. \tag{4.49}$$

Now we consider a set of self-adjoint linear operators A_r, $(r = 1, \cdots, n)$, where the average of A_r is defined by $\langle k|A_r|k\rangle$ and the dispersion is given by

$$\Delta A_r^2 = \left[\langle k\left|A_r^2\right|k\rangle - \langle k|A_r|k\rangle^2\right]. \tag{4.50}$$

For a given A_r we define a vector $|\alpha_r\rangle$ by [10]

$$|\alpha_r\rangle = \frac{A_r - \langle k|A_r|k\rangle}{(\Delta A_r)}|k\rangle. \tag{4.51}$$

The scalar product $\langle\alpha_r|\alpha_r\rangle = 1$ shows that $|\alpha_r\rangle$ is a unit vector and also from (4.51) it follows that $|\alpha_r\rangle$ is orthogonal to $|k\rangle$;

$$\langle k|\alpha_r\rangle = 0, \quad r = 1, 2\cdots, n. \tag{4.52}$$

If we take any vector $|v\rangle$ which is linear combination of $|\alpha_r\rangle$ s with complex coefficients c_r s, i.e.

$$|v\rangle = \sum_r c_r|\alpha_r\rangle, \tag{4.53}$$

then $\mathcal{F}$ defined as the square of the norm of v is real and positive definite

$$\mathcal{F} = \langle v|v\rangle = \sum_r\sum_s c_s^* c_r\langle\alpha_s|\alpha_r\rangle \geq 0. \tag{4.54}$$

Now if we write

$$c_r = a_r + ib_r, \quad \text{and} \quad \langle\alpha_r|\alpha_s\rangle = f_{rs} - ig_{rs}, \tag{4.55}$$

and note that $|\alpha_r\rangle$ is a unit vector, then we have

$$f_{11} = f_{22} = \cdots = f_{nn} = 1. \tag{4.56}$$

By substituting for c_r and $\langle\alpha_r|\alpha_s\rangle$ in (4.54) we find the following expression for $\mathcal{F}$

$$\mathcal{F} = \sum_r\sum_s \left\{(a_r a_s + b_r b_s)f_{rs} + (a_r b_s - a_s b_r)g_{rs}\right\}, \tag{4.57}$$

where from the definitions of f_{rs} and g_{rs} it is clear that f is a symmetric- and g is an antisymmetric matrix. As Eq. (4.57) shows, $\mathcal{F}$ is a quadratic form of $2n$ real quantities a_r and b_r and it is also positive definite, therefore the eigenvlaues of the $2n \times 2n$ matrix formed from a_r s and b_r s, i.e.

$$\begin{bmatrix} f & g \\ -g & f \end{bmatrix}, \tag{4.58}$$

must all be nonnegative. In other words if λ s are the roots of the determinant

$$D(\lambda) = \begin{vmatrix} f - \lambda I & g \\ -g & f - \lambda I \end{vmatrix} = 0, \tag{4.59}$$

then all of the eigenvalues λ are nonnegative. We can express f_{rs} and g_{rs} in terms of the mean values of A_r and A_s using the scalar product $\langle\alpha_r|\alpha_s\rangle$ in (4.55).

The matrix elements of f and g are

$$f_{rs} = \frac{1}{2}\left(\frac{\langle k|A_r A_s + A_s A_r|k\rangle - 2\langle k|A_r|k\rangle\langle k|A_s|k\rangle}{(\Delta A_r)(\Delta A_s)}\right), \tag{4.60}$$

and

$$g_{rs} = \frac{i}{2}\left(\frac{\langle k|A_rA_s - A_sA_r|k\rangle}{(\Delta A_r)(\Delta A_s)}\right), \tag{4.61}$$

and these quantities will be used to find the uncertainty relations for two or more noncommuting observables.

First let us consider the uncertainty relation for a pair of self-adjoint operators A_1 and A_2. In this case the determinant (4.59) becomes

$$D(\lambda) = \begin{vmatrix} 1-\lambda & f_{12} & 0 & g_{12} \\ f_{21} & 1-\lambda & g_{21} & 0 \\ 0 & -g_{12} & 1-\lambda & f_{12} \\ -g_{12} & 0 & f_{21} & 1-\lambda \end{vmatrix}. \tag{4.62}$$

Noting that $f_{12} = f_{21}$ and $g_{12} = -g_{21}$ we can simplify (4.62)

$$D(\lambda) = \left[(1-\lambda)^2 - f_{12}^2 - g_{12}^2\right]^2. \tag{4.63}$$

Thus the eigenvalues $\lambda_\pm$ are the roots of $D(\lambda) = 0$ and are given by

$$\lambda_\pm = 1 \pm \left(f_{12}^2 + g_{12}^2\right)^{\frac{1}{2}} = 1 \pm |\langle\alpha_1|\alpha_2\rangle|, \tag{4.64}$$

and since $\lambda_\pm$ must be nonnegative, we have

$$f_{12}^2 + g_{12}^2 \le 1. \tag{4.65}$$

By substituting from (4.60) and (4.61) we find the uncertainty $\Delta A_1 \Delta A_2$;

$$\begin{aligned} \Delta A_1^2 \Delta A_2^2 \ &\ge\ \frac{1}{4}\left\{\langle k|A_1A_2 + A_2A_1|k\rangle - 2\langle k|A_1|k\rangle\langle k|A_2|k\rangle\right\}^2 \\ &+\ \frac{1}{4}\left(i\langle k|[A_1,\ A_2]|k\rangle\right)^2. \end{aligned} \tag{4.66}$$

This stronger form of the uncertainty relation was first derived by Schrödinger [11],[12]. The usual form of the Heisenberg inequality which is the weaker form results if we ignore the first term on the right-hand side of (4.66), i.e.

$$\Delta A_1^2 \Delta A_2^2 \ge \frac{1}{4}\left(i\langle k|[A_1,\ A_2]|k\rangle\right)^2, \tag{4.67}$$

or

$$\Delta A_1 \Delta A_2 \ge \frac{1}{2}|\langle k|[A_1,\ A_2]|k\rangle|, \tag{4.68}$$

An interesting feature of this approach is that we can determine the uncertainty arising from the measurement of three or more observables at the same time. For instance if we choose three operators A_1, A_2 and A_3 and follow the same argument as the one mentioned above, we obtain the Heisenberg weaker inequality

$$3(\Delta A_1)^2(\Delta A_2)^2(\Delta A_3^2) \ge \frac{1}{4}\sum_c (\Delta A_1)^2\left(i\langle k|\,[A_2,\ A_3]\,|k\rangle\right)^2, \tag{4.69}$$

where $\sum_c$ means a cyclic sum for 1, 2 and 3 [10].

Minimum Uncertainty Products — Returning back to Eq. (4.66) we observe that the uncertainty product $\Delta A_1 \Delta A_2$ assumes its minimum value if the first term on the right-hand side of (4.66) is zero. We can write this term as

$$\langle k|A_1A_2 + A_2A_1|k\rangle - 2\langle k|A_1|k\rangle\langle k|A_2|k\rangle = \langle k|\bar{A}_1\bar{A}_2 + \bar{A}_2\bar{A}_1|k\rangle, \tag{4.70}$$

where

$$\bar{A}_i = A_i - \langle k|A_i|k\rangle, \quad i = 1, 2. \tag{4.71}$$

The expression (4.70) vanishes provided that for a specific state $|k\rangle$

$$\bar{A}_1|k\rangle = \beta\bar{A}_2|k\rangle, \tag{4.72}$$

where β which is a complex constant satisfies

$$\left\langle k\left|\bar{A}_1\bar{A}_2 + \bar{A}_2\bar{A}_1\right|k\right\rangle = (\beta^* + \beta)\left\langle k\left|\bar{A}_2^2\right|k\right\rangle = 0, \tag{4.73}$$

i.e. if β is a pure imaginary constant. Replacing $\bar{A}_1$ and $\bar{A}_2$ by A_1 and A_2, Eq. (4.72) can be written as

$$(A_1 - \mu)|k\rangle = i\gamma(A_2 - \nu)|k\rangle, \tag{4.74}$$

where μ, ν and γ are all constants.

For the two fundamental conjugate quantities $A_1 = p$ and $A_2 = q$ the Heisenberg uncertainty relation (4.66) in its weaker form becomes

$$(\Delta p)^2(\Delta q)^2 \geq \frac{1}{4}(i\langle k|[p,\ q]|k\rangle)^2 = \frac{\hbar^2}{4}. \tag{4.75}$$

Equation (4.74) for the minimum uncertainty can also be written as

$$(\bar{A}_1 - i\gamma\bar{A}_2)|k\rangle = \eta|k\rangle. \tag{4.76}$$

Now suppose that the commutator $\bar{A}_1$ and $\bar{A}_2$ is given by

$$\left[\bar{A}_1,\ \bar{A}_2\right] = iC, \tag{4.77}$$

where C is an operator, then from (4.68) it follows that

$$\langle\bar{A}_1^2\rangle\langle\bar{A}_2^2\rangle \geq \frac{1}{4}\langle C\rangle^2. \tag{4.78}$$

We have also the inequality

$$\left[\langle\bar{A}_1^2\rangle^{\frac{1}{2}} - \gamma\langle\bar{A}_2^2\rangle^{\frac{1}{2}}\right]^2 \geq 0, \tag{4.79}$$

from which we get

$$\begin{aligned} f(\gamma) &= \langle\bar{A}_1^2\rangle + \gamma^2\langle\bar{A}_2^2\rangle - 2\gamma\left[\langle\bar{A}_1^2\rangle\langle\bar{A}_2^2\rangle\right]^{\frac{1}{2}} \\ &\geq \langle\bar{A}_1^2\rangle + \gamma^2\langle\bar{A}_2^2\rangle - \gamma\langle C\rangle \geq 0. \end{aligned} \tag{4.80}$$

The function $f(\gamma)$ is quadratic in γ and has a minimum at

$$\frac{df(\gamma)}{d\gamma} = 2\gamma\langle\bar{A}_2^2\rangle - \langle C\rangle = 0. \tag{4.81}$$

Thus for the minimum uncertainty product we get

$$\langle\bar{A}_2^2\rangle = \frac{\langle C\rangle}{2\gamma}, \tag{4.82}$$

and from (4.78) (with the equality sign) we find

$$\langle\bar{A}_1^2\rangle = \frac{\gamma}{2}\langle C\rangle. \tag{4.83}$$

These two relations (4.82) and (4.83) are important in investigating the classical limit of the hydrogen atom Sec. 9.10.

Two Examples of the Position-Momentum Uncertainty Products — For exactly solvable potentials, the wave function is known analytically, and for these we can calculate the uncertainties for all stationary states [13],[14]. Here we consider two well-known cases and for others we refer the reader to the paper of Nieto [13].

Uncertainty Relation for the Harmonic Oscillator — From the exact normalized wave function for the n-th stationary state Eqs. (8.92),(8.93) we calculate the expectation values of x, p, x^2 and p^2;

$$\langle\psi_n(x)|x|\psi_n(x)\rangle = 0, \tag{4.84}$$

$$\langle\psi_n(x)|p|\psi_n(x)\rangle = 0, \tag{4.85}$$

$$\langle\psi_n(x)\left|x^2\right|\psi_n(x)\rangle = \frac{\hbar}{m\omega}\left(n+\frac{1}{2}\right), \tag{4.86}$$

$$\langle\psi_n(x)\left|p^2\right|\psi_n(x)\rangle = \hbar m\omega\left(n+\frac{1}{2}\right). \tag{4.87}$$

Thus

$$(\Delta x)_n^2 = \frac{\hbar}{m\omega}\left(n+\frac{1}{2}\right), \tag{4.88}$$

and

$$(\Delta p)_n^2 = m\omega\hbar\left(n+\frac{1}{2}\right). \tag{4.89}$$

From thee relations we get the product of the uncertainties

$$(\Delta p)_n(\Delta x)_n = \hbar\omega\left(n+\frac{1}{2}\right). \tag{4.90}$$

Uncertainty Relations for the Hydrogen Atom — The complete normalized wave function for the hydrogen atom is given by Eq. (9.221). Using these wave functions we find the matrix elements of x, p_x, x^2, p_x^2, $y\cdots p_z^2$:

$$\langle\psi_{n,\ell,m}(\mathbf{r})|x|\psi_{n,\ell,m}(\mathbf{r})\rangle = 0, \tag{4.91}$$

$$\langle\psi_{n,\ell,m}(\mathbf{r})|p_x|\psi_{n,\ell,m}(\mathbf{r})\rangle = 0, \tag{4.92}$$

and similar relations for y, z, p_y and p_z. For x^2, $\cdots p_x^2$ etc. we have the matrix elements

$$\begin{aligned}\langle\psi_{n,\ell,m}(\mathbf{r})\left|x^2\right|\psi_{n,\ell,m}(\mathbf{r})\rangle &= \langle\psi_{n,\ell,m}(\mathbf{r})\left|y^2\right|\psi_{n,\ell,m}(\mathbf{r})\rangle \\ &= \frac{2n^2a_0^2\left(\ell^2+m^2+\ell-1\right)\left(5n^2-3\ell^2-3\ell+1\right)}{(2\ell-1)(2\ell+3)},\end{aligned} \tag{4.93}$$

$$\langle\psi_{n,\ell,m}(\mathbf{r})\left|z^2\right|\psi_{n,\ell,m}(\mathbf{r})\rangle = \frac{2n^2a_0^2\left(2\ell^2-2m^2+2\ell-1\right)\left(5n^2-3\ell^2-3\ell+1\right)}{(2\ell-1)(2\ell+3)}, \tag{4.94}$$

$$\begin{aligned}\langle\psi_{n,\ell,m}(\mathbf{r})\left|p_x^2\right|\psi_{n,\ell,m}(\mathbf{r})\rangle &= \langle\psi_{n,\ell,m}(\mathbf{r})\left|p_y^2\right|\psi_{n,\ell,m}(\mathbf{r})\rangle \\ &= \frac{\hbar^2}{4n^2a_0^2}\frac{\left(\ell^2+m^2+\ell-1\right)}{(2\ell-1)(2\ell+3)},\end{aligned} \tag{4.95}$$

and

$$\langle\psi_{n,\ell,m}(\mathbf{r})\left|p_z^2\right|\psi_{n,\ell,m}(\mathbf{r})\rangle = \frac{\hbar^2}{4n^2a_0^2}\frac{\left(2\ell^2-2m^2+2\ell-1\right)}{(2\ell-1)(2\ell+3)}. \tag{4.96}$$

A more interesting result can be found by averaging $(\Delta x)_{n,\ell,m}$ over ℓ and m. In this way we find the product of the uncertainties as a function of the principal quantum number n.

$$\overline{(\Delta x)^2} = \frac{\sum_{\ell=0}^{n-1}\sum_{m=-\ell}^{\ell}(\Delta x)^2_{n,\ell,m}}{\sum_{\ell=0}^{n-1}\sum_{m=-\ell}^{\ell}1} = \frac{a_0^2}{12}\left(7n^2+5\right), \tag{4.97}$$

and

$$\overline{(\Delta p)^2} = \frac{1}{3}\left(\frac{\hbar}{a_0}\right)^2, \tag{4.98}$$

Thus the uncertainty relation averaged over ℓ and m can be written as

$$\sqrt{\overline{(\Delta x)^2}\ \overline{(\Delta p)^2}} = \frac{\hbar}{6}\sqrt{7n^2+5}. \tag{4.99}$$

4.2 Application of the Uncertainty Principle for Calculating Bound State Energies

For simple systems we can find the binding energies with the help of the position-momentum uncertainty. Here we will consider the one-dimensional motions but the method can be extended and applied to three-dimensional systems with spherically symmetric potentials, $V(\mathbf{r}) = V(r)$ [15].

We write the expectation value of the energy as

$$\langle E\rangle = \left\langle \frac{p^2}{2m}\right\rangle + \langle V(x)\rangle, \tag{4.100}$$

where in the following discussion we assume that $V(x)$ is an even function of x. For a general state we know that $\langle x\rangle = \langle p\rangle = 0$ and thus we replace $\langle p^2\rangle$ by $(\Delta p)^2$ and $\langle x^2\rangle$ by $(\Delta x)^2$. Now we take the uncertainty relation as an equality and write $\Delta p = \frac{\hbar}{\Delta x}$ for the ground state and $\Delta p_n = \frac{n\hbar}{\Delta x_n}$ for the n-th excited state. These uncertainties are exact for the harmonic oscillator, but we assume that they are good approximation for other potentials.

For a potential of the form $V(x) = \lambda x^{2j}$, where j is an integer, from Eq. (4.100) we get the following approximate form for E_n

$$E_n = \frac{n^2\hbar^2}{2m(\Delta x)^2} + \lambda(\Delta x)^{2j}. \tag{4.101}$$

Now we choose Δx so that E_n assumes its minimum value. By setting the derivative $\frac{dE_n}{d(\Delta x)}$ equal to zero we obtain expressions for Δx as well E_n;

$$\Delta x_n^{(0)} = \left(\frac{n^2\hbar^2}{2mj\lambda}\right)^{\frac{1}{2j+2}}, \tag{4.102}$$

and

$$E_n^{(0)} = \left(\frac{n^2\hbar^2}{2m\lambda}\right)^{\frac{j}{j+1}} (2j\lambda)^{\frac{1}{j+1}}\left(1+\frac{1}{j}\right), \quad n = 0,\ 1,\ 2\cdots \tag{4.103}$$

The first term in (4.103) is the kinetic energy $T_n^{(0)}$ and the second term is the contribution of the potential energy $V_n^{(0)}$, hence we have

$$T_n^{(0)} = jV_n^{(0)}. \tag{4.104}$$

This result is in agreement with the classical virial theorem [16],

$$2\langle T\rangle = \left\langle x\frac{dV(x)}{dx}\right\rangle = 2j\langle V(x)\rangle. \tag{4.105}$$

In particular for a quartic oscillator we have $j = 2$ and (4.103) reduces to

$$E_n^{(0)} = \frac{3}{4}\left(\frac{4\hbar^4\lambda}{m^2}\right)^{\frac{1}{3}} n^{\frac{4}{3}} \approx 1.1905\left(\frac{\hbar^4\lambda}{m^2}\right)^{\frac{1}{3}} n^{\frac{4}{3}}, \quad n = 1,\ 2,\ 3\cdots. \tag{4.106}$$

If we calculate these eigenvalues using the WKB approximation (Sec. 12.1) we find

$$E_{n'} = \frac{3^{\frac{4}{3}}\pi^2}{\left(\Gamma\left(\frac{1}{4}\right)\right)^{\frac{8}{3}}}\left(\frac{\lambda\hbar^4}{m^2}\right)^{\frac{1}{3}}\left(n'+\frac{1}{2}\right)^{\frac{4}{3}}$$

$$\approx \quad 1.376\left(\frac{\lambda\hbar^4}{m^2}\right)^{\frac{1}{3}}\left(n'+\frac{1}{2}\right)^{\frac{4}{3}}, n'=0,\ 1,\ 2,\cdots. \tag{4.107}$$

These results should be compared with the more accurate estimates of the energy eigenvalue [17].

4.3 Time-Energy Uncertainty Relation

Unlike the $\Delta p\Delta q$ uncertainty that we have studied so far, the time-energy uncertainty $\Delta t\Delta E \geq \frac{1}{2}\hbar$ has been and remains to be a controversial relation, not only in its formulation, but also in its range of validity and interpretation [18],[19]. This uncertainty like the position-momentum uncertainty was considered in the classic paper of Heisenberg for the first time [9].

Starting from the Heienberg equation of motion (4.10) we can write a differential equation for the mean value of the observable A,

$$\frac{d}{dt}\langle A\rangle = \left\langle\frac{\partial A}{\partial t}\right\rangle + \frac{1}{i\hbar}\langle[A,\ H]\rangle. \tag{4.108}$$

We can use this equation to derive a particular form of time-energy uncertainty referred to as Mandelstam-Tamm time-energy uncertainty [18]–[20]. Let us consider a system where H does not explicitly depend on time and let A denote a time-independent observable. If $|k\rangle$ represent the states of a system at the time t, then setting $A_1 = A$ and $A_2 = H$ in (4.68) we find

$$\Delta A\Delta E \geq \frac{1}{2}\langle k|[A,\ H]|k\rangle, \tag{4.109}$$

where $\Delta E = \Delta H$. From Eqs. (4.108) and (4.109) we obtain

$$\frac{\Delta A}{\left|\frac{d\langle k|A|k\rangle}{dt}\right|}\Delta E \geq \frac{1}{2}\hbar. \tag{4.110}$$

Now introducing $\Delta\tau_A$ by

$$\Delta\tau_A = \left|\frac{d\langle k|A|k\rangle}{dt}\right|^{-1}\Delta A \tag{4.111}$$

we find the inequality

$$\Delta\tau_A\Delta E \geq \frac{1}{2}\hbar. \tag{4.112}$$

Thus $\Delta_A\tau$ is the time characteristic of the evolution of the observable A, i.e. the time required for the center $\langle k|A|k\rangle$ of this distribution to be displaced by an amount equal to its width ΔA [18]. Let us emphasize that this uncertainty is

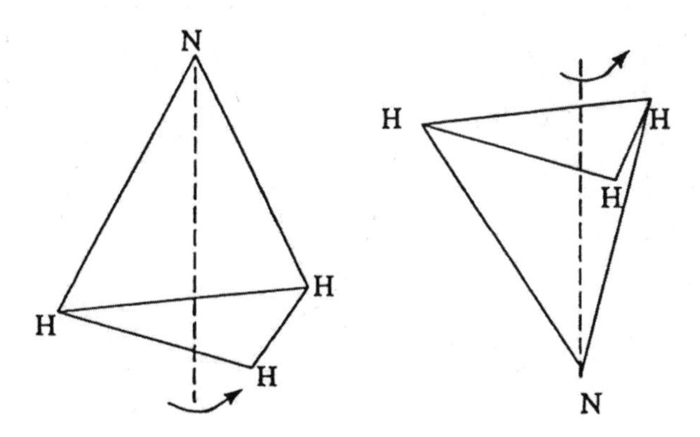

Figure 4.1: The NH_3 molecule where the three hydrogen atoms form an equilateral triangle. The motion of the nitrogen atom is along the x-axis which is shown by the dashed line.

about the time intrinsic to the system and its evolution and is not the dispersion of a dynamical variable t. We should also note that $\langle k|A|k\rangle$ and $\frac{d\langle k|A|k\rangle k}{dt}$ are instantaneous expectation values and in general can depend on time. In principle we can determine the shortest time $\Delta\tau_{A_j}$ among the different observables A_j of a given system, but this $\Delta\tau_{A_j}$ cannot be regarded as the characteristic time of evolution of the system itself.

As an example of the application of this uncertainty, let us consider the case of a two level system such as ammonia molecule NH_3. In this molecule the three hydrogen atoms are at the three equilateral triangle and the nitrogen can move up and down along the axis of the molecule [21]. The potential in which the nitrogen atom moves has the shape of a double-well potential with the two lowest states E_0 and E_1 close together and far from others, E_2, $E_3\cdots$ [21]. The location of H and N atoms are shown in Fig. 4.1. In this case the motion of the nitrogen atom can be described by the time-dependent wave function

$$\psi(x,t) = \psi_1(x)\exp\left(\frac{-iE_1t}{\hbar}\right) + \psi_2(x)\left(\frac{-iE_2t}{\hbar}\right), \tag{4.113}$$

where the axis of the molecule to be chosen as the x axis. Thus the probability distribution is

$$\begin{aligned} P(x,t) &= |\psi(x,t)|^2 \\ &= |\psi_1(x)|^2 + |\psi_2(x)|^2 + 2\mathrm{Re}\left[\psi_1(x)\psi_2(x)e^{\frac{i(E_1-E_2)t}{\hbar}}\right], \end{aligned} \tag{4.114}$$

and this distribution oscillates between the two wave functions shown in Fig. 4.2. $|\psi^R(x)|^2 = |\psi_1(x) - \psi_2(x)|^2$ and $|\psi^L(x)|^2 = |\psi_1(x) + \psi_2(x)|^2$. The period

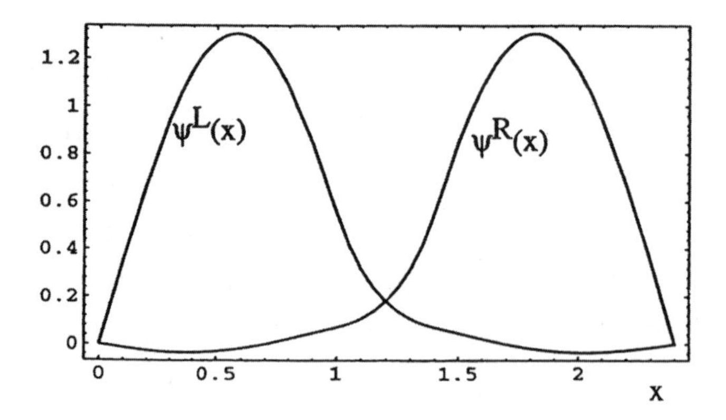

Figure 4.2: The sum of the ground and first excited state and the difference between these two states for the ammonia molecule.

of the oscillation can be found from Eq. (4.114) and is given by

$$\Delta\tau = \frac{\hbar}{|E_1 - E_2|}. \tag{4.115}$$

Thus $\Delta\tau$ represents the time for the transition of the system from one state to the other, or the transition from one well to the other. If we measure the probability distribution at two different times t_1 and t_2 and if $\Delta t = |t_2 - t_1| \ll \Delta\tau$, then we will find the same probability distribution. However the physical properties of the system will be modified over a time interval Δt provided that

$$\Delta t \Delta E \geq \hbar, \tag{4.116}$$

where $\Delta E = |E_1 - E_2|$ and $\Delta t = |t_1 - t_2|$.

An important result of the Mandelstam-Tamm version of the time-energy uncertainty relation is that it gives us a bound on the rate of decay of an unstable state [22]. For this we consider an initial state $|\phi_0\rangle$ which is not an eigenstate of the total Hamiltonian of the system H, but is expressible as an integral over the continuous eigenstates of H. The state of the system after the time t is given by

$$|\phi(t)\rangle = \exp\left(\frac{-iHt}{\hbar}\right)|\phi_0\rangle. \tag{4.117}$$

Using this result we calculate the probability that the system remains in the initial state $|\phi_0\rangle$ after a time t [21]

$$\mathcal{P}(t) = |\langle\phi_0|\phi(t)\rangle|^2. \tag{4.118}$$

From (4.117) and (4.118) we can easily deduce that $\mathcal{P}(t)$ is an even function of t. Next we choose the operator A to be

$$A(t) = |\phi(t)\rangle\langle\phi(t)|. \tag{4.119}$$

By substituting $A(t)$ in (4.110) and using the definition (4.118) we find

$$\frac{[\mathcal{P}(t)(1-\mathcal{P}(t))]^{\frac{1}{2}}}{\left|\frac{d\mathcal{P}(t)}{dt}\right|} \geq \frac{\hbar}{2\Delta E}. \tag{4.120}$$

A rearrangement of the inequality (4.120) gives us

$$\left|\frac{d\mathcal{P}(t)}{dt}\right| \leq \left(\frac{2\Delta E}{\hbar}\right)[\mathcal{P}(t)(1-\mathcal{P}(t))]^{\frac{1}{2}}. \tag{4.121}$$

We observe that the right-hand side of (4.121) takes its maximum value for $\mathcal{P}(t_h) = \frac{1}{2}$ when

$$\left|\frac{d\mathcal{P}}{dt}\right|_{t=t_h} \leq \frac{\Delta E}{\hbar}. \tag{4.122}$$

This time, t_h, which is half-life of a decaying system has the property that only for $t = t_h$ Eq. (4.122) can become an equality.

Another important conclusion that we can draw from (4.122) is that no unstable quantum system can decay completely within a time $\frac{\hbar}{\Delta E}$ [22].
By integrating (4.121) we obtain

$$t \geq \frac{\hbar}{\Delta E}\arccos\sqrt{\mathcal{P}(t)}, \tag{4.123}$$

where we have used the initial condition $\mathcal{P}(0) = 1$. By solving (4.123) for $\mathcal{P}(t)$ we obtain the inequality

$$\mathcal{P}(t) \geq \cos^2\left(\frac{\Delta E t}{\hbar}\right), \quad 0 \leq t \leq \frac{\pi\hbar}{2\Delta E}. \tag{4.124}$$

From Eq. (4.124) we find

$$\mathcal{P}(t_h) = \frac{1}{2} \geq \cos^2\left(\frac{\Delta E\, t_h}{\hbar}\right), \quad 0 \leq t_h \leq \frac{\pi\hbar}{2\Delta E}, \tag{4.125}$$

and by solving this inequality we get

$$\Delta E\, t_h \geq \frac{\pi\hbar}{4}, \quad 0 \leq t_h \leq \frac{\pi\hbar}{2\Delta E}. \tag{4.126}$$

On the other hand for values of $t_h > \frac{\pi\hbar}{2E}$, the inequality becomes

$$\mathcal{P}(t_h) \geq 0 > 1 - \frac{2\Delta E t}{\pi\hbar}, \quad t > \frac{\pi\hbar}{2\Delta E}, \tag{4.127}$$

or thus

$$\Delta E\, t_h > \frac{\pi\hbar}{4} \quad \text{or,} \quad t_h > \frac{\pi\hbar}{2\Delta E}, \tag{4.128}$$

Combining (4.126) and (4.128) we have the following inequality for a decaying quantum system [22]

$$\Delta E\, t_h > \frac{\pi\hbar}{4}. \tag{4.129}$$

As it can be seen from (4.129) the half-life, t_h, and the energy dispersion, ΔE are related.

Wigner's Time-Energy Uncertainty Relation — A different formulation of time-energy relation is due to Wigner and in its simplest form it can expressed as follows [23]:

Let $\psi(x,t)$ be the time-dependent wave function describing a system with the Hamiltonian H. We expand $\psi(x,t)$ in terms of the eigenfunctions of time-independent Hamiltonian H;

$$\psi(x,t) = \int \psi(x,E) \exp\left(\frac{-iEt}{\hbar}\right) dE. \tag{4.130}$$

The uncertainties Δt and ΔE for time and energy are defined by

$$(\Delta t)^2 = \frac{\int_0^\infty (t-\bar{t})^2 |\psi(x,t)|^2 dt}{\int_0^\infty |\psi(x,t)|^2 dt}, \tag{4.131}$$

and

$$(\Delta E)^2 = \frac{\int_0^\infty \left(E-\bar{E}\right)^2 |\psi(x,E)|^2 dE}{\int_0^\infty |\psi(x,E)|^2 dE}, \tag{4.132}$$

where $\bar{t}$ and $\bar{E}$ refer to the mean values of t and E. Here it is assumed that the denominators are finite and not necessarily normalized. From these two relations we obtain

$$\Delta t(x) \Delta E(x) \geq \frac{\hbar}{2}. \tag{4.133}$$

For the simple case of $\bar{E} = \bar{t} = 0$ we can derive (4.133) in the following way: Assuming that $\psi(x,t)$ and $\psi(x,E)$ are not normalized we write

$$N = \int |\psi(x,t)|^2 dt = \frac{1}{2\pi} \int |\psi(x,E)|^2 dE, \tag{4.134}$$

where the second term is found from Parseval relation noting that $\psi(x,E)$ is the Fourier transform of $\psi(x,t)$. Now we write the Schwarz inequality as

$$\left| \int t\psi(x,t) \frac{\partial \psi(x,t)}{\partial t} dt \right|^2 \leq \int t^2 |\psi(x,t)|^2 dt \int \left| \frac{\partial \psi(x,t)}{\partial t} \right|^2 dt \tag{4.135}$$

By integrating the left-hand side of (4.135) by parts we get

$$\int t\psi(x,t) \frac{\partial \psi(x,t)}{\partial t} dt = -\frac{1}{2} \int |\psi(x,t)|^2 dt = -\frac{N}{2}. \tag{4.136}$$

Again from Parseval relation we have

$$\hbar^2 \int \left| \frac{\partial \psi(x,t)}{\partial t} \right|^2 dt = \frac{1}{2\pi} \int E^2 |\psi(x,E)|^2 dE. \tag{4.137}$$

Now if we define $(\Delta\tau)^2$ and $(\Delta E)^2$ by Eqs. (4.131) and (4.132) then (4.135) gives us the uncertainty relation (4.133). Since the energy spectrum has a lower bound therefore Wigner concludes that the equality in (4.133) is not permissible [23]. We note that this uncertainty relates the spreads of energy and time at a fixed point x. In the case of position-momentum the uncertainty $\Delta p(t)\Delta x(t)$ is given at an instant of time and t changes in one direction (increasing t only), whereas in $\Delta\tau(x)\Delta E(x)$, x can change in both directions i.e. from $-\infty$ to $+\infty$, or from $+\infty$ to $-\infty$.

4.4 Uncertainty Relations for Angular Momentum-Angle Variables

The result that we found for two Hermitian operators A_1 and A_2, Eq. (4.66) is valid whenever both A_1 and A_2 are self-adjoint in the domains over which they are defined, e.g. $-\infty < x < +\infty$. However for the pair of variables, the angular momentum L_z and its conjugate ϕ, where

$$\phi = \tan^{-1}\left(\frac{y}{x}\right), \tag{4.138}$$

and

$$L_z = xp_y - yp_x = -i\frac{\partial}{\partial\phi}, \quad \hbar = 1, \tag{4.139}$$

there is a problem since ϕ in (4.138) is defined modulo 2π. Thus L_z as is given by (4.139) is self-adjoint, i.e.

$$\langle L_z f|g\rangle = \langle f|L_z g\rangle, \tag{4.140}$$

only if f and g are periodic functions with period 2π. If for the moment, we assume that ϕ and L_z are conjugate operators

$$[\phi,\ L_z] = i, \tag{4.141}$$

and choose the eigenfunctions of L_z to be $|\ell, m\rangle$ (see Sec. 9.1) [24]

$$L_z|\ell, m\rangle = m|\ell, m\rangle, \quad m = -\ell \cdots \ell, \tag{4.142}$$

then

$$\langle \ell, m'\,|[\phi,\ L_z]|\,\ell, m\rangle = i\,\langle \ell, m'|\ell, m\rangle\,. \tag{4.143}$$

Using the self-adjointness of L_z we operate to the left to obtain

$$(m - m') \langle \ell, m'|\phi|\ell, m\rangle = i\delta_{m,m'}. \tag{4.144}$$

In this relation if we set $m = m'$ we find that it cannot be correct.

We can introduce a periodic function of the coordinate ϕ by defining $\Phi(\phi)$ to be [24],[25]

$$\Phi(\phi) = \phi - 2\pi \sum_{n=0}^{\infty} \theta[\phi - (2n+1)\pi] + 2\pi \sum_{n=0}^{\infty} \theta[-\phi - (2n+1)\pi], \tag{4.145}$$

where $\theta(z)$ is the step function. This type of periodic function will also be used later to describe the problem of a particle bouncing between two rigid walls Sec. 7.6. If we choose $\Phi(\phi)$ instead of ϕ, the commutation relation will take the form

$$[\Phi(\phi),\ L_z] = i\left\{1 - 2\pi \sum_{n=-\infty}^{\infty} \delta[\phi - (2n+1)\pi]\right\}. \tag{4.146}$$

Rather than using $\Phi(\phi)$, we can use $\sin\phi$ and $\cos\phi$, both periodic functions of ϕ with period 2π, to calculate the commutator. The commutation relations for these functions of ϕ and L_z are:

$$[\sin\phi,\ L_z] = i\cos\phi, \tag{4.147}$$

and

$$[\cos\phi,\ L_z] = -i\sin\phi. \tag{4.148}$$

These results follow from the Dirac's rule of association, Eq.(3.121). Denoting the uncertainties (or the dispersions) by $(\Delta L_z)^2$, $(\Delta\sin\phi)^2$ and $(\Delta\cos\phi)^2$, we have according to (4.66) for self-adjoint operator L_z;

$$(\Delta L_z)^2(\Delta\sin\phi)^2 \geq \frac{1}{4}\langle\cos\phi\rangle^2, \tag{4.149}$$

$$(\Delta L_z)^2(\Delta\cos\phi)^2 \geq \frac{1}{4}\langle\sin\phi\rangle^2. \tag{4.150}$$

By adding (4.149) to (4.150) we find an uncertainty relation which is symmetric in $\cos\phi$ and $\sin\phi$;

$$\frac{(\Delta L_z)^2\left[(\Delta\cos\phi)^2 + (\Delta\sin\phi)^2\right]}{\langle\sin\phi\rangle^2 + \langle\cos\phi\rangle^2} \geq \frac{1}{4}. \tag{4.151}$$

Let us now inquire about the conditions under which we can get the standard form of the uncertainty relation

$$(\Delta L_z)^2(\Delta\phi)^2 \geq \frac{1}{4}. \tag{4.152}$$

Consider the probability distribution $P(\phi)$ which is sharply peaked at $\phi = \phi_0$, and is symmetrical about this point. We expand $\cos\phi$ and $\sin\phi$ about $\phi - \phi_0$;

$$\cos\phi = \cos\phi_0 \left[1 - \frac{1}{2}(\delta\phi)^2\right] - \sin\phi_0 \delta\phi, \tag{4.153}$$

$$\sin\phi = \sin\phi_0 \left[1 - \frac{1}{2}(\delta\phi)^2\right] + \cos\phi_0 \delta\phi, \tag{4.154}$$

where

$$\delta\phi = \phi - \phi_0. \tag{4.155}$$

We choose $\delta\phi$ to be much smaller than 2π. Using (4.153),(4.154) we calculate different ϕ-dependent terms in (4.151) remembering that odd power terms in ϕ will not contribute when averaged over $P(\phi)$;

$$\langle\cos\phi\rangle = \cos\phi_0 \left[1 - \frac{1}{2}\left\langle(\delta\phi)^2\right\rangle\right], \tag{4.156}$$

$$\langle\sin\phi\rangle = \sin\phi_0 \left[1 - \frac{1}{2}\left\langle(\delta\phi)^2\right\rangle\right], \tag{4.157}$$

$$\left\langle\cos^2\phi\right\rangle = \cos^2\phi_0 \left[1 - \frac{1}{2}\left\langle(\delta\phi)^2\right\rangle\right] + \sin^2\phi_0 \left\langle(\delta\phi)^2\right\rangle, \tag{4.158}$$

and

$$\left\langle\sin^2\phi\right\rangle = \sin^2\phi_0 \left[1 - \frac{1}{2}\left\langle(\delta\phi)^2\right\rangle\right] + \cos^2\phi_0 \left\langle(\delta\phi)^2\right\rangle. \tag{4.159}$$

Since $\delta\phi$ is small we keep terms up to second order and this gives us

$$(\Delta\cos\phi)^2 = \left\langle\cos^2\phi\right\rangle - \langle\cos\phi\rangle^2 = \sin^2\phi_0 \left\langle(\delta\phi)^2\right\rangle, \tag{4.160}$$

$$(\Delta\sin\phi)^2 = \left\langle\sin^2\phi\right\rangle - \langle\sin\phi\rangle^2 = \cos^2\phi_0 \left\langle(\delta\phi)^2\right\rangle. \tag{4.161}$$

By substituting (4.157)–(4.161) in (4.151) we obtain

$$(\Delta L_z)^2 \left\langle(\delta\phi)^2\right\rangle \geq \frac{1}{4}. \tag{4.162}$$

Thus if we identify $(\Delta\phi)$ by

$$\left\langle(\delta\phi)^2\right\rangle = (\Delta\phi)^2, \tag{4.163}$$

we obtain (4.152). This inequality is of the same form as the uncertainty for $(\Delta A_1)^2(\Delta A_2)^2$ found from (4.66).

We can ask why it is impossible to measure L_z and ϕ at the same time. A possible explanation can be the following [26]. Suppose we measure the total angular momentum of a system and find that it is zero, i.e. the system is in S state. What we have measured is the symmetry of the system under rotation, and we have found that the system is spherically symmetric. But this implies that we cannot measure the conjugate angle since to a spherically symmetric system we cannot assign a privileged angle.

4.5 Local Heisenberg Inequalities

Next we want to study simple generalizations of the uncertainty principle [27]–[30]. We start with an analysis of the local Heisenberg inequalities. Let x and p denote the position and the momentum of a particle, and let $f(x)$ be some arbitrary function of x, then from the commutator

$$[p,\ f(x)] = -if'(x), \quad \hbar = 1, \tag{4.164}$$

we obtain

$$\Delta p\,\Delta f \geq \frac{1}{2}\left|\langle f'(x)\rangle\right|, \tag{4.165}$$

Now we will study two specific cases:

(1) - Let $f(x)$ be a step function

$$f(x) = \theta(x - x_0) = \begin{cases} 0 & x < x_0 \\ 1 & x > x_0 \end{cases}, \tag{4.166}$$

then by calculating the average value of $f(x)$ we get

$$\langle f^2(x)\rangle = \langle f(x)\rangle = \int_{x_0}^{\infty} |\psi(x)|^2 dx = 1 - \mathcal{P}(x_0), \tag{4.167}$$

where $\psi(x)$ is the wave function of the particle and $\mathcal{P}(x_0)$ is the probability of the localization on the half-axis $(-\infty < x \leq x_0)$. From the definition of Δf it follows that

$$\begin{aligned}(\Delta f)^2 &= \langle (f(x))^2\rangle - |\langle f(x)\rangle|^2 \\ &= (1 - \mathcal{P}(x_0)) - (1 - \mathcal{P}(x_0))^2 = \mathcal{P}(x_0)(1 - \mathcal{P}(x_0)).\end{aligned} \tag{4.168}$$

In addition since $f(x) = \theta(x - x_0)$, we have

$$\langle f'(x)\rangle = \langle \delta(x - x_0)\rangle = |\psi(x_0)|^2. \tag{4.169}$$

Substituting these results in Eq. (4.165) we find

$$|\psi(x)|^2 \leq 2\Delta p\left\{\mathcal{P}(x)\left[1 - \mathcal{P}(x)\right]\right\}^{\frac{1}{2}}. \tag{4.170}$$

This expression relates the probability density at the point x to the probability itself and to the momentum uncertainty. Using the fact that the maximum of $\mathcal{P}(x)\,(1 - \mathcal{P}(x))$ is at $\mathcal{P}(x) = \frac{1}{2}$, we have

$$\mathcal{P}(x)\left[1 - \mathcal{P}(x)\right] \leq \frac{1}{4}, \tag{4.171}$$

and thus we conclude that

$$|\psi(x)|^2 \leq \Delta p. \tag{4.172}$$

From either (4.170) or (4.172) it follows that for the normalized wave function $\psi(x)$ to be large at a certain point, it must be narrow enough so that its momentum width is necessarily large. We also observe that the upper bound $|\psi(x)|^2 = \Delta p$ can only be reached for a unique value of x such that $\mathcal{P}(x) = 1$.

(2) - For the second example let us consider the case where

$$f(x) = \frac{1}{x}. \tag{4.173}$$

Then in the inequality

$$\Delta A_1 \Delta A_2 \geq \frac{1}{2}|\langle [A_1,\ A_2]\rangle|, \tag{4.174}$$

if we choose $A_1 = p$ and $A_2 = x^{-1}$ we have

$$(\Delta p)^2 \left[\left\langle \frac{1}{x^2}\right\rangle - \left(\left\langle \frac{1}{x}\right\rangle\right)^2\right] \geq \frac{1}{4}\left\langle \frac{1}{x}\right\rangle^2. \tag{4.175}$$

Now assuming that both $\langle x^{-2}\rangle$ and $\langle x^{-1}\rangle$ are finite, then from the fact that $\frac{a^2}{a-b} \geq 4b$ for $a \geq b$ we find that

$$\Delta p \geq \left\langle \frac{1}{x}\right\rangle. \tag{4.176}$$

Applying this result to the problem of hydrogen atom by setting $x = r$ and considering the Hamiltonian

$$H = \frac{1}{2m}p_r^2 - \frac{e^2}{r}, \tag{4.177}$$

we get

$$\begin{aligned}\langle H\rangle &= \frac{1}{2m}\langle p_r^2\rangle - e^2\left\langle \frac{1}{r}\right\rangle \geq \frac{1}{2m}\left\langle \frac{1}{r}\right\rangle^2 - e^2\left\langle \frac{1}{r}\right\rangle \\ &= \left(\frac{1}{\sqrt{2m}}\left\langle \frac{1}{r}\right\rangle - \sqrt{\frac{m}{2}}e^2\right)^2 - \frac{m}{2}e^4 \geq -\frac{m}{2}e^4.\end{aligned} \tag{4.178}$$

This result shows that in the hydrogen atom the Hamiltonian has a lower bound and therefore the atom is stable.

Uncertainty Relations for Measurements at Different Times — In the derivation of the general form of uncertainty principle, Eq. (4.68), we assumed that both A_1 and A_2 are Hermitian operators corresponding to two observable quantities measured at a given time, e.g. $p(t)$ and $x(t)$. Now we want to obtain an expression for the uncertainty when the position of the particle x is measured at time t while its momentum p is measured at time $t + \delta t$. In the Heisenberg picture we can write

$$p(t+\delta t) = \exp\left(\frac{iH\delta t}{\hbar}\right)p(t)\exp\left(\frac{-iH\delta t}{\hbar}\right), \tag{4.179}$$

where the Hamiltonian is assumed to be time-independent

$$H = \frac{1}{2m}p^2 + V(x). \tag{4.180}$$

Expanding (4.179) using the Baker–Campbell–Hausdorff formula (3.77) we have

$$p(t+\delta t) = p(t) - \frac{i\delta t}{\hbar}[H,\, p] - \frac{(\delta t)^2}{2!\,\hbar^2}[H,\, [H,\, p]] + \cdots. \tag{4.181}$$

This relation can be simplified if we calculate the commutators

$$[H,\, p] = [V(x),\, p] = i\hbar V'(x). \tag{4.182}$$

and

$$\frac{1}{2}[H,\, [H,\, p]\,] = \frac{\hbar}{2}\left[\frac{p^2}{2m},\, iV'(x)\right] = \frac{\hbar^2}{4m}\left(pV'' + V''p\right). \tag{4.183}$$

Substituting these in (4.181) we get

$$[x(t),\, p(t+\delta t)] = i\hbar\left\{1 - \frac{(\delta t)^2}{2m}V''(x) + \mathcal{O}\left(\delta t\right)^3\right\}. \tag{4.184}$$

Thus the uncertainty relation according to (4.68) is

$$\Delta x(t)\Delta p(t+\delta t) \geq \frac{\hbar}{2}\left\{1 - \frac{(\delta t)^2}{2m}\left\langle V''\right\rangle_t + \mathcal{O}\left(\delta t\right)^3\right\}. \tag{4.185}$$

One of the interesting features of this relation is that it shows that for a nonconvex potential, the product of the uncertainties of the position and momentum at different times can be less than $\frac{\hbar}{2}$ [29].

The same method can be used to obtain momentum-momentum and position-position uncertainties. Thus from the commutator

$$[p(t),\, p(t+\delta t)] = i\hbar\delta t V'' + \mathcal{O}\left(\delta t\right)^3, \tag{4.186}$$

we find the uncertainty relation

$$\Delta p(t)\Delta p(t+\delta t) \geq \frac{\hbar}{2}\left|\delta t\left\langle V''(x)\right\rangle_t\right| + \mathcal{O}\left(\delta t\right)^3. \tag{4.187}$$

In order to determine $\Delta x(t)\Delta x(t+\delta t)$ we write the analogue of (4.182) and (4.183) for $x(t)$,

$$[H,\, x] = -i\frac{\hbar}{m}p. \tag{4.188}$$

and

$$\frac{1}{2}[H,\, [H,\, x]\,] = \frac{\hbar^2}{2m}V'(x). \tag{4.189}$$

Then from

$$\begin{aligned} x(t+\delta t) &= \exp\left(\frac{iH\delta t}{\hbar}\right) x(t) \exp\left(\frac{-iH\delta t}{\hbar}\right) \\ &= x(t) + \frac{\delta t}{m} p(t) - \frac{(\delta t)^2}{2m} V'(x) + \mathcal{O}\left(\delta t\right)^3, \end{aligned} \tag{4.190}$$

we obtain

$$[x(t),\ x(t+\delta t)] = \frac{i\hbar}{m}\delta t + \mathcal{O}\left(\delta t\right)^3, \tag{4.191}$$

and thus

$$\Delta x(t)\Delta x(t+\delta t) \geq \frac{\hbar}{2}\frac{|\delta t|}{m} + \mathcal{O}\left(\delta t\right)^3. \tag{4.192}$$

This relation shows the well-known fact of the spreading of the wave packet for very short times [29].

For the problem of harmonic oscillator we can find that these three inequalities are independent of the magnitude of δt. For instance from

$$p(t+\delta t) = \cos(\omega\delta t)\ p(t) - m\omega\sin(\omega\delta t)\ x(t), \tag{4.193}$$

which is the exact solution of Heisenberg's equations for the harmonic oscillator we obtain the uncertainty relation

$$\Delta x(t)\Delta p(t+\delta t) \geq \frac{\hbar}{2}|\cos(\omega\delta t)|. \tag{4.194}$$

Similar results can be found for $\Delta x(t)\Delta x(t+\delta t)$ and for $\Delta p(t)\Delta p(t+\delta t)$ [29].

Correlation of Position and Momentum — In probability theory and in statistics the extent to which two random variables vary together is measured by their covariance. Related to the covariance of two variables is their correlation which measures their degree of independence. This is found by dividing the covariance of two variables by the product of their standard deviation. Denoting the variables by A_1 and A_2 we write the correlation $\rho(A_1, A_2)$ as

$$\rho(A_1, A_2) = \frac{\mathrm{cov}(A_1, A_2)}{\Delta A_1 \Delta A_2}. \tag{4.195}$$

In this relation "cov" denotes the covariance of A_1 and A_2 and is given by

$$\begin{aligned} \mathrm{cov}(A_1, A_2) &= \langle (A_1 - \langle A_1\rangle)(A_2 - \langle A_2\rangle)\rangle \\ &= \langle A_1 A_2\rangle - \langle A_1\rangle\langle A_2\rangle, \end{aligned} \tag{4.196}$$

where $\langle A_1\rangle$ and $\langle A_2\rangle$ are the mean values of A_1 and A_2 respectively and dispersion ΔA_r is defined by

$$\Delta A_r = \left(\langle A_r^2\rangle - \langle A_r\rangle^2\right)^{\frac{1}{2}}, \quad r = 1,\ 2. \tag{4.197}$$

From the Schwarz inequality it follows that

$$|\rho(A_1, A_2)| \le 1. \tag{4.198}$$

Now let us consider the correlation coefficient for two physical observables of the system. For this we replace the operator A_r by A'_r, where

$$A'_r = A_r - \langle k|A_r|k\rangle 1, \quad r = 1,\ 2, \tag{4.199}$$

and 1 is the unit operator (or matrix). In this way the uncertainty ΔA_r defined by (4.50) takes the simpler form of

$$\Delta A_r = \left[\langle k|A'_r|k\rangle\right]^{\frac{1}{2}}. \tag{4.200}$$

In general the two operators A_1 and A_2 do not commute with each other, therefore we replace A_1A_2 by the symmetrized operator $\frac{1}{2}(A_1A_2 + A_2A_1)$ which is Hermitian. But as we have seen in Chapter 3 this is one of the possible forms of constructing a Hermitian operator. Therefore our definition of covariance for two quantum mechanical operators is not unique. For the two operators we define the covariance in analogy with the definition (4.196)

$$\begin{aligned}
\mathrm{cov}_k\langle A_1A_2\rangle &= \frac{1}{2}\langle k|(A_1A_2 + A_2A_1|k\rangle - \langle k|A_1|k\rangle\langle k|A_2|k\rangle \\
&= \frac{1}{2}\langle k\,|(A'_1A'_2 + A'_2A'_1)|\,k\rangle .
\end{aligned} \tag{4.201}$$

From this we find the quantum correlation function for the state $|k\rangle$ to be given by [30]

$$\rho_k(A_1, A_2) = \frac{\mathrm{cov}(A_1, A_2)}{\Delta A_1 \Delta A_2} = \frac{1}{2}\left(\frac{\langle k\,|(A'_1A'_2 + A'_2A'_1)|\,k\rangle}{\sqrt{\langle k\,|{A'_1}^2|\,k\rangle\,\langle k\,|{A'_2}^2|\,k\rangle}}\right). \tag{4.202}$$

It should be noted that if A_1 and A_2 are incompatible observables then they cannot be measured simultaneously and $\mathrm{cov}_k(A_1, A_2)$ cannot be found from a set of measurements of A_1 and A_2 separately. At the same time the Hermitian operator $A'_1A'_2 + A'_2A'_1$ may not be self-adjoint or have self-adjoint extension, and therefore may not be an observable of the system as we have seen in Chapter 3.

Next we replace the expectation value of the anticommutator $A'_1A'_2+A'_2A'_1$ using the generalized Heisenberg's uncertainty relation (4.66) and we find a simple expression for the quantum correlation function in terms of A_1 and A_2

$$\rho_k^2(A_1, A_2) \le 1 - \left[\frac{|\langle k|[A_1,\ A_2]|k\rangle|}{2\Delta A_1 \Delta A_2}\right]^2. \tag{4.203}$$

From this inequality we conclude that when $[A_1,\ A_2] \ne 0$, $\rho_k(A_1, A_2)$ does not reach unity which is its classical bound (4.198). Thus in quantum mechanics there is a decrease in the correlation between any two dynamical variables as compared to the classical one, and this is particularly pronounced when the uncertainties ΔA_1 and ΔA_2 are small [30].

Let us now determine the correlation coefficient $\rho_k(x,p)$ for a one-dimensional motion. In this case (4.203) reduces to

$$\rho_k^2(x,p) \le 1 - \frac{\hbar^2}{4(\Delta x)^2(\Delta p)^2}. \tag{4.204}$$

An immediate consequence of this result is that for a minimum uncertainty, $\Delta p \Delta q = \frac{\hbar}{2}$, Eq. (4.75), $\rho_k^2(x,p) = 0$, i.e. there is no correlation between x and p. On the other hand for a normalized time-dependent Gaussian wave packet of the form

$$\psi(x,t) = \frac{1}{\pi^{\frac{1}{4}}} e^{-i\phi(t)} \sqrt{a}\sqrt{|\alpha(t)|} e^{ik_0 x} \exp\left[-\frac{1}{2}\alpha(t)\left(x - \frac{\hbar k_0 t}{m}\right)\right], \tag{4.205}$$

where $\alpha(t)$ which is the time-dependent phase is given by

$$\alpha(t) = \left[a^2 + \frac{i\hbar t}{m}\right]^{-\frac{1}{2}}, \tag{4.206}$$

we find the uncertainties to be (with $\hbar = 1$ written explicitly)

$$\begin{cases} \Delta x = \dfrac{a}{\sqrt{2}}\left(1 + \dfrac{\hbar^2 t^2}{m^2 a^4}\right)^{\frac{1}{2}} \\ \Delta p = \dfrac{\hbar}{\sqrt{2}a} \end{cases}. \tag{4.207}$$

Here the correlation coefficient is given by

$$\rho_k(x,p) = \frac{\hbar t}{\sqrt{m^2 a^4 + \hbar^2 t^2}}, \tag{4.208}$$

which shows that the correlation is zero at $t = 0$ and approaches unity as t becomes large. Why in the case of a time-dependent wave packet we have such a change in the value of the correlation coefficient? To explain this result we note that the wave packet is composed of an infinite number of components each corresponding to a different momentum in its Fourier transform. The tail of the wave packet, which is the fastest moving part, is associated with high Fourier components, whereas the part close to the center of the wave packet is associated with Fourier components with slowest momenta. Thus there is a correlation between different spatial parts and the Fourier components representing momenta.

We can also calculate the correlation coefficient for a particle in a stationary state for any potential $V(x)$. Starting with the Hamiltonian

$$H = \frac{1}{2m}p^2 + V(x), \tag{4.209}$$

with the eigenstate $|n\rangle$ we find

$$\langle n|xp + px|n\rangle = \frac{im}{\hbar}\langle n\left[H,\ x^2\right]|n\rangle = 0. \tag{4.210}$$

Also we know that for a stationary state

$$\langle n|p|n\rangle = \frac{im}{\hbar}\langle n\left[H,\ x\right]|n\rangle = 0. \tag{4.211}$$

Thus from (4.201) it follows that

$$\mathrm{cov}_n(x,p) = 0, \tag{4.212}$$

and consequently

$$\rho_n(x,p) = 0. \tag{4.213}$$

Thus there is no correlation between x and p for this motion, a result which is independent of the shape of the potential.

Finally let us calculate the correlation function for the components of the angular momentum of a particle. Denoting the eigenstates of the angular momentum by $|k\rangle = |j,m\rangle$ (Chapter 9), then from Eq. (9.54) we have

$$\langle j,m|M_xM_y + M_yM_x|j,m\rangle = \frac{1}{2i}\left\langle j,m\left|M_+^2 - M_-^2\right|j,m\right\rangle = 0. \tag{4.214}$$

Choosing A_1 and A_2 to be M_x and M_y respectively, then from Eqs. (4.203) and (4.214) it follows that

$$\rho_{j,m}(M_x, M_y) = 0. \tag{4.215}$$

Thus we conclude that the correlation coefficient for angular momentum operator is nonzero only for certain superposition of eigenstats [31].

4.6 The Correspondence Principle

We have observed the similarity between Heisenberg's equations of motion and the Hamilton canonical equation. Now let us inquire how the results found from these two theories are related to each other. In particular we want to know whether we can recover the classical description of a motion from the corresponding quantum mechanical problem or not, assuming that the quantum theory contains the classical mechanics as a limiting case. By a close examination of various works one finds that there are at least three distinct ways of getting classical equations from quantum formulation [32]:

The Planck Limit — As early as 1906 Planck observed that his black body radiation formula reduces to Rayleigh–Jean result in the limit of $\hbar \to 0$ [33],[34]

This limit may not be well-defined for every quantum mechanical operator. For instance consider the operators $P_\pm$ defined by [35]

$$P_\pm = \frac{1}{\hbar^2}\mathbf{M}^2 \pm \frac{1}{\hbar}M_z, \tag{4.216}$$

where $\mathbf{M}^2$ and M_z (square and the z-component of angular momentum) are defined by Eqs. (9.74) and (9.75). The operators $\mathbf{M}^2$ and M_z correspond to the classical components of momentum in spherical polar coordinates $(p_r,\ p_\theta,\ p_\phi)$ by

$$M_z = p_\phi, \quad \text{and} \quad \mathbf{M}^2 = p_\theta^2 + \frac{p_\phi^2}{\sin^2\theta}. \tag{4.217}$$

Let us also denote the conjugates of P_+ and P_- operators by $Q_\pm$. Clearly the operators P_- and P_+ do not have classical limits as $\hbar \to 0$, even though they are well-defined operators with the eigenvalues

$$P_\pm|\ell, m\rangle = [\ell(\ell+1) \pm m]\,|\ell, m\rangle, \quad -\ell \le m \le \ell, \tag{4.218}$$

where ℓ is a positive integer or zero. We note that all of the eigenvalues of P_+ and P_- are nondegenerate and nonnegative. Since $[P_+,\ P_-] = 0$, we can use a representation in which both of these operators are diagonal. If here there were connections between the classical Poisson bracket and the quantum commutator, then $[P_+,\ Q_-] = 0$ and thus in this representation Q_- would be diagonal too. However this is not compatible with the requirement that P_- and Q_- are conjugates [35].

Bohr Correspondence Principle — According to Bohr, the prediction of quantum mechanics approaches classical limit as the quantum number of bound systems become large, i.e. the quantum number n tends to infinity. For systems such as a particle in a box with rigid walls Sec. 7.5 or a rigid rotator in the limit of large quantum numbers the classical motion is not recovered. For instance the frequencies of the quantum spectrum $\omega_{mn} = (E_m - E_n)/\hbar$ do not approach the continuous frequencies of the classical spectrum. For a detailed account of the classical limit of quantum mechanics the reader is referred to the papers of Makowski [32] and Liboff [34].

Correspondence Principle and the Asymptotic Form of the Probability Density for the Harmonic Oscillator — While the frequencies of the quantum oscillators do not approach the classical frequencies, we can ask whether quantum probabilities approach the classical probabilities in the limit of large quantum numbers. To investigate this point we consider the example of a simple harmonic oscillator. From the normalized wave function of the simple harmonic oscillator, (see Eq. (8.92)), we can find the probability density $\rho_n(x) = |\psi_n(x)|^2$. If we want to calculate this density in the limit of $n \to \infty$, we can write $\psi_n(x)$ in terms of the Weber function or the confluent hypergeometric function of the third kind [36]

$$\psi_n(x) = D_n(\sqrt{2}\beta x) = 2^{\frac{n}{2}} \exp\left[-\frac{1}{2}\beta^2 x^2 + i\pi n\right] U_2\left(-\frac{1}{2}n, \frac{1}{2}, \beta^2 x^2\right). \tag{4.219}$$

Introducing the dimensionless variable $z = \beta x$ we can write $U_2\left(-\frac{1}{2}n, \frac{1}{2}, \beta^2 x^2\right)$ as a contour integral, and then ψ_n becomes

$$\psi_n(x) = \frac{\beta}{\sqrt{2^n n!\sqrt{\pi}}} 2^n \frac{\Gamma\left(1+\frac{1}{2}n\right)}{2\pi i} \exp\left(-\frac{1}{2}z^2\right) \times \oint \exp\left[z^2 t - \left(\frac{1}{2}n+1\right)\ln t + \left(\frac{1}{2}n - \frac{1}{2}\right)\ln(1-t)\right] dt, \tag{4.220}$$

where the contour in the complex t-plane consists of two straight lines one going from $-\infty$ to about $t = 0$ in the positive direction and there via a semicircle joining another line going from about $t = 0$ to $-\infty$. The saddle point for the integral is where the derivative of the expression

$$z^2 t - \left(\frac{1}{2}n+1\right)\ln t + \left(\frac{1}{2}n - \frac{1}{2}\right)\ln(1-t), \tag{4.221}$$

is zero, viz, at the point

$$z^2 = \frac{\frac{1}{2}n + 1 - \frac{3}{2}t}{t(1-t)}. \tag{4.222}$$

We are interested in the limit when $2n + 1 > z^2$, and when this is the case the two saddle points are symmetrically located above or below the real axis, and we get a cosine factor for the integral. Thus the wave function for $n \gg 1$ becomes

$$\psi_n(x) \approx \frac{\sqrt{\frac{2\beta}{\pi}}}{\left(2n - \beta^2 x^2\right)^{\frac{1}{4}}} \cos\left[\frac{1}{2}\beta x\sqrt{2n - \beta^2 x^2} - \frac{1}{2}\pi n + n\sin^{-1}\left(\frac{\beta x}{\sqrt{2n}}\right)\right], \tag{4.223}$$

and this is valid when $z^2 < 2n$.

By calculating $|\psi_n(x)|^2$ and simplifying we find the probability density for large n values

$$\rho_n(x) \approx \frac{2}{\pi\left(A_n^2 - x^2\right)} \cos^2\left[\frac{1}{2}\beta^2 x\sqrt{A_n^2 - x^2} - \frac{1}{2}\pi n + \frac{1}{2}\beta^2 A_n^2 \sin^{-1}\left(\frac{x}{A_n}\right)\right], \tag{4.224}$$

where $x < A_n$ and

$$A_n^2 = \frac{2n+1}{\beta^2} = \frac{2E_n}{m\omega^2}. \tag{4.225}$$

As Eq. (4.224) shows for large n, $\rho_n(x)$ has a large number of nodes , and that the separation between these nodes is approximately $2\pi/\left(\beta^2 A_n\right)$. In addition this same equation for $\rho_n(x)$ reduces to the classical expression for probability, viz,

$$\rho_{cl}(x) = \frac{1}{\pi\sqrt{A^2 - x^2}}, \tag{4.226}$$

where $A = \sqrt{\frac{2E}{m}}$ is the classical amplitude of motion. That is $\rho_{cl}(x)$ is not the limit of $\rho_n(x)$ as n tend to infinity (which is the Bohr correspondence principle), but it is the average of $\rho_n(x)$ over several wavelengths of oscillation. In other words we get the classical limit if we replace $\cos^2$ in (4.224) by its average value which is $\frac{1}{2}$. For other interesting examples of deviation from the Bohr correspondence principle see [37].

Heisenberg's Correspondence Principle — This principle can be stated in the following way:

Let $g(t)$ be a classical quantity related to a bound motion and $\mathcal{G}(t)$ be its corresponding quantum mechanical operator in the Heisenberg picture, then [7],[38]

$$\langle n|\mathcal{G}(t)|n+s\rangle = \langle n|\mathcal{G}|n+s\rangle e^{\frac{i}{\hbar}(E_{n+s}-E_n)t} \approx \mathcal{G}_s(n)e^{is\omega(n)t}, \tag{4.227}$$

where $\mathcal{G}_s(n)$ is the s-th Fourier component of the classical variable $g(t)$ and $\omega(n)$ is the classical frequency when the energy of the particle is E_n. The connection between the Fourier components of the classical motion and the quantum mechanical eigenvalues, as is formulated by Halpern will be discussed in detail in Chapter 7.

Ehrenfest Theorem — If quantum mechanics is to be a more general theory of motion than classical mechanics, it must contain classical mechanics as a limiting case. Ehrenfest's relations express a formal connection between the time-dependence of the mean value of the observables and the Hamilton equations of motion of classical dynamics [39]. To show this connection we use the time-dependent Schrödinger equation and its complex conjugate to eliminate partial derivatives of $|\psi\rangle$ and $\langle\psi|$ with respect to time from the total time derivative of the expectation value of an arbitrary self-adjoint operator A;

$$\begin{aligned}\frac{d\langle A\rangle}{dt} &= \frac{d}{dt}\langle\psi|A|\psi\rangle = \left\langle \psi\,|A|\,\frac{\partial}{\partial t}\psi\right\rangle + \left\langle\left(\frac{\partial\psi}{\partial t}\right)|A|\,\psi\right\rangle \\ &= \frac{1}{i\hbar}\langle\psi|AH|\psi\rangle - \frac{1}{i\hbar}\langle\psi|HA|\psi\rangle + \left\langle\frac{\partial A}{\partial t}\right\rangle \\ &= \frac{1}{i\hbar}\langle\psi|[H,\ A]|\psi\rangle + \left\langle\frac{\partial A}{\partial t}\right\rangle.\end{aligned} \tag{4.228}$$

This relation shows that any quantum mechanical average obeys the corresponding classical equations of motion.

If we apply (4.228) to the observables $\mathbf{r}$ and $\mathbf{p}$ when the Hamiltonian of the system is given by

$$H = \frac{1}{2m}\mathbf{p}^2 + V(\mathbf{r}), \tag{4.229}$$

then we find

$$\frac{d\langle\mathbf{r}\rangle}{dt} = \frac{1}{m}\langle\mathbf{p}\rangle, \tag{4.230}$$

and

$$\frac{d\langle\mathbf{p}\rangle}{dt} = -\langle\nabla V(\mathbf{r})\rangle. \tag{4.231}$$

These two equations are referred to as Ehrenfest's theorem. The right-hand side of (4.231) is the average of the gradient potential and not the gradient of the potential at $\langle r\rangle$. Consider the motion in one dimension and suppose $V(x)$ is a slowly varying function of x, then we can expand $\frac{dV(x)}{dx}$ around $x = \langle x\rangle$;

$$\begin{aligned}\frac{dV(x)}{dx} &= \frac{dV(\langle x\rangle)}{d\langle x\rangle} + \frac{d^2V(\langle x\rangle)}{d\langle x\rangle^2}(x - \langle x\rangle) \\ &+ \frac{1}{2}\frac{d^3V(\langle x\rangle)}{d\langle x\rangle^3}(x - \langle x\rangle)^2 + \cdots. \end{aligned} \tag{4.232}$$

Now noting that $\langle x - \langle x\rangle\rangle = 0$, and by substituting (4.232) in (4.231) we find

$$\frac{d\langle p_x\rangle}{dt} = -\frac{dV(\langle x\rangle)}{d\langle x\rangle} - \frac{\sigma_x^2}{2}\left(\frac{d^3V(\langle x\rangle)}{d\langle x\rangle^3}\right) + \cdots, \tag{4.233}$$

where $\sigma_x = |(x - \langle x\rangle)|$ is the width of the wave packet. If the higher derivatives of $V(x)$ are much smaller than the first derivative, then we can ignore the higher order terms, and we have the classical equations of motion.

4.7 Determination of the State of a System

In this section we examine the question of determination of the wave function of a state from the measurement of probabilities and currents. To this end we consider the connection between the coordinate and momentum probabilities.

The ket $|\psi\rangle$ is a vector in Hilbert space and is a symbol of all possible functional values. As we have seen earlier The coordinate component of this ket is $\psi(\mathbf{r}) = \langle\mathbf{r}|\psi\rangle$ and represents the functional value of $|\psi\rangle$ at $\mathbf{r}$. For the simple case of free particle the momentum eigenfunction is a solution of

$$-i\hbar\nabla\psi_{\mathbf{p}}(\mathbf{r}) = \mathbf{p}\psi_{\mathbf{p}}(\mathbf{r}), \tag{4.234}$$

where $\mathbf{p}$ is the momentum eigenvalue and has the eigenfunction

$$\psi_{\mathbf{p}}(\mathbf{r}) = \langle\mathbf{p}|\mathbf{r}\rangle = \frac{1}{(2\pi\hbar)^{\frac{3}{2}}}\exp\left(\frac{i\mathbf{p}\cdot\mathbf{r}}{\hbar}\right). \tag{4.235}$$

This eigenfunction is normalized in such a way that

$$\frac{1}{(2\pi\hbar)^3}\int \psi_{\mathbf{p}}^*(\mathbf{r})\psi_{\mathbf{p}'}(\mathbf{r})d^3r = \delta\left(\mathbf{p} - \mathbf{p}'\right). \tag{4.236}$$

A physical state is represented by a normalizable state vector and corresponds to a wave packet. For example a particle localized around $\mathbf{r}_0$ can be represented by

$$|\psi\rangle = \int f(\mathbf{r} - \mathbf{r}')|\mathbf{r}\rangle\, d^3r, \tag{4.237}$$

where f is a function which is peaked around $\mathbf{r}_0$. The normalization condition on $|\psi\rangle$ is

$$\langle|\psi\rangle = \int |f(\mathbf{r})^2|; d^3r = 1. \tag{4.238}$$

The idea of the representation of a vector in Hilbert space and its relation to other representation of the same vector can be generalized in the following way:
Suppose a state is specified in q representation, i.e.

$$|\psi\rangle = \sum_q \psi(q)|q\rangle, \quad \text{with} \quad \psi(q) = \langle q|\psi\rangle, \tag{4.239}$$

where q denotes all labels collectively. Now if $|\psi\rangle$ is specified in another representation say p then similar to (4.239) we have

$$|\psi\rangle = \sum_p \phi(p)|p\rangle, \quad \text{with} \quad \phi(q) = \langle p|\psi\rangle, \tag{4.240}$$

where $\phi(p)$ and $\psi(q)$ are related to each other. Thus by replacing $|\psi\rangle$ from (4.239) in (4.240) we find

$$\phi(p) = \sum_q \langle p|q\rangle\psi(q). \tag{4.241}$$

The transition from one representation to another is achieved by means of a unitary transformation. To show this we write (4.241) as

$$\phi(p) = U(p, q)\psi(q), \tag{4.242}$$

where $U(p, q)$ is an operator with the matrix elements

$$U(p, q) = \langle p|q\rangle. \tag{4.243}$$

The inverse of the transformation (4.242) can be written as

$$|\psi\rangle = U^{-1}(q, p)|\phi\rangle, \tag{4.244}$$

where

$$\begin{aligned} U^{-1}(q,p) &= \langle q|p\rangle = \langle p|q\rangle^* = U^*(p, q) \\ &= U^\dagger(q, p). \end{aligned} \tag{4.245}$$

This last relation shows that U is a unitary transformation.

For the special and important case where q is the coordinate and p is the momentum of the particle (4.241) becomes

$$\phi(p) = \frac{1}{(2\pi\hbar)^{\frac{1}{2}}} \int_{-\infty}^{\infty} \exp\left(\frac{ipq}{\hbar}\right) \psi(q) dq. \tag{4.246}$$

This is the one-dimensional form of the relation between the coordinate wave function $\psi(q)$ and the momentum wave function $\phi(p)$. According to Born's interpretation

$$\mathcal{Q}(q)dq = |\psi(q)|^2 dq, \tag{4.247}$$

measures the probability of finding the particle in the interval q and $q + dq$ and

$$\mathcal{P}(p)dp = |\phi(p)|^2 dp, \tag{4.248}$$

measures the probability of the particle having a momentum between p and $p + dp$. If the state depends on time then

$$\mathcal{Q}(q,t)dq = |\psi(q,t)|^2 dq, \tag{4.249}$$

measures the probability of finding the particle at time t in an infinitesimal region dq about the coordinate q.

From the definition of $\mathcal{Q}(q) = |\psi(q)|^2$ it is clear that in general from the knowledge of $Q(q)$, $\psi(q)$ cannot be determined, since the latter is a complex quantity. As we will see later in this section we can measure both $\mathcal{Q}(q)$ and $\mathcal{P}(p)$, therefore we can ask whether by knowing these two quantities the two complex functions $\psi(q)$ and $\phi(p)$, related to each other by Eq. (4.246), can be found uniquely or not [40],[41]. The answer to this question given by Bargmann is no [41]. Bargmann's argument is as follows:

Let us assume that $\phi(p)$ be an even function of p

$$\phi(p) = \phi(-p), \tag{4.250}$$

and let us define $K(q,p)$ by (we set $\hbar = 1$)

$$K(q,p) = \frac{1}{\sqrt{2}} e^{ipq}. \tag{4.251}$$

Then

$$\psi(q) = \int_{-\infty}^{\infty} K(q,p)\phi(p)dp. \tag{4.252}$$

Next we observe that

$$K(q,p) = K^*(q,-p), \tag{4.253}$$

and

$$\begin{aligned}
\psi(-q) &= \int_{-\infty}^{\infty} K(-q,p)\phi(p)dp \\
&= -\int_{-\infty}^{\infty} K(-q,-p)\phi(-p)dp \\
&= \int_{-\infty}^{\infty} K(q,p)\phi(p)dp = \psi(q).
\end{aligned} \tag{4.254}$$

Similarly $\psi^*(q)$ is the Fourier transform of $\phi^*(q)$;

$$\begin{aligned}\psi^*(q) &= \int_{-\infty}^{\infty} K^*(q,p)\phi^*(p)dp \\ &= -\int_{-\infty}^{\infty} K^*(q,-p)\phi^*(-p)dp \\ &= \int_{-\infty}^{\infty} K(q,p)\phi^*(p)dp.\end{aligned} \tag{4.255}$$

Thus as Eq. (4.254) shows the wave function $\psi(q)$ is an even function of q. Therefore $\mathcal{Q}(q)$ and $\mathcal{P}(p)$ can be obtained from the pair (ψ, ϕ) as well as the pair (ψ^*, ϕ^*). These two couples, however, lead to different probability distribution $\mathcal{R}(u)$ of any other entity u.

For the three-dimensional case the probability of finding the particle in the volume d^3r is

$$\mathcal{Q}(\mathbf{r})d^3r = |\psi(\mathbf{r})|^2 d^3r, \tag{4.256}$$

and in momentum space the corresponding probability is $\mathcal{P}(\mathbf{p})d^3p$ where

$$\mathcal{P}(\mathbf{p}) = \left|\int \psi(\mathbf{r}) \exp\left(-\frac{i\mathbf{p}\cdot\mathbf{r}}{\hbar}\right) d^3r\right|^2. \tag{4.257}$$

Here also the knowledge of $\mathcal{Q}(\mathbf{r})$ and $\mathcal{P}(\mathbf{p})$ will not be sufficient to determine $\psi(\mathbf{r})$. For instance if we take

$$\langle \mathbf{r}|\psi\rangle = \psi(\mathbf{r}) = R_l(r) P_l^{m'}(\cos\theta) e^{-im'\phi}, \tag{4.258}$$

then clearly $\mathcal{Q}(\mathbf{r})$ and $\mathcal{P}(\mathbf{p})$ are independent of the sign of m', and $\psi(\mathbf{r})$ cannot be uniquely determined from (4.256) and (4.257).

However it has been shown that for an ensemble of particles in a given state we can find $\psi(x)$ by placing these particles in an appropriate potential fields and then measuring the distribution of the coordinates and momenta at different times. These times can be arbitrarily close to each other [42].
Also if in addition to $\mathcal{Q}(\mathbf{r})$ the current density $\mathbf{j}(\mathbf{r})$ is known

$$\mathbf{j}(\mathbf{r}) = \frac{\hbar}{2im}\left[\psi^*(\mathbf{r})\nabla\psi(\mathbf{r}) - \psi(\mathbf{r})\nabla\psi^*(\mathbf{r})\right], \tag{4.259}$$

then one can find $\psi(\mathbf{r})$ apart from a constant phase factor [43]–[45].

Measurement of the Probability Density — Suppose that we want to measure $\rho(\mathbf{r}, t)$ at the time t_M, and let us assume that ρ depends only on the radial distance r. We repeat the measurement over an ensemble of a large number of similarly prepared states. At $t = t_M$ we turn on a strong but short range potential $U(r)$ which admits only a single bound sate and at the same time turn off the actual potential of the system $V(r)$. The potential $U_1(r)$ has a bound state $\phi_1(r - r_M)$ and the continuum of states which we denote by

$\phi_n(r - r_M)$. At a later time t, the wave function $\psi(r,t)$ can be expanded in terms of the eigenfunctions $\phi_n(r - r_M)$ and their corresponding energies E_n

$$\psi(r,t) = \sum_n c_n \phi_n(r - r_n) \exp\left[-\frac{i}{\hbar}E_n(t - t_m)\right], \tag{4.260}$$

where $\sum_n$ indicates sum over the discrete and integration over the continuous spectrum of the unbound states. After a short time all the continuum states spread out and therefore can be ignored in the integration over n, and so only the term $n = 1$ survives in (4.260). The probability amplitude for the bound state is given by

$$c_1 = \int \phi^*(r - r_M)\psi(r, t_M)d^3r, \tag{4.261}$$

and the probability of trapping the particle in $U(r)$ is

$$\mathcal{Q} = |c_1|^2. \tag{4.262}$$

If the bound state wave function of the potential $U(r)$ has a short range compared with the distance over which $\psi(r, t_M)$ is appreciable, then we can write (4.262) as

$$\mathcal{Q} \approx |\psi(r_M, t_M)|^2, \tag{4.263}$$

and if ψ is properly normalized we have

$$d\mathcal{Q}(r_M) \approx |\psi(r_M, t_M)|^2 d^3r = \rho(r_M, t_M)d^3r. \tag{4.264}$$

Thus we can determine the probability density $\rho(r_M, t_M)$ by catching the particle in a short-range potential [46].

Regarding the measurement of $\mathcal{P}(p)$, Heisenberg in his well-known book mentions that the momentum distribution of an atomic electron can be found by suddenly turning off the atomic potential and allowing the electron wave function to spread. From the distribution of times of arrival of electrons at a distant detector we can obtain the momentum distribution [7].

Measurement of the Probability Current — Writing the complex wave function $\psi(\mathbf{r})$ as

$$\psi(\mathbf{r}) = R(\mathbf{r}) \exp\left(i\frac{S(\mathbf{r})}{\hbar}\right), \tag{4.265}$$

where $R(\mathbf{r})$ and $S(\mathbf{r})$ are real functions of $\mathbf{r}$, we find that $\rho(\mathbf{r}) = R^2(\mathbf{r})$ and $\mathbf{j} = \frac{1}{m}\rho(\mathbf{r})\nabla S(\mathbf{r})$. Thus $\nabla S(\mathbf{r})$ may be regarded as the average momentum of the particle when it is at $\mathbf{r}$.

To determine $\rho(\mathbf{r})$ operationally we can use a short-range potential which admits a single bound state (see Chapter 20). We assume that another particle is the source of this potential, and that this particle is centered to within an accuracy of $\Delta\mathbf{r}$ of the point $\mathbf{r}_0$, where we want to measure the current. According to the uncertainty principle, the uncertainty in the momentum of the source particle is greater than $(\hbar/\Delta\mathbf{r})$, however we can carry out the position measurement in such a way that $\langle\mathbf{p}\rangle$ is zero. Suppose that we have prepared the system so that the initial wave function is real, i.e. for $t < 0$, $\psi(\mathbf{r}) = R(\mathbf{r})$. Now we turn on the potential

$$U(\mathbf{r}, t) = S(\mathbf{r})\delta(t). \tag{4.266}$$

and by integrating the time-dependent Schrödinger equation

$$i\hbar\frac{\partial\psi(\mathbf{r}, t)}{\partial t} = U(\mathbf{r}, t)\psi(\mathbf{r}, t) = S(\mathbf{r})\delta(t)\psi(\mathbf{r}, t), \tag{4.267}$$

with the boundary condition $\psi(\mathbf{r}, t = -\epsilon) = R(\mathbf{r})$ we find

$$\psi(\mathbf{r}, t) = R(\mathbf{r}) \exp\left[i\frac{S(\mathbf{r})}{\hbar}\right]. \tag{4.268}$$

Thus whereas for $t < 0$, the average momentum was zero at each point, at $t = 0$ the impulse $\nabla S(\mathbf{r})\delta(t)$ produces an average momentum $\nabla S(\mathbf{r})$ for $t > 0$ [44].

Bibliography

[1] For a very clear and detaied examination of these postulates and their implications see C. Cohen-Tannoudji, B. Diu and F. Laloë, *Quantum Mechanics*, Vol. I, (John Wiley & Sons, New York, 1977), Chapter III.

[2] M.A. Morrison, *Understanding Quantum Physics, A Users Manual*, (Prentice Hall, Englewood Cliffs, 1990).

[3] L.D. Landau and E.M. Lifshitz, *Quantum Mechanics, Non-Relativistic Theory*, (Pergamon Press, London, 1958), p. 35.

[4] D. ter Haar, Theory and application of density matrix, Rep. Prog. Phys. 24, 304 (1961).

[5] R. Courant and D. Hilbert, *Methods of Mathematical Physics*, Vol. I, (Interscience, New York, 1953), p. 23.

[6] S.A. Fulling, What is the time derivative of a quantum mechanical observable? Ann. Phys. 165, 315 (1984).

[7] W. Heisenberg, Üeber den anschaulichen Inhalt der quantentheoretischen Kinematik and Mechanik, Z. Physik 43, 172 (1927), English translation in J.A. Wheeler and W.H. Zurek, *Quantum Theory of Measurement*, (Princeton University Press, 1983), pp. 62-84.

[8] P.S. Laplace, *A Philosophical Essay on Probabilities*, (Dover, New York, 1951).

[9] W. Heisenberg, *Physikalische Prinzipien der Quantentheorie*, (Hirzel, Leipzig, 1930)), English translation by C. Eckart and F.C. Hoyt, *The Physical Principles of Quantum Theory*, (University of Chicago Press, Chicago, 1930).

[10] J.L. Synge, Geometrical approach to the Heisenberg uncertainty relation and its generalization, Proc. Roy. Soc. London, 325, 151 (1971).

[11] E. Schrödinger, The uncertainty principle, Abh. Press. Akad. Wiss. 19, 296 (1930).

[12] A. Messiah, *Quantum Mechanics*, Vol. I, (North-Holland, Amsterdam, 1970), p. 142.

[13] M.M. Nieto, Position-momentum uncertainty products for exactly solvable potentials, Phys. Rev. A 20, 700 (1979).

[14] B. Remak, Zwei Beispiele zur Heisenbergschen Unsicherheitsrelation bei gebundenen Teilchen, Z. Physik, 69, 332 (1931).

[15] H.A. Gersch and C.H. Braden, Approximate energy levels and sizes of bound quantum systems, Am. J. Phys. 50, 53 (1982).

[16] H. Goldstein, C. Poole and J. Safco, *Classical Mechanics*, Third Edition, (Addison-Wesley, San Francisco, 2002).

[17] F.T. Hioie and E.W. Montroll, Quantum theory of aharmonic oscillator with positive quartic anharmonicity, J. Math. Phys. 16, 1945 (1975).

[18] For a detailed discussion of different versions of time-energy uncertainty principle see P. Busch's article "The time-energy uncertainty relation" in *Time in Quantum Mechanics*, edited by J.G. Muga, R. Sala Mayato and I.L. Egusquiza (Springer, Berlin, 2002).

[19] M. Bauer and P.A. Mello, The time-energy uncertainty relation, Ann. Phys. 111, 38 (1978).

[20] L. Mandelstam and I. Tamm, The uncertainty relation between energy and time in nonrelativistic quantum mechanics, J. Phys. U.S.S.R. 9, 249 (1945).

[21] M. Razavy, *Quantum Theory of Tunneling*, (World Scientific, Singapore, 2003), p. 497.

[22] K. Bhattacharya, Quantum decay and the Mandelstam–Tamm time-energy inequality, J. Phys. A 16, 2993 (1983).

[23] E.P. Wigner, On the time-energy uncertainty relation, in *Aspects of Quantum Theory*, edited by A. Salam and E.P. Wigner, (Cambridge University Press, London, 1972).

[24] P. Carruthers and M.M. Nieto, Phase and angle variables in quantum mechanics, Rev. Mod. Phys. 40, 411 (1968).

[25] D. Judge and J.T. Lewis, On the commutator $[L_z\ \phi]_-$, Phys. Lett. 5, 190 (1963).

[26] A. Gamba, Einstein–Podolski–Rosen paradox, hidden variables, Bell's inequalities and all that, Am. J. Phys. 55, 295 (1987).

[27] J.-M. Lévy-Leblond, Local Heisenberg inequality, Phys. Lett. 111 A, 353 (1985).

[28] W.G. Faris, Inequalities and uncertainty principles, J. Math. Phys. 19, 461 (1978).

[29] J.-M. Lévy-Leblond, Uncertainty relations for nonsimultaneous measurements, Am. J. Phys. 40, 899 (1978).

[30] J.-M. Lévy-Leblond, Correlation of quantum theory properties and the generalized Heisenberg inequality, Am. J. Phys. 54, 135 (1986).

[31] R.A. Campos, Quantum correlation coefficient for angular momentum and spin, Phys. Lett. A 256, 141 (1999).

[32] A.J. Makowski, A brief survey of various formulations of the correspondence principle, Eur. J. Phys. 27, 1133 (2006).

[33] M. Planck, *Theory of Heat Radiation*, (Dover New York, 1959), translated from the original German published in 1906.

[34] R.L. Liboff, The correspondence principle revisited, Phys. Today, 2, 50 (1984).

[35] A. Peres, *Quantum Theory: Concepts and Methods*, (Kluwer Academic Publishers, Dordrecht 1993), p. 301.

[36] P.M. Morse and H. Feshbach, *Methods of Theoretical Physics*, Part I, (McGraw-Hill, New York, 1953), p. 1643.

[37] G.G. Cabrera and M. Kiwi, Large quantum-number states and the correspondence principle, Phys. Rev. A 36, 2995 (1987).

[38] Q.H. Liu and B. Hu, The hydrogen atom's quantum to classical correspondence in Heisenberg's correspondence principle, J. Phys. A 34, 5713 (2001).

[39] P. Ehrenfest, Bemerkung über die angenäherte Gültigkeit der klassischen Mechanik innerhalb der Quantenmechanik, Z. Phys. 45, 455 (1927).

[40] W. Pauli, *General Principles of Quantum Mechanics*, traslated by P. Achuthan and K. Venkatesan, (Springer, Berlin, 1980).

[41] See H. Reichenbach, *Philosophical Foundations of Quantum Mechanics*, (University of California Press, Berkeley, 1965).

[42] V. Ya. Kreinovich, On the problem of recovering the ψ function in nonrelativistic quantum mechanics, Teo. Mat. Fiz. 28, 56 (1976).

[43] W. Gale, E. Guth and G.T. Trammell, Determination of Quantum States by Measurement, Phys. Rev. 165, 1434 (1968).

[44] G.T. Trammell, Momentum after position, Phys. Today, 22, 9 October (1969).

[45] M. Pavičić, When do position and momentum distribution determine the quantum mechanical state? Phys. Lett. A 118, 5 (1986).

[46] W.E. Lamb, An operational interpretation of nonrelativistic quantum mechanics, Phys. Today, 22, 23, (1969).

Chapter 5

Equations of Motion, Hamiltonian Operator and the Commutation Relations

In classical mechanics the second law of motion together with the initial conditions and constraints completely determine the motion. The equations of motion for most conservative systems with holonomic constraints and for a small group of dissipative systems can be derived from a Lagrangian or a Hamiltonian [1],[2].

In quantum mechanics we have the following set of rules for the operators replacing the classical dynamical variables:

(1) - The generator of the infinitesimal time evolution operator of a system is given by a Hamiltonian

$$H = \sum_{i=1}^{N} \frac{1}{2m_i} p_i^2 + V(q_1, \cdots, q_N), \tag{5.1}$$

where $V(q_1, \cdots, q_N)$ is the potential function.

(2) - The time derivative of any operator function of p_i s and q_i s is found from the Heisenberg equation given by Eq. (4.10) or if F does not depend explicitly on time then

$$i\hbar \frac{dF(p_i(t), q_i(t))}{dt} = [F(p_i(t), q_i(t)), H]. \tag{5.2}$$

(3) - The operators p_i and q_i satisfy the canonical commutation relations

$$[p_i, q_j] = -i\hbar\delta_{ij}, \quad [p_i, p_j] = 0, \quad [q_i, q_j] = 0, \quad i, j = 1, \cdots N. \tag{5.3}$$

(4) - The position and momentum operators satisfy the second law of motion

$$\dot{q}_j = \frac{1}{m_j}\dot{p}_j, \quad \dot{p}_j = -\frac{\partial V(q_1, \cdots, q_N)}{q_j}, \quad j = 1, \cdots N. \tag{5.4}$$

If we take Eqs. (5.1)–(5.3) as fundamental relations, then Eq. (5.4) follows.

5.1 Schwinger's Action Principle and Heisenberg's equations of Motion

We can formally derive the quantized equations of motion from the action-integral operator S exactly as in classical mechanics [3]–[8]. As we have seen earlier $\varepsilon G(t)$ can be regarded as the generator of the classical transformation. In quantum theory the operator form of ε
$G(t)$ will be the generator of a unitary transformation. In particular if we choose

$$\varepsilon G(t) = -\frac{H\delta t}{\hbar}, \tag{5.5}$$

where now $\varepsilon G(t)$ is a dimensionless operator, we have the analogue of Eq. (1.150)

$$-\delta\Lambda = -\frac{i}{\hbar}[H,\ \Lambda]\delta t. \tag{5.6}$$

Again as in classical dynamics for purely time development of a system $\delta\Lambda = -\dot{\Lambda}\delta t$ and thus Eq. (5.6) becomes

$$\dot{\Lambda} = \frac{i}{\hbar}[H,\ \Lambda]. \tag{5.7}$$

This is the Heisenberg equation of motion for the dynamical operator $\Lambda(p_j, q_j)$.

Next let us consider the spatial variation δq_j when t is fixed, i.e. $\delta t = 0$. Let

$$\varepsilon G = \sum_j p_j \delta q_j, \tag{5.8}$$

be the generator of the transformation. Now if in (3.73) we set $A = q_k$ we find

$$\delta q_k = \frac{i}{\hbar}\left[\sum_j p_j \delta q_j,\ q_k\right]. \tag{5.9}$$

This transformation only produces an infinitesimal change in the coordinate operator, therefore

$$\delta p_k = \frac{i}{\hbar}\left[\sum_j p_j \delta q_j,\ p_k\right] = 0. \tag{5.10}$$

Using the operator identity

$$[uv,\ w] = [u,\ w]v + u[v,\ w], \tag{5.11}$$

we can write (5.9) as

$$\delta q_k = \frac{i}{\hbar}\sum_j [p_j,\ q_k]\delta q_j + \frac{i}{\hbar}\sum_j p_j[\delta q_j, q_k]. \tag{5.12}$$

A simple solution of (5.12) is

$$[p_j,\ q_k] = -i\hbar\delta_{jk}, \quad [\delta q_j,\ q_k] = 0. \tag{5.13}$$

If we choose δq_j to be equal to εq_j then the second equation in (5.13) implies that

$$[q_j,\ q_k] = 0, \quad \text{for all } j \text{ and } k. \tag{5.14}$$

Next let us expand (5.10) using (5.11) to obtain

$$\frac{i}{\hbar}\sum_j [p_j,\ p_k]\delta q_j + \frac{i}{\hbar}\sum_j p_j[\delta q_j,\ p_k] = 0. \tag{5.15}$$

With $\delta q_j = \varepsilon q_j$, this last equation becomes

$$\frac{i}{\hbar}\sum_j [p_j,\ p_k]\varepsilon q_j - \varepsilon p_k = 0, \tag{5.16}$$

where we have used Eq. (5.13) to simplify the result. Equation (5.16) must be valid for any state and in particular if we choose the state $|\phi\rangle$ so that $p_k|\phi\rangle = 0$, then we find

$$[p_j,\ p_k] = 0. \tag{5.17}$$

Thus we have shown that from the Schwinger action principle in quantum mechanics we can derive the Heisenberg equations of motion and also the commutation relations (5.13),(5.14) and (5.17). It should be pointed out that a more general form of the commutation relation $[p_j,\ q_k]$ can be directly derived from the equations of motion and the Heisenberg equation (5.7).

Let us also note that this formulation of the Schwinger action principle yields the correct result when the commutator

$$[q_j,\ \dot{q}_k]\,, \tag{5.18}$$

is a *c*-number, i.e. it is equal $\left(\frac{i\hbar}{m}\delta_{jk}\right)$. As we have seen before, for a general curvilinear coordinate, the Hamiltonian is a complicated quadratic function of p_j s, and thus the commutator (1.174) will depend on q_k and is a *q*-number. For these and also the problems related to dissipative forces depending on velocity, the action integral corresponding to the classical Hamilton's principle should be modified. An explicit form of admissible *q*-number variation for the Hamiltonian operator of the general form

$$H(q_i,\ p_i,\ t) = T + V(q_i, t), \tag{5.19}$$

where T is given by (3.216) can be found which reduces to the Schwinger action when T becomes independent of q_i [9],[10].

5.2 Nonuniqueness of the Commutation Relations

The solution (5.13) for the fundamental commutator was obtained under the assumption that the commutator $[p_j,\ q_k]$ is a c-number. There we noted that the Schwinger method has to be modified if the commutator happens to be a q-number.

Wigner studied the possibility of deriving (5.3) from (5.1), (5.2) and (5.4) for a one-dimensional harmonic oscillator and found that the commutator (5.3) is not the only form of the commutation relation which can be derived from these equations [11]. Following Wigner we consider a particle of unit mass with the Hamiltonian

$$H = \frac{1}{2}\left(q^2 + p^2\right), \tag{5.20}$$

where for the sake of simplicity we have set $\hbar = \omega = 1$.
From Eq. (5.4) we have

$$p = \dot{q} = i[H, q], \tag{5.21}$$

and

$$\dot{p} = -q = i[H, p]. \tag{5.22}$$

Let us consider a representation in which H is diagonal. We note that because of the positive-definiteness of the Hamiltonian (5.20) the diagonal elements are positive. If we denote these diagonal elements by $E_0, E_1 \cdots$, then for the matrix elements of p and q we find

$$p_{n,m} = i(E_n - E_m)q_{n,m}, \tag{5.23}$$

and

$$q_{n,m} = -i(E_n - E_m)p_{n,m}. \tag{5.24}$$

From these equations it follows that

$$q_{n,m} = (E_n - E_m)^2 q_{n,m}. \tag{5.25}$$

Thus $q_{n,m}$ will be nonzero only if

$$E_n - E_m = \pm 1. \tag{5.26}$$

Also from (5.23) we have $p_{n,m} = 0$ when $q_{n,m} = 0$.

Equation(5.26) shows that E_n s form an arithmetical series

$$E_n = E_0 + n. \tag{5.27}$$

Now if E_n appears in the diagonal elements of H, then by means of a unitary transformation we can decompose any system of matrices in which E_n occurs repeatedly without changing H. In this way we can assume that all of the eigenvalues E_n in (5.27) are simple. From (5.25) it is clear that only the elements $q_{n,n+1}$ and $q_{n+1,n}$ are nonzero. Since q as an operator is Hermitian, therefore we can take its matrix elements to be real numbers. But from (5.23) and (5.26) it follows that all $p_{n,m}$ s are pure imaginary

$$p_{n,n+1} = -iq_{n,n+1} = -iq_{n+1,n}, \tag{5.28}$$

and

$$p_{n+1,n} = iq_{n+1,n} = -p_{n,n+1}. \tag{5.29}$$

To find the numerical values of $q_{n,n+1}$ and $q_{n,n+1}$ we calculate the elements of H which is a diagonal matrix from (5.20),

$$\begin{cases} E_n = E_0 n = q_{n-1,n}^2 + q_{n,n+1}^2 & n \neq 0 \\ \\ E_0 = q_{0,1}^2 \end{cases} . \tag{5.30}$$

From this relation we can find successive $q_{n,n+1}$ s from (5.30)

$$\begin{cases} q_{n,n+1} = \left(E_0 + \frac{n}{2}\right)^{\frac{1}{2}} & \text{for even } n \\ \\ q_{n,n+1} = \left(\frac{n}{2} + \frac{1}{2}\right)^{\frac{1}{2}} & \text{for odd } n \end{cases} . \tag{5.31}$$

Having obtained $q_{n,n+1}$ we can calculate $p_{n,n+1}$ from (5.29) and thus we get all the matrix elements of q and p. Substituting these in the expression

$$(pq)_{n,m} - (qp)_{n,m}, \tag{5.32}$$

we find that the commutator is diagonal and that its diagonal matrix elements are

$$-2iq_{0,1}^2, -2i\left(q_{1,2}^2 - q_{0,1}^2\right), -2i\left(q_{2,3}^2 - q_{1,2}^2\right)\cdots. \tag{5.33}$$

Hence we have the result

$$([p,q] + i)^2 = -\left(2E_0 - 1\right)^2. \tag{5.34}$$

If we choose $E_0 = \frac{1}{2}$, that is the ground state energy of the harmonic oscillator, then we have the standard commutation relation given by (5.3). However as (5.34) indicates the commutation relation depends on the parameter E_0 and is not unique. This arbitrariness in the commutation relation for the harmonic oscillator can be removed if one requires that the expansion of an arbitrary wave function, $\phi(x)$ in terms of the energy wave functions converges absolutely and uniformly to $\phi(x)$ [12]. Also Pauli has pointed out that if one requires that the fundamental commutation relation be independent of the particular choice of energy, then one gets a unique expression for $[q, \dot{q}]$ [13]. It can also be shown that for any potential of the form $V(x) = Aq^{2n+1} + B$ with n an integer the Hamiltonian $H = \frac{p^2}{2m} + V(q)$ and Eqs. (5.2) and (5.4) imply the commutation relation (5.3)[11]–[14].

From the general form of the commutation relation between p and q for the harmonic oscillator, we can find the commutator between the creation and annihilation operators, $a^\dagger$ and a by defining these operators by

$$a^\dagger = \frac{1}{\sqrt{2}}(q - ip), \tag{5.35}$$

and

$$a = \frac{1}{\sqrt{2}}(q + ip). \tag{5.36}$$

To find the matrix elements of these operators, we first note that in this formulation we have set $\hbar = \omega = 1$, and in these units E_0 is a dimensionless quantity which we denote by N_0. Therefore for the general form of the commutator $[p,\ q]$ from (5.31) and (5.23) we get

$$a_{n,n+1} = a^\dagger_{n+1,n} = \begin{cases} (2N_0 + n)^{\frac{1}{2}} & \text{for } n \text{ even} \\ (n+1)^{\frac{1}{2}} & \text{for } n \text{ odd} \end{cases}. \tag{5.37}$$

Using these we can determine the matrix elements of the commutator $[a, a^\dagger]$;

$$\langle n \left| [a, a^\dagger] \right| n' \rangle = \delta_{n,n'} \begin{cases} 2N_0 & \text{for } n \text{ even} \\ 2(1 - N_0) & \text{for } n \text{ odd} \end{cases}. \tag{5.38}$$

Commutation relations of the type (5.38) have been used to study the quantum field theory of particles obeying parastatistics (i.e. neither bosons nor fermions) [15],[16]. We observe that in this case the fundamental commutation relation is not given by two fold commutator, but by a three fold commutator of the general form

$$\left[a_j, \left[a_j^\dagger, a_k\right]\right] = 2\delta_{ij} a_k. \tag{5.39}$$

Here it is possible to bring ν particles into a single state; and this is called parastatistics of degree ν. In addition to quantum field theory, parastatistics have had applications in atomic nuclei, where for integer spin it is called parabosons and for half-integer spin parafermions [15]–[18].

Assuming that the commutation relation must be independent of the energy, we have the standard result

$$\langle n \left| [a,\ a^\dagger] \right| n' \rangle = \delta_{n,n'}, \quad \text{or} \quad [a,\ a^\dagger] = 1, \tag{5.40}$$

and the energy eigenvalues are given by (5.27) i.e. by

$$E_n = \left(n + \frac{1}{2}\right)\hbar\omega, \tag{5.41}$$

where ω is the angular frequency of the oscillator.

For a more general one-dimensional problem having the Hamiltonian $H = \frac{1}{2m}p^2 + V\left(q^2\right)$ with the equations of motion

$$\dot{q} = \frac{p}{m}, \quad \dot{p} = -\frac{\partial V\left(q^2\right)}{\partial q}, \tag{5.42}$$

we can find the general commutation relation of the form

$$qp - pq = i(1 + 2\beta), \quad \hbar = 1 \tag{5.43}$$

where we have set $\hbar = 1$ and β is independent of time and satisfies the following conditions [19]

$$[\beta, p]_+ \equiv \beta p + p\beta = 0, \tag{5.44}$$

and

$$[\beta q + q\beta, p]_+ = 0, \tag{5.45}$$

and the symbol $[\,,]_+$ denotes the anticommutator.

$$[a,\ b]_+ \equiv ab + ba. \tag{5.46}$$

One way of obtaining the commutation relation of the type (5.43) is by introducing the Pauli matrices (Chapter 8)

$$\sigma_x = \begin{bmatrix} 0 & 1 \\ 1 & 0 \end{bmatrix}, \quad \sigma_y \begin{bmatrix} 0 & -i \\ i & 0 \end{bmatrix}, \quad \sigma_z = \begin{bmatrix} 1 & 0 \\ 0 & -1 \end{bmatrix}, \tag{5.47}$$

and considering the three-dimensional motion of a particle with coordinate $\mathbf{q} = (x, y, z)$ and momentum $\mathbf{p} = (p_x, p_y, p_z)$ where the components of $\mathbf{q}$ and $\mathbf{p}$ satisfy the standard commutation relations

$$q_i p_j - p_j q_i = i\delta_{ij}. \tag{5.48}$$

Now let us introduce q and p in the following way:

$$q = \mathbf{q} \cdot \boldsymbol{\sigma} = x\sigma_x + y\sigma_y + z\sigma_z, \tag{5.49}$$

$$p = \mathbf{p} \cdot \boldsymbol{\sigma} = p_x\sigma_x + p_y\sigma_y + p_z\sigma_z. \tag{5.50}$$

These relations imply that

$$q^2 = \mathbf{q} \cdot \mathbf{q}, \quad \text{and} \quad p^2 = \mathbf{p} \cdot \mathbf{p}. \tag{5.51}$$

Next we define β by

$$\beta = (\mathbf{q} \wedge \mathbf{p}) \cdot \boldsymbol{\sigma} + 1 = (yp_z - zp_y) + (zp_x - xp_y)\sigma_y + (xp_y - yp_x)\sigma_z + 1, \tag{5.52}$$

where $\mathbf{q} \wedge \mathbf{p}$ is dimensionless ($\hbar = 1$). By substituting q and p from Eqs. (5.49) and (5.50) in the commutator $[q, p]$ we find

$$[q, p] \equiv qp - pq = i(1 + 2\beta), \tag{5.53}$$

and

$$q\beta + \beta q = p\beta + \beta p = 0. \tag{5.54}$$

For more recent works on the determination of the commutation relation from the Heisenberg equations of motion see [20]–[26].

5.3 First Integrals of Motion

As we observed in Sec. 1.3, we can construct an infinite number of classical Hamiltonians, all acceptable as generators of motion in coordinate space, or in phase space. Here we want to inquire whether it is possible to obtain similar results in quantum mechanics, i.e. to find a general form of the Hamiltonian (5.1) given the Heisenberg equation (5.2) and the canonical commutation relation (5.3) [25]-[30]. Again we will confine our attention to the one-dimensional motion and we set $\hbar = 1$. Not to be confused with the proper Hamiltonian which is at the same time the energy operator we denote this general form of the Hamiltonian by $K(p,q)$. Then from Eq. (5.2) we have

$$\dot{q} = i[K,q], \quad \text{and} \quad \ddot{q} = i\left[K,\dot{q}\right], \tag{5.55}$$

or

$$\ddot{q} = \frac{1}{m}F(q) = -\left[K,[K,q]\right]. \tag{5.56}$$

The last equation is the quantum analogue of the classical equation (1.49) or (1.131), and like these classical equations, Eq. (5.56) is quadratic in the Hamiltonian $K(p,q)$. Given the force law $F(q)$, the most general solution of (5.56) for $K(p,q)$ is not known. However we can verify that $K(p,q) \equiv H(p,q)$ where

$$H = \frac{1}{2m}p^2 + V(q), \quad F(q) = -\frac{\partial V(q)}{\partial q}, \tag{5.57}$$

is a solution of (5.56).

If in addition to (5.56) we require $K(p,q)$ to satisfy the double commutator

$$[q,[q,H]\,] = -\frac{1}{m}, \tag{5.58}$$

i.e. assume that $p = m\dot{q}$, then $K(p,q) \equiv H(p,q)$, and we have, apart from a multiplicative constant, a unique generator of motion.

A different solution can be found if we choose $K(p,q)$ to be of the form of a product of two operators

$$K(p,q) = R(p)S(q). \tag{5.59}$$

At the first sight (5.59) seems unacceptable since $K(p,q)$ is not a Hermitian operator. But let us first find a solution of this type and then investigate how the resulting solution can be made Hermitian. Substituting (5.59) in (5.56) we obtain

$$\ddot{q} = i\left(R(p)S(q)\frac{dR(p)}{dp} - \frac{dR(p)}{dp}S(q)R(p)\right)S(q). \tag{5.60}$$

Assuming that $S(q)$ can be expanded as a finite or infinite sum of powers of q, i.e.

$$S(q) = \sum_{k=0}^{\infty} c_q q^k, \tag{5.61}$$

the operator $S\left(i\frac{d}{dp}\right)R(p)$ can be written as [31]

$$S\left(i\frac{d}{dp}\right)R(p) = \sum_{n=0}^{\infty}\frac{(i)^n}{n!}\frac{d^nR(p)}{dp^n}\frac{d^nS(q)}{dq^n}. \tag{5.62}$$

Similarly we have

$$S(q)\frac{dR(p)}{dp} = \sum_{n=0}^{\infty}\frac{(i)^n}{n!}\frac{d^{n+1}R(p)}{dp^{n+1}}\frac{d^nS(q)}{dq^n}. \tag{5.63}$$

By substituting (5.62) and (5.63) in (5.60) we find

$$\ddot{q} = \frac{1}{m}F(q) = \sum_{n=0}^{\infty}\frac{(i)^{n+1}}{n!}\left[R(p)\frac{d^{n+1}R(p)}{dp^{n+1}} - \frac{dR(p)}{dp}\frac{d^nR(p)}{dp^n}\right]\frac{d^nS(q)}{dq^n}S(q). \tag{5.64}$$

Since $F(q)$ is a real function we have the following possibilities:

(a) - The term in the square bracket in (5.64) is independent of p, viz,

$$\begin{cases}\left[R(p)\frac{d^{n+1}R(p)}{dp^{n+1}} - \frac{dR(p)}{dp}\frac{d^nR(p)}{dp^n}\right] = C^{-n-1} & \text{for } n \text{ odd integer} \\ \qquad\qquad\qquad = 0 & \text{for } n \text{ even integer}\end{cases} \tag{5.65}$$

The solution of this equation is given by

$$R(p) = \cosh\left(\frac{p}{C} + D\right), \tag{5.66}$$

where C and D are constants. From (5.64) we find that the expression for the force is

$$F(q) = m\sum_{n=0}^{\infty}\frac{(-1)^{n+1}}{(2n+1)!}\frac{d^{2n+1}S(q)}{dq^{2n+1}}\frac{S(q)}{C^{2n+2}}. \tag{5.67}$$

Thus if $F(q)$ is known then $S(q)$ can be found from (5.67). However it is easier to assume a form for $S(q)$ and determine the q-dependence of the force from (5.67). For instance if we choose $S(q) = C\omega q$, then $F = -m\omega^2 q$, a result which is the quantum analogue of (1.61) for this force. Now if we take $S(q)$ to be

$$S(q) = \left(\frac{CF_0}{m}\right)^{\frac{1}{2}}\exp\left(\frac{\mu q}{2}\right), \tag{5.68}$$

where μ is a constant, we find

$$F = F_0e^{-q}\sin\left(\frac{\mu}{2C}\right). \tag{5.69}$$

(b) - For the motion of a free particle, $F(q) = 0$, then

$$R(p)\frac{d^{n+1}R(p)}{dp^{n+1}} - \frac{dR(p)}{dp}\frac{d^nR(p)}{dp^n} = 0, \tag{5.70}$$

for all values of n. The solution of (5.69) is $R(p) = \exp\left(\frac{p}{p_0}\right)$, and therefore

$$K(p, q) = \exp\left(\frac{p}{p_0}\right) S(q), \tag{5.71}$$

with p_0 a constant and $S(q)$ an arbitrary function of q, $K(p, q)$ is the generator of motion. The other possibility is to take $S(q)$ to be a constant. Then as (5.64) shows that the force $F(q)$ vanishes for any function $R(p)$ and in particular for $K(p, q) = H(p) = \frac{p^2}{2m}$.

Once $K(p, q) = R(p)S(q)$ has been determined we have to see whether it is possible to construct a Hermitian operator from $K(p, q)$ or not, and that this operator satisfies (5.56).

In the case of harmonic oscillator where

$$K(p, q) = C\omega \cosh\left(\frac{p}{C} + D\right) q, \tag{5.72}$$

we can find a Hermitian operator by symmetrizing $K(p, q)$, and this symmetrization preserves its essential property, i.e. Eq. (5.56).

For separable $K(p, q)$ operator that we have assumed, Eq. (5.59), the relation between the velocity and the momentum operators is complicated, viz,

$$\dot{q} = \frac{1}{i\hbar}[q, K(p, q)] = \frac{1}{i\hbar}[q, R(p)S(q)] = \frac{dR(p)}{dp} S(q). \tag{5.73}$$

We can try other forms of $K(p, q)$ to see whether it is possible to obtain other first integrals of motion in such a way that not only (5.56) is satisfied, but in addition the commutator

$$[\,[q, K(p, q)], q] = 1, \tag{5.74}$$

remains valid. Again the general solution of this problem is not known, but we can consider a special case where the classical Lagrangian function is given by [32],[33]

$$L\left(\dot{q}, q\right) = \dot{q}\int^{\dot{q}} \frac{D(y)}{y^2} dy - V(q). \tag{5.75}$$

The canonical momentum p is found from (5.75) and is given by

$$p = \int^{\dot{q}} \frac{D(y)}{y^2} dy + \frac{D\left(\dot{q}\right)}{\dot{q}} = \int^{\dot{q}} \frac{1}{y}\frac{dD(y)}{dy} dy. \tag{5.76}$$

This equation can be inverted to give us $\dot{q}$ as a function of p. But unlike (5.73) $\dot{q}$ will be only a function of p. Since the Hamiltonian function which is derived from $L\left(q, \dot{q}\right)$ is the sum of two terms one depending on p and the other on q $K(p, q)$ is of the form

$$K(p, q) = D\left[\dot{q}(p)\right] + V(q), \tag{5.77}$$

and there is no ambiguity of ordering of q and p in $K(p,q)$ for the construction of a Hermitian operator.

Once we have the operator $K(p,q)$ we find the time derivative of q (writing $\hbar$ explicitly in the equations of motion);

$$\dot{q} = \frac{i}{\hbar}[K(p,q),q] = \frac{i}{\hbar}\left[D\left(\dot{q}\right),q\right] = \frac{d}{dp}D\left(\dot{q}\right) = \frac{d\dot{q}}{dp}\frac{dD\left(\dot{q}\right)}{d\dot{q}}, \tag{5.78}$$

a relation which can also be verified using (5.76). Next we consider the equation of motion

$$\ddot{q} = \frac{i}{\hbar}\left[K(p,q),\dot{q}\right] = \frac{i}{\hbar}\left[V(q),\dot{q}\right]. \tag{5.79}$$

and assume that $V(q)$ can be expanded as a power series of q;

$$V(q) = \sum_{n=}^{\infty} c_n q^n, \tag{5.80}$$

where c_n s are the coefficients of expansion. Now for any function of p, say $F(p)$, we have

$$\frac{i}{\hbar}[q,F(p)] = -\frac{dF(p)}{dp}, \tag{5.81}$$

therefore

$$\frac{i}{\hbar}\left[q,\dot{q}\right] = -\frac{d\dot{q}}{dp} = -\frac{\dot{q}}{\left(\frac{dD(\dot{q})}{d\dot{q}}\right)} \equiv -J\left(\dot{q}\right). \tag{5.82}$$

Thus Eq. (5.82) can be used to define $J\left(\dot{q}\right)$. With the help of $J\left(\dot{q}\right)$ we can determine the commutator $[q^n,\dot{q}]$,

$$\frac{i}{\hbar}\left[q^n,\dot{q}\right] = \frac{i}{\hbar}\sum_{k=0}^{n-1} q^k\left[q,\dot{q}\right]q^{n-1-k} = -\sum_{k=0}^{n-1} q^k J\left(\dot{q}\right)q^{n-1-k}. \tag{5.83}$$

From (5.80) and (5.83) we find

$$\frac{i}{\hbar}\left[V(q),\dot{q}\right] = -\hat{O}\left[J\left(\dot{q}\right)\frac{dV(q)}{dq}\right], \tag{5.84}$$

where we have used the notation

$$\hat{O}\left[J\left(\dot{q}\right)\frac{dq^n}{dq}\right] = \sum_{k=0}^{n-1} q^k J\left(\dot{q}\right)q^{n-1-k}, \tag{5.85}$$

to define the quantum mechanical ordering. Thus from (5.79), (5.80) and (5.83) we find the equation of motion to be

$$\ddot{q} = -\hat{O}\left[J\left(\dot{q}\right)\frac{dV(q)}{dq}\right]. \tag{5.86}$$

This results shows that any Hamiltonian of the form (5.77) which satisfies the commutators Eqs. (5.56) and (5.74), generate an equation of motion in which the potential energy is velocity-dependent [33]. If we choose $J(\dot{q}) = 1$, (5.86) reduces to the standard Heisenberg equation. However if we choose $J(\dot{q}) = \frac{\dot{q}}{p_0}$, then (5.86) can be written as

$$\ddot{q} + \frac{1}{p_0}\left(\frac{dV(q)}{dt}\right) = 0. \tag{5.87}$$

In this case from (5.82) we find

$$\dot{q} = \pm v_0 \exp\left(\frac{p}{p_0}\right), \tag{5.88}$$

where v_0 is a constant. The Hamiltonian $K(p,q)$ can be obtained by integrating (5.87);

$$K(p,q) = p_0\dot{q} + V(q) = \pm p_0 v_0 \exp\left(\frac{p}{p_0}\right) + V(q). \tag{5.89}$$

While this first integral generates both (5.55) and (5.74) it should be rejected for violating time-reversal and Galilean invariances.

Bibliography

[1] P. Havas, The range of application of the Lagrange formalism, Nuovo Cimento. 5, 363 (1957).

[2] M. Razavy, *Classical and Quantum Dissipative Systems*, (Imperial College Press, 2005).

[3] J. Schwinger, The theory of quantized fields, Phys. Rev. 82, 914 (1951).

[4] J. Schwinger, *Quantum Kinetics and Dynamics*, (W.A. Benjamin, New York, 1970), Chapter 3.

[5] J. Schwinger, *Quantum Mechanics: Symbolism of Atomic Measurement*, edited by B-G Englert, (Springer, Berlin 2001), Chapter 5.

[6] R.J. Finkelstein, *Nonrelativistic Mechanics*, (W.A. Benjamin, New York 1973), Chapter 3.

[7] E.B. Manoukian, *Quantum Theory: A Wide Spectrum*, (Springer, Dordrecht, The Netherlands, 2006).

[8] P. Roman, *Advanced Quantum Theory, An Outline of the Fundamental Ideas*, (Addison-Wesley, Reading, 1965), Chapter 1.

[9] M.Z. Shaharir, The modified Hamilton-Schwinger action principle, J. Phys. A7, 553 (1974).

[10] H.A. Cohen and M.Z. Shaharir, The action principle in quantum mechanics, Int. J. Theo. Phys. 11, 289 (1974).

[11] E.P. Wigner, Do the equations of motion determine the quantum mechanical commutation relations? Phys. Rev. 77, 711 (1950).

[12] L.M. Yang, A note on the quantum rule of the harmonic oscillator, Phys. Rev. 84, 788 (1951).

[13] See foontnote 30 in A. Pais and G.E. Uhlenbeck's paper, On the field theory with non-localized action, Phys. Rev. 79, 145 (1950).

[14] C.R. Putnam, The quantum-mechanical equations of motion and commutation relations, Phys. Rev. 83, 1047 (1951).

[15] Y. Ohnuki and S. Kamefuchi, *Quantum Field Theory and Parastatistics*, (Springer, Berlin, 1982).

[16] Y. Ohnuki and S. Kamefuchi, On the wave-mechanical representation of a Bose-like oscillator, J. Math. Phys. 19, 67 (1978).

[17] H.J. Fink, W. B. Müller and W. Greiner, Parastatistics as an effective description of complex particles (nucleus-nucleus collisions and muonic atoms), J. Phys. G. 3, 1119 (1977).

[18] R.Y. Cusson, Examples of parastatistics, Ann. Phys. 55, 22 (1969).

[19] H.S. Green, *Matrix Mechanics*, (P. Noordhoff, Netherlands, 1965), p. 41.

[20] M. Arik, N.M. Atakishiyev and K.B. Wolf, Quantum algebraic structures compatible with the harmonic oscillator Newton equation, J. Phys. A 32, L371 (1999).

[21] T.D. Palev and N.I. Stoilova, Wigner quantum oscillators, J. Phys. A 27, 977 (1994).

[22] J. Cislo and J. Lopuszanski, To what extent do the classical equations of motion determine the quantization scheme? J. Math. Phys. 42, 5163 (2001).

[23] R.C. King, T.D. Palev, N.I. Stoilova and J.N. Van der Jeugt, A noncommutative n-particle 3D Wigner quantum oscillator, J. Phys. A. 36, 11999 (2003).

[24] V.V. Dodonov, V.I. Man'ko and V.D. Skarzhinsky, The inverse problem of variational calculus and the nonuniqueness of the quantization of classical systems, Hadronic J. 4, 1734 (1981).

[25] V.I. Man'kov, G. Marmo, E.C.G. Sudarshan and F. Zaccaria, Wigner's problem and alternative commutation relation in quantum mechanics, Int. J. Mod. Phys. B 11, 1281 (1997).

[26] P. Blasiak, Alternative Hamiltonians and Wigner quantization, J. Optics, B 5, S245 (2003).

[27] A. Degasperis, Newton equivalent Hamiltonians for the harmonic oscillator, Ann. Phys. (NY), 29392 (2001).

[28] A. Horzela, On alternative Hamiltonians, Czech. J. Phys. 52, 1239 (2002).

[29] F. Calogero and A. Degasperis, On the quantization of Newton-equivalent Hamiltonians, Am. J. Phys. 72, 1202 (2004).

[30] M. Razavy, q-equivalent Hamiltonian operators, Can. J. Phys. 50, 2037 (1972).

[31] M. Born and N. Wiener, A new formulation of the laws of quantisation of periodic and aperiodic phenomena, J. Math. Phys. 5, 84 (1926).

[32] S. Okubo, Does equation of motion determine commutation relation? Phys. Rev. D22, 919 (1980).

[33] S. Okubo, Canonical quantization of some dissipative systems and nonuniqueness of Lagrangian, Phys. Rev. A 23, 2776 (1981).

Chapter 6

Symmetries and Conservation Laws

We have discussed the possibility of constructing some first integrals of motion in addition to the proper Hamiltonian from the equation of motion and the canonical commutation relation for one-dimensional conservative systems. As we mentioned earlier, even for a free particle, any differentiable function of momentum can be regarded as a first integral. But as we have seen in classical mechanics the requirement of invariance under Galilean transformation eliminates all but the Hamiltonian (or Lagrangian) which is quadratic in velocity. Thus it seems that Galilean invariance, invariance under time-reversal transformation for conservative systems, and other conservation laws may be used to limit the number of quantum mechanical first integrals which can be regarded as generators of motion. To study the limitations that these symmetries and invariances place on the first integral we consider the most general form of transformation of space-time displacements for a system of n interacting particles. Thus let us consider the infinitesimal translation in space by $\delta\boldsymbol{\epsilon}$ and the change in time by δt, then we can write the generator as

$$\delta G = \delta\boldsymbol{\epsilon}\cdot\mathbf{P} + \delta\boldsymbol{\omega}\cdot\mathbf{J} + \delta\mathbf{v}\cdot\mathbf{N} - \delta t H \tag{6.1}$$

where $\delta\boldsymbol{\epsilon}, \delta\boldsymbol{\omega}, \delta\mathbf{v}$ and δt are all infinitesimals, $\mathbf{P}$ is total momentum, $\mathbf{J}$ the total angular momentum, $\mathbf{N}$ is the boost and H is the Hamiltonian.

For the system of n interacting particles these quantities are defined by

$$\mathbf{P} = \sum_{i=1}^{n}\mathbf{p}_i, \tag{6.2}$$

$$\mathbf{J} = \sum_{i=1}^{n} (\mathbf{r}_i \wedge \mathbf{p}_i), \tag{6.3}$$

$$\mathbf{N} = \sum_{i=1}^{n} (\mathbf{p}_i t - m_i \mathbf{r}_i) = \mathbf{P}t - M\mathbf{R}, \tag{6.4}$$

and

$$H = \sum_{i=1}^{n} \frac{p_i^2}{2m_i} + V(\mathbf{r}_1, \cdots \mathbf{r}_n). \tag{6.5}$$

In Eq. (6.4) M is the total mass of the system is

$$M = \sum_{i=1}^{n} m_i, \tag{6.6}$$

and $\mathbf{R}$ is the center of mass coordinate

$$\mathbf{R} = \sum_{i=1}^{n} \frac{m_i}{M} \mathbf{r}_i. \tag{6.7}$$

6.1 Galilean Invariance

For a system of interacting particles we will show that the requirement of the Galilean invariance restricts the dependence of the first integral (or the generator of motion) to a quadratic dependence on the momentum of the center of mass.

Consider two observers O and O' the latter moving relative to O with velocity $\delta\mathbf{v}$, and let us denote the state of an arbitrary but isolated system of n particles as described by the two observers O and O' by $|0, t\rangle$ and by $|\delta\mathbf{v}, t\rangle$ respectively. Setting $\hbar = 1$, the unitary operator for a pure Galilean transformation corresponding to the classical equation (1.155) in this case is [1]

$$\Gamma(\delta\mathbf{v}, t) = \exp\{i\delta\mathbf{v} \cdot (M\mathbf{R} - t\mathbf{P})\} \quad \hbar = 1. \tag{6.8}$$

We first determine the development of the state of the n-particle system for the time δt as seen by the observer O'. If the Hamiltonian of the system is H, then the state of the system at the time $t + \delta t$ is given by

$$|\delta\mathbf{v}, t + \delta t\rangle = (H\delta t + i)|\delta\mathbf{v}, t\rangle = (H\delta t + i)\Gamma(\delta\mathbf{v}, t)|0, t\rangle. \tag{6.9}$$

Since in the nonrelativistic mechanics time is absolute, the same description by the observer O' can be obtained by transforming the state of the system in O at the time $t + \delta t$ to O'. This is done by first operating by $(H\delta t + i)$ on $|0, t\rangle$ and then by $\Gamma(\delta\mathbf{v}, t + \delta t)$, i.e.

$$|\delta\mathbf{v}, t + \delta t\rangle = \Gamma(\delta\mathbf{v}, t + \delta t)(H\delta t + i)|0, t\rangle. \tag{6.10}$$

By equating (6.9) and (6.10) we find

$$\delta t(H\Gamma - \Gamma H) = i\left\{\Gamma(\delta\mathbf{v}, t + \delta t) - \Gamma(\delta\mathbf{v}, t)\right\}. \tag{6.11}$$

Now if we substitute for Γ from (6.8) and assume that $\delta\mathbf{v}$ is infinitesimal we will get

$$[\mathbf{R}, H] - t[\mathbf{P}, H] = i\mathbf{P}. \tag{6.12}$$

This relation shows that for an isolated system, the total Hamiltonian, H, cannot depend on the center of mass coordinate and that H has to be a quadratic function of the center of mass momentum $\mathbf{P}$. Thus we conclude that if we write the Hamiltonian as the sum

$$H = \frac{\mathbf{P}^2}{2M} + K\left(p_i^{(r)}, q_i^{(r)}\right), \tag{6.13}$$

where $p_i^{(r)}$ and $r_i^{(r)}$ refer to the relative momenta and coordinates of the particles, then the Galilean invariance is preserved.

6.2 Wave Equation and the Galilean Transformation

In the following section for the sake of simplicity we consider a one-dimensional motion and we will write the Planck's constant explicitly in the equations. The general proof for three-dimensional case is given in a number of books [1].

Let us consider two inertial frames $S(x,t)$ and $S'(x',t')$, with $S'(x',t')$ moving relative to S with a speed v. The Galilean transformation in this case is

$$\begin{cases} x = x' + vt' \\ t = t'. \end{cases} \tag{6.14}$$

The Schrödinger equation in the frame S is given by

$$-\frac{\hbar^2}{2m}\left(\frac{\partial^2\psi}{\partial x^2}\right) + V\psi = i\hbar\frac{\partial\psi}{\partial t}, \tag{6.15}$$

and in S' by

$$-\frac{\hbar^2}{2m}\left(\frac{\partial^2\psi'}{\partial x'^2}\right) + V'\psi' = i\hbar\frac{\partial\psi'}{\partial t'}. \tag{6.16}$$

Here we impose the following condition on the potential

$$V'(x - vt, t) = V(x, t). \tag{6.17}$$

We want to show that $\psi(x,t)$ and $\psi'(x',t')$ are related to each other by a multiplicative phase factor of the form

$$\psi'(x',t') = e^{-i\chi(x,t)}\psi(x,t), \tag{6.18}$$

i.e. the probability density is the same in S and S' and does not depend on the choice of the system of reference therefore

$$|\psi'(x',t')|^2 = |\psi(x-vt,t)|^2 = |\psi(x,t)|^2 . \tag{6.19}$$

By substituting (6.18) in (6.16) and by introducing the independent variables x and t we find

$$\begin{aligned} & -\frac{\hbar^2}{2m}\left(\frac{\partial^2\psi}{\partial x^2}\right) + i\hbar\left(\frac{\hbar}{m}\frac{\partial\chi}{\partial x} - v\right)\frac{\partial\psi}{\partial x} \\ + & \left[V(x,t) + \frac{i\hbar^2}{2m}\frac{\partial^2\chi}{\partial x^2} + \frac{\hbar^2}{2m}\left(\frac{\partial\chi}{\partial x}\right)^2 - \hbar v\frac{\partial\chi}{\partial x} - \hbar\frac{\partial\chi}{\partial t}\right]\psi \\ = & \; i\hbar\frac{\partial\psi}{\partial t}. \end{aligned} \tag{6.20}$$

This equation should be identical to Eq. (6.15), and for this to be true it is necessary that χ satisfies the following relations:

$$\frac{\hbar}{m}\frac{\partial\chi}{\partial x} - v = 0, \tag{6.21}$$

and

$$\frac{i\hbar^2}{2m}\frac{\partial^2\chi}{\partial x^2} + \frac{\hbar^2}{2m}\left(\frac{\partial\chi}{\partial x}\right)^2 - \hbar v\frac{\partial\chi}{\partial x} - \hbar\frac{\partial\chi}{\partial t} = 0. \tag{6.22}$$

By integrating Eq. (6.21) first and then substituting the result in (6.22) we obtain the function $\chi(x,t)$;

$$\chi(x,t) = \frac{mv}{\hbar}x - \frac{mv^2}{2\hbar}t, \tag{6.23}$$

where we have omitted an arbitrary constant of integration in χ. From Eqs. (6.18) and (6.23) we find $\psi(x,t)$ in terms of $\psi'(x-vt,t)$;

$$\psi(x,t) = \exp\left[\frac{i}{\hbar}\left(mvx - \frac{m}{2}v^2 t\right)\right]\psi'(x-vt,t), \tag{6.24}$$

and this shows that the absolute value of these two wave functions are equal.

The momentum space wave function $\phi(p,t)$ which is defined by the Fourier transform

$$\phi(p,t) = \frac{1}{\sqrt{2\pi\hbar}}\int_{-\infty}^{+\infty}\psi(x,t)\exp\left(\frac{-ipx}{\hbar}\right)dx. \tag{6.25}$$

in the system S can be found in S' by substituting for $\psi(x,t)$ from (6.24);

$$\begin{aligned}\phi(p,t) &= \exp\left[\frac{i}{\hbar}\left(\frac{m}{2}v^2 - pv\right)t\right]\frac{1}{\sqrt{2\pi\hbar}} \\ &\times \int_{-\infty}^{+\infty}\psi'(x',t')\exp\left[\frac{-i}{\hbar}(p-mv)x'\right]dx' \\ &= \exp\left[\frac{i}{\hbar}\left(\frac{1}{2}mv^2 - pv\right)t\right]\phi'(p-mv,t). \end{aligned} \tag{6.26}$$

We note that in S' the momentum is shifted by mv as is expected. Again for the probability density in momentum space we have

$$|\phi(p,t)|^2 = |\phi'(p-mv,t)|^2. \tag{6.27}$$

6.3 Decay Problem in Nonrelativistic Quantum Mechanics and Mass Superselection Rule

Let us study one of the consequences of the invariance of the wave equation under the Galilean transformation. Following the original argument due to Bargmann, let us consider a sequence of the transformation $S_1 \cdots S_4$ defined by two transformations and two Galilean boosts [2],[3]. The first, S_1, is a translation of the coordinate by a distance a with the generator

$$T = \exp\left(\frac{-ip_x a}{\hbar}\right), \tag{6.28}$$

where p_x is the x component of momentum. The second, S_2, is a boost by velocity v. This is followed by the third, S_3, which is a translation by $-a$. Finally the transformation S_4 is a boost by velocity $-v$. Since in the nonrelativistic theory time is not affected by the transformation we have the spatial coordinates

$$x_1 = x - a, \quad x_2 = x_1 - vt, \quad x_3 = x_2 + a, \quad \text{and} \quad x_4 = x, \tag{6.29}$$

i.e. after all these transformations we end up in the original system. Now after the first and the third translations we have the wave functions $\psi_1(x_1)$ and $\psi_3(x_3)$ where

$$\psi_1(x_1) = \psi(x), \quad \psi_3(x_3) = \psi_2(x_2). \tag{6.30}$$

The two boosts will change the wave functions to $\psi_2(x_2)$ and $\psi_4(x_4)$;

$$\psi_2(x_2) = \exp\left[-\frac{im}{\hbar}\left(vx_1 - \frac{v^2t}{2}\right)\right]\psi_1(x_1). \tag{6.31}$$

and

$$\psi_4(x_4) = \exp\left[-\frac{im}{\hbar}\left(-vx_3 - \frac{v^2t}{2}\right)\right]\psi_3(x_3). \tag{6.32}$$

By combining Eqs. (6.30), (6.31) and (6.32) we obtain $\psi_4(x_4)$ in terms of $\psi(x)$;

$$\psi_4(x_4) = \exp\left(\frac{im}{\hbar}va\right)\psi(x). \tag{6.33}$$

For a state with a definite mass (or a mass eigenstate) the phase factor $\exp\left(\frac{im}{\hbar}va\right)$ is of no consequence. However for the superposition of two or many states with different masses, such as the case of the decay of an unstable nucleus, this causes a problem. Let W denote the operator defined by its action on the wave function changing the phase of ψ_m by

$$W\psi_m = \exp\left(-\frac{im}{\hbar}va\right)\psi_m. \tag{6.34}$$

where m is the mass of the particle. This operator, W, as we have seen earlier is expressible in terms of the operators T and Γ

$$W = T^\dagger(a)\Gamma^\dagger(v,t)T(a)\Gamma(v,t). \tag{6.35}$$

Consider the action of W on the linear superposition of two states of different masses m_1 and m_2. For such a superposition, the result of operating by W is

$$W(\alpha_1\psi_{m_1} + \alpha_2\psi_{m_2}) = \exp\left(\frac{-im_1av}{\hbar}\right)\left[\alpha_1\psi_{m_1} + \alpha_2\left(\frac{-i\Delta mav}{\hbar}\right)\psi_{m_2}\right], \tag{6.36}$$

where

$$\Delta m = m_2 - m_1. \tag{6.37}$$

Since W maps the point x into itself and does not change the time, the superposition of states with different masses before and after the action of W must gives us the same physical state. Now as (6.36) shows there is an observables phase shift arising from the action of W. Since this is unacceptable result, according to Bargmann, we should demand that no observable be allowed that have matrix elements between states of different masses. This implies the existence of a superselection rule which forbids transitions between states of different masses. This view has been challenged by Greenberger who has observed that in nuclear physics, for a decaying system, the law of conservation of momentum cannot be maintained unless the change in the rest mass energy is also taken into account even when the velocities of the decaying products are small compared to c the velocity of light [3].

Let us consider a particle of mass M at rest decaying into another particle of mass m and a photon. For instance M can be the mass of an atom in an excited state and m its mass when it is in the ground state. In this case the conservation of momentum along the x axis in the rest frame S of the excited atom gives us

$$mv = \hbar k, \tag{6.38}$$

where k is the wave number of the emitted photon. If we look at the same decay in the system S' which is moving along the negative y axis with velocity u, then in S' the momentum conservation in the y direction gives us

$$Mu = mu + \frac{\hbar k u}{c}. \tag{6.39}$$

This result shows that

$$M = m + \frac{\hbar\omega}{c^2}, \tag{6.40}$$

i.e. as the atom decays, its momentum along the y direction decreases, because the photon carries some momentum. However here the y component of the velocity has not changed, therefore the particle must have lost mass [3].

Now let us see how this idea can be used to resolve the problem associated with superselection rule. We define the relativistic boost by $L(v)$ where L represent the Lorentz transformation [1],[3]. The action of this relativistic boost operator on the wave function can be expressed as

$$L(v)\psi_m(x,t) = \psi_m\left(x+vt,\ t+\frac{vx}{c^2}\right) + \mathcal{O}\left(\frac{v^2}{c^2}\right). \tag{6.41}$$

Thus in this case both x and t have been transformed according to the rules of the Lorentz transformation. Next we define the analogue of W, Eq. (6.35), with $\Gamma(v)$ being replaced by $L(v)$.

$$W_L\psi_m(x,t) = T^\dagger(a)L^\dagger(v)T(a)L(v)\psi_m(x,t) \tag{6.42}$$

Operating with the operator W_L on $\psi_m(x,t)$ we get

$$T^\dagger(a)L^\dagger(v)\psi_m\left(x+a+vt,\ t+\frac{v(x+a)}{c^2}\right) = \psi_m\left(x,\ t+\frac{va}{c^2}\right). \tag{6.43}$$

But the time translation operator is $\exp\left(\frac{iHt}{\hbar}\right)$, thus we can write $W_L\psi_m(x,t)$ as

$$W_L\psi_m(x,t) = \psi_m(x,\ t+\tau) = \exp\left(\frac{iH\tau}{\hbar}\right)\psi_m(x,t), \tag{6.44}$$

where

$$\tau = \frac{av}{c^2}. \tag{6.45}$$

We are interested in the first order correction in powers of $\frac{v}{c}$ and to this order the Hamiltonian is

$$H = mc^2 + \mathcal{O}\left(v^2\right). \tag{6.46}$$

From Eqs. (6.44) and (6.46) it follows that the operator W_L produces the same phase shift when it acts on $\psi_m(x,t)$ as W does, provided that the terms proportional to $\left(\frac{v}{c}\right)^2$ and higher order terms are ignored. We can attribute this change of phase to the change in time introduced by the Lorentz transformation. Thus the superposition of the two state with masses m_1 and m_2 takes the form

$$\begin{aligned} & W_L\left[\alpha_1\psi_{m_1}(x,t)+\alpha_2\psi_{m_2}(x,t)\right] \\ = & \left[\exp\left(-\frac{ivaE_1}{\hbar c^2}\right)\left\{\alpha_1\psi_{m_1}(x,t)+\alpha_2\exp\left(-\frac{iva\Delta E}{\hbar c^2}\right)\psi_{m_2}(x,t)\right\}\right], \end{aligned} \tag{6.47}$$

where $\Delta E=(m_2-m_1)c^2$ is the energy difference between the two states when both are at rest [3].

6.4 Time-Reversal Invariance

For conservative systems the time-reversal requirement also imposes certain constraints on the Hamiltonian or other first integrals of motion. In classical dynamics we know that if we reverse the momenta of a system of particles after it has evolved for a time δt, we find the same motion (i.e. the same coordinates and momenta) as the one obtained by reversing the momenta at the beginning and then allowing the system to evolve backwards for the time δt.

Let θ be the quantum mechanical time reversal transformation. If the initial state is $|\psi\rangle$ then

$$\theta\exp\left(-iH\delta t\right)|\psi\rangle, \tag{6.48}$$

is the state obtained by reversing the momenta at $t=0$ and allowing the system to evolve for the time δt. We should get the same state by first reversing the momenta at $t=0$, $\hat{\theta}|\psi\rangle$ and then moving the system backwards for δt to get

$$\exp\left(\frac{-iH(-\delta t)}{\hbar}\right)\theta|\psi\rangle. \tag{6.49}$$

These two states must be the same. Equating (6.48) and (6.49) gives us the condition that the anticommutator of H and θ has to be zero

$$H\theta+\theta H=0. \tag{6.50}$$

Just as in classical mechanics the time-reversal invariance changes the sign of momentum $\mathbf{p}$, but leaves the position of the particle $\mathbf{r}$ unchanged;

$$\mathbf{r}'=\theta\,\mathbf{r}\,\theta^{-1}=\mathbf{r}, \tag{6.51}$$

$$\mathbf{p}'=\theta\,\mathbf{p}\,\theta^{-1}=-\mathbf{p}. \tag{6.52}$$

Under this transformation the orbital angular momentum $\mathbf{L}$ of the particle as well as its spin angular momentum will also change sign [4],[5];

$$\mathbf{L}'=\theta\,\mathbf{L}\,\theta^{-1}=-\mathbf{L},\quad \text{or}\quad \boldsymbol{\sigma}'=\theta\boldsymbol{\sigma}\,\theta^{-1}=-\boldsymbol{\sigma} \tag{6.53}$$

$$\mathbf{S}' = \theta\, \mathbf{S}\, \theta^{-1} = -\mathbf{S}. \tag{6.54}$$

For example Eq. (6.53) follows from the definition of $\mathbf{L}$

$$\mathbf{L}' = \theta\, (\mathbf{r} \wedge \mathbf{p})\, \theta^{-1} = -\mathbf{r} \wedge \mathbf{p} = -\mathbf{L}. \tag{6.55}$$

A very important result concerning the invariance of fundamental commutation relation under the time-reversal transformation can be obtained by noting that

$$\left[p'_j,\, x'_k\right] = \theta\, \left[p_j,\, x_k\right]\, \theta^{-1} = \theta\, (-i\hbar\delta_{jk})\, \theta^{-1} = \left[-p_j,\, x_k\right] = i\hbar\delta_{kj}. \tag{6.56}$$

From this result it follows that

$$\theta\, i\, \theta^{-1} = -i, \tag{6.57}$$

and therefore

$$\theta\, a|\psi(t)\rangle = a^*\theta|\psi(t)\rangle. \tag{6.58}$$

Now a double reversal of time will have no physical effect, i.e. if θ^2 is applied to a state $|\psi(t)\rangle$ we get the same state back with a possible phase factor

$$\theta^2|\psi(t)\rangle = \beta|\psi(t)\rangle, \tag{6.59}$$

where $\beta = e^{i\phi}$, and ϕ is a real phase. From this result it follows that

$$\begin{aligned} \beta\theta|\psi(t)\rangle &= \beta^2\beta|\psi(t)\rangle = \theta\theta^2|\psi(t)\rangle \\ &= \theta\beta|\psi(t)\rangle = \beta^*\theta|\psi(t)\rangle. \end{aligned} \tag{6.60}$$

Thus $\beta = \beta^*$ and consequently $\beta = 1$, or $\beta = -1$ depending on the nature of the system. Now by examining the expression

$$\langle\theta\psi(t)|\psi(t)\rangle = \left\langle\theta\psi(t)|\theta^2\psi(t)\right\rangle = \beta\,\langle\theta\psi(t)|\psi(t)\rangle\,, \tag{6.61}$$

we find that if $\beta = -1$, then

$$\langle\theta\; \psi(t)|\psi(t)\rangle = 0. \tag{6.62}$$

From this result we conclude that for systems where $\beta = -1$ the time-reversed states are orthogonal to the original states.

Now we want to consider a system composed of n spin $\frac{1}{2}$ particles and examine the result of the time-reversal transformation. We start with the θ transformation for a single particle and then generalize the result to a system of n particles. For a single particle we write θ as a product of two operators;

$$\theta = UK, \tag{6.63}$$

where U is a linear transformation and K is an operator which takes any complex number z into its complex conjugate

$$KzK^{-1} = z^*, \tag{6.64}$$

Since θ is the product of a unitary operator and the complex conjugation operator, then we call it an antiunitary operator. When K acts on the spin of the system it affects only the y component of $\boldsymbol{\sigma}$, i.e.

$$K\sigma_x K^{-1} = \sigma_x, \quad K\sigma_y K^{-1} = -\sigma_y, \quad K\sigma_z K^{-1} = \sigma_z. \tag{6.65}$$

On the other hand for the unitary operator U we have

$$U\sigma_x U^{-1} = -\sigma_x, \quad U\sigma_y U^{-1} = \sigma_y, \quad U\sigma_z U^{-1} = -\sigma_z, \tag{6.66}$$

or

$$U\sigma_x + \sigma_x U = 0, \quad U\sigma_y - \sigma_y U = 0, \quad U\sigma_z + \sigma_z U = 0. \tag{6.67}$$

This follows from Eqs. (6.54), (6.63) and (6.64). Equation (6.67) shows that U which is unitary must be equal to σ_y which is itself a unitary operator. Thus for a single particle Eq. (6.63) becomes

$$\theta = \sigma_y K. \tag{6.68}$$

Now for a system with n particles each of spin $\frac{1}{2}$ we have

$$\theta = \sigma_y^{(1)}\sigma_y^{(2)}\cdots\sigma_y^{(n)}K, \tag{6.69}$$

where $\sigma_y^{(j)}$ is the y components of the spin of the j-th particle. From this result we obtain θ^2

$$\theta^2 = \sigma_y^{(1)}\sigma_y^{(2)}\cdots\sigma_y^{(n)}K\sigma_y^{(1)}\sigma_y^{(2)}\cdots\sigma_y^{(n)}K = (-1)^n 1, \tag{6.70}$$

where 1 is the identity operator. This equation implies that if there are even number of spin $\frac{1}{2}$ particles in the system then we are dealing with an even state, otherwise the state is odd. As we have seen earlier under time reversal of transformation for an odd number of spin $\frac{1}{2}$ particles, a state $|\psi\rangle$ is transformed into an orthogonal state. From this result we obtain Kramer's theorem. This theorem states that if a system is composed of an odd number of spin $\frac{1}{2}$ particles and has a Hamiltonian which is invariant under the time reversal transformation, then all its stationary states are degenerate. This is the case when a spin $\frac{1}{2}$ particle is experiencing a central field of force plus an external electric field. However the theorem also implies that when the Hamiltonian is not invariant under time reversal, e.g. a system in a magnetic field, then each of the levels will split into a number of lines. This is because the magnetic field changes sign under θ transformation.

We have experimental evidence showing that strong forces are invariant under time-reversal transformation [5]. This invariance restricts the type of potentials which are allowed in the theory of nuclear forces, for example in the nucleon-nucleon interaction. Let $\boldsymbol{\sigma}_1$ and $\boldsymbol{\sigma}_2$ denote the spin of the two nucleons and $\mathbf{L}$ denote their orbital angular momentum in the center of mass frame, then a scalar potential of the form

$$V(r) = W(r)\,(\boldsymbol{\sigma}_1 \wedge \boldsymbol{\sigma}_2)\cdot\mathbf{L}, \tag{6.71}$$

is not an admissible potential since it is not invariant under time-reversal transformation [6].

6.5 Parity of a State

In addition to the infinitesimal displacement of the coordinates given by $\delta\boldsymbol{\epsilon}$ in Eq. (6.1) and infinitesimal rotation with the generator $\delta\boldsymbol{\omega}\cdot\mathbf{J}$ in the same equation, we have the invariance of the Hamiltonian under the parity transformation. This invariance does not have a classical analogue. For a closed system the inversion transformation which is a simultaneous change of sign of all coordinates, leave the Hamiltonian unchanged, and this leads to a conservation law.

Let P represents the inversion transformation

$$P\begin{bmatrix}1\\1\\1\end{bmatrix}=\begin{bmatrix}-1\\-1\\-1\end{bmatrix}, \tag{6.72}$$

i.e. the P transformation changes the right-handed coordinate system into a left-handed one or vice-versa. Under this transformation the coordinate $\mathbf{x}$ changes to $-\mathbf{x}$ and the momentum $\mathbf{p}$ changes $-\mathbf{p}$, so that the commutation relation $[x_j, p_k]=i\delta_{jk}$ remains unchanged.

The parity operator thus defined commutes with the Hamiltonian

$$[P,\ H]=0, \tag{6.73}$$

and with the components of the angular momentum operator

$$[P,\ L_x]=[P,\ L_y]=[P,\ L_z]=0,\quad \left[P,\ \mathbf{L}^2\right]=0. \tag{6.74}$$

The reason that L_i commutes with P is that both $\mathbf{r}$ and $\mathbf{p}$ change sign under the action of P, therefore

$$P\mathbf{L}=P(\mathbf{r}\wedge\mathbf{p})=\mathbf{r}\wedge\mathbf{p}=\mathbf{L}. \tag{6.75}$$

From the definition (6.72) it is clear that

$$P(P(\mathbf{r}))=P(-\mathbf{r})=\mathbf{r}. \tag{6.76}$$

Thus

$$P^2=1, \tag{6.77}$$

where 1 is the unit 3×3 matrix. This last equation implies that the eigenvalues of P are either $+1$ or -1, i.e. when P operates on the wave function $\psi(\mathbf{r})$ we either get either $+\psi(\mathbf{r})$ or $-\psi(\mathbf{r})$. In the first case we have an even state and in the second case an odd state.

From Eq. (6.74) we infer that for a closed two-body system with the angular quantum numbers ℓ and m, we have also a state of definite parity, either odd or even. The invariance under parity transformation also implies that all states with the same ℓ but different m have the same parity. To show this we start with the relation

$$(L_x+iL_y)P-P(L_x+iL_y)=0, \tag{6.78}$$

which follows directly from (6.74), then take its matrix element:

$$\begin{aligned} & \langle n,\ell,m|L_x+iL_y|n,\ell,\ m-1\rangle\,\langle n,\ell\ m-1|P|n',\ell,\ m-1\rangle \\ - & \langle n,\ell,m|P|n',\ell,m\rangle\,\langle n',\ell\, m|L_x+iL_y|n',\ell,\ m-1\rangle=0 \end{aligned} \tag{6.79}$$

Again since the matrix elements of L_x+iL_y are independent of n or n', therefore we have

$$\langle n\,\ell,\ m-1|P|n',\ell\ m-1\rangle=\langle n,\ell,m|P|n',\ell,m\rangle\,, \tag{6.80}$$

or the matrix elements of P are independent of m, i.e. the states of different m have the same parity.

For the two body problem with central forces (Chapter 9), where we use spherical coordinates, the parity transformation can be implemented by the coordinate transformation

$$r\to r,\quad \theta\to\pi-\theta,\quad \phi\to\pi+\phi. \tag{6.81}$$

As an application of the invariance under parity and time-reversal transformations we will consider the limitations that they place on the form of the potential [6]. The the most general phenomenological potential acting on a particle with momentum $\mathbf{p}$, orbital angular momentum $\mathbf{L}$ and spin $\mathbf{S}$, which is linear in these dynamical variables is of the form

$$\begin{aligned} V(\mathbf{p},\mathbf{L},\mathbf{S}) &= (\mathbf{p}\cdot\mathbf{L})V_{pL}(r)+(\mathbf{p}\cdot\mathbf{S})V_{pS}(r)+(\mathbf{L}\cdot\mathbf{S})V_{LS}(r) \\ &+ [\mathbf{p}\cdot(\mathbf{L}\wedge\mathbf{S})]\,V_{pLS}(r)+[\mathbf{S}\cdot(\mathbf{p}\wedge\mathbf{L})]\,V_{SpL}(r) \\ &+ [\mathbf{L}\cdot(\mathbf{p}\wedge\mathbf{S})]\,V_{LpS}(r). \end{aligned} \tag{6.82}$$

All these terms commute with the rotation operator with the generator $\mathbf{L}+\mathbf{S}$. The requirement of the time reversal transformation means that the last three terms must be set equal to zero. Also the condition that the Hamiltonian should be invariant under parity transformation requires that $V_{pL}=V_{pS}=0$. Thus the only remaining term will be the spin-orbit potential $V_{LS}(r)$.

6.6 Permutation Symmetry

In the early days of quantum mechanics Heisenberg realized the importance of the permutation symmetry in the description of a many-particle system [8]–[10]. A two-boson state where each particle is characterized by a complete set of quantum numbers ν but otherwise are indistinguishable can be obtained from the vacuum state by the action of the creation operator $a^\dagger(\nu)$ [10]

$$|1_{\nu_1},1_{\nu_2}\rangle=a^\dagger(\nu_1)a^\dagger(\nu_2)|0\rangle. \tag{6.83}$$

By indistinguishability we mean that we cannot associate any other label in addition to ν to the particle. Since the bosons ν_1 and ν_2 are indistinguishable the state (6.83) is the same as the state

$$|1_{\nu_2}, 1_{\nu_1}\rangle = a^\dagger(\nu_2)a^\dagger(\nu_1)|0\rangle, \tag{6.84}$$

i.e. there is a unitary operator U_{12} which connects these two states

$$|1_{\nu_2}, 1_{\nu_1}\rangle = U_{12}|1_{\nu_1}, 1_{\nu_2}\rangle = \left(U_{12}a^\dagger(\nu_1)a^\dagger(\nu_2)U_{12}^{-1}\right) U_{12}|0\rangle. \tag{6.85}$$

Thus we have

$$U_{12}a^\dagger(\nu_1)a^\dagger(\nu_2)U_{12}^{-1}) = a^\dagger(\nu_2)a^\dagger(\nu_1), \tag{6.86}$$

where we have assumed that the vacuum state is invariant under U_{12} transformation, $U_{12}|0\rangle = |0\rangle$. By applying U_{12} twice we get the original state, i.e.

$$U_{12}^2|1_{\nu_1}, 1_{\nu_2}\rangle = |1_{\nu_2}, 1_{\nu_1}\rangle. \tag{6.87}$$

From (6.87) we conclude that the eigenvalues of U_{12} are $+1$ and -1, where the first one corresponds to the symmetric and the second one to the antisymmetric permutations of ν_1 and ν_2;

$$U_{12}|1_{\nu_1}, 1_{\nu_2}\rangle = +|1_{\nu_2}, 1_{\nu_1}\rangle, \tag{6.88}$$

and

$$U_{12}|1_{\nu_1}, 1_{\nu_2}\rangle = -|1_{\nu_2}, 1_{\nu_1}\rangle. \tag{6.89}$$

The symmetric case is for the interchange of two bosons since from the commutation relation

$$\left[a^\dagger(\nu_1),\ a^\dagger(\nu_2)\right] = 0, \tag{6.90}$$

it follows that

$$U_{12}|1_{\nu_1}, 1_{\nu_2}\rangle = a^\dagger(\nu_2)a^\dagger(\nu_1)|0\rangle = a^\dagger(\nu_1)a^\dagger(\nu_2)|0\rangle = |1_{\nu_2}, 1_{\nu_1}\rangle. \tag{6.91}$$

On the other hand for the permutation of two fermions we have the antisymmetric case. Here we use the anticommutator

$$\left[b^\dagger(\nu_1),\ b^\dagger(\nu_2)\right]_+ = 0, \tag{6.92}$$

to get

$$U_{12}|1_{\nu_1}, 1_{\nu_2}\rangle = b^\dagger(\nu_2)b^\dagger(\nu_1)|0\rangle = -b^\dagger(\nu_1)b^\dagger(\nu_2)|0\rangle = -|1_{\nu_2}, 1_{\nu_1}\rangle. \tag{6.93}$$

If we want to express these as single particle states, for the bosons we have the symmetric form

$$|1_{\nu_1}, 1_{\nu_2}\rangle_S = \frac{1}{\sqrt{2}}|1_{\nu_1}\rangle_1|1_{\nu_2}\rangle_2 + \frac{1}{\sqrt{2}}|1_{\nu_2}\rangle_1|1_{\nu_1}\rangle_2, \tag{6.94}$$

where the subscript S indicates the symmetric state. For fermions we write the antisymmetric form (denoted by subscript A)

$$|1_{\nu_1}, 1_{\nu_2}\rangle_A = \frac{1}{\sqrt{2}}|1_{\nu_1}\rangle_1|1_{\nu_2}\rangle_2 - \frac{1}{\sqrt{2}}|1_{\nu_2}\rangle_1|1_{\nu_1}\rangle_2. \tag{6.95}$$

The perturbation symmetry can be extended to a many-boson or many-fermion problem using the same argument that we had earlier for a two-body indistinguishable particles. In the case of bosons the state $|n_{\nu_1}\cdots n_{\nu_r}\cdots\rangle$ must be symmetric under any permutation $\mathcal{P}$ of the particles

$$\mathcal{P}|n_{\nu_1}\cdots n_{\nu_r}\cdots\rangle = |n_{\nu_1}\cdots n_{\nu_r}\cdots\rangle. \tag{6.96}$$

For a many-fermion state $|n_{\nu_1}\cdots n_{\nu_r}\cdots\rangle$ is antisymmetric under permutation

$$\mathcal{P}|n_{\nu_1}\cdots n_{\nu_r}\cdots\rangle = \delta_{\mathcal{P}}|n_{\nu_1}\cdots n_{\nu_r}\cdots\rangle, \tag{6.97}$$

where

$$\delta_{\mathcal{P}} = \begin{cases} +1 & \text{for even } \mathcal{P} \\ -1 & \text{for odd } \mathcal{P} \end{cases}. \tag{6.98}$$

In general the Hamiltonian of the system commutes with U_{12},

$$HU_{12} - U_{12}H = 0, \tag{6.99}$$

and then the symmetry or antisymmetry of the state is preserved in time, that is, the eigenvalues $+1$ or -1 will be constants of motion.

Symmetric and Antisymmetric Wave Functions — One of the results of the permutation symmetry of a system composed of fermions is the Pauli exclusion principle. This principle states that in a system consisting of identical particles with half-integer spin $\frac{1}{2}$, $\frac{3}{2}\cdots$, the wave function must be antisymmetric, and that these particles obey the Fermi-Dirac statistics. On the other hand the wave function describing systems of identical particles with integer spin 0, 1, $2\cdots$ must be symmetric, and these particles called bosons obey the Bose-Einstein statistics.

Writing the two-particle wave function corresponding to the ket (6.95) in the position space we have

$$\psi_A(1,2) = \frac{1}{\sqrt{2}}\left[\psi_{\nu_1}(\mathbf{r}_1)\psi_{\nu_2}(\mathbf{r}_2) - \psi_{\nu_2}(\mathbf{r}_1)\psi_{\nu_1}(\mathbf{r}_2)\right]. \tag{6.100}$$

Now if we set $\nu_1 = \nu_2$ in $\psi(1,2)$ we obtain zero for all values of $\mathbf{r}_1$ and $\mathbf{r}_2$. This result is an expression of the Pauli exclusion principle, i.e. the two fermions cannot occupy the same individual state. For example a three particle wave function, properly symmetrized, can be written as

$$\begin{aligned} \psi_A(1,2,3) &= \frac{1}{\sqrt{6}}\left[\psi(1,2,3) - \psi(2,1,3) + \psi(2,3,1)\right. \\ &- \left.\psi(3,2,1) + \psi(3,1,2) - \psi(1,3,2)\right]. \end{aligned} \tag{6.101}$$

In the case of identical bosons we have the symmetric wave function

$$\begin{aligned} \psi_S(1,2,3) &= \frac{1}{\sqrt{6}}\left[\psi(1,2,3) + \psi(2,1,3) + \psi(2,3,1)\right. \\ &+ \left.\psi(3,2,1) + \psi(3,1,2) + \psi(1,3,2)\right]. \end{aligned} \tag{6.102}$$

An important case of $\psi_A(1,2,3)$ is the wave function for N fermions in a constant common potential when they are not interacting with each other. In this case the total Hamiltonian is the sum of Hamiltonians for individual particles

$$H = \sum_{j=1}^{N} H_j = \sum_{j=1}^{N} \left\{ -\frac{\hbar^2}{2m}\nabla_j^2 + V(\mathbf{r}_j) \right\}. \tag{6.103}$$

Since the motion is separable, for the j-th particle we have

$$H_j\phi(\mathbf{r}_j) = E_j\phi(\mathbf{r}_j), \tag{6.104}$$

and then from (6.103) it follows that the total energy of the system is

$$E = E_1 + E_2 + \cdots + E_N. \tag{6.105}$$

Now if there are only two particles in the system then the antisymmetrized wave function will be

$$\psi_A(\mathbf{r}_1, \mathbf{r}_2) = \frac{1}{\sqrt{2}}[\phi_{E_1}(\mathbf{r}_1)\phi_{E_2}(\mathbf{r}_2) - \phi_{E_1}(\mathbf{r}_2)\phi_{E_2}(\mathbf{r}_1)]. \tag{6.106}$$

This method can easily be generalized to N particles by expressing the total wave function as a determinant (Slater determinant) [7]

$$\psi_A(\mathbf{r}_1, \cdots, \mathbf{r}_N) = \frac{1}{\sqrt{N!}} \begin{vmatrix} \phi_{E_1}(\mathbf{r}_1) & \phi_{E_1}(\mathbf{r}_2) & \cdots & \phi_{E_1}(\mathbf{r}_N) \\ \phi_{E_2}(\mathbf{r}_1) & \phi_{E_2}(\mathbf{r}_2) & \cdots & \phi_{E_2}(\mathbf{r}_N) \\ \cdots & \cdots & \cdots & \cdots \\ \phi_{E_N}(\mathbf{r}_1) & \phi_{E_N}(\mathbf{r}_2) & \cdots & \phi_{E_N}(\mathbf{r}_N) \end{vmatrix}. \tag{6.107}$$

In Chapter 18 we will discuss the theory of N interacting fermions and its application to the pairing forces.

6.7 Lattice Translation

A very important problem in solid state physics is the motion of an electron in a chain of regularly spaced positive ions. The potential function for such a motion is invariant under the translation by a length equal to the lattice spacing a, i.e.

$$V(x + a) = V(x). \tag{6.108}$$

If we denote the translation operator (translation along the x-axis) by $\mathrm{T}(a)$, then

$$\mathrm{T}^\dagger(a)V(x)\mathrm{T}(a) = V(x + a) = V(x). \tag{6.109}$$

The explicit operator form of $\mathrm{T}(a)$ is given in terms of the momentum operator p;

$$\mathrm{T}(a) = \sum_{n=0}^{\infty} \frac{1}{n!} a^n \frac{d^n}{dx^n} = \sum_{n=0}^{\infty} \frac{1}{n!} \left(\frac{ia}{\hbar}\right)^n p^n. \tag{6.110}$$

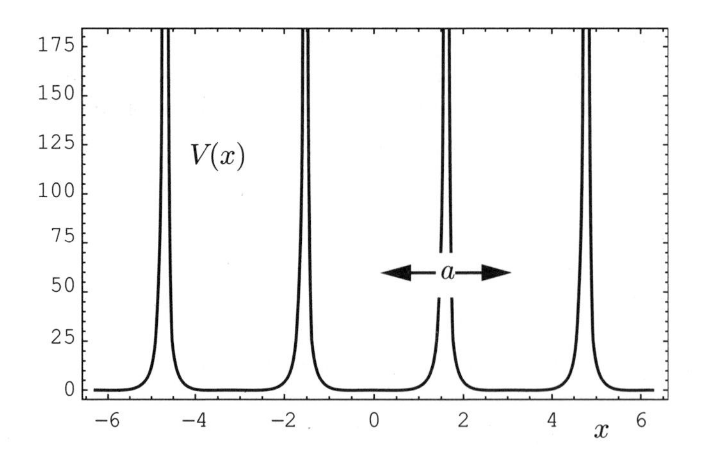

Figure 6.1: A one-dimensional periodic potential with periodicity a.

Noting that the kinetic energy part of the Hamiltonian commutes with $\mathrm{T}(a)$, and that the potential energy also satisfies (6.109), we have

$$\mathrm{T}^{\dagger}(a)H\mathrm{T}(a) = H. \tag{6.111}$$

Since the translation operator is a unitary operator, from Eq. (6.111) it follows that

$$[H,\ \mathrm{T}(a)] = 0. \tag{6.112}$$

The vanishing of this commutator means that we can simultaneously diagonalize H and $\mathrm{T}(a)$. But the unitary operator $\mathrm{T}(a)$ is not Hermitian, therefore its eigenvalue is a complex number of modulus 1.

Let $|n\rangle$ denote the eigenvector which diagonalizes the Hamiltonian

$$H|n\rangle = E_0|n\rangle, \tag{6.113}$$

and at the same time shows that the particle is localized in the n-th site. The wave function for this situation will be completely within this site provided that the height of the barrier between the adjacent lattice sites is infinitely high as is shown in Fig. 6.1. The ket $|n\rangle$ however is is not an eigenket of the lattice translation operator $\mathrm{T}(a)$, because if $\mathrm{T}(a)$ is applied to $|n\rangle$ the adjacent site would be obtained

$$\mathrm{T}(a)|n\rangle = |n+1\rangle. \tag{6.114}$$

Since $|n\rangle$ is not an eigenstate of H and $\mathrm{T}(a)$ at the same time, we try to find a simultaneous eigenstate for these two operators. To this end we consider a linear combination of the form

$$|\alpha\rangle = \sum_{n=-\infty}^{+\infty} e^{in\alpha}|n\rangle. \tag{6.115}$$

where α is a real parameter which can take values between $-\pi$ and $+\pi$. The fact that $|\alpha\rangle$ is an eigenstate of H follows from the relation

$$H|\alpha\rangle = \sum_{n=-\infty}^{+\infty} E_0\, e^{in\alpha}|n\rangle = E_0|\alpha\rangle. \tag{6.116}$$

This result shows that the eigenvalue of H is independent of α. The ket $|\alpha\rangle$ is also an eigenstate of $\mathrm{T}(a)$ since

$$\begin{aligned}\mathrm{T}(a)|\alpha\rangle &= \sum_{n=-\infty}^{+\infty} e^{in\alpha}|n+1\rangle = \sum_{n=-\infty}^{+\infty} e^{i(n-1)\alpha}|n\rangle \\ &= e^{-i\alpha}|n\rangle. \end{aligned} \tag{6.117}$$

Thus $|\alpha\rangle$ which is the simultaneous eigenket of H and $\mathrm{T}(a)$ depends on the continuous parameter α.

Now let us consider this symmetry as is reflected in the form of the wave function. The coordinate representation of the eigenstate $|\alpha\rangle$ can be obtained from $\langle x|\alpha\rangle$. The wave function for the lattice-translated state $\mathrm{T}(a)|\alpha\rangle$ is found from

$$\langle x|\mathrm{T}(a)|\alpha\rangle = \langle x-a|\alpha\rangle, \tag{6.118}$$

where here $\mathrm{T}(a)$ has acted to the left, i.e. on $\langle x|$. On the other hand if in (6.118) $\mathrm{T}(a)$ acts to the right we get

$$\langle x|\mathrm{T}|(a)|\alpha\rangle = e^{-i\alpha}\langle x|\alpha\rangle. \tag{6.119}$$

By equating (6.118) and (6.119) we find

$$\langle x-a|\alpha\rangle = e^{-i\alpha}\langle x|\alpha\rangle. \tag{6.120}$$

Now let us define the function $u_k(x)$ by

$$\psi(x) \equiv \langle x|\alpha\rangle = e^{-ikx}u_k(x), \tag{6.121}$$

where $\alpha = ka$. Since for $-\infty \le x \le +\infty$ the wave function has to be well-defined therefore k must be a real parameter. The function $u_k(x)$ is periodic, as can easily be verified by substituting for $\langle x-a|\alpha\rangle$ and $\langle x|\alpha\rangle$ in (6.120),

$$e^{ik(x-a)}u_k(x-a) = e^{ikx}u_k(x)e^{-ikx}. \tag{6.122}$$

This important result, known as Bloch's theorem, shows that $|\alpha\rangle$ which is an eigenstate of $\mathrm{T}(a)$ can be expressed as the product of a plane wave and a periodic function $u_k(x)$ with periodicity a. In the mathematical literature this symmetry property of the wave function which is a solution of the Schrödinger equation with periodic potential is called "Flouqet theorem" [11],[12].

We also note that k and $k+\frac{2\pi n}{a}$ both give us the same eigenvalue e^{ika}. Therefore for k in the interval

$$-\frac{\pi}{a} < k \le \frac{\pi}{a}, \tag{6.123}$$

we can have all of the eigenvalues of T(a).

The technique that we used to find the symmetry of the wave function and the properties of the lattice translation operator T can be generalized to two- or three-dimensional systems in straight forward way [13].

6.8 Classical and Quantum Integrability

In classical dynamics a system is called completely integrable if for a system of N degrees of freedom, in addition to the Hamiltonian, there are $N-1$ globally defined functions whose mutual Poisson bracket is zero, i.e. [14]–[18]

$$\{I_j,\ I_k\} = 0, \quad j,k = 1,2,\cdots,N-1, \tag{6.124}$$

and

$$\{I_j,\ H\} = 0, \quad j,k = 1,2,\cdots,N-1. \tag{6.125}$$

Since H which is a member of the set $\{I_j\}$, satisfies (6.125) therefore all $\{I_j\}$ s are constants of motion. In classical mechanics complete integrability means that in principle the motion can be determined by quadrature [19].

Following the classical definition of an integrable system we call a quantum system of N degrees of freedom integrable if there are $N-1$ independent global operators which commute with each other and with the Hamiltonian. That is we replace the Poisson brackets by commutators and require that the commutator for any pair of the Hermitian first integrals of motion must vanish. If we follow the Dirac rule of association, Chapter 3, then the classical integrability always imply quantum integrability. But as we have seen Dirac's rule for more complicated dynamical variables leads to inconsistencies, and a close examination shows that in general the quantum integrability does not follow from the classical one. For instance, let us consider the following Hamiltonian describing a motion with two degrees of freedom [20]:

$$H = \frac{1}{2}p_1^2 + \frac{1}{2}p_2^2 + a\left[\left(q_1^2 + b\right)q_2^{-\frac{2}{3}} + \frac{3}{4}q_2^{\frac{4}{3}}\right]. \tag{6.126}$$

For this two-dimensional motion we have a second conserved quantity I which is given by

$$I = 2p_1^3 + 3p_1p_2^2 + 3ap_1\left[2\left(q_1^2 + b\right)q_2^{-\frac{2}{3}} - 3q_2^{\frac{4}{3}}\right] + 18ap_2q_1q_2^{\frac{1}{3}}. \tag{6.127}$$

There is no ambiguity in constructing a Hermitian operator corresponding to the classical Hamiltonian H. For constructing a Hermitian operator for I Eq. (6.127) we observe that the last two terms of I are linear in p_1 and p_2 and therefore we can make the operator $\hat{I}$ Hermitian by symmetrization, that is we can write it as

$$\begin{aligned}\hat{I} &= 2p_1^3 + 3p_1p_2^2 + \frac{3}{2}ap_1\left[2\left(q_1^2+b\right)q_2^{-\frac{2}{3}}\right] \\ &+ \frac{3}{2}a\left[2\left(q_1^2+b\right)q_2^{-\frac{2}{3}}p_1\right] - 9aq_2^{\frac{4}{3}}p_1 + 9aq_2\left(p_2q_2^{\frac{1}{3}} + q_2^{\frac{1}{3}}p_2\right), \end{aligned} \tag{6.128}$$

where $p_1 = -i\frac{\partial}{\partial q_1}$ and $p_2 = -i\frac{\partial}{\partial q_2}$. Now by calculating the commutator $\left[H,\ \hat{I}\right]$ we find that it does not vanish and therefore the condition for complete integrability is not satisfied [14]–[17].

Another simple example of a two-dimensional motion where a classically conserved quantity does not have a corresponding quantum first integral will be discussed in connection with the question of classical versus quantum degeneracy.

6.9 Classical and Quantum Mechanical Degeneracies

An accepted view in quantum theory is that the classical degeneracy of a motion implies its quantum degeneracy, as the correspondence principle requires [21]. As we have already noted the example of anisotropic two-dimensional oscillator may be an exception to this rule. Now let us study this question in some detail. For this purpose again we consider the anisotropic two-dimensional oscillator with the Hamiltonian given by (1.221), where we know that the classical motion is degenerate. This example is separable we can write the Schrödinger equation for the motion and find its eigenvalues. Since the potential has a discontinuous derivative at $x = 0$ the corresponding Heisenberg equation is not easy to solve. From the separability of the Hamiltonian it follows that the total energy of the system can be written as [22]

$$E = E(\nu_1) + E(\nu_2), \tag{6.129}$$

where $E(\nu_i)$ is the eigenvalue obtained by solving the Schrödinger equation

$$-\frac{\hbar^2}{2m}\frac{d^2\psi_i(x)}{d\,x^2} + \frac{1}{2}m\omega^2x^2\left[\frac{\theta(x)}{(1+\lambda_i)^2} + \frac{\theta(-x)}{(1-\lambda_i)^2}\right]\psi_i(x) = E(\nu_i)\psi_i(x), \quad i = 1, 2. \tag{6.130}$$

The general solution of (6.130) which is finite at $x = \pm\infty$ is given by [23]

$$\psi_{\nu_i}(x) = \begin{cases} ND_{\nu_i}\left[-\sqrt{\frac{2m\omega}{(1-\lambda_i)\hbar}}\ x\right] & \text{for } x \le 0 \\ ND_{\nu_i}\left[+\sqrt{\frac{2m\omega}{(1+\lambda_i)\hbar}}\ x\right] & \text{for } x \ge 0 \end{cases} \tag{6.131}$$

where N is the normalization constant and $D_{\nu_i}(z)$ is the parabolic cylinder function

$$D_{\nu_i}(z) = 2^{\frac{\nu_i}{2}} e^{\frac{-z^2}{4}} \left[\frac{\Gamma\left(\frac{1}{2}\right)}{\Gamma\left(\frac{1-\nu_i}{2}\right)} {}_1F_1\left(-\frac{\nu_i}{2};\frac{1}{2};\frac{z^2}{2}\right) + \frac{z}{\sqrt{2}}\frac{\Gamma\left(\frac{1}{2}\right)}{\Gamma\left(-\frac{\nu_i}{2}\right)} {}_1F_1\left(\frac{1-\nu_i}{2};\frac{3}{2};\frac{z^2}{2}\right)\right]. \tag{6.132}$$

The function ${}_1F_1$ is the confluent hypergeometric (or Kummer) function [24].

At $x = 0$ the two parts of ψ_{ν_i} should join smoothly

$$\sqrt{1-\lambda_i}\frac{D'_{\nu_i}(0)}{D_{\nu_i}(0)} = \sqrt{1+\lambda_i}\frac{D'_{\beta_i}(0)}{D_{\beta_i}(0)}, \quad i = 1, 2. \tag{6.133}$$

This gives us the eigenvalue equation

$$1 + \sqrt{\frac{1-\lambda_i}{1+\lambda_i}}\left[\frac{\Gamma\left(\frac{1-\nu_i}{2}\right)\Gamma\left(\frac{-\beta_i}{2}\right)}{\Gamma\left(\frac{1-\beta_i}{2}\right)\Gamma\left(\frac{-\nu_i}{2}\right)}\right] = 0, \tag{6.134}$$

where β_i and ν_i are related to each other by

$$\left(\beta_i + \frac{1}{2}\right)(1+\lambda_i) = \left(\nu_i + \frac{1}{2}\right)(1-\lambda_i). \tag{6.135}$$

The energy eigenvalues are related to ν_1 and ν_2 the roots of the transcendental Eq. (6.134) by

$$E(\nu_1, \nu_2) = \hbar\omega\left[\frac{\nu_1 + \frac{1}{2}}{(1+\lambda_1)} + \frac{\nu_2 + \frac{1}{2}}{(1+\lambda_2)}\right]. \tag{6.136}$$

These energy levels are nondegenerate, so that the symmetry of the classical motion associated with the closed orbit does not carry over to quantum mechanics.

Now let us consider the asymptotic form of the eigenvalues given by Eq. (6.134). For this we first rewrite (6.134) by changing the Γ functions with negative argument in terms of Γ functions with positive argument using the relation [24]

$$\Gamma(1-z)\Gamma(z) = \frac{\pi}{\sin(\pi z)}. \tag{6.137}$$

Then we replace $\Gamma(z)$ s by their asymptotic expansion for large z, [24]

$$\Gamma(z) \to z^{z-\frac{1}{2}} e^{-z}\sqrt{2\pi}, \tag{6.138}$$

to get [23]

$$\frac{\sin\left[\frac{1}{2}\pi(\beta_i - \nu_i)\right] + \sin\left[\frac{1}{2}\pi(\beta_i + \nu_i)\right]}{\sin\left[\frac{1}{2}\pi(\beta_i - \nu_i)\right] - \sin\left[\frac{1}{2}\pi(\beta_i + \nu_i)\right]} = 1. \tag{6.139}$$

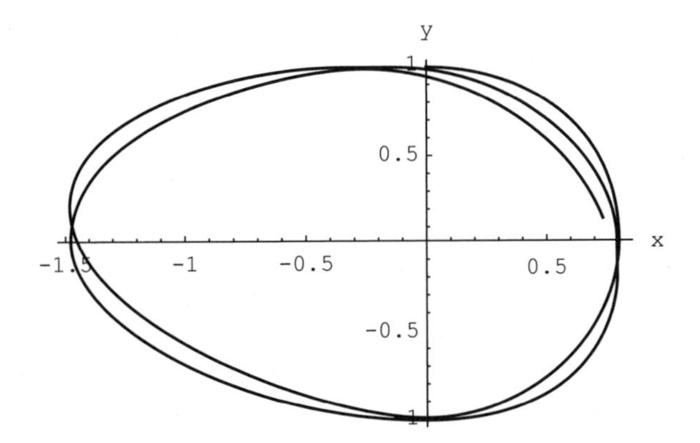

Figure 6.2: The classical limit of a degenerate quantum system given by the Hamiltonian (6.142) shows no sign of degeneracy. The orbit is precessing due to the additional term $\Delta v(x)$ in the Hamiltonian.

This equation is satisfied whenever $(\beta_i + \nu_i)$ is an even integer. Combining this result with (6.135) we find that

$$\nu_i + \frac{1}{2} \to n_i(1 + \lambda_i), \quad n_i \text{ an integer.} \tag{6.140}$$

Substituting this result in (6.136) we have the standard eigenvalues for two-dimensional oscillator

$$E(n_1, n_2) = \hbar\omega(n_1 + n_2), \tag{6.141}$$

i.e. for large quantum numbers we recover the degeneracy associated with closed orbits as is required by the Bohr correspondence principle.

Having shown the loss of degeneracy caused by quantization, we can also demonstrate by an example that the converse is also true. That is there are quantum mechanical degenerate systems that will have nondegenerate classical limits. For this let us consider the Hamiltonian operator [25],[26]

$$H = \frac{p_x^2}{2m} + \frac{p_y^2}{2m} + \frac{1}{2}m\omega^2\left(x^2 + y^2\right) + \frac{\hbar^2}{2m}\Delta v(x), \tag{6.142}$$

where

$$\begin{aligned} \Delta v(x) &= -2g\frac{d}{dx}\left[\frac{e^{-\alpha^2 x^2}}{1 + g\int_{-\infty}^{x} e^{-\alpha^2 z^2} dz}\right] \\ &= -2g\frac{d}{dx}\left[\frac{e^{-\alpha^2 x^2}}{1 + g\left(\sqrt{\frac{\pi}{\alpha}}\left\{1 + erf(\alpha x)\right\}\right)}\right], \end{aligned} \tag{6.143}$$

where g is a constant and $\alpha^2 = \frac{m\omega}{\hbar}$. The energy eigenvalues of this Hamiltonian is given by

$$E(n_1, n_2) = (n_1 + n_2)\hbar\omega, \tag{6.144}$$

where n_1 and n_2 are integers. Thus the term $\frac{\hbar^2}{2m}\Delta v(x)$ changes the wave functions but not the eigenvalues of H [25].

In the classical limit we can derive the canonical equations of motion from (6.142) and solve for $x(t)$ and $y(t)$, and thus determine the orbit. We know that in the absence of $\Delta v(x)$ the orbit will be an ellipse with two axes of symmetry and the conserved quantities are given by (1.194)–(1.196). But because of the addition of $\Delta v(x)$ the classical orbit calculated from $x(t)$ and $y(t)$ does not close on itself (see Fig. 6.2).

Bibliography

[1] K. Gottfried and T-M Yan, *Quantum Mechanics: Fundamentals*, (Springer, New York, 2003).

[2] V. Bargmann, On the unitary ray representation of continuous groups, Ann. Math. 59, 1 (1954).

[3] D.M. Greenberger, Inadequacy of the usual Galilean transformation in quantum mechanics, Phys. Rev. Lett, 87, 100405 (2001).

[4] R. Sachs, *The Physics of Time Reversal*, (The University of Chicago Press, 1987).

[5] E.M. Henley, Parity and time-reversal invariance in nuclear physics, Annual Rev. Nucl. Science, 19, 367 (1969).

[6] S. Okubo and R.E. Marshak, Velocity-dependence of the two nucleon interaction, Ann. Phys. 4, 166 (1958).

[7] J.C. Slater, The theory of complex spectra, Phys. Rev. 34, 1293 (1929).

[8] W. Heisenberg, Mehrkörperproblem und Resonanz in der Quantunmechanik, Z. Physik 38, 411 (1926).

[9] W. Heisenberg, Über die spectra von atomsystemen mit zwei electron, Z. Phys. 39, 499 (1926), reprinted in W. Heisenberg, *Collected Works*, Series A1, (Springer, Berlin, 1985)

[10] F.A. Kaempffer, *Concepts in Quantum Mechanics*, (Academic Press, New York, 1965).

[11] D. Zwillinger, *Hanbook of Differential Equations*, (Academic Press, Boston, 1989) p. 404.

[12] A.A. Cottey, Floquet's theorem and band theory in one-dimension, Am. J. Phys. 39, 1235 (1971).

[13] J. Callaway, *Energy Band Theory*, (Academic Press, 1964), p. 102.

[14] J. Hietarinta, Quantum integrability is not a trivial consequence of classical integrability, Phys. Lett. A 93, 55 (1982).

[15] J. Hietarinta, Classical versus quantum integrability, J. Math. Phys. 25, 1833 (1984).

[16] J. Hietarinta, Direct methods for the search of the 2nd invariants, Phys. Rep. 147, 87 (1987).

[17] J. Hietarinta, Solvability in quantum mechanics and classically superfluous invariants, J. Phys. A 22 L143 (1989).

[18] M. Razavy, Does the integrability of a classical Hamiltonian implies its quantum integrability, Phys. Lett. A 118, 387 (1986).

[19] V.I. Arnold, *Mathematical Methods of Classical Mechanics*, (Springer Verlag, Berlin, 1980).

[20] C.R. Holt, Construction of new integrable Hamiltonians in two degrees of freedom, J. Math. Phys. 23, 1037 (1982).

[21] See for instance, R.J. Finkelstein, *Nonrelativistic Mechanics*, (W.A. Benjamin, New York, 1973).

[22] J.F. Marko and M. Razavy, Dynamical symmetries of a class of action equivalent Hamiltonians, Lett. Nuovo Cimento 40, 533 (1984).

[23] G. Ghosh and R.W. Hasse, Inequivalence of the classes of quantum and classical harmonic potentials: Proof by example, Phys. Rev. D 24, 1027 (1981).

[24] I.S. Gradshteyn and I.M. Ryzhik, *Tables of Integrals, Series, and Products*, Fourth Edition, (Academic Press, New York, 1965), p. 1064.

[25] P.B. Abraham and H.E. Moses, Changes in potential due to changes in the point spectrum: Anharmonic oscillator with exact solution, Phys. Rev. A 23, 2088 (1981).

[26] M. Razavy, Some counterexamples in quantum mechanics, Iranian J. Science and Technology, A 25, 329 (2001).

Chapter 7

Bound State Energies for One-Dimensional Problems

As we noticed in Chapter 2, by a careful examination of the problem of anharmonic oscillator, Born, Heisenberg and Jordan were led to the discovery of matrix mechanics. In this and later chapters we want to follow up the development of the matrix mechanics from those early days to the present time. Here we must emphasize that as a practical tool the Heisenberg matrix mechanics is not a simple and straightforward method of solving quantum mechanical problems. In particular for the approximate solution of nonseparable systems, motions with insufficient symmetries or cases where the boundary conditions are complicated, the solution of the wave equation is definitely preferable. However this method gives us a better insight about the connection between the Hamiltonian operator, the equations of motion and the commutation relation, a subject that we discussed for the harmonic oscillator in Sec. 5.2. Furthermore we observe that even for simple systems such as the motion of a particle in a box, we find the presence of quantum forces proportional to $\hbar$, $\hbar^2 \cdots$. As we will in this chapter see there have been attempts by Klein *et al* [1]–[8], by Halpern, [9] and others to use Heisenberg's formulation to determine the eigenvalues of simple systems.

The methods that we will be considered here, in principle, can yield exact values. Approximate techniques based on matrix mechanics which will be useful in more complicated systems will be considered later.

First let us consider some general properties for one-dimensional motion. The ground state wave function has no nodes for the range of coordinates that the particle is allowed to move. However this result may not be true for the wave function of several electrons in their ground state [10]. If $|\psi_0\rangle$ denotes the ground state wave function which we can choose to be real without any nodes, then it follows that the wave functions for the excited states $|\psi_n\rangle$ must have nodes, otherwise $\langle\psi_0|\psi_n\rangle$ cannot vanish. The fact that $|\psi_0\rangle$ has no nodes implies that the ground state cannot be degenerate. To show this result let us suppose that $\left|\psi_0^{(1)}\right\rangle$ and $\left|\psi_0^{(2)}\right\rangle$ are two different states corresponding to the same eigenvalue E_0. Since the wave equation is linear, it follows that if C_1 and C_2 are arbitrary constants, then $C_1\left|\psi_0^{(1)}\right\rangle + C_2\left|\psi_0^{(2)}\right\rangle$ will also be an eigenstate with the eigenvalue E_0. By choosing C_1 and C_2 judiciously we can make $|\psi\rangle$ zero at any given point in space, i.e. we have an eigenfunction with a node contrary to our original assumption.

7.1 Klein's Method

While studying certain problems related to the collective motion in nuclear physics, Klein and collaborators discovered that the Heisenberg equations together with canonical commutation relation can be used to calculate the energy eigenvalues. They applied this method to find the eigenvalues of the confining quartic potential [1]–[8]

$$V(x) = \frac{1}{2}x^2 + \frac{1}{4}\lambda x^4. \tag{7.1}$$

Later the same method was used to calculate the energy levels of a particle moving in the double well potential [11]

$$V(x) = -\frac{1}{2}x^2 + \frac{1}{4}\lambda x^4, \tag{7.2}$$

and other one-dimensional systems. But before discussing these and other problems we want to show that the solution of the equation of motion and the commutation relation for the matrix elements $\langle j|x|n\rangle$ are compatible with the diagonal form of the Hamiltonian. This, in a way, is similar to what we discussed regarding the connection between the Hamiltonian, the equations of motion and the commutation relation in Sec. 5.2. First we note that for a particle of unit mass under the action of the potential $V(x)$ the Hamiltonian is

$$H = \frac{p^2}{2} + V(x), \tag{7.3}$$

and we have the commutation relation,

$$[x, H] = ip. \tag{7.4}$$

From this relation it follows that

$$p^2 = [x, H][H, x]. \tag{7.5}$$

The equation of motion obtained from (7.3) is given by

$$[\,[x, H], H] = V'(x), \tag{7.6}$$

where prime denotes the derivative with respect to x. We can rewrite (7.6) also in the following two different forms:

$$x\,[\,[x, H], H] + [\,[x, H], H]\,x = 2xV'(x). \tag{7.7}$$

and

$$p\,[\,[x, H], H] + [\,[x, H], H]\,p = pV'(x) + V'(x)p. \tag{7.8}$$

Let us use a representation in which H is diagonal, i.e.

$$H|n\rangle = E_n|n\rangle. \tag{7.9}$$

In this representation from (7.5) we obtain the result

$$\langle j\,|p^2|\,n\rangle = -\sum_k (E_k - E_j)(E_n - E_k)\langle j|x|k\rangle\langle k|x|n\rangle, \tag{7.10}$$

and from (7.4) we have

$$(E_{n'} - E_n)\,\langle n\,|x|\,n'\rangle = i\,\langle n\,|p|\,n'\rangle\,. \tag{7.11}$$

We also write Eq. (7.7) in terms of the matrix elements

$$\sum_k \left\{(E_k - E_j)^2 + (E_n - E_k)^2\right\}\langle j|x|k\rangle\langle k|x|n\rangle = 2\,\langle j|xV'(x)|n\rangle\,. \tag{7.12}$$

By multiplying (7.10) by two and adding it to (7.12) we obtain

$$\sum_k (2E_k - E_j - E_n)^2\langle j|x|k\rangle\langle k|x|j\rangle = 2\,\langle j\,|p^2|\,n\rangle + 2\,\langle j|xV'(x)|n\rangle\,. \tag{7.13}$$

Similarly the matrix elements of (7.8) satisfy the following relation

$$\begin{aligned} &(E_n - E_j)\sum_k (2E_k - E_j - E_n)^2\langle j|x|k\rangle\langle k|x|n\rangle \\ = \;& -2i\,\langle j\,|V'(x)p + pV'(x)|\,n\rangle + 2(E_n - E_j)\,\langle j|(xV'(x)|n\rangle\,, \quad n \neq j, \end{aligned} \tag{7.14}$$

where we have used (7.11) to simplify (7.14).
But

$$\begin{aligned} &i\,\langle j\,|V'(x)p + pV'(x)|\,n\rangle = 2\langle j|[V(x), H]|n\rangle \\ = \;& 2(E_n - E_j)\langle j|V(x)|n\rangle, \quad n \neq j, \end{aligned} \tag{7.15}$$

and therefore (7.13) reduces to

$$\sum_k (2E_k - E_j - E_n)^2 \langle j|x|k\rangle\langle k|x|j\rangle = -4\,\langle j\,|V(x)|\,n\rangle$$
$$+\quad 2\,\langle j|xV'(x)|n\rangle\,, \quad n \neq j. \tag{7.16}$$

Now we eliminate $\langle j|xV'(x)|n\rangle$ between (7.13) and (7.16) to get

$$\frac{1}{2}\,\langle j\,|p^2|\,n\rangle + \langle j\,|V(x)|\,n\rangle = \langle j|H|n\rangle = 0, \quad n \neq j, \tag{7.17}$$

that is H is a diagonal matrix [5] as we assumed in (7.9), and this shows the consistency of this formulation for one-dimensional problems.

7.2 The Anharmonic Oscillator

In the early works on quantum mechanics the anharmonic oscillator with cubic nonlinearity was used as a simple system to illustrate the method of quantization. But here we consider a potential with quartic nonlinearity. The Heisenberg equations of motion and the commutation relation for the problem of anharmonic oscillator with $V(x)$ given by (7.1) are:

$$[x, H] = ip, \tag{7.18}$$

$$[p, H] = -ix - i\lambda x^3, \tag{7.19}$$

and

$$[x, p] = i. \tag{7.20}$$

Since this is a one-dimensional motion with a simple interaction, the eigenvalues will be nondegenerate. Here again we will use a representation of matrices where H is diagonal, and we denote the eigenstates of H by $|n\rangle$, therefore

$$\langle n\,|H|\,n'\rangle = 0, \quad \text{if} \quad n \neq n'. \tag{7.21}$$

The commutator (7.20) in this representation becomes

$$\sum_{n''} \left(\langle n\,|x|\,n''\rangle\,\langle n''\,|p|\,n'\rangle - \langle n\,|p|\,n''\rangle\,\langle n''\,|x|\,n'\rangle\right) = i\delta_{nn'}. \tag{7.22}$$

We also write (7.19) in its matrix form

$$(E_{n'} - E_n)\,\langle n\,|p|\,n'\rangle = -i\,\langle n\,|x|\,n'\rangle$$
$$-\quad i\lambda \sum_{n''}\sum_{n'''} \left(\langle n\,|x|\,n''\rangle\,\langle n''\,|x|\,n'''\rangle\,\langle n'''\,|x|\,n'\rangle\right). \tag{7.23}$$

Now we eliminate $(E_{n'} - E_n)$ between (7.11) and (7.23) to get the matrix elements of p in terms of the matrix elements of x;

$$\begin{aligned} \langle n|p|n'\rangle^2 &= -\langle n|x|n'\rangle^2 \\ &- \lambda \langle n|x|n'\rangle \sum_{n''}\sum_{n'''} (\langle n|x|n''\rangle \langle n''|x|n'''\rangle \langle n'''|x|n'\rangle). \end{aligned} \tag{7.24}$$

At this point we make the assumption that only a finite number of terms contribute significantly to the sums in Eqs. (7.22) and (7.24), i.e. as $|n-n'|$, $|n-n''|$, or $|n'-n''|$ become large, the matrix elements become small and negligible.

To simplify the problem further, we note that in the potentials (7.1) and (7.2) only even powers of x appears, therefore the energy eigenstates have definite parities which alternate, viz,

$$\langle n|x|n'\rangle = \langle n|p|n'\rangle = 0, \quad |n-n'| = \text{ even integer.} \tag{7.25}$$

In addition because of the time-reversal invariance we have the following relations between the matrix elements

$$\langle n|x|n'\rangle = \langle n'|x|n\rangle, \tag{7.26}$$

and

$$\langle n|p|n'\rangle = -\langle n'|p|n\rangle. \tag{7.27}$$

Now we change our notation and write these equations in a way that the conditions (7.26) and (7.27) take a simple form. Let us introduce two integers I and J by [1]

$$n = 2I - 2, \quad \text{and} \quad n' = 2J - 1, \tag{7.28}$$

and set

$$X(I,J) = \langle n|x|n'\rangle = \langle n'|x|n\rangle, \tag{7.29}$$

$$Y(I,J) = -i\langle n|p|n'\rangle = i\langle n'|p|n\rangle. \tag{7.30}$$

Thus I refers to even and J refers to odd parity states. The commutation relation (7.22) now divides into two sets of equations

$$C_E(I,I') \equiv \sum_J [X(I,J)Y(I',J) + Y(I,J)X(I',J)] + \delta_{II'} = 0, \tag{7.31}$$

and

$$C_O(J,J') \equiv \sum_I [X(I,J)Y(I,J') + Y(I,J)X(I,J')] - \delta_{JJ'} = 0. \tag{7.32}$$

In terms of the states I and J, Eq. (7.24) becomes

$$\begin{aligned} E_M(I,J) &\equiv -Y(I,J)^2 + X(I,J)^2 \\ &+ \lambda X(I,J) \sum_{I'}^{\nu+1}\sum_{J'}^{\nu} X(I,J')X(I',J')X(I',J) = 0. \end{aligned} \tag{7.33}$$

In these equations the sum over I, J, I', and J' are all from one to infinity, but since these sums are convergent, we have written them in truncated form. Thus in the ν-th order of approximation we have $2\nu+1$ states, $\nu+1$ of which have even and ν have odd parities. If we consider the set of $2\nu(\nu+1)$ unknowns

$$X(I,J), \quad Y(I,J), \quad 1 \le J \le \nu, \quad 1 \le I \le \nu+1, \tag{7.34}$$

then by determining these unknowns, we find the $2\nu+1$ low-lying energy eigenvalues.

Having decided on the number of unknowns, $2\nu(\nu+1)$, let us examine the number of equations. In every order of approximation we have more equations than unknowns, viz, the total number of equations is $(2\nu+1)(\nu+1)$. Since the major truncation errors come from the matrix elements close to boundary states, we omit the terms $I = \nu+1, \;\; I' = 1, 2 \cdots \nu+1$ from the commutation relation for even terms, i.e. from $C_E(I, I')$.

For the case of anharmonic oscillator, a simple method for the numerical solution of Eqs. (7.31)–(7.33) was devised by Klein and collaborators. In this method we consider the set of nonlinear equations

$$G_i(x_1 \cdots x_M) = 0, \quad i = 1, \cdots M. \tag{7.35}$$

Assuming that for $\lambda = \lambda_0$ the approximate solution $x_j^{(0)}$ is known and that

$$G_i(x_1^{(0)} \cdots x_M^{(0)}) = -B_i, \quad i = 1, \cdots M, \tag{7.36}$$

then from (7.35) and (7.36) it follows that for the difference $\delta x_j^{(0)} = x_j - x_j^{(0)}$, we have

$$\sum_j \left(\frac{\partial G_i}{\partial x_j^{(0)}}\right) \delta x_j^{(0)} = \sum_j A_{ij} \delta x_j^{(0)} \approx B_i, \quad i = 1, \cdots M. \tag{7.37}$$

Thus to the first order the solution to Eq. (7.35) is given by

$$x_j^{(1)} = x_j^{(0)} + \delta x_j^{(0)}. \tag{7.38}$$

This process can be repeated and we find $x_j^{(2)}, x_j^{(3)}$ etc. until the desired accuracy is reached, i.e. the difference $x_j^{(n+1)} - x_j^{(n)}$ is as small as we want. In this way we have calculated the matrix elements $X(I,J)$ and $Y(I,J)$ given in TABLE I for the parameter $\lambda = 1$.

TABLE I: Matrix elements of the coordinate and momentum (for momentum, times $(-i)$), in the first, second and third order approximation $\nu = 1$, $\nu = 2$ and $\nu = 3$. Here the value of $\lambda = 1$ has been used in the calculation.

	$\nu = 1$	$\nu = 2$	$\nu = 3$
X(1,1)	0.5911	0.5953	0.5953
Y(1,1)	−0.8459	−0.8365	−0.8365
X(2,1)	0.8359	0.7711	0.7710
Y(2,1)	1.1963	1.2880	1.2893
X(1,2)		0.0200	0.0201
Y(1,2)		−0.0993	−0.0994
X(2,2)		0.8883	0.8944
Y(2,2)		−1.6809	−1.6631
X(3,1)		0.0302	0.0272
Y(3,1)		0.1596	0.1509
Y(3,2)		1.8540	1.9937
X(1,3)			0.0007
Y(1,3)			−0.0064
X(2,3)			0.0323
X(3,3)			1.0674
Y(33)			−2.3265
X(4,1)			0.0011
Y(4,1)			0.0105
X(4,2)			0.0410
Y(4,2)			0.2506
X(4,3)			1.2445
Y(4,3)			2.4024

Having obtained the elements $X(I,J)$ and $Y(I,J)$ we can calculate the ground state energy from $\langle 0|H|0\rangle$, where the diagonal elements of H are obtained from the expectation value H;

$$\begin{aligned}\langle n|H|n\rangle &= \frac{1}{2}\sum_{n'}\left(\langle n|p|n'\rangle\langle n'|p|n\rangle + \langle n|x|n'\rangle\langle n'|x|n\rangle\right)\\ &+ \frac{\lambda}{4}\sum_{n',n'',n'''}\left(\langle n|x|n'\rangle\langle n'|x|n''\rangle\langle n''|x|n'''\rangle\langle n'''|x|n\rangle\right).\end{aligned} \tag{7.39}$$

We then find the ground state energy to be

$$\begin{aligned} E(0) &= \langle 0|H|0\rangle = \frac{1}{2}\sum_{J}^{\nu}\left(Y(1,J)^2 + X(1,J)^2\right) \\ &+ \frac{\lambda}{4}\sum_{J=1}^{\nu}\sum_{I=1}^{\nu+1}\sum_{J'=1}^{\nu} X(1,J)X(I,J)X(I,J')X(1,J'). \end{aligned} \tag{7.40}$$

Once $E(0)$ has been obtained we can determine the odd-parity-states energies from (7.39) with the result that

$$E(2J-1) = E(0) - \frac{Y(1,J)}{X(1,J)}, \quad J = 1, \cdots \nu. \tag{7.41}$$

For the even-parity states we find a similar relation:

$$E(2J-2) = E(1) + \frac{Y(I,1)}{X(I,1)}, \quad I = 2, \cdots \nu + 1. \tag{7.42}$$

The energies for the lowest level of anharmonic oscillator are tabulated in TABLE II.

TABLE II: The energies of the low-lying states of the anharmonic oscillator ($\frac{1}{2}x^2 + \frac{1}{4}x^4$) calculated from the Heisenberg equations are shown in this table for three orders of approximation. The last column shows the result of the sixth order approximation which is very close to the exact result.

	$\nu = 1$	$\nu = 2$ A	$\nu = 3$	$\nu = 4$	$E(n)$
$n = 0$	0.6240	0.6209	0.6209	0.6209	0.6209
1	2.0551	2.0261	2.0260	20260	20260
2	3.4863	3.6965	3.6984	3.6984	3.6984
3		5.5887	5.5577	5.5576	5.5576
4		7.3115	7.5659	7.5684	7.5684
5				9.7455	9.7092
6				11.6760	11.9645

7.3 The Double-Well Potential

For the double-well potential, Eq. (7.2), we have similar relations except for a change of sign of the terms quadratic in $\langle n|x|n'\rangle$ and quadratic in $X(I,J)$ in Eqs. (7.33), (7.39) and (7.40). Thus for this case we have

$$\begin{aligned} E_M(I,J) &\equiv Y(I,J)^2 + X(I,J)^2 \\ &- \lambda X(I,J)\sum_{I'}^{\nu+1}\sum_{J'}^{\nu} X(I,J')X(I',J')X(I',J) = 0, \end{aligned} \tag{7.43}$$

and

$$\begin{aligned} E(0) &= \frac{1}{2}\sum_{J}^{\nu}\left(Y(1,J)^2 - X(1,J)^2\right) \\ &+ \frac{\lambda}{4}\sum_{J=1}^{\nu}\sum_{I=1}^{\nu+1}\sum_{J'=1}^{\nu} X(1,J)X(I,J)X(I,J')X(1,J'). \end{aligned} \tag{7.44}$$

In the first order of approximation, $\nu = 1$, and there are four unknowns, $X(1,1), Y(1,1), X(2,1)$ and $Y(2,1)$, and four equations for $E_M(1,1)$, $E_M(2,1)$, $C_E(1,1)$ and $C_O(1,1)$. By eliminating $Y(1,1), X(2,1)$ and $Y(2,1)$ from this set we find that $X(1,1)$ is a solution of the equation

$$\left[1 + 4X(1,1)^2 - 12\lambda X(1,1)^6\right]\left[1 + 4X(1,1)^4 + 4\lambda X(1,1)^6\right] = 0. \tag{7.45}$$

Of course we do not expect that in the first order of calculation the result will be close to the exact result (see TABLE III). Also let us note that the solution of (7.45) is not unique, and we can change the signs of $X(I,J)$ s and $Y(I,J)$ s and we get a new set of solutions.

Among the possible sets of solutions only two, ν=2 A and ν=2 B are shown in this table. Here rather than using the numerical technique that we discussed earlier for solving these equations, we try to find different roots for the set of equations $C_E(I,I'), C_O(J,J')$ and $E_M(I,J)$ without assuming any starting set of solutions for these equations. In this way for $\nu = 2$ we get different solutions, two of which are shown in TABLE III, and are denoted by $2A$ and $2B$. The first one, $2A$, is close to the solution of the first order calculation $\nu = 1$, therefore it is this one that we choose for the calculation of the eigenvalues [11]. The eigenvalues obtained for $2A$ are shown in TABLE IV.

TABLE III: Matrix elements of the coordinate and momentum (for momentum, times (-i)), in the first, and second approximation $\nu = 1$, and $\nu = 2$ for the double-well potential, $(-\frac{1}{2}x^2 + \frac{1}{4}x^4)$.

	$\nu = 1$	$\nu = 2$ A	$\nu = 2$ B
X(1,1)	0.7619	0.8129	0.0749
Y(1,1)	−0.6562	−0.5907	−0.2642
X(2,1)	1.0776	0.8764	−0.9665
Y(2,1)	0.9280	1.0896	1.5054
X(1,2)		0.0749	−0.8129
Y(1,2)		−0.2642	0.5907
X(2,2)		0.9665	0.8764
Y(2,2)		−1.5054	1.0896
X(3,1)		0.0788	1.2542
Y(3,1)		0.3198	1.5745
X(3,2)		1.2542	−0.0788
Y(3,2)		1.5745	−0.3198

TABLE IV: The energies of the low-lying states of the double-well potential $(-\frac{1}{2}x^2 + \frac{1}{4}x^4)$ calculated from the Heisenberg equations are shown in this table for three orders of approximation. In the last column the same energies are calculated accurately using finite difference method and are shown for comparison.

	$\nu = 1$	$\nu = 2$ A	$\nu = 3$	$E(n)$
n=0	0.1776	0.1474	0.1474	0.1465
1	1.0389	0.8741	0.8741	0.8672
2	1.9001	2.1173	2.1173	2.0197
3		3.6748	3.5316	3.5456
4		4.9303	5.1777	5.1544
5			6.8729	6.8964
6			8.5001	8.7472

Here we can ask whether these other solutions, i.e. those not obtainable by the iterative method have any physical significance or not. Should these solutions be rejected as being unphysical and if so on what grounds? At present we do not have answers to these questions.

7.4 Chasman's Method

A method of solving Heisenberg's equations which is very similar to Klein's but applicable to the potentials of the type $V(x) = \frac{\lambda}{n}x^n$ with n an even integer has been suggested by Chasman [12]. Writing the Hamiltonian as

$$H = \frac{1}{2}p^2 + \frac{1}{n}\lambda x^n, \quad (n \text{ an even integer}), \tag{7.46}$$

we have the Heisenberg equations

$$-ip = xH - Hx, \tag{7.47}$$

and

$$i\lambda x^{n-1} = pH - Hp. \tag{7.48}$$

By multiplying (7.47) by p from left and (7.48) by x from right and adding the results we have

$$[xp, H] = -ip^2 + i\lambda x^n. \tag{7.49}$$

The diagonal elements of the left-hand side of (7.49) are zero, therefore

$$\left(p^2\right)_{j,j} = \lambda \left(x^n\right)_{j,j}, \tag{7.50}$$

which is just the virial theorem, (see also Eq. (12.104)). If we take the diagonal elements of H, Eq. (7.46), and combine it with (7.50) we get

$$E_j = \frac{2+n}{2n}\left(p^2\right)_{j,j}, \tag{7.51}$$

where we have used a representation in which H is diagonal, $\langle j|H|k\rangle = E_j\delta_{k,j}$. Now we consider the commutator

$$[x, [x,\ H]\,] = -1, \tag{7.52}$$

and take the diagonal matrix elements of this equation to find

$$\sum_k x_{j,k}\left(E_k - E_j\right)x_{k,j} = \frac{1}{2}. \tag{7.53}$$

Next we square Eq. (7.47) and again find its (j, j) element

$$\sum_k x_{j,k}\left(E_k - E_j\right)^2 x_{k,j} = \left(p^2\right)_{j,j} = \frac{2n}{2+n}E_j. \tag{7.54}$$

Equations (7.53) and (7.54) may be used to calculate higher energy eigenvalues in terms of the ground state energy. To see how this method works, let us consider the simple case of the harmonic oscillator, i.e. when $n = 2$. Then as we have seen earlier, Eq. (5.25), the nonvanishing off-diagonal elements of x

are $x_{j,j\pm1}$. If we choose $j = 0$ in (7.53), $x_{0,k}$ is zero unless $k = 1$. Thus (7.53) and (7.54) reduce to

$$x_{0,1}(E_1 - E_0)x_{1,0} = \frac{1}{2}, \tag{7.55}$$

and

$$x_{0,1}(E_1 - E_0)^2 x_{1,0} = E_0, \tag{7.56}$$

and from these we get

$$E_1 = 3E_0. \tag{7.57}$$

So starting from E_0 we can find successive E_j s in terms of E_0.

When $n \neq 2$, then as a first order approximation, just as in Klein's method we choose

$$x_{j,k} = 0, \quad \text{for} \quad k \neq j \pm 1, \tag{7.58}$$

and use these matrix elements in (7.53) and (7.54) to calculate energies to the first order. Having obtained the first order energies, we can calculate the matrix elements $x_{j,j\pm3}$ from these energies. We note that the matrix elements $x_{i,i\pm2k}$ do not contribute to the sums. By substituting for p from Eq. (7.47) in the commutation relation $[x,\ p] = i$ and taking the $(j, j+2)$ matrix element of the resulting expression we obtain

$$\sum_{k=j\pm1,j\pm3} (E_{j+2} - 2E_k + E_j)\, x_{j,k} x_{k,j+2} = 0. \tag{7.59}$$

Next we set $j = 0$ and substitute for $x_{j,j+1}$ and E_j calculated to the first order from (7.53) and (7.54) we get the ratios

$$\frac{x_{j,j\pm3}\,(E_{j\pm3} - E_j)\, x_{j\pm3,j}}{x_{j,j+1}(E_{j+1} - E_j)x_{j+1,j}}. \tag{7.60}$$

We substitute these ratios in (7.53) and (7.54) and obtain $x_{j,j\pm1}$ and E_j to the second order, this time by summing over $k = j \pm 1$, $k = j \pm 3$.

For the third order approximation we use the equations

$$\sum_{k=j\pm1,j\pm3,j\pm5} (E_{j+2} - 2E_k + E_j)\, x_{j,k} x_{k,j\pm2} = 0, \tag{7.61}$$

and

$$\sum_{k=j\pm1,j\pm3,j\pm5} (E_{j+4} - 2E_k + E_j)\, x_{j,k} x_{k,j\pm4} = 0. \tag{7.62}$$

Since (7.61) and (7.62) are coupled we must iterate them until $x_{j,j+3}$ and $x_{j,j+5}$ matrix elements do not change.

The ground state energy is determined by combining Eqs. (7.47) and (7.48) to find

$$x_{j,k}(E_k - E_j)^2 = \lambda \left(x^{n-1}\right)_{j,k}, \qquad (7.63)$$

and then express all the quantities in this equation in terms of E_0.

In TABLE V the results of the first and second order calculation are shown and are compared with the exact result for the quartic harmonic oscillator $V(x) = x^4$.

TABLE V: The energy eigenvalues of the quartic anharmonic oscillator $V(x) = x^4$ in units of E_0 [12].

	$E_j^{(1)}$	$E_j^{(2)}$	Exact
$j = 0$	1	1	1
$j = 1$	3.66	3.58	3.583
$j = 2$	7.21	7.02	7.031
$j = 3$	11.25	10.97	10.980
$j = 4$	15.71	15.31	15.333
$j = 5$	20.51	20.01	20.024
$j = 6$	25.62	24.99	25.010
$j = 7$	30.99	30.24	30.259
$j = 8$	36.61	35.72	35.747
$j = 9$	42.45	41.43	41.454

7.5 Heisenberg's Equations of Motion for Impulsive Forces

While the analytical determination of the wave function and eigenvalues for the problem of a particle confined in a box in wave mechanics is elementary and simple, a close examination shows that the details of such a motion are complicated. For instance the definition of the time-evolution operator for the motion in a box, and the possibility of different self-adjoint extension of the operators need careful examination [13],[14]. For instance, it is clear that the one possible self-adjoint extension where a particle striking one wall and appearing at the other wall, cannot correspond to a classical motion of a particle bouncing between two walls.

Heisenberg's Equations for Motion Between Fixed Walls — Let us study the classical motion of a ball which is bouncing between two rigid walls located at $x = 0$ and $x = 1$ respectively. If we denote the initial position and momentum of the particle by x_0 and p_0 respectively, then at a later time, t, the classical position of the particle can be expanded as a Fourier series;

$$x_c(t) = \frac{1}{2} - \sum_{n=-\infty}^{\infty}{}' \left[\frac{1-(-1)^n}{(n\pi)^2}\right] \exp[in\pi s(t)], \tag{7.64}$$

where the prime on summation means that the term with $n = 0$ must be excluded, and $s(t)$ is given by the function

$$s(t) = \left(\frac{p_0 t}{m} + x_0\right). \tag{7.65}$$

Next we want to find the quantum version of the classical expression (7.64). We note that there are different ways of ordering the operators x_0 and p_0 as we have seen in Chapter 3. For instance if we utilize Weyl–McCoy's rule of ordering, we can write (7.65) as an operator equation for the position of the particle

$$\begin{aligned} x(t) &= \mathcal{O}_W\{x_c\} = \frac{1}{2} - \sum_{n=-\infty}^{\infty}{}' \left[\frac{1-(-1)^n}{(n\pi)^2}\right] \mathcal{O}_W[\exp(in\pi s(t))] \\ &= \frac{1}{2} - \sum_{n=-\infty}^{\infty}{}' \left[\frac{1-(-1)^n}{(n\pi)^2}\right] \exp\left[\left(\frac{-i\hbar}{2}\right)\left(\frac{\partial^2}{\partial x_0 \partial p_0}\right)\right] \\ &\times \left\{\exp(in\pi x_0)\exp\left(\frac{in\pi p_0 t}{m}\right)\right\} \end{aligned} \tag{7.66}$$

where $\mathcal{O}_W$ denotes the Weyl-ordered product. In this equation the exponential differential operator acts only on the product in the curly brackets, and the order of terms must be preserved in differentiation. Using the commutation relation

$$[x_0, p_0] = i\hbar, \tag{7.67}$$

we obtain

$$\begin{aligned} x(t) = \mathcal{O}_W\{x_c(t)\} &= \frac{1}{2} - \sum_{n=-\infty}^{\infty}{}' \left[\frac{1-(-1)^n}{(n\pi)^2}\right] \exp\left(\frac{in^2\pi^2\hbar t}{2m}\right) \\ &\times \left[\exp(in\pi x_0)\exp\left(\frac{inp_0 t}{m}\right)\right]. \end{aligned} \tag{7.68}$$

Once the commutation relation (7.67) is assumed then from (7.67) and (7.68) it follows that

$$[x(t), p(t)] = i\hbar, \tag{7.69}$$

$$i\hbar\left(\frac{dx(t)}{dt}\right) = [x(t), H], \tag{7.70}$$

and

$$i\hbar\left(\frac{dp(t)}{dt}\right) = [p(t), H], \tag{7.71}$$

where $p(t)$ is the mechanical momentum operator

$$p(t) = m\frac{dx(t)}{dt}. \tag{7.72}$$

For instance we obtain (7.70) by noting that the Hamiltonian H is given by $\frac{p^2}{2m}$ and then use the identity

$$[A, B^n] = [A, B]\, B^{n-1} + B\,[A, B]\, B^{n-2} + \cdots + B^{n-1}\,[A, B] \tag{7.73}$$

to get

$$\begin{aligned} [x(t), H] &= -\sum_{n=-\infty}^{\infty\,\prime}\left[\frac{1-(-1)^n}{(n\pi)^2}\right]\mathcal{O}_W\left\{\sum_{j=0}^{\infty}\left(\frac{(in\pi)^j}{j!}\right)\left[s(t)^j, H\right]\right\} \\ &= -\left(\frac{i\hbar}{m}\right)\sum_{n=-\infty}^{\infty\,\prime}\left[\frac{1-(-1)^n}{(n\pi)^2}\right]\mathcal{O}_W\left\{\sum_{k=0}^{j-1} s(t)^k p_0 s(t)^{j-k-1}\right\}. \end{aligned} \tag{7.74}$$

Also by differentiating (7.68) we find

$$\begin{aligned} i\hbar\left(\frac{dx(t)}{dt}\right) &= i\hbar\frac{p(t)}{m} = -i\hbar\sum_{n=-\infty}^{\infty\,\prime}\left[\frac{1-(-1)^n}{(n\pi)^2}\right]\sum_{j=0}^{\infty}\left[\frac{(in\pi)^j}{j!}\right] \\ &\times\ \mathcal{O}_W\left\{\sum_{k=0}^{j-1} s(t)^k p_0 s(t)^{j-k-1}\right\}. \end{aligned} \tag{7.75}$$

Now if we compare (7.74) and (7.75) we obtain (7.70). Similarly we can verify the correctness of commutation relation (7.69).

Let us also investigate the validity of (7.71) for the rate of change of momentum of the particle. If we multiply (7.75) by m and differentiate it with respect to t we have

$$\begin{aligned} \left(\frac{dp(t)}{dt}\right) &= \sum_{n=-\infty}^{\infty\,\prime}\left[1-(-1)^n\right]\mathcal{O}_W\left\{\frac{\exp(in\pi s(t))p_0^2}{2m}\right\} \\ &+ \frac{\hbar}{m}\sum_{n=-\infty}^{\infty\,\prime}\left[1-(-1)^n\right](n\pi)\mathcal{O}_W\left\{\exp(in\pi s(t))p_0\right\} \\ &+ \frac{\hbar^2}{4m}\sum_{n=-\infty}^{\infty\,\prime}\left[1-(-1)^n\right](n\pi)^2\mathcal{O}_W\left\{\exp(in\pi s(t))\right\}. \end{aligned} \tag{7.76}$$

Now we define $\Theta(s(t))$ by

$$\Theta(s(t)) = -i\mathcal{O}_W \left\{ \sum_{n=-\infty}^{\infty}{}' \left[\frac{1-(-1)^n}{(n\pi)} \right] \exp(in\pi s(t)) \right\}, \tag{7.77}$$

and we observe that

$$\Theta(s(0)) = \Upsilon(x_0)\Upsilon(1-x_0), \quad 0 \le x_0 \le 1, \tag{7.78}$$

where $\Upsilon(x_0)$ is given by

$$\Upsilon(x_0) = \begin{cases} -1 & \text{for } x_0 < 0 \\ 0 & \text{for } x_0 = 0 \\ +1 & \text{for } x_0 > 0 \end{cases}. \tag{7.79}$$

By differentiating $\Theta(s(t))$ with respect to x_0 we obtain

$$\begin{aligned} \frac{d\Theta(s(t))}{dx_0} &= \Delta(s(t)) = \Delta(x_0, p_0, t) \\ &= 2\sum_{j=-\infty}^{\infty} \{\delta[s(t) - 2j] - \delta[(2j+1) - s(t)]\}. \end{aligned} \tag{7.80}$$

Thus the initial value of $\frac{dp(t)}{dt}$, Eq. (7.76), can be written as

$$\begin{aligned} &\left(\frac{dp(t)}{dt}\right)_{t=0} \\ &= -\frac{1}{m}\left[\frac{1}{2}\Delta(t=0)p_0^2 - i\hbar\left(\frac{d\Delta(t=0)}{dx_0}\right)p_0 - \frac{\hbar^2}{4}\left(\frac{d^2\Delta(t=0)}{d\,x_0^2}\right)\right]. \end{aligned} \tag{7.81}$$

As we note here the initial force has terms proportional to $\hbar$ and $\hbar^2$ and the same is true about $\left(\frac{dp(t)}{dt}\right)$. So we arrive at the interesting result that the impulsive forces in quantum mechanics have terms not found in their classical counterpart.

Let us note that while in classical dynamics most of the forces of constraint can be imposed on the motion without difficulty in quantum theory this is not the case and it is not possible to idealize stiff potentials by constraints [15].

7.6 Motion of a Wave Packet

Having determined the time dependence of the position and momentum operators for a particle in a box, we can now study the time development of the wave packet associated with this motion.

Let $\Psi(x)$ represents a narrow wave packet which is localized within the box. We want to calculate the expectation value of $x(t)$ and $p(t)$ with this wave packet, i.e. $\langle\Psi(x)|x(t)|\Psi(x)\rangle$ and $\langle\Psi(x)|p(t)|\Psi(x)\rangle$. It is convenient to expand $\Psi(x)$ in terms of the Fourier series (or the complete set of eigenfunctions) $\psi_n(x) = \sqrt{2}\sin(n\pi x)$

$$\Psi(x) = \sqrt{2}\sum_{n=0}^{\infty} C_n \sin(n\pi x), \tag{7.82}$$

and then find $\langle n|x(t)|j\rangle$. For this we first calculate

$$I(x,t) = \exp\left(\frac{ik\pi p_0 t}{m}\right)\sin(n\pi x), \tag{7.83}$$

where k is an integer. By expanding the exponential and separating the result into odd and even states we obtain

$$I(x,t) = \cos\left(\frac{\hbar t k n\pi^2}{m}\right)\sin(n\pi x) + \sin\left(\frac{\hbar t k n\pi^2}{m}\right)\cos(n\pi x). \tag{7.84}$$

Using this result we calculate $\langle j|x(t)|n\rangle$,

$$\begin{aligned}\langle j|x(t)|n\rangle &= \frac{1}{2}\delta_{j,n} - 2\sum_{k=-\infty}^{\infty}{}' \left(\frac{1-(-1)^k}{\pi^2 k^2}\right)\exp\left(\frac{i\hbar t k^2\pi^2}{2m}\right)\\ \times & \int_0^1 e^{ik\pi x}\left\{\cos\left(\frac{\hbar t k n\pi^2}{m}\right)\sin(j\pi x)\sin(n\pi x)\right.\\ + & \left.\sin\left(\frac{\hbar t k n\pi^2}{m}\right)\sin(j\pi x)\cos(n\pi x)\right\}dx.\end{aligned} \tag{7.85}$$

The integration in (7.85) can be carried out, and with further simplification we obtain

$$\begin{aligned}\langle j|x(t)|n\rangle &= \frac{1}{2}\delta_{j,n} + \left(\frac{1-(-1)^{j+n}}{\pi^2}\right)\\ \times & \left[\frac{1}{(j+n)^2} - \frac{1}{(j-n)^2}\right]\exp\left[\frac{i\hbar t\left(j^2-n^2\right)\pi^2}{2m}\right].\end{aligned} \tag{7.86}$$

From this expression we can determine the position of the center of the wave packet $\Psi(x)$ as a function of time;

$$\langle x(t)\rangle = \frac{1}{2}\sum_{n=0}^{\infty}\sum_{j=0}^{\infty} C_n C_j \left(\langle j|x(t)|n\rangle + \langle n|x(t)|j\rangle\right). \tag{7.87}$$

The velocity of the center of wave packet can be found by differentiating $\langle x(t)\rangle$;

$$\frac{1}{m}\langle p(t)\rangle = \frac{d}{dt}\langle x(t)\rangle. \tag{7.88}$$

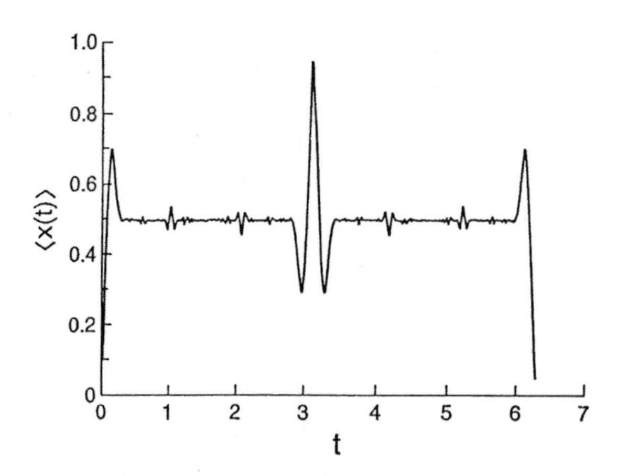

Figure 7.1: The motion of the center of wave packet as a function of time (N=10). This time is measured in units of $\left(\frac{2m}{\pi\hbar}\right)$.

To illustrate this type of bouncing motion we choose the wave packet

$$\Psi(x) = \begin{cases} \sqrt{2N}\sin(N\pi x) & 0 \le x \le \frac{1}{N} \\ 0 & x > \frac{1}{N} \end{cases}, \tag{7.89}$$

and from it we can find C_n and thus calculate $\langle n|x(t)|j\rangle$. The results for $\langle x(t)\rangle$ and for $\frac{d}{dt}\langle x(t)\rangle$ are shown in Figs. 7.1 and 7.2. In Fig. 7.2 we observe how the direction of velocity changes upon the collision with the well. This motion is periodic with a period of $\frac{4m}{\pi\hbar}$ independent of N.

The correspondence between the classical periodicity and the quantum mechanical periodicity for this motion has been discussed at length by Styer [16].

As we mentioned earlier, there are different self-adjoint extensions of this problem. The case where the particle strikes one well and appears at the other is such an example. Again this is a periodic motion and we can write the classical position of the particle as a Fourier series;

$$x_c(t) == \frac{1}{2} - \frac{1}{2\pi i}\sum_{n=-\infty}^{\infty\,\prime}\left(\frac{1}{n}\right)\exp[2in\pi s(t)]. \tag{7.90}$$

We note that this classical motion is discontinuous at $s(t) = 1, 2, \cdots$ and velocity does not change sign.

Using the Weyl–McCoy rule of ordering discussed earlier we find the quantum operator $x(t)$ as

$$\begin{aligned} x(t) &= \frac{1}{2} - \frac{1}{2\pi i}\sum_{n=-\infty}^{\infty\,\prime}\left(\frac{1}{n}\right)\exp\left[\left(-\frac{i\hbar}{2}\right)\frac{\partial^2}{\partial x_0 \partial p_0}\right] \\ &\times \left\{\exp(2\pi i n\pi x_0)\exp\left(\frac{inp_0 t}{m}\right)\right\}. \end{aligned} \tag{7.91}$$

7.7 Heisenberg's and Newton's Equations of Motion

We have already seen that Heisenberg's Correspondence principle indicates a connection between the Fourier transform of the classical solution and the quantal matrix elements for large quantum numbers Sec. 4.5. We also know of the formal similarity between the equations of motion of Heisenberg and of Newton.

This formal similarity for a one-dimensional motion of a particle of mass m subject to a force $F(x)$ can be seen in the equation of motion

$$m\frac{d^2x}{d\,t^2} = F(x). \tag{7.92}$$

But whereas in classical dynamics x is a c-number in quantum mechanics it is a matrix, or an operator. In this section we want to show that a solution of Newton's equation of motion can be regarded as the generating function for

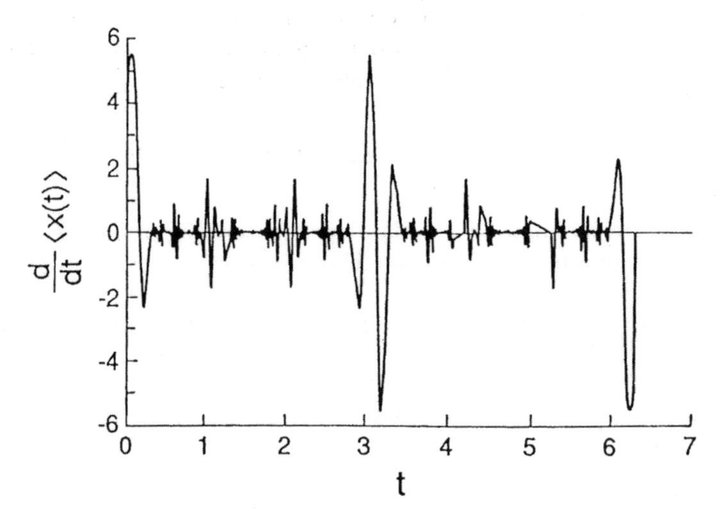

Figure 7.2: The velocity of the center of wave packet as a function of time (N=10). The velocity is in units of $\left(\frac{\pi\hbar}{2m}\right)$.

the matrix elements of the coordinate operator in matrix mechanics [9]. As we have seen in Klein's approach, the solution of the quantum mechanical problem amounts to the determination of a pair of matrices p and x such that

(a) - the commutation relations are satisfied, and

(b) - the Hamiltonian operator written in terms of p and x is diagonal.

Here again we will consider polynomial potentials and as in Klein's approach we make the following assumptions:

(1) - That the matrix elements of p and x decrease rapidly as the distance from the main diagonal increases. Let us note that for a simple harmonic oscillator the elements p_{kj} and x_{kj} become zero for $|k-j|>1$.

(2) - That the matrix elements of p and x vary slowly as we move parallel to the principal diagonal of each of the matrices. Again for harmonic oscillator the change from one element of the matrix, (k,j), to the next is slow and the ratio of these two elements is given by $\left(\frac{k+j+1}{k+j}\right)$. With the help of these assumptions we can simplify the matrix product considerably.

If we denote the energies of the k and j states by E_k and E_j respectively, then the time derivative of the matrix element x_{kj} is given by

$$m\left(\frac{d^2x}{d\,t^2}\right)_{kj} = -m\left(\frac{E_k-E_j}{\hbar}\right)^2 x_{kj}. \tag{7.93}$$

By substituting (7.93) in (7.92) we find

$$-m\left(\frac{E_k-E_j}{\hbar}\right)^2 x_{kj} = [F(x)]_{kj}\,. \tag{7.94}$$

This is a set of coupled nonlinear equations for the matrix element x_{kj}. For a polynomial potential, $F(x)$ is also a polynomial, and if we assume that $F(x) = x^n$ then the right hand side of (7.94) can be written as

$$(x^n)_{jk} = \sum_{l_1}\sum_{l_2}\cdots\sum_{l_{n-1}} x_{jl_1}x_{l_1l_2}\cdots x_{l_{n-1}k}. \tag{7.95}$$

Now according to assumption (1) the dependence of x_{jk} on j and k can be written in terms of a function ξ in the following way:

$$x_{kj} = [\xi(j+k)]_{j-k} + [\xi(j+k)]_{k-j} \approx \xi_{j-k}+\xi_{k-j} = \sum_r \xi_r\left(\delta_{k,j+r}+\delta_{k,j-r}\right). \tag{7.96}$$

If we substitute (7.96) in (7.95) we find

$$(x^n)_{kj} = \sum_r\left(\delta_{k,j+r}+\delta_{k,j-r}\right)\sum_{l_1l_2\cdots l_n}\left(\xi_{l_1}\xi_{l_2}\cdots\xi_{l_n}\right)\delta_{r,\,\pm l_1\pm l_2\cdots l_n}, \tag{7.97}$$

where the sum is over all r and over 2^n possible permutations of signs. Equation (7.97) is not exact and there are terms which we have omitted [9]. We can write

(7.97) in the simpler form of

$$[F(x)]_{jk} = (x^n)_{jk} = \sum_r \beta_r \left(\delta_{k,\, j+r} + \delta_{k,\, j-r}\right), \tag{7.98}$$

where β_r denotes the sum over l_i s in (7.97).

Next we introduce a function $g(\theta)$ by

$$g(\theta) = \sum_r \xi_r \left(e^{ir\theta} + e^{-ir\theta}\right). \tag{7.99}$$

By comparing (7.96), (7.98) and (7.99) we find the following relation

$$F[g(\theta)] = \sum_r \beta_r \left(e^{ir\theta} + e^{-ir\theta}\right). \tag{7.100}$$

Note that the energy difference $E_k - E_j$ depends on $(k-j)$ as well as $(k+j)$. If we write $E_k - E_j$ as

$$(E_k - E_j) = (k-j)\Delta E(k,j), \tag{7.101}$$

and for the moment assume that ΔE is independent of k and j then (7.94) becomes

$$-m(k-j)^2 \left(\frac{\Delta E}{\hbar}\right)^2 x_{kj} = [F(x)]_{kj}. \tag{7.102}$$

In the next step we replace x_{kj}, $[F(x)]_{kj}$ and $(k-j)$ in terms of ξ_r, β_r and r^2;

$$-mr^2 \left(\frac{\Delta E}{\hbar}\right)^2 \xi_r = \beta_r. \tag{7.103}$$

By multiplying (7.103) by $(e^{ir\theta} + e^{-ir\theta})$ and summing over r we find that $g(\theta)$ is the solution of the differental equation

$$m\left(\frac{\Delta E}{\hbar}\right)^2 \frac{d^2 g(\theta)}{d\,\theta^2} = F(g). \tag{7.104}$$

Changing θ to t where

$$t = \left(\frac{\hbar}{\Delta E}\right)\theta, \tag{7.105}$$

we obtain Newton's equation of motion

$$m\frac{d^2 g}{d\,t^2} = F(g). \tag{7.106}$$

This result shows that the solution of the classical equation of motion, $g(\theta)$, if expanded as a Fourier series, Eq. (7.99), yields the coefficients ξ_r and these are related to the matrix elements x_{kj} by Eq. (7.96) (see also Heisenberg's

Correspondence principle Sec. 4.6. Of course we arrive at this result if the approximations that we have made are valid. In particular the assumption that $\Delta E(k, j)$ in (7.101) is independent of k and j is crucial one in this derivation.

If we multiply (7.106) by $\frac{dg}{dt}$ and integrate we obtain the law of conservation of energy;

$$\frac{m}{2}\left(\frac{dg(t)}{dt}\right)^2 + V[g(t)] = \frac{m}{2}\left(\frac{dg(t)}{dt}\right)^2_{t=0} + V[g(0)] = H(t) = H(0). \quad (7.107)$$

Since $H(t) = H(0)$, the Fourier coefficient of $H(t)$ is $H(0)$ and this according to the above argument should be regarded as the matrix element of H, i.e.

$$H_{kj} = H(0)\delta_{kj}. \quad (7.108)$$

Hence the classical conservation of energy is equivalent to the diagonalization of the Hamiltonian [9] (See Chapter 3 and also Sec. 12.8).

While the method of calculating the Fourier coefficients is applicable to the equation of motion, it cannot be used in the case of the commutators, since we are dealing with the small differences of large numbers. If we want to approximate the commutation relation we sum the matrix elements in the following way:

We start with the commutator

$$[H,\, [H,\, x]\,] = \frac{\hbar}{mi}\,[p,\, x] = -\frac{\hbar^2}{2m}. \quad (7.109)$$

The diagonal elements of (7.109) are given by

$$\sum_k (E_k - E_j)(x_{kj})^2 = \frac{\hbar^2}{2m}, \quad j = 1,\, 2, \cdots, \quad (7.110)$$

where k is a nonnegative integer. We rewrite (7.110) in the form

$$\sum_{r\geq -j} (E_{j+r} - E_j)(x_{j+r,j})^2 = \frac{\hbar^2}{2m}, \quad j = 1,\, 2, \cdots. \quad (7.111)$$

In (7.111) all the terms with $r > 0$ are positive and for $r < 0$ are negative. If we sum the two sides of (7.111) from $j = 0$ to $j = J$, then some of the negative terms cancel some of the positive terms. Let us consider this part of the problem in detail.

If S_1 and S_2 represent the partial sums

$$S_1 = \sum_{j=0}^{J}\sum_{r=1}^{\infty}(E_{j+r} - E_j)(x_{j+r,j})^2 > 0, \quad (7.112)$$

$$S_2 = \sum_{j=0}^{J}\sum_{r=-j}^{0}(E_{j+r} - E_j)(x_{j+r,j})^2 < 0, \quad (7.113)$$

so that from (7.111) we find

$$S_1 + S_2 = \frac{\hbar^2}{2m}\sum_{j+1}^{J} 1 = \frac{\hbar^2}{2m}(J+1). \tag{7.114}$$

We change the sign of r in S_2 and write it as

$$S_2 = \sum_{j=0}^{J}\sum_{r=0}^{j}(E_{j-r} - E_j)(x_{j,j-r})^2, \tag{7.115}$$

and we also interchange the j and r sums and use symmetry of x_{kj} to get

$$S_2 = \sum_{r=0}^{J}\sum_{j=r}^{J}(E_{j-r} - E_j)(x_{r-j,j})^2. \tag{7.116}$$

Now in this expression the sum over j runs from $j = 0$ to $J - r$, thus

$$S_2 = -\sum_{r=0}^{J}\sum_{j=0}^{J-r}(E_{j+r} - E_j)(x_{j,j+r})^2. \tag{7.117}$$

By adding S_1 and S_2 we find

$$S_1 + S_2 = \sum_{j=0}^{J}\sum_{r=J-j+1}^{\infty}(E_{j+r} - E_j)(x_{j,j+r})^2 = \frac{\hbar^2}{2m}(J+1). \tag{7.118}$$

The approximate method that we discussed earlier if applied to Eq. (7.117) yields the following result

$$\sum_{j=0}^{J}\sum_{r=J-j+1}^{\infty} r\Delta E\xi_r^2 = \left(\frac{\hbar^2}{2m}\right)(J+1). \tag{7.119}$$

By inverting the order of summation in (7.119) we obtain

$$\sum_{r=0}^{\infty} r\Delta E\xi_r^2 \sum_{j=J-r+1}^{J} 1 = \frac{\hbar^2}{2m}(J+1), \tag{7.120}$$

or by carrying out the second sum we find

$$(\Delta E)\sum_{r} r^2\xi_r^2 = \frac{\hbar^2}{2m}(J+1). \tag{7.121}$$

Let us note that (7.121) implies that:

(a) - ΔE be independent of J, the index parallel to the diagonal and

(b) - because of the dependence of ΔE on J ξ_r s also pick up some of the

previously neglected matrix elements.
Now for a given $F(x)$ we solve (7.106) and determine $g(t)$. Then using (7.105) we write the solution as $g(\theta)$ and from the Fourier coefficients of $g(\theta)$ we find ξ_r according to Eq. (7.99). Finally by substituting ξ_r in (7.121) we find ΔE.

Next let us consider two examples where it is easy to apply this method:
(1) - First let us solve the problem of simple the harmonic oscillator

$$\frac{d^2 x}{d\,t^2} + \omega^2 x = 0. \tag{7.122}$$

using this method. According (7.106) $g(t)$ satisfies (7.122), i.e.

$$g(t) = \xi_1 \left(e^{i\omega t} + e^{-i\omega t}\right), \tag{7.123}$$

or

$$g(\theta) = \xi_1 \left[\exp\left(\frac{i\hbar\omega}{\Delta E}\theta\right) + \exp\left(-\frac{i\hbar\omega}{\Delta E}\theta\right)\right]. \tag{7.124}$$

Thus in the Fourier expansion of $g(\theta)$ there is one coefficient ξ_1 and the matrix elements of x according to (7.96) are

$$\xi_1 = x_{j,j+1} = x_{j+1,j}. \tag{7.125}$$

From the commutation relation (7.121) we find ξ_1^2 to be

$$\xi_1^2 = \frac{\hbar^2}{2m\Delta E}(j+1). \tag{7.126}$$

Now by comparing (7.99) and (7.124) we obtain

$$\Delta E = \hbar\omega, \tag{7.127}$$

and therefore

$$\xi_1^2 = \frac{\hbar}{2m\omega}(j+1). \tag{7.128}$$

This result together with (7.125) gives us the matrix elements of the harmonic oscillator.
(2) - As a second example let us study the motion of a particle bouncing between two rigid walls separated by a unit length (Sec. 7.5). The classical motion $x_c(t)$ can be written as a Fourier series over positive n values;

$$x_c(t) = g(\theta)$$
$$= \frac{2}{\pi^2}\sum_{n=0}^{\infty}\frac{1}{(2n+1)^2}\left[\exp\left\{i(2n+1)\frac{np_0\hbar}{m\Delta E}\theta\right\} + \exp\left\{-i(2n+1)\frac{np_0\hbar}{m\Delta E}\theta\right\}\right]. \tag{7.129}$$

Comparing this relation with Eq. (7.99) we obtain ξ_n;

$$\xi_n = \frac{2}{\pi^2(2n+1)^2}. \tag{7.130}$$

Also from the commutation relation (7.121) we have

$$\sum_n (2n+1)^2 \left[\frac{2}{\pi^2(2n+1)^2}\right]^2 = \frac{\hbar^2}{2m\Delta E}(j+1). \tag{7.131}$$

By carrying out the summation in (7.131) we get the simple relation

$$\left(\frac{2}{\pi^2}\right)^2 \frac{\pi^2}{8} = \frac{\hbar^2}{2m\Delta E}(j+1), \tag{7.132}$$

from which we conclude that

$$\Delta E = \frac{\pi^2\hbar^2}{4m}(j+1). \tag{7.133}$$

Thus the matrix elements of x and p are given by

$$x_{j,j+2n} = p_{j,j+2n} = 0, \tag{7.134}$$

$$x_{j,j+2n+1} = \frac{2}{\pi^2(2n+1)^2}, \tag{7.135}$$

and

$$p_{j,j+2n+1} = \frac{2i\hbar}{(2n+1)}(j+1). \tag{7.136}$$

The last relation is found by calculating $\left(\frac{dx}{dt}\right)_{j,j+2n+1}$;

$$\begin{aligned} p_{j,j+2n+1} &= -\frac{mi}{\hbar}\left(E_j - E_{j+2n+1}\right) x_{j,j+2n+1} = \frac{mi}{\hbar}(2n+1)\Delta E\, x_{j,j+2n+1} \\ &= \frac{2i\hbar(j+1)}{(2n+1)}. \end{aligned} \tag{7.137}$$

For large integer j these approximate results agree with the exact result found from Eqs. (7.86) and (7.88) for $|x_{j,j+2n+1}|$ and $|p_{j,j+2n+1}|$:

$$|x_{j,j+2n+1}| = \frac{2}{\pi^2}\left[\frac{1}{(2n+1)^2} - \frac{1}{(2j+2n+1)^2}\right], \tag{7.138}$$

and

$$|p_{j,j+2n+1}| = 4i\hbar\left[\frac{j(j+2n+1)}{(2j+2n+1)(2j+1)}\right]. \tag{7.139}$$

This method can be generalized to systems with N degrees of freedom, but the results are complicated [9],[17].

Nonclassical Forces — The quantization of the motion of a particle bouncing between two rigid walls shows us that there are forces at the boundaries which are absent in the classical problem and that these forces are proportional to $\hbar$ and $\hbar^2$. Similar nonclassical forces appear in the description of the quantum mechanics of a free particle in D dimensions [18]. The number of dimensions are one, two or three i.e. the dimensions for a single particle in the coordinate space. It can also be considered as a hyperspace of the quantum state of a single hypothetical particle having the configuration space of two or more non-interacting (say N) particles in one, two or three dimensions. If d denotes the dimensions of the space, then $D = Nd$. Here for the sake of simplicity we consider only S wave for the state of the hypothetical particle, i.e. we assume that the dynamical variable is $\hat{r}$, where

$$\hat{r} = \sqrt{\sum_{n=1}^{N} x_n^2}. \tag{7.140}$$

We will consider the non-relativistic motion of a free particle (or a hypothetical particle) in D dimensions for which the Hamiltonian is just the kinetic energy

$$H = \frac{\mathbf{p}_D^2}{m} = -\frac{\hbar^2}{2m}\nabla_D^2, \tag{7.141}$$

where m denotes the mass and $\mathbf{p}_D$ the momentum of the particle. The Heisenberg equations of motion for the coordinate operator $\hat{r}$ defined by (7.140) is

$$m\frac{d\hat{r}}{dt} = -\frac{i\hbar}{2}\left[\nabla_D^2, \hat{r}\right]. \tag{7.142}$$

In order to investigate the behavior of the derivatives of r at the origin, we replace r by

$$r_\epsilon = \sqrt{\epsilon^2 + r^2}, \tag{7.143}$$

and at the end we let ϵ go to zero.

By calculating the commutator in (7.142) we find

$$\begin{aligned} m\frac{d\hat{r}}{dt} &= -i\hbar\left(\frac{D-1}{2r} + \frac{\partial}{\partial r}\right) - i\hbar\mathcal{D}(\hat{r}) \\ &= -i\hbar\frac{1}{r^{\frac{D-1}{2}}}\frac{\partial}{\partial r}r^{\frac{D-1}{2}} - i\hbar\mathcal{D}(\hat{r}), \end{aligned} \tag{7.144}$$

where

$$\mathcal{D}(r) = \lim_{\epsilon\to 0}\frac{1}{2}\left[\frac{\epsilon^2}{(\epsilon^2 + r^2)^{\frac{3}{2}}}\right]. \tag{7.145}$$

In the same way we calculate the second derivative of $\hat{r}$;

$$\frac{d^2\hat{r}}{dt^2} = -\frac{1}{\hbar^2}[H, [H, \hat{r}]] = -\frac{\hbar^2}{4m^2}[\nabla_D^2, [\nabla_D^2, \hat{r}]]. \tag{7.146}$$

From this relation we can find the forces acting on any partial wave ℓ

$$m\frac{d^2\hat{r}}{dt^2} = -\frac{\partial V(\hat{r})}{\partial \hat{r}} + F_q(\hat{r}) + F_\ell(\hat{r}), \tag{7.147}$$

where

$$V(r) = \frac{\hbar^2}{2m}\frac{(D-1)(D-3)}{4r^2}, \tag{7.148}$$

$$F_q(\hat{r}) = -\frac{\hbar^2}{2m}\left\{2\frac{D-1}{r}\frac{\partial \mathcal{D}}{\partial r} + \frac{\partial^2 \mathcal{D}}{\partial \hat{r}^2} + \frac{4}{r^{D-1}}\frac{\partial}{\partial r}\left(r^{D-1}\mathcal{D}\frac{\partial}{\partial r}\right)\right\}, \tag{7.149}$$

and

$$F_\ell(r) = \lim_{\epsilon\to 0}\left\{\frac{1}{m}\frac{1}{\sqrt{\epsilon^2+r^2}}\frac{\mathbf{L}^2}{r^2}\right\}. \tag{7.150}$$

In this last expression $\mathbf{L}$ is the angular momentum operator (Sec. 9.1). For the s wave only the last term, $F_\ell(r)$, vanishes. From Eq. (7.147) it is clear that for any partial wave there are three components to this nonclassical force. The sum of the first two components $F_\ell(r) - \frac{\partial V(r)}{\partial r}$ is nonzero for $r \neq 0$. The force $-\frac{\partial V(r)}{\partial r}$ vanishes for one- or three-dimensional spaces and is negative for $D = 2$, and $F_\ell(r)$ is the usual centrifugal force and the term. The nonclassical force $F_q(\hat{r})$ depends on the behavior of the wave function and its derivatives at the origin and as can be seen from the coefficient of $\frac{\hbar^2}{2m}$ in (7.149) it has a purely quantum-mechanical origin.

Bibliography

[1] C.-T. Li, A. Klein and F. Krejs, Matrix mechanics as a practical tool in quantum theory: The anharmonic oscillator, Phys. Rev. D 12, 2311 (1975).

[2] A. Klein and H.A. Weldon, Equation of motion, variational principles, and WKB approximations in quantum mechanics and quantum field theory, Phys. Rev. D 17, 1009 (1978).

[3] A. Klein and C.-T. Li, Variational principles and the Heisenberg matrix mechanics, Physica, 96 A, 254 (1979).

[4] A. Klein, C.-T. Li and M. Vassanji, Variational principles for particles and fields in Heisenberg matrix mechanics, J. Math. Phys. 21, 2521 (1980).

[5] W.R. Greenberg, A. Klein and C.-T. Li, The invariant tori and Heisenberg matrix mechanics- A new window on the quantum-classical correspondence, Phys. Rev. Lett. 75, 1244 (1995).

[6] W.R. Greenberg, A. Klein and C.-T. Li, The invariant tori and the Heisenberg matrix mechanics: A new window on the quantum-classical correspondence, Phys. Rep. 264, 167 (1996).

[7] C.-T. Li and A. Klein, Some structural and numerical aspects of Heisenberg matrix mechanics with applications to one-dimensional systems, Fizika, 22,67 (1990).

[8] C.-T. Li, G.H. Suen and A. Klein, Some structural and numerical aspects of Heisenberg matrix mechanics with applications to nonpolynomial potentials, Phys. Rev. A 63, 052110 (2001).

[9] F.R. Halpern, Relation between the solution of Newton's and Heisenberg's equations of motion, Phys. Rev. D 18, 1894 (1978).

[10] L.D. Landau and E.M. Lifshitz, *Quantum Mechanics, Non-Relativistic Theory*, (Pergamon Press, London, 1958), footnote on page 214.

[11] M. Razavy, Energy levels and matrix elements of a double-well potential calculated from the Heisenberg equations of motion, Nuovo Cimento. 116B, 317 (2001).

[12] R.R. Chasman, Eigenvalue problem in matrix mechanics, J. Math. Phys. 2, 733 (1961).

[13] A.S. Wightman, in *Cargese Lectures in Theoretical Physics*, edited by M. Levy, (Gordon and Birch, New York, 1964), p. 262.

[14] W.A. Atkinson and M. Razavy, Impulsive forces and the Heisenberg equations of motion for a particle in a box, Can. J. Phys. 71, 380 (1993).

[15] N.G. van Kampen and J.J. Lodder, Constraints, Am. J. Phys. 52, 419 (1984).

[16] D.F. Styer, Quantum revivals versus classical periodicity in an infinite square well, Am. J. Phys. 69, 56 (2001).

[17] F.R. Halpern, General techniques for single and coupled quantum anharmonic oscillators, J. Math. Phys. 16, 2469 (1975).

[18] I. Bialynicki-Birula, M.A. Cirone, J.P. Dahl, T.H. Seligman, F. Straub and W.P. Schleich, Quantum fictitious forces, Fortschr. Phys. 50, 599 (2002).

Chapter 8

Exactly Solvable Potentials, Supersymmetry and Shape Invariance

The aim of this chapter is to give an introduction to a group of solvable potential mostly found using the factorization method. This method introduced earlier to solve the harmonic oscillator problem. For these solvable problems we can determine all of the energy eigenvalues, and these in turn can be used to obtain the time development of any dynamical variable that we wish to study. We will also consider the theory of supersymmetry applied to one-dimensional quantum mechanics and its relation to the shape invariant potentials.

Time Dependence of the Operators — The time dependence of any operator A which does not depend explicitly on time can be expressed in terms of the energy eigenvalues and the matrix elements of A at a given time, e.g. at $t = 0$. Our starting point is the Heisenberg equation

$$i\hbar\frac{dA}{dt} = [A, H], \tag{8.1}$$

where we have assumed that that $\frac{\partial H}{\partial t} = 0$ (see Eq. (4.11)). This equation can be solved for $A(t)$ if the initial operator $A(0)$ is known

$$A(t) = \exp\left(-\frac{iHt}{\hbar}\right) A(0) \exp\left(\frac{iHt}{\hbar}\right), \tag{8.2}$$

For instance to find the matrix $q(t)$ as a function of time, we substitute $q(0)$ for $A(0)$ in (8.2) and determine $q(t)$. This can be easily done if we use a representation in which H is diagonal. Thus the matrix elements of $A(t)$ obtained from

Eq. (8.2) are

$$\langle j|A(t)|k\rangle = \left\langle j\left|e^{\frac{-iHt}{\hbar}}A(0)e^{\frac{iHt}{\hbar}}\right|k\right\rangle = \exp\left[-\frac{i}{\hbar}(E_j - E_k)t\right]\langle j|(A(0)|k\rangle. \quad (8.3)$$

Once we have found $\langle j|A(0)|k\rangle$, E_k and E_j, then we have the complete solution of the problem.

8.1 Energy Spectrum of the Two-Dimensional Harmonic Oscillator

As we noted in the classical discussion of this problem there is a conserved symmetric tensor T_{ij} given by (1.198). The corresponding Heisenberg operators for T_{ij} are found by replacing p_i and q_i by the corresponding operators satisfying the canonical commutation relations. Here we will write down the operators corresponding to the group $\{G_i\}$, Eqs. (1.194)–(1.196).

$$G_1 = -\frac{i\hbar}{2}\left(q_1\frac{\partial}{\partial q_2} - q_2\frac{\partial}{\partial q_1}\right), \quad (8.4)$$

$$G_2 = \frac{1}{\omega}\left(-\frac{\hbar^2}{2m}\frac{\partial^2}{\partial q_1 \partial q_2} + \frac{1}{2}m\omega^2 q_1 q_2\right), \quad (8.5)$$

and

$$G_3 = -\frac{1}{2\omega}\left[\left(-\frac{\hbar^2}{2m}\frac{\partial^2}{\partial q_1^2} + \frac{1}{2}m\omega^2 q_1^2\right) - \left(-\frac{\hbar^2}{2m}\frac{\partial^2}{\partial q_2^2} + \frac{1}{2}m\omega^2 q_2^2\right)\right]. \quad (8.6)$$

These operators satisfy the commutation rules for the three-dimensional rotation group

$$[G_i, G_j] = -i\hbar\sum_k \varepsilon_{ijk}G_k. \quad (8.7)$$

We can also relate these G_i s to the Hamiltonian operator by

$$H^2 - \frac{\hbar^2\omega^2}{2} = 2\omega^2\left(G_1^2 + G_2^2\right) + \frac{1}{4}\left[(H + 2\omega G_3)^2 + (H - 2\omega G_3)^2\right] \quad (8.8)$$

or

$$H^2 = \hbar^2\omega^2 + 4\omega^2\left(G_1^2 + G_2^2 + G_3^2\right). \quad (8.9)$$

The eigenvalues of $\mathbf{G}^2$ are identical to those of $\mathbf{L}^2$, i.e. $\hbar^2 n(n+1)$, (Sec. 9.2), and therefore the eigenvalues of H^2 are

$$\left\langle n\left|H^2\right|n\right\rangle = \hbar^2\omega^2 + 4\omega^2 n(n+1) = \hbar^2\omega^2(2n+1)^2, \quad n = 0, 1, \cdots. \quad (8.10)$$

(The connection between the eigenvalues of the two-dimensional isotropic harmonic oscillator and the orbital angular momentum will be discussed in Sec. 9.5). From this we conclude that the matrix H is a diagonal matrix with the elements

$$\langle n|H|n\rangle = \hbar\omega(2n+1), \quad n = 0, 1, \cdots, \tag{8.11}$$

and that these are the energy levels of an isotropic two-dimensional oscillator.

8.2 Exactly Solvable Potentials Obtained from Heisenberg's Equation

Soon after the discovery of wave mechanics there were attempts to find exact solutions for the Schrödinger equation with different potential functions. Apart from the well-known problems of the harmonic oscillator and the hydrogen atom (or Kepler problem), other potentials for specific physical situations were found for which the eigenvalues and the wave function could be obtained exactly. The first requirement for exact solvability is to find a curvilinear coordinate system in which the wave equation can be separated into a number of ordinary differential equations. For velocity independent forces, we know that there are eleven coordinate systems were the Laplacian is separable [1]. From these eleven coordinate systems, three are very important in quantum mechanics. Cartesian system for the problems of harmonic oscillator and free particle, spherical polar coordinates for a large number of problems having spherical symmetry, and parabolic coordinates for solving the Stark effect and also scattering by the Coulomb force. By separating the variables in these systems we obtain a set of three ordinary differential equations each of the the second order. We call these problems solvable if all three equations can be solved analytically.

Once the ordinary differential equation is obtained, then with the choice of proper potential functions, we can find the eigenfunctions and the eigenvalues. Among these potentials we have the Eckart potential [2] for quantum tunneling, the Morse potential [3] for molecular vibration and the Kronig-Penny potential for the motion of an electron in a crystal lattice [4]. A very general approach to determine the form of exactly potentials for the Schrödinger equation was found by Manning [5]. Recent works on the role of supersymmetry in quantum field theory has renewed interest in solvable potentials, particularly those that can be found by factorization method.

To obtain solvable potentials for one-dimensional motion we start with the basic equations, viz, the canonical commutation relation (7.20) and the equation of motion (7.6), and for simplicity we take the mass of particle m as the unit of mass. When written in the representation in which the Hamiltonian H is diagonal these equations become

$$\sum_k (2E_k - E_n - E_j) \langle n|x|k\rangle\langle k|x|j\rangle = \delta_{nj}, \tag{8.12}$$

and

$$(E_j - E_n)^2 \langle n|x|j\rangle = \langle n|V'|j\rangle. \tag{8.13}$$

If the potential is not a polynomial then in general we want to change the position coordinate from x to a function of x, say $f(x)$. When this is done then the equation for the commutator

$$[x\,[x, H]] = -1 \tag{8.14}$$

changes to

$$[\,[f(x), H], f(x)] = (f'(x))^2 \tag{8.15}$$

and the equation of motion (7.6) becomes

$$[\,[f(x), H], H] = -\left(f''(x)H + Hf''(x)\right) + 2f''(x)V(x) + f'(x)V'(x) - \frac{1}{4}f^{(4)}(x). \tag{8.16}$$

Here primes denote derivatives with respect to x and $f^{(4)}(x)$ denotes the fourth derivative of $f(x)$.

We can find a set of exactly solvable potentials if the double commutator (8.16) reduces to a linear expression in $f(x)$. For this reduction we set [6]

$$2f''(x)V(x) + f'(x)V'(x) = \alpha f(x) + \beta, \tag{8.17}$$

and

$$f''(x) = \mu f(x) + \nu, \tag{8.18}$$

where α, β, μ and ν are all constants. Equation (8.18) is a condition that we impose on $f(x)$ so that $f^{(4)}(x)$ can be absorbed in $f''(x)$. Once these equations are satisfied then Eq. (8.17) becomes a differential equation for $V(x)$. By integrating (8.17) and (8.18) we find that

$$V(x) = \frac{1}{2\left(f'(x)\right)^2}\left(\alpha f^2(x) + 2\beta f(x) + \gamma\right), \tag{8.19}$$

with

$$\left(f'(x)\right)^2 = \mu f^2(x) + 2\nu f(x) + \sigma. \tag{8.20}$$

In these relations γ and σ are the integration constants. By substituting for $V(x)$, $f'(x), f''(x)$ and $f^{(4)}(x)$ in (8.16) we obtain the following expression for the double commutator which is linear in $f(x)$;

$$[\,[f(x), H], H] = \left(\alpha - \frac{1}{4}\mu^2\right) f(x) - \mu[f(x)H + Hf(x)] - 2\nu H + \beta - \frac{1}{4}\mu\nu. \tag{8.21}$$

Similarly we need the double commutator $[\,[f(x),H],f(x)]$ for the determination of the solvable potentials. This can be derived from the Hamiltonian

$$H=\frac{1}{2}p^2+\frac{1}{2\left(f'(x)\right)^2}\left(\alpha f^2(x)+2\beta f(x)+\gamma\right), \tag{8.22}$$

and is given by

$$[\,[f(x),H],f(x)]=\left(f'(x)\right)^2=\mu f^2(x)+2\nu f(x)+\sigma. \tag{8.23}$$

To obtain the general form of the solvable potentials we first integrate (8.23) and solve for $f(x)$ to get

$$f(x)=\frac{1}{4\mu}e^{\sqrt{\mu}\,(x+c)}+\left(\frac{\nu^2}{\mu}-\sigma\right)e^{-\sqrt{\mu}\,(x+c)}-\frac{\nu}{\mu}. \tag{8.24}$$

Substituting this result in (8.22) we find the Hamiltonian and the potential.

Some of the exactly solvable potentials obtained in this way are as follows [6]:

(1) - If we choose $\mu=\sigma=0, \nu=2$ in (8.20) and $c=0$, then $f(x)=x^2$. Now with $\alpha=4\omega^2$, $\beta=\gamma=0$ we find the harmonic oscillator Hamiltonian.

(2) - With the same $f(x)$ and α as in (1) but with $\gamma=4\hbar^2\ell(\ell+1)$ we have

$$V(x)=\frac{1}{2}\omega^2x^2+\frac{\hbar^2\ell(\ell+1)}{2x^2},\quad \hbar=1, \tag{8.25}$$

which is the effective potential for a three-dimensional isotropic harmonic oscillator.

We can determine the matrix elements of the function $f(x)$ and the eigenvalues of the Hamiltonian (8.22) when the potential is given by the general form (8.19). Here in addition to the commutators (8.22) and (8.23) we need the following anticommutator

$$\begin{aligned}\left[f'(x)^2,H\right]_+ &= [f(x),H][H,f(x)]-\frac{1}{2}f'(x)f'''(x)\\ &- \frac{3}{4}\left(f''(x)\right)^2+2\left(f'(x)\right)^2V(x)\\ &= [f(x),H][H,f(x)]+\left(\alpha-\frac{5}{4}\mu^2\right)f^2(x)\\ &+ 2\left(\beta-\frac{5}{4}\mu\nu\right)f(x)+\gamma-\frac{3}{4}\nu^2-\frac{1}{2}\mu\sigma.\end{aligned} \tag{8.26}$$

Now we calculate the matrix elements of (8.21) between the states $\langle n|$ and $|n'\rangle$, i.e. between the eigenstates of the Hamiltonian (8.22)

$$\begin{aligned}&\left[(E_{n'}-E_n)^2+\mu(E_{n'}+E_n)-\left(\alpha-\frac{1}{4}\mu^2\right)\right]f_{n,n'}\\ &= \delta_{n,n'}\left(\beta-\frac{1}{4}\mu\nu-2\nu E_n\right).\end{aligned} \tag{8.27}$$

Because (8.21) is linear in $f(x)$, as in the case of harmonic oscillator, the non-vanishing matrix elements are the diagonal elements, $n = n'$, and those of neighboring states $n' = n \pm 1$. From (8.27) it follows that for $n' = n + 1$ we have

$$(E_{n+1} - E_n)^2 + \mu(E_{n+1} + E_n) = \alpha - \frac{1}{4}\mu^2, \tag{8.28}$$

which gives us a recursion relation between the eigenvalues, but not each individual eigenvalue. In addition the diagonal element, $f_{n,n}$, is given in terms of E_n by

$$f_{n,n} = \frac{\beta - \frac{1}{4}\mu\nu - 2\nu E_n}{2\mu E_n - \alpha + \frac{1}{4}\mu^2}. \tag{8.29}$$

Noting that only $f_{n,n}$ and $f_{n,n\pm 1}$ are nonzero, from the matrix elements of the commutator (8.23) we find that

$$[2(E_{n+1} - E_n) - \mu]\, f_{n,n+1}^2 = [2(E_n - E_{n-1}) + \mu]\, f_{n-1,n}^2 + \mu f_{n,n}^2 + 2\nu f_{n,n} + \sigma. \tag{8.30}$$

For the determination of E_n s we also need the diagonal elements of (8.26), viz,

$$\begin{aligned} & 2E_n\left[\mu f_{n,n}^2 + 2\nu f_{n,n} + \sigma\right] \\ = \; & (E_{n+1} - E_n)^2 f_{n,n+1}^2 + (E_n - E_{n-1})^2 f_{n-1,n}^2 + \left(\alpha - \frac{5}{4}\mu^2\right) f_{n,n}^2 \\ + \; & 2\left(\beta - \frac{5}{4}\mu\nu\right) f_{n,n} + \gamma - \frac{3}{4}\nu^2 - \frac{1}{2}\mu\sigma, \end{aligned} \tag{8.31}$$

where for $(f'(x))^2$ we have substituted from (8.20) and where

$$\left(f^2\right)_{n,n} = \langle n\,|f^2|\,n\rangle = f_{n,n+1}^2 + f_{n,n-1}^2 + f_{n,n}^2. \tag{8.32}$$

By eliminating $f_{n,n}, f_{n,n\pm1}, E_{n-1}$ and E_{n+1} between equations (8.28)-(8.31) we find the energy eigenvalue E_n given in terms of the parameters of the potential;

$$E_n = -\frac{1}{2\mu}\left\{\Gamma(\alpha,\beta,\gamma,\mu,\nu,\sigma) - \left(n + \frac{1}{2}\right)^2 \mu\right\}^2 + \frac{\alpha}{2\mu}. \tag{8.33}$$

In this expression n is an integer which for $\mu > 0$ must be between zero and $\left(\frac{\Gamma}{\mu} - \frac{1}{2}\right)$, and $\Gamma(\alpha,\beta,\gamma,\mu,\nu,\sigma)$ is a function of the parameters of the potential [6]

$$\begin{aligned} & \Gamma(\alpha,\beta,\gamma,\mu,\nu,\sigma) \\ = \; & \left\{\frac{1}{2}\left(\alpha + \frac{\mu^2}{4} - \frac{\mu\gamma'}{\sigma'}\right) \pm \frac{1}{2}\left[\left(\alpha + \frac{\mu^2}{4} - \frac{\mu\gamma'}{\sigma'}\right)^2 + \frac{4\mu\left(\beta'\right)^2}{\sigma'}\right]^{\frac{1}{2}}\right\}^{\frac{1}{2}}, \end{aligned} \tag{8.34}$$

with β', γ' and σ' given by

$$\beta' = \beta - \frac{\alpha\nu}{\mu}, \tag{8.35}$$

$$\gamma' = \gamma - \frac{2\beta\nu}{\mu} + \frac{\alpha\nu^2}{\mu^2}, \tag{8.36}$$

and

$$\sigma' = \sigma - \frac{\nu^2}{\mu}. \tag{8.37}$$

Next we choose the parameters of $V(x)$ in (8.19) as follows:

$$\alpha = \nu = \sigma = 0, \quad \text{and} \quad \gamma = \frac{\lambda}{8\mu}, \quad \beta = -\frac{\lambda\mu}{4}, \tag{8.38}$$

then (8.19) reduces to the Morse potential [3]

$$V(x) = \lambda\left(e^{-2\sqrt{\mu}\,(x+c)} - 2e^{-\sqrt{\mu}\,(x+c)}\right). \tag{8.39}$$

Since we want to take the limit of $\sigma \to 0$, we choose the positive square root of the term inside the curly bracket in (8.34) to get a real result. In the limit of $\sigma \to 0$ and $\nu \to 0$, we obtain $\Gamma = \sqrt{2\lambda\mu}$ and then from (8.33) for the bound states of the Morse potential we find the eigenvalues

$$E_n = -\lambda + \sqrt{2\lambda\mu}\left(n + \frac{1}{2}\right) - \frac{1}{2}\mu\left(n + \frac{1}{2}\right)^2, \quad n = 0, 1, 2 \cdots. \tag{8.40}$$

If we want to write these eigenvalues for a particle of mass m then E_n becomes

$$E_n = -\lambda + \sqrt{\frac{2\lambda\mu}{m}}\left(n + \frac{1}{2}\right) - \frac{1}{2m}\mu\left(n + \frac{1}{2}\right)^2, \quad n = 0, 1, 2 \cdots. \tag{8.41}$$

We can also find the eigenvalues of the Pöschl-Teller potential as a special case of (8.19). To get this potential we choose the following set of parameters

$$c = \alpha = \beta = 0, \quad \frac{1}{4\mu} = \left(\frac{\nu^2}{4\mu} - \sigma\right), \quad \gamma = -\frac{\lambda}{8\mu}, \tag{8.42}$$

and with these parameters the potential $V(x)$ becomes

$$V(x) = \frac{-\lambda\mu}{\cosh^2(\sqrt{\mu}x)}. \tag{8.43}$$

The eigenvalues calculated from (8.33) are given by

$$E_n = -\frac{\mu}{2}\left[\sqrt{2\lambda + \frac{1}{4}} - \left(n + \frac{1}{2}\right)\right]^2, \quad n = 0, 1, \cdots, N. \tag{8.44}$$

where N is an integer smaller but closest to $\left(\sqrt{2\lambda + \frac{1}{4}} - \frac{1}{2}\right)$.

8.3 Creation and Annihilation Operators

The assumption that we made to simplify Eq. (8.16), i.e. those given by Eqs. (8.17) and (8.18), can be slightly generalized and be written in terms of time-dependent creation and annihilation operators. This generalization also helps us to relate the present method to the factorization technique which will be discussed later in this chapter.

As we argued earlier we want to reduce (8.16) to a relation where the double commutator becomes a linear function of $f(x)$. The most general form for the right side of (8.16) in this case is [7],[8];

$$[H,\ [H,\ f(x)]] = f(x)R_0(H) + [H,\ f(x)]R_1(H) + R_{-1}(H), \tag{8.45}$$

where R_0, R_1 and R_{-1} are all functions of the Hamiltonian. These functions are related to the parameters of the potential $V(x)$, Eq. (8.19), by

$$R_0(H) = -2\mu H + \left(\alpha - \frac{\mu^2}{4}\right), \tag{8.46}$$

$$R_1(H) = -\mu, \tag{8.47}$$

and

$$R_{-1}(H) = -2\nu H + \left(\beta - \frac{\mu\nu}{4}\right). \tag{8.48}$$

Since (8.45) is also linear in $f(x)$, we expect the potential to be of the general form given by (8.19). However now we are interested in the time evolution of $f(x)$ and in the time-dependent operators corresponding to the creation and annihilation operators. To this end we introduce the notation

$$\underbrace{[H,\ [H,\ [H,\ \cdots,\ [H,}_{n} f(x)]\cdots]\,] = (\mathcal{C}H)^n f(x), \tag{8.49}$$

and consider the Heisenberg equation

$$\begin{aligned} e^{iHt}f(x)e^{-iHt} &= \sum_{n=0}^{\infty}\frac{(it)^n}{n!}(\mathcal{C}H)^n f(x) \\ &= [H,\ f(x)]\left(\frac{e^{i\alpha_+(H)t} - e^{i\alpha_-(H)t}}{\alpha_+(H)-\alpha_-(H)}\right) - \frac{R_{-1}(H)}{R_0(H)} \\ &+ \left(f(x) + \frac{R_{-1}(H)}{R_0(H)}\right)\left(\frac{\alpha_+(H)e^{i\alpha_-(H)t} - \alpha_-(H)e^{i\alpha_+(H)t}}{\alpha_+(H)-\alpha_-(H)}\right). \end{aligned} \tag{8.50}$$

In this relation $\alpha_+(H)$ and $\alpha_-(H)$ are frequency like quantities and are given by

$$\alpha_\pm(H) = \frac{1}{2}\left\{R_1(H) \pm \sqrt{(R_1(H))^2 + 4R_0(H)}\right\}. \tag{8.51}$$

As the right-hand side of Eq. (8.50) has two different time-dependent factors and also a constant term, the left-hand side can have three nonvanishing matrix elements. Therefore

$$\left\langle \phi_j \left| e^{iHt} f(x) e^{-iHt} \right| \phi_n \right\rangle, \tag{8.52}$$

is zero unless $j = n$ or $j = n \pm 1$. Hence the eigenvalues must satisfy the conditions

$$E_{n+1} - E_n = \alpha_+(E_n), \tag{8.53}$$

and

$$E_{n-1} - E_n = \alpha_-(E_n). \tag{8.54}$$

These equations will give us the same eigenvalues as (8.28). If we add a constant number to the potential so that the ground state energy becomes zero, $E_0 = 0$, then we can determine $E_1,\ E_2, \cdots$ from Eq. (8.53).

Now let us write (8.50) in terms of positive and negative frequency parts;

$$\begin{aligned} e^{iHt} f(x) e^{-iHt} &= a^+(H, f(x)) e^{i\alpha_+(H)t} \\ &+ a^-(H, f(x)) e^{i\alpha_-(H)t} - \frac{R_{-1}(H)}{R_0(H)}, \end{aligned} \tag{8.55}$$

where the creation and annihilation operators which we denote by $a^+(H, f(x))$ and $a^-(H, f(x))$ are defined by

$$a^+(H, f(x)) = \frac{[H,\ f(x)] - \left(f(x) + \frac{R_{-1}(H)}{R_0(H)}\right)\alpha_-(H)}{\alpha_+(H) - \alpha_-(H)}, \tag{8.56}$$

and

$$a^-(H, f(x)) = \frac{-[H,\ f(x)] + \left(f(x) + \frac{R_{-1}(H)}{R_0(H)}\right)\alpha_+(H)}{\alpha_+(H) - \alpha_-(H)}. \tag{8.57}$$

The action of annihilation operator on $\phi_n(x)$ gives us

$$\begin{aligned} a^-(H, f(x))\phi_n(x) &= \frac{-1}{E_{n+1} - E_{n-1}} \\ &\times \left\{[H,\ f(x)] + (E_n - E_{n+1}) f(x) + \frac{R_{-1}(H)}{E_{n-1} - E_n}\right\}\phi_n(x). \end{aligned} \tag{8.58}$$

Similarly for the creation operator we find

$$\begin{aligned} a^+(H, f(x))\phi_n(x) &= \frac{1}{E_{n+1} - E_{n-1}} \\ &\times \left\{[H,\ f(x)] + (E_n - E_{n-1}) f(x) + \frac{R_{-1}(H)}{E_{n+1} - E_n}\right\}\phi_n(x). \end{aligned} \tag{8.59}$$

In particular for $n = 0$ and $E_0 = 0$, therefore the action of $a^-(H, f(x))$ on ϕ_0 should yield zero,

$$-[H,\ f(x)]\phi_0(x) + \left(f(x)\alpha_+(0) - \frac{R_{-1}(0)}{\alpha_-(0)}\right)\phi_0(x) = 0. \tag{8.60}$$

This is the equation which defines the ground state wave function up to a normalization constant.

A Simple Example — Consider the motion of a particle of unit mass in the Pöschl–Teller potential

$$V(x) = \frac{g(g-1)}{2\sin^2 x} - \frac{g^2}{2}, \quad 0 < x < \pi \tag{8.61}$$

This potential like (8.43) is a special case of the general form given by (8.19). It is found by setting $\alpha = \beta = 0, \gamma = g(g-1)$ and $f(x) = \cos x$ in (8.19), and then adding the constant term $-\frac{g^2}{2}$ to it. The commutators $[H,\ f(x)]$ and $[H,\ [H,\ f(x)]]$ for this potential are

$$[H,\ f(x)] = i\sin xp + \frac{1}{2}\cos x, \tag{8.62}$$

and

$$[H,\ [H,\ f(x)]] = \cos x\left(2H + g^2 - \frac{1}{4}\right) + [H,\ \cos x]. \tag{8.63}$$

Comparing (8.63) with (8.45) we find

$$R_0(H) = \left(2H + g^2 - \frac{1}{4}\right), \tag{8.64}$$

$$R_1(H) = 1, \tag{8.65}$$

$$R_{-1}(H) = 0. \tag{8.66}$$

Substituting these in (8.51) we obtain $\alpha_\pm(H)$;

$$\alpha_\pm(H) = \frac{1}{2} \pm \sqrt{2H + g^2}. \tag{8.67}$$

From Eqs. (8.56), (8.57), (8.62) and (8.67) we find the creation and annihilation operators $a^\pm$;

$$\begin{aligned} a'^{\pm} &= a^\pm\left(2\sqrt{2H+g^2}\right) = \pm i\sin x\, p + \cos x\sqrt{2H+g^2} \\ &= \cos x\sqrt{2H+g^2} \pm \sin x\frac{d}{dx}. \end{aligned} \tag{8.68}$$

We note that a'^{+} is not the Hermitian adjoint of a'^{-} and for this reason we have not used $a^\dagger$ and a to denote these operators.

The eigenvalues for this problem can be found from Eqs. (8.54) and (8.67) and remembering that we have set $E_0 = 0$. The recurrence relation (8.54) gives us

$$E_n = n\left(\frac{n}{2} + g\right), \quad n = 0,\ 1,\ 2\cdots. \tag{8.69}$$

Now let us consider the action of the $a'^{\,-}$ and $a'^{\,+}$ on $\phi_n(x)$;

$$\begin{aligned} a'^{\,-}\phi_n(x) &= -\sin x\frac{d\phi_n(x)}{dx} + (n+g)\cos x\,\phi_n(x) \\ &= \left(n + g - \frac{1}{2}\right)\phi_{n-1}, \end{aligned} \tag{8.70}$$

and

$$\begin{aligned} a'^{\,+}\phi_n(x) &= \sin x\frac{d\phi_n(x)}{dx} + (n+g)\cos x\,\phi_n(x) \\ &= \frac{2(n+1)(n+2g)}{2n+2g+1}\phi_{n+1}. \end{aligned} \tag{8.71}$$

These recursion relations determine the wave function to within a normalization constant,

$$\phi_n(x) = (\sin x)^g P_n^{\left(g-\frac{1}{2},\, g-\frac{1}{2}\right)}(\cos x), \tag{8.72}$$

where $P_n^{\beta,\beta}$ is the Jacobi polynomial [9].

8.4 Determination of the Eigenvalues by Factorization Method

Let A be a Hermitian operator which is independent of time, and we want to find its eigenvalues. For the ground state we write $A = A_1$ and we express A_1 as

$$A_1 = a_1^\dagger a_1 + \lambda_1, \tag{8.73}$$

where $a_1^\dagger$ is the Hermitian conjugate of a_1 and λ_1 is a number multiplies by a unit operator. If we can factorize A_1 in more than one way, then we choose the way which gives us the largest λ_1.

Eigenvalues and Eigenfunctions of a Simple Harmonic Oscillator — The simplest problem that we can solve by factorization method is that of the harmonic oscillator. Let us start with the simple example where

$$A = H = \frac{1}{2m}\left(p^2 + m^2\omega^2 q^2\right), \tag{8.74}$$

that is the problem of simple harmonic oscillator. In this case we write (8.73) as

$$\begin{aligned} A_1 &= \frac{1}{2m}\left(p^2 + m^2\omega^2 q^2\right) = \frac{1}{2m}(p + im\omega q)(p - im\omega q) - \frac{i\omega}{2}(qp - pq) \\ &= \frac{1}{2m}(p + im\omega q)(p - im\omega q) + \frac{\hbar\omega}{2}, \end{aligned} \tag{8.75}$$

i.e. we have chosen

$$a_1^\dagger = \frac{1}{\sqrt{2}}(p + im\omega q), \quad \text{and} \quad a_1 = \frac{1}{\sqrt{2}}(p - im\omega q). \tag{8.76}$$

Next we define A_2 by

$$A_2 = a_1 a_1^\dagger + \lambda_1, \tag{8.77}$$

and try to write it as

$$A_2 = a_2^\dagger a_2 + \lambda_2, \tag{8.78}$$

where λ_2 is again a number. Using the definitions of a_1 and $a_1^\dagger$ we can write A_2 as

$$\begin{aligned} A_2 &= \frac{1}{2m}(p - im\omega q)(p + im\omega q) + \frac{\hbar\omega}{2} \\ &= \frac{1}{2m}(p + im\omega q)(p - im\omega q) + \frac{3\hbar\omega}{2}. \end{aligned} \tag{8.79}$$

This process can be continued and in this way we can determine all of the eigenvalues of H, viz,

$$\lambda_n = \left(n + \frac{1}{2}\right)\hbar\omega. \tag{8.80}$$

Let us denote the eigenvector of A by $|\psi\rangle$ and its corresponding eigenvalue by λ, i.e. $A|\psi\rangle = \lambda|\psi\rangle$, and define a state $|\phi_n\rangle$ by

$$|\phi_n\rangle = a_n a_{n-1} \cdots a_2 a_1 |\psi\rangle. \tag{8.81}$$

Then

$$\langle\phi_1|\phi_1\rangle = \left\langle \psi \left| a_1^\dagger a_1 \right| \psi \right\rangle = (\lambda - \lambda_1). \tag{8.82}$$

The left-hand side of (8.82) is positive definite, therefore $\lambda - \lambda_1 \geq 0$, so λ cannot be less than λ_1. From the way that we have defined different A_j s it follows that

$$A_{j+1} a_j = \left(a_j a_j^\dagger + \lambda_j\right) a_j = a_j \left(a_j^\dagger a_j + \lambda_j\right) = a_j A_j. \tag{8.83}$$

Using this result and the definition of $|\phi_2\rangle$ we have

$$\begin{aligned} \langle\phi_2|\phi_2\rangle &= \left\langle \psi \left| a_1^\dagger a_2^\dagger a_2 a_1 \right| \psi \right\rangle = \left\langle \psi \left| a_1^\dagger (A_2 - \lambda_2) a_1 \right| \psi \right\rangle \\ &= \left\langle \psi \left| a_1^\dagger a_1 (A_1 - \lambda_2) \right| \psi \right\rangle = (\lambda - \lambda_2)\langle\phi_1|\phi_1\rangle = (\lambda - \lambda_2)(\lambda - \lambda_1). \end{aligned} \tag{8.84}$$

Again this relation shows that $\lambda \geq \lambda_2$ unless $\lambda = \lambda_1$. Continuing this process we find

$$(\lambda - \lambda_n)(\lambda - \lambda_{n-1}) \cdots (\lambda - \lambda_1) \geq 0. \tag{8.85}$$

This relation indicates that either $\lambda \geq \lambda_n$ or otherwise we have the relation

$$(\lambda - \lambda_{n-1}) \cdots (\lambda - \lambda_1) = 0. \tag{8.86}$$

Thus λ must be equal to one of the λ_j s or be greater than all of them. Noting that $\lambda_{j+1} \geq \lambda_j$, if the sequence $\{\lambda_j\}$ is unbound then λ must coincide with one of the λ_j s. On the other hand if the sequence $\{\lambda_j\}$ is bound, and if λ^{max} is its upper bound, then λ can have one of the values λ_j or an unrestricted value not less than λ^{max} [19].

Eigenfunctions of the Simple Harmonic Oscillator — We have already discussed the problem of determination of the eigenvalues for the harmonic oscillator using the factorization method. In order to obtain the eigenfunctions for this problem we start with the ground state of the system. This state is defined as the eigenfunction of the annihilation operator

$$a_1\psi_0(q) = \frac{1}{\sqrt{2}}\left(-i\hbar\frac{d}{dq} - im\omega q\right)\psi_0(q) = 0. \tag{8.87}$$

Solving the differential equation (8.87) we have

$$\psi_0(q) = N_0 \exp\left(-\frac{1}{2}\beta^2 q^2\right), \tag{8.88}$$

where $\beta = \sqrt{\frac{m\omega}{\hbar}}$. The factor N_0 is the normalization constant obtained from

$$\int_{-\infty}^{\infty} |\psi_0(q)|^2 dq = 1, \tag{8.89}$$

and is given by $N_0 = \left(\frac{m\omega}{\pi\hbar}\right)^{\frac{1}{4}}$. The eigenfunction for the first excited state is found by applying $a_2 = a_1^\dagger$ to $\psi_0(q)$;

$$\psi_1(q) = a_2\psi_0(q) = \frac{1}{\sqrt{2}}\left(-i\hbar\frac{d}{dq} + im\omega q\right)\psi_0(q). \tag{8.90}$$

This equation gives us $|\psi_1\rangle$, which after being normalized can be written as

$$\psi_1(q) = \sqrt{\frac{2m\omega}{\hbar}}\left(\frac{m\omega}{\pi\hbar}\right)^{\frac{1}{4}} q\exp\left(-\frac{1}{2}\beta^2 q^2\right) = \sqrt{\frac{2}{\pi}}\beta^{-\frac{3}{2}} q\exp\left(-\frac{1}{2}\beta^2 q^2\right). \tag{8.91}$$

We can continue this process and find the n-th excited state wave function

$$\psi_n(q) = N_n H_n(\alpha q)\exp\left(-\frac{1}{2}\beta^2 q^2\right), \tag{8.92}$$

where $H_n(\beta q)$ is the Hermite polynomial of degree n and N_n, the normalization constant is

$$N_n = \frac{1}{\sqrt{2^n n!}}\left(\frac{m\omega}{\pi\hbar}\right)^{\frac{1}{4}}. \tag{8.93}$$

Finally we can find the time-dependence of the operators $a_1(t)$ and $a_1^\dagger(t)$ by differentiating them with respect to time. Then using the Heisenberg equations of motion

$$\dot{p}(t) = m\omega q(t), \quad \dot{q}(t) = \frac{1}{m}p(t), \tag{8.94}$$

we obtain

$$\dot{a}_1(t) = \frac{1}{\sqrt{2}}(\dot{p}(t) - im\omega\dot{q}(t)) = -i\omega a_1(t). \quad (8.95)$$

Thus the time-dependence of $a_1(t)$ and $a_1^\dagger(t)$ are given by

$$a_1(t) = a_1(0)e^{-i\omega t}, \quad (8.96)$$

and

$$a_1^\dagger(t) = a_1^\dagger(0)e^{i\omega t}, \quad (8.97)$$

Eigenvalues for the Motion of a Particle in a Box — As a second example let us calculate the energy eigenvalues for a particle trapped in a box of unit length $-\frac{1}{2} \le x \le \frac{1}{2}$. For this case the Hamiltonian is simple $H = \frac{p^2}{2m}$, but we must find suitable raising and lowering operators $a_j^\dagger$ and a_j so that we can satisfy the boundary conditions. Let us define a_j by

$$a_j = \frac{1}{\sqrt{2m}}[p + i\gamma_j \tan(k_j x)]. \quad (8.98)$$

Since the motion of the particle is confined to $-\frac{1}{2} \le x \le \frac{1}{2}$, a_j can become infinite only at the boundaries, and thus $|k_j| \le \pi$.

According to our formulation $a_1^\dagger a_1$, apart from the number λ_1, must be the Hamiltonian, Eq. (8.73). But first we calculate $a_j^\dagger a_j$ as well as $a_j a_j^\dagger$;

$$\begin{aligned} a_j^\dagger a_j &= \frac{1}{2m}\left\{p^2 + \gamma_j^2 \tan^2(k_j x) + i\gamma_j[p,\ \tan(k_j x)]\right\} \\ &= \frac{1}{2m}\left\{p^2 + \gamma_j^2 \tan^2(k_j x) + \hbar k_j \gamma_j \sec^2(k_j x)\right\} \\ &= \frac{1}{2m}\left\{p^2 + \hbar k_j \gamma_j + \gamma_j(\gamma_j + \hbar k_j)\tan^2(k_j x)\right\}, \end{aligned} \quad (8.99)$$

and

$$a_j a_j^\dagger = \frac{1}{2m}\left\{p^2 - \hbar k_j \gamma_j + \gamma_j(\gamma_j - \hbar k_j)\tan^2(k_j x)\right\}. \quad (8.100)$$

For $j = 1$, Eq. (8.99) should reduce to the Hamiltonian plus a number c_1, and for this to be true we must have

$$\gamma_1(\gamma_1 + \hbar k_1) = 0, \quad (8.101)$$

and

$$\gamma_1 \hbar k_1 + \lambda_1 = 0. \quad (8.102)$$

Solving these equations for γ_1 and λ_1 we find

$$\gamma_1 = -\hbar k_1, \quad \text{and} \quad \lambda_1 = -\gamma_1 \hbar k_1 = (\hbar k_1)^2. \quad (8.103)$$

The maximum value of λ_1 is found when k_1 assumes its maximum value which is π, and thus the ground state energy is $\lambda_1 = (\pi\hbar)^2$.

To calculate the energy of other states we use the relation

$$a_{j+1}^\dagger a_{j+1} + \lambda_{j+1} = a_j a_j^\dagger + \lambda_j, \quad (8.104)$$

which shows that

$$k_{j+1} = k_j = \cdots = k_1 = \pi, \tag{8.105}$$

and

$$\gamma_{j+1}(\gamma_{j+1} + \hbar k_{j+1}) = \gamma_j(\gamma_j - \hbar k_j). \tag{8.106}$$

Thus the eigenvalues satisfy the relation

$$\lambda_{j+1} + \gamma_{j+1}\hbar k_{j+1} = \lambda_j - \gamma_j \hbar k_j, \tag{8.107}$$

or

$$\lambda_{j+1} - \gamma_{j+1}^2 = \lambda_j - \gamma_j^2 = \cdots = \lambda_1 - \gamma_1^2 = 0. \tag{8.108}$$

Equation (8.108) shows us that $\lambda_j = \gamma_j^2$ and also from (8.106)

$$\gamma_{j+1}(\gamma_{j+1} + \pi\hbar) = \gamma_j(\gamma_j - \pi\hbar). \tag{8.109}$$

Solving (8.109) for γ_{j+1} we find that either $\gamma_{j+1} = -\gamma_j$ or

$$\gamma_{j+1} = \gamma_j - \hbar\pi = \cdots = \gamma_1 - j\hbar\pi = -(j+1)\hbar\pi. \tag{8.110}$$

The first solution, i.e. $\gamma_{j+1} = -\gamma_j$ is unacceptable and thus we get

$$\lambda_j = \gamma_j^2 = j^2\hbar^2\pi^2. \tag{8.111}$$

To obtain the wave function we start with the ground state. Since this is the lowest state and we cannot produce any lower state, therefore $a_1|\phi_0\rangle = 0$, where $|\phi_0\rangle$ denotes the ground sate. To find the wave function in coordinate space we use a_1 as defined by (8.98) and write

$$\left[-i\hbar\frac{d}{dx} - i\hbar\pi\tan(\pi x)\right]\phi_0(x) = 0, \tag{8.112}$$

where in (8.112) we have substituted for λ and k_1 from (8.105) and (8.110). Integrating (8.112) we find

$$\phi_0(x) = N_0\cos(\pi x), \tag{8.113}$$

where N_0 is the normalization constant.

Wave functions for the excited states can be found by noting that if we write

$$\phi_j(x) = \cos^j(\pi x)\phi_0(x), \tag{8.114}$$

then $a_j\phi_{j-1}(x) = 0$, and with the help of this relation we can find other wave functions, $\psi_j(x)$, $j = 1,\ 2\cdots$ from (8.81) or from

$$\psi_j(x) = a_1^* a_2^* \cdots a_{j-1}^* \phi_{j-1}(x). \tag{8.115}$$

8.5 A General Method for Factorization

The method that we outlined in the last section to obtain the eigenvalues and eigenfunctions can be extended to other solvable potentials [10]–[15].

Let us consider a self-adjoint operator H depending on a parameter j. We want to determine the eigenvalues of H which is of the form $[\lambda - L(j)]$;

$$H|\lambda, j\rangle = [\lambda - L(j)]|\lambda, j\rangle. \tag{8.116}$$

We assume that there exists a pair of mutually adjoint operators $\beta^+(j)$ and $\beta^-(j)$ such that

$$\left[H, \beta^+(j)\right] = H\beta^+(j) - \beta^+(j)H = \beta^+(j), \tag{8.117}$$

and

$$\left[H, \beta^-(j)\right] = H\beta^-(j) - \beta^-(j)H = -\beta^-(j). \tag{8.118}$$

From Eqs. (8.116)–(8.118) we find

$$H\left(\beta^+(j)|\lambda, j\rangle\right) = [\lambda - L(j+1)]\left(\beta^+(j)|\lambda, j\rangle\right), \tag{8.119}$$

and

$$H\left(\beta^-(j)|\lambda, j\rangle\right) = [\lambda - L(j-1)]\left(\beta^-(j)|\lambda, j\rangle\right). \tag{8.120}$$

We may interpret these results as

$$\beta^+(j)|\lambda, j\rangle \sim |\lambda, j+1\rangle, \tag{8.121}$$

and

$$\beta^-(j)|\lambda, j\rangle \sim |\lambda, j-1\rangle, \tag{8.122}$$

i.e. $\beta^+(j)$ and $\beta^-(j)$ are the ladder operators changing the eigenstates of H upwards or downwards. We note that these ladder operators are not self-adjoint operators.

For local potentials the Hamiltonian operator is a quadratic function of the momentum of the particle and the corresponding wave equation is a second order differential equation. In this method of factorization we write the Hamiltonian as the product of two operators, each linear in momentum, one being a raising and the other a lowering operator. The two example that we have seen earlier illustrates how this technique can be utilized to obtain the eigenvalues and the eigenfunctions of a number of problems.

For exactly solvable potentials it is more convenient to work with real raising and lowering operators. To this end we introduce $\beta^+ = ia$, and $\beta^- = -ia^\dagger$ and we write the Hamiltonian in the simple form of

$$H = -\frac{d^2}{d\,q^2} + V(q, j), \tag{8.123}$$

where we have used the units $\hbar = 2m = 1$. The constant j which is appearing in the potential is assumed to be an integer. Under certain conditions to be discussed later, Eq. (8.123), can be factorized and can be written as

$$H = \beta^-(j)\beta^+(j), \tag{8.124}$$

or as

$$H = \beta^+(j+1)\beta^-(j+1), \tag{8.125}$$

where β^+ and β^- are raising and lowering operators. They are defined by

$$\beta^\pm(j) = W(q,j) \pm \frac{d}{dq}. \tag{8.126}$$

The function $W(q,j)$ which depends on the form of the potential will be determined later.

Denoting the eigenfunction of H corresponding to the eigenvalue λ by $|\lambda, q\rangle$ we want $\beta^+(j)$ and $\beta^-(j)$ to have the properties discussed earlier, (8.121) and (8.122), and are properly normalized,

$$|\lambda, j+1\rangle = \beta^-|\lambda, j\rangle, \tag{8.127}$$

$$|\lambda, j-1\rangle = \beta^+|\lambda, j\rangle. \tag{8.128}$$

These relations show that if $|\lambda, j\rangle$ is known we can find two other eigenfunctions $|\lambda, j+1\rangle$ and $|\lambda, j-1\rangle$, with the help of Eqs. (8.127) and (8.128). Repeating this process we can determine a ladder of solutions for a fixed λ. Since H defined by Eq. (8.116) can be expressed either as (8.124) or (8.125), therefore we have

$$\beta^-(j)\beta^+(j)|\lambda, j\rangle = [\lambda - L(j)]|\lambda, j\rangle, \tag{8.129}$$

or

$$\beta^+(j+1)\beta^-(j+1)|\lambda, j\rangle = [\lambda - L(j+1)]|\lambda, j\rangle. \tag{8.130}$$

Substituting for $\beta^-(j)$, $\beta^+(j)$, $\beta^+(j+1)$ and $\beta^-(j+1)$ from (8.126) and subtracting (8.129) from (8.130) we find

$$W^2(q,j+1) - W^2(q,j) + \frac{dW(q,j+1)}{dq} + \frac{dW(q,j)}{dq} = [L(j) - L(j+1)]. \tag{8.131}$$

This condition is both necessary and sufficient for the factorization of the Hamiltonian (8.123).

Next we will consider the relationship between the two functions $W(q,j)$ and $L(j)$ and the potential function $V(q,j)$ in the Hamiltonian. This is found by substituting for $\beta^-(j)$ and $\beta^+(j)$ or for $\beta^-(j+1)$ and $\beta^+(j+1)$ from (8.126) and comparing the result with (8.123);

$$\begin{aligned} V(q,j) &= W^2(q,j) - \frac{d}{dq}W(q,j) + L(j) \\ &= W^2(q,j+1) + \frac{d}{dq}W(q,j+1) + L(j+1). \end{aligned} \tag{8.132}$$

Thus for a given $V(q,j)$ we can determine $W(q,j)$ and $L(j)$ by solving (8.132), which means solving a Riccati equation for $W(q,j)$ (or $W(q,j+1)$). If the ground state wave function for the Hamiltonian (8.123), $\phi_0(q,j)$, is known then $W(q,j)$ can be obtained from

$$W(q,j) = -\frac{d}{dq}\ln\phi_0(q,j). \tag{8.133}$$

Note that in (8.133), $\phi_0(q,j)$ has no nodes, therefore $W(q,j)$ is well-defined. A more convenient way is to assume a simple dependence of $W(q,j)$ on j (or on q) and then find $V(q,j)$ from (8.132). Using this method we can obtain a table of $V(q,j)$ s and the corresponding $W(q,j)$ s and $L(j)$ s. Let us consider the application of the factorization method in some detail:

(i) - The simplest case is to assume that $W(q,j)$ is independent of q, then $L(j) = -W^2(j)$ and $V(q,j)$ will be zero. This corresponds to the motion of a free particle where the wave function is a trigonometric function.

(ii) - The next simple case is the one in which $W(q,j)$ is a linear function of j;

$$W(q,j) = v(q) + jw(q). \tag{8.134}$$

By substituting (8.134) in (8.131) we observe that for $L(j)$ to be independent of q, $v(q)$ and $w(q)$ must satisfy the following relations:

$$\frac{dw(q)}{dq} + w^2(q) = A, \tag{8.135}$$

and

$$\frac{dv(q)}{dq} + v(q)w(q) = B, \tag{8.136}$$

where A and B are constants.

(iii) - The third possible choice of $W(q,j)$ for which makes $L(j)$ independent of q is when $W(q,j)$ is proportional to $\frac{1}{j}$

$$W(q,j) = \frac{1}{j}y(q) + jw(q), \tag{8.137}$$

where again from Eq. (8.131) it follows that

$$\frac{dw(q)}{dq} + w^2(q) = A', \quad y(q) = B'. \tag{8.138}$$

In these equations A' and B' are constants.

By solving Eqs. (8.135), (8.136) and (8.138) with different values of the constants A, B, A' and B', including zero we find the following forms for $W(q,j)$ and $L(j)$:

(a) - When $L(j)$ is a linear function of j we have two possible forms for

$W(q,j)$.

(1) - For $L(j) = -b\left(2j - \frac{1}{2}\right)$, $W(q,j)$ found from (8.134) is

$$W(q,j) = (j+c)\frac{1}{q} + \frac{1}{2}bq, \tag{8.139}$$

and the potential $V(q,j)$ is

$$V(q,j) = -\frac{1}{4}b^2q^2 + b(j-c) - \frac{(j+c)(j+c+1)}{q^2}. \tag{8.140}$$

(2) - If we choose $L(j) = -2bj$, then

$$W(q,j) = bq + d, \tag{8.141}$$

and is independent of j. Substituting these in (8.132) gives us the potential

$$V(q,j) = -(bq+d)^2 + b(2j+1). \tag{8.142}$$

(b) - Now let us examine those cases where $L(j)$ is a quadratic function of j:

(3) - If we choose $L(j) = b^2(j+c)^2$ we observe that the right-hand side of (8.131) is $[-2b^2(j+c) - b^2]$ and is linear in j. Therefore we take $W(q,j)$ to be of the form (8.134) and solve (8.135) and (8.136) to find the general form of $W(q,j)$ compatible with this $L(j)$;

$$W(q,j) = (j+c)b\cot[b(q+\theta)] + d\csc[b(q+\theta)], \tag{8.143}$$

where d and θ are constants. The potential derived from this $W(q,j)$ is

$$V(q,j) = \frac{\left\{b^2(j+c)(j+c+1) + d^2 + 2bd\left(j+c+\frac{1}{2}\right)\cos[b(q+\theta)]\right\}}{\sin^2[b(q+\theta)]}. \tag{8.144}$$

(4) - When

$$L(j) = -(j+c)^2, \tag{8.145}$$

again $W(q,j)$ is linear in j. Thus by solving (8.135) and (8.136) we find $W(q,j)$ to be

$$W(q,j) = -j - c + de^q, \tag{8.146}$$

and the corresponding potential is

$$V(q,j) = d^2e^{2q} - 2d\left(j+c+\frac{1}{2}\right)e^q. \tag{8.147}$$

With the same $L(j)$, Eq. (8.145), but with different constants A and B in (8.135) and (8.136) we get a different $W(q,j)$ which is still a linear function of j.

(5) - Choosing $W(q,j)$ to be

$$W(q,j) = (j+c)b\coth[b(q+\alpha)], \tag{8.148}$$

then the solvable potential is of the form

$$V(q,j)=\frac{b^2(j+c)(1+c+j)}{\sinh^2[b(q+\alpha)]}. \tag{8.149}$$

(6) - Alternatively we can choose

$$W(q,j)=(j+c)b\tanh[b(q+\alpha)], \tag{8.150}$$

with the resulting potential

$$V(q,j)=-\frac{b^2(j+c)(1+c+j)}{\cosh^2[b(q+\alpha)]}. \tag{8.151}$$

(c) - When $L(j)$ is inversely proportional to j^2, $L(j)=-\frac{1}{j^2}$, then $W(q,j)$ is a solution of (8.137)

$$W(q,j)=\frac{B'}{j}+jw(q), \tag{8.152}$$

where $w(q)$ is a solution of (8.138). Setting A' in (8.138) equal to zero and $B'=-1$ we find $w(q)=q^{-1}$ and

$$W(q,j)=\left(\frac{j}{q}-\frac{1}{j}\right). \tag{8.153}$$

The potential in this case is that of the Kepler problem (or hydrogen atom)

$$V(q,j)=\frac{2}{q}-\frac{j(j+1)}{q^2}. \tag{8.154}$$

We can generalize this problem by assuming that $L(j)=-(q+c)^{-1}$, and then the corresponding $W(q,j)$ and the potential are

$$W(q,j)=\left(\frac{(j+c)}{q}-\frac{1}{(j+c)}\right), \tag{8.155}$$

and

$$V(q,j)=\frac{2}{q}-\frac{(j+c)(j+c+1)}{q^2}. \tag{8.156}$$

respectively. A number of other local potentials for which the eigenvalues and eigenvectors can be found are discussed in detail by Infeld and Hull [13]-[14].

Now that we have found analytical expressions for raising and lowering operators, we want to determine the eigenvalues and the corresponding wave functions for these solvable potentials.

When $V(q,j)$ is such that the Eq. (8.132) is satisfied and in addition we have $L(j+1)>L(j)$, then as we will see the eigenvalues will be independent of j. Let us consider the scalar product

$$\begin{aligned}\langle\lambda_n,j+1|\lambda_n,j+1\rangle &= \langle\lambda_n,j|\beta^+(j+1)\beta^-(j+1)|\lambda_n,j\rangle\\ &= (\lambda_n-L(j+1))\langle\lambda_n,j|\lambda_n,j\rangle,\end{aligned} \tag{8.157}$$

where we have used Eqs. (8.125) and (8.130) to obtain (8.157). Since both of the scalar products on the left and on the right of (8.157) are positive, $(\lambda_n - L(j+1))$ cannot be negative. Now $L(j+1)$ increases with increasing j, consequently there can be no eigenfunction for j larger than a certain maximum number n, i.e.

$$\lambda_n = L(n+1), \quad n = j, j+1, j+2, \cdots, \tag{8.158}$$

and since $L(j)$ is known therefore λ_n can be determined. From Eq. (8.157) it follows that for $j = n$, the ket $|\lambda_n, n+1\rangle$ has to be zero or

$$\beta^-(n+1)|\lambda_n, n\rangle = \left(W(q, n+1) - \frac{d}{dq}\right)|\lambda_n, n\rangle = 0. \tag{8.159}$$

Denoting the wave function corresponding to the state $|\lambda_n, n\rangle$ by $\phi_n(q, n)$ we have

$$\beta^-(n+1)\phi_n(q, n) = \left(W(q, n+1) - \frac{d}{dq}\right)\phi_n(q, n) = 0. \tag{8.160}$$

We can integrate (8.160) to find $\phi_n(q, n)$;

$$\phi_n(q, n) = \frac{e^{\int_a^q W(q', n+1)dq'}}{\left\{\int_a^b \exp\left[2\int_a^b W(q', n+1)dq'\right]dq\right\}^{\frac{1}{2}}}. \tag{8.161}$$

Here we have normalized the wave function assuming that the coordinate q can change continuously from a to be b (a and b can be finite or infinite) .

For other j states having the same eigenvalue, we can use the raising operator $\beta^+(j+1), \beta^+(j+2), \cdots$ and operate on $\phi_n(q, n)$;

$$\phi_n(q, j) = \beta^+(j+1)\beta^+(j+2)\cdots\beta^+(n)\phi_n(q, n). \tag{8.162}$$

The wave function $\phi_n(q, j)$ obtained in this way is

$$\phi_n(q, j) = \frac{1}{\sqrt{L(n+1) - L(j+1)}}\left[W(q, j+1) + \frac{d}{dx}\right]\phi_n(q, j+1). \tag{8.163}$$

We note that the wave functions belonging to the same j but with two different eigenvalues are orthogonal to each other;

$$\int_a^b \phi_n(q, j)\phi_{n'}(q, j)dq = \delta_{nn'}. \tag{8.164}$$

If $L(j+1) < L(j)$, then

$$\lambda_n = L(n), \quad n = j, j-1, j-2, \cdots. \tag{8.165}$$

Here as in the other case we can calculate $\phi_n(q, n)$ and then in turn find $\phi_n(q, j)$;

$$\phi_n(q, n) = \frac{e^{-\int_a^q W(q', n)dq'}}{\left\{\int_a^b \exp\left[-2\int_a^b W(q', n)dq'\right]dq\right\}^{\frac{1}{2}}}, \tag{8.166}$$

and

$$\phi_n(q,j) = \frac{1}{\sqrt{L(n) - L(j)}}\left[W(q,j) - \frac{d}{dx}\right]\phi_n(q, j-1). \tag{8.167}$$

Pöschl–Teller Potential — Consider the general form of the Pöschl–Teller potential given by

$$V(q,j) = \left\{\frac{b^2(j+c)(1+c+j)}{\sinh^2[b(q+\alpha)]} - \frac{b^2(j-c)(1-c+j)}{\cosh^2[b(q+\alpha)]}\right\}, \tag{8.168}$$

where as in (8.149) and (8.150) we have

$$W(q,j) = (j-c)b\tanh[b(q+\alpha)] + (j+c)b\coth[b(q+\alpha)], \tag{8.169}$$

and $L(j) = -4b^2j^2$. For this potential the energy eigenvalues are:

$$\lambda_n = -4b^2(j-n)^2, \quad n = 0, 1, \cdots j-1. \tag{8.170}$$

We can write the wave function Using the results given in Eqs. (8.161) and (8.167) we can write the wave function as

$$\phi_n(q,n) = \left[\frac{2b\Gamma\left(n - c + \frac{1}{2}\right)}{\Gamma\left(-n - c + \frac{1}{2}\right)\Gamma(2n)}\right]^{\frac{1}{2}} \sinh^{-n-c}[b(q+\alpha)]\cosh^{-n+c}[b(q+\alpha)], \tag{8.171}$$

and

$$\phi_n(q,j) = \frac{1}{2b[(j+n)(j-n)]^{\frac{1}{2}}}\left[W(q,j) - \frac{d}{dq}\right]\phi_n(q,j-1), \tag{8.172}$$

provided that $c < \frac{1}{2} - j$ [13].
For more recent work on the factorization method see [16],[17].

8.6 Supersymmetry and Superpotential

The method of factorization which we studied earlier has an interesting application in supersymmetric quantum mechanics and allows one to find a large number of exactly solvable potentials. Let us consider a one-dimensional problem with the Hamiltonian

$$H_1 = \beta^-\beta^+ = -\frac{d^2}{d\,q^2} + V_1(q), \tag{8.173}$$

where we have set $\hbar = 2m = 1$. As we have seen before the raising and lowering operators are β^+ and β^-;

$$\beta^+ = W(q) + \frac{d}{dq}, \quad \beta^- = W(q) - \frac{d}{dq}. \tag{8.174}$$

This is a special case of a factorizable Hamiltonian since H_1 and $\beta^{\pm}$ do not depend on the parameter j as we assumed in Eqs. (8.123) and (8.126). Substituting for β^+ and β^- from (8.174) in (8.173) we find that the potential $V(q)$ is related to $W(q)$ by

$$V_1(q) = W^2(q) - \frac{dW(q)}{dq}. \tag{8.175}$$

In supersymmetric quantum mechanics $W(q)$ is known as superpotential [16]. Next we define the Hamiltonian H_2 by reversing the order of β^+ and β^-;

$$H_2 = \beta^+\beta^- = -\frac{d^2}{d\,q^2} + V_2(q), \tag{8.176}$$

where now from (8.174) it follows that $V_2(q)$ is given by

$$V_2(q) = W^2(q) + \frac{dW(q)}{dq}. \tag{8.177}$$

The two potentials $V_1(q)$ and $V_2(q)$ are known as supersymmetric partner potentials.

If the energy eigenvalues of H_1 denoted by $E_n^{(1)}$ are known, then the eigenvalues of H_2 will be related to these $E_n^{(1)}$ s. To find this relationship we consider the eigenvalue equation

$$H_1|1,n\rangle = \beta^-\beta^+|1,n\rangle = E_n^{(1)}|1,n\rangle. \tag{8.178}$$

By multiplying (8.178) by β^+ we have

$$\beta^+\beta^-\left(\beta^+|1,n\rangle\right) = E_n^{(1)}\left(\beta^+|1,n\rangle\right), \tag{8.179}$$

or if $H_2 = \beta^+\beta^-$, then

$$H_2\left(\beta^+|1,n\rangle\right) = E_n^{(1)}\left(\beta^+|1,n\rangle\right). \tag{8.180}$$

Now we observe that the eigenvalue equation for H_2 is

$$H_2|2,n\rangle = \beta^+\beta^-|2,n\rangle = E_n^{(2)}|2,n\rangle. \tag{8.181}$$

The same argument as the one used in getting (8.180) gives us

$$H_2\left(\beta^-|2,n\rangle\right) = \beta^-\beta^+\left(\beta^-|2,n\rangle\right) = E_n^{(2)}\left(\beta^+|2,n\rangle\right). \tag{8.182}$$

By adding a constant to the potential we make $E_0^{(1)} = 0$. Then from Eqs. (8.178)–(8.180) we find that the eigenvalues and the eigenstates of H_1 and H_2 are related to each other by

$$E_n^{(2)} = E_{n+1}^{(1)}, \quad n = 1, 2, \cdots, \tag{8.183}$$

$$|2,n\rangle = \frac{1}{\sqrt{E_{n+1}^{(1)}}}\left(\beta^+|1,n+1\rangle\right), \tag{8.184}$$

and

$$|1, n+1\rangle = \frac{1}{\sqrt{E_n^{(2)}}}\left(\beta^-|2,n\rangle\right). \tag{8.185}$$

Equation (8.184) shows that β^- converts $|1, n+1\rangle$ to $|2, n\rangle$ and (8.185) shows that β^+ converts $|2, n\rangle$ to $|1, n+1\rangle$. At the same time the action of β^- annihilates an extra state with the energy $E_0^{(1)}$.

To establish a relation between the degeneracy of the spectra of these two Hamiltonians and the conserved quantities we consider the supersymmetric Hamiltonian

$$H = \begin{bmatrix} H_1 & 0 \\ 0 & H_2 \end{bmatrix}, \tag{8.186}$$

and instead of β^+ and β^- we define the operators Q^+ and Q^- by

$$Q^- = \begin{bmatrix} 0 & \beta^- \\ 0 & 0 \end{bmatrix}, \quad \text{and} \quad Q^+ = \begin{bmatrix} 0 & 0 \\ \beta^+ & 0 \end{bmatrix}. \tag{8.187}$$

The operators Q^- and Q^+ are called supercharge operators . They commute with the Hamiltonian H, and thus are constants of motion and are responsible for the degeneracies in the spectra of H_1 and H_2

$$\left[H, Q^-\right] = \left[H, Q^+\right] = 0. \tag{8.188}$$

In addition they have the important property that they satisfy the following anti-commutation relations

$$\left[Q^-, Q^+\right]_+ = H, \tag{8.189}$$

and

$$\left[Q^-, Q^-\right]_+ = \left[Q^+, Q^+\right]_+ = 0. \tag{8.190}$$

The fact that we have both commutation and anti-commutation relations indicates that the operators Q^+ and Q^- change the bosonic degrees of freedom corresponding to the commutation relation into fermionic degrees of freedom connected with the anti-commutation relation. To illustrate this point let us consider the case where H is the Hamiltonian of a harmonic oscillator of mass $m = \frac{1}{2}$;

$$H = p^2 + \frac{1}{4}q^2. \tag{8.191}$$

As we have seen earlier, Eq. (8.126), β^+ and β^- for this case are

$$\beta^\pm = W(q) \pm \frac{d}{dq} = \frac{1}{2}q \pm \frac{d}{dq}, \tag{8.192}$$

and they satisfy the commutation relation

$$\left[\beta^+, \beta^-\right] = 1. \tag{8.193}$$

If we define the number operator N by $N = \beta^-\beta^+$, then we can relate H to N

$$N = \beta^-\beta^+ = -\frac{d^2}{d\,q^2} + \frac{1}{4}q^2 - \frac{1}{2} = H - \frac{1}{2}. \tag{8.194}$$

This operator, N, satisfy the following commutation relations:

$$\left[N, \beta^+\right] = -\beta^+, \quad \left[N, \beta^-\right] = \beta^-. \tag{8.195}$$

We also define the fermionic creation and annihilation operators by

$$\sigma_+ = \begin{bmatrix} 0 & 1 \\ 0 & 0 \end{bmatrix}, \tag{8.196}$$

and

$$\sigma_- = \begin{bmatrix} 0 & 0 \\ 1 & 0 \end{bmatrix}. \tag{8.197}$$

These operators obey the anti-commutation relations

$$[\sigma_+, \sigma_-]_+ = 1, \quad [\sigma_-, \sigma_-]_+ = 0, \quad [\sigma_+, \sigma_+]_+ = 0, \tag{8.198}$$

and in addition we have the commutator

$$[\sigma_+, \sigma_-] = \sigma_3 = \begin{bmatrix} 1 & 0 \\ 0 & -1 \end{bmatrix}. \tag{8.199}$$

With the help of $\boldsymbol{\sigma}$ matrices the supersymmetric Hamiltonian H for this case can be written as a 2×2 matrix

$$H = Q^-Q^+ + Q^+Q^- = \left(-\frac{d^2}{d\,q^2} + \frac{1}{4}q^2\right)1 - \frac{1}{2}\sigma_3, \tag{8.200}$$

where Q^- and Q^+ are defined by (8.187) and 1 is a unit 2×2 matrix.

We can express the action of $\beta^-, \beta^+, \sigma^-$ and σ^+ operators on the state specified by the number of bosons and fermions (Fock space). To this end let us define the number operator of fermions, N_f, by the relation

$$N_f = \sigma_-\sigma_+. \tag{8.201}$$

If we square the operator N_f we find

$$N_f^2 = \sigma_-\sigma_+\sigma_-\sigma_+ = \sigma_-(1 - \sigma_-\sigma_+)\sigma_+ = \sigma_-\sigma_+ = N_f, \tag{8.202}$$

since as Eq. (8.198) shows $(\sigma_+)^2 = (\sigma_-)^2 = 0$. If $|n_f\rangle$ denotes the eigenstate of N_f then

$$N_f^2|n_f\rangle = n_f^2|n_f\rangle = N_f|n_f\rangle = n_f|n_f\rangle. \tag{8.203}$$

Thus

$$n_f^2 = n_f, \quad \text{or} \quad n_f = 0, \quad \text{or} \quad 1, \tag{8.204}$$

i.e. we have either one particle in a given state or none. In general the action of $\beta^-, \beta^+, \sigma_-$ and σ_+ on the state with n_b bosons and n_f fermions can be written as

$$\beta^+|n_b, n_f\rangle = |n_b - 1, n_f\rangle, \tag{8.205}$$

$$\beta^-|n_b, n_f\rangle = |n_b + 1, n_f\rangle, \tag{8.206}$$

$$\sigma_+|n_b, n_f\rangle = |n_b, n_f - 1\rangle, \tag{8.207}$$

and

$$\sigma_-|n_b, n_f\rangle = |n_b, n_f + 1\rangle. \tag{8.208}$$

We observe that the operators Q^- and Q^+ acting on these states change a boson into a fermion and vice versa without changing the energy of the state

$$Q^-|n_b, n_f\rangle = |n_b - 1, n_f + 1\rangle, \tag{8.209}$$

and

$$Q^+|n_b, n_f\rangle = |n_b + 1, n_f - 1\rangle. \tag{8.210}$$

8.7 Shape Invariant Potentials

If a pair of supersymmetric partner potentials $V_1(q)$ and $V_2(q)$ defined by (8.175) and (8.177) are similar in shape and differ only in the parameters appearing in them, then we call these potentials "shape invariant". Thus $V_1(q)$ and $V_2(q)$ are shape invariant pairs provided

$$V_1(q, a) = V_2(q, a') + R(a), \tag{8.211}$$

where $R(a)$ is independent of q and a denotes a single parameter or a set of parameters. For well-known analytically solvable potentials the parameter a' is related to a by a simple translation

$$a' = a - c. \tag{8.212}$$

However we can get other solvable potentials if we choose $a' = ca_1$ [16],[17].

Harmonic Oscillator — As an example of shape invariant potentials let us consider $W(x, \omega) = \frac{1}{2}\omega q - b$, then from (8.175) and (8.177) we have

$$V_{1,2}(q, \omega) = W^2(q, \omega) \pm \frac{dW(q, \omega)}{dq} = \frac{1}{4}\omega^2\left(q - \frac{2b}{\omega}\right)^2 \pm \frac{1}{2}\omega. \tag{8.213}$$

Therefore the shifted harmonic oscillator potential is shape invariant and for this potential $R = \omega$.

The Morse Potential — As a second example consider the Morse potential for which the superpotential is

$$W(q, A) = A - \sqrt{\lambda}e^{-\sqrt{\mu}\,q}. \tag{8.214}$$

From $W(q, A)$ we calculate the partners

$$V_1(q, A) = \lambda e^{-2\sqrt{\mu}\, q} - \sqrt{\lambda}(2A + \sqrt{\mu})e^{-\sqrt{\mu}\, q} + A^2, \tag{8.215}$$

and

$$V_2(q, A) = \lambda e^{-2\sqrt{\mu}\, q} - \sqrt{\lambda}(2A - \sqrt{\mu})e^{-\sqrt{\mu}\, q} + A^2, \tag{8.216}$$

By choosing

$$A = \sqrt{\lambda} - \frac{1}{2}\sqrt{\mu}, \tag{8.217}$$

$V_1(x, A)$ takes the form of the Morse potential, Eq. (8.39), apart from an additive constant. The shape invariance follows from the fact that

$$V_2(q, A) = V_1(q, A - \sqrt{\mu}) + A^2 - (A - \sqrt{\mu})^2. \tag{8.218}$$

Calculating the Eigenvalues for Shape Invariant Potentials — The shape invariance of the partner potentials can be used to find the complete spectrum of the Hamiltonian algebraically. For this we can construct a series of Hamiltonians H_n where n is an integer in such a way that H_n will have the same set of eigenvalues as H_1 except that the first $(n-1)$ levels of H_1 will be missing in H_n. By repeated application of the shape invariance condition (8.211) we obtain

$$H_{n+1} = -\frac{d^2}{d\,x^2} + V_1(x, a_{n+1}) + \sum_{j=1}^{n} R(a_j), \tag{8.219}$$

where in this relation

$$a_j = a_0 - jc. \tag{8.220}$$

If in Eq. (8.219) we substitute for $R(a_n)$ from (8.211) i.e.

$$R(a_n) = V_2(x, a_n) - V_1(x, a_{n+1}), \tag{8.221}$$

we find

$$H_{n+1} = -\frac{d^2}{d\,x^2} + V_2(x, a_n) + \sum_{j=1}^{n-1} R(a_j). \tag{8.222}$$

We note that H_n and H_{n+1} are supersymmetric partner Hamiltonians, and have identical bound states spectra except for the ground state of H_n. From (8.219) and the fact that $E_0^{(1)} = 0$ it follows that the ground state for H_n is given by

$$\begin{cases} E_n^{(1)} = \sum_{j=1}^{n} R(a_j), & n > 1 \\ E_0^{(0)} = 0 \end{cases}. \tag{8.223}$$

Thus by successive application of raising and lowering operators we can find the complete spectrum for any solvable potential.

For a harmonic oscillator the spectrum according to (8.80) is very simple $E_n = \left(n+\frac{1}{2}\right)\omega$. For the Morse potential, $R(\mu) = A^2 - (A - \sqrt{\mu})^2$ and therefore

$$E_n^{(1)} = A^2 - (A - n\sqrt{\mu})^2 = \frac{\mu}{4} - \sqrt{\lambda\mu} + 2\sqrt{\lambda\mu}\left(n + \frac{1}{2}\right) - \mu\left(n + \frac{1}{2}\right)^2. \quad (8.224)$$

This result agrees with the one found earlier Eq. (8.40) for the Morse potential provided that in the latter equation we set $m = \frac{1}{2}$ and calculate $E_n - E_0$. Note that in the present formulation we have set the ground state energy equal to zero.

Generalized Hulthén Potential — Finally let us apply this method of determination of eigenvalues to the generalized Hulthén potential. This central potential has the form

$$V(r) = -\frac{V_0}{e^{\alpha r} - \beta}, \quad (8.225)$$

where α and β are constants. If we choose $\beta = 1$ then (8.225) becomes the standard Hulthén potential which is used as a simple model for short range nucleon-nucleon interaction [18]–[20].
The superpotential for (8.225) is

$$W(r) = \frac{A}{e^{\alpha r} - \beta} + B, \quad (8.226)$$

where

$$B = \frac{V_0 - \beta\alpha^2}{2\beta\alpha}, \quad (8.227)$$

From $W(r)$ we find the radial dependence of $V_1(r, A)$ and $V_2(r, A)$;

$$\begin{aligned} V_{1,2}(r, A) &= W^2(r) \mp \frac{dW(r)}{dr} \\ &= \frac{\left(\frac{A^2}{\beta} \pm A\alpha\right) e^{\alpha r}}{(e^{\alpha r} - \beta)^2} - \frac{V_0}{(e^{\alpha r} - \beta)} + B^2. \end{aligned} \quad (8.228)$$

The shape invariance of the potential $V_1(r, A)$ is evident from the relation

$$V_2(r, A) = V_1(r, A - \beta\alpha) + R(A), \quad (8.229)$$

where from (8.228) and (8.229) we get

$$R(A) = \left(\frac{A}{2\beta} - \frac{V_0}{2A}\right)^2 - \left(\frac{A - \alpha\beta}{2\beta} - \frac{V_0}{2(A - \alpha\beta)}\right)^2. \quad (8.230)$$

The energy spectrum of the Hamiltonian

$$H = -\frac{d^2}{d\,r^2} + V_1(r, A), \quad (8.231)$$

is obtained by the repeated application of the shape invariance condition and is given by

$$\begin{cases} E_n^{(1)} = \left(\frac{A}{2\beta} - \frac{V_0}{2A}\right)^2 - \left(\frac{A-n\alpha\beta}{2\beta} - \frac{V_0}{2(A-n\alpha\beta)}\right)^2 \\ E_0^{(1)} = 0 \end{cases}. \tag{8.232}$$

To calculate the ground state we observe that

$$V_1(r, A = -\beta\alpha) = V(r) - E, \tag{8.233}$$

and this gives us

$$E_0 = -B^2 = -\left(\frac{V_0 - \beta\alpha^2}{2\beta\alpha}\right)^2. \tag{8.234}$$

For the Hulthén potential $\beta = 1$, and the ground state is given by [19]

$$E_0 = -\frac{1}{4}\left(\alpha - \frac{1}{\alpha}V_0\right)^2. \tag{8.235}$$

If we add (8.234) to $E_n^{(1)}$, Eq. (8.232), we find the complete energy spectrum of the generalized Hulthén potential to be

$$E_n = -\frac{1}{4}\left[(n+1)^2\alpha^2 + \frac{V_0^2}{\alpha^2\beta^2(n+1)^2} - \frac{2V_0}{\beta}\right], \quad n = 0,\ 1,\ 2\cdots. \tag{8.236}$$

The Eckart Potential — This potential can be derived from the superpotential

$$W(x, A) = -A\coth(\gamma x) + \frac{B}{A}, \tag{8.237}$$

and has the form

$$V_1(x, A) = W^2(x, A) - \frac{dW(x, A)}{dx} = \left(\frac{B}{A} - A\coth(\gamma x)\right)^2 - \gamma A\left(\frac{1}{\sinh(\gamma x)}\right)^2. \tag{8.238}$$

The partner potential $V_2(x)$ obtained from $W(x)$ is

$$V_2(x, A) = W^2(x, A) + \frac{dW(x, A)}{dx} = \left(\frac{B}{A} - A\coth(\gamma x)\right)^2 + \gamma A\left(\frac{1}{\sinh(\gamma x)}\right)^2. \tag{8.239}$$

The shape invariance of the Eckart potential can be established through the transformation

$$R(A) = V_2(x, A) - V_1(x, A+\gamma) = A^2 - (A+\gamma)^2 - \frac{B^2}{(A+\gamma)^2} + \left(\frac{B}{A}\right)^2. \tag{8.240}$$

Thus for the complete discrete energy spectrum we have

$$E_n = A^2 - (A+n\gamma)^2 - \left(\frac{B}{A+n\gamma}\right)^2 + \left(\frac{B}{A}\right)^2. \tag{8.241}$$

A complete list of other solvable potentials is given in references [16] and [17].

In addition to the completely solvable potentials we have quasi-solvable potentials for which only a finite number of eigenstates and eigenvalues can be found analytically [21].

One-Dimensional Motion with Continuous Spectra — So far we have considered potentials having discrete energy levels. The same method outlined in this section can be applied to the one-dimensional problems with continuous spectra. Here the potentials $V_1(q)$ and $V_2(q)$ are finite as $x \to -\infty$ or as $x \to +\infty$ or both. Denoting the asymptotic form of $W(q)$ by

$$W(q \to \pm\infty) = W_{\pm}, \tag{8.242}$$

we have

$$V_{1,2} \to W_{\pm}^2, \quad \text{as} \quad q \to \pm\infty. \tag{8.243}$$

Suppose that a plane wave e^{ikx} is incident from $x \to -\infty$, then we get reflected waves $R_1(k)e^{-ikx}$ and R_2e^{-ikx} for the two partner potentials $V_1(q)$ and $V_2(q)$, as well as transmitted waves $T_1(k)e^{ik'x}$ and $T_2(k)e^{ik'x}$. Thus the asymptotic form of the wave functions are

$$\psi_{1,2}(k, q \to -\infty) \to e^{ikx} + R_{1,2}(k)\, e^{-ikx}, \tag{8.244}$$

and

$$\psi_{1,2}(k, q \to +\infty) \to T_{1,2}(k)\, e^{-ik'x}, \tag{8.245}$$

where k and k' are related to $W_{\pm}$ by

$$k = \sqrt{(E - W_-^2)}, \quad \text{and} \quad k' = \sqrt{(E - W_+^2)}. \tag{8.246}$$

Here E is the energy associated with the incoming wave, $E = k^2$. As in the case of discrete spectra, H_1 and H_2 have the same energy. Now by applying the operators β^- and β^+, Eqs. (8.127) and (8.128), to the continuum wave functions of H_1 and H_2 we find two equations

$$e^{ikx} + R_1(k)e^{-ikx} = \mathcal{N}\left[(W_- - ik)e^{ikx} + (W_- + ik)R_2(k)e^{-ikx}\right], \tag{8.247}$$

and

$$T_1(k)e^{ik'x} = \mathcal{N}\left[(W_+ - ik')T_2(k)e^{ik'}\right], \tag{8.248}$$

where $\mathcal{N}$ is the normalization constant. We solve these equations for the reflection and transmission amplitudes for the potential $V_1(q)$:

$$R_1(k) = \left(\frac{W_- + ik}{W_- - ik}\right) R_2(k), \tag{8.249}$$

and

$$T_1(k) = \left(\frac{W_- - ik'}{W_- - ik}\right) T_2(k). \tag{8.250}$$

From these expressions for $R_1(k)$ and $T_1(k)$ we deduce the following results:

(a) - The two potentials $V_1(q)$ and $V_2(q)$ have identical reflection and transmission coefficients

$$|R_1(k)|^2 = |R_2(k)|^2, \quad \text{and} \quad |T_1(k)|^2 = |T_2(k)|^2. \tag{8.251}$$

(b) - If $W_+ = W_-$, then $k = k'$ and $T_1(k) = T_2(k)$.

(c) - When $W_- = 0$, we have $R_1(k) = -R_2(k)$.

Let us consider the following example where the superpotential is given by

$$W(q) = A\tanh(\alpha q). \tag{8.252}$$

Then

$$V_1(q) = A^2 - \frac{A(A+\alpha)}{\cosh^2(\alpha q)}, \tag{8.253}$$

$$V_2(q) = A^2 - \frac{A(A-\alpha)}{\cosh^2(\alpha q)}. \tag{8.254}$$

Now if we choose $A = \alpha$, then $V_2(q) = A^2$ is a constant potential and the reflection coefficient for this potential, $R_2(k)$, is zero for all k. Therefore from (8.249) it follows that $R_1(k) \equiv 0$, and that

$$V_1(q) = A^2\left(1 - \frac{2}{\cosh^2(\alpha q)}\right), \tag{8.255}$$

is a reflectionless potential.

8.8 Solvable Examples of Periodic Potentials

In Chapter 6 we discussed the lattice translation operator and showed that this operator commutes with the Hamiltonian of the system. We also found how the symmetry associated with the periodicity is reflected in the translational symmetry of the wave function, Eq. (6.122). The eigenvalues and the eigenfunction of these potentials for a few cases can be found exactly from the solution of the Schrödinger equation. Among them we find the following potentials:

(1) - In a crystal we can assume that the potential takes the form of an infinite series of equidistant rectangular barriers we can solve the wave equation. In the limiting case such a series of barriers reduces to an infinite array of δ-functions located at the sites of the positive ions. In this limit we have

$$V(x) = \sum_{n=-\infty}^{\infty} V_0\delta(x - na). \tag{8.256}$$

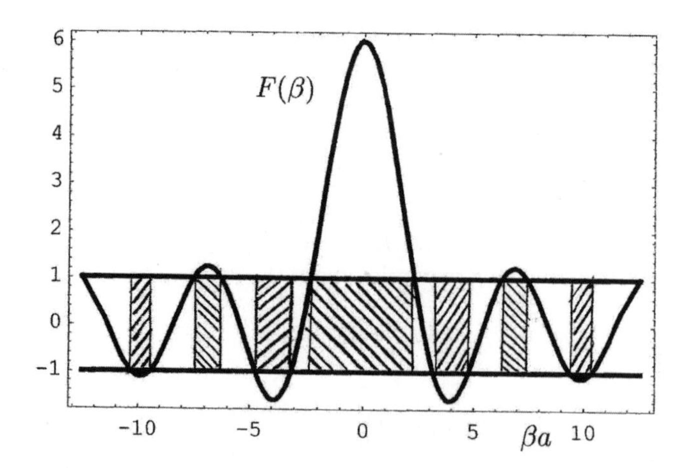

Figure 8.1: The function $F(\beta)$, Eq. (8.265), is plotted as a function of the dimensionless quantity βa. The energy of the electron is proportional to β^2.

To find the eigenvalue equation for this model, which is called the Kronig-Penney model, we solve the wave equation for $\psi(x)$. Noting that in the interval $0 < x < a$ the potential is zero and thus $\psi(x)$ is given by

$$\psi(x) = Ae^{i\beta x} + Be^{-i\beta x}. \tag{8.257}$$

In this relation $\hbar\beta = \sqrt{2mE}$, E being the energy of the electron and A and B are constants. From (6.121) we find $u_k(x)$ to be

$$u_k(x) = Ae^{i(\beta-k)x} + Be^{-i(\beta+k)x}. \tag{8.258}$$

The periodicity condition (6.122) means that for $x = a - \varepsilon$ we have

$$u_k(a-\varepsilon) = u_k(-\varepsilon), \tag{8.259}$$

where ε is a very small positive number. The continuity of the wave function at $x = 0$, viz,

$$u_k(\varepsilon) = u_k(-\varepsilon), \tag{8.260}$$

and Eq. (8.259) gives us a relation between A and B

$$A + B = Ae^{i(\beta-k)a} + Be^{-i(\beta+k)a}. \tag{8.261}$$

To find a second boundary condition we observe that by integrating the Schrödinger equation with the potential (8.256) from $-\varepsilon$ to $+\varepsilon$ we have

$$\frac{\hbar^2}{2m}\left[\left(\frac{d\psi}{dx}\right)_{x=\varepsilon} - \left(\frac{d\psi}{dx}\right)_{x=-\varepsilon}\right] = V_0\psi(0). \tag{8.262}$$

Substituting for ψ from (8.257), this last condition reduces to

$$A - B - Ae^{i(\beta-k)a} + Be^{-i(\beta+k)a} = \frac{2mV_0}{i\beta\hbar^2}(A+B). \tag{8.263}$$

The two homogeneous equations (8.261) and (8.262) can have a nontrivial solution only if the determinant of the coefficients of A and B vanish, i.e. if

$$\cos ka = \cos\beta a + \left(\frac{mV_0 a}{\hbar^2}\right)\frac{\sin\beta a}{\beta a}. \tag{8.264}$$

The quantity $\frac{mV_0 a}{\hbar^2}$ is a dimensionless constant which is a measure of the strength of the potential. If we denote the right-hand side of (8.264) by $F(\beta)$;

$$F(\beta) = \cos\beta a + \left(\frac{mV_0 a}{\hbar^2}\right)\frac{\sin\beta a}{\beta a}, \tag{8.265}$$

then the acceptable solution of the problem are those parts of βa axis which satisfies the condition $-1 \le F(\beta) \le 1$ (see Fig. 8.1).

From the solution of the Kronig-Penney model we find the following results: If the potential barrier between adjacent sites is strong, then the energy bands are narrow and far apart. This is the case of crystals where the electrons are tightly bound to the ions and the wave function is well localized within a given site. On the other hand if the barrier is weak, the energy bands are wide and spaced close to each other. This is typically the situation for alkali metals with weakly bound electrons.

(2) - For an oscillating potential of the form

$$V(x) = V_0 \cos\left(\frac{\pi x}{a}\right), \tag{8.266}$$

the Schrödinger equation reduces to the Mathieu equation which is exactly solvable [3], [22].

(3) - A periodic potential for which the wave equation can be transformed to the differential equation for the hypergeometric function and for which the energy band can be calculated exactly is [23]

$$V(x) = \frac{V_0}{\sin^2\left(\frac{\pi x}{a}\right)}. \tag{8.267}$$

(4) - For some applications in condensed matter physics it is useful to consider the energy band structure of an electron moving in a one-dimensional periodic potential consisting of a lattice with two different rectangular barriers and wells per site. A lattice with this kind of periodicity has been studied in reference [24].

Bibliography

[1] For a complete list of the eleven coordinates in which to wave equation (or Helmholz equation) is separable see P.M. Morse and H. Feshbach, *Methods of Theoretical Physics*, Part I, (McGraw-Hill, New York, 1953) p. 655.

[2] C. Eckart, The penetration of a potential barrier by electrons, Phys. Rev. 35, 1303 (1930).

[3] P.M. Morse, Diatomic molecules according to wave mechanics, II, Vibrational levels, Phys. Rev. 34, 57 (1929).

[4] R. de L. Kronig and W.G. Penney, Quantum mechanics of electrons in crystal lattices, Proc. R. Soc. (London) A 130, 499 (1931).

[5] M.F. Manning, Exact solutions of the Schrödinger equation, Phys. Rev. 48, 161 (1935).

[6] C.-T. Li, Nontrivial exactly solvable potentials with linear equations of motion, Chinese J. Phys. 32, 309 (1994).

[7] S. Odake and R. Sasaki, Exact solution in the Heiseberg picture and annihilation and creation operators, Phys. Lett B 641, 112 (2006).

[8] S. Odake and R. Sasaki, Unified theory of annihilation-creation operators for solvable ("discrete") quantum mechanics, J. Math. Phys. 47, 102102 (2006).

[9] M. Abramowitz and I.A. Stegun, *Handbook of Mathematical Functions*, (Dover, New York, 1965), p. 773.

[10] E. Schrödinger, Eigenvalues and eigenfunctions, Proc. Roy. Irish Acad. 46, 9 (1940).

[11] E. Schrödinger, Further studies on solving eigenvalue problems by factorization, Proc. Roy. Irish Acad. 46, 183 (1941).

[12] E. Schrödinger, The factorization of the hypergeometric equation, Proc. Roy. Irish Acad. 47, 53 (1941).

[13] L. Infeld and T.E. Hull, The factorization method, Rev. Mod. Phys. 23, 21 (1951).

[14] A detailed account of the factorization method for solvable potentials can be found in S-H Dong's *Factorization Method in Quantum Mechanics*, (Springer, Dordrecht, The Netherlands, 2007).

[15] S. Kuru and I. Negro, Factorization of one-dimensional classical systems, Ann. Phys. 323, 413 (2008).

[16] F. Cooper, A. Khare and U. Sukhatme, Supersymmetry and quantum mechanics, Phys. Rep. 251, 267 (1995).

[17] See for instancce, F. Cooper, A. Khare and U. Sukhatme, *Supersymmetry in Quantum Mechanics*, (World Scientific, Singapore, 2001) Chapter 3.

[18] L. Hulthén, On the characteristic solutions of the Schrödinger deutron equation, Ark. Mat. Astron. Fys. 28 A, 5, (1942), 12pp.

[19] H.S. Green, *Matrix Mechanics*, (P. Noordhoff, Netherlands, 1965).

[20] G. Chen, Shape invariance and the supersymmetric WKB approximation for the generalized Helthén potential, Physica Scripta, 69, 257 (2004).

[21] V.M. Tkachuk, Quasi-exactly solvable potentials with two known eigenstates, Phys. Lett. A 245, 177 (1998).

[22] F. Bloch, Quantum mechanics of electrons in crystal lattices, Z. Phys. 52, 555 (1928).

[23] F.L. Scarf, New soluble energy band problem, Phys. Rev. 112, 1137 (1958).

[24] H.X. Jiang and J.Y. Lin, Band structure of a periodic potential with two wells and two barriers per period, Am. J. Phys. 55, 462 (1987).

Chapter 9

The Two-Body Problem

In classical mechanics, a system composed of two particles interacting by a potential which depends only on the relative coordinate of the two particles is separable and can be solved and the solution is given by a definite integral [1]. This separability is carried over to quantum mechanics where the problem is reduced to that of the motion of a particle in a field of force. As we will see the angular part is completely solvable and the solution for the radial part of some potentials can be found by factorization method.

We start by writing the Hamiltonian for the two-body problem as

$$H = \frac{\mathbf{p}_1^2}{2m_1} + \frac{\mathbf{p}_2^2}{2m_2} + V(|\mathbf{r}_1 - \mathbf{r}_2|), \tag{9.1}$$

where m_1 and m_2 are the masses of the two particles, $\mathbf{p}_1$ and $\mathbf{p}_2$ represent their momenta and $\mathbf{r}_1$ and $\mathbf{r}_2$ their coordinates.

We can separate the translational part of the Hamiltonian by introducing the center of mass and relative coordinates;

$$(m_1 + m_2)\mathbf{R} = m_1\mathbf{r}_1 + m_2\mathbf{r}_2, \qquad \mathbf{r} = \mathbf{r}_1 - \mathbf{r}_2, \tag{9.2}$$

and the center of mass and relative momenta

$$\mathbf{P} = \mathbf{p}_1 + \mathbf{p}_2, \qquad \mathbf{p} = \frac{m_2\mathbf{p}_1 - m_1\mathbf{p}_2}{m_1 + m_2}. \tag{9.3}$$

Substituting for $\mathbf{p}_1$, $\mathbf{p}_2$ and $|\mathbf{r}_1 - \mathbf{r}_2|$ from (9.2) and (9.3) in (9.1) and simplifying the result we find

$$H = \frac{\mathbf{P}^2}{2M} + \frac{\mathbf{p}^2}{2m} + V(r). \tag{9.4}$$

Here M is the total mass, $M = m_1 + m_2$, and m is the reduced mass

$$m = \frac{m_1 m_2}{m_1 + m_2}. \tag{9.5}$$

Equation (9.4) which is identical to its classical counterpart shows that **P** is a cyclic operator since it commutes with **p** and with **r**, and the Hamiltonian is independent of the center of mass coordinate **R**. Thus

$$i\hbar\frac{d\mathbf{P}}{dt} = [\mathbf{P}, H] = 0, \tag{9.6}$$

and **P** is a constant of motion. For this reason $\frac{\mathbf{P}^2}{2M}$ in the Hamiltonian (9.4) which has positive eigenvalues can be replaced by a c-number multiplied by a unit operator.

The second term in H, i.e. $\frac{\mathbf{p}^2}{2m}$, is the kinetic energy of the internal degrees of freedom and can be split further into rotational and vibrational kinetic energies. To this end we define the radial momentum p_r conjugate to r by

$$rp_r = \mathbf{r}\cdot\mathbf{p} - i\hbar, \tag{9.7}$$

where $-i\hbar$ is added to $\mathbf{r}\cdot\mathbf{p}$ to make p_r Hermitian [2]. Next we observe that

$$r\mathbf{p} - \mathbf{p}r = -i\hbar(r\nabla - \nabla r) = i\hbar\frac{\mathbf{r}}{r}. \tag{9.8}$$

Thus from (9.7) and (9.8) we have

$$rp_r - p_r r = \frac{1}{r}(r\mathbf{r}\cdot\mathbf{p} - \mathbf{r}\cdot\mathbf{p}r) = i\hbar. \tag{9.9}$$

Now taking the Hermitian conjugate of the operator relation (9.7) we get

$$p_r^\dagger r = \mathbf{p}\cdot\mathbf{r} + i\hbar = (\mathbf{r}\cdot\mathbf{p} - 3i\hbar) + i\hbar = rp_r - i\hbar = p_r r. \tag{9.10}$$

Here we have used

$$(AB)^\dagger = B^\dagger A^\dagger, \tag{9.11}$$

and

$$\mathbf{r}\cdot\mathbf{p} - \mathbf{p}\cdot\mathbf{r} = 3i\hbar, \tag{9.12}$$

where both **r** and **p** are assumed to be Hermitian operators.

The fact that $p_r^\dagger r = p_r r$ does not prove that that p_r is self-adjoint. Thus if we choose $\phi(r)$ and $\psi(r)$ to be two square integrable functions, then for the self-adjointness of p_r we must have

$$\langle\phi(r)|p_r\psi(r)\rangle = \langle p_r\phi(r)|\psi(r)\rangle. \tag{9.13}$$

We note that from (9.7) that we can write p_r as

$$p_r = -i\hbar\left(\frac{1}{r}\frac{d}{dr}\right)r. \tag{9.14}$$

Using the differential form of p_r as given by (9.14) and integrating by parts we find that [3]

$$p_r^\dagger = p_r - \frac{i\hbar}{2}\delta(r). \tag{9.15}$$

Here we have used the integral

$$\int_0^\infty f(r)\delta(r) = \frac{1}{2}f(0), \tag{9.16}$$

noting that $r = 0$ is the end point of integration [4].

A different way of investigating the self-adjointness of p_r is to find its deficiency indices. In Sec. 3.9 we have seen that the deficiency indices of the operator p_r given by (9.14) is obtained by solving the differential equation [5]

$$p_r f_\pm(r) = \pm i\hbar f_\pm(r), \tag{9.17}$$

with the result that

$$f_\pm(r) = \frac{A}{r}e^{\mp r}. \tag{9.18}$$

From this result it follows that while $f_+(r)$ is square integrable for the range $0 \le r < \infty$, and the function $f_-(r)$ is not, and thus the deficiency indices are $(1, 0)$. We conclude that p_r is not an observable, but as we will see it is a useful operator for calculating the eigenvalues[5],[6].

Our aim is to express $\mathbf{p}^2$ in terms of p_r^2 and for this we decompose $\mathbf{p}$ into components parallel and perpendicular to $\mathbf{r}$. This can be done with the well-known formula from vector calculus (see Sec. 3.3)

$$\mathbf{p} = \frac{1}{r^2}[\mathbf{r}(\mathbf{r}\cdot\mathbf{p}) - \mathbf{r}\wedge(\mathbf{r}\wedge\mathbf{p})]. \tag{9.19}$$

Next we find the scalar product of $\mathbf{p}$ in (9.19) with $\mathbf{p}$ multiplying from the left;

$$\mathbf{p}^2 = \mathbf{p}\cdot\frac{1}{r^2}[\mathbf{r}(\mathbf{r}\cdot\mathbf{p}) - \mathbf{r}\wedge(\mathbf{r}\wedge\mathbf{p})]. \tag{9.20}$$

We also have

$$\mathbf{p}\frac{1}{r^2} - \frac{1}{r^2}\mathbf{p} = \frac{2i\hbar\mathbf{r}}{r^4}. \tag{9.21}$$

Thus from (9.20) and (9.21) we find

$$\mathbf{p}^2 = \frac{1}{r^2}[(\mathbf{p}\cdot\mathbf{r})(\mathbf{r}\cdot\mathbf{p}) + 2i\hbar\mathbf{r}\cdot\mathbf{p} - (\mathbf{p}\wedge\mathbf{r})\cdot(\mathbf{r}\wedge\mathbf{p})]. \tag{9.22}$$

Using the definition of $\mathbf{L}$

$$\mathbf{L} = (\mathbf{r}\wedge\mathbf{p}) = -(\mathbf{p}\wedge\mathbf{r}), \tag{9.23}$$

we can write (9.22) as

$$\mathbf{p}^2 = \frac{1}{r^2}\left[(\mathbf{r}\cdot\mathbf{p} - i\hbar)\,\mathbf{r}\cdot\mathbf{p} + \mathbf{L}^2\right]. \tag{9.24}$$

In addition we have

$$\begin{aligned}\frac{1}{r^2}(\mathbf{r}\cdot\mathbf{p} - i\hbar)\mathbf{r}\cdot\mathbf{p} &= \frac{1}{r}p_r(rp_r + i\hbar)\\ &= \frac{1}{r}(rp_r - i\hbar)p_r + i\hbar\frac{1}{r}p_r = p_r^2.\end{aligned} \tag{9.25}$$

Finally from (9.24) and (9.25) we obtain

$$\mathbf{p}^2 = p_r^2 + \frac{\mathbf{L}^2}{r^2}, \tag{9.26}$$

where

$$\mathbf{L} = \mathbf{r} \wedge \mathbf{p}, \tag{9.27}$$

is the angular momentum about the center of mass. The reason that we can write the second term in (9.26) without specifying the order of $\mathbf{L}^2$ and r^{-2} is that these two operators commute. In fact $\mathbf{L}^2$ commutes with $\mathbf{p}^2$ and with any function of $\mathbf{r}^2$ (see Sec. 3.3).

9.1 The Angular Momentum Operator

The theoretical framework for calculating the eigenvalues, specifically those of the orbital angular momentum using the algebraic technique and raising and lowering operators was fully developed in one of the very first papers on matrix mechanics [11]. The angular momentum $\mathbf{L}$ which is defined by (9.27) obeys the commutation relation

$$\mathbf{L} \wedge \mathbf{L} = i\hbar \mathbf{L}. \tag{9.28}$$

To show this we write $\mathbf{L}$ in terms of its components

$$L_x = yp_z - zp_y, \tag{9.29}$$

$$L_y = zp_x - xp_z, \tag{9.30}$$

$$L_z = xp_y - yp_x, \tag{9.31}$$

and by direct calculation we find the following commutators;

$$[L_x, L_y] = i\hbar L_z, \tag{9.32}$$

$$[L_y, L_z] = i\hbar L_x, \tag{9.33}$$

$$[L_z, L_x] = i\hbar L_y. \tag{9.34}$$

We can write (9.32)–(9.34) in a compact form as

$$[L_i, L_j] = i\hbar \sum_k \varepsilon_{ijk} L_k, \tag{9.35}$$

where ε_{ijk} is the completely antisymmetric tensor defined by Eqs. (1.32) and (1.33). Equation (9.28) is the vector version of (9.32)–(9.34).

Since $x^\dagger = x, y^\dagger = y \cdots p_x^\dagger = p_x \cdots$ and $p_z^\dagger = p_z$, we have

$$L_z^\dagger = (xp_y)^\dagger - (yp_x)^\dagger = xp_y - yp_x = L_z. \tag{9.36}$$

Similarly we find

$$L_x^\dagger = L_x, \quad L_y^\dagger = L_y. \tag{9.37}$$

The fact that L_x, L_y and L_z do not commute means that they are not simultaneously measurable. The only exception to this is when the eigenvalue of $\mathbf{L}^2$ is zero, then all of the components have simultaneously zero eigenvalues.

We can choose one of the three components of $\mathbf{L}$, say L_z, and measure it simultaneously with $\mathbf{L}^2$. Now we will show that $\mathbf{L}^2$ commutes with L_z. Consider the commutators

$$\begin{aligned}\left[L_z, L_x^2\right] &= [L_z, L_x]L_x + L_x[L_z, L_x] \\ &= i\hbar(L_yL_x + L_xL_y),\end{aligned} \tag{9.38}$$

and

$$\begin{aligned}\left[L_z, L_y^2\right] &= [L_z, L_y]L_y + L_y[L_z, L_y] \\ &= -i\hbar(L_xL_y + L_yL_x),\end{aligned} \tag{9.39}$$

where in calculating (9.38) and (9.39) we have used the relation

$$[A, BC] = [A, B]C + B[A, C]. \tag{9.40}$$

By adding (9.38) to (9.39) and noting that $[L_z, L_z^2] = 0$ we have

$$[L_z, \mathbf{L}^2] = 0. \tag{9.41}$$

Next we want to show that any component of the angular momentum operator, L_i, commutes with $\mathbf{r}^2, \mathbf{p}^2$ and $\mathbf{r}\cdot\mathbf{p}$ (see also Sec. 3.3). For this we observe that the angular momentum commutator (9.28) will not change if we make one of the two following replacements [2]:

(a) - If we replace $\mathbf{r}$ by $i\mathbf{p}$ and $\mathbf{p}$ by $i\mathbf{r}$ simultaneously.

(b) - If we replace $\mathbf{r}$ by $\frac{1}{2}(\mathbf{r}+\mathbf{p})$ and at the same time $\mathbf{p}$ by $(\mathbf{p}-\mathbf{r})$.

Now let us proceed with the proof that L_i satisfies the following commutators (see also Sec. 3.5)

$$\left[L_i,\ \mathbf{r}^2\right] = \left[L_i,\ \mathbf{p}^2\right] = [L_i,\ \mathbf{r}\cdot\mathbf{p}] = 0. \tag{9.42}$$

Using the fundamental commutation relations:

$$[x,\ p_x] = [y,\ p_y] = [z,\ p_z] = i\hbar, \tag{9.43}$$

and with the other commutators for x, y, z, p_x, p_y and p_z being equal to zero we can calculate the following commutators:

$$\left[L_z,\ x^2\right] = [L_z,\ x]x + x[L_z,\ x] = 2i\hbar xy, \tag{9.44}$$

$$\left[L_z,\ y^2\right] = [L_z,\ y]y + y[L_z,\ x] = -2i\hbar xy, \tag{9.45}$$

$$\left[L_z,\ z^2\right] = 0. \tag{9.46}$$

By adding these three relations we get

$$\left[L_z,\ \mathbf{r}^2\right] = 0. \tag{9.47}$$

Similarly we can show that

$$\left[L_x,\ \mathbf{r}^2\right] = \left[L_y,\ \mathbf{r}^2\right] = 0. \tag{9.48}$$

Now if we make the replacement given by (a) and use the same argument that we have just made, we reach the conclusion that

$$\left[L_x,\ \mathbf{p}^2\right] = \left[L_y,\ \mathbf{p}^2\right] = \left[L_z,\ \mathbf{p}^2\right] = 0. \tag{9.49}$$

Finally if we make the substitution suggested in (b) we obtain

$$\left[L_x,\ (\mathbf{r}+\mathbf{p})^2\right] = 0. \tag{9.50}$$

From this together with $\left[L_z, \mathbf{r}^2\right] = \left[L_z, \mathbf{p}^2\right] = 0$ and $\mathbf{r}\cdot\mathbf{p} - \mathbf{p}\cdot\mathbf{r} = 3i\hbar$, we get

$$[L_x,\ (\mathbf{r}\cdot\mathbf{p})] = 0. \tag{9.51}$$

We can summarize our result in this way: If $F\left(\mathbf{r}^2, \mathbf{p}^2, \mathbf{r}\cdot\mathbf{p}\right)$ is any scalar constructed from the operators shown in its argument, then every component of the angular momentum vector commutes with F.

9.2 Determination of the Angular Momentum Eigenvalues

As we noticed before L_z and $\mathbf{L}^2$ commute, and both are self-adjoint operators and thus observables. We will use the factorization technique to find the eigenvalues of L_z and $\mathbf{L}^2$. Since the same method can be used to obtain the eigenvalues of the spin operator, we use the symbol $\mathbf{M}$ for a general operator satisfying the fundamental commutation relation (9.28),

$$\mathbf{M}\wedge\mathbf{M} = i\hbar\mathbf{M}. \tag{9.52}$$

Now let us introduce two non-Hermitian operators $M_\pm$ by

$$M_\pm = M_x \pm iM_y. \tag{9.53}$$

These are the raising and lowering operators as we will see later.

From (9.52) and (9.53) it follows that

$$M_xM_y + M_yM_x = \frac{1}{2i}\left(M_+^2 - M_-^2\right), \tag{9.54}$$

$$[M_\pm,\ \mathbf{M}^2] = 0, \tag{9.55}$$

and

$$\begin{aligned}[M_z,\ M_\pm] &= [M_z,\ M_x] \pm i[M_z,\ M_y] = i\hbar M_y \pm i(-i\hbar M_x) \\ &= \pm(M_x \pm iM_y) = \pm\hbar M_\pm.\end{aligned} \tag{9.56}$$

In addition from (9.53) we find

$$M_\pm M_\mp = M_x^2 + M_y^2 \pm \hbar M_z. \tag{9.57}$$

We denote the eigenvalues of M_z and $\mathbf{M}^2$ by $m\hbar$ and $\nu\hbar^2$ respectively, i.e.

$$\mathbf{M}^2|m,\nu\rangle = \nu\hbar^2|m,\nu\rangle, \tag{9.58}$$

and

$$M_z|m,\nu\rangle = m\hbar|m,\nu\rangle. \tag{9.59}$$

From the condition $\mathbf{M}^2 \geq M_z^2$, it follows that

$$\nu \geq m^2. \tag{9.60}$$

Now we operate on (9.59) with M_+;

$$M_+M_z|m,\nu\rangle = m\hbar M_+|m,\nu\rangle = (M_zM_+ - \hbar M_+)|m,\nu\rangle, \tag{9.61}$$

where we have substituted for M_+M_z on the left-hand side of (9.61) from the commutator (9.56). We can rewrite Eq. (9.61) as

$$M_z(M_+|m,\nu\rangle) = (m+1)\hbar(M_+|m,\nu\rangle). \tag{9.62}$$

This relation shows that $(M_+|m,\nu\rangle)$ is an eigenfunction of M_z with the eigenvalue $(m+1)\hbar$. Since $\mathbf{M}^2$ commutes with M_z, the ket $(M_+|m,\nu\rangle)$ is also an eigenstate of $\mathbf{M}^2$. As (9.60) shows ν has to be larger or equal to m^2. Thus there must be a maximum eigenvalue of M_z otherwise the inequality (9.60) will be violated. If $|\mu,\nu\rangle$ is the eigenstate corresponding to the largest eigenvalue of M_z, then we have $M_+|\mu,\nu\rangle = 0$, or $M_-M_+|\mu,\nu\rangle = 0$,

$$M_-M_+|\mu,\nu\rangle = \left(\mathbf{M}^2 - M_z^2 - \hbar M_z\right)|\mu,\nu\rangle = 0. \tag{9.63}$$

Noting that $|\mu,\nu\rangle$ is an eigenstate of both $\mathbf{M}^2$ and M_z, from (9.63) it follows that

$$\nu - \left(\mu^2 + \mu\right) = 0. \tag{9.64}$$

Let us now investigate that action of the lowering operator M_- on (9.59) when m is equal to μ

$$M_zM_-|\mu,\nu\rangle = (\mu - 1)\hbar M_-|\mu,\nu\rangle. \tag{9.65}$$

This relation is found exactly as we obtained (9.62) but with M_- rather than M_+. If we repeat this operation n times we find

$$M_z[(M_-)^n|\mu,\nu\rangle] = (\mu - n)\hbar[(M_-)^n|\mu,\nu\rangle]. \tag{9.66}$$

Now by choosing n large enough we can violate (9.60) unless for some integer $n = N$

$$(M_-)^N|\mu,\nu\rangle = 0. \tag{9.67}$$

We can operate on the last relation (9.67) with M_+M_- and use (9.57) by writing it as

$$M_+M_- = \mathbf{M}^2 - M_z^2 + \hbar M_z, \tag{9.68}$$

to get

$$\left(\mathbf{M}^2 - M_z^2 + \hbar M_z\right)(M_-)^N|\mu,\nu\rangle = 0. \tag{9.69}$$

Noting that $(M_-)^N|\mu,\nu\rangle$ is an eigenstate of M_z and $\mathbf{M}^2$ therefore we have the following relation between the eigenvalues;

$$\left[\nu - (\mu - N)^2 + (\mu - N)\right] = 0, \tag{9.70}$$

where μ is the largest eigenvalue of M_z. By substituting for ν from (9.64) in (9.70) we find

$$\mu = \frac{N}{2} \equiv j, \tag{9.71}$$

where N is zero or is an integer. Therefore j values are restricted to

$$j = 0,\ \frac{1}{2},\ 1,\ \frac{3}{2},\cdots. \tag{9.72}$$

From Eqs. (9.64) and (9.71) we find ν

$$\nu = j(j+1). \tag{9.73}$$

Thus we have

$$\mathbf{M}^2|m,j\rangle = \hbar^2 j(j+1)|m,j\rangle, \tag{9.74}$$

and

$$M_z|m,j\rangle = \hbar m|m,j\rangle, \quad m = -j,\ -j+1\cdots j-1,\ j. \tag{9.75}$$

Matrix Elements of M_x and M_y — From the fact that $\mathbf{M}^2$ commutes with M_+

$$\left[\mathbf{M}^2,\ M_+\right] = 0, \tag{9.76}$$

we find

$$\mathbf{M}^2\left(M_+|m,j\rangle\right) = M_+\left(\mathbf{M}^2|m,j\rangle\right) = \hbar^2 j(j+1)M_+|m,j\rangle. \tag{9.77}$$

But as Eq. (9.62) shows while the eigenvalue of $\mathbf{M}^2$ has not changed, by the action of M_+ the eigenvlaue of M_z has increased by $\hbar$. Let us assume that

$|m,j\rangle$ is normalized, but $M_+|m,j\rangle$ is not. To normalize this ket we determine its norm;

$$\left\langle m,j\left|M_+^\dagger M_+\right|m,j\right\rangle = \langle m,j|M_-M_+|m,j\rangle$$
$$= \left\langle m,j\left|\mathbf{M}^2 - M_z(M_z+\hbar)\right|m,j\right\rangle = \hbar^2\left[j(j+1)-m(m-1)\right]. \quad (9.78)$$

Since $|m,j\rangle$ is normalized therefore

$$|m+1,j\rangle = \left[(j-m)(j+m+1)\hbar^2\right]^{-\frac{1}{2}} M_+|m,j\rangle, \quad (9.79)$$

is the normalized eigenvector. From this and the corresponding relation for $|m-1,j\rangle$ we find the matrix elements of M_+ and M_-;

$$\langle j,m-1|M_+|j,m\rangle = [j(j+1)-m(m+1)]^{\frac{1}{2}}\,\hbar = [(j-m)(j+m+1)]^{\frac{1}{2}}\,\hbar, \quad (9.80)$$

and

$$\langle j,m-1|M_-|j,m\rangle = [j(j+1)-m(m-1)]^{\frac{1}{2}}\,\hbar = [(j+m)(j-m+1)]^{\frac{1}{2}}\,\hbar. \quad (9.81)$$

In this representation where both $\mathbf{M}^2$ and M_z are diagonal, we have a discrete sequences of values for quantum numbers j and m. It is convenient to write these matrices as an infinite set of finite matrices of the form

$$M_\alpha = \begin{bmatrix} M_\alpha^{(1)} & 0 & 0 & \cdots \\ 0 & M_\alpha^{(2)} & 0 & \cdots \\ 0 & 0 & M_\alpha^{(3)} & \cdots \\ \cdot & \cdot & \cdot & \cdots \end{bmatrix}, \quad (9.82)$$

where each $M_\alpha^{(2j+1)}$ represented by a given value of j and is a $(2j+1)\times(2j+1)$ matrix.

Suppressing the superscript $(2j+1)$ for all $M_\alpha^{(2j+1)}$, we note that for $j=0$, $\mathbf{M}^2$ and M_z are represented by null matrices of unit rank.

For $j=\frac{1}{2}$ we obtain

$$M_x = \frac{\hbar}{2}\begin{bmatrix} 0 & 1 \\ 1 & 0 \end{bmatrix}, \quad M_y = \frac{\hbar}{2}\begin{bmatrix} 0 & -i \\ i & 0 \end{bmatrix}, \quad (9.83)$$

$$M_z = \frac{\hbar}{2}\begin{bmatrix} 1 & 0 \\ 0 & -1 \end{bmatrix}, \quad \mathbf{M}^2 = \frac{3\hbar^2}{4}\begin{bmatrix} 1 & 0 \\ 0 & 1 \end{bmatrix}. \quad (9.84)$$

This case will be discussed in detail later.

For the next one, $j=1$, the matrices are given by

$$M_x = \frac{\hbar}{\sqrt{2}}\begin{bmatrix} 0 & 1 & 0 \\ 1 & 0 & 1 \\ 0 & 1 & 0 \end{bmatrix}, \quad M_y = \frac{\hbar}{\sqrt{2}}\begin{bmatrix} 0 & -i & 0 \\ i & 0 & -i \\ 0 & i & 0 \end{bmatrix}, \quad (9.85)$$

$$M_z = \hbar \begin{bmatrix} 1 & 0 & 0 \\ 0 & 0 & 0 \\ 0 & 0 & -1 \end{bmatrix}, \quad \mathbf{M}^2 = 2\hbar^2 \begin{bmatrix} 1 & 0 & 0 \\ 0 & 1 & 0 \\ 0 & 0 & 1 \end{bmatrix}. \tag{9.86}$$

Similar matrices can be found for $j = 2, 3, \cdots$.

We are thus led to the conclusion that the eigenvalues of $\mathbf{M}^2$ are determined by quantum numbers $j = 0, 1, 2\cdots$, as is given in Eq. (9.74). Now to each value of the quantum number j, there corresponds $(2j+1)$ eigenfunctions $|m, j\rangle$, $m = -j, -j+1, \cdots j$, Eq. (9.75). The fact that the eigenvalues of $\mathbf{M}^2$ are not dependent on m (in other words the degeneracy of the eigenvalues) is associated with the invariance of the motion in a central potential under rotations.

9.3 Matrix Elements of Scalars and Vectors and the Selection Rules

For a two-particle system interacting with a central potential, the angular momentum vector which we define by Eqs. (9.32)–(9.34) commutes with any scalar function of the operators $\mathbf{p}$ and $\mathbf{r}$. This is because any rotation of the coordinate system leaves scalar quantities unchanged. If we denote the scalar quantity by f, we have

$$[f, L_x] = [f, L_y] = [f, L_z] = 0. \tag{9.87}$$

Since f commutes with L_z and $\mathbf{L}^2$, therefore the matrix form of f will be diagonal in a representation where L_z and $\mathbf{L}^2$ are diagonal matrices. If $|\ell, m\rangle$ is an eigenstate of L_z and $\mathbf{L}^2$, and n denotes all the remaining quantum numbers which define the state of the system, then we want to show that the matrix elements

$$\langle n', \ell, m|f|n\, \ell, m\rangle \tag{9.88}$$

are independent of m. To prove this result we first note that from (9.87) we get

$$f(L_x + iL_y) - (L_x + iL_y)f = 0. \tag{9.89}$$

Now we find the matrix element of this operator with the states $\langle n, \ell, m|$ and $|n', \ell, m-1\rangle$;

$$\begin{aligned} &\langle n, \ell, m|f|n', \ell, m\rangle \langle n', \ell, m|(L_x + iL_y)|n', \ell, m-1\rangle \\ - \;&\langle n, \ell, m|(L_x + iL_y)|n, \ell, m-1\rangle \langle n, \ell, m-1|f|n', \ell, m-1\rangle = 0, \end{aligned} \tag{9.90}$$

where we have used the fact that $L_x + iL_y$ has nonzero elements only when $n, \ell, m \to n, \ell, m-1$. We observe that the matrix elements involving $L_x + iL_y$ do not depend on n or n', therefore (9.90) yields the result that

$$\langle n, \ell, m|f|n', \ell, m\rangle = \langle n, \ell, m-1|f|n', \ell, m-1\rangle, \tag{9.91}$$

or that the above matrix element will be the same for all allowed m values [7]. From this it follows that the nonzero elements of f are

$$\langle n,\ell,m|f|n',\ell,m\rangle = \langle n,\ell|f|n',\ell\rangle\,, \tag{9.92}$$

and are independent of m. Noting that the Hamiltonian H is a scalar we can draw an interesting conclusion from (9.92), that the energy of stationary states of H are independent of m, and that the energy levels are $(2\ell+1)$ fold degenerate. Now we consider the same two-body system and we assume that there is a real vector $\mathbf{K}$ which corresponds to a physical quantity. An infinitesimal rotation about the i-th axis will change the components of $\mathbf{K}$ and the new components will be linear combinations of the old ones. Thus the commutator of L_i with the operator $\mathbf{K}$ will produce components of $\mathbf{K}$. For instance if we choose $\mathbf{K}$ to be the radius vector $\mathbf{r}$, then

$$[L_i,\ r_j] = i\hbar\sum_k \varepsilon_{ijk}r_k. \tag{9.93}$$

On the other hand if $\mathbf{K}=\mathbf{p}$, then

$$[L_i,\ p_j] = i\hbar\sum_k \varepsilon_{ijk}p_k. \tag{9.94}$$

In general we have the commutator

$$[L_i,\ K_j] = i\hbar\sum_k \varepsilon_{ijk}K_k, \tag{9.95}$$

a result that we will use later to solve the problem of the hydrogen atom [7].

It is convenient to set $\hbar=1$ in the following derivation, and keeping this in mind from (9.95) we find

$$[L_x^2,\ K_x] = 0, \tag{9.96}$$

$$\begin{aligned}\left[L_y^2,\ K_x\right] &= L_x[L_y,\ K_x]+[L_y,\ K_x]L_y = -i(L_yK_z+K_zL_y)\\ &= -2iL_yK_z - K_x,\end{aligned} \tag{9.97}$$

and

$$\left[L_z^2,\ K_x\right] = 2iL_zK_y - K_x. \tag{9.98}$$

By adding these three relations we obtain

$$\left[\mathbf{L}^2,\ K_x\right] = 2i(L_zK_y - L_yK_z) - 2K_x. \tag{9.99}$$

Having found the commutator $\left[\mathbf{L}^2,\ K_x\right]$ we calculate the double commutator $\left[\mathbf{L}^2,\ \left[\mathbf{L}^2,\ K_x\right]\ \right]$ in two different ways [7]:

(1) - We can write $\mathbf{L}^2 = L_x^2+L_y^2+L_z^2$ and for each term calculate

$$\left[L_i^2,\ \left[\mathbf{L}^2,\ K_x\right]\right] = \left[L_i^2,\ 2i(L_zK_y - L_yK_z) - 2K_x\right], \tag{9.100}$$

using the commutation relations for different components of $\mathbf{L}$ and $\mathbf{K}$ and then adding up the results we find

$$\left[\mathbf{L}^2,\ \left[\mathbf{L}^2,\ K_x\right]\right] = 2\left(\mathbf{L}^2 K_x + K_x \mathbf{L}^2\right) - 4L_x(\mathbf{L}\cdot\mathbf{K}). \tag{9.101}$$

(2) - We can also write the double commutator as

$$\left[\mathbf{L}^2,\ \left[\mathbf{L}^2,\ K_x\right]\right] = \mathbf{L}^4 K_x - 2\mathbf{L}^2 K_x \mathbf{L}^2 + K_x \mathbf{L}^4. \tag{9.102}$$

By equating (9.101) and (9.102) we obtain

$$\mathbf{L}^4 K_x - 2\mathbf{L}^2 K_x \mathbf{L}^2 + K_x \mathbf{L}^4 = 2\left(\mathbf{L}^2 K_x + K_x \mathbf{L}^2\right) - 4L_x(\mathbf{L}\cdot\mathbf{K}). \tag{9.103}$$

The matrix element corresponding to the transition $n, \ell, m \to n', \ell', n'$ $(\ell \neq \ell')$ of Eq. (9.103) is

$$\begin{aligned} &\left[\ell^2(\ell+1)^2 - 2\ell(\ell+1)\ell'\left(\ell'+1\right) + \ell'^{\,2}\left(\ell'+1\right)^2\right]\langle n,\ell,m|K_x|n',\ell',m'\rangle \\ -\ &2\left[\ell(\ell+1) + \ell'\left(\ell'+1\right)\right]\langle n,\ell,m|K_x|n',\ell',m'\rangle = 0. \end{aligned} \tag{9.104}$$

We arrive at this result by noting that

$$\langle n,\ell,m|L_x(\mathbf{L}\cdot\mathbf{K})|n',\ell',m'\rangle = 0, \quad \ell \neq \ell'. \tag{9.105}$$

This follows from the fact that the matrix representing the scalar $\mathbf{L}\cdot\mathbf{K}$ is diagonal with respect to m and ℓ, and the matrix L_x is diagonal with respect to ℓ and n. By rearranging the terms in (9.104) we find

$$\left[\left(\ell+\ell'+1\right)^2 - 1\right]\left[\left(\ell-\ell'\right)^2 - 1\right]\langle n,\ell,m|K_x|n',\ell',m'\rangle = 0, \quad \ell \neq \ell'. \tag{9.106}$$

The first bracket in (9.106) cannot be zero, since ℓ and ℓ' are equal or greater than zero and $\ell \neq \ell'$. Therefore we have the condition

$$\ell' - \ell = \pm 1, \tag{9.107}$$

which makes the second bracket in (9.106) zero or otherwise $\langle n,\ell,m|K_x|n',\ell',m'\rangle$ must vanish. In addition to (9.107) the case where

$$\ell' - \ell = 0, \tag{9.108}$$

also allows for the nonzero matrix element of K_x [7],[11]. If we replace K_x either by K_y and K_z we find the same results.

The rules (9.107) and (9.108) are called selection rules. In addition to these rules any transition between two states with $\ell = 0$ is forbidden. This can easily be shown by taking the matrix elements of $[L_y,\ K_z] = iK_x$;

$$\langle n,0,0|L_y K_z - K_z L_y|n',0,0\rangle = i\,\langle n,0,0|K_x|n',0,0\rangle\,. \tag{9.109}$$

The left-hand side of (9.109) is zero since L_y is diagonal with the eigenvalue $\ell = 0$. Thus we have

$$\langle n,0,0|K_x|n',0,0\rangle = 0. \tag{9.110}$$

Similar selection rules can be found with respect to the quantum number m [7].

9.4 Spin Angular Momentum

Some particles such as electrons and protons have intrinsic angular momentum in addition to the orbital angular momentum **L** which is associated with their motion. The spin may be compared to the angular momentum of a rigid body about its center of mass, but it arises from the internal degrees of freedom of the particle.

Let us denote the components of the spin angular momentum of a particle by the Hermitian operator **S** with the components S_x, S_y and S_z. These components satisfy the fundamental commutation relation

$$\mathbf{S} \wedge \mathbf{S} = i\hbar\mathbf{S}. \tag{9.111}$$

A trivial solution of (9.111) is found when all of the components of **S** are zero. This is the case of spin zero particle such as π-meson. But particles like electrons and protons are spin $\frac{1}{2}$ particles. This means that the eigenvalues of the components of **S** along a given direction can take on the values $\pm\frac{\hbar}{2}$. Thus for any direction **n**, there are two eigenstates $|\mathbf{n}\uparrow\rangle$ and $|\mathbf{n}\downarrow\rangle$ corresponding to the eigenvalues $\frac{1}{2}\hbar$ and $-\frac{1}{2}\hbar$. We can choose the eigenstates of S_z as the basis for spin space. Once this choice is made the operator S_z can be written in this basis as

$$S_z = \begin{bmatrix} \langle\mathbf{k}\uparrow|S_z|\mathbf{k}\uparrow\rangle & \langle\mathbf{k}\uparrow|S_z|\mathbf{k}\downarrow\rangle \\ \langle\mathbf{k}\downarrow|S_z|\mathbf{k}\uparrow\rangle & \langle\mathbf{k}\downarrow|S_z|\mathbf{k}\downarrow\rangle \end{bmatrix}, \tag{9.112}$$

where **k** is the unit vector in the z-direction. Now for this representation we have

$$S_z|\mathbf{k}\uparrow\rangle = \frac{1}{2}\hbar|\mathbf{k}\uparrow\rangle; \quad \text{and} \quad S_z|\mathbf{k}\downarrow\rangle = -\frac{1}{2}\hbar|\mathbf{k}\downarrow\rangle, \tag{9.113}$$

and therefore the matrix (9.112) becomes

$$S_z = \frac{\hbar}{2}\begin{bmatrix} 1 & 0 \\ 0 & -1 \end{bmatrix}. \tag{9.114}$$

The other two components, S_x and S_y, are also Hermitian 2×2 matrices and these can be written as

$$S_x = \frac{\hbar}{2}\begin{bmatrix} 0 & 1 \\ 1 & 0 \end{bmatrix}, \quad \text{and} \quad S_y = \frac{\hbar}{2}\begin{bmatrix} 0 & -i \\ i & 0 \end{bmatrix}, \tag{9.115}$$

(see Eqs. (9.83) and (9.84)). We can easily verify that these three matrices satisfy the fundamental commutation relation (9.111).
The eigenstates $|\uparrow,\mathbf{k}\rangle$ and $|\downarrow,\mathbf{k}\rangle$ obtained from (9.114) are given by

$$|\uparrow,\mathbf{k}\rangle = \begin{bmatrix} 1 \\ 0 \end{bmatrix}, \quad |\downarrow,\mathbf{k}\rangle = \begin{bmatrix} 0 \\ 1 \end{bmatrix}. \tag{9.116}$$

Generally the spin operator **S** is expressed in terms of Pauli matrices σ_x, σ_y and σ_z, where

$$\mathbf{S} = \frac{\hbar}{2}\boldsymbol{\sigma}, \tag{9.117}$$

and

$$\sigma_x = \begin{bmatrix} 0 & 1 \\ 1 & 0 \end{bmatrix}, \quad \sigma_y = \begin{bmatrix} 0 & -i \\ i & 0 \end{bmatrix} \quad \sigma_z = \begin{bmatrix} 1 & 0 \\ 0 & -1 \end{bmatrix}. \tag{9.118}$$

Some of the properties of the Pauli spin matrices are as follows [8]:

(a) - They obey the commutation relation

$$\sigma_i \sigma_j - \sigma_j \sigma_i = 2i \sum_k \varepsilon_{ijk} \sigma_k, \tag{9.119}$$

which is the same as (9.111) but written in terms of σ.

(b) - The square of each σ_i is a unit matrix

$$\sigma_x^2 = \sigma_y^2 = \sigma_z^2 = \begin{bmatrix} 1 & 0 \\ 0 & 1 \end{bmatrix} = 1. \tag{9.120}$$

(c) - By direct calculation we can show that when $i \neq j$ the Pauli matrices satisfy the anti-commutation relation

$$\sigma_i \sigma_j + \sigma_j \sigma_i = 0. \tag{9.121}$$

We can combine (9.120) and (9.121) into a single relation

$$\sigma_i \sigma_j + \sigma_j \sigma_i = 2\delta_{ij}. \tag{9.122}$$

(d) - We can also combine (9.119) and (9.122) and write

$$\sigma_i \sigma_j = \delta_{ij} + i \sum_k \varepsilon_{ijk} \sigma_k. \tag{9.123}$$

(e) - If $\mathbf{A}$ and $\mathbf{B}$ are two c-number vectors we have

$$\begin{aligned} (\mathbf{A} \cdot \boldsymbol{\sigma})(\mathbf{B} \cdot \boldsymbol{\sigma}) &= \sum_{ij} A_i B_j \sigma_i \sigma_j = \sum_{ij} A_i B_j (\delta_{ij} + i \sum_k \varepsilon_{ijk} \sigma_k) \\ &= \mathbf{A} \cdot \mathbf{B} + i(\mathbf{A} \wedge \mathbf{B}) \cdot \boldsymbol{\sigma}. \end{aligned} \tag{9.124}$$

(f) - From (9.117) and (9.120) we find $\mathbf{S}^2$ to be

$$\mathbf{S}^2 = \frac{\hbar^2}{4} \left(\sigma_x^2 + \sigma_y^2 + \sigma_z^2 \right) = \frac{3\hbar^3}{4}. \tag{9.125}$$

(g) - The total angular momentum $\mathbf{J}$ of a spinning particle is the sum of its orbital angular momentum and its spin angular momentum

$$\mathbf{J} = \mathbf{L} + \mathbf{S}. \tag{9.126}$$

9.5 Angular Momentum Eigenvalues Determined from the Eigenvalues of Two Uncoupled Oscillators

An elegant way of finding the eigenvalues and eigenvectors of angular momentum in terms of the creation and annihilation operators for the uncoupled oscillators has been derived by Schwinger [9].

Let us consider two simple harmonic oscillators for which the creation and annihilation operators are denoted by $a_+^\dagger$, a_+, $a_-^\dagger$ and a_-. The number operators for $(+)$ and $(-)$ oscillators are

$$N_+ = a_+^\dagger a_+, \quad N_- = a_-^\dagger a_-, \tag{9.127}$$

and these operators satisfy the commutation relations

$$\left[a_\pm,\, a_\pm^\dagger\right] = 1, \quad [N_\pm,\, a_\pm] = -a_\pm, \quad \left[N_\pm,\, a_\pm^\dagger\right] = a_\pm^\dagger, \tag{9.128}$$

where in each case we have to take all plus signs or all minus signs. Since all these oscillators are uncoupled we have

$$\left[a_+,\, a_-^\dagger\right] = \left[a_-,\, a_+^\dagger\right] = 0, \tag{9.129}$$

and so forth. The two number operators N_+ and N_- commute, a result that simply follows from (9.129). Hence they can be diagonalized simultaneously. Let $|n_+, n_-\rangle$ represent a state where the $(+)$ oscillator has an eigenvalue n_+ and the oscillator $(-)$ has an eiganvalue n_-, then we have

$$N_+|n_+, n_-\rangle = n_+|n_+, n_-\rangle, \quad N_-|n_+, n_-\rangle = n_-|n_+, n_-\rangle. \tag{9.130}$$

The action of the creation and annihilation operators on these states will change the numbers n_+ and n_-, i.e.

$$a_+^\dagger|n_+, n_-\rangle = \sqrt{n_+ + 1}\ |n_+ + 1, n_-\rangle, \tag{9.131}$$

$$a_-^\dagger|n_+, n_-\rangle = \sqrt{n_- + 1}\ |n_+, n_- + 1\rangle, \tag{9.132}$$

$$a_+|n_+, n_-\rangle = \sqrt{n_+}\ |n_+ - 1, n_-\rangle, \tag{9.133}$$

and

$$a_-|n_+, n_-\rangle = \sqrt{n_-}\ |n_+, n_- - 1\rangle. \tag{9.134}$$

Next we define M_+, M_- and M_z by

$$M_+ = \hbar a_+^\dagger a_-, \quad M_- = \hbar a_-^\dagger a_+, \tag{9.135}$$

and

$$M_z = \frac{\hbar}{2}\left(a_+^\dagger a_+ - a_-^\dagger a_-\right) = \frac{\hbar}{2}\left(N_+ - N_-\right). \tag{9.136}$$

From the commutation relations of the operators $a_+^\dagger \cdots a_-$, N_+, N_- we can verify that $M_\pm$ and M_z satisfy the angular momentum commutation relations;

$$[M_z, \ M_\pm] = \pm\hbar M_\pm, \tag{9.137}$$

and

$$[M_+, \ M_-] = 2\hbar M_z. \tag{9.138}$$

Moreover if N denotes the sum of N_+ and N_-

$$N = N_+ + N_- = a_+^\dagger a_+ + a_-^\dagger a_-, \tag{9.139}$$

then we have

$$\mathbf{M}^2 = M_z^2 + \frac{1}{2}\left(M_+M_- + M_-M_+\right) = \frac{\hbar^2}{2}N\left(\frac{N}{2}+1\right). \tag{9.140}$$

Now using Eqs. (9.131)–(9.134) we determine the action of the operators M_+, M_- and M_z on $|n_+, n_-\rangle$

$$M_+|n_+, n_-\rangle = \hbar a_+^\dagger a_-|n_+, n_-\rangle = \hbar\sqrt{n_-(n_+ + 1)}\ |n_+ + 1, n_- - 1\rangle \tag{9.141}$$

$$M_-|n_+, n_-\rangle = \hbar a_-^\dagger a_+|n_+, n_-\rangle = \hbar\sqrt{n_+(n_- + 1)}\ |n_+ - 1, n_- + 1\rangle \tag{9.142}$$

and

$$M_z|n_+, n_-\rangle = \frac{\hbar}{2}(N_+ - N_-)|n_+, n_-\rangle = \frac{\hbar}{2}(n_+ - n_-)|n_+, n_-\rangle. \tag{9.143}$$

We note that $|n_+, n_-\rangle$ is an eigenstate of the operator M_z. To write these in the notation of Sec. 9.2 we replace n_+ and n_- by

$$n_+ \to \ell + m, \quad n_- \to \ell - m, \tag{9.144}$$

then

$$M_+|\ell + m,\ \ell - m\rangle = \hbar\sqrt{(\ell - m)(\ell + m + 1)}\ |\ell + m + 1,\ \ell + m - 1\rangle, \tag{9.145}$$

$$M_-|\ell + m,\ \ell - m\rangle = \hbar\sqrt{(\ell + m)(\ell - m + 1)}\ |\ell + m - 1,\ \ell - m + 1\rangle, \tag{9.146}$$

$$M_z|\ell + m,\ \ell - m\rangle = m\hbar\ |\ell + m,\ \ell - m\rangle, \tag{9.147}$$

and

$$\mathbf{M}^2\ |\ell + m,\ \ell - m\rangle = \hbar^2\ell(\ell + 1)\ |\ell + m,\ \ell - m\rangle. \tag{9.148}$$

A physical picture of the connection between (+) and (−) oscillators and the eigenvalues of angular momentum has been discussed by Sakurai [10].

An object of high ℓ can be visualized as a collection of spin $\frac{1}{2}$ particles, $\ell + m$ of them with spin up and $\ell - m$ of them with spin down. Thus if we have 2ℓ spin $\frac{1}{2}$ particles we can add them up in different ways to get states with angular momenta $\ell,\ \ell - 1,\ \ell - 2, \cdots$. If we have just two particles each with a spin $\frac{1}{2}$ then we can have either a system of $\ell = 0$ or $\ell = 1$. The problem with this interpretation is that

(a) - we have to start with an even number of particles, 2ℓ, and

(b) - the (+) and (−) particles are in fact bosons and not fermions.

9.6 Rotations in Coordinate Space and in Spin Space

If the state of a particle in a given coordinate system is described by $|\mathbf{r}\rangle$, and in a rotated system the state is $|\mathbf{r}'\rangle$, then these two are related to each other by a unitary transformation T. Let us consider the simple case of a rotation about the z axis by an angle φ, where the rotated coordinates are related to the original coordinates by a set of linear transformation;

$$\begin{cases} x' = x\cos\varphi + y\sin\varphi, & x = x'\cos\varphi - y'\sin\varphi \\ y' = -x\sin\varphi + y\cos\varphi, & y = x'\sin\varphi + y'\cos\varphi\,. \\ z' = z \end{cases} \tag{9.149}$$

From these relations we find the following derivatives

$$\frac{\partial x}{\partial\varphi} = -y, \quad \frac{\partial y}{\partial\varphi} = x, \quad \frac{\partial z}{\partial\varphi} = 0. \tag{9.150}$$

Now let us expand the transformed state $|\mathbf{r}'\rangle$ as a power series in φ

$$|\mathbf{r}'\rangle = |x',\, y',\, z'\rangle = |x,\, y,\, z\rangle + \frac{\varphi}{1!}\left[\frac{d|\mathbf{r}\rangle}{d\,\varphi}\right]_{\varphi=0} + \frac{\varphi^2}{2!}\left[\frac{d^2|\mathbf{r}\rangle}{d^2\varphi^2}\right]_{\varphi=0} + \cdots. \tag{9.151}$$

The coefficient of φ in (9.151) can be calculated in terms of the derivatives;

$$\left[\frac{\partial\,|\mathbf{r}\rangle}{\partial\varphi}\right]_{\varphi=0} = \frac{\partial\,|\mathbf{r}\rangle}{\partial x}\frac{\partial x}{\partial\varphi} + \frac{\partial\,|\mathbf{r}\rangle}{\partial y}\frac{\partial y}{\partial\varphi}$$

$$= \left(x\frac{\partial}{\partial y} - y\frac{\partial}{\partial x}\right)|\mathbf{r}\rangle = \frac{i}{\hbar}L_z|\mathbf{r}\rangle, \tag{9.152}$$

where L_z is the z component of the angular momentum operator. In a similar way we can determine the coefficients φ^2, φ^3 etc. The resulting infinite series can be summed up as an exponential operator

$$|\mathbf{r}'\rangle = \sum_{n=0}^{\infty}\left(\frac{\varphi^n}{n!}\right)\left(\frac{iL_z}{\hbar}\right)^n|\mathbf{r}\rangle = e^{\frac{i\varphi L_z}{\hbar}}\,|\mathbf{r}\rangle\,. \tag{9.153}$$

The result for the rotation about the z-axis can be generalized to a rotation about a given axis $\mathbf{n}$, and in this case the general form of the unitary operator T for this rotation is

$$T = \exp\left(\frac{i(\mathbf{n}\cdot\mathbf{L})\phi}{\hbar}\right), \tag{9.154}$$

where ϕ is the angle of rotation.

Rotations and Spin Space — The average value of the spin must transform as a vector, therefore we expect that the averages of the components of spin transform according to

$$\bar{\sigma}_x = \sin\vartheta\cos\varphi, \quad \bar{\sigma}_y = \sin\vartheta\sin\varphi, \quad \bar{\sigma}_z = \cos\varphi. \tag{9.155}$$

Let us assume that the average value of σ_z in a state with spin up amplitude α and spin down amplitude β, i.e.

$$|a\rangle = \begin{bmatrix} \alpha \\ \beta \end{bmatrix}, \tag{9.156}$$

is $\cos\vartheta$. Then

$$\begin{aligned} \bar{\sigma}_z &= \langle a|\sigma_z|a\rangle = [\alpha^*,\ \beta^*] \begin{bmatrix} 1 & 0 \\ 0 & -1 \end{bmatrix} \begin{bmatrix} \alpha \\ \beta \end{bmatrix} \\ &= |\alpha|^2 - |\beta|^2 = \cos\vartheta. \end{aligned} \tag{9.157}$$

To this relation we add the normalization condition

$$\langle a|a\rangle = |\alpha|^2 + |\beta|^2 = 1, \tag{9.158}$$

By solving Eqs. (9.157) and (9.158) for α and β we find two solutions for $|a\rangle$ which we denote by $|a_1\rangle$ and $|a_2\rangle$;

$$|a_1\rangle = \begin{bmatrix} \cos\frac{\vartheta}{2} \\ e^{i\chi_1}\sin\frac{\vartheta}{2} \end{bmatrix}, \quad \text{and} \quad |a_2\rangle = \begin{bmatrix} -e^{i\chi_2}\sin\frac{\vartheta}{2} \\ \cos\frac{\vartheta}{2} \end{bmatrix}, \tag{9.159}$$

where χ_1 and χ_2 are real phases to be determined. Again for the state $|a\rangle$ we find $\langle a_i\,|\sigma_x|\,a_i\rangle$ and $\langle a_i\,|\sigma_y|\,a_i\rangle$, $i = 1,\ 2$ and equate them with $\bar{\sigma}_x$ and $\bar{\sigma}_y$, of Eq. (9.155).

$$\langle a_i|\sigma_x|a_i\rangle = \sin\vartheta\cos\varphi, \tag{9.160}$$

and

$$\langle a_i|\sigma_y|a_i\rangle = \sin\vartheta\sin\varphi. \tag{9.161}$$

By substituting for $|a_1\rangle$ and $|a_2\rangle$ from (9.159) in (9.160) and (9.161) we find that $\chi_1 = \chi_2 = \varphi$. Therefore $|a_1\rangle$ and $|a_2\rangle$ will be given in terms of ϑ and φ by the matrices

$$|a_1\rangle = \begin{bmatrix} \cos\frac{\vartheta}{2} \\ e^{i\varphi}\sin\frac{\vartheta}{2} \end{bmatrix}, \quad \text{and} \quad |a_2\rangle = \begin{bmatrix} -e^{-i\varphi}\sin\frac{\vartheta}{2} \\ \cos\frac{\vartheta}{2} \end{bmatrix}. \tag{9.162}$$

For $\vartheta = 0$, $|a_1\rangle$ becomes spin up state $|\uparrow \mathbf{k}\rangle$ and $|a_2\rangle$ becomes spin down state $|\downarrow \mathbf{k}\rangle$.

Suppose that a spinor is in the state given by $|a_1\rangle$, Eq. (9.159), then a rotation by an angle ϕ about the z axis changes $|a_1\rangle$ to $|a_1'\rangle$. In analogy with rotation in coordinate space, Eq. (9.152), we expect that the generator of such a transformation to be

$$T_\phi = e^{\frac{i\phi\sigma_3}{2}}, \tag{9.163}$$

i.e. the transformed state $|a_1'\rangle$ be given by

$$|a_1'\rangle = T_\phi|a_1\rangle. \tag{9.164}$$

Noting that

$$\sigma_3 = \begin{bmatrix} 1 & 0 \\ 0 & -1 \end{bmatrix}, \quad \text{and} \quad \sigma_3^2 = \begin{bmatrix} 1 & 0 \\ 0 & 1 \end{bmatrix} = 1, \tag{9.165}$$

we find the expansion of T_ϕ in powers of ϕ;

$$\begin{aligned} T_\phi &= \left[1 + \frac{1}{2!}\left(\frac{i\phi}{2}\right)^2 + \cdots\right] 1 + \left[\left(\frac{i\phi}{2}\right) + \frac{1}{3!}\left(\frac{i\phi}{2}\right)^3 + \cdots\right]\sigma_3 \\ &= \cos\left(\frac{\phi}{2}\right) 1 + i\sin\left(\frac{\phi}{2}\right)\sigma_3 = \begin{bmatrix} e^{\frac{i\phi}{2}} & 0 \\ 0 & e^{\frac{-i\phi}{2}} \end{bmatrix}. \end{aligned} \tag{9.166}$$

Thus

$$|a_1'\rangle = e^{\frac{i\phi}{2}} \begin{bmatrix} \cos\frac{\vartheta}{2} \\ e^{i(\varphi-\phi)}\sin\frac{\vartheta}{2} \end{bmatrix}, \tag{9.167}$$

which, apart from the physically unobservable phase factor, is the same as $|a_1\rangle$ except for the change in azimuthal angle $\varphi \to \varphi - \phi$ as is expected.

9.7 Motion of a Particle Inside a Sphere

The first problem that we will study is the three-dimensional motion of a particle confined inside a sphere of radius R without the presence of any other force. The Hamiltonian for this motion is simple;

$$H = \frac{1}{2m}\left[p_r^2 + \frac{\mathbf{L}^2}{r^2}\right] = \frac{1}{2m}\left[p_r^2 + \frac{\ell(\ell+1)\hbar^2}{r^2}\right]. \tag{9.168}$$

As in the case of the motion of a particle in a box, (Sec. 8.4), we choose the lowering operator a_j as

$$a_j = \frac{1}{\sqrt{2m}}\left[p_r + i\gamma_j \cot\left(\frac{\pi r}{R}\right)\right], \tag{9.169}$$

and $a_j^\dagger$ as the Hermitian adjoint of a_j. For zero angular momentum, $\ell = 0$, Eq. (9.168) has exactly the same form as the Hamiltonian for a particle in a box, $H = \frac{p^2}{2m}$. In this special case, ($\ell = 0$), we can use the method of factorization outlined Sec. 8.4 to find the eigenvalues,

$$\gamma_j = \left(\frac{j\pi\hbar}{R}\right)^2. \tag{9.170}$$

When $l \neq 0$ we use the operator

$$\alpha_j = \frac{1}{\sqrt{2m}}\left[p_r + \frac{i\hbar j}{r}\right]. \tag{9.171}$$

The Hamiltonian (9.168) can be written in terms of this operator and $\alpha_j^\dagger$ as

$$H = \alpha_j \alpha_j^\dagger. \tag{9.172}$$

Then the eigenvalue equation is

$$H\psi_\ell^{(j)}(r) = E_\ell^{(j)}\psi_\ell^{(j)}, \tag{9.173}$$

where $E_\ell^{(j)}$ is the energy eigenvalue.

The two operators $\alpha_j^\dagger$ and α_j have the following additional properties:

$$\alpha_1^\dagger \alpha_1 = \frac{p_r^2}{2m} \tag{9.174}$$

and

$$\alpha_{j+1}^\dagger \alpha_{j+1} = \alpha_j \alpha_j^\dagger. \tag{9.175}$$

These results can be derived directly from the definition of α_j, Eq. (9.171), and its Hermitian conjugate.

Next we determine $\phi_l^{(1)}(r)$ which is the solution of $a_l|\phi_l\rangle = 0$ or

$$a_l\phi_l(r) = \frac{1}{\sqrt{2m}}\left[-\frac{i\hbar}{r}\frac{d}{dr}(r\phi_l(r)) + i\gamma_l \cot(k_l r)\phi_l(r)\right] = 0. \tag{9.176}$$

If we substitute for a_j and $a_j^\dagger$ from (9.169) in the Hamiltonian (9.168) and set the coefficient of $\cot^2(k_l r)$ equal to zero we find $\gamma_j = \hbar k_j$ and

$$\psi_l^{(1)} = \alpha_l \alpha_{l-1} \cdots \alpha_1 \phi_l, \tag{9.177}$$

with the energy eigenvalues

$$E_\ell^{(j)} = \frac{\hbar^2\left(k_\ell^{(j)}\right)^2}{2m}. \tag{9.178}$$

Now by eliminating p_r between (9.169) and (9.171) we have

$$\alpha_j = a_j + \frac{ij\hbar}{r} - i\hbar k_j \cot(k_j r). \tag{9.179}$$

From (9.177) and (9.179) we find $\psi_1^{(1)}(r)$;

$$\begin{aligned}\psi_1^{(1)}(r) &= \alpha_1\phi_1(r) = a_1\phi_1(r) + i\hbar\left(\frac{1}{r} - k_1\cot(k_1 r)\right)\phi_1(r)\\ &= i\hbar\left(\frac{1}{r} - k_1\cot(k_1 r)\right)\phi_1(r).\end{aligned} \tag{9.180}$$

The general form of eigenvector $\psi_l^{(1)}(r)$ is found from (9.177) turns out to be

$$\psi_\ell^{(1)} = (i\hbar k_1)^l g_\ell(k_\ell r)\phi_\ell(r), \tag{9.181}$$

where $g_l(kr)$ is obtained from the recurrence relation

$$g_{\ell+1}(x) = (\ell+1)\frac{g_\ell(x)}{x} - \left(\frac{dg_\ell(x)}{dx} + g_\ell(x)\cot(x)\right). \tag{9.182}$$

In the range of

$$0 < x < k_\ell R, \tag{9.183}$$

the term in the parenthesis in (9.182) varies between $-\infty$ and ∞ and at the boundary $\cot(k_\ell R)$ becomes infinite. So at this point the coefficient of $\cot(k_\ell R)$, i.e. $g_\ell(k_\ell R)$ has to be zero. The smallest root of $g_\ell(k_\ell R) = 0$ gives us the lowest eigenvalue $k_\ell^{(1)}$ with the smallest energy $E_\ell^{(1)}$ given by (9.178). We note that Eq. (9.182) in the absence of the last term, $g_\ell(x)\cot(x)$, is the recurrence relation for the spherical Bessel function of the the order ℓ;

$$j_{\ell+1}(x) = (\ell+1)\frac{j_\ell(x)}{x} - \frac{dj_\ell(x)}{dx}. \tag{9.184}$$

9.8 The Hydrogen Atom

We used the factorization method to find the eigenvalues of the Kepler problem, Eq. (8.154). Now let us apply the same method to a system composed of a positive charge $-Ze$ interacting with an electron of charge e. The Hamiltonian (9.4) in this case is

$$H = \frac{\mathbf{P}^2}{2M} + \frac{p_r^2}{2m} + \frac{\mathbf{L}^2}{2mr^2} - \frac{Ze^2}{r}. \tag{9.185}$$

Denoting the eigenvalues of the operators $\mathbf{L}^2$ and $\frac{\mathbf{P}^2}{2M}$ by $\hbar^2 l(l+1)$ and $E_{c.m.}$ respectively we have

$$H - E_{c.m.} = H_r = \frac{p_r^2}{2m} + \frac{\hbar^2\ell(\ell+1)}{2mr^2} - \frac{Ze^2}{r}. \tag{9.186}$$

As we have seen in the case of one-dimensional motion we can write H_r as

$$H_r = a_1^\dagger a_1 + \lambda_1, \tag{9.187}$$

where

$$a_j = \frac{1}{\sqrt{2}}\left[p_r + i\left(\gamma_j + \frac{B_j}{r}\right)\right], \tag{9.188}$$

and γ_j and B_j are real numbers to be determined. Similarly we can define the operators A_{j+1} and A_j by

$$A_{j+1} = a_j a_j^\dagger + \lambda_j, \tag{9.189}$$

and

$$A_j = a_j^\dagger a_j + \lambda_j. \tag{9.190}$$

Next we calculate $a_j^\dagger a_j$

$$\begin{aligned} a_j^\dagger a_j &= \frac{1}{2m}\left\{p_r - i\left(\gamma_j + \frac{B_j}{r}\right)\right\}\left\{p_r + i\left(\gamma_j + \frac{B_j}{r}\right)\right\} \\ &= \frac{1}{2m}\left\{p_r^2 + \left(\gamma_j + \frac{B_j}{r}\right)^2 + iB_j\left[p_r, \frac{1}{r}\right]\right\} \\ &= \frac{1}{2m}\left\{p_r^2 + \gamma_j^2 + \frac{2\gamma_j B_j}{r} + \frac{\left(B_j^2 - B_j\hbar\right)}{r^2}\right\}, \end{aligned} \tag{9.191}$$

and in a similar way we calculate $a_j a_j^\dagger$

$$\begin{aligned} a_j a_j^\dagger &= \frac{1}{2m}\left\{p_r^2 + \left(\gamma_j + \frac{B_j}{r}\right)^2 - iB_j\left[p_r, \frac{1}{r}\right]\right\} \\ &= \frac{1}{2m}\left\{p_r^2 + \gamma_j^2 + \frac{2\gamma_j B_j}{r} + \frac{\left(B_j^2 + B_j\hbar\right)}{r^2}\right\}. \end{aligned} \tag{9.192}$$

By comparing $a_1^\dagger a_1$ obtained from (9.191) with the same operator found from (9.186) and (9.187) we find

$$\gamma_1 B_1 = -mZe^2, \quad \text{and} \quad B_1(a_1 - \hbar) = \hbar^2\ell(\ell+1), \tag{9.193}$$

with

$$\lambda_1 = -\frac{\gamma_1^2}{2m}. \tag{9.194}$$

These equation have two sets of solutions:
We can either choose

$$B_1 = -\ell\hbar, \quad \gamma_1 = \frac{mZe^2}{\hbar\ell}, \quad \text{and} \quad \lambda_1 = -\frac{\left(mZe^2\right)^2}{2m\hbar^2\ell^2}, \tag{9.195}$$

or

$$B_1 = (\ell+1)\hbar, \quad \gamma_1 = -\frac{mZe^2}{\hbar(\ell+1)}, \quad \text{and} \quad \lambda_1 = -\frac{\left(mZe^2\right)^2}{2m\hbar^2(\ell+1)^2}. \tag{9.196}$$

The second set gives us a larger eigenvalue and therefore we choose (9.196) rather than (9.195). Also from (9.189) and (9.190) we have

$$a_{j+1}^\dagger a_{j+1} + \lambda_{j+1} = a_j a_j^\dagger + \lambda_j. \tag{9.197}$$

Similarly for the general j we choose

$$\gamma_{j+1}B_{j+1} = \gamma_j B_j, \quad \text{and} \quad B_{j+1}(B_{j+1} - \hbar) = B_j(B_j + \hbar), \tag{9.198}$$

with the corresponding eigenvalue λ_j satisfying the relation

$$\lambda_{j+1} + \left(\frac{\gamma_{j+1}^2}{2m}\right) = \lambda_j + \left(\frac{\gamma_j^2}{2m}\right). \tag{9.199}$$

Again for a fixed ℓ we choose the solution which maximizes the eigenvalue;

$$B_{j+1} = B_j + \hbar = \cdots = B_1 + j\hbar = (\ell + 1 + j)\hbar. \tag{9.200}$$

Solving for B_j, γ_j and λ_j we find

$$B_j = (\ell + j)\hbar, \quad \gamma_j = -\frac{mZe^2}{(\ell + j)\hbar}, \quad \text{and} \quad \lambda_j = -\frac{m^2 Z^2 e^4}{2m(\ell + j)^2\hbar^2}. \tag{9.201}$$

For a hydrogen atom at rest $\mathbf{P}^2 = 0$ and the energy eigenvalues are

$$E_{l,j} = -\frac{mZ^2e^4}{2(\ell + j)^2\hbar^2}, \tag{9.202}$$

where ℓ is a nonnegative integer and j is a positive integer.

Hydrogen Atom Eigenstates — The wave function for the Coulomb potential can be found from Eq. (8.81) or from

$$\psi^{(j)}(r) = a_1^* a_2^* \cdots a_{j-1}^* \phi^{(j)}(r). \tag{9.203}$$

The lowest state for a given j is a solution of

$$a_j \phi^{(j-1)} = \frac{1}{\sqrt{2m}} \left\{ p_r + i \left(\gamma_j + \frac{B_j}{r} \right) \right\} \phi^{(j-1)}(r) = 0, \tag{9.204}$$

where γ_j and B_j are given by Eq. (9.201) and

$$p_r = -i\frac{\hbar}{r}\frac{\partial}{\partial r} r, \tag{9.205}$$

is the radial momentum operator. Now by integrating this first order differential equation we find $\phi^{(j-1)}(r)$;

$$r\phi^{(j-1)}(r) = N_{\ell j}\, r^{\ell + j} \exp\left[-\frac{mZe^2}{(\ell + j)\hbar^2} r \right]. \tag{9.206}$$

For a given ℓ this wave function has no nodes, i.e. it is the wave function of the ground state which is found for $j = 1$ in (9.203). If we write $n = \ell + j$ with $j = 1$ or $\ell = n - 1$ then

$$r\phi_n^{(\ell)}(r) = r\phi_n^{(n-1)}(r) = N_n r^n \exp\left[-\frac{Zr}{na_0} \right]. \tag{9.207}$$

In Eq. (9.207) a_0 is the Bohr radius, $a_0 = \frac{\hbar^2}{me^2}$ represents the nodeless wave function for the principal quantum number n.

Having obtained $\phi_n^{(n-1)}(r)$, we can construct

$$\phi_n^{(\ell=n-2)}(r),\ \phi_n^{(\ell=n-3)}(r),\ \cdots, \tag{9.208}$$

by the following method:

Let us consider the hydrogen atom for which $Z=1$. Then the normalized $\phi_n^{(n-1)}(r)$ can be written as

$$\phi_n^{(n-1)}(r) = \left(\frac{2}{na_0}\right)^{\frac{3}{2}}\left(\frac{1}{\sqrt{2n(2n-1)!}}\right)\left(\frac{2r}{na_0}\right)^{n-1}\exp\left[-\frac{r}{na_0}\right]. \tag{9.209}$$

Next we introduce the raising and lowering operators by

$$a_n^\ell(r) = \frac{n\ell}{\sqrt{n^2-\ell^2}}\left[\frac{1}{\ell} + a_0\left(\frac{d}{dr} - \frac{\ell-1}{r}\right)\right], \tag{9.210}$$

and

$$a_n^{\ell\,*}(r) = \frac{n\ell}{\sqrt{n^2-\ell^2}}\left[\frac{1}{\ell} - a_0\left(\frac{d}{dr} + \frac{\ell+1}{r}\right)\right]. \tag{9.211}$$

The lowering operator $a_n^{\ell\,*}(r)$ reduces the angular momentum quantum number ℓ by one unit [12]–[15]. Then we have

$$\phi_n^{(\ell-1)}(r) = a_n^{\ell\,*}(r)\phi_n^{(\ell)}, \tag{9.212}$$

and

$$\phi_n^{(\ell-2)}(r) = a_n^{\ell-1\,*}(r)\phi_n^{(\ell-1)} = a_n^{\ell-1\,*}(r)b_n^{\ell-1\,*}(r)\phi_n^{(\ell)}. \tag{9.213}$$

This lowering of the ℓ s can be continued up to the point where ℓ becomes zero. In this way we can generate all of the wave functions of the hydrogen atom.

As an example consider the normalized $\phi_2^{(1)}(r)$ which is given by

$$\phi_2^1(r) = \frac{1}{\sqrt{4!}}\left(\frac{1}{a_0}\right)^{\frac{3}{2}}\frac{r}{a_0}e^{-\frac{r}{2a_0}}, \tag{9.214}$$

and the lowering operator which according to (9.211) is

$$a_2^{1\,*}(r) = \frac{2}{\sqrt{3}}\left[1 - a_0\left(\frac{2}{r} + \frac{d}{dr}\right)\right]. \tag{9.215}$$

From these expressions we find $\phi_2^{(0)}(r)$

$$\phi_2^{(0)} = a_2^{1\,*}(r)\phi_2^{(1)}(r) = \frac{1}{2\sqrt{2}a_0^{\frac{3}{2}}}\left(-2+\frac{r}{a_0}\right)\frac{r^2}{a_0^2}e^{-\frac{r}{3a_0}}. \tag{9.216}$$

Likewise for $n = 3$, from (9.209) we obtain

$$\phi_3^{(2)}(r) = \left(\frac{2}{3a_0}\right)^{\frac{3}{2}} \frac{1}{\sqrt{6}} \left(\frac{2r}{3a_0}\right)^2 e^{-\frac{r}{3a_0}}, \tag{9.217}$$

and from (9.211) we have

$$a_3^{2\,*}(r) = \frac{6}{\sqrt{5}}\left[\frac{1}{2} - a_0\left(\frac{d}{dr} + \frac{3}{r}\right)\right]. \tag{9.218}$$

Thus

$$\phi_3^1(r) = a_3^{2\,*}(r)\phi_3^{(2)}(r) = \frac{4}{81\sqrt{6}a_0^{\frac{3}{2}}}\left(\frac{r^2}{a_0^2} - \frac{6r}{a_0}\right) e^{-\frac{r}{3a_0}}, \tag{9.219}$$

and from $\phi_3^1(r)$ and $a_3^{1\,*}(r)$ we can calculate $\phi_3^{(0)}(r)$,

$$\phi_3^0(r) = a_3^{1\,*}(r)\phi_3^{(2)}(r) = \frac{2}{81\sqrt{3}a_0^{\frac{3}{2}}}\left[27 - 18\frac{r}{a_0} + \frac{2r^2}{a_0^2}\right] e^{-\frac{r}{3a_0}}. \tag{9.220}$$

Here the phase of the wave function is given by $(-1)^{n-\ell-1}$ whereas this phase is 1 in the standard textbooks [7].

The complete wave function which includes the angular part can be written as

$$\psi_{n\ell m}(r,\theta,\phi) = \phi_n^{\ell}(r) Y_{\ell,m}(\theta,\phi), \tag{9.221}$$

where $Y_{\ell m}(\theta,\phi)$ is the spherical harmonics;

$$Y_{\ell m}(\theta,\phi) = \sqrt{\frac{(2\ell+1)(\ell-m)!}{4\pi(\ell+1)!}} P_\ell^m(\cos\theta) e^{im\phi}. \tag{9.222}$$

9.9 Calculation of the Energy Eigenvalues Using the Runge–Lenz Vector

This method is a version of the elegant solution of the hydrogen atom problem found by Pauli within the framework of the matrix mechanics before the discovery of wave mechanics [16], [17].

In the classical Kepler problem the orbit of a particle attracted by a potential $\frac{-e^2}{r}$ to the center of force close on themselves, i.e. the orbit does not precess. The reason for this is the fact that the period associated with the angular motion, i.e. the time that the polar angle θ changes from zero to 2π is the same (or in general an integral multiple) of the period that the radial distance goes from its minimum value to the maximum value and back. The non-precessing orbit is also related to the conservation of the Runge–Lenz vector (see Sec. 1.10)[18]. Let $\mathbf{R}$ which is defined by

$$\mathbf{R} = \frac{1}{m}\mathbf{p} \wedge \mathbf{L} - \frac{e^2}{r}\mathbf{r}, \tag{9.223}$$

denote the Runge–Lenz vector. This vector, $\mathbf{R}$, lies in the plane of motion of the particle and is directed along the semi-major axis of the ellipse. Since $\mathbf{R}$ is constant, the major axis of the ellipse is fixed in space and cannot precess. The quantum mechanical operator form of the Runge–Lenz vector can be found by writing for $\mathbf{p}$ and $\mathbf{L}$ their operator forms and symmetrizing the result to get a Hermitian $\mathbf{R}$

$$\mathbf{R} = \frac{1}{2m}(\mathbf{p} \wedge \mathbf{L} - \mathbf{L} \wedge \mathbf{p}) - \frac{Ze^2}{r}\mathbf{r}. \tag{9.224}$$

Note that this is the only Hermitian operator for $\mathbf{R}$ that we can construct. This $\mathbf{R}$ operator commutes with either of the Hamiltonians (9.185) or (9.186) and therefore is a constant of motion,

$$[\,\mathbf{R},\ H_r] = 0, \tag{9.225}$$

In addition to this property the $\mathbf{R}$ operator is (a) - perpendicular to $\mathbf{L}$,

$$\mathbf{R} \cdot \mathbf{L} = \mathbf{L} \cdot \mathbf{R} = 0, \tag{9.226}$$

and (b) - the square of its magnitude can be expressed in terms of the Hamiltonian [8]

$$\mathbf{R}^2 = \left(Ze^2\right)^2 + \frac{2H_1\left(\mathbf{L}^2 + \hbar^2\right)}{m}. \tag{9.227}$$

For bound states of the hydrogen atom, the eigenvalues of H_r are negative numbers. When this is the case it is more convenient to work with the vector operator $\mathbf{K}$ rather than $\mathbf{R}$ where

$$\mathbf{K} = \sqrt{\frac{-m}{2H_r}}\mathbf{R}. \tag{9.228}$$

This operator, $\mathbf{K}$, is Hermitian provided that it acts on the bound state eigenstates. The vector $\mathbf{R}$ does not commute with itself and we have the following commutators for its components

$$[R_i,\ R_j] = i\hbar\left(\frac{-2H_r}{m}\right)\sum_k \varepsilon_{ijk}L_k, \tag{9.229}$$

where ε_{ijk} is the totally antisymmetric matrix introduced earlier, with its nonzero elements given by (1.32) and (1.33). When we replace $\mathbf{R}$ by $\mathbf{K}$, Eq. (9.229) becomes

$$[K_i,\ K_j] = i\hbar\sum_k \varepsilon_{ijk}L_k. \tag{9.230}$$

In addition to (9.230) we also have the commutator of $\mathbf{K}$ with $\mathbf{L}$, viz,

$$[K_i,\ L_j] = i\hbar \sum_k \varepsilon_{ijk} K_k. \tag{9.231}$$

Equations (9.229) and (9.231) can be derived by writing for R_i, K_i and L_i in terms of x, y, z, p_x, p_y and p_z and use the commutators of position and momentum to simplify the result, exactly as was done for the commutators of the components of $\mathbf{L}$.

By solving Eq. (9.227) for H_r, noting that $\mathbf{K}$ and $\mathbf{L}$ are constants of motion we obtain

$$H_r = -\frac{mZ^2e^4}{2\left(\mathbf{K}^2 + \mathbf{L}^2 + \hbar^2\right)}. \tag{9.232}$$

Since $\mathbf{K}$ and $\mathbf{L}$ do not commute with each other we introduce two new operators $\mathbf{M}$ and $\mathbf{N}$ by [8]

$$\mathbf{M} = \frac{1}{2}(\mathbf{L} + \mathbf{K}), \tag{9.233}$$

and

$$\mathbf{N} = \frac{1}{2}(\mathbf{L} - \mathbf{K}). \tag{9.234}$$

The commutation relations for $\mathbf{M}$ and $\mathbf{N}$ found from (9.35), (9.230) and (9.231) are

$$[M_i,\ M_j] = i\hbar \sum_k \varepsilon_{ijk} M_k, \tag{9.235}$$

$$[N_i,\ N_j] = i\hbar \sum_k \varepsilon_{ijk} N_k, \tag{9.236}$$

and

$$[M_i,\ N_j] = 0. \tag{9.237}$$

These relations show that $\mathbf{M}$ and $\mathbf{N}$ commute with each other and that they satisfy the commutation relations for the components of angular momentum. Using the operators $\mathbf{M}$ and $\mathbf{N}$, we can diagonalize the Hamiltonian together with $\mathbf{M}^2$, M_z, $\mathbf{N}^2$ and N_z since they all commute with each other. Let us denote the eigenstates of these four commuting operators by $|\mathcal{M}, \mathcal{N}, \mu, \nu\rangle$, then

$$\mathbf{M}^2|\mathcal{M}, \mathcal{N}, \mu, \nu\rangle = \hbar^2\mathcal{M}(\mathcal{M}+1)|\mathcal{M}, \mathcal{N}, \mu, \nu\rangle, \tag{9.238}$$

$$\mathbf{N}^2|\mathcal{M}, \mathcal{N}, \mu, \nu\rangle = \hbar^2\mathcal{N}(\mathcal{N}+1)|\mathcal{M}, \mathcal{N}, \mu, \nu\rangle, \tag{9.239}$$

$$M_z|\mathcal{M}, \mathcal{N}, \mu, \nu\rangle = \hbar\mu|\mathcal{M}, \mathcal{N}, \mu, \nu\rangle, \tag{9.240}$$

and

$$N_z|\mathcal{M}, \mathcal{N}, \mu, \nu\rangle = \hbar\nu|\mathcal{M}, \mathcal{N}, \mu, \nu\rangle. \tag{9.241}$$

The eigenvalues $\mathcal{M}^2$ and $\mathcal{N}^2$ are the same as those of $\mathbf{M}^2$ discussed earlier with no restriction for the eigenvalues to be integers

$$\mathcal{M},\ \mathcal{N} = 0,\ \frac{1}{2},\ 1,\ \frac{3}{2},\ 2, \cdots, \tag{9.242}$$

and μ and ν have the same eigenvalues as those of M_z;

$$\mu = -\mathcal{M}, \quad -\mathcal{M}+1, \quad \cdots \mathcal{M}-1, \quad \mathcal{M}, \tag{9.243}$$

$$\nu = -\mathcal{N}, \quad -\mathcal{N}+1, \quad \cdots \mathcal{N}-1, \quad \mathcal{N}. \tag{9.244}$$

Now from Eq. (9.226) and (9.228) we find that

$$\mathbf{K}\cdot\mathbf{L} = \mathbf{L}\cdot\mathbf{K} = 0. \tag{9.245}$$

Using these relations we have

$$(\mathbf{K}+\mathbf{L})^2 - (\mathbf{K}-\mathbf{L})^2 = \frac{1}{4}\left(\mathbf{M}^2 - \mathbf{N}^2\right) = 0. \tag{9.246}$$

Therefore for the eigenstate $|\mathcal{M},\mathcal{N},\mu,\nu\rangle$ we get

$$\mathbf{M}^2|\mathcal{M},\mathcal{N},\mu,\nu\rangle = \mathbf{N}^2|\mathcal{M},\mathcal{N},\mu,\nu\rangle, \tag{9.247}$$

or

$$\mathcal{M} = \mathcal{N}. \tag{9.248}$$

This result shows that the eigenvalues corresponding to the eigenstate $|\mathcal{M},\mathcal{N}=\mathcal{M},\mu,\nu\rangle$ is given by

$$H_r|\mathcal{M},(\mathcal{N}=\mathcal{M}),\mu,\nu\rangle = -\left[\frac{mZ^2e^4}{2\hbar^2(2\mathcal{M}+1)^2}\right]|\mathcal{M},(\mathcal{N}=\mathcal{M}),\mu,\nu\rangle, \tag{9.249}$$

where

$$n = 2\mathcal{M}+1 = 1, \quad 2, \quad 3, \cdots. \tag{9.250}$$

Let us examine the state $|\mathcal{M},(\mathcal{N}=\mathcal{M}),\mu,\nu\rangle$. This is an eigenstate of $L_z = M_z+N_z$ with the eigenvalue $\frac{1}{2}\hbar(\mu+\nu)$ which is $\hbar$ times an integer. However this state is not an eigenstate of $\mathbf{L}^2$ because it is a linear combination of the hydrogen atom states (denoted by $|n,\ell,m\rangle$) with fixed n and m but different ℓ values. Now we want to know about the degeneracy of the system, i.e. the number of states with different ℓ and m values having the same energy. We note that the energy eigenvalue depends only on $\mathcal{M}=\mathcal{N}$, Eq. (9.249). Thus for a fixed $\mathcal{M}$ there are $2\mathcal{M}+1$ ν values and for fixed $\mathcal{N}=\mathcal{M}$ there are $(2\mathcal{M}+1)$ ν values. Altogether there are

$$(2\mathcal{M}+1)^2 = n^2, \tag{9.251}$$

different levels corresponding to the same energy eigenvalue E_n

$$E_n = -\frac{mZ^2e^4}{2\hbar^2n^2}, \quad n = 1,2,3\cdots. \tag{9.252}$$

Two-Dimensional Kepler Problem — In Sec. 1.9 we considered the classical formulation of this problem. The Hamiltonian of the system has the simple form of

$$H = \frac{\mathbf{p}^2}{2m} - \frac{Ze^2}{\rho}, \tag{9.253}$$

where $\rho = \sqrt{q_1^2 + q_2^2}$. There we found that three conserved quantities are associated with this Hamiltonian, the Runge–Lenz vector $(R_1,\ R_2)$ and the z component of angular momentum which we denoted by L_3. Following Pauli's method we can solve the quantum mechanical version of this motion for the bound states. For the sake of simplicity we use units where $Ze^2 = m = \hbar = 1$. We first construct Hermitian operators for

$$G_1 = \frac{R_1}{\sqrt{-2E}}, \qquad G_2 = \frac{R_2}{\sqrt{-2E}}, \tag{9.254}$$

by symmetrization method and then replace the Poisson brackets (1.206)-(1.208) by the commutators

$$[L_3,\ G_1] = iG_2, \tag{9.255}$$

$$[L_3,\ G_2] = -iG_1, \tag{9.256}$$

and

$$[G_1,\ G_2] = iL_3. \tag{9.257}$$

Now we construct a three-dimensional vector operator $\mathbf{J}$;

$$\mathbf{J} = \mathbf{G} + \mathbf{L}, \tag{9.258}$$

where $\mathbf{L} = L_3\mathbf{k}$. The components of $\mathbf{J}$ satisfy the commutation relations of the angular momentum

$$\mathbf{J} \wedge \mathbf{J} = i\mathbf{J}. \tag{9.259}$$

We observe that $\mathbf{G}\cdot\mathbf{L} + \mathbf{L}\cdot\mathbf{G} = 0$, therefore

$$\mathbf{J}^2 = (\mathbf{G} + \mathbf{L})^2 = \mathbf{G}^2 + \mathbf{L}^2. \tag{9.260}$$

Next we find the square of the Runge–Lenz vector

$$\begin{aligned} 4\mathbf{R}^2 &= \left[(\mathbf{p}\wedge\mathbf{L}) - (\mathbf{L}\wedge\mathbf{p}) - \frac{2\boldsymbol{\rho}}{\rho}\right]^2 \\ &= \left\{[2(\mathbf{p}\wedge\mathbf{L}) - i\mathbf{p}]^2 - \frac{2\boldsymbol{\rho}}{\rho}\cdot[2(\mathbf{p}\wedge\mathbf{L}) - i\mathbf{p}] \right. \\ &- \left. \frac{2}{\rho}[2(\mathbf{p}\wedge\mathbf{L}) - i\mathbf{p}]\cdot\boldsymbol{\rho} + 4\right\}. \end{aligned} \tag{9.261}$$

Expanding the first term on the left-hand side of (9.261) and simplifying the result we get

$$\begin{aligned} [(2\mathbf{p}\wedge\mathbf{L}) - i\mathbf{p}]^2 &= 4(\mathbf{p}\wedge\mathbf{L})^2 - 2i\mathbf{p}\cdot(\mathbf{p}\wedge\mathbf{L}) - 2i(\mathbf{p}\wedge\mathbf{L})\cdot\mathbf{p} - \mathbf{p}^2 \\ &= \mathbf{p}^2\left(4\mathbf{L}^2 + 1\right). \end{aligned} \tag{9.262}$$

Similarly we expand other terms in Eq. (9.261)

$$\frac{2\boldsymbol{\rho}}{\rho}\cdot[2(\mathbf{p}\wedge\mathbf{L}) - i\mathbf{p}] + \frac{2}{\rho}[2(\mathbf{p}\wedge\mathbf{L}) - i\mathbf{p}]\cdot\boldsymbol{\rho} = \frac{2}{\rho}\left(4\mathbf{L}^2 + 1\right). \tag{9.263}$$

By substituting (9.262) and (9.263) in (9.261) and factoring $\left(4\mathbf{L}^2+1\right)$ we find

$$4\mathbf{R}^2 = \left(\mathbf{p}^2 - \frac{2}{\rho}\right)\left(4\mathbf{L}^2+1\right) + 4, \tag{9.264}$$

or

$$4\mathbf{R}^2 = -8E\mathbf{G}^2 = 2H\left(4\mathbf{L}^2+1\right) + 4. \tag{9.265}$$

Since $\left[H,\ \mathbf{J}^2\right] = 0$, we can diagonalize H and $\mathbf{J}^2$ simultaneously, knowing that the eigenvalues of $\mathbf{J}^2$ are $j(j+1)$ where j is an integer. Thus from (9.265) we obtain

$$j(j+1) = -\left(\frac{1}{4} + \frac{1}{E}\right). \tag{9.266}$$

Now by identifying j with the principal quantum number l_n we find the eigenvalues to be [19]

$$E_n = -\frac{1}{\left(l_n + \frac{1}{2}\right)^2}, \quad l_n = 1,\ 2, \cdots. \tag{9.267}$$

The complete normalized wave function obtained either from the factorization method or directly by solving the Schrödinger equation with the Coulomb potential in polar coordinates is given by

$$\psi_{n,\ell}(\rho,\phi) = \sqrt{\frac{\gamma^3(n-|\ell|)!}{\pi(n+|\ell|)!}}(2\gamma\rho)^{|\ell|}e^{-\gamma\rho}L^{2|\ell|}_{n-|\ell|}(2\gamma\rho)e^{i\ell\phi}, \tag{9.268}$$

where $E = -\gamma^2$ and $L^{2|\ell|}_{n-|\ell|}(2\gamma\rho)$ are the Laguerre polynomials.

9.10 Classical Limit of Hydrogen Atom

As we observed in our study of the correspondence principle, a basic question in quantum theory is the way that classical mechanics can be viewed as a limit of quantum theory. Classically the Kepler problem can be formulated as a motion in a plane containing the Runge–Lenz vector and is perpendicular to the angular momentum vector of the particle, and in this plane we have a well-defined elliptic orbit. Since in the classical limit we have a two-dimensional motion, it is convenient to start with the problem of hydrogen atom in two dimensions, the one that we solved in the preceding section.

The uncertainty $\Delta R_1 \Delta R_2$ obtained from Eq. (9.254) and the commutation relation (9.257) is given by

$$\Delta R_1 \Delta R_2 \geq \frac{1}{2}|\langle -2HL_3\rangle|. \tag{9.269}$$

To get the minimum uncertainty, i.e. satisfying the equality sign in (9.269) we find the solution of the eigenvalue equation (see Eq. (4.76)) [20]

$$(R_1 + i\xi R_2)\psi = \eta\psi, \tag{9.270}$$

where ξ is a real parameter and η is an eigenvalue of the non-Hermitian operator $R_1 + i\xi R_2$. Since both R_1 and R_2 commute with H, therefore we can find the eigenfunctions $\psi_n^\xi(\rho, \phi)$ which diagonalize H and at the same time satisfies (9.270). The solution of the eigenvalue equation (9.270) gives us the minimum uncertainty product ΔR_1 and ΔR_2, which according to (4.82) and (4.83) are:

$$(\Delta R_1)^2 = -E_n \xi \langle L_3 \rangle, \tag{9.271}$$

$$(\Delta R_2)^2 = -\frac{E_n}{\xi} \langle L_3 \rangle, \tag{9.272}$$

and these uncertainties satisfy (9.269) with the equality sign. To generate other eigenvalues we introduce the raising and lowering operators $A^\pm$ by

$$A^\pm = \pm \frac{1}{\sqrt{-2H}} (\xi R_1 + i R_2) - \sqrt{1-\xi^2}\, L_3, \tag{9.273}$$

for $0 \le \xi \le 1$. By applying these operators m times we find

$$\left(A^\pm\right)^m \psi_n = \left[\eta \pm m \left(-2E_n \sqrt{1-\xi^2}\right)^{\frac{1}{2}}\right] \psi_n, \tag{9.274}$$

where m is an integer satisfying the condition

$$m = -\ell_n,\ -\ell_n + 1, \cdots \ell_n - 1,\ \ell_n. \tag{9.275}$$

The eigenvalue η is real and is related to the mean value of the eccentricity ε of the elliptic orbit (see also Sec. 12.8)

$$\varepsilon = m \left[-2E\left(1-\xi^2\right)\right]^{\frac{1}{2}}. \tag{9.276}$$

We are particularly interested in the eigenstate which corresponds to the maximum eigenvalue $m = \ell_n$. This eigenstate satisfies the condition

$$A^+ \psi_n^\xi = 0. \tag{9.277}$$

Now for large quantum numbers ℓ_n from equation (9.276) it follows that the eccentricity ε depends only on ξ and is independent of ℓ_n. In fact in this limit Eq. (9.276) shows that

$$\varepsilon = \sqrt{1-\xi^2} = \left(1 - 2E_n \langle L_z \rangle^2\right)^{\frac{1}{2}}. \tag{9.278}$$

We can compare this result with the classical expression for the eccentricity of the orbit given in terms of the semi-major and semi-minor axes, a and b;

$$\varepsilon_c^2 = 1 - \frac{b^2}{a^2} = 1 - \frac{2|E|\mathbf{L}^2}{m\left(Ze^2\right)^2}, \tag{9.279}$$

or

$$\varepsilon_c^2 = 1 - 2|E|\mathbf{L}^2, \tag{9.280}$$

in the units that we have chosen.

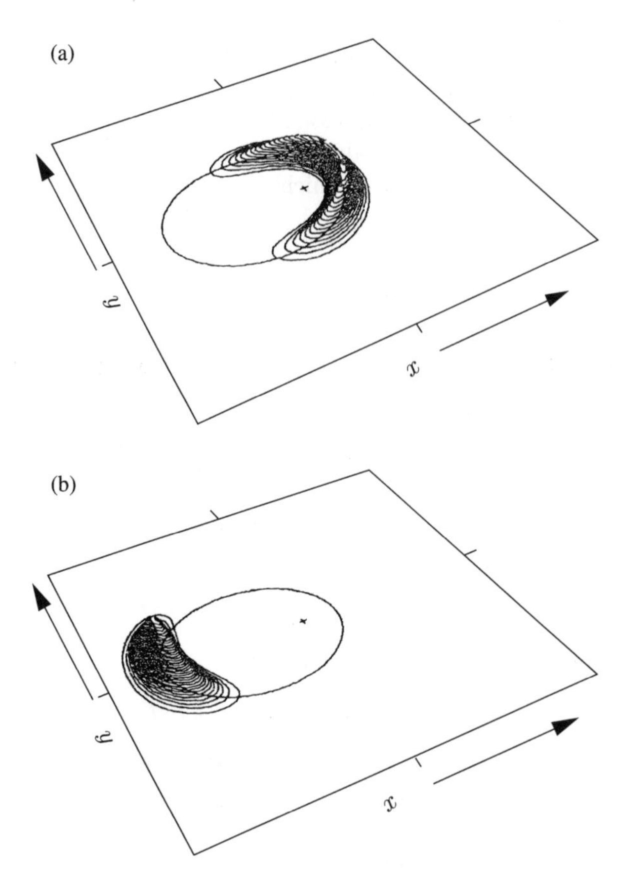

Figure 9.1: The time-dependent wave packet obtained from Eq. (9.284) is shown for two different times (a) for $t = 0$ and (b) for $t = \frac{1}{2}T$ where T is the Kepler period for this orbit $T = 2\pi\ell_0^3$. The classical orbit for this motion is also shown in both (a) and (b). For this calculation the eccentricity is chosen to be $\varepsilon = 0.6$ and the average angular momentum is assumed to be $\langle L_3\rangle = 32$. Other parameters used in the calculation are $\ell_0 = 40$ and $\sigma^2 = 3.0$ [20].

Returning to the wave function $\psi_n^\xi(\rho,\phi)$ we note that a general linear superposition of these states for large quantum numbers also minimizes the uncertainty (9.269) and therefore has minimal fluctuations in R_1 and R_2. In order to solve Eqs. (9.270) and (9.277) we expand $\psi_n^\xi(\rho,\phi)$ in terms of the eigenfunctions of the Coulomb Hamiltonian and the angular momentum L_3;

$$\psi_n^\xi(\rho,\phi) = \sum_{\ell=-\ell_n}^{\ell_n} C_{n,\ell}^\xi\, \psi_{n,\ell}(\rho,\phi), \tag{9.281}$$

where the coefficients $C^{\xi}_{n,\ell}$ are given by [20]

$$C^{\xi}_{n,\ell} = \frac{1}{2^{\ell_n}}\left[\frac{(2\ell_n)!}{(\ell_n+\ell)!(\ell_n-\ell)!}\right]^{\frac{1}{2}}\left(1-\xi^2\right)^{\frac{\ell_n}{2}}\left(\frac{1+\xi}{1-\xi}\right)^{\frac{\ell}{2}}. \tag{9.282}$$

When ℓ_n is large, say around 40, the coefficient $C^{\xi}_{n,\ell}$ given by (9.282) can be approximated by a Gaussian function of ℓ

$$C^{\xi}_{n,\ell} \approx \left[\frac{\pi}{2}\ell_n\left(1-\xi^2\right)\right]^{-\frac{1}{4}}\exp\left[-\frac{(\ell-\xi\ell_n)^2}{\ell_n\left(1-\xi^2\right)}\right]. \tag{9.283}$$

The wave function (9.281) which minimizes $\Delta R_1 \Delta R_2$ has a spatial probability distribution which is peaked about the Kepler orbit having the eccentricity ε

To find the time evolution of this wave packet we superimpose the energy eigenstates $\psi^{\xi}_{n,\ell}(\rho,\phi)$ with time-dependent factor $e^{-iE_n t}$, i.e.

$$\psi^{\xi}(\rho,\phi,t) = \sum_n a_n \psi^{\xi}_n(\rho,\phi)\exp(-iE_n t), \tag{9.284}$$

where the coefficient a_n is sharply peaked about a fixed principal quantum numbers ℓ_0. For instance we can choose

$$a_n = \left(2\pi\sigma^2\right)^{-\frac{1}{4}}\exp\left[-\frac{(\ell_n-\ell_0)^2}{4\sigma^2}\right], \tag{9.285}$$

and calculate $\psi^{\xi}(\rho,\phi,t)$ from (9.284) and then observe the wave packet $|\psi^{\xi}(\rho,\phi,t)|^2$ at different times.

In Fig. 9.1 this wave packet is plotted at two different times. The original wave packet is shown in (a) and the wave packet after a time equal to half of the Kepler period is displayed in (b). The wave packet moves around the elliptic orbit with a period of $T = \frac{2\pi Z e^2\sqrt{m}}{(-2E)^{\frac{3}{2}}}$. Writing this period in units that we are using and noting that $E = -\frac{1}{2\ell_0^2}$, the period becomes $T = 2\pi\ell_0^3$. Thus the wave packet starts its motion at the perihelion, Fig. 9.1 (a), and then it slows down, contracts, and becomes steeper as it reaches aphelion Fig. 9.1 (b). As it returns to perihelion it speed up and spreads faster [20]. This motion of the wave packet is counterclockwise. For the motion in the opposite direction we can either choose ξ to be a negative quantity or keep ξ positive but choose the condition

$$A_-\psi^{\xi}_n = 0, \tag{9.286}$$

instead of (9.277).

The spreading of the wave packet as it moves around the orbit is due to the initial uncertainty in the position and momentum as is required by the Heisenberg principle (Sec. 4.5). But there is an additional and important quantum interference effect which happens when the head of the wave packet catches up with its tail. This causes a nonuniform varying amplitude of the wave packet along the ellipse [20].

9.11 Self-Adjoint Ladder Operator

The ladder operators that we have seen so far are not self-adjoint. This is because the adjoint of either of the ladder operators changes the eigenstate in the opposite sense. It is possible to formulate the factorization method in such a way that the ladder operators are self-adjoint [22],[23].

Let M be the self-adjoint operator that we want to determine its eigenvalues and β_j^+ and β_j^- be the non-self-adjoint raising and lowering operators for M. We define $\widehat{M}$ and $\widehat{B}_j$ as the following operators:

$$\widehat{M} = \begin{bmatrix} M & 0 \\ 0 & -M \end{bmatrix}, \quad \widehat{B}_j = \begin{bmatrix} 0 & \beta_j^- \\ \beta_j^+ & 0 \end{bmatrix}. \tag{9.287}$$

Both of these operators are self-adjoint and satisfy the anti-commutation relation

$$\left[\widehat{M},\, \widehat{B}_j\right]_+ = \widehat{M}\widehat{B}_j + \widehat{B}_j\widehat{M} = -\widehat{B}_j. \tag{9.288}$$

The operator $\widehat{M}$ acts on a pair of states each of which can be represented by a 1×2 matrix (spinor form)

$$|\lambda, j\rangle^a = \begin{bmatrix} |\lambda, j\rangle \\ 0 \end{bmatrix}, \quad \text{and} \quad |\lambda, j\rangle^b = \begin{bmatrix} 0 \\ |\lambda, j\rangle \end{bmatrix}. \tag{9.289}$$

That is

$$\widehat{M}|\lambda, j\rangle^a = [\lambda - L(j)]|\lambda, j\rangle^a \tag{9.290}$$

and

$$\widehat{M}|\lambda, j\rangle^b = -[\lambda - L(j)]|\lambda, j\rangle^b \tag{9.291}$$

Upon the action of the matrix $\widehat{B}$ on $|\lambda, j\rangle^a$ and $|\lambda, j+1\rangle^b$ states we get

$$\widehat{B}|\lambda, j\rangle^a = |\lambda, j+1\rangle^b \tag{9.292}$$

and

$$\widehat{B}|\lambda, j+1\rangle^b = |\lambda, j\rangle^a \tag{9.293}$$

Thus the ladder operator $\widehat{B}$ changes the eigenstates from one subspace to the other and at the same time changes the eigenvalues. We observe that in this formulation instead of having one set of eigenstates from one subspace and a pair of mutually adjoint ladder operators, we have one ladder operator and two subspaces.

Let us consider the example of orbital angular momentum where we choose M to be the z component of the angular momentum, $M = L_z$, and β^+ and β^- are given by

$$\beta^\pm = L_x \pm iL_y. \tag{9.294}$$

Substituting for M and $\beta^{\pm}$ in (9.287) we have

$$\widehat{M} = \begin{bmatrix} L_z & 0 \\ 0 & -L_z \end{bmatrix} = \sigma_z L_z, \tag{9.295}$$

and

$$\widehat{B} = \begin{bmatrix} 0 & (L_x - iL_y) \\ (L_x + iL_y) & 0 \end{bmatrix} = \sigma_x L_x + \sigma_y L_y, \tag{9.296}$$

where σ_x, σ_y and σ_z are the Pauli spin matrices, Eq. (9.118).

9.12 Self-Adjoint Ladder Operator for Angular Momentum

We now formulate the problem of determination of the eigenvalues of angular momentum in a space of n dimensions in terms of the self-adjoint ladder operators. In the following sections, for convenience, we set $\hbar = 1$ and we define $\hat{L}_{ij}$ by

$$\hat{L}_{ij} = x_i p_j - x_j p_i, \quad i = 1, 2 \cdots, n. \tag{9.297}$$

These $\hat{L}_{ij}$ s are obvious generalization of L_x, L_y and L_z operators for the three-dimensional space. The operator $\hat{L}_{ij}$ has the following properties:

$$\hat{L}_{ij} = -\hat{L}_{ji}, \tag{9.298}$$

$$\hat{L}_{ij} = \hat{L}_{ij}^{\dagger}, \quad \text{(self-adjointness)}, \tag{9.299}$$

$$\left[\hat{L}_{ij}, \hat{L}_{ik}\right] = i\hat{L}_{jk}, \quad \hbar = 1 \tag{9.300}$$

$$\left[\hat{L}_{ij}, \hat{L}_{kl}\right] = 0, \tag{9.301}$$

and

$$\hat{L}_{ij}\hat{L}_{kl} + \hat{L}_{ki}\hat{L}_{jl} + \hat{L}_{jk}\hat{L}_{il} = 0, \tag{9.302}$$

where i, j, k and l are all different integers, and they all run from 1 to n.

The set $\{\hat{L}\}$ form the elements of a Lie algebra with a single Casimir invariant [22]

$$\hat{L}_n^2 = \sum_{i<j}^{n} \hat{L}_{ij}^2. \tag{9.303}$$

From the properties of $\hat{L}_{ij}$ we can show that $\hat{L}_m^2$, $m = 2, 3 \cdots, n$ form a commuting set of operators and that the eigenvalues of $\hat{L}_n^2$ can be determined from the irreducible representations of the Lie group. The eigenvalues of $\hat{L}_n^2$ can also be found from the equation

$$\hat{L}_n^2|\psi\rangle = \text{constant }|\psi\rangle, \tag{9.304}$$

by separating (9.304) in generalized spherical polar coordinates (hyperspherical coordinates)

$$\begin{cases} x_1 = r\cos\phi_1, \\ x_2 = r\sin\phi_1\cos\phi_2 \\ \vdots \\ x_{n-1} = r\sin\phi_1\sin\phi_2\cdots\cos\theta \\ x_n = r\sin\phi_1\sin\phi_2\cdots\sin\phi_{n-2}\sin\theta \end{cases}. \tag{9.305}$$

However in the present case we are dealing with the representation in which the commuting sets $\hat{L}_m^2$, $m = 2, 3\cdots n$ are simultaneously diagonal. Here we can solve the problem by induction. That is we assume that the solution is known in n space and the we find in $(n+1)$ space. For this we can use a ladder operator for $\hat{L}_n^2$ which is self-adjoint of the type that we have discussed.

9.13 Generalized Spin Operators

For the construction of a self-adjoint ladder operator for $\hat{L}_n^2$ we first need to consider a generalized form of spin operators for n dimensions.

Let us define σ_{ij} operators $i, j = 1, 2, \cdots, n$ by the following set of relations

$$\sigma_{ij} = -\sigma_{ji}, \quad \sigma_{ij} = \sigma_{ij}^\dagger, \tag{9.306}$$

$$\sigma_{ij}^2 = 1, \quad (1 \text{ is the unit operator}), \tag{9.307}$$

$$\sigma_{lj}\sigma_{lk} = i\sigma_{jk}, \quad l, j, k \text{ are all different}, \tag{9.308}$$

$$[\sigma_{ij}, \sigma_{kl}] = 0, \quad i, j, k, l \text{ are all different}. \tag{9.309}$$

These generalized spin matrices are related to the matrices α_j defined by Eq. (9.122), i.e.

$$\alpha_i\alpha_j + \alpha_j\alpha_i = 2\delta_{ij}, \tag{9.310}$$

and the relation is

$$\sigma_{ij} = -\left(\frac{i}{2}\right)[\alpha_i, \alpha_j]. \tag{9.311}$$

From the aforementioned properties it follows that

$$[\sigma_{ij}, \sigma_{ik}]_+ = 0, \quad i, j, k \text{ are all different}. \tag{9.312}$$

Since the spin and orbital angular momentum are independent operators we have

$$[\sigma_{ij}, \hat{L}_{kl}] = 0, \quad \text{for all } i, j, k, l. \tag{9.313}$$

An important result concerning σ_{ij} and $\hat{L}_{kl}$ matrices can be stated by the following result

$$\sum_{ijk}^{n\,\prime} \left(\sigma_{ij}\hat{L}_{ij}\right)\left(\sigma_{kn+1}\hat{L}_{kn+1}\right) = 0, \tag{9.314}$$

where the prime on the summation sign means that i, j, k are all different integers. For the proof of this result see the paper of Joseph [22]. Also as in the case of angular momentum in three dimensions we have

$$p_i\hat{L}_{jk} + p_j\hat{L}_{ki} + p_k\hat{L}_{ij} = 0, \tag{9.315}$$

and

$$x_i\hat{L}_{jk} + x_j\hat{L}_{ki} + x_k\hat{L}_{ij} = 0, \tag{9.316}$$

when i, j and k are all different integers. From these results and Eq. (9.314) we obtain

$$\sum_{ijk}^{n\,\prime} \left(\sigma_{in+1}p_i\right)\left(\sigma_{jk}\hat{L}_{jk}\right) = 0, \tag{9.317}$$

and

$$\sum_{ijk}^{n\,\prime} \left(\sigma_{in+1}x_i\right)\left(\sigma_{jk}\hat{L}_{jk}\right) = 0. \tag{9.318}$$

Now we can find an explicit form for the ladder operator.

9.14 The Ladder Operator

The operator L_{n+1} defined by

$$L_{n+1} = \sum_{i=1}^{n} \sigma_{in+1}\hat{L}_{in+1}, \tag{9.319}$$

is a self-adjoint ladder operator for the eigenstates of $\hat{L}_n^2$ if and only if they are also eigenstates of the self-adjoint operator

$$\mathcal{L}_n = \sum_{m=2}^{n} L_m. \tag{9.320}$$

This result follows from Eqs. (9.297), (9.299), (9.306) and (9.313).

Next we consider the anti-commutation relation

$$\left[L_{n+1}, \left(\mathcal{L}_n + \frac{1}{2}(n-1)\right)\right]_+ = 0. \tag{9.321}$$

This result can be proven from the anticommutator

$$\left[\left(\sum_{ij}^{n}\sigma_{ij}\hat{L}_{ij}\right),\left(\sum_{k}^{n}\sigma_{kn+1}\hat{L}_{kn+1}\right)\right]_{+}$$
$$= \sum_{ij}^{n}\left[\left(\sigma_{ij}\hat{L}_{ij}\right),\left(\sigma_{in+1}\hat{L}_{in+1}+\sigma_{jn+1}\hat{L}_{jn+1}\right)\right]_{+}, \tag{9.322}$$

which follows from (9.314).

Now if we substitute from (9.300), (9.301) and (9.308) in (9.322) we find

$$\left[\left(\sum_{ij}^{n}\sigma_{ij}\hat{L}_{ij}\right),\left(\sum_{k}^{n}\sigma_{kn+1}\hat{L}_{k}\right)\right]_{+} = -2(n-1)\sum_{j=1}^{n}\sigma_{jn+1}\hat{L}_{jn+1}. \tag{9.323}$$

Using the definitions of L_{n+1} and $\mathcal{L}_n$ we observe that (9.323) reduces to (9.321).

Next we want to express the total angular momentum $\hat{L}_n^2$ in terms of $\mathcal{L}_n$. To this end we square both sides of Eq. (9.319) and use the properties of σ_{ij} s and L_i s to simplify the result (note that we have set $\hbar = 1$);

$$L_{n+1}^2 = \hat{L}_{n+1}^2 - \hat{L}_n^2 - \mathcal{L}_n. \tag{9.324}$$

We can also obtain another expression for L_{n+1}^2 between Eqs. (9.319) and (9.320);

$$L_{n+1}^2 = \mathcal{L}_{n+1}(\mathcal{L}_{n+1}+n-1) - \mathcal{L}_n(\mathcal{L}_n+n-1). \tag{9.325}$$

By eliminating L_{n+1}^2 from (9.324) and (9.325) we find

$$\hat{L}_{n+1}^2 - \mathcal{L}_{n+1}(\mathcal{L}_{n+1}+n-1) = \hat{L}_n^2 - \mathcal{L}_n(\mathcal{L}_n+n-2). \tag{9.326}$$

Now for $n = 2$ we have

$$\mathcal{L}_2^2 = L_2^2 = \left(\sigma_{12}\hat{L}_{12}^2\right)^2 = \hat{L}_{12}^2 = \hat{L}_2^2, \tag{9.327}$$

therefore by induction from Eq. (9.326) we get

$$\hat{L}_n^2 - \mathcal{L}_n(\mathcal{L}_n+n-2) = 0. \tag{9.328}$$

Since $\mathcal{L}_n$ is self-adjoint we can choose a representation in which it is diagonal, and since $\mathcal{L}_n$ commutes with $\hat{L}_n^2$ we can diagonalize these two operators simultaneously. The common eigenstates for $\mathcal{L}$ and $\hat{L}^2$ are doubly degenerate with respect to $\hat{L}^2$. As in the case of three-dimensional space we write

$$\mathcal{L}_n|\ell_n\rangle^a = \ell_n|\ell_n\rangle^a, \tag{9.329}$$

and

$$\mathcal{L}_n|\ell_n\rangle^b = -(\ell_n+n-2)|\ell_n\rangle^b. \tag{9.330}$$

From these solutions we conclude that

$$L_n^2|\ell_n\rangle = \ell_n(\ell_n + n - 2)|\ell_n\rangle, \tag{9.331}$$

for both a and b subspaces.

Finally let us write the ladder operators for this problem

$$\mathcal{L}_n\left(L_{n+1}|\ell_n\rangle^a\right) = -(\ell_n + n - 1)\left(L_{n+1}|\ell_n\rangle^a\right), \tag{9.332}$$

$$\mathcal{L}_{n+1}\left(L_{n+1}|\ell_{n+1}\rangle^b\right) = \ell_n\left(L_{n+1}|\ell_n\rangle^b\right). \tag{9.333}$$

These relations show that

$$L_{n+1}|\ell_n^a\rangle \sim |\ell_{n+1}\rangle^b, \tag{9.334}$$

and

$$L_{n+1}|\ell_{n+1}\rangle^b \sim |\ell_n\rangle^a. \tag{9.335}$$

The matrix elements of L_{n+1} can be obtained from (9.324). Apart from a phase factor they are

$$^b\langle\ell_n|L_{n+1}|\ell_{n+1}\rangle^a = \left[(\ell_{n+1} - \ell_n)(\ell_{n+1} + \ell_n + n - 1)\right]^{\frac{1}{2}}. \tag{9.336}$$

An interesting result of this rather long derivation is that it shows that except in the three-dimensional space where ℓ_2 and ℓ_3 may be half-integers, the ℓ_n s are integers and they satisfy the following relation

$$\ell_n \geq \ell_{n-1} \geq \ell_{n-2} \cdots \geq \ell_2. \tag{9.337}$$

Here we assume, without the loss of generality, that all ℓ_m s are positive. In order to justify this assumption we observe that since L_{m+1}^2 is positive definite

$$M_m = \ell_{m+1} - \ell_m, \tag{9.338}$$

is compatible with the stepping procedure in ℓ_m only if M_m is a positive integer. Then this guarantees that ℓ_{m+1} is an upper bound to the ladder. There is also a lower bound preventing the generation of negative ℓ_n values. Consider the case of $m = 2$, then from (9.335) it follows that for the lowering operator

$$L_3|\ell_2\rangle^b \sim |\ell_2 - 1\rangle^a, \tag{9.339}$$

for

$$0 < \ell_2 < 1, \tag{9.340}$$

can be interpreted as

$$L_3|\ell_2\rangle^b \sim |1 - \ell_2\rangle^b. \tag{9.341}$$

When the inequality (9.340) is satisfied we have a new set of eigenvalues $(1-\ell_2)$, $(2-\ell_2)$ and $(3-\ell_2)$ which must have ℓ_3 as an upper bound. If we combine these with (9.338) we find that ℓ_2 and ℓ_3 may have integers or half-integer values, a result which is true when $n = 3$.

Bibliography

[1] H. Goldstein, C. Poole and J. Safco, *Classical Mechanics*, Third Edition (Addison-Wesley, San Francisco, 2002).

[2] H.S. Green, *Matrix Mechanics*, (P. Noordhoff, Netherlands, 1965).

[3] R. L. Liboff, I. Nebenzahl and H.H. Fleischmann, On the radial momentum operator, Am. J. Phys. 41, 976, (1973).

[4] B. Friedman, *Principles and Techniques of Applied Mathematics*, (John Wiley & Sons, New York, 1957), p. 154.

[5] A.Z. Capri, *Nonrelativistic Quantum Mechanics*, Third Edition, (World Scientific, 2002), p. 209.

[6] G. Paz, The non-self-adjointness of the radial operator in n dimensions, J. Phys. A 35, 3727 (2002).

[7] L.D. Landau and E.M. Lifshitz, *Quantum Mechanics, Non-relativistic Theory*, (Pergamon Press, London, 1958).

[8] G. Baym, *Lectures on Quantum Mechanics*, (W.A. Benjamin, Reading, 1969).

[9] J. Schwinger in L.C. Biedenharn and H. van Dam, *Quantum Theory of Angular Momentum*, (Academic Press, New York, 1965), p. 229.

[10] J.J. Sakurai, *Modern Quantum Mechanics*, (Addison-Wesley, Reading, 1994), p. 217.

[11] M. Born, W. Heisenberg and P. Jordan, Zur Quantenmechanic II, Z. Phys. 35, 557 (1926), translated in B.L. van der Waerden, *Sources of Quantum Mechanics*, (Dover Publications, New York, 1967), p. 321.

[12] L. Infeld and T.E. Hull, The factorization method, Rev. Mod. Phys. 23, 21 (1951).

[13] Y.F. Liu, Y.A. Lei and J.Y. Zeng, Factorization of the radial Schrödinger equation and four kinds of the raising and lowering operators of hydrogen atoms and isotropic harmonic oscillators, Phys. Lett. A 231, 9 (1997)

[14] B-W. Xu and F-M. Kong, Factorization of the radial Schrödinger equation of the hydrogen atom, Phys. Lett. A 259, 212 (1999).

[15] J.D. Hey, On the determination of radial matrix elements for high-n transitions in hydrogenic atoms and ions, J. Phys. B, 39, 2641 (2006).

[16] W. Pauli, Uber das Wasserstoffspektrum vom Standpunkt der neuen Quantenmechanik, Z. Phys. 36, 336 (1926), translated in *Sources of Quantum Mechanics*, edited by B.L. van der Waarden, (North-Holland, Amsterdam, 1967), p. 387 ff.

[17] M. Taketani and M. Nagasaki, *The Formation and Logic of Quantum Mechanics*, Vol. III (World Scientific, 2001), p. 336.

[18] H. Goldstein, C. Poole and J. Safco, *Classical Mechanics*, Third Edition (Addison-Wesley, San Francisco, 2002).

[19] For obtaining the eigenvalues from the wave equation see X.L. Yang, S.H. Guo, F.T. Chan, K.W. Wong and W.Y. Ching, Analytic solution of a two-dimensional hydrogen atom. I. Nonrelativistic theory, Phys. Rev. A 43, 1186 (1991).

[20] M. Nauenberg, Quantum wave packet on Kepler orbits, Phys. Rev. A 40, 1133 (1989).

[21] A wave packet solution for the hydrogen atom for large principal quantum number can be found in L.S. Brown, Classical limit of the hydrogen atom, Am. J. Phys. 41, 525 (1973).

[22] A. Joseph, Self-Adjoint Ladder Operators (I), Rev. Mod. Phys. 39, 829 (1967).

[23] C.A. Coulson and A. Joseph, Self-Adjoint Ladder Operators (II), Rev. Mod. Phys. 39, 838 (1967).

Chapter 10

Methods of Integration of Heisenberg's Equations of Motion

In Chapter 7 we studied methods of solving the one-dimensional problem and of finding the eigenvalues of a particle in confining potentials. There, by solving a set of nonlinear algebraic equations resulting from the Heisenberg equations of motion we found the matrix elements and the eigenvalues for these problems. Now we want to consider the integration of the operator differential equations of motion for a wider class of potentials and thus determine the motion of a wave packet or the tunneling of a particle through a potential barrier. The main motivation for studying the solution to the known problems in wave mechanics by the more difficult operator method lies in the hope that these techniques may give us a better insight to the much more complicated but in a way similar problems of quantum field theory.

10.1 Discrete-Time Formulation of the Heisenberg's Equations of Motion

A practical method of integration the operator equations of the motion advanced by Bender and collaborators is to replace these equations by finite-difference equations. This formulation can be used for the determination of the eigenvalues of confining potentials and for one-dimensional quantum tunneling [1]–[8].

Setting $\hbar = m = 1$, and for simplicity considering one-dimensional motion of the form

$$\frac{dq(t)}{dt} = \frac{1}{i}\left[q(t),\ H\right] = p(t), \tag{10.1}$$

and

$$\frac{dp(t)}{dt} = \frac{1}{i}\left[p(t),\ H\right] = -\frac{\partial V(q)}{\partial q} = F(q(t)), \tag{10.2}$$

the object is to integrate these equations. To this end we approximate these with a special form of difference equations so as to preserve the equal time commutation relation $[p,\ q] = -i$. Let us replace $p(t)$ and $q(t)$ by $p_j = p(jh)$ and $q_j = q(jh)$ where h is the time step, $t = jh$ and j is an integer, and these operators are defined at these times. Next we introduce an auxiliary momentum $\pi_{j+\frac{1}{2}}$ defined at half-integral time steps. Now we can write the finite difference approximation to Eqs. (10.1) and (10.2);

$$\frac{q_{j+1} - q_j}{h} = \pi_{j+\frac{1}{2}}, \tag{10.3}$$

and

$$\frac{\pi_{j+\frac{1}{2}} - \pi_{j-\frac{1}{2}}}{h} = F(q_j). \tag{10.4}$$

The canonical momentum operator p_j is related to $\pi_{j\pm\frac{1}{2}}$ by

$$p_j = \frac{1}{2}(\pi_{j+\frac{1}{2}} + \pi_{j-\frac{1}{2}}). \tag{10.5}$$

Now by eliminating the auxiliary momentum $\pi_{j\pm\frac{1}{2}}$ in Eqs. (10.3) and (10.4) we obtain the coupled difference equations

$$q_{j+1} = q_j + h\left[p_j + \frac{1}{2}hF(q_j)\right], \tag{10.6}$$

and

$$p_{j+1} = p_j + \frac{h}{2}\left[F(q_j) + F(q_{j+1})\right]. \tag{10.7}$$

The operators p_j and q_j defined in this way satisfy the canonical commutation relation at the discrete points $j = 0,\ 1,\ 2,\ \cdots$

$$[q_j,\ p_j] = i, \tag{10.8}$$

while other commutators vanish. To show that Eq. (10.8) is true for any point j, let us assume that (10.8) is true for a given j and then prove that it is true for $j+1$. This can be demonstrated by noting that p_{j+1} and q_{j+1} are related to p_j and q_j by the unitary evolution operator $U_j(h)$;

$$q_{j+1} = U_j^\dagger(h) q_j U_j(h), \tag{10.9}$$

$$p_{j+1} = U_j^\dagger(h) p_j U_j(h), \tag{10.10}$$

where $U_j(h)$ is given by the discrete form of the evolution operator $U(t) = U(jh)$

$$U_j(h) = \exp\left[-i\frac{h}{2}V(q_j)\right]\exp\left[-i\frac{h}{2}p_j^2\right]\exp\left[-i\frac{h}{2}V(q_j)\right]. \tag{10.11}$$

First let us consider two transformations

$$\exp\left[\frac{ih}{2}p_j^2\right]q_j\exp\left[-\frac{ih}{2}p_j^2\right] = q_j + hp_j, \tag{10.12}$$

and

$$\exp\left[\frac{ih}{2}V(q_j)\right]p_j\exp\left[-\frac{ih}{2}V(q_j)\right] = p_j + \frac{1}{2}hF(q_j), \tag{10.13}$$

These equations enable us to calculate q_{j+1}, Eq. (10.9),

$$\begin{aligned} q_{j+1} &= U_j^\dagger q_j U_j = \exp\left[\frac{ih}{2}V(q_j)\right][q_j + hp_j]\exp\left[-\frac{ih}{2}V(q_j)\right] \\ &= q_j + h\left[p_j + \frac{1}{2}hF(q_j)\right], \end{aligned} \tag{10.14}$$

and the result is the same as (10.6). Similarly for p_{j+1} we obtain (10.7). Since p_{j+1} and q_{j+1} are related by unitary transformations to p_j and q_j, the equal time commutation relation will remain unchanged [9].

Accuracy of Finite-Difference Equations — To test the accuracy of the finite-difference equations (10.6) and (10.7) we will use them to find the long time behavior of an exactly solvable problem [10]. The solvable problem is given by the quadratic Hamiltonian

$$H = \frac{1}{2}\left(p^2 - q^2\right), \tag{10.15}$$

which generates the equations of motion

$$\frac{dq}{dt} = p, \qquad \frac{dp}{dt} = q. \tag{10.16}$$

The operator equations (10.16) can be solved exactly in terms of the initial operators q_0 and p_0;

$$q(t) = q_0\cosh t + p_0\sinh t, \tag{10.17}$$

and

$$q(t) = q_0\sinh t + p_0\cosh t. \tag{10.18}$$

The corresponding difference equations are found from (10.6) and (10.7) which after rearrangement can be written as

$$q_{n+1} = \frac{1}{2}[f(h) + f(-h)]q_n + \frac{1}{2\sqrt{1+\frac{h^2}{4}}}[f(h) - f(-h)]p_n, \tag{10.19}$$

and

$$p_{n+1} = \frac{1}{2}\sqrt{1+\frac{h^2}{4}}\,[f(h) - f(-h)]q_n + \frac{1}{2}[f(h) + f(-h)]p_n, \tag{10.20}$$

where

$$f(h) = 1 + h\sqrt{1+\frac{h^2}{4}} + \frac{1}{2}\,h^2. \tag{10.21}$$

By iterating Eqs. (10.19) and (10.20) we find

$$q_N = \frac{1}{2}\,\left\{[f(h)]^N + [f(-h)]^N\right\} q_0 + \frac{1}{2\sqrt{1+\frac{h^2}{4}}}\,\left\{[f(h)]^N - [f(-h)]^N\right\} p_0, \tag{10.22}$$

and a similar relation for p_N. For the initial condition we choose a state given by a Gaussian wave packet centered at the origin, then

$$\langle q_0\rangle = \langle p_0\rangle = 0, \tag{10.23}$$

and

$$\langle q_0^2\rangle = \langle p_0^2\rangle = \frac{1}{2}. \tag{10.24}$$

Now with these initial conditions we first obtain $\langle q^2(t)\rangle$ from Eqs. (10.17) and (10.18);

$$\langle q^2(t)\rangle = \frac{1}{2} + \sinh^2 t = \frac{1}{2} + t^2 + \frac{1}{3}t^4 + \mathcal{O}\left(t^6\right). \tag{10.25}$$

Using the finite difference equations, (10.19) and (10.20), with $t = hN$ we find [10]

$$\begin{aligned}\langle q_N^2\rangle &= \left[\frac{1}{4}\left\{[f(h)]^{2N} + [f(-h)]^{2N}\right\}\right.\\ &+ \left.\frac{h^2}{32}\left\{[f(h)]^N + [f(-h)]^N\right\}^2\right]\left(1+\frac{h^2}{4}\right)^{-1}.\end{aligned} \tag{10.26}$$

To compare this approximate form with the exact result when t is large, let us choose $t = 10$, then from (10.25) we find

$$\langle q^2(10)\rangle = 1.2191 \times 10^8. \tag{10.27}$$

Now we choose $N = 100$ and $h = 0.1$ and calculate $\langle q_{100}^2\rangle$ from (10.26);

$$\langle q_{100}^2\rangle = 1.2014 \times 10^8. \tag{10.28}$$

By comparing these results we conclude that finite-difference approximation gives reasonable results for long times.

For small t we substitute $h = \frac{t}{N}$ and expand $\langle q_N^2\rangle$ in powers of t to find

$$\langle q_N^2\rangle = \frac{1}{2} + t^2 + \frac{1}{3}\left(1 - \frac{5}{8N^2}\right)t^4 + \cdots. \tag{10.29}$$

When N is large (e.g. $N \approx 100$) (10.29) is a good approximation when it is compared to the expansion of $\langle q^2(t)\rangle$ given in (10.25).

10.2 Quantum Tunneling Using Discrete-Time Formulation

The finite difference equations (10.6) and (10.7) may be used to determine tunneling in a double-well potential, e.g. for the potential

$$V(q) = \frac{4q^2(q-\beta)^2}{\beta^2}. \tag{10.30}$$

In the neighborhood of $q = 0$ this potential behaves like a harmonic oscillator $V(q) \approx 4q^2$, and for the initial state we choose a state given by a Gaussian wave function located in the left well

$$\psi(q) \to \left(\frac{\omega}{\pi}\right)^{\frac{1}{4}} \exp\left(-\frac{\omega q^2}{2}\right), \tag{10.31}$$

where in the present case $\omega = \sqrt{8}$. Now if $\beta^2 > 2\omega$, then the height of the potential barrier is higher than the energy associated with the wave packet and we have tunneling between the two wells.

Since the potential (10.30) is nonlinear in q, successive iterations of q_{n+1} and p_{n+1} will have powers of q_0 and p_0. Thus for calculating the matrix elements of $\langle q_0^n\rangle$ we have to truncate the infinite matrix elements of

$$\langle j|q_0|k\rangle = \frac{1}{\sqrt{2\omega}}\left(\sqrt{j}\,\delta_{j,k+1} + \sqrt{k}\,\delta_{k,j+1}\right), \tag{10.32}$$

and

$$\langle j|p_0|k\rangle = i\sqrt{\frac{\omega}{2}}\left(\sqrt{j}\,\delta_{j,k+1} - \sqrt{k}\,\delta_{k,j+1}\right), \tag{10.33}$$

and replace them by a finite but large $D \times D$ matrices. That is to assume that j and k are finite and $1 \le j,\ k \le D$ where $D = \frac{1}{2}\left(3^N + 1\right)$ [7].

Bender and collaborators have used the finite-difference equations (10.6) and (10.7) with the parameters $\beta = 2.5$, $D = 32$ and $h = 0.008$ and calculated the expectation value $\langle 0|q(t)|0\rangle$ using the Gaussian wave packet (10.31).

The oscillation of $\langle 0|q(t)|0\rangle$ as a function of time is shown in Fig. 10.1. The period of oscillations is approximately given by

$$T = \frac{2\pi}{E_1 - E_0}, \tag{10.34}$$

where E_0 and E_1 are the lowest and the first excited state eigenvalues of the trapped particle [14]. If we calculate the two lowest eigenvalues for the potential (10.30) we find $E_0 = 1.03359$ and $E_1 = 1.44203$, and these give us a period of about 15.3838.

In this finite basis we write Eqs. (10.6) and (10.7) in terms of their matrix elements and the potential $V'(x) = -F(x)$:

$$\langle j|x_{n+1}|k\rangle = \langle j|x_n|k\rangle + h\langle j|p_n|k\rangle - \frac{1}{2}h^2\langle j|V_n'|k\rangle, \tag{10.35}$$

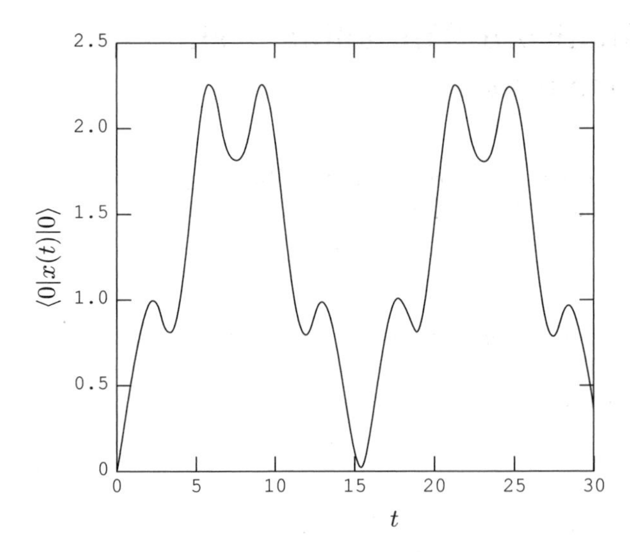

Figure 10.1: Expectation value of $q(t)$ with a Gaussian wave packet is displayed as a function of time. In this calculation p_0 and q_0 are approximated by 32×32 matrices and $\beta = 2.5$ and a step size $h = 0.008$ have been used to obtain this plot [7].

$$\langle j|p_{n+1}|k\rangle = \langle j|p_n|k\rangle - \frac{1}{2}h\left\{\langle j|V'_{n+1}|k\rangle + \langle j|V'_n|k\rangle\right\}, \tag{10.36}$$

where $V'_n = V'(x_n)$, and $1 \le j,\ k \le D$. One way of carrying out this numerical integration is to introduce a unitary matrix $\langle j|\mathcal{T}_n|\alpha\rangle$ at each time step to diagonalize the coordinate matrix $\langle j|x|k\rangle$ [11], [12]

$$(x(t_n))_\alpha \delta_{\alpha,k} = \sum_{j=1}^{D} \langle\alpha|\mathcal{T}(t_n)|j\rangle\langle j|x(t_n)|k\rangle. \tag{10.37}$$

Here x_α refers to the grid point and this together with the matrix $\langle\alpha|\mathcal{T}_n|j\rangle$ are obtained from Eq. (10.37). Having found $\langle\alpha|\mathcal{T}_n|j\rangle$ we construct the force matrix by writing

$$\langle j|V'_n|k\rangle = \sum_{k=1}^{D} \langle j|\mathcal{T}(t_n)|\alpha\rangle V'[x_\alpha(t_n)]\langle\alpha|\mathcal{T}(t_n)|k\rangle. \tag{10.38}$$

Once $\langle j|V'_n|k\rangle$ is determined, we can advance to the next lattice time $h(n+1)$ with the help of Eqs. (10.35) and (10.36).

Tunneling in an Infinite Domain — So far we have studied the solution of Heisenberg's equations for tunneling in a confining double-well potential. Now we want to consider an extension of this method for the solution of the problem of tunneling in an infinite domain. Here again the discrete form of the Heisenberg equations are given by (10.6) and (10.7). Now for the initial position operator we use an orthonormal plane wave basis set which is of the form

$$\langle x|j\rangle = \frac{1}{\sqrt{2L_0}} e^{ik_j x}. \tag{10.39}$$

Here the wave number (or momentum) k_j is given by

$$k_j = k_{min} + (j-1)\Delta k, \quad j = 1,\ 2 \cdots D, \tag{10.40}$$

where

$$\Delta k = \frac{k_{max} - k_{min}}{D-1}, \tag{10.41}$$

and $2L_0 = \frac{2\pi}{\Delta k}$ is the total length of the coordinate grid at $t = 0$. This coordinate grid expands as the wave packet moves, i.e. $L = L(t)$. In terms of this plane wave basis set we find the matrix elements of x_0 and p_0:

$$\begin{aligned} \langle j|x_0|n\rangle &= 12L_0 \int_{-L_0}^{L_0} e^{-ik_j x}\, x\, e^{ik_n x} dx \\ &= \begin{cases} \dfrac{i(-1)^{n-j+1}}{(k_n - k_j)} & j \neq n \\ 0 & n = j \end{cases}, \end{aligned} \tag{10.42}$$

and

$$\langle j|p_0|k\rangle = \frac{1}{2L_0} \int_{-L_0}^{L_0} e^{-ik_j x} \left(-i\frac{\partial}{\partial x}\right) e^{ik_n x} dx = k_j \delta_{jn}. \tag{10.43}$$

Now we want to examine the motion of a wave packet as it tunnels through a barrier with finite width. If $|\Psi(t)\rangle$ denotes the initial state of the system, then from the definition of $\langle\alpha|\mathcal{T}(t_n)|j\rangle$ it follows that the time evolution of the position operator can be written as

$$\langle x|\Psi(t)\rangle = \sum_{j=1}^{D} \langle\alpha|\mathcal{T}(t_n)|j\rangle^\dagger \langle j|\Psi(t_0)\rangle, \tag{10.44}$$

where $\mathcal{T}(t_n)$ is defined by (10.37). In order to calculate the transmission and reflection probabilities $|T|^2$ and $R|^2$, we first find the transmitted and reflected wave packets

$$\Psi_T(x,t) = \lim \Psi(x,t), \quad \text{as} \quad t \to \infty,\ x \to \infty, \tag{10.45}$$

and

$$\Psi_R(x,t) = \lim \Psi(x,t), \quad \text{as} \quad t \to \infty,\ x \to -\infty. \tag{10.46}$$

From these wave packets we calculate these probabilities:

$$|T|^2 = \int |\Psi_T(x,t)|^2 dx, \quad t \to \infty, \tag{10.47}$$

and

$$|R|^2 = \int |\Psi_R(x,t)|^2 dx, \quad t \to \infty. \tag{10.48}$$

Alternatively we can project the wave packet onto the desired momentum or energy states:

$$|T(p_\alpha)|^2 = \frac{1}{N}\left(\langle +|p_\alpha||\Psi(t)\rangle\right)^2, \tag{10.49}$$

and

$$|R(p_\alpha)|^2 = \frac{1}{N}\left(\langle -p_\alpha||\Psi(t)\rangle\right)^2, \tag{10.50}$$

where N is the normalization constant such that $|T(p_\alpha)|^2 + |R(p_\alpha)|^2 = 1$ is satisfied. These projected states are calculated from the initial state by

$$\langle p_\alpha|\Psi(t)\rangle = \sum_{j=1}^{D}\langle j|p_\alpha\rangle\langle j|\Psi(t_0)\rangle. \tag{10.51}$$

For the initial state we choose a Gaussian wave packet in momentum space;

$$\langle j|\Psi(t_0)\rangle = \langle k_j|\Psi(t_0)\rangle = \frac{1}{\sqrt{N}}\exp\left[-\frac{\sigma^2}{2}(k_j - k_0)^2\right]e^{ik_jx_0}, \tag{10.52}$$

where N is the normalization constant and σ^2 is the width of the wave packet. From Eqs. (10.44) and (10.52) we can calculate $|\langle x|\Psi(t)\rangle|^2$.

This method has been used to calculate the transmission (or reflection) coefficient for tunneling through a symmetric from of the Eckart potential;

$$V(x) = \frac{V_0}{\cosh^2\left(\frac{x}{a}\right)}. \tag{10.53}$$

Choosing the parameters of the potential to be $V_0 = 0.425$ eV, $a = 1.0$ a.u. (atomic unit), and those of the wave packet to be $\sigma = 1.0$ a.u. and $k_0 = \sqrt{2mE_0}$ with $E_0 = 0.35$ eV, one can follow the motion of the wave packet (10.52), when at $t = 0$ it is centered to the left of the barrier. At a later time this wave packet passes through the barrier and then it divides, with a larger wave packet reflected and a smaller transmitted (see [12]).

We can examine the accuracy of the finite difference method by comparing the reflection probabilities $|R(k_0)|^2$ obtained using this method with the analytic solution which is known. The results of numerical calculation obtained from Eqs. (10.35) and (10.36) agree very well with the exact results.

10.3 Determination of Eigenvalues from Finite-Difference Equations

A different approach for calculating the energy eigenvalues of a particle in a confining potential is to work with the finite difference equations (10.6) and (10.7). Bender and Green have suggested the use of matrix representation for

q_n, $n = 0,\ 1,\ 2 \cdots N$ to find the energy spectrum [13]. Noting that the matrix element $A_n = \langle 0|q_n(t)|1\rangle$ form a numerical time sequence we can compute its discrete Fourier transform by writing

$$\tilde{A}_m = \frac{1}{N+1}\sum_{n=0}^{N} A_n \exp\left(\frac{2\pi i m n}{N+1}\right). \tag{10.54}$$

The coefficient A_n can be expressed as

$$\begin{aligned} A_n = \langle 0|q(t)|1\rangle &= \sum_j\sum_k \langle 0|E_j\rangle\langle E_j|q(t)|E_k\rangle\langle E_k|1\rangle \\ &= \sum_j\sum_k a_{ij}\exp[i(E_j - E_k)t], \end{aligned} \tag{10.55}$$

where

$$a_{jk} = \langle 0|E_j\rangle\langle E_j|q(0)|E_k\rangle\langle E_k|1\rangle. \tag{10.56}$$

In the energy representation $\langle E_j|q(0)|E_k\rangle$ is zero unless j is even and k is odd. If we substitute (10.55) in (10.54) and set $t = nh$ we find that A_m contains the sum

$$\sum_{n=0}^{N} \exp\left[inh(E_j - E_k)\right]\exp\left(\frac{2\pi i m n}{N+1}\right). \tag{10.57}$$

For large N this sum has a spike at the points where

$$h(E_j - E_k) + \frac{2\pi m}{N+1} = 0,\ 2\pi,\ \cdots. \tag{10.58}$$

Thus if we plot $\left|\tilde{A}_m\right|^2$, which is a real positive quantity, versus m then we observe sudden jumps at

$$E_{odd} - E_{even} = \frac{2\pi m}{(N+1)h}, \tag{10.59}$$

when measured from the right axis and at

$$E_{even} - E_{odd} = \frac{2\pi(N+1-m)}{(N+1)h}, \tag{10.60}$$

when measured from the left axis. The approximate values of level spacing found by this method are compared with the exact values in TABLE VI.

TABLE VI: Exact and approximate level spacing obtained for the quartic potential $V(q) = 0.885q^4$, using the parameters $N = 1000$ and $D = 11$ for the calculation. The approximate values are found from the position of the spikes in the plot of $\left|\tilde{A}_m\right|^2$ versus m. These are found from Eq. (10.58) [13].

Level spacing	Exact	Approximate
$E_1 - E_0$	1.728	1.674
$E_2 - E_1$	2.142	2.218
$E_3 - E_2$	2.537	2.595
$E_4 - E_3$	2.790	2.846
$E_5 - E_4$	3.000	3.097
$E_6 - E_5$	3.210	3.306

10.4 Systems with Several Degrees of Freedom

The finite difference method of solving the Heisenberg equations of motion can be extended to systems with more than one degree of freedom. Here we will study this formulation as it is applied to a system with two degrees of freedom. Let us assume that a particle of unit mass moves in a potential $V(x, y)$ and that the Hamiltonian for the motion is given by

$$H = \frac{1}{2}\left(p^2 + \wp^2\right) + V(x, y), \tag{10.61}$$

where p and $\wp$ are the momenta conjugate to the coordinates x and y. In this two-dimensional motion we have to verify that there is no operator ordering problem when $V(x, y)$ is written in terms of finite differences. In addition we have to show that the independent degrees of freedom (p, x) and $(\wp, y)$ remain independent at every discrete lattice point. That is the difference equations advancing the operators by a unit time step should preserve the commutation relation between the independent pairs of operators (x, y), (x, p), $(x, \wp)$ etc.

The Heisenberg equations obtained from the Hamiltonian (10.61) are

$$\dot{x} = p, \quad \dot{y} = \wp, \quad \dot{p} = -\frac{\partial V}{\partial x}[x, y], \quad \dot{\wp} = -\frac{\partial V}{\partial y}[x, y]. \tag{10.62}$$

These operators satisfy the canonical commutation relations

$$[x(t),\ p(t)] = [y(t),\ \wp(t)] = i, \tag{10.63}$$

with other equal time commutators like $[x(t), \wp(t)]$ etc. being equal to zero.

Now we approximate the dynamical operators x, p, y and $\wp$ by linear polynomials on the interval $0 \leq t \leq h$ and denote the operators at $t = 0$ and $t = h$ by $(x_1, p_1, y_1, \wp_1)$ and $(x_2, p_2, y_2, \wp_2)$ respectively. Next we introduce linear finite elements for these operators by

$$x(t) = \left(1 - \frac{t}{h}\right) x_1 + \left(\frac{t}{h}\right) x_2, \tag{10.64}$$

and

$$p(t) = \left(1 - \frac{t}{h}\right) p_1 + \left(\frac{t}{h}\right) p_2, \tag{10.65}$$

with similar relations for $y(t)$ and $\wp(t)$. The difference equations corresponding to the operator differential equations (10.62) are found by replacing the time derivatives with the differences divided by h, and by writing the operators on the right-hand side of these equations by the corresponding operators at the midpoint of the interval, $t = \frac{h}{2}$. By these replacements we obtain the following difference equations:

$$\frac{1}{h}(x_2 - x_1) = \frac{1}{2}(p_1 + p_2), \tag{10.66}$$

$$\frac{1}{h}(y_2 - y_1) = \frac{1}{2}(\wp_1 + \wp_2),, \tag{10.67}$$

$$\frac{1}{h}(p_2 - p_1) = -\frac{\partial V}{\partial x}\left[\frac{1}{2}(x_1 + x_2),\ \frac{1}{2}(y_1 + y_2)\right], \tag{10.68}$$

and

$$\frac{1}{h}(\wp_2 - \wp_1) = -\frac{\partial V}{\partial y}\left[\frac{1}{2}(x_1 + x_2),\ \frac{1}{2}(y_1 + y_2)\right]. \tag{10.69}$$

Here the brackets are used to show the arguments of the partial derivatives. These four equations give us x_2, y_2, p_2, $\wp_2$ at $t = h$ in terms of x_1, y_1, p_1, $\wp_1$ at $t = 0$. We observe that in the exact equation (10.62) there is no ambiguity of ordering of x and y since both of these operators are defined at t and that their equal time commutator is zero. However in Eqs. (10.68) and (10.69) the pairs (x_1, y_2) and (x_2, y_1) are defined at different times and that their commutator, in general, does not vanish. Therefore we must first show that there is no problem of ordering in these equations. Let us introduce two new operators σ and τ by the relations

$$\sigma = \frac{1}{2}(x_1 + x_2), \quad \tau = \frac{1}{2}(y_1 + y_2). \tag{10.70}$$

Now if σ and τ commute with each other then there is no operator-ordering ambiguity. To demonstrate this we solve (10.66) and (10.67) for p_2 and $\wp_2$:

$$p_2 = -p_1 + \frac{2}{h}(x_2 - x_1), \tag{10.71}$$

and

$$\wp_2 = -\wp_1 + \frac{2}{h}(y_2 - y_1). \tag{10.72}$$

Then we substitute for p_2 and $\wp_2$, from Eqs. (10.71) and (10.72) in (10.68) and (10.69). The resulting operator equations can be written as

$$\alpha = \frac{\partial V}{\partial \sigma}[\sigma, \tau] + \frac{4\sigma}{h^2}, \tag{10.73}$$

and

$$\beta = \frac{\partial V}{\partial \tau}[\sigma, \tau] + \frac{4\tau}{h^2}, \tag{10.74}$$

where in these relations α and β are defined by

$$\alpha = \frac{2p_1}{h} + \frac{4x_1}{h^2}, \tag{10.75}$$

and

$$\beta = \frac{2\wp_1}{h} + \frac{4y_1}{h^2}. \tag{10.76}$$

We note that α and β are defined in terms of the initial operators and these commute with each other. The set of nonlinear equations (10.73) and (10.74) can, in principle, be solved for the unknowns σ and τ, and the solution can be written as

$$\sigma = \sigma(\alpha, \beta), \quad \text{and} \quad \tau = \tau(\alpha, \beta). \tag{10.77}$$

Since α and β commute and σ and τ are functions of these operators, they also commute. Thus we conclude that there is no problem of ordering in Eqs. (10.68) and (10.69). Having established this fact, we now find the solutions for p_2, x_2, $\wp_2$ and y_2:

$$p_2 = -p_1 - \frac{4x_1}{h} + \frac{4}{h}\sigma(\alpha, \beta), \tag{10.78}$$

$$x_2 = -x_1 + 2\sigma(\alpha, \beta), \tag{10.79}$$

$$\wp_2 = -\wp_1 - \frac{4y_1}{h} + \frac{4}{h}\tau(\alpha, \beta), \tag{10.80}$$

and

$$y_2 = -y_1 + 2\tau(\alpha, \beta). \tag{10.81}$$

Using these solutions we can verify the constancy of the commutation relation in time. Thus for the two conjugate variables x_2 and p_2 we have

$$\begin{aligned} [x_2, p_2] &= \left[-x_1 + 2\sigma(\alpha, \beta),\ -p_1 - \frac{4x_1}{h} + \frac{4}{h}\sigma(\alpha, \beta)\right] \\ &= [x_1, p_1] - \frac{4}{h}[x_1,\ \sigma(\alpha, \beta)] + 2\left[p_1 + \frac{4x_1}{h},\ \sigma(\alpha, \beta)\right] \\ &= i + [h\alpha,\ \sigma(\alpha, \beta)] = i, \end{aligned} \tag{10.82}$$

and a similar result for $[y_2,\ \wp_2]$. For the commutators of two coordinates or two momenta the proof is more tedious. For instance for $[x_2,\ y_2]$ we have

$$\begin{aligned} [x_2, y_2] &= [-x_1 + 2\sigma(\alpha, \beta),\ -y_1 + 2\tau(\alpha, \beta)] \\ &= -2[x_1,\ \tau(\alpha, \beta)] - 2[\sigma(\alpha, \beta),\ y_1] \\ &= \frac{4i}{h}\left[\frac{\partial \sigma}{\partial \beta} - \frac{\partial \tau}{\partial \alpha}\right]. \end{aligned} \tag{10.83}$$

This commutator is zero provided that

$$\frac{\partial \sigma}{\partial \beta} - \frac{\partial \tau}{\partial \alpha} = 0. \tag{10.84}$$

To show that (10.84) is true we start with the definition of α and β, Eqs. (10.76) and (10.77) and differentiate them with respect to α and β, and in this way we find the following results:

$$\frac{\partial^2 V}{\partial \sigma^2}\frac{\partial \sigma}{\partial \alpha} + \frac{\partial^2 V}{\partial \sigma \partial \tau}\frac{\partial \tau}{\partial \alpha} + \frac{4}{h^2}\frac{\partial \sigma}{\partial \alpha} = 1, \tag{10.85}$$

$$\frac{\partial^2 V}{\partial \sigma^2}\frac{\partial \sigma}{\partial \beta} + \frac{\partial^2 V}{\partial \sigma \partial \tau}\frac{\partial \tau}{\partial \beta} + \frac{4}{h^2}\frac{\partial \sigma}{\partial \beta} = 0, \tag{10.86}$$

$$\frac{\partial^2 V}{\partial \sigma \partial \tau}\frac{\partial \sigma}{\partial \alpha} + \frac{\partial^2 V}{\partial \tau^2}\frac{\partial \tau}{\partial \alpha} + \frac{4}{h^2}\frac{\partial \tau}{\partial \alpha} = 0, \tag{10.87}$$

and

$$\frac{\partial^2 V}{\partial \sigma \partial \tau}\frac{\partial \sigma}{\partial \beta} + \frac{\partial^2 V}{\partial \tau^2}\frac{\partial \tau}{\partial \beta} + \frac{4}{h^2}\frac{\partial \tau}{\partial \beta} = 1. \tag{10.88}$$

Now we solve this set of equations for $\frac{\partial \sigma}{\partial \beta}$ and $\frac{\partial \tau}{\partial \alpha}$;

$$\frac{\partial \sigma}{\partial \beta} = \frac{\partial \tau}{\partial \alpha} = \frac{\frac{\partial^2 V}{\partial \sigma \partial \tau}}{\frac{\partial^2 V}{\partial \sigma \partial \tau} - \left[\frac{\partial^2 V}{\partial \sigma^2} + \frac{4}{h^2}\right]\left[\frac{\partial^2 V}{\partial \tau^2} + \frac{4}{h^2}\right]}. \tag{10.89}$$

This result shows the validity of (10.84) and therefore the fact that $[x_2,\ y_2] = 0$. In the same way we can prove the vanishing of other commutators.

The unitary transfer operator which advances the dynamical operators $p_j,\ x_j,\ \cdots$ by one time step must have the following properties:

$$p_{j+1} = U_j p_j U_j^{-1}, \qquad x_{j+1} = U_j x_j U_j^{-1}, \tag{10.90}$$

$$\wp_{j+1} = U_j \wp_j U_j^{-1}, \qquad y_{j+1} = U_j y_j U_j^{-1}. \tag{10.91}$$

This unitary operator can be derived just like the operator U_j, Eq. (10.11) and is given by

$$U_j = \exp\left[\frac{ih}{4}\left(p_j^2 + \wp_j^2\right)\right]\exp[ihA_j(x_j, y_j)]\exp\left[\frac{ih}{4}\left(p_j^2 + \wp_j^2\right)\right], \tag{10.92}$$

where

$$\begin{aligned} A_j(x_j, y_j) &= \frac{2}{h^2}\left[x_j - \sigma\left(\frac{4x_j}{h^2},\ \frac{4y_j}{h^2}\right)\right]^2 + \frac{2}{h^2}\left[y_j - \tau\left(\frac{4x_j}{h^2},\ \frac{4y_j}{h^2}\right)\right]^2 \\ &+ V\left[\sigma\left(\frac{4x_j}{h^2},\ \frac{4y_j}{h^2}\right),\ \tau\left(\frac{4x_j}{h^2},\ \frac{4y_j}{h^2}\right)\right]. \end{aligned} \tag{10.93}$$

A Second Way of Formulating Finite Difference Equations for Systems with Two Degrees of Freedom — For a system with several degrees of freedom we can derive difference equations exactly as we found Eqs. (10.6) and (10.7). The two-dimensional analogue of these equations are:

$$x_2 = x_1 + hp_1 - \frac{1}{2}h^2\frac{\partial V}{\partial x}[x_1, y_1], \tag{10.94}$$

$$p_2 = p_1 - \frac{h}{2}\left(\frac{\partial V}{\partial x}[x_1, y_1] + \frac{\partial V}{\partial x}[x_2, y_2]\right), \tag{10.95}$$

$$y_2 = y_1 + h\wp_1 - \frac{1}{2}h^2\frac{\partial V}{\partial y}[x_1, y_1], \tag{10.96}$$

and

$$\wp_2 = \wp_1 - \frac{h}{2}\left(\frac{\partial V}{\partial y}[x_1, y_1] + \frac{\partial V}{\partial y}[x_2, y_2]\right). \tag{10.97}$$

These equations preserve the canonical commutation relations and are accurate through order h^2. The unitary transfer operator in this case is

$$U_j = \exp\left[\frac{ih}{2}V(x_j, y_j)\right]\exp\left[\frac{ih}{2}\left(p_j^2 + \wp_j^2\right)\right]\exp\left[\frac{ih}{2}V(x_j, y_j)\right], \tag{10.98}$$

and this U_j leaves the canonical commutation relations unchanged [4].

10.5 Weyl-Ordered Polynomials and Bender–Dunne Algebra

In Sec. 1.7 we studied the time development of a classical quantity $u(q(t), p(t))$ and showed that it can be written as a power series in Δt, where the coefficients of expansion are given in terms of multiple Poisson brackets of u with the Hamiltonian H evaluated at $t = 0$, Eq. (1.137). In quantum theory we can solve the Heisenberg equations of motion

$$i\frac{dq(t)}{dt} = [q(t), H], \quad i\frac{dp(t)}{dt} = [p(t), H], \tag{10.99}$$

as an initial value problem, i.e. express the operator $q(t)$ as

$$q(t) = \sum_{m,n} c_{m,n}(t)p^m(0)q^n(0), \tag{10.100}$$

where $q(0)$ and $p(0)$ are the initial coordinate and momentum operators. Since $p^m(0)q^n(0)$ is not a Hermitian operator we replace it with a Weyl-ordered operator $T_{m,n}(0)$, (Sec. 3.6), which is Hermitian and is defined by [15]

$$T_{m,n}(t) = \left(\frac{1}{2}\right)^n \sum_{k=0}^{n} \frac{n!}{(n-k)!k!}q^k(t)p^m(t)q^{n-k}(t). \tag{10.101}$$

In terms of these $T_{m,n}(t)$ operators the Heisenberg equations of motion (10.99) can be written as

$$i\frac{dT_{0,1}(t)}{dt} = [T_{0,1},\ H], \tag{10.102}$$

and

$$i\frac{dT_{1,0}(t)}{dt} = [T_{1,0},\ H], \tag{10.103}$$

Noting that we have set $\hbar = 1$. We can use the fundamental commutation relation $[q(t), p(t)] = i$ to rewrite $T_{m,n}(t)$ as

$$T_{m,n}(t) = \left(\frac{1}{2}\right)^m \sum_{j=0}^{m} \frac{m!}{(m-j)!j!} p^j(t) q^n(t) p^{m-j}(t). \tag{10.104}$$

We can use either (10.101) or (10.104) for the integration of Eqs. (10.103). The set $T_{m,n}$ can be generalized to the cases where either m or n is negative. Thus if $m < 0$ and $n \geq 0$ we use (10.101) and for $n < 0$ and $m \geq 0$ we use (10.104). The basis elements of $T_{m,n}(t)$ form an algebra (Bender–Dunne algebra) which is closed under multiplication [15]. From the definition (10.101) or (10.104) it follows that the product $T_{m,n}(t)T_{r,s}(t)$ can be written in terms of $T_{j,k}(t)$ by rearranging the orders of $p(t)$ and $q(t)$;

$$\begin{aligned} T_{m,n}T_{r,s} &= \sum_{j=0}^{\infty} \frac{\left(\frac{i}{2}\right)^j}{j!} \sum_{k=0}^{j} (-1)^{j-k} \binom{j}{k} \frac{n!}{(n-k)!} \frac{m!}{(m+k-j)!} \\ &\times \frac{r!}{(r-k)!} \frac{s!}{(s+k-j)!} T_{m+r-j,\ n+s-j}, \end{aligned} \tag{10.105}$$

where m, n, r and s are all positive integers. Since the two sides of (10.105) are defined for the same time, we have suppressed the time dependence of the T operators. One can generalize (10.105) to the cases where one or more of these subscripts are negative;

$$\begin{aligned} T_{m,n}T_{r,s} &= \sum_{j=0}^{\infty} \frac{\left(\frac{i}{2}\right)^j}{j!} \sum_{k=0}^{j} (-1)^{j-k} \binom{j}{k} \\ &\times \frac{\Gamma(n+1)\Gamma(m+1)\Gamma(r+1)\Gamma(s+1)}{\Gamma(n-k+1)\Gamma(m+k-j+1)\Gamma(r-k+1)\Gamma(s+k-j+1)} \\ &\times T_{m+r-j,\ n+s-j}. \end{aligned} \tag{10.106}$$

From Eq. (10.106) we can determine the commutation and anti-commutation relations between $T_{m,n}$ and $T_{r,s}$ and express the result as a linear expression in $T_{j,k}$;

$$[T_{m,n},\ T_{r,s}] = 2\sum_{j=0}^{\infty} \frac{\left(\frac{i}{2}\right)^{2j+1}}{(2j+1)!} \sum_{k=0}^{2j+1} (-1)^k \binom{2j+1}{k}$$

$$\times \quad \frac{\Gamma(n+1)\Gamma(m+1)\Gamma(r+1)\Gamma(s+1)}{\Gamma(m-k+1)\Gamma(n+k-2j)\Gamma(r+k-2j)\Gamma(s-k+1)}$$
$$\times \quad T_{m+r-2j-1,\, n+s-2j-1}, \qquad (10.107)$$

and

$$[T_{m,n},\, T_{r,s}]_+ = 2\sum_{j=0}^{\infty}\frac{\left(\frac{i}{2}\right)^{2j}}{(2j)!}\sum_{k=0}^{2j}(-1)^k\binom{2j}{k}\times$$
$$\times \quad \frac{\Gamma(n+1)\Gamma(m+1)\Gamma(r+1)\Gamma(s+1)}{\Gamma(m-k+1)\Gamma(n+k-2j+1)\Gamma(r+k-2j+1)\Gamma(s-k+1)}$$
$$\times \quad T_{m+r-2j,\, n+s-2j}. \qquad (10.108)$$

The commutation relation (10.107) is particularly useful in the integration of the operator differential equations. Now we observe that the Hamiltonian can be written as a finite or infinite sum of $T_{m,n}$ s. For instance if we consider the motion of a particle in a quartic potential where

$$V(q) = \frac{1}{2}q^2 + \frac{\lambda}{4}q^4, \qquad (10.109)$$

then the Hamiltonian can be expressed as

$$H = \frac{1}{2}p^2 + \frac{1}{2}q^2 + \frac{1}{4}\lambda q^4 = \frac{1}{2}T_{2,0} + \frac{1}{2}T_{0,2} + \frac{\lambda}{4}T_{0,4}. \qquad (10.110)$$

For this Hamiltonian the time development of $T_{m,n}$ is obtained from the Heisenberg equation

$$i\frac{dT_{m,n}}{dt} = [T_{m,n},\, H] = \left[T_{m,n},\, \frac{1}{2}T_{2,0} + \frac{1}{2}T_{0,2} + \frac{\lambda}{4}T_{0,4}\right]$$
$$= i\left\{nT_{m+1,n-1} - m\left(T_{m-1,n+1} + \lambda T_{m-1,n+3}\right)\right\}. \qquad (10.111)$$

This relation shows that $T_{0,1}(t) = q(t)$ is coupled to all other $T_{j,k}$ s. Only for $\lambda = 0$, $T_{0,1}$ and $T_{1,0}$ are coupled to each other and to no other element of the $T_{j,k}$ set. In this special case we have two coupled first order operator equation to be solved;

$$\frac{dT_{0,1}}{dt} = T_{1,0}, \qquad (10.112)$$

and

$$\frac{dT_{1,0}}{dt} = -T_{0,1}. \qquad (10.113)$$

Thus in the case of harmonic oscillator we find

$$T_{0,1}(t) = T_{0,1}(0)\cos t + T_{1,0}(0)\sin t, \qquad (10.114)$$

$$T_{1,0}(t) = T_{1,0}(0)\cos t - T_{0,1}(0)\sin t. \qquad (10.115)$$

Connection with a Tensor Operator in $SU(2)$ Group — In Sec. 9.5 we studied the relation between the creation and annihilation operators for the harmonic oscillator and the raising and lowering operators of angular momentum. Now we want to consider a similar connection between the polynomials $\mathcal{P}_j^m(q,p)$ defined as components of a tensor operator and Weyl-ordered products $T_{m,n}(p,q)$. These $\mathcal{P}_j^m(q,p)$ are expressible in terms of hypergeometric function of argument -1 [17],[18].

Again setting $\hbar = 1$, let us consider the $SU(2)$ generators given by

$$M_+ = -\frac{1}{2}q^2, \tag{10.116}$$

$$M_- = -\frac{1}{2}p^2, \tag{10.117}$$

and

$$M_z = \frac{i}{4}(qp + pq). \tag{10.118}$$

These operators satisfy the commutation relations (9.137) and (9.138) for angular momentum. From q and p we can construct a tensor operator with components $\mathcal{P}_j^m(q,p)$, where $j = 0,\ \frac{1}{2},\ 1, \cdots$ and $m = -j,\ -j+1,\ \cdots j$. These $\mathcal{P}_j^m(q,p)$ operators can be defined as polynomials in q and p;

$$\mathcal{P}_j^m(q,p) = 2^m\sqrt{\frac{(j-m)!}{(j+m)!}}\sum_{s=0}^{j-m}\frac{(-ip)^{j-m-s}q^{j+m}(-ip)^s}{s!(j-m-s)!}. \tag{10.119}$$

Alternatively we can define $\mathcal{P}_j^m(q,p)$ by means of its commutator with $M_\pm$ and M_z;

$$\left[M_\pm,\ \mathcal{P}_j^m\right] = \sqrt{(j\mp m)(j\pm m+1)}\ \mathcal{P}_j^{m\pm 1}, \tag{10.120}$$

$$\left[M_z,\ \mathcal{P}_j^m\right] = m\mathcal{P}_j^m. \tag{10.121}$$

These relations can be derived by substituting Eqs. (10.116)–(10.119) in the right-hand sides of Eqs. (10.120) and (10.121).

Equation (10.119) shows that $\mathcal{P}_j^m(q,p)$ is essentially the same as Weyl polynomials $T_{m,n}(p,q)$ and the two are related to each other by

$$T_{j-m,\ j+m}(p,q) = \frac{(i)^{-j+m}}{2^j}\sqrt{(j-m)!(j+m)!}\ \mathcal{P}_j^m(q,-ip). \tag{10.122}$$

We can also write $T_{m,n}$ in terms of qp if we introduce the operator $\hat{Z}$ by

$$\hat{Z} = iqp, \tag{10.123}$$

and note that this $\hat{Z}$ operator satisfies the commutation relations

$$\left[\hat{Z},\ q\right] = q, \quad \text{and} \quad \left[\hat{Z},\ p\right] = -p. \tag{10.124}$$

We can then generalize Eq. (10.123) for any positive integer n by induction, and thus we find

$$q^n p^n = (-i)^n \frac{\hat{Z}!}{\left(\hat{Z}-n\right)!}, \quad \text{and} \quad p^n q^n = (-i)^n \frac{\left(\hat{Z}+n\right)!}{\left(\hat{Z}\right)!}. \tag{10.125}$$

We can also use these forms when n is a negative integer. By grouping the factors of q and p in a special way, viz, by writing

$$p^j q^n p^{m-j} = \left(p^j q^j\right)\left(q^{n-j} p^{n-j}\right) p^{m-n}, \tag{10.126}$$

we can express these factors in terms of $\hat{Z}$. Now by substituting for the first two factors in (10.126) in terms of $\hat{Z}$, we can write $T_{m,n}$ as a hypergeometric function of argument (-1);

$$T_{m,n}(p,q) = \frac{(-i)^n}{2^m} \frac{\hat{Z}!}{\left(\hat{Z}-n\right)!}\, {}_2F_1\left[\hat{Z}+1,\ -m;\ \hat{Z}-n+1,\ -1\right] p^{m-n}. \tag{10.127}$$

We note that the order of parameters in ${}_2F_1$ is immaterial since $\hat{Z}$ commutes with all of the other parameters. By using the formula

$$p^j f\left(\hat{Z}\right) = f\left(\hat{Z}+j\right) p^j, \tag{10.128}$$

when j is a positive or negative integer we can calculate commutators involving $T_{m,n}$, e.g. we have

$$\begin{aligned}\left[T_{m,n},\ p^j\right] = \frac{(-i)^n}{2^m} &\left\{ \frac{\hat{Z}!}{\left(\hat{Z}-n\right)!}\, {}_2F_1\left[\hat{Z}+1,\ -m;\ \hat{Z}-n+1,\ -1\right] \right. \\ - &\left. \frac{\left(\hat{Z}+j\right)!}{\left(\hat{Z}-n+j\right)!}\, {}_2F_1\left[\hat{Z}+1+j,\ -m;\ \hat{Z}-n+1+j,\ -1\right] \right\} p^{m-n+j}.\end{aligned} \tag{10.129}$$

The two hypergeometric functions multiplied by p^{m-n+j} in Eq. (10.129) can be expressed again in terms of $T_{m,n}$ s. A trivial result of (10.129) is that

$$\left[T_{m,0},\ p^j\right] = 0, \tag{10.130}$$

and another simple result is given by

$$\left[T_{0,2},\ p^2\right] = 2\hat{Z} = 4iqp + 2. \tag{10.131}$$

10.6 Integration of the Operator Differential Equations

A different technique for integrating the equations of motion is to make use of the basis set $T_{m,n}$ [15]. In this approach we first try to determine the operator which is conjugate to the Hamiltonian [16]. This operator which will be denoted by $\Theta(q,p)$ may not be self-adjoint, but nonetheless it is useful since it has a very simple dependence on time. Since Θ is conjugate to H, it satisfies the commutation relation

$$-i\left[\Theta(p,q),\ H(p,q)\right] = \frac{d\Theta(p,q)}{dt} = 1. \tag{10.132}$$

This definition does not give us a unique operator, since we can add any function of H to it without changing Eq. (10.132).

Equation (10.132) can be trivially integrated with the result that

$$\Theta(p(t),q(t)) = \Theta(p(0),q(0)) + t. \tag{10.133}$$

That is Θ is a first integral of motion which depends linearly on t. Noting that the Hamiltonian H which does not depend on explicitly on time is a constant of motion

$$H(p(t),q(t)) = H(p(0),q(0)), \tag{10.134}$$

we conclude that (10.133) and (10.134) form a set of two equations for the two unknowns $p(t)$ and $q(t)$, and thus we can solve them to find $p(t)$ and $q(t)$ separately.

Motion of a Particle in a Constant Field — Let us examine this idea on a very simple case, viz, the quantum problem of the motion of a particle in a constant field of force, where

$$H = \frac{1}{2}p^2 + gq = \frac{1}{2}T_{2,0}(p,q) + gT_{0,1}(p,q), \tag{10.135}$$

and g is a constant. This problem was originally solved using matrix mechanics by Born and Wiener [19]. We expand $\Theta(p,q)$ in terms of the basis set of $T_{m,n}$ s;

$$\Theta(p,q) = \sum_{m,n} \alpha_{m,n} T_{m,n}(p,q), \tag{10.136}$$

and we substitute (10.135) and (10.136) in (10.132) to get

$$\sum_{m,n} \alpha_{m,n}\left[T_{m,n},\ \frac{1}{2}T_{2,0} + gT_{0,1}\right] = i, \tag{10.137}$$

where we have suppressed the dependence of $T_{m,n}$ on p and q. Now from the commutator (10.107) we find

$$\left[T_{0,1},\ T_{m,n}\right] = imT_{m-1,n}, \tag{10.138}$$

and

$$[T_{2,0},\ T_{m,n}] = -2inT_{m+1,n-1}. \tag{10.139}$$

Using these commutators we simplify (10.137)

$$\sum_{m,n} \alpha_{m,n} \left(nT_{m+1,n-1} - gmT_{m-1,n}\right) = 1. \tag{10.140}$$

Now the right hand side of (10.140) is equal to $T_{0,0}$ therefore

$$\sum_{m,n} \left[(n+1)\alpha_{m-1,n+1} - g(m+1)\alpha_{m+1,n}\right] T_{m,n} = T_{0,0}, \tag{10.141}$$

and hence $\alpha_{m,n}$ is a solution of the linear difference equation

$$(n+1)\alpha_{m-1,n+1} - g(m+1)\alpha_{m+1,n} = \delta_{m,0}\delta_{n,0}. \tag{10.142}$$

The simplest solution of (10.142) is

$$\alpha_{1,0} = -\frac{1}{g}, \quad \alpha_{m,n} = 0 \ \text{ for } \ m \neq 1, \ \ n \neq 0. \tag{10.143}$$

Thus

$$\Theta = -\frac{1}{g}p + t, \quad \text{or} \quad p = g(t - \Theta). \tag{10.144}$$

By substituting for p in the Hamiltonian (10.135) we find the coordinate operator $q(t)$;

$$q(t) = \frac{1}{g}\left\{H - \frac{1}{2}g^2(t\,1 - \Theta)^2\right\}, \tag{10.145}$$

where H and Θ are constant operators and 1 is the unit operator.

Conjugate Operator for the Hamiltonian of a Simple Harmonic Oscillator — We now consider using this technique to solve the problem of the simple harmonic oscillator with the Hamiltonian

$$H = \frac{1}{2}\left(p^2 + q^2\right) = \frac{1}{2}(T_{2,0} + T_{0,2}). \tag{10.146}$$

By writing $\Theta(p, q)$ as a Weyl-ordered operator, Eq. (10.136), and substituting (10.146) and (10.136) in (10.132) we find that

$$\sum_{m,n} \alpha_{m,n} \left[T_{m,n},\ \frac{1}{2}(T_{2,0} + T_{0,2})\right] = i. \tag{10.147}$$

This relation can be simplified with the help of the commutation relation (10.107)

$$\sum_{m,n} \alpha_{m,n} \left(nT_{m+1,n-1} - mT_{m-1,n+1}\right) = 1 = T_{0,0}. \tag{10.148}$$

Setting the coefficients of $T_{m,n}$ equal to zero we find that $\alpha_{m,n}$ satisfies the linear difference equation

$$(n+1)\alpha_{m-1,n+1} - (m+1)\alpha_{m+1,n-1} = \delta_{m,0}\delta_{n,0}. \tag{10.149}$$

To find the simplest solution of this equation we first note that we cannot set all $\alpha_{m,n}$ s equal to zero because of the inhomogeneous term in (10.149). To construct this solution we start with Eq. (10.149) and with $m = n = 0$. Since the difference equation relates $\alpha_{m-1,n+1}$ to $\alpha_{m+1,n-1}$ and on the integer planar lattice, the line joining the points $(m-1, n+1)$, $(m+1, n-1)$ is parallel to the diagonal line $(-1, 1)$, $(1, -1)$, and that this line passes through $m = n = 0$, we can obtain a set of solution by putting all $\alpha_{m,n}$ equal to zero except for $\alpha_{-2m-1,2m+1}$, $m = 0,\ 1,\ 2\cdots$. For this special solution we get

$$\begin{cases} \alpha_{-2m-1,2m+1} = \dfrac{(-1)^m}{2m+1}, \quad m = 0,\ 1, 2\cdots \\ \alpha_{mn} = 0, \quad \text{otherwise} \end{cases}. \tag{10.150}$$

If we substitute these in (10.136) we obtain

$$\Theta = \sum_{m=0}^{\infty} \frac{(-1)^m}{2m+1} T_{-2m-1,2m+1}. \tag{10.151}$$

This is in fact the Weyl-ordered form of the classical function

$$\theta = \tan^{-1}\left(\frac{q}{p}\right), \tag{10.152}$$

which is conjugate to the classical Hamiltonian $H = \frac{1}{2}\left(p^2 + q^2\right)$.

Integration of the Heisenberg Equation for a Quartic Anharmonic Oscillator — The next problem that we want to consider is the motion of a particle in a potential where the Hamiltonian is

$$H = \frac{1}{2}p^2 + \frac{1}{4}q^4 = \frac{1}{2}T_{2,0} + \frac{1}{4}T_{0,4}. \tag{10.153}$$

By substituting this Hamiltonian in (10.132), with Θ as given in (10.136) and simplifying the result with the help of (10.107) we find the analogue of (10.148)

$$\sum_{m,n} \alpha_{m,n}\left[nT_{m+1,n-1} - mT_{m-1,n+3} + \frac{1}{4}m(m-1)(m-2)T_{m-3,n+1}\right] = 1. \tag{10.154}$$

From this equation we deduce the partial difference equation for $\alpha_{m,n}$;

$$\begin{aligned} &(n+1)\alpha_{m-1,n+1} - (m+1)\alpha_{m+1,n-3} \\ +\ &\frac{1}{4}(m+1)(m+2)(m+3)\alpha_{m+3,n-1} = \delta_{m,0}\delta_{n,0}. \end{aligned} \tag{10.155}$$

For integers $m < 0$ and $n > 0$ we can find successive nonzero values of $\alpha_{m,n}$ by noting that for $m = n = 0$, $\alpha_{-1,1} = 1$. Then difference equation equation shows that for $m = -2$, $n = 4$ we have $\alpha_{-3,5} = -\frac{1}{5}$. Similarly if we choose $m = -4$ and $n = 2$ we find $\alpha_{-3,5} = \frac{1}{2}$, and this infinite sequence continues. We can simplify this partial difference equation by two successive transformations [15]. First we introduce two independent variables M and N by

$$M = -\frac{1}{6}(n + 2m), \quad N = \frac{1}{6}(n - m), \tag{10.156}$$

and replace $\alpha_{m,n}$ by $A_{M,N}$ where

$$A_{M,N} = \alpha_{-2N-2M-1,\, 4N-2N+1} = \alpha_{m-1,n+1}. \tag{10.157}$$

The constraints $m < 0$ and $n > 0$, for the nontrivial solution of $\alpha_{m,n}$ shows that $A_{M,N}$ must vanish for $M < 0$, $N < 0$. BY changing $\alpha_{m,n}$ to $A_{M,N}$ we find that Eq. (10.155) transforms Eq. (10.155) to a first order partial difference equation

$$\begin{aligned} & (4N - 2M + 1)A_{M,N} + (2N + 2M - 1)A_{M,N-1} \\ - & \frac{1}{4}(2N + 2M - 1)(2N + 2M - 2)(2N + 2M - 3)A_{M-1,N-1} \\ = & \delta_{M,0}\delta_{N,0}. \end{aligned} \tag{10.158}$$

In the second transformation we change the dependent variables $A_{M,N}$ to $B_{M,N}$ by defining $B_{M,N}$ as

$$B_{M,N} = \frac{\Gamma\left(\frac{1}{2}\right)}{2^N \Gamma\left(M + N + \frac{1}{2}\right)} A_{M,N}. \tag{10.159}$$

Replacing $A_{M,N}$ s by $B_{M,N}$ s in (10.158) yields a linear partial difference equation of first order;

$$\begin{aligned} & (4N - 2M + 1)B_{M,N} + B_{M,N-1} \\ - & (N + M - 1)B_{M-1,N-1} = \delta_{M,0}\delta_{N,0}. \end{aligned} \tag{10.160}$$

The solution of this difference equation can be written in the form of a triangular matrix. The first few matrix elements of $B_{M,N}$ are tabulated in TABLE VII.

TABLE VII: Nonzero Values of $B_{M,N}$ defined as the solution of Eq. (10.160) [15].

	$M=0$	$M=1$	$M=2$	$M=3$	$M=4$	$M=5$
$N=0$	1					
$N=1$	$-\frac{1}{5}$	$\frac{1}{3}$				
$N=2$	$\frac{1}{45}$	$-\frac{11}{105}$	$\frac{1}{5}$			
$N=3$	$-\frac{1}{585}$	$\frac{6}{385}$	$-\frac{13}{189}$	$\frac{1}{7}$		
$N=4$	$\frac{1}{9945}$	$-\frac{202}{135135}$	$\frac{305}{27027}$	$-\frac{5}{99}$	$\frac{1}{9}$	
$N=5$	$-\frac{1}{208845}$	$\frac{353}{3357585}$	$-\frac{23}{19305}$	$\frac{-100}{11583}$	$-\frac{17}{429}$	$\frac{1}{11}$

The operator $\Theta(p,q)$ which is given as an expansion in $T_{m,n}$ with coefficients $\alpha_{m,n}$ can also be written as an expansion in terms $T_{m,n}$ but now with the coefficients $B_{m,n}$. This is found by replacing m and n by M and N and transforming $\alpha_{m,n}$ to $B_{m,n}$. The final form of $\Theta(p,q)$ is the infinite series

$$\Theta(p,q) = \sum_{N=0}^{\infty}\sum_{M=0}^{N} 2^N \frac{\Gamma\left(M+N+\frac{1}{2}\right)}{\Gamma\left(\frac{1}{2}\right)} B_{M,N}\, T_{-2N-2M-1,4N-2N+1}(p,q). \tag{10.161}$$

The coefficients $B_{m,n}$ can be obtained from a rather complicated generating function [15].

10.7 Iterative Solution for Polynomial Potentials

An interesting method for solving the operator equation of motion which can be used when the potential is a polynomial in q is as follows [23]:

We start with the Hamiltonian of a particle of unit mass;

$$H(p,q) = \frac{1}{2}p^2 + \frac{1}{2}q^2 + \sum_{k=2}^{K} \frac{1}{k+1}\lambda_k q^{k+1}. \tag{10.162}$$

When we set $\lambda_k = 0$ for all k s then we have the harmonic oscillator problem for which the operator solution is given by (10.114) and (10.115). We write this solution in the compact matrix form of

$$\begin{bmatrix} q(t) \\ p(t) \end{bmatrix} = \exp(Jt) \begin{bmatrix} T_{0,1} \\ T_{1,0} \end{bmatrix}, \tag{10.163}$$

where

$$J = \begin{bmatrix} 0 & 1 \\ -1 & 0 \end{bmatrix}, \tag{10.164}$$

and $T_{0,1} = q(0)$ and $T_{1,0} = p(0)$ are the operators at $t = 0$. The equation of motion for the complete system derived from (10.162) are

$$\frac{dq(t)}{dt} = p(t), \tag{10.165}$$

$$\frac{dp(t)}{dt} = -q(t) - \sum_{k=2}^{K} \lambda_k q(t)^k. \tag{10.166}$$

These first order differential equations can be converted into a matrix integral equation

$$\begin{bmatrix} q(t) \\ p(t) \end{bmatrix} = \exp(Jt) \begin{bmatrix} T_{0,1} \\ T_{1,0} \end{bmatrix} - \sum_{k=2}^{K} \lambda_k \int_0^t e^{J(t-\tau)} \begin{bmatrix} 0 \\ [q(\tau)]^k \end{bmatrix} d\tau. \tag{10.167}$$

In particular if we want to find q and p as functions of time from (10.167) we obtain

$$q(t) = T_{0,1} \cos t + T_{1,0} \sin t - \sum_{k=2}^{K} \lambda_k \int_0^t \sin(t-\tau) q^k(\tau) d\tau, \tag{10.168}$$

and

$$p(t) = -T_{0,1} \sin t + T_{1,0} \cos t - \sum_{k=2}^{K} \lambda_k \int_0^t \cos(t-\tau) q^k(\tau) d\tau. \tag{10.169}$$

For instance for the cubic potential

$$V(x) = \frac{1}{2}x^2 + \frac{\lambda}{3}x^3, \tag{10.170}$$

we have

$$\lambda_2 = \lambda, \quad \lambda_3 = \lambda_4 = \cdots = 0. \tag{10.171}$$

Thus the operators $q(t)$ and $p(t)$ satisfy the following integral equations

$$q(t) = T_{0,1} \cos t + T_{1,0} \sin t - \lambda \int_0^t \sin(t-x) q^2(x) dx, \tag{10.172}$$

and

$$p(t) = -T_{0,1} \sin t + T_{1,0} \cos t - \lambda \int_0^t \cos(t-x) q^2(x) dx. \tag{10.173}$$

Assuming that for small values of λ and for short times the iterative solution converges, we write

$$q(t) = q_0(t) + \lambda q_1(t) + \lambda^2 q_2(t) + \cdots. \tag{10.174}$$

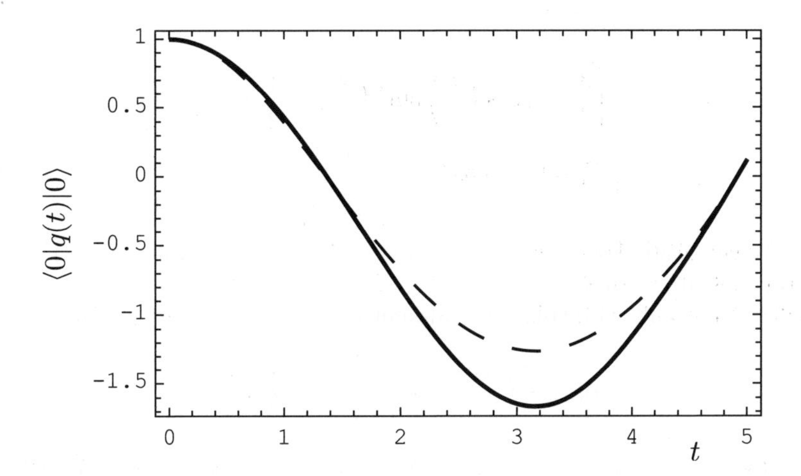

Figure 10.2: The expectation value of the position operator $\langle 0|q(t)|0\rangle$ (the solid curve) and the classical position of the particle (the dashed curve) are plotted as functions of time. The parameters used in the calculation are: $\lambda = 0.4$, $\nu = 1$, and $q_0 = 1$.

Substituting (10.172) in (10.168) and equating different powers of λ we find

$$q_0(t) = T_{0,1}\cos t + T_{1,0}\sin t, \tag{10.175}$$

$$q_1(t) = -\int_0^t \sin(t-x)q_0^2(x)dx, \tag{10.176}$$

$$q_2(t) = \int_0^t \sin(t-x)\left[q_0(x)q_1(t) + q_1(x)q_0(t)\right]dx, \tag{10.177}$$

and so on. For instance for $q_1(t)$ we obtain

$$\begin{aligned} q_1(t) &= -\frac{2}{3}\sin^2\left(\frac{t}{2}\right) \\ &\times \left[T_{0,2}\,(2+\cos t) + 2T_{2,0}\,\sin^2\left(\frac{t}{2}\right) + 2T_{1,1}\,\sin t\right], \end{aligned} \tag{10.178}$$

and a much longer expression for $q_2(t)$ in terms of $T_{3,0},\ T_{2,1}\cdots$. Substituting for $q_0(t), q_1(t)\cdots$ in (10.174) we get an expression for the operator $q(t)$ in terms of $T_{m,n}$ s.

Similarly for $p(t)$ we write

$$p(t) = p_0(t) + \lambda p_1(t) + \lambda^2 p_2(t) + \cdots, \tag{10.179}$$

where

$$p_0(t) = -T_{0,1}\sin t + T_{1,0}\cos t, \tag{10.180}$$

and

$$\begin{aligned} p_1(t) &= -\left[\frac{8}{3}T_{0,2}\cos\left(\frac{t}{2}\right)\sin^3\left(\frac{t}{2}\right) + \frac{1}{3}T_{2,0}(\sin t + \sin 2t)\right. \\ &- \left.\frac{4}{3}T_{1,1}(1 + 2\cos t)\sin^2\left(\frac{t}{2}\right)\right]. \end{aligned} \quad (10.181)$$

We can study this motion by taking the expectation value of $q(t)$ with an arbitrary localized wave packet.
Consider a normalized displaced Gaussian wave packet $\phi(q)$ given by

$$\phi(q) = \left(\frac{\nu}{\pi}\right)^{\frac{1}{4}}\exp\left[-\frac{1}{2}\nu(q - q_0)^2\right]. \quad (10.182)$$

Using this wave packet we calculate the expectation values of $T_{m,n}$ s for the first few m s and n s;

$$\begin{aligned} &\langle 0|T_{0,0}|0\rangle = 1, \quad \langle 0|T_{0,1}|0\rangle = q_0, \quad \langle 0|T_{1,0}|0\rangle = 0, \\ &\langle 0|T_{1,1}|0\rangle = 0, \qquad \langle 0|T_{0,2}|0\rangle = \frac{1}{2\nu}\left(1 + 2\nu q_0^2\right), \\ &\langle 0|T_{2,0}|0\rangle = \frac{1}{2}\nu, \cdots. \end{aligned} \quad (10.183)$$

Thus

$$\langle 0|q(t)|0\rangle = q_0\cos t - \frac{2\lambda}{3}\sin^2\left(\frac{t}{2}\right)\left[\frac{1}{2\nu}\left(1 + 2\nu q_0^2\right) + \frac{1}{2\nu}\sin^2\left(\frac{t}{2}\right)\right]. \quad (10.184)$$

In a similar way we can find the time-dependence of the momentum operator which in this case is

$$\begin{aligned} p(t) &= T_{1,0}\cos t - T_{0,1}\sin t \\ &+ \lambda\left\{\frac{8}{3}T_{0,2}\cos\left(\frac{t}{2}\right)\sin^3\left(\frac{t}{2}\right) + \frac{1}{3}T_{2,0}(\sin t + \sin 2t)\right. \\ &- \left.\frac{4}{3}T_{1,1}(1 + 2\cos t)\sin^2\left(\frac{t}{2}\right)\right\}. \end{aligned} \quad (10.185)$$

From Eq. (10.185) we calculate the expectation value of $\langle 0|p(t)|0\rangle$ with the Gaussian wave packet (10.182). In Fig. 10.2 the expectation value of the position operator calculated from (10.184) is plotted as a function of time. For comparison the classical position is also shown. Because of the approximate nature of this calculation the expectation values of the commutation relation and that of the Hamiltonian of the particle will be time-dependent, i.e.

$$\langle 0|[q(t),\, p(t)]|0\rangle = i\left\{1 + \lambda c_1(t) + \lambda^2 c_2(t) + \cdots\right\}, \quad (10.186)$$

and

$$\left\langle 0 \left| \frac{1}{2}p^2 + \frac{1}{2}q^2 - \frac{\lambda}{3}q^3 \right| 0 \right\rangle = \left\{e_0 + \lambda e_1(t) + \lambda^2 e_2(t) + \cdots\right\}, \tag{10.187}$$

where c_1, $c_2 \cdots$ and e_0, e_1, $e_2 \cdots$ are functions of time. In the next section we will consider a direct method of integration of equations of motion which leaves these expectation values, over a finite range of time, unchanged.

10.8 Another Numerical Method for the Integration of the Equations of Motion

In our study of the iterative method of solution of the Heisenberg equations of motion we noted that except for very short times both the energy and the commutation relation between $p(t)$ and $q(t)$ will not remain constant particularly if the strength of the interaction, λ, is large.

A different approach is to consider the position operator of the particle as a function of time and try to determine the coefficients of the expansion of $q(t)$ in terms of the set $T_{m,n}(0)$ by numerical integration [20],[21]. Let $f(q) = -\frac{\partial V(q)}{\partial q}$ represent the force which acts on the particle, then from Taylor expansion we have

$$q(\Delta t) = T_{0,1} + \frac{\Delta t}{1!}T_{1,0} + \frac{(\Delta t)^2}{2!}f(T_{0,1}) + \frac{1}{2}\frac{(\Delta t)^3}{3!}\left[T_{1,0}f'(T_{0,1}) + f'(T_{0,1})T_{1,0}\right] + \cdots, \tag{10.188}$$

where prime denotes derivative with respect to q. We can write a relation similar to (10.188) for $p(\Delta t)$.

In general if t_j denotes the time $j\Delta t$ with j an integer, then $q(t_{j+1})$ and $p(t_{j+1})$ are expressible in terms of $T_{m,n}(t_j)$;

$$\begin{aligned} q(t_{j+1}) &= T_{0,1}(t_j) + \frac{\Delta t}{1!}T_{1,0}(t_j) + \frac{(\Delta t)^2}{2!}f(T_{0,1}(t_j)) \\ &+ \frac{1}{2}\frac{(\Delta t)^3}{3!}\left[T_{1,0}(t_j)\, f'(T_{0,1}(t_j)) + f'(T_{0,1}(t_j))\, T_{1,0}(t_j)\right] + \cdots, \end{aligned} \tag{10.189}$$

and

$$\begin{aligned} p(t_{j+1}) &= T_{1,0}(t_j) + \frac{\Delta t}{1!}f(T_{0,1}(t_j)) \\ &+ \frac{1}{2}\frac{(\Delta t)^2}{2!}\left[T_{1,0}(t_j)\, f'(T_{0,1}(t_j)) + f'(T_{0,1}(t_j))\, T_{1,0}(t_j)\right] + \cdots. \end{aligned} \tag{10.190}$$

These relations show that if $\{T_{m,n}\}$ s are known at t_j then $q(t_{j+1})$ and $p(t_{j+1})$ can be obtained. From $q(t_{j+1})$ and $p(t_{j+1})$ we can calculate $T_{m,n}$ at a later time, i.e. $T_{m,n}(t_{j+1})$ using Eqs. (10.101) and (10.104)

$$T_{m,n}\left[q(t_{j+1}),\ p(t_{j+1})\right] = T_{m,n}(t_{j+1}). \tag{10.191}$$

From Eqs. (10.189) and (10.190) it follows that $T_{m,n}(t_{j+1})$ depends on the products of the elements $\{T_{m,n}(t_j)\}$ and that we can simplify these using the product relation (10.106). In turn the elements of $\{T_{m,n}(t_j)\}$ are given in terms of $\{T_{m,n}(t_{j-1})\}$ and so on. Therefore the result of integration will be given as a series in $T_{m,n}(0)$.

An alternative way of considering this problem is to find the equation of motion of $T_{m,n}(t)$ for the Hamiltonian of the problem. For instance let us consider the case where the potential is cubic in q and the Hamiltonian is

$$\begin{aligned} H &= \frac{1}{2}p^2 + \frac{1}{2}q^2 - \frac{1}{3}\lambda q^3 \\ &= \frac{1}{2}T_{2,0}(t) + \frac{1}{2}T_{0,2}(t) - \frac{\lambda}{3}T_{0,3}(t). \end{aligned} \tag{10.192}$$

The Heisenberg equation for $T_{m,n}$ is

$$i\frac{dT_{m,n}}{dt} = [T_{m,n},\ H] = \frac{1}{2}\left[T_{m,n}, T_{2,0}\right] + \frac{1}{2}[T_{m,n},\ T_{0,2}] - \frac{\lambda}{3}\left[T_{m,n}, T_{0,3}\right]. \tag{10.193}$$

We can evaluate the commutators on the right hand side of (10.193) using Eq. (10.107);

$$\frac{dT_{m,n}}{dt} = nT_{m+1,n-1} - mT_{m-1,n+1} + 3\lambda m T_{m-1,n+2} - \frac{\lambda m!}{12(m-3)!}T_{m-3,n}. \tag{10.194}$$

From this last relation we find the operator equations for $p = T_{1,0}$ and $q = T_{0,1}$;

$$\frac{dT_{0,1}}{dt} = T_{1,0}, \tag{10.195}$$

and

$$\frac{dT_{1,0}}{dt} = -T_{0,1} + 3\lambda T_{0,2}. \tag{10.196}$$

We note that $T_{1,0}$ is coupled to $T_{0,2}$ and $T_{0,2}$ to other $T_{m,n}$ s. Therefore we have an infinite set of coupled equations. In general $T_{m,n}(\Delta t)$ can be expanded in a Taylor series like (10.188);

$$T_{m,n}(\Delta t) = T_{m,n}(0) + \frac{\Delta t}{1!}\left(\frac{dT_{m,n}}{dt}\right)_{t=0} + \frac{(\Delta t)^2}{2!}\left(\frac{d^2T_{m,n}}{d\,t^2}\right)_{t=0} + \cdots. \tag{10.197}$$

Then from (10.193) it follows that $\left(\frac{dT_{m,n}}{dt}\right)_{t=0}$ can be expressed as $T_{m+1,n-1}(0)$ and other elements of the basis set. The final result can be written as an operator equation

$$T_{m,n}(N\Delta t) = \sum_{j,k} C_{m,n;j,k}(N\Delta t)T_{j,k}(0), \tag{10.198}$$

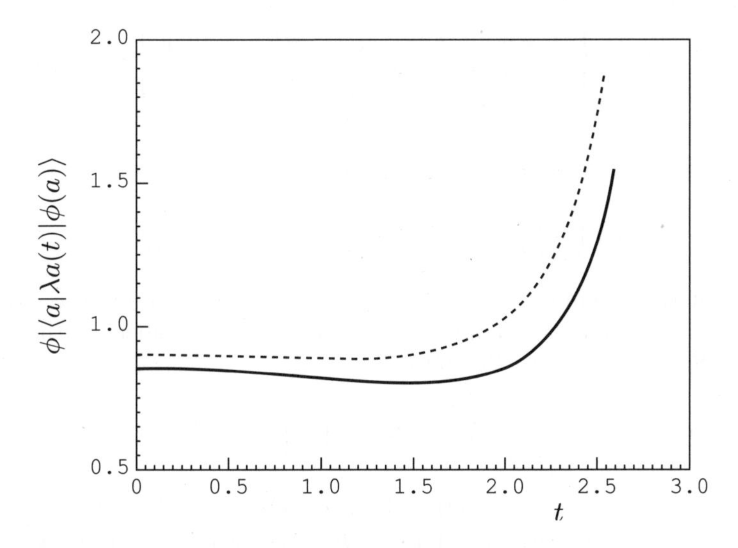

Figure 10.3: The expectation value of the position operator multiplied by λ, $\langle\phi(q)|\lambda q(t)|\phi(q)\rangle$ is plotted as a function of time. Here the particle is trapped in the trough of a cubic potential $V(q) = \frac{1}{2}q^2 - \frac{1}{3}\lambda q^3$. The solid line shows the result for $\lambda q_0 = 0.85$ and the dotted line is for $\lambda q_0 = 0.9$. In the first case the energy of the wave packet is slightly below and in the second slightly above the maximum height of the barrier. Both are calculated for $\lambda = \sqrt{0.1}$.

where the coefficients $C_{m,n;j,k}(N\Delta t)$ are determined numerically for a given $N\Delta t$.

From (10.197) we can calculate the expectation value of

$$\langle T_{0,1}\rangle = \langle\phi(q)|q(t)|\phi(q)\rangle, \tag{10.199}$$

with the wave packet $\phi(q)$ given by a displaced Gaussian function (10.182). The time-dependence of $\langle T_{0,1}\rangle$ is shown in Fig. 10.3. The expectation value of the momentum operator $p(t) = T_{1,0}(t)$ can be obtained in a similar way, and this is displayed in Fig. 10.4.

A similar formulation is possible when the potential is a polynomial in $\left(\frac{1}{q}\right)$. For instance if we are dealing with the radial problem with a potential of the form

$$V(r) = \sum_{k=1}^{K} \frac{A_k}{r^k}, \tag{10.200}$$

then we can determine the equation of motion for $T_{m,n}$ as before. In this case we replace $T_{m,n}(p,q)$ by $S_{m,n}(p_r, r)$ where p_r is the radial momentum operator, (9.205), and $S_{m,n}(t)$ is defined as

$$S_{m,n}(t) = \frac{1}{2^n}\sum_{k=0}^{n} \frac{n!}{k!(n-k)!} r^k \left(-i\frac{\partial}{r\partial r}\right)^m r^{(n-k)}$$

$$= \frac{1}{r}\left\{\frac{1}{2^n}\sum_{k=0}^{n}\frac{n!}{k!(n-k)!}r^k\left(-i\frac{\partial}{\partial r}\right)^m r^{(n-k)}\right\}r = \frac{1}{r}T_{m,n}\,r, \tag{10.201}$$

where we have simplified $S_{m,n}(t)$ using the result

$$\left(-\frac{i}{r}\frac{\partial}{\partial r}r\right)\left(-\frac{i}{r}\frac{\partial}{\partial r}r\right)\cdots\left(-\frac{i}{r}\frac{\partial}{\partial r}r\right) = \frac{(-i)^m}{r}\frac{\partial^m}{\partial r^m}r \tag{10.202}$$

Noting that the commutation relation $[p_r,\ r]$ has the same form as $[p,\ q]$, we find that the commutation relation between the two members of the set, say $S_{m,n}$ and $S_{r,s}$ is similar to (10.107) i.e.

$$\begin{aligned}[S_{m,n},\ S_{r,s}] &= 2\sum_{j=0}^{\infty}\frac{\left(\frac{i}{2}\right)^{2j+1}}{(2j+1)!}\sum_{k=0}^{2j+1}(-1)^k\binom{2j+1}{k} \\ &\times \frac{\Gamma(n+1)\Gamma(m+1)\Gamma(r+1)\Gamma(s+1)}{\Gamma(m-k+1)\Gamma(n+k-2j)\Gamma(r+k-2j)\Gamma(s-k+1)} \\ &\times S_{m+r-2j-1,\ n+s-2j-1}.\end{aligned} \tag{10.203}$$

For the potential (10.200) the Hamiltonian is

$$H = \frac{1}{2}S_{2,0} + \sum_{k=1}^{K}\frac{A_k}{r^k}, \tag{10.204}$$

and from this Hamiltonian and the commutation relation we can find $\frac{dS_{m,n}}{dt}$;

$$\begin{aligned}i\frac{dS_{m,n}}{dt} &= [S_{m,n},\ H] = inS_{m+1,\ n-1} \\ &+ 2\sum_{j=0}^{\infty}\left(\frac{i}{2}\right)^{2j+1}\frac{m!}{(m-2j-1)!(2j+1)!} \\ &\times \left(\sum_{k=1}^{K}A_k\frac{(k+2j)!}{(k-1)!}S_{m-2j-1,\ n-k-2j-1}\right).\end{aligned} \tag{10.205}$$

For the radial part of the Kepler problem (or the hydrogen atom) where

$$V(r) = -\frac{\kappa}{r} + \frac{l(l+1)}{2r^2} \tag{10.206}$$

Eq. (10.207) reduces to

$$\begin{aligned}i\frac{dS_{m,n}}{dt} &= [S_{m,n},\ H] = inS_{m+1,\ n-1} \\ &+ 2\sum_{j=0}^{\infty}\left(\frac{i}{2}\right)^{2j+1}\frac{m!}{(m-2j-1)!} \\ &\times \{-\kappa S_{m-2j-1,\ n-2j-2} + 2(j+1)l(l+1)S_{m-2j-1,\ n-2j-3}\}.\end{aligned} \tag{10.207}$$

Equation (10.205) is valid for positive as well as negative integers n. But if we consider the Taylor expansion of $S_{0,1}(\Delta t)$ which is similar to (10.197) and express the derivatives

$$\left(\frac{dS_{0,1}(t)}{dt}\right)_{t=0}, \left(\frac{d^2 S_{0,1}(t)}{d\,t^2}\right)_{t=0}, \cdots \tag{10.208}$$

in terms of the basis set $S_{m,n}(t)$ we have

$$\begin{aligned} S_{0,1}(\Delta t) &= S_{0,1}(0) + \frac{\Delta t}{1!} S_{1,0}(0) + \frac{(\Delta t)^2}{2!} \sum_k k A_k S_{0,-(k+1)}(0) \\ &- \frac{(\Delta t)^3}{3!} \sum_k k(k+1) A_k S_{1,-(k+2)}(0) - \frac{(\Delta t)^4}{4!} \sum_k k(k+1) A_k \\ &\times \left[-(k+2) S_{2,-(k+3)}(0) + \sum_j j A_j S_{0,\,-(k+j+3)}(0) \right] + \cdots. \end{aligned} \tag{10.209}$$

This relation shows that only $S_{m,n}$ s with $n \le 1$ are needed for calculating the time development of the radial coordinate $S_{0,1}(\Delta t) = r(\Delta t)$.

10.9 Motion of a Wave Packet

Let us assume that the operator equation is integrated and the relation (10.198) is found. Then the position of the particle as a function of time can be determined from the expectation value $\langle \phi(q) | T_{0,1}(t) | \phi(q) \rangle$, where $\phi(q)$ is a localized wave packet. Similarly the momentum associated with the motion of the particle is obtained from $\langle \phi(q) | T_{1,0}(t) | \phi(q) \rangle$. We have already seen an example of such a motion for a particle bouncing between two rigid walls, Sec. 7.5, where we chose a sinusoidal wave packet confined to a small region between the wells. When $T_{0,1}(t)$ can be written as a sum involving $T_{m,n}(0)$ s with positive n, then it is convenient to choose a displaced Gaussian (10.182) for the wave packet.

With this wave packet the integrals involving $T_{m,n}$ s for the first few terms of expansion of $T_{0,1}(t)$ can be done analytically. In particular we observe that for a simple harmonic oscillator we can solve (10.195),(10.196) with $\lambda = 0$ and find $T_{0,1}$;

$$T_{0,1}(t) = T_{0,1}(0) \cos t + T_{1,0}(0) \sin t. \tag{10.210}$$

The expectation value of $T_{0,1}(t)$ with the wavepacket (10.182) gives us

$$\langle \phi(q) | T_{0,1}(t) | \phi(q) \rangle = q_0 \cos t, \tag{10.211}$$

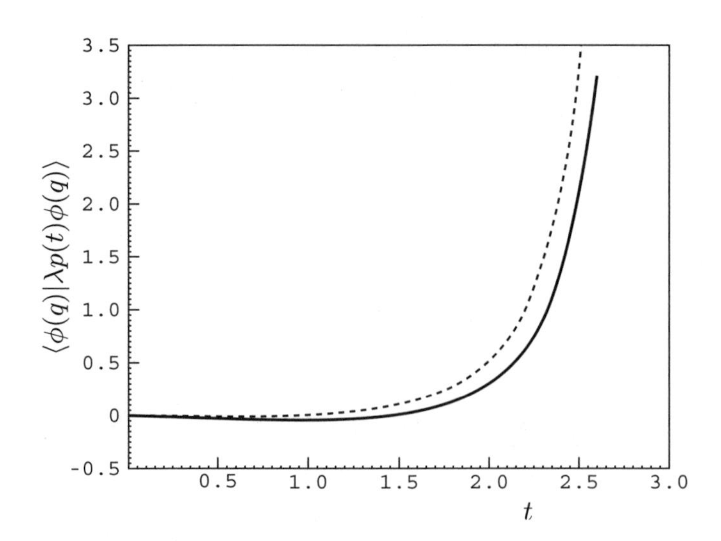

Figure 10.4: The expectation value of the momentum operator for a particle trapped in the potential $V(q) = \frac{1}{2}q^2 - \frac{1}{3}\lambda q^3$ as a function of time.

i.e. the center of the wave packet oscillates about the origin with the frequency 2π and the amplitude q_0.

When λ is not zero we assume that the wavepacket at $t = 0$ is somewhere within the potential well i.e. $\lambda q_0 < 1$. Then after a time t it tunnels through the barrier, which in this case has a maximum height of $\frac{1}{6\lambda^2}$ and goes to $+\infty$. We can use Eqs. (10.195),(10.196) to determine $\langle\phi(q)|T_{0,1}(t)|\phi(q)\rangle$ and $\langle\phi(q)|T_{1,0}(t)|\phi(q)\rangle$ as functions of time. Figures 10.3 and 10.4 show these expectation values for early times $t < 2.5$, for two different values of the initial displacement q_0 (t is dimensionless). Due to the coupling between different $T_{m,n}$ s the accuracy of the numerical integration cannot be maintained for longer times. This can be seen by evaluating

$$C = \langle\phi(q)|[q,\, p]|\phi(q)\rangle, \tag{10.212}$$

and

$$E = \left\langle \phi(q) \left| -\frac{1}{2}\frac{d^2}{d\,q^2} + \frac{1}{2}q^2 - \frac{\lambda}{3}q^3 \right| \phi(q) \right\rangle, \tag{10.213}$$

at every step of integration. For $\lambda = \sqrt{0.1}$ after a time of 2.5 (dimensionless) both C and E will start changing with time.

Since we have used a wave packet to describe the motion of a particle we need to differentiate between the passage over the barrier and quantum tunneling. But we must note that this distinction is not sharply defined. For the tunneling problem we choose q_0 such that the expectation value of the energy Eq. (10.213) is less than the maximum height of the barrier. Alternatively if we start with the initial values $q(0) = q_0$ and $p(0) = 0$ and solve the classical equation of motion $\ddot{q} = -\frac{\partial V(q)}{\partial q}$, then the classical motion will have two turning

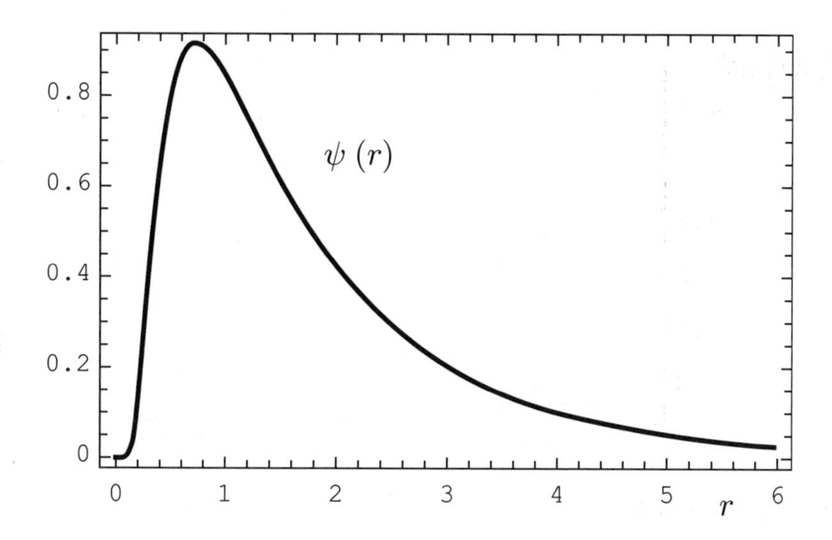

Figure 10.5: The normalized wave packet (10.217) is shown as a function of r.

points. That is the classical limit of quantum tunneling results in bounded oscillation about the equilibrium point $q = q_0$.

The shifted Gaussian wave packet is suitable when the potential is a polynomial in q. If we are considering an inverse power-law potential such as (10.200) then the wave packet should go to zero as $r \to 0$ fast enough so that the matrix elements $\langle \psi(r)|S_{m,-n}(t)|\psi(r)\rangle$ remains finite for all n. Thus

$$\lim \frac{1}{r^n}\left(r\psi(r)\right) \to 0, \quad \text{as} \quad r \to 0 \quad \text{for all } n. \tag{10.214}$$

If we calculate the matrix elements of $S_{m,-n}$ with the wave packet $\psi(r)$ we find that

$$\langle \psi(r)|S_{2m+1,-n}|\psi(r)\rangle = 0, \tag{10.215}$$

and

$$\begin{aligned}\langle \psi(r)|S_{2m,-n}|\psi(r)\rangle &= \frac{(-1)^m}{2^m}\sum_{j=0}^{2m}\frac{(2m)!}{(2m-j)!j!}(-1)^j \\ &\times \int_0^\infty \frac{\partial^j(r\psi)}{\partial r^j}\frac{1}{r^n}\frac{\partial^{2m-j}(r\psi)}{\partial r^{2m-j}}dr.\end{aligned} \tag{10.216}$$

Thus all the matrix elements of $S_{m,n}$ are real.

A simple and analytically tractable wave packet which satisfies (10.214), shown in Fig. 10.5, is

$$r\psi(r) = N\exp\left[-\frac{1}{2}\left(\frac{a}{r} + br\right)\right], \tag{10.217}$$

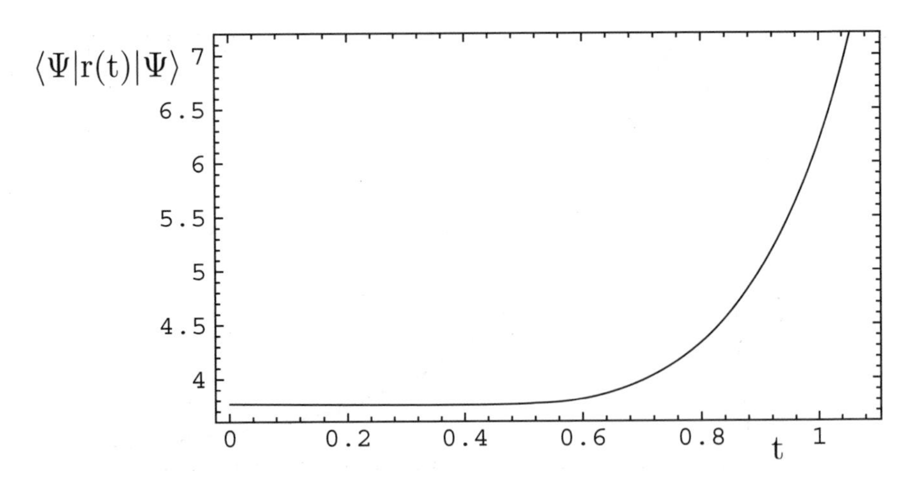

Figure 10.6: The position of the center of mass of the wave packet plotted as a function of time.

where a and b are constants. The normalization constant N is given in terms of a and b

$$N = \frac{1}{\sqrt{2}}\left(\frac{b}{a}\right)^{\frac{1}{4}} \frac{1}{\sqrt{K_1(2\sqrt{ab})}}. \tag{10.218}$$

Here $K_1(x)$ is the modified Bessel function of order 1 [25]. The center of this wave packet is located at r_0 where

$$r_0 = \left(\frac{a}{b}\right)^{\frac{1}{2}} \frac{K_2(2\sqrt{ab})}{K_1(2\sqrt{ab})}. \tag{10.219}$$

Using this wave packet we can also find the matrix elements of $S_{0,n}$;

$$\langle\psi|S_{0,-n}|\psi\rangle = \left(\frac{b}{a}\right)^{\frac{n}{2}} \frac{K_{n-1}(2\sqrt{ab})}{K_1(2\sqrt{ab})}. \tag{10.220}$$

For a specific example we choose $V(r)$, Eq. (10.200), to be the sum of three terms [22]:

$$V(r) = \frac{3.877}{r} - \frac{15.488}{r^2} + \frac{9.0349}{r^3}. \tag{10.221}$$

The coefficients of $V(r)$ are chosen in such a way that the wave packet trapped within the potential can tunnel and escape to infinity. The position of the center of mass of the wave packet as a function of time is shown in Fig. 10.6.

In order to compare the result of the solution of the Heisenberg equation for $\langle\psi(r)|r(t)|\psi(r)\rangle$ and the solution of the time-dependent Schrödinger equation $\langle\psi(r,t)|r|\psi(r,t)\rangle$ we choose a Rydberg wave packet [26]. The Rydberg wave packets are formed from the superposition of many eigenstates belonging to different principle quantum numbers n. The wave packets thus obtained from the

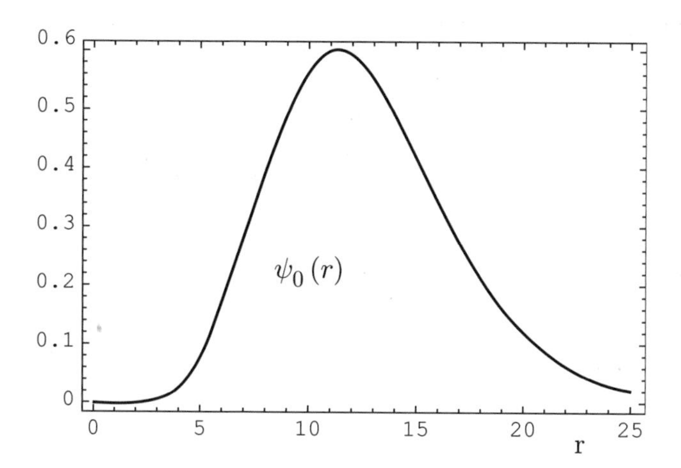

Figure 10.7: A Rydberg wave packet $\psi_{n,\ell=0}(r)$ found by superimposing different S waves.

hydrogen atom wave functions are of great interest in studying the classical limit of the problem of interaction between atoms and the external electromagnetic field [26]. An example of such a wave packet found from superposition a number of S waves is displayed in Fig. 10.7.

Let us denote the radial wave function of a hydrogen-like atom by $u_{n,\ell}(r)$ where

$$u_{n,\ell}(r) = \left[\left(\frac{2Z}{n}\right)^3 \frac{(n-\ell-1)!}{2n(n+l)!}\right]^{\frac{1}{2}} \exp\left(-\frac{Zr}{n}\right)\left(\frac{2Zr}{n}\right)^\ell L_{n-\ell-1}^{2\ell+1}\left(\frac{2Zr}{n}\right), \quad n = 1, 2, \cdots, \tag{10.222}$$

and Z is the nuclear charge. In this equation $L_{n-\ell-1}^{2\ell+1}$ denotes the associated Laguerre polynomial and r is measured in units of the Bohr radius $a_0 = \frac{\hbar^2}{Me^2}$ [27]. To construct Rydberg wave packet we fix ℓ and write

$$\psi_\ell(r) = \sum_n C_n u_{n,\ell}(r). \tag{10.223}$$

Now we want to impose the condition

$$\lim \psi_\ell(r) \to r^K \quad \text{as} \quad r \to 0, \tag{10.224}$$

and for this we expand the right hand side of (10.223) in powers of r and set the coefficients of r^s, $s = 0, 1, \cdots K-1$ equal to zero and the coefficient of r^K equal to unity. In this way we get a set of $K+1$ linear equations for C_n s used in Eq. (10.223). Figure 10.7 shows such a Rydberg wave packet obtained for S wave by adding five waves with $n = 1, \cdots, 5$.
We choose the potential in this case to be the Coulomb potential

$$V(r) = -\frac{Z}{r} = -\frac{3}{r}, \tag{10.225}$$

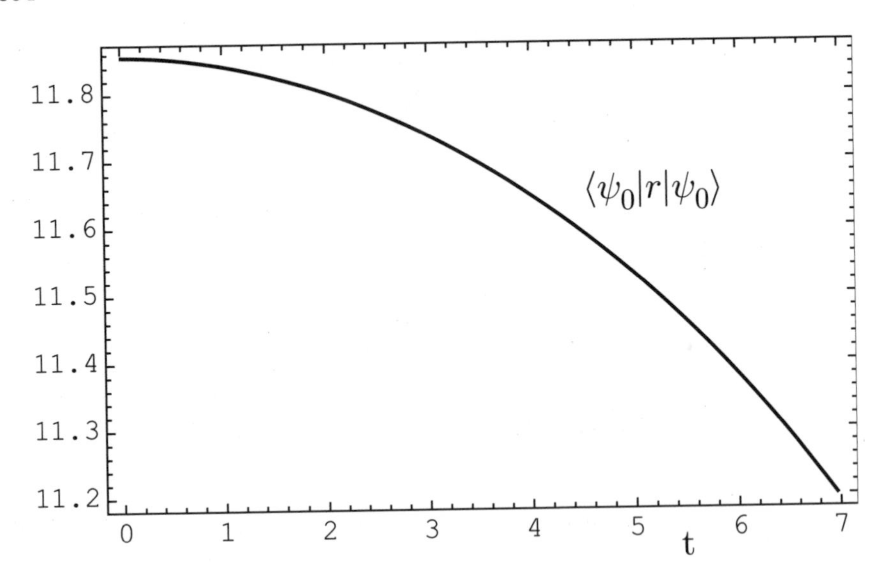

Figure 10.8: The position of the center of Rydberg wave packet as a function of time when the potential is the Coulomb potential. For the range of t shown in the figure, the result of integration of the Heisenberg equation overlaps with the result found from the solution of the wave equation.

in atomic units. The calculation is done exactly as before. However now we can compare our result for the motion of the center of the Rydberg wave packet with a similar calculation using the Schrödinger equation. In the latter case the motion of the center of the wave packet is expressible in terms of $u_{n,\ell}(r)$ as

$$\langle\psi_\ell(r,t)|r|\psi_\ell(r,t)\rangle = \sum_{n,j}^{K} C_n C_j^* \exp[i(E_j - E_n)t]\langle u_{j,\ell}(r)|r|u_{n,\ell}(r)\rangle. \quad (10.226)$$

By comparing this result with

$$\langle\psi_\ell(r)|S_{0,1}(t)|\psi_\ell(r)\rangle \quad (10.227)$$

found by using the Rydberg wave packet and the equation of motion for the operator $S_{0,1}(t)$, we observe that the two curves overlap for $0 \le t \le 7$ (dimensionless units) (see Fig. 10.8).

We can apply this method to other systems as well. For instance consider the Morse potential;

$$V(x) = V_0\left[\exp\left(-2\sqrt{\mu}\,(x - x_0)\right) - 2\exp\left(-\sqrt{\mu}\,(x - x_0)\right)\right]. \quad (10.228)$$

If we change x to q where q is defined by

$$q = \exp\left(-\sqrt{\mu}\,(x - x_0)\right), \quad (10.229)$$

then the Hamiltonian becomes

$$H = \mu \left[-\frac{1}{2}\left(q^2 \frac{d^2}{d\,q^2} + q\frac{d}{dq} \right) + B_0(q^2 - q) \right], \tag{10.230}$$

where $B_0 = \frac{V_0}{\mu}$ is a dimensionless constant. Next we write H in terms of the basis set $\{T_{m,n}\}$;

$$H = \mu \left[\frac{1}{2}\left(T_{0,2}T_{2,0} - iT_{1,1} \right) + B_0(T_{0,2} - 2T_{0,1}) \right]. \tag{10.231}$$

With the help of the product formula (10.106) we can write H as an operator which is linear in $T_{m,n}$ s,

$$H = \mu \left[\frac{1}{2}T_{2,2} + iT_{0,1}T_{1,0} + B_0(T_{0,2} - 2T_{0,1}) \right]. \tag{10.232}$$

Using this Hamiltonian we can find the equations of motion for $T_{m,n}(t)$;

$$\begin{aligned} & i\frac{dT_{m,n}}{dt} = [T_{m,n},\ H] = \mu\left\{ (i/4)\, mn(m-n)T_{m-1,\,n-1} \right. \\ + \quad & \left. i(m-n)T_{m,n} + i(n-m)T_{m+1,\,n+1} + 2mB_0\left(T_{m-1,n} - T_{m-1,\,n+1} \right) \right\}. \end{aligned} \tag{10.233}$$

Again we observe that we need to determine $T_{m,n}$ for positive integers, only. Thus the final result will be of the form (10.198).

Bibliography

[1] C.M. Bender and D.H. Sharp, Solution of the operator field equation by the method of finite elements, Phys. Rev. Lett. 50, 1535 (1985).

[2] C.M. Bender, K.A. Milton, D.H. Sharp, L.M. Simmons, Jr., and R. Stong, Discrete-time quantum mechanics, Phys. Rev. D 32, 1476 (1985).

[3] C.M. Bender, Finite element approximation in quantum theory, Physica 124 A, 91 (1984).

[4] C.M. Bender, K.A. Milton, S.S. Pinsky, and L.M. Simmons, Jr., Discrete-time quantum mechanics II, Systems with several degrees of freedom, Phys. Rev. D 33, 1692 (1986).

[5] C.M. Bender, L.R. Mead, S.S. Pinsky, Resolution of the operator-ordering problem by the method of finite elements, Phys. Rev. Lett. 56, 2445 (1986).

[6] C.M. Bender, L.M. Simmons, Jr., and R. Strong, Matrix methods in discrete-time quantum mechanics, Phys. Rev. 33, 2362 (1986).

[7] C.M. Bender, F. Cooper, J.E. O'Dell and L.M. Simmons, Jr., Quantum tunneling using discrete-time operator difference equations, Phys. Rev. Lett. 55, 901 (1985).

[8] C.M. Bender, F. Cooper, V.P. Gutschick and M.M. Nieto, Simple approach to tunneling using the method of finite elements, Phys. Rev. D 32, 1486 (1985).

[9] V. Moncrief, Finite-difference approach to solving operator equations of motion in quantum theory, Phys. Rev. D 82, 2485 (1983).

[10] F. Cooper, K.A. Milton and L.M. Simmons, Jr., Quantum roll: A study of the long-time behavior of the finite element method, Phys. Rev. D 32, 2056 (1985).

[11] O.A. Sharafeddin, Finite-difference Heisenberg approach: an approximate treatment of many-particle systems, energy level separations and the time-dependent Schrödinger wavefunction, Chem. Phys. Lett. 247, 470 (1995).

[12] O.A. Sharafeddin, Finite-difference Heisenberg approach: The treatment of the unbound states and the elimination of artificial boundary reflections, J. Chem. Phys. 105, 1084 (1996).

[13] C.M. Bender and M.L. Green, Accurate determination of spectra in discrete-time quantum mechanics, Phys. Rev. D 34, 3255 (1986).

[14] M. Razavy, *Quantum Theory of Tunneling*, (World Scientific, Singapore, 2003).

[15] C.M. Bender and G.V. Dunne, Exact solutions to operator differential equations, Phys. Rev. D 40, 2739 (1989).

[16] M. Razavy, Quantum-mechanical conjugate of the Hamiltonian operator, Nuovo Cim. 63 B, 271 (1969).

[17] M.A. Lohe, L.C. Biedenharn and J.D. Louck, Tensor operator formulation of Weyl-ordered polynomials, Phys. Rev. D 43, 617 (1991).

[18] L.C. Biedenharn and J.D. Louck, An intrinsically self-conjugate boson structure: The symplecton, Ann. Phys. 63, 459 (1971).

[19] M. Born and N. Wiener, A new formulation of the laws of quantization of periodic and aperiodic phenomena, J. Math. and Phys. 5, 84 (1926).

[20] M. Kamela and M. Razavy, Integration of the Heisenberg equation of motion for quantum tunneling, Phys. Rev. A 45, 2695 (1992).

[21] M. Hron and M. Razavy, Integration of the Heisenberg equations of motion for quartic potential, Phys. Rev. A 51, 4365 (1995).

[22] M. Hron and M. Razavy, Integration of the Heisenberg equations for inverse power-law potentials, Phys. Rev. A 54, 3801 (1996).

[23] M. Znojil, Polynomial oscillators in Heisenbrg picture, Czech. J. Phys. 41, 201 (1991).

[24] Z-B Chen and H-X Lu, Solving operator differential equations in terms of the Weyl-ordered polynomials, Mod. Phys. Lett. A 17, 2009 (2002).

[25] I.S. Gradshteyn and I.M. Ryzhik, *Tables of Integrals, Series, and Products*, Fourth Edition, (Academic Press,1965), p. 952.

[26] G. Alber and P. Zoller, Laser excitation of electronic wave packets in rydberg atoms, Phys. Rep. 199, 231 (1991).

[27] See for instance: G. Baym, *Lecturs on Quantum Mechanics*, (W.A. Benjamin, Reading, 1976), p.173.

Chapter 11

Perturbation Theory

In their pioneering work on matrix mechanics, Born, Heisenberg and Jordan among other original contributions also developed a systematic method of approximate calculation of the eigenvalues based on perturbation theory [1]. In "three men's paper" this perturbation theory was formulated in the following way:

Let us write the Hamiltonian of the system which is not explicitly time-dependent as

$$H = H_0(p,q) + \lambda H_1(p,q) + \lambda^2 H_2(p,q) + \cdots, \tag{11.1}$$

where λ is a small dimensionless parameter. We assume that the solution for the unperturbed Hamiltonian H_0 is known and we use a representation in which $H_0(p,q)$ is diagonal, i.e.

$$\langle n|H_0(p_0,q_0)|j\rangle = E_j^{(0)}\delta_{nj} \tag{11.2}$$

where p_0 and q_0 are matrices which make H_0 diagonal. Here $p_0 = \lim p$ and $q_0 = \lim p$ as $\lambda \to 0$.

In order to diagonalize $H(p,q)$ we choose a unitary transformation U such that

$$p = Up_0U^{-1}, \quad \text{and} \quad q = Uq_0U^{-1}, \tag{11.3}$$

and then the Hamiltonian

$$H(p,q) = UH(p_0,q_0)U^{-1}, \tag{11.4}$$

becomes a diagonal matrix

$$\langle n|H|j\rangle = E_j\delta_{nj}. \tag{11.5}$$

To determine the form of the transformation, U, we write it as a power series in λ

$$U = 1 + \lambda U_1 + \lambda^2 U_2 + \cdots, \tag{11.6}$$

with its inverse given by

$$U^{-1} = 1 - \lambda U_1 + \lambda^2\left(U_1^2 - U_2\right) + \cdots. \tag{11.7}$$

Next we substitute (11.1), (11.6) and (11.7) in (11.4) and equate different powers of λ and we obtain the following set of equations:

$$\langle n|H_0(p_0, q_0)|j\rangle = E_j^{(0)}\delta_{nj}, \tag{11.8}$$

$$\langle n|U_1 H_0 - H_0 U_1 + H_1|j\rangle = E_j^{(1)}\delta_{nj}, \tag{11.9}$$

$$\langle n|U_2 H_0 - H_0 U_2 + H_0 U_1^2 - U_1 H_0 U_1 + U_1 H_1 - H_1 U_1 + H_2|j\rangle = E_j^{(2)}\delta_{nj}, \tag{11.10}$$

....

$$\langle n|U_r H_0 - H_0 U_r + F_r(H_0 \cdots H_r,\ U_0 \cdots U_{r-1})|j\rangle = E_j^{(n)}\delta_{nj}. \tag{11.11}$$

In these relations all of the operators are functions of p_0 and q_0. The operator F_r is found by collecting all the contributions coming from $H_0, \cdots H_r; U_0, \cdots U_{r-1}$ as can be seen from (11.9) and (11.10). From the property $UU^{-1} = U\left(U^*\right)^t = 1$ and the expansion (11.6) it follows that in the first order of approximation

$$U_1 + \left(U_1^*\right)^t = 0. \tag{11.12}$$

Also to this order we have for the correction to the energy,

$$E_n^{(1)} = \langle n|H_1|n\rangle. \tag{11.13}$$

The correction to the operator U to the first order is

$$\langle n|U|j\rangle = \frac{\langle n|H_1|j\rangle}{\hbar\omega_{nj}^{(0)}}(1 - \delta_{nj}), \tag{11.14}$$

where

$$\hbar\omega_{nj}^{(0)} = E_n^{(0)} - E_j^{(0)} = -\hbar\omega_{jn}^{(0)}. \tag{11.15}$$

From the Hermiticity of H_1 it follows that (11.14) satisfies (11.12).

The second order correction to the energy eigenvalues can be found from (11.10) and is given by

$$E_n^{(2)} = \langle n|H_2|n\rangle + \sum_{k\neq n}\frac{\langle n|H_1 k\rangle\langle k|H_1|n\rangle}{\hbar\omega_{nk}^{(0)}}. \tag{11.16}$$

These results for the first and second order perturbation corrections obtained here are identical to those found from wave mechanics [2]. In the same way we can calculate higher order corrections to the energy. Furthermore we can determine matrix p (and or q) as a power series in λ which can be done by expanding either of these operators as

$$p = p_0 + \lambda p_1 + \lambda^2 p_2 + \cdots, \tag{11.17}$$

and

$$q = q_0 + \lambda q_1 + \lambda^2 q_2 + \cdots. \tag{11.18}$$

Then using (11.3) with the expanded form of U in powers of λ given by (11.6), we obtain to the first order in λ

$$p_1 = U_1 p_0 - p_0 U_1, \tag{11.19}$$

and

$$q_1 = U_1 q_0 - q_0 U_1. \tag{11.20}$$

Now we substitute for U_1 from (11.14) to get

$$\langle n|p_1|j\rangle = \frac{1}{2}\sum_k{}' \left(\frac{\langle n|H_1|k\rangle\langle k|p_0|j\rangle}{\hbar\omega_{nk}^{(0)}} - \frac{\langle n|p_0|k\rangle\langle k|H_1|j\rangle}{\hbar\omega_{kj}^{(0)}}\right), \tag{11.21}$$

and

$$\langle n|q_1|j\rangle = \frac{1}{2}\sum_k{}' \left(\frac{\langle n|H_1|k\rangle\langle k|q_0|j\rangle}{\hbar\omega_{nk}^{(0)}} - \frac{\langle n|q_0|k\rangle\langle k|H_1|j\rangle}{\hbar\omega_{kj}^{(0)}}\right), \tag{11.22}$$

where the prime on the summation sign indicates that the terms with vanishing denominator are to be excluded from the sum.

We can calculate higher order corrections to the energy eigenvalue E_n for the state $|n\rangle$ with the result that

$$E_n = E_n^{(0)} + \lambda E_n^{(1)} + \lambda^2 E_n^{(2)} + \cdots. \tag{11.23}$$

If the Hamiltonian H in (11.1) consists of H_0 and the perturbation $\lambda H_1 = \lambda V(q)$, the latter being a function of the coordinate(s), then from (11.16) it is clear that E_n can be expressed as a power series in λ. However the rate of convergence of the power series (11.23) for a general perturbation $\lambda V(q)$ cannot be determined. Now let us examine the magnitudes of the corrections $E_n^{(2)}$, $E_n^{(3)}$ etc. in the series (11.23). As we have seen earlier for $H_2 = 0$ and $H_1(q) = V(q)$, $E_n^{(2)}$ is given by

$$E_n^{(2)} = \sum_k{}' \frac{|\langle n|V|k\rangle|^2}{\hbar\omega_{nk}^{(0)}} = \sum_k{}' \frac{|\langle n|V|k\rangle|^2}{E_n^{(0)} - E_k^{(0)}}. \tag{11.24}$$

If we calculate the next term in perturbation series we find [2]

$$\begin{aligned} E_n^{(3)} &= \sum_k{}' \sum_j{}' \frac{\langle n|V|j\rangle\langle j|V|k\rangle\langle k|V|n\rangle}{\hbar^2 \omega_{jn}^{(0)} \omega_{kn}^{(0)}} \\ &- \langle n|V|n\rangle \sum_j{}' \frac{|\langle n|V|j\rangle|^2}{\hbar^2 \left(\omega_{nj}^{(0)}\right)^2}. \end{aligned} \tag{11.25}$$

By comparing different terms in the expansion (11.23) we observe that the perturbation series is essentially an expansion in powers of the dimensionless number $\frac{\lambda\langle n|V|j\rangle}{\hbar\omega_{nj}^{(0)}}$, i.e. for fast convergence the matrix element $\langle n|V|j\rangle$ must be smaller than the level spacing $\hbar\omega_{nj}^{(0)} = E_n^{(0)} - E_j^{(0)}$.

Perturbative Calculation of the Energy Levels of a General Anharmonic Oscillator — As an example let us calculate the approximate energy levels of the anharmonic oscillator with the Hamiltonian

$$\begin{aligned} H &= \frac{1}{2}p^2 + \frac{1}{2}\omega^2 x^2 + \frac{1}{3}\lambda_3 x^3 + \frac{1}{4}\lambda_4 x^4 \\ &= H_0 + H_1 + H_2, \end{aligned} \tag{11.26}$$

where in (11.26) H_0 is the Hamiltonian for the harmonic oscillator, H_1 is the cubic and H_2 is the quartic term. To the first order of perturbation the matrix elements $\langle n\left|x^3\right|n\rangle$ (where $|n\rangle$ is the harmonic oscillator state) is zero. Thus

$$\langle n|H|n\rangle = \hbar\omega\left(n+\frac{1}{2}\right) + \frac{3}{8}\left(\frac{\lambda_4}{4}\right)\left(n^2+n+\frac{1}{2}\right). \tag{11.27}$$

In the second order perturbation, both H_1 and H_2 contribute. Using the matrix elements of $\langle n\left|x^3\right|j\rangle$ and $\langle n\left|x^4\right|j\rangle$ shown in TABLE XI (Chapter 12) we find $E_n^{(2)}$ from (11.13)

$$\begin{aligned} E_n^{(2)} &= -\left(\frac{\lambda_3}{3}\right)^2\left(\frac{\hbar}{\omega}\right)^3\left(\frac{15}{4\hbar\omega}\right)\left(n^2+n+\frac{11}{30}\right) \\ &- \left(\frac{\lambda_4}{4}\right)^2\left(\frac{\hbar}{\omega}\right)^4\left(\frac{1}{8\hbar\omega}\right)\left(34n^3+51n^2+59n+21\right). \end{aligned} \tag{11.28}$$

Therefore the energy levels of the anharmonic oscillator (11.26) to the second order are given by [3]

$$\begin{aligned} E_n &\approx \hbar\omega\left(n+\frac{1}{2}\right) - \frac{1}{\hbar\omega}\left(\frac{\lambda_3}{3}\right)^2\left(\frac{\hbar}{\omega}\right)^3 A_n^{(2)} + \left(\frac{\lambda_4}{4}\right)\left(\frac{\hbar}{\omega}\right)^2 B_n^{(1)} \\ &- \left(\frac{\lambda_4}{4}\right)^2\left(\frac{\hbar}{\omega}\right)^4 B_n^{(2)}, \end{aligned} \tag{11.29}$$

where

$$A_n^{(2)} = \frac{15}{4}\left(n^2 + n + \frac{11}{30}\right), \tag{11.30}$$

$$B_n^{(1)} = \frac{3}{4}\left(2n^2 + 2n + 1\right), \tag{11.31}$$

and

$$B_n^{(2)} = \frac{1}{8}\left(34n^3 + 51n^2 + 59n + 21\right). \tag{11.32}$$

11.1 Perturbation Theory Applied to the Problem of a Quartic Oscillator

Earlier in Chapter 7 we studied the problem of the quartic anharmonic oscillator with the Hamiltonian

$$H = \frac{1}{2}p^2 + \frac{1}{2}\omega^2 q^2 + \frac{1}{4}\lambda q^4, \tag{11.33}$$

in connection with the Klein's approximate calculation of the eigenvalues. In the following discussion we set $\omega = 1$. First we want to examine the perturbation technique as is applied to this problem. For this we examine the classical solution of the equation of motion

$$\ddot{q}(t) + q(t) + \lambda q^3(t) = 0, \tag{11.34}$$

found by using the standard form of the perturbation theory [5],[6].

Assuming that λ is a small parameter the classical perturbation expansion of (11.34) is

$$q(t) = \sum_{n=0}^{\infty} \lambda^n q_n(t), \tag{11.35}$$

where $q_n(t)$ s are the coefficients of expansion of $q(t)$. For the determination of $q(t)$ as a function of time, we assume that the initial conditions are

$$q(0) = 1, \quad \dot{q}(0) = 0. \tag{11.36}$$

By substituting (11.35) in (11.34) and setting the coefficients of λ^0 and λ equal to zero we find

$$\ddot{q}_0(t) + q_0(t) = 0, \tag{11.37}$$

and

$$\ddot{q}_1(t) + q_1(t) = -q_0^3(t). \tag{11.38}$$

Now by solving Eq. (11.37) with the boundary conditions (11.36) we obtain

$$q_0(t) = \cos t, \tag{11.39}$$

and by substituting q_0 in (11.38) we get the solution of the forced harmonic oscillator

$$q_1(t) = \frac{1}{32}\left[\cos(3t) - \cos t - 12t\sin t\right]. \tag{11.40}$$

Combining (11.35) and (11.39) we find $q(t)$ to the first order in λ

$$q(t) \approx \left\{\cos t + \frac{\lambda}{32}\left[\cos(3t) - \cos t - 12\, t\sin t\right]\right\}. \tag{11.41}$$

An examination of this result shows that even in the first order perturbation, the solution becomes unbounded for times larger than $\frac{1}{\lambda}$ no matter how small the parameter λ is. For higher orders of approximate calculation the problem of boundedness becomes worse. Clearly these spurious solutions found in the perturbation expansion of $q(t)$ is absent from the exact solution which is bounded and is periodic. Thus the presence of the nonlinear term $q^3(t)$ in (11.34) introduces many time scales whereas for $\lambda = 0$ we have just one time scale $T = 2\pi$.

Classical Anharmonic Oscillator — In the classical description of this motion we can avoid the presence of the so called secular term, $12t\sin t$. This term introduces an error which grows linearly in time and makes the solution unacceptable. One possible method is the following:

We consider a canonical transformation

$$(p(t),\ q(t)) \to (P(t),\ Q(t)), \tag{11.42}$$

in such a way that the Hamiltonian to the order λ has the form [4]

$$H = H_0 + \alpha\lambda H_0^2 + \mathcal{O}\left(\lambda^2\right). \tag{11.43}$$

In this relation H_0 is given by

$$H_0 = \frac{1}{2}\left(P^2(t) + Q^2(t)\right). \tag{11.44}$$

and α is a constant to be determined. Noting that under time-reversal $q(t)$ remains unchanged whereas $p(t)$ changes sign, we write $q(t)$ and $p(t)$ as a combination of linear and cubic terms;

$$q(t) = Q(t) - \frac{3\lambda}{32}\left[3P^2(t)Q(t) + \frac{5}{3}Q^3(t)\right] + \mathcal{O}\left(\lambda^2\right), \tag{11.45}$$

and

$$p(t) = P(t) + \frac{3\lambda}{32}\left[5P(t)Q^2(t) + P^3(t)\right] + \mathcal{O}\left(\lambda^2\right). \tag{11.46}$$

The constant coefficients in (11.45) and (11.46) are obtained in such a way that if we substitute these equations in (11.33) we recover Eq. (11.43). For this case the parameter α in (11.43) must be equal to $\frac{3}{8}$. As we have seen earlier, Sec. 1.3, the Hamiltonian (11.43) up to the order λ^2 is pq-equivalent to H_0, therefore $P(t)$ and $Q(t)$ are sinusoidal functions of time. For the initial condition on $Q(t)$

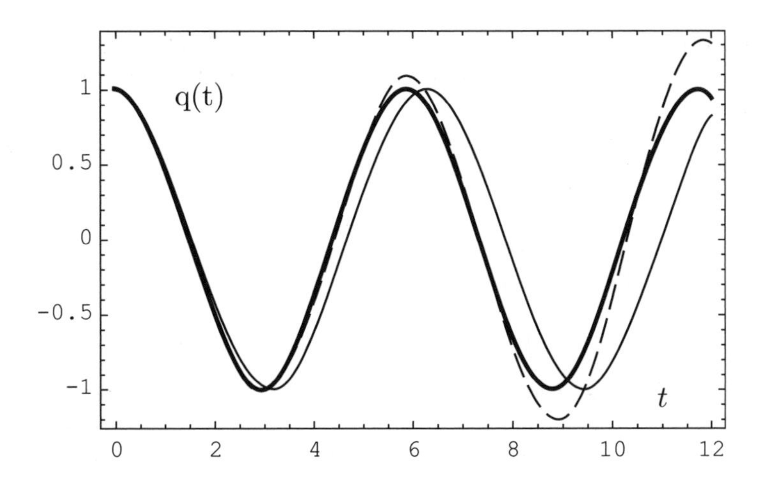

Figure 11.1: The result found from perturbation, Eq. (11.41), for anharmonic oscillator (dashed line) is compared with the exact result (thick solid line) and with the approximate calculation, Eq. (11.45) (thin solid line). These values have been calculated for $\lambda = 0.2$.

and $P(t)$, we choose $Q(0)$ and $P(0)$ so that $q(0) = 1$ and $p(0) = 0$. Thus we have

$$Q(t) = Q_0 \cos(\omega t), \tag{11.47}$$

$$P(t) = P_0 \sin(\omega t), \tag{11.48}$$

where ω is given by

$$\omega = 1 + \frac{3}{8}\lambda, \tag{11.49}$$

and Q_0 is a solution of

$$\frac{5\lambda}{32} Q_0^3 - Q_0 + 1 = 0. \tag{11.50}$$

Substituting (11.47) and (11.48) in (11.45) and (11.46) we find the classical solution of the anharmonic oscillator to the first order in λ. In Fig. 11.1 the results found from two different approximate calculations are compared with the exact calculation. Now for solving the same problem in quantum theory we use a multiple-scale perturbation theory which can be formulated in the following way:

We assume the existence of a number of time scales $(t, \tau, \tau_1 \cdots)$ for the nonlinear system, e.g. cubic nonlinearity [5]–[7]. For the sake of simplicity we consider only two times t and $\tau = \lambda t$ and assume that these two are independent variables. The displacement $q(t)$ will now depend on t as well as τ. Let us consider the perturbation solution of Eq. (11.34) which we write as an expansion in powers of λ

$$q(t) = q_0(t, \tau) + \lambda q_1(t, \tau) + \mathcal{O}\left(\lambda^2\right). \tag{11.51}$$

By substituting (11.51) in (11.34) then using the chain rule and the fact that $\frac{d\tau}{dt} = \lambda$ we find that $q_0(t,\tau)$ and $q_1(t,\tau)$ are the solutions of the partial differential equations

$$\frac{\partial^2 q_0}{\partial t^2} + q_0 = 0, \tag{11.52}$$

$$\frac{\partial^2 q_1}{\partial t^2} + q_1 = -q_0^3 - 2\frac{\partial^2 q_0}{\partial t \partial \tau}. \tag{11.53}$$

Equation (11.52) has the general solution

$$q_0(t,\tau) = \mathcal{A}(\tau)\cos t + \mathcal{B}(\tau)\sin t, \tag{11.54}$$

where the amplitudes $\mathcal{A}$ and $\mathcal{B}$ are functions of τ. Next we substitute (11.54) in (11.53) and set the coefficients of $\sin t$ and $\cos t$ equal to zero. These are the two terms responsible for the resonant coupling between q_0 and q_1. In this way we get two coupled equations for $\mathcal{A}(\tau)$ and $\mathcal{B}(\tau)$;

$$2\frac{d\mathcal{B}}{d\tau} = -\frac{3}{4}\mathcal{A}\left(\mathcal{A}^2 + \mathcal{B}^2\right), \tag{11.55}$$

and

$$2\frac{d\mathcal{A}}{d\tau} = \frac{3}{4}\mathcal{B}\left(\mathcal{A}^2 + \mathcal{B}^2\right). \tag{11.56}$$

By multiplying (11.55) by $\mathcal{B}$ and (11.56) by $\mathcal{A}$ and adding the results we obtain

$$\frac{d\mathcal{C}(\tau)}{d\tau} = 0, \tag{11.57}$$

where

$$\mathcal{C}(\tau) = \frac{1}{2}\left(\mathcal{A}^2(\tau) + \mathcal{B}^2(\tau)\right). \tag{11.58}$$

As Eq. (11.57) shows $\mathcal{C}(\tau)$ is a constant, therefore we set $\mathcal{C}(\tau) = \mathcal{C}(0)$. Next we substitute from (11.58) in (11.55) and (11.56) and we find two linear coupled equations

$$\frac{d\mathcal{B}}{d\tau} = -\frac{3}{4}C(0)\mathcal{A}, \tag{11.59}$$

and

$$\frac{d\mathcal{A}}{d\tau} = \frac{3}{4}C(0)\mathcal{B}. \tag{11.60}$$

Imposing the initial conditions (11.36) we obtain the initial values $\mathcal{A}(0) = 1$ and $\mathcal{B}(0) = 0$. This means that $\mathcal{C}(0) = \frac{1}{2}$.

By solving (11.59) and (11.60) with this value of $C(0)$ and by substituting the result in Eq. (11.51) we find that to this order of perturbation we have

$$q(t) = \cos\left[\left(1 + \frac{3}{8}\lambda\right)t\right]. \tag{11.61}$$

Noting that $q(t)$ calculated in this way is correct to the order of λ^2, this result would be valid for $t \ll \frac{1}{\lambda^2}$.

Now let us consider the quantum-mechanical solution of this problem which can be found from the Heisenberg equations of motion. The equations of motion for the nonlinear oscillator (11.34) are

$$\begin{cases} \dot{p} + x + \lambda x^3 = 0 \\ \\ p = \dot{x} \end{cases}, \tag{11.62}$$

and these equations are subject to the initial conditions

$$x(0) = x_0, \quad \text{and} \quad p(0) = p_0, \tag{11.63}$$

where x_0 and p_0 are the initial operators satisfying the commutation relation $[x_0,\, p_0] = i\hbar$.
As in the classical case we expand the operator $x(t)$ in powers of λ;

$$x(t) = X(t,\tau) = X_0(t,\tau) + \lambda X_1(t,\tau) + \mathcal{O}\left(\lambda^2\right), \tag{11.64}$$

where X_0 and X_1 are operator-valued functions. Now by substituting for $x(t)$ from (11.64) in (11.62) and equating the coefficients of λ^0 and λ we obtain two partial differential equations for the operators X_0 and X_1 and these are the analogues of Eqs. (11.52) and (11.53);

$$\frac{\partial^2 X_0}{\partial t^2} + X_0 = 0, \tag{11.65}$$

$$\frac{\partial^2 X_1}{\partial t^2} + X_1 = -X_0^3 - 2\frac{\partial^2 X_0}{\partial t \partial \tau}. \tag{11.66}$$

Again we observe that (11.65) is homogeneous and linear equation with the general solution

$$X_0(t,\tau) = \hat{A}(\tau)\cos t + \hat{B}(\tau)\sin t, \tag{11.67}$$

and

$$p(t,\tau) = \frac{dx(t)}{dt} = -\hat{A}(\tau)\sin t + \hat{B}(\tau)\cos t + \mathcal{O}(\lambda). \tag{11.68}$$

From the canonical commutation relation $[x(t),\, p(t)] = i\hbar$ and Eqs. (11.67) and (11.68) we find that the operators $\hat{A}(\tau)$ and $\hat{B}(\tau)$ satisfy the relation

$$\left[\hat{A}(\tau),\, \hat{B}(\tau)\right] = i\hbar. \tag{11.69}$$

Also by imposing the initial conditions (11.63) on the solutions (11.67) and (11.68) we get

$$\hat{A}(0) = x_0, \quad \text{and} \quad \hat{B}(0) = p_0, \tag{11.70}$$

to the order λ. The quantum resonance coupling between $X_0(t,\tau)$ and $X_1(t,\tau)$ can be removed in the same way as in the classical problem, that is by substituting for $X_0(t,\tau)$ from (11.65) in (11.66) and setting the coefficients of $\sin t$ and $\cos t$ equal to zero. Since $\hat{A}$ and $\hat{B}$ are operators we have to preserve their orders in the expansion of $X_0(t,\tau)$. The results are similar to Eqs. (11.55) and (11.56)

$$8\frac{d\hat{B}}{d\tau} = -3\hat{A}^3 - \hat{B}\hat{A}\hat{B} - \hat{B}\hat{B}\hat{A} - \hat{A}\hat{B}\hat{B}, \tag{11.71}$$

and

$$8\frac{d\hat{A}}{d\tau} = 3\hat{B}^3 + \hat{A}\hat{B}\hat{A} + \hat{A}\hat{A}\hat{B} + \hat{B}\hat{A}\hat{A}. \tag{11.72}$$

Following the method that we used to solve the classical equations, we multiply Eq. (11.71) once from the left and then from the right by $\hat{B}$ and Eq. (11.72) from left and then right by $\hat{A}$ and add the resulting four equations thus found to get

$$\frac{d}{d\tau}\hat{H}(\tau) = 0, \tag{11.73}$$

where

$$\hat{H}(\tau) = \frac{1}{2}\left[\hat{A}^2(\tau) + \hat{B}^2(\tau)\right]. \tag{11.74}$$

Now Eq. (11.73) shows that $\hat{H}$ is independent of τ, therefore

$$\hat{H}(\tau) = \hat{H}(0) = \frac{1}{2}\left(x_0^2 + p_0^2\right). \tag{11.75}$$

From the definition of $\hat{H}(\tau)$, Eq. (11.74), and the commutation relation (11.69) we can write (11.72) and (11.71) as

$$\frac{d\hat{B}}{d\tau} = -\frac{3}{8}\left(\hat{H}\hat{A} + \hat{A}\hat{H}\right), \tag{11.76}$$

and

$$\frac{d\hat{A}}{d\tau} = \frac{3}{8}\left(\hat{H}\hat{B} + \hat{B}\hat{H}\right). \tag{11.77}$$

Assuming for the moment that $\hat{H}$ can be replaced by its classical value $C(0)$, then we can write the solutions of (11.76) and (11.77) as

$$\hat{A}(\tau) = x_0 \cos\left(\frac{3}{4}C(0)\tau\right) + p_0 \sin\left(\frac{3}{4}C(0)\tau\right), \tag{11.78}$$

and

$$\hat{B}(\tau) = p_0 \cos\left(\frac{3}{4}C(0)\tau\right) - x_0 \sin\left(\frac{3}{4}C(0)\tau\right). \tag{11.79}$$

The complete quantum mechanical result is obtained when we replace (11.78) and (11.79) by the Weyl-ordered products of the $\hat{A}$ and $\hat{B}$ operators with $C(0)$ being replaced by $\hat{H}$:

$$\hat{A}(\tau) = \mathcal{O}_W\left[x_0 \cos\left(\frac{3}{4}\hat{H}\tau\right) + p_0 \sin\left(\frac{3}{4}\hat{H}\tau\right)\right], \tag{11.80}$$

and

$$\hat{B}(\tau) = \mathcal{O}_W \left[p_0 \cos\left(\frac{3}{4}\hat{H}\tau\right) - x_0 \sin\left(\frac{3}{4}\hat{H}\tau\right)\right]. \tag{11.81}$$

(For Weyl-ordered operators see Sec. 3.6).

Let us consider the way that we can find the ordered products in Eqs. (11.80) and (11.81). For this we expand $f\left(\hat{H}\tau\right)$ as a Taylor series in powers of $\hat{H}\tau$, where f can be cosine, sine or an exponential of the argument $\hat{H}\tau$. A typical term after the expansion will depend on $\hat{H}^n$ and we will use Weyl ordering for each term, i.e.

$$\mathcal{O}_W\left(x_0\hat{H}^n\right) = \frac{1}{2^n}\sum_{n}^{n}\left(\begin{array}{c} n \\ j \end{array}\right)\hat{H}^j x_0 \hat{H}^{n-j}. \tag{11.82}$$

Now for each power of τ we reorder the operators so that they becomes symmetric with respect to x_0, i.e. x_0 appears in front and at the end of the expanded form:

$$\mathcal{Q}_W(x_0) = x_0 = \frac{1}{2}(x_0 + x_0), \tag{11.83}$$

$$\mathcal{Q}_W\left(x_0\hat{H}\right) = \frac{1}{2}\left(x_0\hat{H} + \hat{H}x_0\right) = \frac{\hbar}{2}\left[x_0\left(\frac{\hat{H}}{\hbar}\right) + \left(\frac{\hat{H}}{\hbar}\right)x_0\right], \tag{11.84}$$

$$\begin{aligned} \mathcal{Q}_W\left(x_0\hat{H}^2\right) &= \frac{1}{4}\left(x_0\hat{H}^2 + 2\hat{H}x_0\hat{H} + \hat{H}^2x_0\right) \\ &= \frac{\hbar^2}{2}\left[x_0\left(\frac{\hat{H}^2}{\hbar^2} - \frac{1}{4}\right) + \left(\frac{\hat{H}^2}{\hbar^2} - \frac{1}{4}\right)x_0\right], \end{aligned} \tag{11.85}$$

$$\mathcal{Q}_W\left(x_0\hat{H}^3\right) = \frac{\hbar^3}{2}\left[x_0\left(\frac{\hat{H}^3}{\hbar^3} - \frac{3}{4}\frac{\hat{H}}{\hbar}\right) + \left(\frac{\hat{H}^3}{\hbar^3} - \frac{3}{4}\frac{\hat{H}}{\hbar}\right)x_0\right], \tag{11.86}$$

and so on.

The general form of $\mathcal{O}_W\left(q_0\hat{H}^n\right)$ is given by

$$\mathcal{O}_W\left(x_0\hat{H}^n\right) = \frac{\hbar^n}{2}\left[x_0E_n\left(\frac{\hat{H}}{\hbar} - \frac{1}{2}\right) + E_n\left(\frac{\hat{H}}{\hbar} - \frac{1}{2}\right)x_0\right], \tag{11.87}$$

where E_n s are Euler's polynomials shifted by $\frac{1}{2}$ [5],[8].

Using the well-known generating function defining the Euler's polynomials i.e.

$$\frac{2\exp\left[\left(\frac{\hat{H}}{\hbar} + \frac{1}{2}\right)\left(\frac{\tau\hbar}{4}\right)\right]}{e^{\frac{\hbar\tau}{4}} + 1} = \sum_{n=0}^{\infty}\frac{\left(\frac{\hbar\tau}{4}\right)^n}{n!}E_n\left(\frac{\hat{H}}{\hbar} + \frac{1}{2}\right), \quad |\hbar\tau| < 4\pi \tag{11.88}$$

we can write the Weyl-ordered product in a compact form

$$\mathcal{O}_W\left(x_0 e^{\frac{\hat{H}\tau}{4}}\right) = \frac{x_0 e^{\frac{\hat{H}\tau}{4}} + e^{\frac{\hat{H}\tau}{4}} x_0}{2\cosh\left(\frac{\tau\hbar}{8}\right)}. \tag{11.89}$$

From the complex exponential form of (11.89) we obtain

$$\mathcal{O}_W\left[x_0 \cos\left(\frac{3}{4}\hat{H}\tau\right)\right] = \frac{x_0 \cos\left(\frac{3}{4}\hat{H}\tau\right) + \cos\left(\frac{3}{4}\hat{H}\tau\right) x_0}{2\cosh\left(\frac{3\tau\hbar}{8}\right)}. \tag{11.90}$$

and

$$\mathcal{O}_W\left[x_0 \sin\left(\frac{3}{4}\hat{H}\tau\right)\right] = \frac{x_0 \sin\left(\frac{3}{4}\hat{H}\tau\right) + \sin\left(\frac{3}{4}\hat{H}\tau\right) x_0}{2\cosh\left(\frac{3\tau\hbar}{8}\right)}. \tag{11.91}$$

By substituting these ordered operators in $\hat{A}(\tau)$ and $\hat{B}(\tau)$, Eqs. (11.80) and (11.81) and then calculating $X_0(t,\tau)$ from Eq. (11.67) we find that

$$\begin{aligned} X_0(t,\tau) &= \frac{x_0 \cos\left(t + \frac{3}{4}\hat{H}\tau\right) + \cos\left(t + \frac{3}{4}\hat{H}\tau\right) x_0}{2\cosh\left(\frac{3\lambda\hbar t}{8}\right)} \\ &+ \frac{p_0 \sin\left(t + \frac{3}{4}\hat{H}\tau\right) + \sin\left(t + \frac{3}{4}\hat{H}\tau\right) p_0}{2\cosh\left(\frac{3\lambda\hbar t}{8}\right)} \\ &= \frac{x_0 \cos\left[\left(1 + \frac{3}{4}\hat{H}\lambda\right)t\right] + \cos\left[\left(1 + \frac{3}{4}\hat{H}\lambda\right)t\right] x_0}{2\cosh\left(\frac{3\lambda\hbar t}{8}\right)} \\ &+ \frac{p_0 \sin\left[\left(1 + \frac{3}{4}\hat{H}\lambda\right)t\right] + \sin\left[\left(1 + \frac{3}{4}\hat{H}\lambda\right)t\right] p_0}{2\cosh\left(\frac{3\lambda\hbar t}{8}\right)}. \end{aligned} \tag{11.92}$$

This expression reduces to the classical result, Eq. (11.61), in the limit of $\hbar \to 0$ provided that we set $p_0 = 0$, $x_0 = 1$ and the classical value of $\hat{H} = C(0) = \frac{1}{2}$.

Having found the time-dependence of the position operator $x(t)$, we can determine the energy level spacing for this anharmonic oscillator. We note that as (11.75) shows $\hat{H}$ is the Hamiltonian for a harmonic oscillator, therefore if $|n\rangle$ represents the eigenstate of $\hat{H}$ we have

$$\hat{H}|n\rangle = \frac{1}{2}\left(x_0^2 + p_0^2\right)|n\rangle = \left(n + \frac{1}{2}\right)|n\rangle. \tag{11.93}$$

Now we take the expectation value of $X_0(t,\tau)$ between $\langle n-1|$ and $|n\rangle$

$$\begin{aligned} &\langle n-1|X_0(t,\tau)|n\rangle \\ = \; &\langle n-1|x_0|n\rangle \frac{\cos\left(t + \frac{3}{4}\left(n + \frac{1}{2}\right)\hbar\lambda t\right) + \cos\left(t + \frac{3}{4}\left(n - \frac{1}{2}\right)\hbar\lambda t\right)}{2\cosh\left(\frac{3\lambda\hbar t}{8}\right)} \end{aligned}$$

$$
\begin{aligned}
&+ \quad \langle n-1|p_0|n\rangle \frac{\sin\left(t+\frac{3}{4}\left(n+\frac{1}{2}\right)\hbar\lambda t\right)+\sin\left(t+\frac{3}{4}\left(n-\frac{1}{2}\right)\hbar\lambda t\right)}{2\cosh\left(\frac{3\lambda\hbar t}{8}\right)} \\
&= \quad \langle n-1|x_0|n\rangle \cos\left[\left(1+\frac{3}{4}n\hbar\lambda\right)t\right] \\
&+ \quad \langle n-1|p_0|n\rangle \sin\left[\left(1+\frac{3}{4}n\hbar\lambda\right)t\right]. \qquad (11.94)
\end{aligned}
$$

The result shows that the level spacing for the anharmonic oscillator is

$$\Delta E_n = E_n - E_{n-1} = 1 + \frac{3}{4}n\lambda\hbar + \mathcal{O}\left(\lambda^2\right). \qquad (11.95)$$

This result agrees with the corresponding result that we will find from the variational technique, Eq. (12.119), for $E_n - E_{n-1}$.

An extension of this method for the calculation of higher order terms of the anharmonic oscillator has been worked out by Auberson and Capdequi Peyranère [7].

11.2 Degenerate Perturbation Theory

By examining the expressions for $E_n^{(2)}, E_n^{(3)}$, Eqs. (11.24) and (11.25) it is clear that the second and third order corrections to the energy found at the beginning of this chapter diverge. That is the method breaks down whenever $\hbar\omega_{nj}^{(0)} = E_n^{(0)} - E_j^{(0)} = 0$, but the matrix element $\langle n|V|n\rangle$ is not zero. We can bypass this difficulty in the following way:

Let us assume that we have a finite number of states $|n_a\rangle, |n_b\rangle \cdots |n_k\rangle$ all eigenfunctions of H_0 but with the same eigenvalue $E_n^{(0)}$,

$$H_0|n_j\rangle = E_n^{(0)}|n_j\rangle, \quad j = a, b \cdots k. \qquad (11.96)$$

Now if $\langle n_j|V|n_k\rangle \neq 0$ for $j \neq k$, then we cannot use the series (11.23). Instead we start with a state which is a linear combination of the states $|n_j\rangle$ defined by (11.96)

$$|n_\alpha\rangle = \sum_{j=a}^{k} A_{\alpha j}|n_j\rangle, \qquad (11.97)$$

and choose $A_{\alpha j}$ s such that

$$\langle n_\beta|V|n_\alpha\rangle = 0, \quad \text{if} \quad \alpha \neq \beta. \qquad (11.98)$$

But as (11.98) shows the state $|n_\alpha\rangle$ diagonalizes V within that one group of states.

To diagonalize V we determine the eigenvectors $A_{\alpha j}$ of the $k \times k$ matrix

$$\begin{bmatrix} \langle n_a|V|n_a\rangle & \langle n_a|V|n_b\rangle & \cdots\cdots & \langle n_a|V|n_k\rangle \\ \langle n_b|V|n_a\rangle & \langle n_b|V|n_b\rangle & \cdots\cdots & \cdots\cdots \\ \cdots\cdots & \cdots\cdots & \cdots\cdots & \cdots\cdots \\ \langle n_k|V|n_a\rangle & \cdots\cdots & \cdots\cdots & \langle n_k|V|n_k\rangle \end{bmatrix}. \tag{11.99}$$

The eigenvectors $A_{\alpha j}$ are the solution of the homogeneous linear equation

$$\sum_i \langle n_j|V|n_i\rangle A_{\alpha i} = E^{(1)}_{n_\alpha} A_{\alpha j}, \tag{11.100}$$

where $E^{(1)}_{n_\alpha}$ is the eigenvalue and $A_{\alpha j}$ s are the normalized eigenvectors

$$\sum_i |A_{\alpha i}|^2 = 1. \tag{11.101}$$

Once the eigenvectors $A_{\alpha j}$ are found from (11.101) we construct the state $|n_\alpha\rangle$ using (11.97). These eigenvectors form an orthonormal set, i.e.

$$\sum_i A^*_{\beta i} A_{\alpha i} = \delta_{\alpha\beta}. \tag{11.102}$$

Thus by multiplying (11.100) by $A^*_{\beta j}$ and summing over j we obtain

$$\langle n_\beta|V|n_\alpha\rangle = E^{(1)}_{n_\alpha}\delta_{\alpha\beta}. \tag{11.103}$$

The energy $E^{(1)}_{n_\alpha}$ is the first order correction. We can obtain higher order corrections as in the case of nondegenerate systems. For instance the energy eigenvalue for the state $|n_\alpha\rangle$ to the second order is given by

$$E_{n_\alpha} = E^{(0)}_{n_\alpha} + \lambda\langle n_\alpha|V|n_\alpha\rangle + \lambda^2 {\sum_k}' \frac{|\langle k|V|n_\alpha\rangle|^2}{E^{(0)}_n - E^{(0)}_k}, \tag{11.104}$$

where the prime over summation sign means that the sum runs over all states except the k states $|n_\alpha\rangle, |n_\beta\rangle \cdots$.

The Stark Effect — A very simple yet important example of the application of degenerate perturbation theory is provided by the Stark effect. Suppose that a hydrogen atom is placed in a uniform electric field $\mathcal{E}$ which is in the z direction and let us consider the splitting of the fourfold degenerate $n = 2$ levels of the H-atom [11]. Using the spectroscopic notation, these levels are

$$|2S_0\rangle, \quad |2P_1\rangle, \quad |2P_0\rangle, \quad |2P_{-1}\rangle, \tag{11.105}$$

where the first level corresponds to the quantum numbers $\ell = 0,\ m = 0$ and the other three levels have the quantum numbers $\ell = 1$ and $m = 1,\ m = 0$ and

$m = -1$ respectively. Since the perturbation potential in this case is $V(z) = e\mathcal{E}z$ and this perturbation commutes with L_z,

$$[L_z,\ e\mathcal{E}z] = 0, \tag{11.106}$$

thereforeL_z is a conserved quantity. Taking the matrix element of this commutator between $\langle 2P_{-1}|$ and $|2P_1\rangle$ we obtain

$$\begin{aligned}\langle 2P_{-1}|[e\mathcal{E}z,\ L_z]|2P_1\rangle &= \langle 2P_{-1}|e\mathcal{E}z(L_z|2P_1\rangle) - (\langle 2P_{-1}|L_z)e\mathcal{E}z|2P_1\rangle \\ &= 2\hbar\langle 2P_{-1}|e\mathcal{E}z|2P_1\rangle = 0.\end{aligned} \tag{11.107}$$

If we examine other matrix elements of $V = e\mathcal{E}z$ we find that the only nonvanishing off-diagonal matrix elements are $\langle 2S_0|e\mathcal{E}z|2P_0\rangle$ and $\langle 2P_0|e\mathcal{E}z|2S_0\rangle$, i.e. the two states with $m = 0$. These matrix elements are equal and their value can be evaluated with the help of the hydrogen atom wave function

$$\langle 2S_0|e\mathcal{E}z|2P_0\rangle = -3e\mathcal{E}a_0, \tag{11.108}$$

where a_0 is the Bohr radius. For the H-atom because of the symmetry under the parity operation the diagonal elements of V are zero. Thus the matrix (11.99) reduces to

$$\begin{bmatrix} 0 & -3e\mathcal{E}a_0 \\ -3e\mathcal{E}a_0 & 0 \end{bmatrix}. \tag{11.109}$$

For instance if the H atom which is originally in the $|2S_0\rangle$ state is placed in a uniform electric field then the spectral line will split and we get two perturbed eigenstates

$$\frac{1}{\sqrt{2}}\left(|2S_0\rangle + |2P_0\rangle\right), \quad \text{and} \quad \frac{1}{\sqrt{2}}\left(|2S_0\rangle - |2P_0\rangle\right), \tag{11.110}$$

and the energies of these two states are

$$-\frac{e^2}{2a_0}\left(\frac{1}{4} + \frac{6\mathcal{E}}{\frac{e}{a_0^2}}\right), \quad \text{and} \quad -\frac{e^2}{2a_0}\left(\frac{1}{4} - \frac{6\mathcal{E}}{\frac{e}{a_0^2}}\right), \tag{11.111}$$

respectively.

11.3 Almost Degenerate Perturbation Theory

Now let us consider the effect of a small perturbation on almost degenerate levels. By "almost" we mean that the level spacing $E_2^{(0)} - E_1^{(0)}$ for the two levels

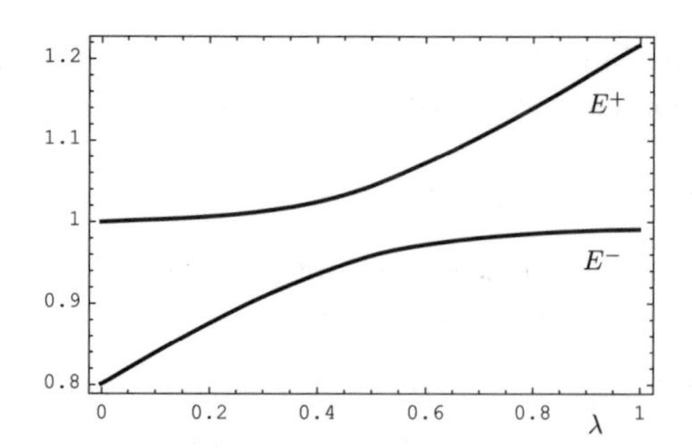

Figure 11.2: Plot of the energy levels E^+ and E^- as functions of the strength of the perturbation λ.

1 and 2 is comparable with the energy shift produced by the perturbation. We write the eigenvalue equations for the two closely spaced levels as

$$H_0 \left|n_1^{(0)}\right\rangle = E_1^{(0)} \left|n_1^{(0)}\right\rangle, \quad H_0 \left|n_2^{(0)}\right\rangle = E_2^{(0)} \left|n_2^{(0)}\right\rangle, \tag{11.112}$$

and assume that other levels of the system are far from these two. Denoting the perturbation by $H_1 = \lambda V$ we write the exact eigenvalue equation as

$$(H_0 + \lambda V)|n\rangle = E_n|n\rangle. \tag{11.113}$$

We also assume that the approximate state of the system $|n\rangle$ is a linear combination of $\left|n_1^{(0)}\right\rangle$ and $\left|n_2^{(0)}\right\rangle$, i.e. we ignore the overlap between $|n\rangle$ and the other levels of the system

$$|n\rangle = c_1 \left|n_1^{(0)}\right\rangle + c_2 \left|n_2^{(0)}\right\rangle. \tag{11.114}$$

By substituting (11.114) in (11.113) and multiplying the result first by $\left\langle n_1^{(0)}\right|$ and then by $\left\langle n_2^{(0)}\right|$ we obtain two coupled equations for c_1 and c_2;

$$\left(\lambda V_{11} - E + E_1^{(0)}\right) c_1 + \lambda V_{12} c_2 = 0, \tag{11.115}$$

and

$$\lambda V_{21} c_1 + \left(\lambda V_{22} - E + E_2^{(0)}\right) c_2 = 0, \tag{11.116}$$

where V_{ij} stands for $\langle i|V|j\rangle$. Setting the determinant of these two homogeneous equations equal to zero we find a quadratic equation for E whose solutions are

$$\begin{aligned} E^{\pm} &= \frac{1}{2}\left(E_1^{(0)} + E_2^{(0)}\right) \\ &+ \frac{1}{2}\left[\lambda(V_{11} + V_{22}) \pm \sqrt{\left(\lambda(V_{11} - V_2) + E_1^{(0)} - E_2^{(0)}\right)^2 + 4|\lambda V_{12}|^2}\right]. \end{aligned} \tag{11.117}$$

Now by changing λ from zero to one, the energies $E^{\pm}$ will change, and these changes are shown in Fig. 11.2. We observe that as we increase λ, E^{-} approaches E^{+} but in general the two energies will not become equal, and the curves shown will not cross each other [10].

11.4 van der Waals Interaction

We can use perturbation theory to derive the leading term of the potential energy between two widely separated atoms. Let us consider two H atoms again. At a large distance R from the first atom, the electric field will be that of a dipole, and the second atom which may be also regarded as a dipole, interacts with the dipole field of the first atom. We take the two protons as two fixed points in space separated by a distance $\mathbf{R}$. The electron in the first atom is at $\mathbf{r}_1$ from the first proton, and the electron in the second atom is at $\mathbf{r}_2$ relative to the second proton (Fig. 11.3). Thus the interaction between the two atoms is the sum of Coulomb potentials between different charges with the total potential energy

$$V = e^2\left[\frac{1}{R} + \frac{1}{|\mathbf{R}+\mathbf{r}_2-\mathbf{r}_1|} - \frac{1}{|\mathbf{R}+\mathbf{r}_2|} - \frac{1}{|\mathbf{R}-\mathbf{r}_1|}\right]. \tag{11.118}$$

For large R we can expand the denominators in (11.118) in powers of $\frac{r_1}{R}$ and $\frac{r_2}{R}$, and keep the leading terms. In this way we find $V(\mathbf{R})$ to be

$$V(\mathbf{R}) = e^2\left[\frac{\mathbf{r}_1\cdot\mathbf{r}_2}{R^3} - \frac{3(\mathbf{r}_1\cdot\mathbf{R})(\mathbf{r}_2\cdot\mathbf{R})}{R^5}\right]. \tag{11.119}$$

It will be convenient to choose $\mathbf{R}$ to be in the direction of the z-axis, then (11.119) reduces to the simple form of

$$V(R) = \frac{e^2}{R^3}\left(x_1x_2 + y_1y_2 - 2z_1z_2\right). \tag{11.120}$$

Assuming that the atom 1 is in the state $|n_1\rangle$ and atom 2 is in the state $|n_2\rangle$, then in the absence of V, the energies of the two atoms is $E_1 + E_2$. Now if we calculate the total energy of the system as a whole then this energy will depend on R and to the first order it is given by

$$E(R) = E_1 + E_2 + \langle n_1, n_2|V|n_1, n_2\rangle. \tag{11.121}$$

The last term in (11.121) represents the potential energy and is dependent on the distance R. The potential energy can be obtained by noting that for the

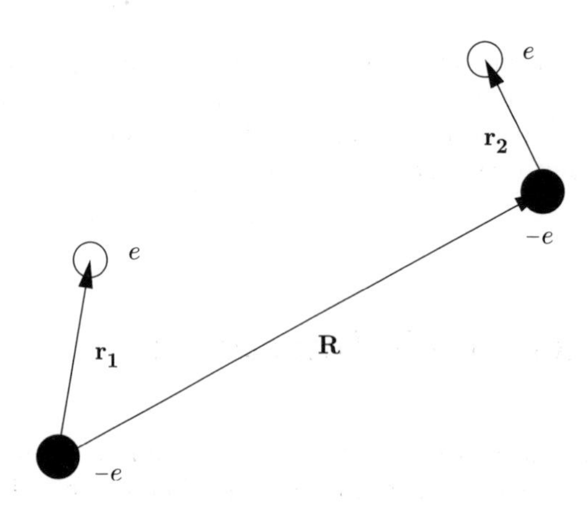

Figure 11.3: Two H atoms separated by a distance R ($R \gg a_0$), where a_0 is the Bohr radius.

unperturbed system $|n_1, n_2\rangle$ can be written as $|n_1\rangle|n_2\rangle$, thus

$$\begin{aligned}\langle n_1, n_2|V|n_1, n_2\rangle &= \frac{e^2}{R^3}\left[\langle n_1|x_1|n_1\rangle\langle n_2|x_2|n_2\rangle\right.\\ &+ \langle n_1|y_1|n_1\rangle\langle n_2|y_2|n_2\rangle\\ &- \left.2\langle n_1|z_1|n_1\rangle\langle n_2|z_2|n_2\rangle\right].\end{aligned} \tag{11.122}$$

If we assume that the two atoms are in their ground states then (11.122) vanishes again because of the reflection symmetry of the wave function which we have discussed earlier. Thus the potential energy shows up as the second order correction to $E(R)$:

$$E^{(2)}(R) = \frac{e^4}{R^6}\sum_{j_1,\, j_2}{}' \frac{|\langle n_1, n_2|(x_1x_2 + y_1y_2 - 2z_1z_2)|j_1, j_2\rangle|^2}{E_{n_1} + E_{n_2} - E_{j_1} - E_{j_2}}. \tag{11.123}$$

Here the prime means that $n_1 = j_1$ and $n_2 = j_2$ must be excluded from the sum. This expression shows that if $|n_1\rangle$ and $|n_2\rangle$ are ground states of the H atom, $E^{(2)}(R)$ is negative, that is the van der Waals force in this case is attractive. We can write $E(R)$ as

$$E(R) = E_{n_1} + E_{n_2} - \frac{e^2}{a_0}\left(\frac{a_0}{R}\right)^6 \zeta, \tag{11.124}$$

where ζ is a dimensionless number

$$\zeta = \frac{e^2}{a_0^5}\sum_{j_1\, j_2}{}' \frac{|\langle n_1 n_2|(x_1x_2 + y_1y_2 - 2z_1z_2)|j_1 j_2\rangle|^2}{E_{n_1} + E_{n_2} - E_{j_1} - E_{j_2}}, \tag{11.125}$$

and has the value 6.5 for the ground state of the H atom [11].

It is worth noting that at larger distances the R^{-6} attraction changes to an attraction proportional to R^{-7} because of the retardation effect [12]. We also note that while the hydrogen atom in its ground state does not have a permanent dipole moment, the presence of the second atom induces a dipole moment in the first. The potential shown in (11.124) which is attractive and drops off as R^{-6} is the asymptotic form of the potential for large R. The short range interaction between the atoms 1 and 2 is much more complicated to calculate, but it is repulsive and generally behaves as R^{-12}.

11.5 Time-Dependent Perturbation Theory

In classical mechanics if $H_0(p,q)$ represents the Hamiltonian for a solvable unperturbed system, and if this system is perturbed by the action of a small time-dependent interaction $\lambda H_1(p,q,t)$ so that the total Hamiltonian is

$$H(p,q,t) = H_0(p,q) + \lambda H_1(p,q,t), \tag{11.126}$$

then we can make a canonical transformation generated by $F_2(q,P,t)$ and change p and q to P and Q;

$$(p,q) \longleftrightarrow (P,Q). \tag{11.127}$$

Under this transformation the total Hamiltonian changes to $K(P,Q,t)$ where [9]

$$K(P,Q,t) = H_0 + \lambda H + \frac{\partial F_2}{\partial t}. \tag{11.128}$$

From this Hamiltonian we find, the equations of motion for P and Q

$$\dot{P} = -\frac{\partial K}{\partial Q}, \quad \text{and} \quad \dot{Q} = \frac{\partial K}{\partial P}. \tag{11.129}$$

In quantum mechanics we can generate the effect of this type of perturbation by a time-dependent unitary transformation. Here we replace every $U(p,q)$ in the formulation shown in the beginning of this chapter by a time-dependent $U(p,q,t)$, and this adds a term $\frac{\hbar}{i}\frac{\partial U}{\partial t}$ to the Hamiltonian. In this way we can be modify the perturbation method for application to the cases where the interactions, λH_1, $\lambda^2 H_2 \cdots$ are dependent on time but H_0 is not. For this we need to replace every term of the form $H_0 U_r - U_r H_0$ by

$$H_0 U_r - U_r H_0 + \frac{\hbar}{i}\frac{\partial U_r}{\partial t}. \tag{11.130}$$

This replacement changes the group of equations (11.9), (11.10) $\cdots$ to the following set

$$\left\langle n \left| U_1 H_0 - H_0 U_1 - \frac{\hbar}{i}\frac{\partial U_1}{\partial t} + H_1 \right| j \right\rangle = E_j^{(1)} \delta_{nj}, \tag{11.131}$$

$$\left\langle n \left| U_2H_0 - H_0U_2 - \frac{\hbar}{i}\frac{\partial U_2}{\partial t} \right| j \right\rangle$$
$$+ \left\langle n \left| \left(H_0U_1 - U_1H_0 + \frac{\hbar}{i}\frac{\partial U_1}{\partial t}\right) + U_1H_1 - H_1U_1 + H_2 \right| j \right\rangle = E_j^{(2)}\delta_{nj}, \tag{11.132}$$

....

Let us apply this form of the time-dependent perturbation theory to the important problem of calculating the matrix elements of p and q for a charged particle in an oscillating electric field where

$$H_1 = e\mathcal{E}q_0 \cos\omega_0 t. \tag{11.133}$$

To the first order the operators p and q are:

$$p = p_0 + \lambda(U_1p_0 - p_0U_1), \tag{11.134}$$

and

$$q = q_0 + \lambda(U_1q_0 - q_0U_1). \tag{11.135}$$

If we denote the time dependence of the matrix elements such as p, q or H by

$$\langle n|p(t)|j\rangle = \langle n|p(\tau)|j\rangle \exp[i(\omega_{nj} + \tau\omega_0)t], \tag{11.136}$$

then we have

$$\langle n|H_1(\tau = 1)|j\rangle = \langle n|H_1(\tau = -1)|j\rangle = \frac{e\mathcal{E}}{2}\langle n|q_0|j\rangle, \tag{11.137}$$

and

$$\langle n|U_1(\tau = 1)|j\rangle = \frac{e\mathcal{E}}{2\hbar}\frac{\langle n|q_0|j\rangle}{\omega_{nj}^{(0)} + \omega_0}. \tag{11.138}$$

by substituting $\langle n|U_1(1)|j\rangle$ in (11.135) we find $\langle n|q_1(1)|j\rangle$;

$$\langle n|q_1(\tau = 1)|j\rangle = \frac{e\mathcal{E}}{2\hbar}\sum_k\left(\frac{\langle n|q_0|k\rangle\langle k|q_0|j\rangle}{\omega_{nk}^{(0)} + \omega_0} - \frac{\langle n|q_0|k\rangle\langle k|q_0|j\rangle}{\omega_{kj}^{(0)} + \omega_0}\right). \tag{11.139}$$

If q is the position of the particle in rectangular coordinates, then $p = m\dot{q}$ and (11.139) can be written as

$$\langle n|q_1(1)|j\rangle = \frac{e\mathcal{E}}{2\hbar im}\sum_k\left(\frac{\langle n|q_0|k\rangle\langle k|p_0|j\rangle - \langle n|p_0|k\rangle\langle k|q_0|j\rangle}{\left(\omega_{nk}^{(0)} + \omega_0\right)\left(\omega_{kj}^{(0)} + \omega_0\right)}\right). \tag{11.140}$$

similarly for $\langle n|q_1(\tau = -1)|j\rangle$ we have

$$\langle n|q_1(-1)|j\rangle = \frac{e\mathcal{E}}{2\hbar im}\sum_k\left(\frac{\langle n|q_0|k\rangle\langle k|p_0|j\rangle - \langle n|p_0|k\rangle\langle k|q_0|j\rangle}{\left(\omega_{nk}^{(0)} - \omega_0\right)\left(\omega_{kj}^{(0)} - \omega_0\right)}\right). \tag{11.141}$$

11.6 The Adiabatic Approximation

A special case of the time-dependent perturbation theory is the adiabatic approximation where the motion of a system can be separated into a fast moving part and a slow time variation. This method is often used when the Hamiltonian consists of a time independent part H_0 and a small time-dependent part $\lambda H'(t)$

$$H(t) = H_0 + \lambda H'(t). \tag{11.142}$$

The eigenvalue equation will be time-dependent of the form

$$H(t)\psi_n(\mathbf{r},t) = E_n(t)\psi_n(\mathbf{r},t), \tag{11.143}$$

where here t is considered to be a parameter. The presence of the small term $\lambda H'(t)$ in the Hamiltonian modifies the time-dependent wave function from

$$\exp\left(-\frac{i}{\hbar}E_n t\right)\psi_n(\mathbf{r}), \tag{11.144}$$

to

$$\exp\left[-\frac{i}{\hbar}\int_0^t E_n(t')\,dt'\right]\psi_n(\mathbf{r},t), \tag{11.145}$$

where $E_n(t')$ varies slowly with time. We note that (11.145) is not an exact solution of the time-dependent Schrödinger equation

$$i\hbar\frac{\partial\Psi(\mathbf{r},t)}{\partial t} = H(t)\Psi(\mathbf{r},t). \tag{11.146}$$

To obtain $\Psi(\mathbf{r},t)$ in terms of $\psi_n(\mathbf{r},t)$ we can expand $\Psi(\mathbf{r},t)$ in terms of $\psi_n(\mathbf{r},t)$ and determine the coefficients of expansion, assuming that these are slowly varying eigenstates of the Hamiltonian, Eq. (11.143). We find the wave function $\Psi(\mathbf{r},t)$ to be

$$\Psi(\mathbf{r},t) = \sum_n a_n(t)\psi_n(\mathbf{r},t)\exp\left[-\frac{i}{\hbar}\int_0^t E_n(t')\,dt'\right]. \tag{11.147}$$

Further we assume that at each instant t, the eigenfunctions $\{\psi_n(\mathbf{r},t)\}$ form an orthonormal set

$$\int \psi_n^*(\mathbf{r},t)\psi_k(\mathbf{r},t)d^3r = \delta_{nk}. \tag{11.148}$$

Next by substituting (11.147) in (11.146) we find

$$i\hbar\sum_n\left[\frac{da_n(t)}{dt}\psi_n(\mathbf{r},t) + a_n(t)\frac{\partial\psi_n(\mathbf{r},t)}{\partial t} - \frac{i}{\hbar}E_n a_n(t)\psi_n(\mathbf{r},t)\right]$$
$$\times\ \exp\left[-\frac{i}{\hbar}\int_0^t E_n(t')\,dt'\right] = \sum_n a_n(t)H\psi_n(\mathbf{r},t)\exp\left[-\frac{i}{\hbar}\int_0^t E_n(t')\,dt'\right]. \tag{11.149}$$

With the help of Eq. (11.143) we can reduce (11.149) to the following equation

$$\sum_n \left[\frac{da_n(t)}{dt}\psi_n(\mathbf{r},t) + a_n(t)\frac{\partial \psi_n(\mathbf{r},t)}{\partial t}\right] \exp\left[-\frac{i}{\hbar}\int_0^t E_n(t')\,dt'\right] = 0. \quad (11.150)$$

By multiplying (11.150) with

$$\psi_k^*(\mathbf{r},t)\exp\left[\frac{i}{\hbar}\int_0^t E_k(t')\,dt'\right], \quad (11.151)$$

and integrating over $\mathbf{r}$ we obtain

$$\frac{da_k(t)}{dt} = \sum_n \left\{a_n(t)\left\langle \psi_k(\mathbf{r},t)|\frac{\partial}{\partial t}\psi_n(\mathbf{r},t)\right\rangle \exp\left[\frac{i}{\hbar}\int_0^t \left(E_k(t') - E_n(t')\right) dt'\right]\right\}. \quad (11.152)$$

This is the equation that we must solve for $a_k(t)$. Here we need to calculate the scalar product

$$\left\langle \psi_k(\mathbf{r},t)\frac{\partial}{\partial t}\psi_n(\mathbf{r},t)\right\rangle, \quad (11.153)$$

and we do it by differentiating Eq. (11.143) with respect to time ;

$$\frac{\partial H}{\partial t}\psi_n(\mathbf{r},t) + H\frac{\partial \psi_n(\mathbf{r},t)}{\partial t} = \frac{\partial E_n(t)}{\partial t}\psi_n(\mathbf{r},t) + E_n(t)\frac{\partial \psi_n(\mathbf{r},t)}{\partial t}. \quad (11.154)$$

Next we multiply this equation by $\psi_k^*(\mathbf{r},t)$ and integrate over $\mathbf{r}$ to find

$$\left\langle \psi_k(\mathbf{r},t)\left|\frac{\partial H}{\partial t}\right|\psi_n(\mathbf{r},t)\right\rangle + E_k(t)\left\langle \psi_k(\mathbf{r},t)\left|\frac{\partial \psi_n(\mathbf{r},t)}{\partial t}\right.\right\rangle$$
$$= \frac{\partial E_n(t)}{\partial t}\langle \psi_k(\mathbf{r},t)|\psi_n(\mathbf{r},t)\rangle + E_n(t)\left\langle \psi_k(\mathbf{r},t)\left|\frac{\partial \psi_n(\mathbf{r},t)}{\partial t}\right.\right\rangle. \quad (11.155)$$

For $k \neq n$, (11.155) reduces to

$$\left\langle \psi_k(\mathbf{r},t)\left|\frac{\partial \psi_n(\mathbf{r},t)}{\partial t}\right.\right\rangle = \frac{\lambda}{E_n - E_k}\left\langle \psi_k(\mathbf{r},t)\left|\frac{\partial H'}{\partial t}\right|\psi_n(\mathbf{r},t)\right\rangle, \quad (11.156)$$

since $\frac{\partial H}{\partial t} = \frac{\partial H'}{\partial t}$. But when $k = n$ then the normalization condition

$$\langle \psi_n(\mathbf{r},t)|\psi_n(\mathbf{r},t)\rangle = 1, \quad (11.157)$$

gives us

$$\left\langle \psi_n(\mathbf{r},t)|\frac{\partial \psi_n(\mathbf{r},t)}{\partial t}\right\rangle + \left\langle \psi_n(\mathbf{r},t)|\frac{\partial \psi_n(\mathbf{r},t)}{\partial t}\right\rangle^* = 0, \quad (11.158)$$

and hence

$$\left\langle \psi_n(\mathbf{r},t)|\frac{\partial \psi_n(\mathbf{r},t)}{\partial t}\right\rangle = i\alpha_n(t), \quad (11.159)$$

where $\alpha_n(t)$ is a real function of time. Now by a judicious choice of the phase of $\psi_n(\mathbf{r},t)$ we can make the scalar product (11.159) equal to zero. Thus let us write

$$\phi_n(\mathbf{r},t) = \psi_n(\mathbf{r},t)\exp\left[-i\int^t \alpha_n(t')\,dt'\right], \tag{11.160}$$

and substitute for $\psi_n(\mathbf{r},t)$ in (11.159) to find

$$\left\langle \phi_n(\mathbf{r},t)|\frac{\partial\phi_n(\mathbf{r},t)}{\partial t}\right\rangle = 0. \tag{11.161}$$

For many problems where this approximation is used this additional phase

$$\gamma_n(t) = \int^t \alpha(t')\,dt'. \tag{11.162}$$

is of no physical significance and can be ignored but this is not always the case as we will see in the discussion of Berry's phase Sec. 16.13. Setting $\gamma_n(t) = 0$ and replacing the bracket

$$\left\langle \psi_k(\mathbf{r},t)|\frac{\partial\psi_n(\mathbf{r},t)}{\partial t}\right\rangle, \tag{11.163}$$

from (11.156) in (11.152) we arrive at the following exact equation for $a_n(t)$;

$$\begin{aligned}\frac{da_k(t)}{dt} &= -\sum_n \frac{\lambda}{E_n - E_k}\left\langle \psi_k(\mathbf{r},t)\left|\frac{\partial H'(t)}{\partial t}\right|\psi_n(\mathbf{r},t)\right\rangle \\ &\times \exp\left[\frac{i}{\hbar}\int_0^t \left(E_k(t') - E_n(t')\right)dt'\right]a_n(t).\end{aligned} \tag{11.164}$$

At this point we assume that $H'(t)$, $E_n(t)$ and $\psi_k(\mathbf{r},t)$ all slowly varying functions of time, and that the right-hand side of (11.164) is small or that the matrix element

$$\frac{\lambda}{E_n - E_k}\left\langle \psi_k(\mathbf{r},t)\left|\frac{\partial H'(t)}{\partial t}\right|\psi_n(\mathbf{r},t)\right\rangle, \tag{11.165}$$

is small. Let us assume that the initial state of this system is $\psi_j(\mathbf{r},0)$, then by setting $t = 0$ in Eq. (11.147) we get

$$a_n(t) \approx a_j(0) = \delta_{jn}. \tag{11.166}$$

By substituting this value of $a_j(t)$ in the right-hand side of (11.164) and integrating the resulting equation we find

$$\begin{aligned}a_k(t) &= \frac{\lambda\hbar}{i(E_j - E_k)^2}\left\langle \psi_k(\mathbf{r},t)\left|\frac{\partial H'(t)}{\partial t}\right|\psi_j(\mathbf{r},t)\right\rangle \\ &\times \left\{\exp\left[\frac{i}{\hbar}(E_k - E_j)t\right] - 1\right\}, \quad j \neq k.\end{aligned} \tag{11.167}$$

For the validity of this approximation the change in the Hamiltonian over the time $\frac{2\pi\hbar}{E_j - E_k}$ has to be small compared to the level spacing $E_j - E_k$. In other words the matrix element

$$\frac{\lambda\hbar}{i(E_j - E_k)^2}\left\langle \psi_k(\mathbf{r},t)\left|\frac{\partial H'(t)}{\partial t}\right|\psi_j(\mathbf{r},t)\right\rangle, \tag{11.168}$$

must be small.

An Application of the Adiabatic Approximation — As an example of the application of the adiabatic approximation we examine the motion of an harmonic oscillator with the Hamiltonian

$$H = \frac{p^2}{2m} + \frac{1}{2}m\omega^2(q - \xi(t))^2, \tag{11.169}$$

where $\xi(t)$ is a function of time. The exact eigenstates of this Hamiltonian are simply the harmonic oscillator wave functions displaced by $a(t)$, viz,

$$\psi_n(q) = N_n H_n[\alpha(q - \xi(t))] \exp\left[-\frac{1}{2}\alpha^2(q - \xi(t))^2\right], \tag{11.170}$$

where $\alpha = \sqrt{\frac{m\omega}{\hbar}}$. The eigenvalues of $H(t)$ are $E_n = \left(n + \frac{1}{2}\right)\hbar\omega$. Suppose that the oscillator is initially in the ground state, $(n = 0)$, and we want to investigate the condition under which the adiabatic approximation for this problem is valid. For this example we have

$$H'(t) = \frac{1}{2}m\omega^2\left[\xi^2(t) - 2\xi(t)q\right], \tag{11.171}$$

therefore

$$\frac{\partial H'(t)}{\partial t} = -m\omega^2(q - \xi(t))\frac{d\xi(t)}{dt}, \tag{11.172}$$

As we have seen earlier the condition for the validity of this approximation is expressed by the smallness of the fraction (11.168) or in the present case by the term

$$r = \left\langle 1\left|\frac{\partial H'}{\partial t}\right|0\right\rangle = -\frac{d\xi(t)}{dt}\left(\frac{m\omega^2}{\sqrt{2}\alpha}\right), \tag{11.173}$$

being small. Substituting (11.172) in (11.173) we obtain

$$r = \frac{\frac{d\xi(t)}{dt}}{\sqrt{\frac{2\hbar\omega}{m}}}. \tag{11.174}$$

Noting that the denominator in (11.174) is roughly the maximum speed of hypothetical classical oscillator which has only the zero point energy, we observe that r will be small provided that the equilibrium position moves slowly relative to the classical oscillator speed.

11.7 Transition Probability to the First Order

In this section we will study two important cases of the time-dependent perturbation theory. The first case is when the interaction $H'(t)$ is turned on suddenly at $t = 0$,

$$H'(t) = \begin{cases} 0, & t < 0 \\ H', & t > 0 \end{cases}. \tag{11.175}$$

If we substitute from (11.175) in (11.167) and calculate $|a_k(t)|^2$ which is the transition probability we find

$$\begin{aligned} P_{j\to k}(t) &= |a_k(t)|^2 = \left| \frac{\left(e^{\frac{i(E_k-E_j)t}{\hbar}} - 1\right)}{(E_k - E_j)} \langle k\,|H'|\,j\rangle \right|^2 \\ &= \left\{ \frac{\sin\left[\frac{(E_k-E_j)t}{2\hbar}\right]}{\left(\frac{E_k-E_j}{2}\right)} \right\}^2 |\,\langle k\,|H'|\,j\rangle\,|^2, \end{aligned} \tag{11.176}$$

We observe that for very short times $P_{j\to k}(t)$ is proportional to t^2. But as t grows then the term in the curly bracket in (11.176) will have a sharper peak about $E_k \approx E_j$, and the area under this curve grows as $\frac{2\pi t}{\hbar}$. Thus the curly bracket takes the form of

$$\frac{2\pi t}{\hbar}\delta(E_j - E_k). \tag{11.177}$$

This follows from the fact that most of the contributions comes from the central peak $E_k \approx E_j$ and also the total area under the rapidly oscillating function of $\omega_{kj} = \frac{(E_k-E_j)}{\hbar}$ is

$$\int_{-\infty}^{+\infty} \frac{\sin^2\left(\frac{\omega_{kj}t}{2}\right)}{\omega_{kj}^2} = \frac{\pi t}{2}. \tag{11.178}$$

Therefore the transition probability in this limit of large t is

$$P_{j\to k}(t) = \frac{2\pi t}{\hbar}|\,\langle k\,|H'|\,j\rangle\,|^2\delta(E_j - E_k), \tag{11.179}$$

and the transition rate is

$$\Gamma_{j\to k} = \frac{P_{j\to k}(t)}{t} = \frac{2\pi}{\hbar}|\,\langle k\,|H'|\,j\rangle\,|^2\delta(E_j - E_k). \tag{11.180}$$

A very useful form of the transition rate can be found from (11.179) by first summing $P_{j\to k}(t)$ over the final states available for the transition and then dividing by t, i.e. calculating the transition probability per unit time. If $\rho(E_k)$ denotes the density of final states, or the number of states per unit energy interval then the transition probability is

$$\int \rho(E_k)P_{j\to k}(t)dE_k = \frac{2\pi t}{\hbar}\,\langle k\,|H'|\,j\rangle\,|^2\rho(E_k)|_{E_k=E_j}, \tag{11.181}$$

and the transition rate once summed over a continuous group of final states becomes

$$\Gamma = \sum_k \Gamma_{j\to k} = \frac{2\pi}{\hbar}\,|\langle k\,|H'|\,j\rangle\,|^2\rho(E_k)|_{E_k=E_j}. \tag{11.182}$$

Special Case of Oscillating Perturbation — When the Hamiltonian is periodic in time then even a small change in H can produce a substantial change in $a_k(t)$ over a long period of time. Thus in this case the result (11.167) becomes invalid.

Let us consider an oscillatory perturbation that is turned on slowly. This is the situation, for example, when light waves shine on an electron, then it takes sometime to build up the perturbation to its steady state [11]. We assume that in the distant past the system was in the state j. We also assume that the perturbation applied to the system is of the form

$$H'(t) = H'e^{\varepsilon t}\cos(\omega t). \tag{11.183}$$

In this relation $\varepsilon > 0$, H' is a constant and ω is a part of a continuum frequencies avaiable to the system. Once we have done the calculation we set $\varepsilon = 0$. Thus we substitute (11.183) in (11.164), assuming that at $t = -\infty$, $a_k = \delta_{kj}$ and then we integrate the result over t from $-\infty$ to t to find

$$a_k = \frac{e^{\varepsilon t}}{2}\langle k\,|H'|\,j\rangle\left[\frac{e^{\frac{i(E_k-E_j-\hbar\omega)t}{\hbar}}}{E_j - E_k + \hbar\omega + i\hbar\varepsilon} + \frac{e^{\frac{i(E_k-E_j+\hbar\omega)t}{\hbar}}}{E_j - E_k - \hbar\omega + i\hbar\varepsilon}\right]. \tag{11.184}$$

Now we define the transition probability for a system to jump from a state j to another state k after the passage of a time t. For such a jump the transition probability which we denote by $P_{j\to k}(t)$ is given by the absolute value of $a_k(t)$

$$\begin{aligned} P_{j\to k}(t) &= |a_k|^2 = \frac{e^{2\varepsilon t}}{4}|\,\langle k\,|H'|\,j\rangle\,|^2 \\ &\times \left\{\frac{1}{(E_j - E_k + \hbar\omega)^2 + (\hbar\varepsilon)^2} + \frac{1}{(E_j - E_k - \hbar\omega)^2 + (\hbar\varepsilon)^2}\right. \\ &+ \left. 2\mathrm{Re}\left[\frac{e^{-2i\omega t}}{(E_j - E_k + \hbar\omega + i\hbar\varepsilon)(E_j - E_k - \hbar\omega + i\hbar\varepsilon)}\right]\right\}. \end{aligned} \tag{11.185}$$

The last term in (11.185) is due to the interference between the positive and negative frequency parts of (11.183), i.e. $e^{i\omega t}e^{\varepsilon}$ and $e^{-i\omega t}e^{\varepsilon}$. The transition rate $\Gamma_{j\to k}$ is defined as the rate of change of $P_{j\to k}(t)$ with respect to t;

$$\begin{aligned} \Gamma_{j\to k}(t) &= \frac{dP_{j\to k}(t)}{dt} = \frac{e^{2\varepsilon t}}{4}|\,\langle k\,|H'|\,j\rangle\,|^2 \\ &\times \left\{\left[\frac{2\varepsilon}{(E_j - E_k + \hbar\omega)^2 + (\hbar\varepsilon)^2} + \frac{2\varepsilon}{(E_j - E_k - \hbar\omega)^2 + (\hbar\varepsilon)^2}\right](1 - \cos 2\omega t)\right. \\ &+ \left. 2\sin(2\omega t)\left[\frac{E_j - E_k + \hbar\omega}{(E_j - E_k + \hbar\omega)^2 + (\hbar\varepsilon)^2} - \frac{E_j - E_k - \hbar\omega}{(E_j - E_k - \hbar\omega)^2 + (\hbar\varepsilon)^2}\right]\right\}. \end{aligned} \tag{11.186}$$

The first two terms vanish unless $E_j - E_k = \pm\hbar\omega$. The last term will not be zero, however if we average $\Gamma_{j\to k}(t)$ over a few cycles of $H'(t)$ then only the non-oscillating terms will survive. Thus in the limit of $\varepsilon \to 0$ we get

$$\Gamma_{j\to k}(t) = \frac{dP_{j\to k}(t)}{dt} = \frac{\pi}{2\hbar} |\langle k|H'|j\rangle|^2 \left[\delta(E_k - E_j - \hbar\omega) + \delta(E_k - E_j + \hbar\omega)\right]. \tag{11.187}$$

As we assumed earlier ω is a part of different frequencies applied to the system.

We note that the positive (negative) frequency part in the first order of H' is responsible for an increase (decrease) of the energy of the system by an amount $\hbar\omega$.

Bibliography

[1] M. Born, W. Heisenberg and P. Jordan, Zur Quantenmechanik II, Z. Phys. 35, 557 (1926). The English translation of this paper can be found in B.L. van der Waerden, *Sources of Quantum Mechanics*, (North-Holland, Amsterdam, 1967).

[2] See for example L.D. Landau and E.M. Lifshitz, *Quantum Mechanics, Non-Relativistic Theory*, (Pergamon Press, London, 1958), Chapter VI.

[3] S. Flügge, *Practical Quantum Mechanics*, (Springer, New York, 1971), p. 82.

[4] A. Hackl and S. Kehrein, A unitary perturbation theory approach to real-time evolution problems, J. Phys: Condens. Matter 21, 015601 (2009).

[5] C.M. Bender and L.M.A. Bettencourt, Multiple-scale analysis of the quantum anharmonic oscillator, Phys. Rev. Lett. 77, 4114 (1996).

[6] C.M. Bender and L.M.A. Bettencourt, Multiple-scale analysis of the quantum systems, Phys. Rev. D 54, 7710 (1996).

[7] G. Auberson and M.C. Peyranère, Quantum anharmonic oscillator in the Heisenberg picture and multiple scale techniques Phys. Rev. A 65, 032120 (2002).

[8] M. Abramowitz and I.A. Stegun, editors, *Handbook of Mathematical Functions*, (Dover, New York, 1970), p. 804.

[9] H. Goldstein, C. Poole and J. Safco, *Classical Mechanics*, Third Edition (Addison-Wesley, San Francisco, 2002).

[10] D. Park, *Classical Dynamics and Its Quantum Analogues*, Second Edition, (Springer-Verlag, Berlin 1990).

[11] G. Baym, *Lectures on Quantum Mechanics*, (W.A. Benjamin, Reading, 1969), p. 233.

[12] B. Holstein, The van der Walls Interaction, Am. J. Phys. 441, 69 (2001).

Chapter 12

Other Methods of Approximation

In addition to the perturbation theory which has a wide range of applications in quantum mechanics, there are a number of other methods, some derived from the approximate solution of the wave equation while others derived from the matrix or the operator formulation. In this chapter we want to consider methods derived mainly from the Heisenberg equations of motion and are for time-independent problems [1]–[5]. Some of the approximate methods used for the approximate solution to the scattering problem, or time-dependent systems will be reviewed in later chapters.

12.1 WKB Approximation for Bound States

The semi-classical approximation well-known in wave mechanics can also be derived from the Heisenberg equations of motion. In this section we show how the bound state energies can be obtained for a one-dimensional motion of a particle moving in a potential well. We write the Hamiltonian in the following form

$$H = \frac{1}{2}[x, H][H, x] + V(x), \tag{12.1}$$

and also assume that the classical motion in this potential is periodic and that

$$x(t, \omega) = x(t + T, \omega), \tag{12.2}$$

where $T = \frac{2\pi}{\omega}$ is the period of oscillation.

In this semiclassical approximation we consider a method for determination of the matrix elements of the equation of motion for a large quantum number n. The equation of motion for the one-dimensional system is

$$[\,[x, H], H] = \frac{dV(x)}{dx}, \tag{12.3}$$

which in terms of the matrix elements can be written as

$$[E(n \pm \nu) - E_n]^2 \langle n|x|n \pm \nu\rangle = \left\langle n \left| \frac{dV(x)}{dx} \right| n \pm \nu \right\rangle, \quad \nu \text{ an integer.} \tag{12.4}$$

Here we want to calculate the values of $\langle n|x|n\pm\nu\rangle$ approximately when n is large and ν is small. This is done by the expansion of the matrix element $\langle n|x|n \pm \nu\rangle$ in powers of ν and then by referring these matrix elements to a common matrix element $\langle n - \frac{1}{2}\nu|x|n + \frac{1}{2}\nu\rangle$,

$$\langle n|x|n \pm \nu\rangle = \left\langle n - \frac{1}{2}\nu|x|n + \frac{1}{2}\nu \right\rangle \pm \frac{1}{2}\nu\frac{\partial}{\partial n}\left\langle n - \frac{1}{2}\nu|x|n + \frac{1}{2}\nu \right\rangle + \cdots. \tag{12.5}$$

We now observe that for the matrix elements of any operator

$$A(x) = xB(x), \tag{12.6}$$

when n is large we have

$$\begin{aligned} &\langle n|A|n+\nu\rangle + \langle n|A|n-\nu\rangle \\ = &\sum_{\nu_1>0} \langle n|x|n-\nu_1\rangle\langle n-\nu_1|B|n+\nu\rangle \\ + &\sum_{\nu_1>0} \langle n|x|n+\nu_1\rangle\langle n+\nu_1|B|n-\nu\rangle. \end{aligned} \tag{12.7}$$

By expanding the matrix elements of B in (12.7) we get

$$\begin{aligned} &\langle n \mp \nu_1|B|n \pm \nu\rangle = \langle n|B|n \pm (\nu+\nu_1)\rangle \\ \mp \quad &\nu_1\frac{\partial}{\partial n}\left\langle n - \frac{1}{2}(\nu+\nu_1)|B|n + \frac{1}{2}(\nu+\nu_1)\right\rangle. \end{aligned} \tag{12.8}$$

Substituting (12.5) and (12.8) in (12.7) we find that because of the cancelation of the linear terms we have

$$\begin{aligned} &\langle n|A|n+\nu\rangle + \langle n|A|n-\nu\rangle \\ = \quad &2\left\langle n + \frac{1}{2}\nu|A|n - \frac{1}{2}\nu\right\rangle\left[1 + \mathcal{O}\left(\frac{1}{n^2}\right)\right]. \end{aligned} \tag{12.9}$$

To write these relations in a more compact form let us introduce $x_\nu(\bar{n})$ which is defined by

$$x_\nu(\bar{n}) = x_{-\nu}(\bar{n}) = \left\langle n - \frac{1}{2}|x|n + \frac{1}{2}\right\rangle, \tag{12.10}$$

and use this relation to write the equation of motion (12.4) as

$$[\nu\omega(\bar{n})]^2 x_\nu(\bar{n}) = \left(\frac{dV(x)}{dx}\right)_\nu, \tag{12.11}$$

where $\omega(\bar{n})$ is defined by

$$\omega(\bar{n}) = \frac{dE(n)}{dn}. \tag{12.12}$$

If we write the Fourier transform of the classical motion $x(t,\bar{n})$ as

$$x(t,\bar{n}) = \sum_{\nu=-\infty}^{\infty} x_\nu(\bar{n}) \exp[i\nu\omega(\bar{n})t], \tag{12.13}$$

then Eq. (12.11) can be associated with the Fourier transform, and $\left(\frac{dV(x)}{dx}\right)_\nu$ will be the ν-th component of the Fourier transform of the force exerted on the particle. We can also obtain (12.11) from a variational principle [1],[2]. To this end let us start with the Lagrangian $L = p\dot{q} - H$ and write it as

$$\begin{aligned} L(\bar{n}) &= \langle n|\dot{x}^2|n\rangle - \langle n|H|n\rangle = 2\sum_{\nu>0}^{\infty}[\nu\omega(\bar{n})]^2 x_{-\nu}(\bar{n})x_\nu(\bar{n}) \\ &- \sum_{\nu>0}^{\infty}[\nu\omega(\bar{n})]^2 x_{-\nu}(\bar{n})x_\nu(\bar{n}) + V(x(t,\bar{n}))_0. \end{aligned} \tag{12.14}$$

Here $V(x(t,\bar{n}))_0$ is the constant term in the Fourier expansion of the potential. The Euler–Lagrange equation for (12.14) which is

$$\frac{\delta L}{\delta x_{-\nu}(\bar{n})} \equiv \frac{d}{dt}\left(\frac{\partial L(\bar{n})}{\partial \dot{x}_{-\nu}(\bar{n})}\right) - \frac{\partial L(\bar{n})}{\partial x_{-\nu}(\bar{n})} = 0, \tag{12.15}$$

gives us the equation of motion (12.11). In this variation of $L(\bar{n})$, $\omega(\bar{n})$ is held fixed. The next step is to vary $L(\bar{n})$ with respect to n.

$$\begin{aligned} \frac{\delta L}{\delta n} &= \left(\frac{\delta L(\bar{n})}{\delta x_{-\nu}(\bar{n})}\right)\frac{\partial x_{-\nu}(\bar{n})}{\partial n} + \left(\frac{\delta L(\bar{n})}{\delta x_\nu(\bar{n})}\right)\frac{\partial x_\nu(\bar{n})}{\partial n} \\ &+ \frac{dL(\bar{n})}{dn} = \frac{\partial L(\bar{n})}{\partial n} = 0. \end{aligned} \tag{12.16}$$

Thus if we find the partial derivative of $L(\bar{n})$, defined by Eq. (12.14), with respect to $\bar{n}$ and then substitute for $\frac{dE(\bar{n})}{dn}$ from (12.12) we get

$$\frac{d}{dn}\left\{2\sum_{\nu>0}\nu^2\omega(\bar{n})x_{-\nu}(\bar{n})x_\nu(\bar{n})\right\} = 1. \tag{12.17}$$

Integrating this last relation we have

$$S(\bar{n}) = 2\pi\left\{\sum_{-\infty}^{\infty}\nu^2\omega(\bar{n})x_{-\nu}(\bar{n})x_\nu(\bar{n})\right\} = 2\pi\left(n+\frac{1}{2}\right), \tag{12.18}$$

where we have chosen the constant of integration to be $\frac{1}{2}$.

We can easily see that $S(\bar{n})$ is related to the standard phase integral of the WKB approximation. Remembering that the period is given by

$$T(\bar{n}) = 2\pi/\omega(\bar{n}), \tag{12.19}$$

we can write (12.18) as

$$\begin{aligned} S(\bar{n}) &= T(\bar{n}) \sum_{\nu=-\infty}^{\infty} \nu^2\omega^2(\bar{n}) x_{-\nu}(\bar{n}) x_\nu(\bar{n}) \\ &\to \int_0^{T(\bar{n})} [\dot{x}(t,n)]^2\, dt = \oint p\dot{x}dt = \oint pdx. \end{aligned} \tag{12.20}$$

From the equation of motion (12.11) and the quantization rule (12.18) we can find $x_\nu(\bar{n})$ and the energy eigenvalues [1]–[4]. Detailed calculations for the problem of anharmonic oscillator with $V(x) = \frac{1}{2}x^2 + \frac{1}{4}x^4$ show that the energy of the ground state is off by as much as 8%, whereas the matrix elements are accurate to about 1% of their exact values [5].

12.2 Approximate Determination of the Eigenvalues for Nonpolynomial Potentials

If we choose $f(x)$ in such a way that the equation of motion for $f(x)$ and $(f(x)^2)$ and also the equation for the commutation relation are all become polynomials in $f(x)$, then we can use Klein's method for polynomial potentials that we discussed earlier and obtain the energy eigenvalues and the matrix elements $\langle n|f(x)|k\rangle$. Let us assume that we have found such a function $f(x)$ which makes the right-hand sides of (8.15) and (8.16) polynomials in $\xi = f(x)$. By changing the variable from x to ξ, we proceed to take the matrix elements of the above operator equations. We also write $\langle n\left|\xi^2\right|j\rangle$, $\langle n\left|\xi^3\right|j\rangle, \cdots$ as matrix products

$$\langle n\left|\xi^m\right|j\rangle = \sum_{n_1,\cdots,n_{m-1}} \langle n\left|\xi\right|n_1\rangle\langle n_1\left|\xi\right|n_2\rangle\langle n_2\left|\xi\right|n_3\rangle\cdots\langle n_{m-1}\left|\xi\right|j\rangle. \tag{12.21}$$

Using the condition that we applied to the polynomial potentials, viz, the saturation of the infinite sums in (12.21), or in other words the condition

$$|\langle n\left|\xi\right|n\pm 1\rangle| \gg |\langle n\left|\xi\right|n\pm 3\rangle| \gg |\langle n\left|\xi\right|n\pm 5\rangle| \gg \cdots, \tag{12.22}$$

we observe that we can truncate the infinite sum and get a finite number of nonlinear algebraic equations.

To illustrate this technique, let us consider the motion of a particle of unit mass with the Hamiltonian

$$H=\frac{p^2}{2}+V(x)=\frac{p^2}{2}+\lambda\sinh^2 x,\quad \lambda>0. \tag{12.23}$$

We choose

$$\xi=\sinh x, \tag{12.24}$$

and we write Eqs. (8.15) and (8.16) as [4]

$$[[\xi,H],\xi]=1+\xi^2, \tag{12.25}$$

$$[[\xi,H],H]=-(\xi H+H\xi)+\left(2\lambda-\frac{1}{4}\right)+4\lambda\xi^3. \tag{12.26}$$

We also add the equation of motion for ξ^2 to these two;

$$\left[\left[\xi^2,H\right],H\right]=-2\left\{\left(1+2\xi^2\right)H+H\left(1+2\xi^2\right)\right\}+12\lambda\xi^4+(8\lambda-4)\xi^2-2. \tag{12.27}$$

The matrix elements of these three equations are given by

$$\sum_k\left(E_k-E_n-\frac{1}{2}\right)\langle n\,|\xi|\,k\rangle\,\langle k|\xi|n\rangle=\frac{1}{2}, \tag{12.28}$$

$$\left[(E_k-E_n)^2+E_k+E_n-2\lambda+\frac{1}{4}\right]\langle n|\xi|k\rangle-4\lambda\left\langle n\left|\xi^3\right|k\right\rangle=0, \tag{12.29}$$

and

$$-\left[1+2\left\langle n\left|\xi^2\right|n\right\rangle\right]E_n+3\lambda\left\langle n\left|\xi^4\right|n\right\rangle+(2\lambda-1)\left\langle n\left|\xi^2\right|n\right\rangle-\frac{1}{2}=0, \tag{12.30}$$

where (12.30) is written for the diagonal elements of the double commutator (12.26).

An approximate method of solving the coupled set of nonlinear equations (12.28)–(12.30) can be found by keeping all the matrix elements between neighboring eigenstates $\langle n|\xi|n\pm1\rangle$ and ignoring all other matrix elements, e.g. $\langle n|\xi|n\pm3\rangle$. Using this technique we can reduce (12.28)–(12.30) for $k=n+1$ to the following set:

$$\left(\omega_n-\frac{1}{2}\right)\left\langle n\left|\xi^2\right|n+1\right\rangle-\left(\omega_{n-1}+\frac{1}{2}\right)\left\langle n-1\left|\xi^2\right|n\right\rangle-\frac{1}{2}=0, \tag{12.31}$$

$$\begin{aligned}\omega_n^2\ &+\ \omega_n+2E_n-4\lambda\left[\left\langle n-1\left|\xi^2\right|n\right\rangle+\left\langle n\left|\xi^2\right|n+1\right\rangle\right.\\ &+\ \left.\left\langle n+1\left|\xi^2\right|n+2\right\rangle\right]-2\lambda+\frac{1}{4}=0\end{aligned} \tag{12.32}$$

and

$$\begin{aligned}
& 3\lambda\left[\langle n\left|\xi^4\right|n+1\rangle+\langle n\left|\xi^2\right|n+1\rangle\langle n+1\left|\xi^2\right|n+2\rangle\right.\\
+ & 2\langle n-1\left|\xi^2\right|n\rangle\langle n\left|\xi^2\right|n+1\rangle\\
+ & \left.\langle n-1\left|\xi^2\right|n\rangle\langle n-2\left|\xi^2\right|n-1\rangle+\langle n-1\left|\xi^4\right|n\rangle\right]\\
+ & (2\lambda-1)\left[\langle n\left|\xi^2\right|n+1\rangle+\langle n-1\left|\xi^2\right|n\rangle\right]-\frac{1}{2}\\
- & \left\{1+2\left(\langle n\left|\xi^2\right|n+1\rangle+\langle n-1\left|\xi^2\right|n\rangle\right)\right\}E_n=0,
\end{aligned} \tag{12.33}$$

where

$$\omega_n = E_{n+1} - E_n, \tag{12.34}$$

and

$$\langle n\left|\xi^2\right|j\rangle \approx \left(\langle n\left|\xi\right|j\rangle\right)^2, \quad \langle n\left|\xi^4\right|j\rangle \approx \left(\langle n\left|\xi\right|j\rangle\right)^4. \tag{12.35}$$

To introduce a cut-off and make these equations a closed set we make the additional assumption that for large N

$$\langle N+1|\xi|N+2\rangle \approx \langle N|\xi|N+1\rangle. \tag{12.36}$$

This assumption will hardly affect the energies of the low-lying states, but will simplify (12.31)–(12.33) considerably and reduce them to a set of $3(N+1)$ equations for $\langle n|\xi|n+1\rangle$, ω_n and E_n. At the first sight this method seems to be a crude one. In order to get a better idea about the accuracy of this approximation let us consider the case of strong potential where $\lambda = 10$. Then from the solution of Eqs. (12.31)-(12.33) we find the following values for the matrix elements [4]:

$$\langle 0|\xi|1\rangle = 0.337, \quad \langle 1|\xi|2\rangle = 0.4813, \quad \langle 2|\xi|3\rangle = 0.5969. \tag{12.37}$$

These should be compared with the exact results:

$$\langle 0|\xi|1\rangle = 0.3343, \quad \langle 1|\xi|2\rangle = 0.4739, \quad \langle 2|\xi|3\rangle = 0.5821. \tag{12.38}$$

For the energy eigenvalues this approximation yields the following numbers

$$E_0 = 2.309, \quad E_1 = 7.212, \quad E_2 = 12.52. \tag{12.39}$$

whereas the exact energies are

$$E_0 = 2.355, \quad E_1 = 7.283, \quad E_2 = 12.625. \tag{12.40}$$

12.3 Generalization of the Semiclassical Approximation to Systems with N Degrees of Freedom

For a system of N degrees of freedom the eigenvalue equation (7.9) can be written as

$$H|n_1\cdots n_N\rangle = E(n_1\cdots n_N)|n_1\cdots n_N\rangle, \tag{12.41}$$

where n_j is the quantum number for the j-th degree of freedom. Setting the mass of each of the particles equal to one and denoting the coordinates of the N degrees of freedom of the system by $x^{(1)},\cdots x^{(N)}$, we write the Hamiltonian H as

$$H = \sum_{j=1}^{N}\frac{1}{2}\left(p^j\right)^2 + V\left(x^{(1)},\cdots x^{(N)}\right), \tag{12.42}$$

from which we get the Heisenberg equations of motion

$$\ddot{x}^{(j)} = [\,[x^{(j)},\ H],H] = \frac{\partial V}{\partial x^{(j)}},\quad j=1,2\cdots N. \tag{12.43}$$

In the representation in which H is diagonal, Eq. (12.41), the matrix elements of (12.43) can be expressed as [2]

$$\begin{aligned}&[E(n_1,\cdots n_N) - E(n_1+\nu_1,\cdots n_N+\nu_n)]^2\\ \times\ &\left\langle n_1\cdots n_N\left|x^{(j)}\right|n_1+\nu_1\cdots n_N+\nu_N\right\rangle\\ =\ &\left\langle n_1\cdots n_N\left|\frac{\partial V}{\partial x^{(j)}}\right|n_1+\nu_1\cdots n_N+\nu_N\right\rangle.\end{aligned} \tag{12.44}$$

As in the case of $N=1$, we can expand (12.44) for large quantum numbers n_i, $i=1,\cdots N$ in powers of n_i^{-1} about the matrix element

$$x_{\boldsymbol{\nu}}^{(j)} \equiv x_{\boldsymbol{\nu}}^{(j)}(n_1\cdots n_N) = \left\langle n_1-\frac{\nu_1}{2}\cdots n_N-\frac{\nu_N}{2}\left|x^{(j)}\right|n_1+\frac{\nu_1}{2}\cdots n_N+\frac{\nu_N}{2}\right\rangle, \tag{12.45}$$

and keep the first two terms in the expansion. In this way we obtain the classical equation

$$(\mathbf{k}\cdot\boldsymbol{\omega})^2\, x_{\boldsymbol{\nu}}^{(j)}(\mathbf{n}) = \left(\frac{\partial V}{\partial x^{(j)}}\right)_{\boldsymbol{\nu}}, \tag{12.46}$$

where ω_i is defined by

$$\omega_i = \frac{\partial E(\mathbf{n})}{\partial n_i}. \tag{12.47}$$

Equation (12.46) may be regarded as an N-dimensional Fourier transform with N distinct times, i.e.

$$x^{(j)}(\omega_1 t_1,\ \omega_2 t_2\cdots\omega_n t_n,\mathbf{n}) = \sum_{\boldsymbol{\nu}} x_{\boldsymbol{\nu}}^{(j)}(\mathbf{n})\exp\left(-i\sum_l \nu_l\omega_l t_l\right), \tag{12.48}$$

and since ω_l s are constants we can use the dimensionless constant θ_l

$$\theta_l = \omega_l t. \tag{12.49}$$

Therefore (12.48) can be written as

$$x^{(j)}(\boldsymbol{\theta},\, \mathbf{n}) = \sum_{\boldsymbol{\nu}} x^{(j)}_{\boldsymbol{\nu}}(\mathbf{n}) \exp\left(-i\boldsymbol{\nu}\cdot\boldsymbol{\theta}\right). \tag{12.50}$$

By setting all t_l s equal $t_1 = \cdots = t_N = t$, then Eq. (12.48) reduces to the classical expression (1.108) for $A = x^{(j)}$.

We now can consider a Lagrangian formulation for this system as we did for the motion with one degree of freedom. Thus let us start with the Lagrangian averaged over the phases $\boldsymbol{\theta}$;

$$\bar{L} = \frac{1}{(2\pi)^N}\int_0^{2\pi} L\left[\mathbf{x}(\boldsymbol{\theta}),\, \boldsymbol{\omega}\cdot\nabla_{\boldsymbol{\theta}}\mathbf{x}(\boldsymbol{\theta})\right] d\boldsymbol{\theta}, \tag{12.51}$$

where

$$\boldsymbol{\omega}\cdot\nabla_{\boldsymbol{\theta}} = \sum_{i=1}^{N}\frac{\partial}{\partial t_i} \equiv \frac{d}{dt}, \tag{12.52}$$

and $\mathbf{x}(\boldsymbol{\theta})$ is defined by its components $x^{(j)}(\boldsymbol{\theta})$ as is given in (12.50). The equations of motion can be derived from $\bar{L}$ by requiring that

$$\delta\bar{L} = 0, \tag{12.53}$$

provided that the trial functions satisfy the periodicity condition

$$\delta x^{(j)}(\theta_1\cdots\theta_j + 2\pi, \cdots\theta_N) = \delta x^{(j)}(\theta_1\cdots\theta_j, \cdots\theta_N), \tag{12.54}$$

for all j. The variation $\delta\bar{L} = 0$ yields the equations of motion

$$(\boldsymbol{\omega}\cdot\nabla_{\boldsymbol{\theta}})\frac{\partial L}{\partial\dot{x}^{(j)}} = \frac{\partial L}{\partial x^{(j)}}, \tag{12.55}$$

which in view of (12.52) is just the familiar form of the Lagrange equation. Note that in (12.55)

$$\dot{x}^{(j)} = (\boldsymbol{\omega}\cdot\nabla_{\boldsymbol{\theta}})x^{(j)}. \tag{12.56}$$

In addition Eqs. (12.46) and (12.55) are related to each other, the former being the Fourier transform of the latter. This can be seen by substituting (12.50) in (12.51) and averaging over $\boldsymbol{\theta}$.

We can also write the Lagrangian as

$$L = \frac{1}{2}\sum_{j=1}^{N}\dot{x}^{(j)}\dot{x}^{(j)} - V(\mathbf{x}), \tag{12.57}$$

where for $\dot{x}^{(j)}$ we substitute from (12.50). Then by averaging over $\boldsymbol{\theta}$ we obtain (compare with Eq. (12.14) which is for $N = 1$),

$$\bar{L} = \frac{1}{2}\sum_{\boldsymbol{\nu}}(\boldsymbol{\omega}\cdot\boldsymbol{\nu})^2 x_{\boldsymbol{\nu}} x_{-\boldsymbol{\nu}} - V_0. \tag{12.58}$$

In this relation $V_0 \equiv \bar{V}$ is the constant term in the Fourier series expansion of the potential. From (12.58) it follows that

$$\frac{\partial \bar{L}}{\partial x^{(j)}_{-\boldsymbol{\nu}}} = (\boldsymbol{\omega}\cdot\boldsymbol{\nu})^2 x^{(j)}_{\boldsymbol{\nu}} - \frac{\partial V_0}{\partial x^{(j)}_{-\boldsymbol{\nu}}}. \tag{12.59}$$

The last term in (12.59) can be calculated from the Fourier transform of $V\left(x^{(j)}\right)$;

$$\begin{aligned}
\frac{\partial V_0}{\partial x^{(j)}_{-\boldsymbol{\nu}}} &= \frac{\partial}{\partial x^{(j)}_{-\boldsymbol{\nu}}}\frac{1}{(2\pi)^N}\int V\left(x^{(j)}\right) d\boldsymbol{\theta} \\
&= \frac{\partial}{\partial x^{(j)}_{\boldsymbol{\nu}}}\frac{1}{(2\pi)^N}\int V\left(\sum_{\boldsymbol{\mu}} x^{(j)}_{\boldsymbol{\mu}} e^{i\boldsymbol{\mu}\cdot\boldsymbol{\theta}}\right) d\boldsymbol{\theta} \\
&= \frac{1}{(2\pi)^N}\int e^{i\boldsymbol{\nu}\cdot\boldsymbol{\theta}}\frac{\partial V}{\partial x^{(j)}} d\boldsymbol{\theta} = \left(\frac{\partial V}{\partial x^{(j)}}\right)_{\boldsymbol{\nu}}.
\end{aligned} \tag{12.60}$$

If we substitute (12.60) in (12.59) and let

$$\frac{\partial \bar{L}}{\partial x^{(j)}_{-\boldsymbol{\nu}}} = 0, \tag{12.61}$$

we find then Eq. (12.46).

Transformation to Action-Angle Variables — An examination of Eq. (12.46) shows that the solution of the problem can be found if we know the values of the N Fourier components. We can choose the set

$$\mathbf{x_1} \equiv \left(x^{(1)}_{1,0,0\cdots},\ x^{(2)}_{0,1,0\cdots}\ \cdots\ x^{(N)}_{0,0,\cdots 1}\right), \tag{12.62}$$

to represent the solution $\mathbf{x_1}$. The complex elements $x_{0,0\cdots,\,1,\,0,\,\cdots}$ constitute the $2N$ integration constants. If we choose all elements of $\mathbf{x_1}$ to be real, and let $\boldsymbol{\delta}$ denote the set of N real phases $\boldsymbol{\delta} = (\delta_1\ \delta_2 \cdots\ \delta_N)$ then the general solution will be of the form

$$\mathbf{x}_{\boldsymbol{\nu}} \to \mathbf{x}_{\boldsymbol{\nu}} e^{i\boldsymbol{\nu}\cdot\boldsymbol{\delta}}. \tag{12.63}$$

We note that $\bar{L}$ and $\bar{H}$, where the latter is defined by

$$\bar{H} = \frac{1}{2}\sum_{\boldsymbol{\nu}}(\boldsymbol{\nu}\cdot\boldsymbol{\omega})^2 \mathbf{x}_{\boldsymbol{\nu}}\mathbf{x}_{-\boldsymbol{\nu}} + V_0. \tag{12.64}$$

remain unchanged if we make the replacement shown in (12.63), therefore we will only consider the vector $\mathbf{x}_{\boldsymbol{\nu}}$ satisfying

$$\mathbf{x}_{\boldsymbol{\nu}} = \mathbf{x}_{-\boldsymbol{\nu}}, \tag{12.65}$$

in the following discussion. Once $\mathbf{x}_{\mathbf{1}}$, Eq.(12.62), has been determined, we can find the vector $\boldsymbol{\omega}$ in terms of $\mathbf{x}_{\mathbf{1}}$,

$$\boldsymbol{\omega} = \boldsymbol{\omega}(\mathbf{x}_{\mathbf{1}}). \tag{12.66}$$

Similarly we can express $\mathbf{x}_{\boldsymbol{\nu}}$ in terms of $\boldsymbol{\omega}$, i.e.

$$\mathbf{x}_{\boldsymbol{\nu}} = \mathbf{x}_{\boldsymbol{\nu}}(\boldsymbol{\omega}), \quad \text{for all } \boldsymbol{\nu}. \tag{12.67}$$

By substituting $\mathbf{x}_{\boldsymbol{\nu}}(\boldsymbol{\omega})$ in $\bar{L}$ we find

$$\bar{L} = \frac{1}{2}\sum_{\boldsymbol{\nu}}(\boldsymbol{\omega}\cdot\boldsymbol{\nu})^2\mathbf{x}_{\boldsymbol{\nu}}\cdot\mathbf{x}_{\boldsymbol{\nu}} - \bar{V}, \tag{12.68}$$

where the summation is over all $\boldsymbol{\nu}$ with positive and negative components. From the Lagrangian $\bar{L}$ we define the action I_k by

$$I_k = \frac{\delta\bar{L}}{\delta\omega_k}, \tag{12.69}$$

where the δ derivative means that we keep $\omega_j (j \neq k)$ and $\mathbf{x}_{\boldsymbol{\nu}}$ fixed. We also note that

$$\frac{\partial\bar{L}}{\partial\omega_k} = \frac{\delta\bar{L}}{\delta\omega_k} + \sum_{\boldsymbol{\nu}}\frac{\partial\bar{L}}{\partial\mathbf{x}_{\boldsymbol{\nu}}}\frac{\partial\mathbf{x}_{\boldsymbol{\nu}}}{\partial\omega_k} = \frac{\delta\bar{L}}{\delta\omega_k} = I_k. \tag{12.70}$$

By substituting for $\bar{L}$ from (12.68) in (12.69) we obtain

$$I_k = \sum_{\boldsymbol{\nu}}\nu_k(\boldsymbol{\nu}\cdot\boldsymbol{\omega})x_{\boldsymbol{\nu}}x_{\boldsymbol{\nu}} = \frac{1}{(2\pi)^{2N}}\int_0^{2\pi}\mathbf{p}\cdot\left(\frac{\partial\mathbf{x}}{\partial\theta_k}\right)d\boldsymbol{\theta}. \tag{12.71}$$

Now if we multiply (12.71) by ω_k and sum over k we find

$$\sum_k\omega_k I_k = \frac{1}{(2\pi)^N}\int_0^{2\pi}\mathbf{p}\cdot\left(\sum_k\omega_k\frac{\partial\mathbf{x}}{\partial\theta_k}\right)d\boldsymbol{\theta} = \overline{\mathbf{p}\cdot\dot{\mathbf{x}}}, \tag{12.72}$$

where we have used (12.52) to obtain the last term of Eq. (12.72). The average Hamiltonian $\bar{H}$ can be found by averaging (12.64) over the angles to get $H(\mathbf{I})$ which is also expressible as

$$\bar{H}(\mathbf{I}) = \overline{\mathbf{p}\cdot\dot{\mathbf{x}}} - \bar{L}(\boldsymbol{\omega}) = \boldsymbol{\omega}\cdot\mathbf{I} - \bar{L}(\boldsymbol{\omega}), \tag{12.73}$$

and has the familiar form of

$$H(\mathbf{p},\,\mathbf{q}) = \mathbf{p}\cdot\dot{\mathbf{q}} - L\left(\dot{\mathbf{q}},\,\mathbf{q}\right). \tag{12.74}$$

The Hamiltonian $H(\mathbf{I})$ which has been derived from $L(\boldsymbol{\omega})$ by a Legendre transformation satisfies the canonical equations of motion

$$\omega_k = \frac{\partial \bar{H}(\mathbf{I})}{\partial I_k}. \tag{12.75}$$

While in this derivation of $\bar{H}(\mathbf{I})$ we have suppressed the set of quantum numbers $\mathbf{n}$, Eq. (12.75) is related to (12.47). To establish this connection we quantize (12.75) using the Bohr–Sommerfeld quantization rule (12.93).

A Simple Example — Let us consider the motion of a system of anharmonic coupled oscillators given by the Lagrangian

$$L = \frac{1}{2}\left(\dot{x}^2 + \dot{y}^2\right) - \frac{1}{4}\kappa_1 x^4 - \frac{1}{4}\kappa_2 y^4 - \frac{1}{2}\lambda x^2 y^2. \tag{12.76}$$

When $\lambda = 0$, the problem is separable and we can use the WKB approximation to find the energy levels of the system. Alternatively we can use a simple trial function and find the average Lagrangian $\bar{L}$. Thus if we set $\lambda = 0$ and choose

$$x(t) \approx 2x_{10}\cos(\omega_1 t_1), \tag{12.77}$$

and

$$y(t) \approx 2y_{01}\cos(\omega_2 t_2), \tag{12.78}$$

and use these in (12.51), we obtain

$$\bar{L} \approx \bar{L}(\lambda = 0) = \omega_1^2 x_{10}^2 + \omega_2^2 y_{01}^2 - \frac{3}{2}\kappa_1 x_{10}^4 - \frac{3}{2}\kappa_2 y_{01}^4. \tag{12.79}$$

The extremum of $\bar{L}$ yields the following relations

$$x_{10}^2 = \frac{\omega_1^2}{3\kappa_1}, \tag{12.80}$$

and

$$y_{01}^2 = \frac{\omega_2^2}{3\kappa_2}. \tag{12.81}$$

The action variables can be found from (12.70);

$$I_1 = 2\omega_1 x_{10}^2 = \frac{2}{3\kappa_1}\omega_1^3, \tag{12.82}$$

and

$$I_2 = 2\omega_2 y_{01}^2 = \frac{2}{3\kappa_2}\omega_2^3. \tag{12.83}$$

Writing Eq. (12.75) as

$$\omega_k = \frac{\partial E}{\partial I_k}, \quad k = 1,\ 2, \tag{12.84}$$

we observe that from (12.82), (12.83) and (12.84) we have

$$\omega_1 = \frac{\partial E}{\partial I_1} = \left(\frac{3\kappa_1}{2}\right)^{\frac{1}{3}} I_1^{\frac{1}{3}}, \tag{12.85}$$

and

$$\omega_2 = \frac{\partial E}{\partial I_2} = \left(\frac{3\kappa_2}{2}\right)^{\frac{1}{3}} I_2^{\frac{1}{3}}. \tag{12.86}$$

By integrating (12.85) and (12.86) and adding the results we get an expression for the energy

$$E(I_1, I_2) = \frac{1}{2}\left(\frac{3}{2}\right)^{\frac{4}{3}} \left[\kappa_1^{\frac{1}{3}} I_1^{\frac{4}{3}} + \kappa_2^{\frac{1}{3}} I_2^{\frac{4}{3}}\right]. \tag{12.87}$$

Having obtained the classical expression for $E(I_1, I_2)$ we use the Bohr-Sommerfeld quantization rule to calculate the approximate energy levels of the two uncoupled oscillators:

$$E(n_1, n_2) = \frac{1}{2}\left(\frac{3\hbar}{2}\right)^{\frac{4}{3}} \left[\kappa_1^{\frac{1}{3}}\left(n_1 + \frac{1}{2}\right)^{\frac{4}{3}} + \kappa_2^{\frac{1}{3}}\left(n_2 + \frac{1}{2}\right)^{\frac{4}{3}}\right]. \tag{12.88}$$

Next let us determine the shift in the energy levels caused by the mixing term $\left(-\frac{1}{2}\lambda x^2 y^2\right)$ in (12.76) when λ is small, $\lambda \ll \kappa_{1,2}$. Adding the average of this term to $\bar{L}$, Eq. (12.79) we get

$$\bar{L}_\lambda = \bar{L} - 2\lambda x_{10}^2 y_{10}^2. \tag{12.89}$$

In this case the extremum of $\bar{L}_\lambda$ will give us [2]

$$x_{10}^2 = \frac{\omega_1^2}{3\kappa_1} - \frac{2\lambda}{3\kappa_1} y_{01}^2, \tag{12.90}$$

and

$$y_{01}^2 = \frac{\omega_2^2}{3\kappa_2} - \frac{2\lambda}{3\kappa_2} y_{10}^2. \tag{12.91}$$

Following the previous steps, now for $E_\lambda(n_1, n_2)$ we obtain an expression for the quantized energy

$$\begin{aligned} E(n_1, n_2) &= \frac{1}{2}\left(\frac{3\hbar}{2}\right)^{\frac{4}{3}} \left\{\kappa_1^{\frac{1}{3}}\left(n_1 + \frac{1}{2}\right)^{\frac{4}{3}} + \kappa_2^{\frac{1}{3}}\left(n_2 + \frac{1}{2}\right)^{\frac{4}{3}}\right. \\ &+ \left.\frac{4}{9}\frac{\lambda}{\kappa_1\kappa_2}\left[\kappa_1\left(n_1 + \frac{1}{2}\right)\right]^{\frac{2}{3}} \left[\kappa_2\left(n_2 + \frac{1}{2}\right)\right]^{\frac{2}{3}}\right\}. \end{aligned} \tag{12.92}$$

Semiclassical Quantization of Nonseparable Systems — For a separable motion according to the old quantum theory, the Bohr–Sommerfeld quantization condition is

$$I_k = \left(n_k + \frac{\beta_k}{4}\right)\hbar. \tag{12.93}$$

In this relation n_k is a positive integer or zero, and the constant β_k which is called Maslov index is usually an integer [6].

The generalization of the Bohr–Sommerfeld quantization to nonseparable (or non-integrable) systems is originally due to Einstein with modifications by Brillouin and Keller [7]. Einstein proposed the following quantization rule: We find the integral

$$I = \frac{1}{2\pi}\sum_j \oint p_j dq_j, \tag{12.94}$$

over a closed path in phase space that makes an integral number of complete cycles of each of the various periodicities and set this equal to $n\hbar$.

TABLE VIII: Correspondence between Heisenberg's matrix mechanics and the semi-classical quantization. Here we have used the representation where the energy is diagonal.

Quantum mechanical	Semi-classical
Matrix elements of x and p $\langle n\|x\|j\rangle,\ \langle n\|p\|j\rangle$	Fourier components of $x(I,\theta), p(I,\theta)$ $x_k(I), p_k(I)$
Equation of motion $(E_{n+k}-E_n)^2\langle n+k\|x\|n\rangle =$ $\left\langle n+k\left\|\frac{\partial V}{\partial x}\right\|n\right\rangle$	Fourier component of of the equation of motion $(k\omega)^2 x_k = \frac{d\langle V\rangle}{dx_{-k}}$
Commutation relations $\langle\langle n\|[x,\ p]\|n\rangle = i$	Quantization of the action $I = \left\langle p\frac{\partial x}{\partial\theta}\right\rangle = n+\frac{1}{2}$
Diagonal elements of H $E_n = \langle n\|H\|n\rangle$	Energy in terms of action $E(I) = \langle H\rangle$
Angular frequency $\omega_{n+k,\ n} = E_{n+k}-E_n$	Classical frequency $\omega = \frac{dE(I)}{dI}$
Operator $A(x,\ p)$	Dynamical variable $A(x(I,\theta), p(I,\theta))$
Matrix elements $\langle n\|A\|n+k\rangle$	Fourier coefficients $A_k(I) = \frac{1}{2\pi}\int A(I,\theta)\exp(-ik\theta)d\theta$
Hamiltonian matrix is diagonal $\sum_{k\neq 0}\langle n\|H\|n+k\rangle$ $\times\langle n+k\|H\|n\rangle = 0$	Hamiltonian independent of θ $\sum_{k\neq 0} H_k H_{-k} = 0$

Let us apply this method to the familiar problem of two-dimensional harmonic oscillator Eq. (1.213),

$$H = \left(\frac{p_1^2}{2m_1} + \frac{1}{2}m_1\omega_1^2 q_1^2\right) + \left(\frac{p_2^2}{2m_2} + \frac{1}{2}m_2\omega_2^2 q_2^2\right), \tag{12.95}$$

if we integrate over n_1 cycles of q_1 and n_2 cycles of q_2, the integral (12.94) becomes

$$I(n_1, n_2) = \frac{n_1 E_2}{\omega_1} + \frac{n_2(E_1 - E_2)}{\omega_2} = n\hbar, \tag{12.96}$$

where E_2 and $E_1 - E_2$ are the energies of the q_1 and q_2 coordinates respectively. Now according to Einstein (12.96) must be satisfied for all integral values of n_1 and n_2. This can happen only if $\frac{E_2}{\omega_1}$ and $\frac{E_1-E_2}{\omega_2}$ are each integral multiples of $\hbar$. Thus for separable systems this rule reduces to the Bohr–Sommerfeld rule.

12.4 A Variational Method Based on Heisenberg's Equation of Motion

The variational method in quantum mechanics provides a powerful technique for calculating the low-lying energy levels of a bound particle or determining the partial wave phase shifts and the scattering amplitude in scattering theory [8],[9]. In the matrix formulation there has been some work done mainly on the bound state of one-dimensional problems [10],[11]. The interesting outcome of this approach is that the variational method is closely related to the method of finite differences used to solve operator differential equations discussed in Sec. 10.1 [12].

For a one-dimensional motion of a particle of unit mass with the Hamiltonian

$$H = \frac{1}{2}p^2 + V(q), \tag{12.97}$$

we can write the Heisenberg equations of motion as the limit of the finite difference equations;

$$\frac{q_{n+1} - q_n}{\tau} = p_{n+1}, \tag{12.98}$$

and

$$\frac{p_{n+1} - p_n}{\tau} = F(q_n) = -\left(\frac{\partial V(q)}{\partial q}\right)_{q_n}, \tag{12.99}$$

as τ tends to zero. If we compare these with Eqs. (10.6) and (10.7), we observe that for the iterative solution, the latter difference equations are better, but Eqs. (12.98) and (12.99) are simpler for the variational calculation.

Now let us consider an operator $Q(p(t), q(t))$ in the Heisenberg picture and calculate its matrix elements between two eigenvectors $|k\rangle$ and $|j\rangle$ of the Hamiltonian for the system H;

$$\begin{aligned}\langle n|Q(p(t), q(t))|j\rangle &= \exp\left[-\frac{i}{\hbar}(E_j - E_n)t\right]\langle n|Q(p(0),\ q(0))|j\rangle \\ &= \langle n|Q(p(0),\ q(0))|j\rangle\sum_{k=0}^{\infty}\left[-\frac{i(E_j - E_n)t}{\hbar}\right]^k\frac{t^k}{k!}.\end{aligned} \tag{12.100}$$

This equation takes a simple form when $j = n$,

$$\langle n|Q(p(t), q(t))|n\rangle = \langle n|Q(p(0), q(0)|n\rangle. \tag{12.101}$$

In particular for the operator

$$Q(p(t), q(t)) = q(t)p(t), \tag{12.102}$$

we use Eqs. (12.98) and (12.99) to find

$$\begin{aligned}\langle n|q(\tau)p(\tau)|n\rangle &= \langle n|q(0)p(0)|n\rangle + \\ &+ \tau\langle n|(p(0))^2 - q(0)V'(q(0))|n\rangle + \mathcal{O}\left(\tau^2\right),\end{aligned} \tag{12.103}$$

where $\mathcal{O}\left(\tau^2\right)$ denotes terms proportional to τ^2. By substituting from (12.102) in (12.103) we find that to the order τ the following condition must be satisfied,

$$\left\langle n\left|(p(0))^2 - q(0)V'(q(0))\right|n\right\rangle = 0, \tag{12.104}$$

and this is the quantum mechanical statement of the virial theorem (see Secs. 4.2 and 7.4).

Next let us introduce a variational parameter η, and write [10],[11]

$$q(0) \to \eta q, \tag{12.105}$$

and

$$p(0) \to \frac{1}{\eta}p, \tag{12.106}$$

where η is introduced in such a way that the canonical commutation relation $[q(0),\ p(0)]$ remains unchanged. In addition for infinitesimal τ, Eq. (12.104) must be satisfied and thus we find

$$\left\langle n\left|p^2\right|n\right\rangle = \eta_n^3\left\langle n\left|q\frac{\partial V(\eta_n q)}{\partial q}\right|n\right\rangle, \tag{12.107}$$

and this is the variational condition on the virial theorem. Here we have assumed that the parameter η is dependent on the state $|n\rangle$. The condition (12.107) is the same as the one obtained by minimizing the expectation value of the Hamiltonian (12.97);

$$\begin{aligned} & \delta\langle\phi_n(\eta_n)|H|\phi_n(\eta_n)\rangle \\ = & \ \delta\left\{\frac{1}{2}\left\langle\phi_n(\eta_n)\left|p^2\right|\phi_n(\eta_n)\right\rangle + \langle\phi_n(\eta_n)|V(q)|\phi_n(\eta_n)\rangle\right\} = 0, \end{aligned} \tag{12.108}$$

where the normalized variational wave function $\phi_n(\eta_n)$ is given by

$$\phi_n(q) \to \sqrt{\eta_n}\phi_n(\eta_n q). \tag{12.109}$$

This technique allows us to calculate the energies of different levels. For instance let us consider the case of the anharmonic oscillator

$$V(q) = \frac{1}{2}q^2 + \frac{\lambda}{4}q^4, \tag{12.110}$$

for which (12.107) becomes

$$\begin{aligned} \left\langle n\left|p^2\right|n\right\rangle &= \eta_n^3\left\langle n\left|qV'(\eta_n q)\right|n\right\rangle \\ &= \eta_n^4\left\langle n\left|q^2\right|n\right\rangle + \lambda\eta_n^6\left\langle n\left|q^4\right|n\right\rangle. \end{aligned} \tag{12.111}$$

Now we write

$$A_n = \left\langle n\left|p^2\right|n\right\rangle, \quad B_n = \left\langle n\left|q^2\right|n\right\rangle, \tag{12.112}$$

and

$$C_n = \left\langle n\left|q^4\right|n\right\rangle, \tag{12.113}$$

and expand η_n as a power series in λ

$$\eta_n = \left(1 + \epsilon_n\lambda + \delta_n\lambda^2 + \cdots\right). \tag{12.114}$$

Then by equating different powers of λ in (12.111) we find

$$A_n = B_n \quad \epsilon_n = \frac{C_n}{6A_n - 2B_n}, \quad \text{and} \quad \delta_n = -\frac{7}{2}\epsilon_n^2. \tag{12.115}$$

Substituting these in the expression for the diagonal elements of the Hamiltonian

$$\begin{aligned} \langle n|H|n\rangle &= \frac{1}{2}\eta_n^2\left\langle n\left|p^2\right|n\right\rangle + \frac{1}{2\eta_n^2}\left\langle n\left|q^2\right|n\right\rangle + \frac{\lambda}{4\eta_n^4}\left\langle n\left|q^4\right|n\right\rangle \\ &= \frac{1}{2}\left(A_n\eta_n^2 + \frac{B_n}{\eta_n^2}\right) + \frac{\lambda}{4\eta_n^4}C_n. \end{aligned} \tag{12.116}$$

Noting that

$$\left\langle n\left|p^2(0)\right|n\right\rangle = \left\langle n\left|q^2(0)\right|n\right\rangle = \left(n + \frac{1}{2}\right), \tag{12.117}$$

and

$$\langle n\left|q^4(0)\right|n\rangle = \frac{3}{2}\left(n^2+n+\frac{1}{2}\right), \tag{12.118}$$

we find the following expression for E_n in powers of λ [13]

$$\begin{aligned} E_n &= \left(n+\frac{1}{2}\right)+\frac{3}{2}\left(n^2+n+\frac{1}{2}\right)\frac{\lambda}{4} \\ &- \frac{2\left(n^2+n+\frac{1}{2}\right)^2}{n+\frac{1}{2}}\left(\frac{\lambda}{4}\right)^2+\cdots. \end{aligned} \tag{12.119}$$

This variational method can also be used to calculate the level spacing (or the energy differences) for potentials with many bound states. We write the matrix elements of the Heisenberg equations for $p(\tau)$ and $q(\tau)$ where τ is infinitesimal time

$$\langle n\left|q(\tau)\right|k\rangle = \langle n\left|q(0)\right|k\rangle + \tau\langle n\left|p(0)\right|k\rangle + \mathcal{O}\left(\tau^2\right), \tag{12.120}$$

$$\langle n\left|p(\tau)\right|k\rangle = \langle n\left|p(0)\right|k\rangle + \tau\langle n\left|F(q(0))\right|k\rangle + \mathcal{O}\left(\tau^2\right), \tag{12.121}$$

TABLE IX: Energy eigenvalues for a quartic anharmonic oscillator $V(q) = q^4$ (see [10]).

Quantum number n	Exact results	Approximate eigenvalues
0	0.6680	0.6814
1	2.3936	2.4237
2	4.6968	4.6850
3	7.3367	7.2911
4	10.244	10.167
5	13.379	13.267

TABLE X: Energy differences between two adjacent levels $\omega_n = E_{n+1} - E_n$, Eq. (12.124), for a quartic anharmonic oscillator $V(q) = q^4$ [10].

ω_n	Exact results	Approximate values
ω_0	1.7256	1.8171
ω_1	2.3032	2.2894
ω_2	2.6399	2.6207
ω_3	2.9073	2.8845
ω_4	3.1350	3.1072

where $F(q) = -\frac{\partial V(q)}{\partial q}$. Now from Eqs. (12.100), (12.120) and (12.121) it follows that

$$\langle n |p(0)| k\rangle = i\omega_{nk} \langle n |q(0)| k\rangle , \tag{12.122}$$

and

$$\langle n |F(q(0))| k\rangle = i\omega_{nk} \langle n |p(0)| k\rangle , \tag{12.123}$$

where $\omega_{nk} = E_n - E_k$. Now if we introduce the variational parameter η as in (12.105) and (12.106) in Eqs. (12.122) and (12.123) we obtain

$$i\omega_{nk} = \frac{\langle n|p(0)|k\rangle}{\langle n|q(0)|k\rangle} = \frac{\langle n|p|k\rangle}{\eta^2\langle n|q|k\rangle}, \tag{12.124}$$

and

$$i\omega_{nk} = \frac{\langle n|F(q(0))|k\rangle}{\langle n|p(0)|k\rangle} = \eta\frac{\langle n|F(\eta q)|k\rangle}{\langle n|p|k\rangle}. \tag{12.125}$$

By dividing (12.125) by (12.124) we find that η must satisfy the condition

$$\frac{\langle n|p|k\rangle}{\langle n|q|k\rangle} = \eta^3\frac{\langle n|F(\eta q)|k\rangle}{\langle n|p|k\rangle}, \tag{12.126}$$

where η now depends on n as well as k.

The above-mentioned method can be applied to a general quantum-mechanical system with two degrees of freedom whether it is integrable or not, for example to the general form of the Henon-Heiles Hamiltonian [14].

12.5 Raleigh–Ritz Variational Principle

The variational principle is a powerful technique for calculating the ground state energy of a quantum mechanical system. For the formulation of Raleigh-Ritz variational principle we first observe that for a given Hamiltonian H with any number of degrees of freedom and any normalized state $|\psi\rangle$, the expectation value $\langle\psi|H|\psi\rangle$ is always greater or equal to the ground state energy of the system, E_0,

$$E_0 \le \langle\psi|H|\psi\rangle. \tag{12.127}$$

The equality is obtained if $|\psi\rangle$ is the exact ground state wave function. To prove the inequality (12.127), let us expand $|\psi\rangle$ in terms of the eigenstates of H, viz,

$$|\psi\rangle = \sum_n c_n|\psi_n\rangle, \tag{12.128}$$

then

$$H|\psi\rangle = H\sum_n c_n|\psi_n\rangle = \sum_n E_n|\psi_n\rangle, \tag{12.129}$$

or

$$\langle\psi|H|\psi\rangle = \sum_n E_n|c_n|^2. \tag{12.130}$$

Since $E_0 \le E_n$, for all $n > 0$, we have the inequality

$$\langle\psi|H|\psi\rangle \ \ge E_0 \sum_n |c_n|^2 = E_0. \tag{12.131}$$

If $|\psi\rangle$ is not normalized then we write (12.131) as

$$E_0 \le \frac{\langle\psi|H|\psi\rangle}{\langle\psi|\psi\rangle}. \tag{12.132}$$

Thus by finding a trial wave function which minimizes the right-hand side of (12.132), i.e. $\left|\psi^{(0)}\right\rangle$, we find an approximate value for E_0. A simple example of this method is applied for the calculation the ionization energy for helium-like atoms that we will consider in Chapter 17.

For calculating the first excited state of the system, we choose the trial wave function $\left|\psi^{(1)}\right\rangle$ to be normalized and at the same time be orthogonal to the ground state

$$\left\langle\psi^{(0)}|\psi^{(1)}\right\rangle = 0. \tag{12.133}$$

This technique can be generalized in the following way:
If $\left|\psi^{(0)}\right\rangle$, $\left|\psi^{(1)}\right\rangle \cdots$, $\left|\psi^{(n-1)}\right\rangle$ are already determined then

$$\left\langle\psi^{(n)}|\psi^{(n)}\right\rangle = 1, \quad \left\langle\psi^{(n)}|\psi^{(j)}\right\rangle = 0, \quad j = 0,\ 1,\ \cdots, n-1 \tag{12.134}$$

are the conditions which must be satisfied for an acceptable trial function for the n-th excited state. Thus by finding a trial wave function which minimizes the right-hand side of (12.131) we find an approximate value of E_0 [15].

12.6 Tight-Binding Approximation

This approximation which is useful in determining the energy band structure in solids is applicable to the systems where the Hamiltonian is invariant under space translation equal to the lattice spacing.

If the potential between two adjacent sites in a periodic potential is not infinitely strong, then the wave function will not be completely localized in the n-th site and through the mechanism of quantum tunneling it will leak to the neighboring sites. The diagonal elements of H will be the same, independent of the site $|n\rangle$, i.e.

$$\langle n|H|n\rangle = E_0. \tag{12.135}$$

For a finite but high barriers between the sites, the matrix elements of the Hamiltonian between well separated sites will be negligible, and to a good approximation we can ignore all but the matrix elements of the Hamiltonian between the neighboring sites. When this is the case, we use the tight-binding approximation which can be expressed as

$$\langle n' |H| n\rangle \neq 0, \quad \text{for} \quad n' - n = 0,\ \pm 1. \tag{12.136}$$

Because of the translational invariance of the Hamiltonian, $\langle n \pm 1|H|n\rangle$ will be constant independent of $|n\rangle$. Let us denote this matrix element by $-\Delta$,

$$\langle n \pm 1|H|n\rangle = -\Delta. \tag{12.137}$$

Noting that the off diagonal elements of H are not zeros, we have

$$H|n\rangle = E_0|n\rangle - \Delta|n+1\rangle - \Delta|n-1\rangle, \tag{12.138}$$

and this relation shows that $|n\rangle$ is not an eigenstate of H. Now let us consider the eigenstate of $\mathrm{T}(a)$, Sec. 6.7 in this approximation. For this we apply H to the state $|\alpha\rangle$, Eq. (6.115),

$$\begin{aligned} H|\alpha\rangle &= H \sum_{n=-\infty}^{\infty} e^{in\alpha}|n\rangle \\ &= E_0 \sum_{n=-\infty}^{\infty} e^{in\alpha}|n\rangle - \Delta \sum_{n=-\infty}^{\infty} e^{in\alpha}|n+1\rangle - \Delta \sum_{n=-\infty}^{\infty} e^{in\alpha}|n\rangle \\ &= \left\{E_0 \sum_{n=-\infty}^{\infty} e^{in\alpha} - \Delta \sum_{n=-\infty}^{\infty} \left(e^{in\alpha - i\alpha} + e^{in\alpha + i\alpha}\right)\right\}|n\rangle \\ &= (E_0 - 2\Delta\cos\alpha) \sum_{n=-\infty}^{\infty} e^{in\alpha}|n\rangle. \end{aligned} \tag{12.139}$$

As this equation shows in this approximation the energy eigenvalue depends on α and changes continuously between $E_0 - 2\Delta$ and $E_0 + 2\Delta$ [16].

12.7 Heisenberg's Correspondence Principle

The idea that the quantum matrix elements can be approximately determined from the Fourier coefficient of the expansion of classical motion goes back to Heisenberg [17]. In this section we want to discuss the application of Heisenberg's Correspondence principle mentioned earlier Sec. 4.6 for the calculation of the matrix elements of bound states. Let us assume that for a bound classical motion of a particle of unit mass the position is given by $q(t)$. We can expand $q(t)$ as a Fourier series

$$q(t) = \sum_s q_s e^{is\omega t}, \tag{12.140}$$

where s is an integer and ω is the classical frequency which, in general is energy-dependent. Then the Heisenberg form of correspondence Eq. (4.227) gives us the matrix element of the quantum mechanical $q(t)$

$$\langle n|q(t)|n+s\rangle = q_s, \tag{12.141}$$

when n is much larger than s.

Before discussing the general case, let us examine the result that we obtain by applying this principle to the simple harmonic oscillator. By writing

$$q(t) = \sqrt{\frac{2E}{\omega^2}}\cos(\omega t), \tag{12.142}$$

where $E = \frac{1}{2}\omega^2 q_0^2$ is the energy and q_0 is its amplitude, and substituting (12.142) in (12.141) we find that the only non-zero Fourier components are those with $s = \pm 1$. Thus for the off diagonal matrix elements we obtain

$$\langle n|q|n-1\rangle = \langle n|q(t)|n+1\rangle = \sqrt{\frac{\hbar}{2\omega}n}, \tag{12.143}$$

where for E we have substituted $\hbar\omega$. For the exact solution of the harmonic oscillator we have the matrix elements

$$\langle n|q|n+1\rangle = \sqrt{\frac{\hbar}{\omega}}\left(\frac{n+1}{2}\right)^{\frac{1}{2}}, \quad \text{and} \quad \langle n|q|n-1\rangle = \sqrt{\frac{\hbar}{2\omega}}\left(\frac{n}{2}\right)^{\frac{1}{2}}, \tag{12.144}$$

which shows that the amplitudes for the upward and downward transitions are not equal, whereas (12.143) predicts that they are equal. To remedy this defect we modify (12.144) so that for small $\frac{s}{n}$ it gives the same result as (12.144). At the same time we generalize (12.143) so that it can be used for any positive integral power of q. To this end we write [18]

$$\langle n\left|q^k\right|n+s\rangle = \left(\frac{\hbar}{2\omega}n_c\right)^{\frac{k}{2}} \frac{k!}{\left[\frac{1}{2}(k-s)\right]!\left[\frac{1}{2}(k+s)\right]!}. \tag{12.145}$$

Here s is restricted to

$$s = k-2j, \quad j = 0,\ 1, \cdots k, \tag{12.146}$$

and n_c is given by

$$n_c = \left[\frac{(n+s)!}{n!}\right]^{\frac{1}{s}}, \quad s \neq 0 \tag{12.147}$$

As $s \to 0$, by expanding the right-hand side of (12.147) we get

$$n_c = n + \frac{1}{2}, \quad s = 0. \tag{12.148}$$

This version of Heisenberg's Correspondence principle works very well for different matrix elements of the coordinate and momentum. It also satisfies Bohr's

TABLE XI: Matrix elements of q^k and p^k for harmonic oscillator calculated from Heisenberg's Correspondence principle, Eqs. (12.145) and (12.149). These are compared with the exact results.

Matrix element	Correspondence value	Exact value
$\langle n\|q\|n+1\rangle$	$\sqrt{\frac{\hbar}{2\omega}(n+1)}$	$\sqrt{\frac{\hbar}{2\omega}(n+1)}$
$\langle n\|q\|n-1\rangle$	$\sqrt{\frac{\hbar}{2\omega}n}$	$\sqrt{\frac{\hbar}{2\omega}n}$
$\langle n\left\|q^2\right\|n+2\rangle$	$\frac{\hbar}{2\omega}\sqrt{(n+1)(n+2)}$	$\frac{\hbar}{2\omega}\sqrt{(n+1)(n+2)}$
$\langle n\left\|q^2\right\|n\rangle$	$\frac{\hbar}{\omega}\left(n+\frac{1}{2}\right)$	$\frac{\hbar}{\omega}\left(n+\frac{1}{2}\right)$
$\langle n\left\|q^3\right\|n-1\rangle$	$\left(\frac{\hbar}{\omega}\right)^{\frac{3}{2}}\sqrt{\frac{9n^3}{8}}$	$\left(\frac{\hbar}{\omega}\right)^{\frac{3}{2}}\sqrt{\frac{9n^3}{8}}$
$\langle n\left\|q^3\right\|n-3\rangle$	$\left(\frac{\hbar}{\omega}\right)^{\frac{3}{2}}\sqrt{\frac{1}{8}n(n-1)(n-2)}$	$\left(\frac{\hbar}{\omega}\right)^{\frac{3}{2}}\sqrt{\frac{1}{8}n(n-1)(n-2)}$
$\langle n\left\|q^4\right\|n\rangle$	$\frac{3\hbar^2}{2\omega^2}\left(n^2+n+\frac{1}{4}\right)$	$\frac{3\hbar^2}{2\omega^2}\left(n^2+n+\frac{1}{2}\right)$
$\langle n\|p\|n+1\rangle$	$i\sqrt{\frac{\hbar\omega}{2}(n+1)}$	$i\sqrt{\frac{\hbar\omega}{2}(n+1)}$
$\langle n\|p\|n-1\rangle$	$i\sqrt{\frac{\hbar\omega}{2}n}$	$i\sqrt{\frac{\hbar\omega}{2}n}$
$\langle n\left\|p^3\right\|n+3\rangle$	$-i\left(\frac{\hbar\omega}{2}\right)^{\frac{3}{2}}\sqrt{(n+1)(n+2)}$	$-i\left(\frac{\hbar\omega}{2}\right)^{\frac{3}{2}}\sqrt{(n+1)(n+2)}$ $\times\sqrt{(n+3)}$
$\langle n\left\|p^4\right\|n+2\rangle$	$-4\left(\frac{\hbar\omega}{2}\right)^2(n+1)(n+2)$	$-4\left(\frac{\hbar\omega}{2}\right)^2\left(n+\frac{3}{2}\right)\sqrt{(n+1)}$ $\times\sqrt{(n+2)}$
$\langle n\|p^4\|n-2\rangle$	$-4(\frac{\hbar\omega}{2})^2 n(n-1)$	$-4(\frac{\hbar\omega}{2})^2(n-\frac{1}{2})\sqrt{n(n-1)}$

correspondence principle and agrees with quantum mechanical calculation in the limit $\frac{s}{n} \to 0$. The momentum matrix elements can be found by noting the symmetry in position and momentum matrix elements which is the case for the harmonic oscillator. Thus from (12.145) we obtain

$$\left\langle n \left| p^k \right| n+s \right\rangle = i^{2k-s} \left(\frac{\hbar\omega}{2} n_c \right)^{\frac{k}{2}} \frac{k!}{\left[\frac{1}{2}(k-s)\right]! \left[\frac{1}{2}(k+s)\right]!}. \tag{12.149}$$

which apart from the constant factor is the same as (12.145).

From Eqs. (12.145) and (12.149) we find the following simple relation between the matrix elements of q^k and those of p^k;

$$\left\langle n \left| p^k \right| n+s \right\rangle = i^{2k-s} \omega^k \left\langle n \left| q^k \right| n+s \right\rangle. \tag{12.150}$$

TABLE XI shows the matrix elements $\left\langle n \left| q^k \right| n+s \right\rangle$ and $\left\langle n \left| p^k \right| n+s \right\rangle$ for integers k and s. The exact results are also shown for comparison. We observe that for a number of matrix elements, what we find by this method agrees with the exact results. For $k \geq 4$, only the matrix elements of the form $\left\langle n \left| p^k \right| n \pm k \right\rangle$ are identical, but other matrix elements remain close to the exact values.

The Morse Potential — The approximate calculation of the matrix elements of the Morse potential, $V(x)$, defined by (8.39) provides a nontrivial example of the usefulness of this correspondence principle. The classical motion of a particle of unit mass in this potential can be formulated in terms of the action-angle variables I and ω [21],[20];

$$H = I\omega \left(1 - \frac{I\omega}{4\lambda} \right) - \lambda, \tag{12.151}$$

where

$$\omega = \sqrt{2\mu\lambda}. \tag{12.152}$$

If we replace I in (12.151) by $\left(n + \frac{1}{2}\right)\hbar$ (see Eq. (12.93) we find the exact eigenvalues of the Morse potential, Eq. (8.40). The coordinate and momentum of the particle written in terms of the action-angle variable are:

$$q(t) = \frac{1}{\sqrt{\mu}} \ln \left[\frac{1 + \sqrt{\frac{E}{\lambda}} \cos w}{1 - \frac{E}{\lambda}} \right], \tag{12.153}$$

and

$$p(t) = \left[\frac{\sqrt{E} \left(\sqrt{1 - \frac{E}{\lambda}} \right) \sin w}{1 + \sqrt{\frac{E}{\lambda}} \cos w} \right]. \tag{12.154}$$

where E is the energy of the particle and where q and p are periodic functions of time with $w = \omega t$. By expanding q in a Fourier series we find that the coefficients of expansion of q are

$$\langle n|q|n+s\rangle = \frac{-i^s}{\sqrt{\mu}s}\left(\frac{n}{\frac{2}{\hbar}\sqrt{\frac{2\lambda}{\mu}}-n}\right)^{\frac{s}{2}}. \tag{12.155}$$

Similarly an expansion of the periodic function p shows that the coefficients of expansion are:

$$\langle n|p|n+s\rangle = \begin{cases} i^{s-2}\left(\frac{\omega}{\sqrt{\mu}}\right)\left(\frac{n}{\frac{2}{\hbar}\sqrt{\frac{2\lambda}{\mu}}-n}\right)^{\frac{s}{2}} & s \neq 0\,. \\ 0 & s = 0 \end{cases} \tag{12.156}$$

In order to obtain more accurate results for the matrix elements we replace n in (12.155) and (12.156) by n_c, where as before n_c is related to n by (12.147) and (12.148). The exact result found from the solution of the Schrödinger equation is [22]

$$\begin{aligned}\langle n|q|n+s\rangle &= -\frac{1}{\sqrt{\mu}s}\left\{\frac{\left[\left(\frac{1}{\hbar}\sqrt{\frac{2\lambda}{\mu}}-n\right)\left(\frac{1}{\hbar}\sqrt{\frac{2\lambda}{\mu}}-n-s\right)\right]^{\frac{1}{2}}}{\left(\frac{1}{\hbar}\sqrt{\frac{2\lambda}{\mu}}-n-\frac{1}{2}-\frac{s}{2}\right)}\right\} \\ &\times \left[\frac{(n+s)!\,\Gamma\left(\frac{2}{\hbar}\sqrt{\frac{2\lambda}{\mu}}-n-s\right)}{n!\,\Gamma\left(\frac{2}{\hbar}\sqrt{\frac{2\lambda}{\mu}}-n\right)}\right]^{\frac{1}{2}}. \end{aligned} \tag{12.157}$$

TABLE XII: The results found for $Q(n,s)$, Eq. (12.158), from Heisenberg's Correspondence principle are compared to the exact result for the Morse potential [18].

Transition $n \to n+s$	Correspondence value	Exact value
$1 \to 2$	0.020408	0.020406
$3 \to 4$	0.041667	0.041662
$7 \to 8$	0.086957	0.086944
$1 \to 4$	2.6203×10^{-5}	2.6272×10^{-5}
$3 \to 6$	1.397×10^{-4}	1.398×10^{-4}
$5 \to 10$	4.557×10^{-6}	4.575×10^{-6}

In order to compare the result of the approximate calculation based on the correspondence, Eq. (12.155), and the exact result given by (12.157), we first note that the quantity $\frac{1}{\hbar}\sqrt{\frac{2\lambda}{\mu}}$ is a dimensionless number which is a measure of the strength of the potential and thus the number of bound states. From the matrix elements we find the quantity

$$Q(n,s) = \mu s^2 |\langle n|q|n+s\rangle|^2, \tag{12.158}$$

which depends on $\frac{1}{\hbar}\sqrt{\frac{2\lambda}{\mu}}$, n and s. When the potential has a large number of bound states, the approximate and the exact values of $Q(n,s)$ are very close to each other. In TABLE XII, the results are shown for $\frac{1}{\hbar}\sqrt{\frac{2\lambda}{\mu}} = 50$, corresponding to a Morse potential with about 48 bound states [18].

12.8 Bohr and Heisenberg Correspondence and the Frequencies and Intensities of the Emitted Radiation

In classical dynamics the orbit of a particle attracted by an inverse square law of force, e.g. Coulomb force with the potential $V(r) = -\frac{Ze^2}{r}$ is, in general, an ellipse with the equation

$$r = \frac{(1-\varepsilon^2)\,a}{1+\varepsilon\cos\theta}, \tag{12.159}$$

in polar coordinates [23]. The semi-major axis a is inversely proportional to E, the total energy, which is negative

$$a = -\frac{Ze^2}{2E}, \tag{12.160}$$

and the eccentricity is given by

$$\varepsilon = \left[1 - \frac{2|E|}{m}\left(\frac{\ell_c}{Ze^2}\right)^2\right]^{\frac{1}{2}}. \tag{12.161}$$

In this relation m is the reduced mass of the system and l_c is the classical angular momentum. The period of motion which is given by the Kepler's third law is

$$T = \frac{2\pi}{\omega} = \frac{2\pi Z e^2\sqrt{m}}{(-2E)^{\frac{3}{2}}}. \tag{12.162}$$

We can also write the classical motion in terms of the eccentricity anomaly u which is defined by

$$r = a(1-\varepsilon\cos u), \tag{12.163}$$

where both u and r are functions of time. In particular we have

$$u - \varepsilon \sin u = \omega t, \tag{12.164}$$

with ω, the angular frequency, given by (12.162). For the two-dimensional elliptic motion it is convenient to use coordinates $\xi(t)$ and $\eta(t)$ defined by

$$\xi(t) = a(\cos u - \varepsilon), \tag{12.165}$$

and

$$\eta(t) = a\sqrt{1 - \varepsilon^2} \sin u. \tag{12.166}$$

These are parametric equations for the orbit. Thus if we eliminate u between (12.165) and (12.166) we find the equation of orbit in rectangular coordinates:

$$\left(\frac{\xi}{a} + \varepsilon\right)^2 + \frac{\eta^2}{a^2\left(1 - \varepsilon^2\right)} = 1. \tag{12.167}$$

In order to apply Heisenberg's Correspondence principle to the two-dimensional motion we need the coefficients of the Fourier transforms of $\xi(t)$ and $\eta(t)$. If we expand these two functions as

$$\xi(t) = \sum_{-\infty}^{+\infty} \xi_s e^{-is\omega t}. \tag{12.168}$$

and

$$\eta(t) = \sum_{-\infty}^{+\infty} \eta_s e^{-is\omega t}, \tag{12.169}$$

then the coefficients ξ_s and η_s are given by [24]

$$\begin{cases} \xi_s = \frac{a}{s} J_s'(s\varepsilon) e^{-is\delta} & |s| \geq 1 \\ \xi_0 = -\frac{3}{2} a\varepsilon & \mathrm{s}_0 = 0 \end{cases} \tag{12.170}$$

and

$$\begin{cases} \eta_s = \frac{ia\sqrt{1-\varepsilon^2}}{s\varepsilon} J_s(s\varepsilon) e^{-is\delta} & |s| \geq 1 \\ \eta_0 = 0 & s = 0 \end{cases} \tag{12.171}$$

These are found from the inverse Fourier transform of (12.168) and (12.169) and by noting that [26]

$$J_s(z) = \frac{1}{\pi} \int_0^{\pi} \cos(s\theta - z \sin\theta) d\theta. \tag{12.172}$$

Heisenberg's Correspondence Principle Applied to the Quantum Theory of Dipole Transition — In classical electrodynamics, we know that the intensity of dipole radiation per unit time by a particle of charge e located at $\mathbf{r}$ is given by [25]

$$I = \frac{2e^2}{3c^3} \left|\ddot{\mathbf{r}}(t)\right|^2 . \tag{12.173}$$

When the particle is bound and the Hamiltonian is separable we can expand $\mathbf{r}(t)$ in terms of multiple Fourier series

$$\mathbf{r}(t) = \sum_{\mathbf{s}} \mathbf{r_s} \exp(-i\mathbf{s}\cdot\boldsymbol{\theta}), \tag{12.174}$$

where

$$\boldsymbol{\theta} = \boldsymbol{\omega} t. \tag{12.175}$$

The quantities $\mathbf{r_s}$, $\boldsymbol{\theta}$ and $\boldsymbol{\omega}$ are all vectors, each with x, y and z components. Substituting for $\mathbf{r}(t)$ in (12.173) we find the classical intensity to be

$$I = \frac{4e^2}{3c^3}(\mathbf{s}\cdot\boldsymbol{\omega})^4 |\mathbf{r_s}|^2. \tag{12.176}$$

Now we apply Heisenberg's Correspondence principle to the problem of dipole radiation in order to find an approximate quantum expression for the emitted power. In this case the correspondence principle can be stated in the following way [24]:

The mean quantum mechanical power associated with the transition $\mathbf{n} \to \mathbf{n}'$ is approximately equal to the mean emitted power derived by classical theory. This is found from the Fourier components of order $\mathbf{s}$ and $-\mathbf{s}$, where $\mathbf{s} = \mathbf{n} - \mathbf{n}'$.

Next we want to consider the matrix elements of any classical periodic function of $\mathbf{r}$, say $\mathbf{F}^c(\mathbf{r})$. For this we express $\mathbf{r}(t)$ in terms of the action angle variables $\boldsymbol{\theta}$ and $\mathbf{J}$ and write

$$\mathbf{F}^c(\mathbf{r}) = \sum_{\mathbf{s}} \mathbf{F}^c_{\mathbf{s}}(\mathbf{J}) \exp(-i\mathbf{s}\cdot\boldsymbol{\theta}). \tag{12.177}$$

Now according to Heisenberg's Correspondence principle, for the expectation value of the quantum operator $\mathbf{F}(\mathbf{r})$ which corresponds to the classical function $\mathbf{F}(\mathbf{r})$ we have

$$\langle \mathbf{n}' \left|\mathbf{F}(\mathbf{r})\right| \mathbf{n}\rangle \approx \mathbf{F}^c_{\mathbf{s}}(\mathbf{J}). \tag{12.178}$$

In this relation $|\mathbf{n}\rangle$ and $|\mathbf{n}'\rangle$ are the eigenstates of the Hamiltonian for the bound system

$$H_0|\mathbf{n}\rangle = E(\mathbf{n})|\mathbf{n}\rangle. \tag{12.179}$$

The classical expression (12.176) shows that for the quantum mechanical problem of dipole radiation we need to find a mean over one or more of the quantum numbers of the square of the matrix element. We also note that for the problem involving the Coulomb force which is a central force, the energy levels are degenerate and the classical motion is two-dimensional. The mean value in this case means averaging over ℓ and m (see Sec. 9.5). Let us first consider the summation over the quantum number m of the bound system. We have the classical actions J_ℓ and J_m corresponding to quantum numbers ℓ and m, i.e.

$$\frac{J_m}{J_\ell} = \frac{m}{\ell} = \mu = \cos\beta, \tag{12.180}$$

where β is the angle between the z-axis and the angular momentum vector. Thus we can replace the sum over all m by an integral

$$\sum_{m=-\ell}^{\ell} \to 2\ell \int_{-1}^{1} \frac{d\mu}{2}. \tag{12.181}$$

For a vector function of $\mathbf{r}$, say $\mathbf{F}(\mathbf{r})$, we have to evaluate the sum

$$\sum_{m=-\ell}^{\ell} \sum_{m'=-\ell'}^{\ell'} \left|\langle n'\ell'm'\left|\mathbf{F}(\mathbf{r})\right| n\ell m\rangle\right|^2, \tag{12.182}$$

approximately. From the correspondence principle (12.178), this expression can be approximated by

$$\sum_m \sum_{\Delta m} \left(\frac{1}{2\pi}\right)^6 \int\int \exp\left[i\mathbf{s}\cdot(\boldsymbol{\theta}-\boldsymbol{\theta}')\right] \mathbf{F}(\theta,\mathbf{J})\cdot\mathbf{F}^*\left(\boldsymbol{\theta}',\mathbf{J}'\right) d^3\theta d^3\theta', \tag{12.183}$$

where for the vector $\mathbf{s}$ we have written

$$\mathbf{s} = \mathbf{n} - \mathbf{n}' = -(\Delta n, \Delta\ell, \Delta m). \tag{12.184}$$

We can do the summation over Δm in (12.183) with the help of the following relation

$$\frac{1}{2\pi}\sum_{\Delta m} \exp\left[i\Delta m\left(\theta_m - \theta'_m\right)\right] \approx \delta\left(\theta_m - \theta'_m\right). \tag{12.185}$$

Using this relation Eq. (12.183) becomes

$$\frac{\ell}{\pi}\int_0^{2\pi} d\theta_m \int_{-1}^{1}\frac{d\mu}{2}\left|\left(\frac{1}{2\pi}\right)^2 \int\int \exp[-i(\Delta n\,\theta_n + \Delta\ell\,\theta_\ell)]\,\mathbf{F}(\boldsymbol{\theta},\mathbf{J})d\theta_n d\theta_\ell\right|^2. \tag{12.186}$$

For the case of the Coulomb potential, the eigenvalues are degenerate and for a given n the individual (ℓ, m) states are equally populated, therefore instead of (12.182) we need the sum

$$\frac{1}{n^2}\sum_{\ell=0}^{n-1}\sum_{\ell'=0}^{n'-1}\sum_{m=-\ell}^{\ell}\sum_{m'=-\ell'}^{\ell'} \left|\langle n'\ell'm'\left|\mathbf{F}(\mathbf{r})\right| n\ell m\rangle\right|^2. \tag{12.187}$$

To apply Heisenberg's correspondence principle to the hydrogen (or Kepler) problem we write the three-dimensional analogue of (12.141) which is

$$\sum_{n'=1}^{\infty}\sum_{\ell'=0}^{n'-1}\sum_{m'=-\ell'}^{\ell'} \langle n'\,\ell'\,m'\,|q|\,n\,\ell\,m\rangle \exp\left[i(E_{n'}-E_n)\frac{t}{\hbar}\right]$$

$$= \sum_{s=-n+1}^{\infty}\left(\sum_{\ell'=0}^{n+s-1}\sum_{m'=-\ell'}^{\ell'} \langle (n+s)\,\ell'\,m'\,|q|\,n\,\ell\,m\rangle\right)\exp\left[i(E_{n+s}-E_n)\frac{t}{\hbar}\right]. \tag{12.188}$$

Now for large quantum numbers, according to Heisenberg's Correspondence principle, Eq. (12.188) tends to the classical function $q(t)$. For instance let us consider the hydrogen atom with the energy E_n, given by Eq. (9.252). Associated with this motion we have a Keplerian orbit with a semi-major axis $a = n^2 a_0$ where $a_0 = \frac{\hbar^2}{me^2}$ is the Bohr radius. In addition the classical orbit is characterized by the angular momentum $L = \ell\hbar$, its projection on the z-axis, $L_z = m\hbar$, and on the x-axis $L_x = \sqrt{\ell^2 - m^2}\,\hbar$. For this we use Eqs. (12.165), (12.166), (12.168)–(12.171) to obtain [38]

$$\frac{\xi}{a} = -\frac{m}{\ell}\left[-\frac{3}{2}\varepsilon + \sum_{s=-\infty}^{\infty}{}' \frac{1}{s}J_s'(s\varepsilon)e^{is\omega t}\right], \tag{12.189}$$

$$\frac{\eta}{a} = -\frac{\sqrt{1-\varepsilon^2}}{\varepsilon}\sum_{s=-\infty}^{\infty}{}' \frac{i}{s}J_s(s\varepsilon)e^{is\omega t}, \tag{12.190}$$

and

$$\frac{\zeta}{a} = \frac{\sqrt{\ell^2-m^2}}{\ell}\left[-\frac{3}{2}\varepsilon + \sum_{s=-\infty}^{\infty}{}' \frac{i}{s}J_s'(s\varepsilon)e^{is\omega t}\right], \tag{12.191}$$

where prime on $\sum$ indicates the term with $s = 0$ should be omitted from the summation over s.

We can establish the following connection between the exact quantum mechanical matrix elements (12.188) and the classical equations (12.189)–(12.191). Suppose that we replace q by ξ in (12.188) and calculate the right-hand side of (12.188) for n, ℓ and m, all very large integers, then in this limit we recover (12.189), remembering that the selection rule allows a nonzero result only for $\ell' = \ell \pm 1$ and $m' = m \pm 1$ [18]. We find a similar result, Eq. (12.190), when we set $q = \eta$ with the same conditions on ℓ' and m'. Finally if we take q to be ζ in (12.188) and follow the same argument we obtain (12.191) but now for $\ell' = \pm 1$ and $m' = m$ [18].

Quantum Theory of Oscillator Strength — As an application of the Heisenberg correspondece principle we want to consider the semi-classical determination of the oscillator strength. In classical electrodynamics the relation between polarization $\mathbf{P}$ and the applied electric field $\mathbf{E} = \mathbf{E}_0 \sin(\omega t)$ is given by

$$\mathbf{P} = \alpha_p\,\mathbf{E}, \tag{12.192}$$

where

$$\alpha_p = \sum_j \frac{e^2}{m}\left(\frac{\mathcal{F}_j}{\omega_j^2 - \omega^2}\right). \tag{12.193}$$

In this relation m is the mass of the electron, e its charge and $\mathcal{F}_j$ is the oscillator strength which is the number of electrons per atom in the state of angular frequency ω_j.

In quantum theory one can calculate the polarizability using the first order perturbation theory with the result that [29].

$$\alpha_p = \left(\frac{2\omega}{\hbar}\right)\sum_j \frac{|\langle f|\mathbf{p}|i\rangle|^2}{(\omega_f - \omega_i)^2 - \omega^2}, \tag{12.194}$$

where ω_i and ω_f are frequencies associated with the initial and final states of the atom respectively and $|\langle f|\mathbf{p}|i\rangle|^2$ is the expectation value of the dipole moment of the atom,

$$\langle f|\mathbf{p}|i\rangle = e\langle f|\mathbf{r}|i\rangle. \tag{12.195}$$

Note that f and i refer to a set of quantum numbers e.g. $(n\,\ell\,m)$ in the case of Coulomb potential. By comparing (12.193) and (12.194) we obtain the quantum mechanical expression for the sum rule

$$\langle n\,\ell\,m\,|\mathcal{F}|\,n'\,\ell'\,m'\rangle = \frac{2m}{3\hbar}\omega_{n',n}\left[|\langle n'\,\ell'\,m'\,|\mathbf{r}|\,n\,\ell\,m\rangle|^2\right]_{\text{averaged over } m}, \tag{12.196}$$

with $\hbar\omega_{n',n} = E_{n'} - E_n$. In order to determine the right-hand side of (12.196) we will use Eq. (12.186) with $\mathbf{F}$, the action-angle coordinate corresponding to the position coordinate $\mathbf{r}$. Thus we write $\mathbf{r} = (\xi, \eta, \zeta)$, and we express each component in terms of the angle variables $(\theta_n, \theta_\ell, \theta_m)$,

$$\begin{aligned}\xi(\theta_n,\theta_\ell,\theta_m) &= \sum_{s,\Delta\ell,\Delta m} \langle n+s\,\ell+\Delta\ell\,m+\Delta m|\xi|n\,\ell, m\rangle \\ &\times\ \exp[i(s\,\theta_n + \Delta\ell\,\theta_\ell + \Delta m\,\theta_m)],\end{aligned} \tag{12.197}$$

with similar expressions for $\eta(\theta_n,\theta_\ell,\theta_m)$ and $\zeta(\theta_n,\theta_\ell,\theta_m)$. The sele4ction rules mentioned earlier implies that for ξ and η coordinates $\Delta\ell = \pm 1$ and $\Delta m = \pm 1$ and for ζ coordinate $\Delta\ell = \pm 1$ and $\Delta m = 0$. Now from Eqs. (12.168)–(12.171) we find that [18]

$$\begin{aligned}&\langle n\,\ell\,m|\xi|n+s\,\ell+\Delta\ell\,m+\Delta m\rangle \\ &= \frac{\Delta\ell}{4}\left(1+\frac{\Delta m}{\Delta\ell}\frac{m}{\ell}\right)\langle n+s\,\ell+\Delta\ell|R|n\,\ell\rangle,\end{aligned} \tag{12.198}$$

$$\begin{aligned}&\langle n\,\ell\,m|\eta|n+s\,\ell+\Delta\ell\,m+\Delta m\rangle \\ &= -i\Delta m\frac{\Delta\ell}{4}\left(1+\frac{\Delta m}{\Delta\ell}\frac{m}{\ell}\right)\langle n+s\,\ell+\Delta\ell|R|n\,\ell\rangle,\end{aligned} \tag{12.199}$$

and

$$\langle n\ \ell\ m|\zeta|n{+}s\ \ell{+}\Delta\ell\ m{+}\Delta m\rangle = -\frac{i}{2}\sqrt{\left(1-\frac{m^2}{\ell^2}\right)}\langle n{+}s\ \ell{+}\Delta\ell|R|n\ \ell\rangle. \quad (12.200)$$

In these relations the matrix element $\langle n+s\ \ell+\Delta\ell|R|n\ \ell\rangle$ is given by

$$\langle n+s\ \ell+\Delta\ell|R|n\ \ell\rangle$$
$$\approx \left(\frac{n_c^2\Delta\ell}{2s}\right)\left[\left(1+\Delta\ell\frac{l_c}{n_c}\right)J_{s-1}(s\varepsilon)-\left(1-\Delta\ell\frac{l_c}{n_c}\right)J_{s+1}(s\varepsilon)\right], \quad (12.201)$$

where we have substituted $a = n_c^2$ (a in units of the Bohr radius). Here as in the case of the Morse potential n_c is an average of the quantum numbers n and n' which can be chosen either as $n_c = \frac{1}{2}(n+n')$ or for a better approximation as $n_c = \frac{n(n+s)}{n+\frac{1}{2}s}$.

In TABLE XIII we compare the results found by exact quantum mechanical calculation of the dipole moment squared with its approximate value using Heisenberg's Correspondence principle.

TABLE XIII: Exact and approximate values for the dipole moment squared $|\langle n+s\ \ell+\Delta\ell|R|n\ \ell\rangle|^2$ with $n_c = \frac{n(n+s)}{n+\frac{1}{2}s}$ and $\ell_c = \ell$ [18].

Transition $n \to n+s$	Correspondence value	Exact value
$2s-3p$	9.720	9.393
$4s-5p$	73.181	72.553
$6s-7p$	275.25	274.19
$4p-5d$	123.18	121.86
$4d-5f$	200.46	197.83

Bibliography

[1] W.R. Greenberg, A. Klein, I. Zlatev and C.T. Li, From Heisenberg matrix mechanics to semi-classical quantization: Theory and first approximation, Phys. Rev. A 54, 1820 (1996).

[2] A. Klein and C-T. Li, Semiclassical quantization of nonseparable systems, J. Math. Phys. 20, 572 (1979).

[3] W.R. Greenberg, A. Klein and C.-T. Li, The invariant tori and Heisenberg matrix mechanics- A new window on the quantum-classical correspondence, Phys. Rev. Lett. 75, 1244 (1995).

[4] C.-T. Li, G.H. Suen and A. Klein, Some structural and numerical aspects of Heisenberg matrix mechanics with applications to nonpolynomial potentials, Phys. Rev. A 63, 052110 (2001).

[5] C.-T. Li and A. Klein, An improved semi-classical approximation based on Heisenberg's matrix mechanics, Chinese J. Phys. 39, 555 (2001).

[6] M.C. Gutzwiller, *Chaos in Classical and Quantum Mechanics*, (Springer Verlag, New York, 1990), p. 218.

[7] For a readable account of EBK quantization condition see J.B. Keller, Semi-classical Mechanics, SIAM Review, 27, 485 (1985).

[8] See for example: J. Schwinger, *Quantum Mechanics: Symbolism of Atomic Measurement*, Edited by B-G Englert, (Springer, Berlin, 2001), p. 309.

[9] E. Merzbacher, *Quantum Mechanics*, (John Wiley & Sons, New York, 1970) p.169.

[10] L. Vazquez, Relation between two variational methods to calculate the energy levels, Phys. Lett. A 144, 15 (1990).

[11] G. Dattoli, P.L. Ottaviani, A. Torre and L. Vazquez, Evolution operator equations: integration with algebraic and finite difference methods. Applications to physical problems in classical and quantum mechanics and quantum field theory, Rev. Nuovo Cimento, 20, 1(1997).

[12] C.M. Bender and D.H. Sharp, Solution of the operator field equation by the method of finite elements, Phys. Rev. Lett. 50, 1535 (1985).

[13] J.D. de Deus, Variational method for the double-well anharmonic oscillator, Phys. Rev. D 26, 2782 (1982).

[14] M.J. Rodriguez and L. Vazquez in : Lectures notes in mathematics, Vol. 1394. Nonlinear semigroups, partial differential equations and attractors, Edited by T.L. Gill and W.W. Zachary (Springer, Berlin, 1989) p. 131.

[15] This powerful method can be used to find the low-lying energy levels of atomic, molecular or nuclear systems, see for instance W. Schweizer, *Numerical Quantum Dynamics*, (Kluwer Academic Publisher, 2001).

[16] J.J. Sakurai, *Modern Quantum Mechanics*, (Addison-Wesley, Reading, 1994).

[17] W. Heisenberg, *The Physical Principles of the Quantum Theory*, translated by C. Eckart and F.C. Hoyt, (Dover, New York, 1949), p. 116.

[18] P.F. Naccache, Matrix elements and correspondence principle J. Phys. B 5, 1308 (1972).

[19] I.C. Percival and D. Richards, A correspondence principle for strongly coupled states, J. Phys. B 3, 1035 (1970).

[20] R.B. Shirts, Use of the classical Fourier amplitudes as quantum metrix elements: A comparison of Morse oscillator Fourier coefficients with quantum matrix elements, J. Phys. Chem. 91, 2258 (1987).

[21] H. Goldstein, *Classical Mechanics*, Second Edition, (Addison-Wesley Reading, 1980).

[22] G.D. Carney and R.N. Porter, H_3^+: Ab initio calculation of transition vibration spectrum, J. Chem. Phys. 65, 3547 (1976).

[23] See for example, L.D. Landau and E.M. Lifshitz, *Mechanics*, (Pergamon, Oxford, 1960), Chapter III.

[24] I.C. Percival and D. Richards, Collisions of particles and highly excited atoms, Adv. At. Mol. Phys. 11, 1 (1975).

[25] L.D. Landau and I.M. Lifshitz, *The Classical Theory of Fields*, Fourth Edition (Pergamon Press, Oxford, 1975), p. 175.

[26] E.T. Whittaker and G.N. Watson, *A Course of Modern Analysis*, (Cambridge University Press, Cambridge, 1948), p. 362.

[27] Q.H. Liu and B. Hu, The hydrogen atom's quantum-to-classical corresspondence in Heisenberg's corrspondence principle, J. Phys. A 34, 5713 (2001).

[28] G. Baym, *Lectures on Quantum Mechanics*, (W.A. Benjamin, Reading, 1969).

[29] S. Flügge, *Practical Quantum Mechanics*, (Springer-Verlag, New York, 1970), p. 157.

Chapter 13

Quantization of the Classical Equations of Motion with Higher Derivatives

Quantum mechanical systems whose equation of motion involves higher derivatives than the second have been studied in connection with field theories with non-localized action [1] and also in the quantum theory of radiating electron [2]. These equations can be of finite order in derivatives or they can be of infinite order. For such systems the advantage of using the Heisenberg approach, for consistency of the formulation and as a way of determining the spectra, becomes evident.

13.1 Equations of Motion of Finite Order

If we are interested in reversible and conservative motions, then the equation of motion will be of the form

$$F(D)q = 0, \qquad D \equiv \frac{d}{dt}, \tag{13.1}$$

where F is a polynomial of degree $2N$. In this case Eq. (13.1) is derivable from an action principle with the Lagrangian

$$L = -qF(D)q. \tag{13.2}$$

If we can write $F(D)$ as a product

$$F = \prod_{i=1}^{N}\left(1 + \frac{D^2}{\omega_i^2},\right). \tag{13.3}$$

where ω_i s are real and distinct then the solution of (13.1) is a linear combination of oscillators with frequencies ω_i, $i = 1,\ 2,\ \cdots N$. In this case the Hamiltonian will be the same linear combination of the oscillator Hamiltonians and can easily be quantized. This can be done in the following way:

Let us define N coordinates Q_i by

$$Q_i = \prod_{j=1}^{N\ \prime}\left(1 + \frac{D^2}{\omega_j^2}\right) q, \tag{13.4}$$

where prime on the product sign means that the i-th factor should be deleted. From Eqs. (13.1) and (13.4) it follows that

$$\left(D^2 + \omega_i^2\right) Q_i = 0, \quad i = 1, 2\cdots, N. \tag{13.5}$$

That is, the number of independent solutions of (13.1) is the same as (13.5).

Since the classical set (13.5) is equivalent to (13.1) and (13.3) therefore we can replace the Lagrangian (13.2) by its equivalent

$$\tilde{L} = -\sum_{j=1}^{N} \eta_j Q_j \left(D^2 + \omega_j^2\right) Q_j. \tag{13.6}$$

In this expression we have to fix the N constants η_i s in such a way that $L - \tilde{L}$ becomes a total time derivative of a function of q and its derivatives (this is what is meant by the equivalent Lagrangian). Substituting for Q_i from (13.4) in (13.6) we have

$$\begin{aligned}\tilde{L} &= -q\left[\sum_k \eta_k \left\{\prod_j{}' \left(1 + \frac{D^2}{\omega_j^2}\right)\right\}^2 \left(D^2 + \omega_k^2\right)\right] q \\ &+ \quad \text{a total time derivative.}\end{aligned} \tag{13.7}$$

Next we substitute from (13.3) in (13.7) to find

$$\tilde{L} = -qF^2(D)\sum_k \left(\frac{\eta_k \omega_k^2}{1 + \frac{D^2}{\omega_k^2}}\right) q, \tag{13.8}$$

and this is equivalent to L in Eq. (13.2) provided that

$$\sum_k \frac{\eta_k \omega_k^2}{1 + \frac{D^2}{\omega_k^2}} = \frac{1}{F(D)}. \tag{13.9}$$

Since the roots of $F(D)$ are simple, we can find a partial fraction decomposition of F^{-1} where the numerators are D-independent. That is

$$\omega^4 \eta_k = \frac{1}{F'\left(-\omega_k^2\right)}, \tag{13.10}$$

where $F'\left(-\omega_i^2\right)$ is defined by

$$F'\left(-\omega_k^2\right) = \left(\frac{dF}{dD^2}\right)_{D^2=-\omega_k^2}. \tag{13.11}$$

From Eq. (13.10) we obtain the important result about the sign of η_k. Noting that F is a single valued function of its argument, the quantities $F'\left(-\omega_k^2\right)$ and therefore η_k in (13.10) alternate in sign. For instance if $F'\left(-\omega_1^2\right)$ is positive then $F'\left(-\omega_{2k+1}^2\right)$ will be positive and $F'\left(-\omega_{2k}^2\right)$ will be negative. We can find a simple relation between η_j s and ω_j^2 by setting $D = 0$ in (13.9) to get

$$\sum_j \eta_j \omega_j^2 = 1, \tag{13.12}$$

and

$$\sum_k \eta_k \omega_k^{2n} = 0, \qquad n = 2,\ 3 \cdots N. \tag{13.13}$$

The last relation, Eq. (13.13), is found by expanding both sides of (13.9) in powers of D^{-2} and equating the coefficients of different powers of D^{-2} on the two sides.

Returning to the Lagrangian $\tilde{L}$, Eq. (13.6) we observe that

$$\tilde{L} = \sum_{j=1}^{N} \eta_j \left[(DQ_j)^2 - \omega_j^2 Q_j^2\right], \tag{13.14}$$

where we have preformed a partial time integration and we have omitted a total time derivative from $\tilde{L}$. Having found a Lagrangian which is quadratic in DQ_j and Q_j we can write a Hamiltonian for this system;

$$H = \sum_j \left[\frac{P_j^2}{4\eta_j} + \eta_j \omega_j^2 Q_j^2\right]. \tag{13.15}$$

This Hamitonian can be transformed to the standard form of the Hamiltonian for a set of oscillators if we make the contact transformation

$$P_j \to P_j \left(2|\eta_j|\right)^{\frac{1}{2}}, \quad Q_j \to \frac{Q_j}{\left(2|\eta_j|\right)^{\frac{1}{2}}}, \tag{13.16}$$

and note the sign properties of the η_j s. Thus H can be written as the sum of harmonic oscillator Hamiltonians

$$H = \frac{1}{2}\sum_j (-1)^{j-1}\left(P_j^2 + \omega_j^2 Q_j^2\right), \tag{13.17}$$

We can quantize (13.17) and the result is that the eigenvalues of H are given by

$$E_{n_1,\cdots,n_N} = \sum_{j=1}^{N}(-1)^{j-1}\left(n_j + \frac{1}{2}\right)\omega_j, \qquad n_j = 0, 1\cdots. \tag{13.18}$$

As this result shows the total energy of these oscillators in not positive definite.

In recent years there have been a number of attempts to find a Hamiltonian formulation which gives satisfactory quantized solution [3]–[5]. For simplicity let us consider the case of a fourth-order oscillator knowing that this method can be generalized to the case of N oscillators. Using the notation that we introduced earlier we have the equation of motion

$$\left(D^2 + \omega_1^2\right)\left(D^2 + \omega_2^2\right) q = 0, \tag{13.19}$$

or in terms of the $\{Q_i\}$ coordinates we have

$$Q_1 = \left(1 + \frac{D^2}{\omega_2^2}\right) q, \quad Q_2 = \left(1 + \frac{D^2}{\omega_1^2}\right) q, \tag{13.20}$$

where now Q_1 and Q_2 satisfy the equation of motion

$$\left(D^2 + \omega_1^2\right) Q_1 = 0, \quad \left(D^2 + \omega_2^2\right) Q_2 = 0, \tag{13.21}$$

We can write the Hamiltonian for the two oscillators (13.21) as

$$H = \frac{1}{2}\left(P_1^2 + \omega_1^2 Q_1^2\right) + \frac{1}{2}\left(P_2^2 + \omega_2^2 Q_2^2\right), \tag{13.22}$$

and this H differs in the sign of the second term from H given by (13.17). Upon quantization this Hamiltonian will give us positive definite eigenvalues.

13.2 Equation of Motion of Infinite Order

Now let us consider the equations of motion of infinite order in time derivatives. In classical dynamics one way of writing this type of motion is in the form of an integral equation for $q(t)$ [6];

$$\int K\left(t - t'\right) q\left(t'\right) dt' = 0, \tag{13.23}$$

or what is essentially the same form [7]

$$m\ddot{q}(t) = \int K_1(t-t')\,q(t')\,dt'. \tag{13.24}$$

Classically this type of equation arises naturally when we are considering the motion of an extended object or when we are discussing the motion of a radiative electron [2]. The Lagrangian for (13.23) is

$$L = -q(t)\int K(t-t')\,q(t')\,dt', \tag{13.25}$$

and we will use the form of the Lagrangian given in (13.25) to find the Hamiltonian and then proceed with the quantization. To relate this motion with the one given by (13.1) and (13.5) we assume that $K(t)$ can be written as a Fourier integral

$$K(t) = \frac{1}{2\pi}\int_{-\infty}^{+\infty} e^{ikt}F(ik)dk. \tag{13.26}$$

Among all possible forms of $F(ik)$ we restrict our attention to those cases where F is an entire function in the sense of functions of complex variable [1]. Then by the product theorem of Weierstrass we can write $F(D)$ as [8]

$$F(D) = e^{f(D)}\prod_j\left(1+\frac{D^2}{\omega_j^2}\right), \tag{13.27}$$

i.e. we have expressed F in terms of its zeros and an exponential function. This is permitted as long as the sum

$$\sum_j \omega_j^{-2}, \tag{13.28}$$

is convergent. In Eq. (13.27) $f(D)$ is again an entire function. When $f(D)$ is zero then we can formulate the problem as before, only in this case we have an infinite product. As an example consider the case where

$$F(D) = \cosh(\alpha D), \tag{13.29}$$

then using the product representation of $\cosh(\alpha D)$ we have

$$F(D) = \cos(\alpha D) = \prod_{j=1}^{\infty}\left(1+\frac{D^2}{\omega_j^2}\right), \qquad \omega_j = \frac{\pi}{2\alpha}(2j-1). \tag{13.30}$$

If we define Q_j s exactly as before, Eq. (13.4), and define $\tilde{L}$ as in (13.6) with $N\to\infty$ then we have the analogue of (13.9)

$$\frac{1}{\cosh(\alpha D)} = \sum_{k=1}^{\infty}\frac{\eta_k\omega_k^2}{1+\frac{D^2}{\omega_k^2}}. \tag{13.31}$$

Using the partial fraction expansion of $[\cosh(\alpha D)]^{-1}$ we obtain

$$\frac{1}{\cosh(\alpha D)} = \frac{4}{\pi}\sum_{k=1}^{\infty}\frac{(-1)^{(k-1)}}{(2k-1)}\frac{1}{1+\frac{D^2}{\omega_k^2}}. \tag{13.32}$$

Therefore η_k is given by

$$\eta_k = \frac{4}{\pi}\frac{(-1)^{(k-1)}}{(2k-1)\omega_k^2} = \frac{16\alpha^2}{\pi^3}\left(\frac{(-1)^{(k-1)}}{(2k-1)^3}\right), \tag{13.33}$$

and as this relation shows again that η_k s alternates in sign. For the case where F is given by (13.27) the equivalent Lagrangian $\tilde{L}$ can be written in terms of Q_j s as before

$$\tilde{L} = \sum_k \eta_k Q_k e^{f(D)}\left(1+\frac{D^2}{\omega_k^2}\right)Q_k. \tag{13.34}$$

For this system the construction of the Hamiltonian from Lagrangian is not easy. We know of a simpler way of quantizing the system without having a classical Hamiltonian and canonical coordinates and momenta. This method which was first advocated by Heisenberg is as follows [9]:

We start with the equations of motion and we obtain an expression for the energy E in terms of q and its time derivatives and then attribute to E the role of time-displacement operator, i.e.

$$[E,\ g(q,\dot{q},\ddot{q},\cdots)] = -i\dot{g}, \quad \hbar = 1, \tag{13.35}$$

for any function g of $q,\dot{q},\cdots$. To this end we first obtain an expression for the classical energy of the system in terms of $q(t)$ and its derivatives, or as an integral operator acting on $q(t)$.

13.3 Classical Expression for the Energy

The energy is defined as the first integral of motion for the time displacement. For a Lagrangian quadratic in velocity, the energy is defined as

$$E = \dot{q}\frac{\partial L}{\partial \dot{q}} - L. \tag{13.36}$$

We can generalize this definition to the equations of infinite order by writing

$$\begin{aligned} E &= \sum_{n=1}^{\infty}(D^n q)\frac{\delta L}{\delta(D^n q)} - L \\ &= \sum_{n=1}^{\infty} D^n q\left[\sum_{m=0}^{\infty}(-1)^m D^m\left(\frac{\partial L}{\partial(D^{m+n}q)}\right)\right] - L, \end{aligned} \tag{13.37}$$

where $\frac{\delta L}{\delta(D^n q)}$ is the functional derivative of L which is defined by the expression in the bracket in (13.37). Next let us write the Fourier transforms of $q(t)$ and $E(t)$

$$q(t) = \frac{1}{2\pi}\int_{-\infty}^{+\infty} e^{-ikt}R(k)dk, \tag{13.38}$$

$$E(t) = \frac{1}{2\pi}\int_{-\infty}^{+\infty} e^{-ist}\mathcal{E}(s)ds, \tag{13.39}$$

and write $F(D)$ as

$$F(D) = \sum_{0}^{\infty} \lambda_n D^n. \tag{13.40}$$

Now by substituting these in (13.37) we obtain

$$\mathcal{E}(s) = -\frac{1}{2\pi}\sum_{n=1}^{\infty}\sum_{m=0}^{\infty}(-1)^m\lambda_{n+m}\int_{-\infty}^{+\infty}(-ik)^m[-i(s-k)]^n R(k)R(s-k)dk. \tag{13.41}$$

Replacing n by $p-m$, we can first sum over m and then express the summation over p in terms of F. Thus $\mathcal{E}(s)$ takes the form

$$\mathcal{E}(s) = \frac{1}{2\pi s}\int_{-\infty}^{\infty}(k-s)R(k)R(s-k)[F(i(k-s)) - F(ik)]dk. \tag{13.42}$$

The equation of motion in Fourier transform form can be obtained from Eqs. (13.23) and (13.26) and is given by

$$F(-ik)R(k) = 0. \tag{13.43}$$

Thus as a consequence of the equation of motion we have

$$s\mathcal{E}(s) = 0, \tag{13.44}$$

and this implies the conservation of energy.

Let us consider the following simple example where

$$L = -qF_1(D)q, \quad F_1(D) = \frac{1}{2}e^{f(D)}\left(D^2 + \omega^2\right), \tag{13.45}$$

then we have

$$q(t) = A\cos\omega t, \tag{13.46}$$

and

$$R(k) = \pi\left[\delta(k+\omega) + \delta(k-\omega)\right]. \tag{13.47}$$

By substituting (13.47) in (13.43) we find that the latter equation is satisfied. Also from (13.42) it follows that

$$\mathcal{E}(s) = \pi A^2\omega^2 e^{f(i\omega)}\delta(s), \tag{13.48}$$

and the energy is conserved since $s\mathcal{E}(s) = 0$. From (13.39) we have the energy of the system which is real, since f is an even function of its argument, and this energy can be expressed as

$$E = \frac{1}{2}A^2\omega^2 e^{f(i\omega)}. \tag{13.49}$$

13.4 Energy Eigenvalues when the Equation of Motion is of Infinite Order

The method of Heisenberg mentioned earlier will be used here to find the energy eigenvalues when the equation of motion is of infinite order with the Lagrangian (13.45) [1]. For this formulation it is convenient to make use of the Fourier transform of $q(t)$. Thus the time derivative of q is given by the integral of $(-ikR(k))$ and in addition we have the commutator of $R(k)$ and E;

$$[R(k),\ E] = kR(k). \tag{13.50}$$

In the representation where E is diagonal, by taking the matrix elements of the two sides of (13.50), we find

$$(E' - E'' - k)\left\langle E'|R(k)|E''\right\rangle = 0, \tag{13.51}$$

or

$$\left\langle E'|R(k)|E''\right\rangle = \delta\left(E' - E'' - k\right)\left\langle E'|R|E''\right\rangle. \tag{13.52}$$

The equation of motion

$$F_1(-ik)R(k) = 0, \tag{13.53}$$

implies that $R(k)$ is zero except for $k = \pm\omega$. Therefore (13.52) can also be written as

$$\left\langle E'|R(k)|E''\right\rangle = \delta\left(k^2 - \omega^2\right)\delta\left(E' - E'' - k\right)\left\langle E'|R_0|E''\right\rangle. \tag{13.54}$$

Having found the matrix elements of $R(k)$ we can proceed to find the diagonal elements of E. These diagonal elements satisfy the equation

$$\frac{1}{2\pi}\int_{-\infty}^{+\infty}\left\langle E'|\mathcal{E}(s)|E''\right\rangle e^{-ist}ds = E'\delta\left(E' - E''\right), \tag{13.55}$$

where $\mathcal{E}(s)$ is given by (13.42) with F being replaced by F_1. Now if we substitute from (13.54) in (13.42) we get

$$E' = \frac{1}{16\pi^2}e^{f(i\omega)}\left[\left|\left\langle E'|R_0|E' + \omega\right\rangle\right|^2 + \left|\left\langle E'|R_0|E' - \omega\right\rangle\right|^2\right]. \tag{13.56}$$

From this equation we can find the diagonal elements of E as well as the matrix elements $\langle E'|R_0|E' \pm \omega\rangle$. First let us consider the case where $f = 0$, then we have the ordinary harmonic oscillator for which we know the eigenvalues

$$E' = \left(n' + \frac{1}{2}\right)\omega, \tag{13.57}$$

and the matrix elements of R_0 consists of two terms:

$$\langle E'|R_0|E''\rangle = 2\pi\sqrt{2}\left[\sqrt{n'\omega}\,\delta\left(n' - n'' - 1\right) + \sqrt{n''\omega}\,\delta\left(n' - n'' + 1\right)\right]. \tag{13.58}$$

For the equation of motion $F_1(D)q = 0$, we observe that ω is replaced by $\omega e^{f(i\omega)}$, therefore (13.57) becomes

$$E' = \left(n' + \frac{1}{2}\right)\omega e^{f(i\omega)}, \tag{13.59}$$

while $\langle E'|R_0|E''\rangle$ is the same as for ordinary oscillator. Thus we conclude that $\{q_{mn}\}$, the matrix elements of q in the energy representation, for $f \neq 0$ is the same as for $f = 0$.

The same matrix elements of q is obtained if we replace $\left(D^2 + \omega^2\right)$ by $\left(D^2 + \omega^2\right)^k$ where k is an integer. However the presence of the exponential factor $e^{if(D)}$ in Eq. (13.45) has a profound effect if our oscillator is coupled to another system, e.g. to a simple harmonic oscillator with frequency ω_0.
Let us take the Lagrangian of the coupled system to be

$$\begin{aligned} L &= -\frac{1}{2}q\exp\left[\lambda^{2k}\left(D^2+\omega^2\right)^k\right]\left(D^2+\omega^2\right)q \\ &- \frac{1}{2}q_0\left(D^2+\omega_0^2\right)q_0 + \epsilon q q_0, \end{aligned} \tag{13.60}$$

where λ has the dimension of inverse frequency. From this Lagrangian we find the coupled equations of motion to be

$$\begin{cases} \exp\left[\lambda^{2k}\left(D^2+\omega^2\right)^k\right]\left(D^2+\omega^2\right)q = \epsilon q_0 \\ \left(D^2+\omega_0^2\right)q_0 = \epsilon q \end{cases}. \tag{13.61}$$

To determine the normal modes of this coupled system we take

$$q \sim e^{i\nu t}, \quad q_0 \sim e^{i\nu t}, \tag{13.62}$$

and substitute these in (13.61) to find an equation for ν

$$\exp\left[\lambda^{2k}\left(\omega^2-\nu^2\right)^k\right]\left(\omega^2-\nu^2\right)\left(\omega_0^2-\nu^2\right) = \epsilon^2. \tag{13.63}$$

For $\epsilon = 0$ we have the roots of ν at $\pm\omega$ and $\pm\omega_0$, but when $\epsilon \neq 0$ and $\lambda \neq 0$ there are infinity of solutions and the problem of determination of eigenvalues and eigenvectors becomes very complicated.

Bibliography

[1] A. Pais and G.E. Uhlenbeck, On the field theories with non-localized action, Phys. Rev. 79, 145 (1950).

[2] A detailed account of the problem of radiating electron can be found in M. Razavy, *Classical and Quantum Dissipative Systems*, (Imperial College Press, London, 2005), see also F. Rohrlich, *Classical Charged Particles*, 3rd Edition (World Scientific, Singapore, 2007).

[3] P.D. Manneheim and A. Davidson, Dirac quantization of the Pais-Uhlenbeck fourth order oscillator, Phys. Rev. A 71, 0421110 (2005).

[4] K. Bolonek and P. Kosinski, Hamiltonian structure for Pais-Uhlenbeck oscillator, Acta Phys. Polonica, B 36, 2115 (2005).

[5] E.V. Damaskinsky and M.A. Sokolov, Remarks on quantization of Pais-Uhlenbeck oscillators, J. Phys. A 39, 10499 (2006).

[6] R.M. Santilli, *Foundations of Theoretical Mechanics II*, (Springer Verlag, New York, 1983).

[7] W. Pauli, On the Hamiltonian structure of non-local field theories, Nuovo Cimento, 10, 648 (1953).

[8] See for instance, E.T. Whittaker and G.N. Watson, *A Course of Modern Anaysis*, (Cambridge University Press, 1948), p. 137.

[9] W. Heisenberg, Observable magnitudes in the theory of elementary particles III, Z. Physik, 123, 93 (1944).

Chapter 14

Potential Scattering

Quantum scattering is one of the most powerful methods of investigating the structure of atoms, molecules and nuclei. The simplest way of formulating the potential scattering, for short as well as long range forces, is to employ the Schrödinger equation and its integral equation form [1]–[3]. However there is an interesting determinantal formulation due to Schwinger in which discrete energy eigenvalues and their eigenfunctions are used, and in a way this is related to the matrix mechanics [6]. We use the wave equation formulation to define the scattering amplitude and phase shifts and later we will consider the determinantal formulation.

In a typical scattering experiment a narrow beam of monochromatic particles is directed toward a target. The intensity of the incident beam is low enough so that there is no interaction between the particles in the beam. After the incident beam interacts with the target the particles in the beam scatter in all directions and these particles are observed at a great distance from the target. The important quantity measured in a scattering experiment is the differential scattering cross section $\frac{d\sigma}{d\Omega}$ which is defined as the ratio of the scattered particle flux at the detector per unit solid angle and the flux in the incident beam [1]-[3]. If we choose the origin of the polar coordinates as the scattering center (or the center of the target) and the polar axis (z-axis) as the axis of the incident beam then according to the definition we have

$$\frac{d\sigma}{d\Omega} = \frac{I(\theta,\phi)}{I_0}. \tag{14.1}$$

Here I_0 denotes the incident flux and $I(\theta,\phi)\ d\Omega$ is the flux of the scattered particles through the cone subtended by the detector. From this definition it is clear that (a) - $\frac{d\sigma}{d\Omega}$ has the dimension of area and (b) - that the number of particles scattered per unit time into the solid angle $d\Omega$ in the direction (θ,ϕ)

is

$$dN = I_0 \frac{d\sigma}{d\Omega} \, d\Omega. \tag{14.2}$$

The total cross section σ_t is obtained by integrating $d\sigma$ over the solid angle $d\Omega$;

$$\sigma_t = \int \frac{d\sigma}{d\Omega} \, d\Omega = \int \frac{d\sigma}{d\Omega} \, \sin\theta \, d\theta \, d\phi. \tag{14.3}$$

The differential cross section depends on the angles θ and ϕ as well as on the energy of the particles in the incident beam.

We assume that the interaction between a particle in the beam and the target is given by the potential $V(\mathbf{r})$, and that this potential has a short range, i.e. $V(\mathbf{r}) \to 0$ faster than $r^{-2-\epsilon}$, $\epsilon > 0$ as r tends to infinity, or

$$\int V(r) dr < \infty, \tag{14.4}$$

$$\int V(r) r dr < \frac{\hbar^2}{2m} M, \tag{14.5}$$

and

$$\int V(r) r^2 dr < \frac{\hbar^2}{2m} N. \tag{14.6}$$

In these relations M and N are constants.

The Hamiltonian for the system composed of the projectile plus the target can be written as

$$H = H_0 + V(\mathbf{r}) = \frac{\mathbf{p}^2}{2m} + V(\mathbf{r}), \tag{14.7}$$

where

$$m = \frac{m_1 m_2}{m_1 + m_2}, \tag{14.8}$$

is the reduced mass of the system composed of the target and the projectile (incident particle). Thus we need the solution of the equation

$$[H_0 + V(\mathbf{r})] |\psi\rangle = E|\psi\rangle, \tag{14.9}$$

with E which is the total energy of the system being a positive quantity. We can write (14.9) as an integral equation

$$|\psi\rangle = |\mathbf{k}\rangle + GV(\mathbf{r})|\psi\rangle, \tag{14.10}$$

where G is the Green function

$$G = \frac{1}{E - H_0}, \tag{14.11}$$

and $|\mathbf{k}\rangle$ represents an incident plane wave and is a solution of the Schrödinger equation

$$(E - H_0)|\mathbf{k}\rangle = 0. \tag{14.12}$$

Multiplying (14.10) from the left by $(E - H_0)$ we get

$$(E - H_0)|\psi\rangle = (E - H_0)|\mathbf{k}\rangle + V(\mathbf{r})|\psi\rangle = V(\mathbf{r})|\psi\rangle, \tag{14.13}$$

which is Eq. (14.9). Thus Eq. (14.10) is equivalent to (14.9), but in addition it contains the initial state $|\mathbf{k}\rangle$ which is the state of the particle when it is far from the target.

In order to determine the state $|\psi\rangle$ we need to find the Green function G in (14.10). If we work in the momentum representation, we have a simple expression for G:

$$\begin{aligned}\langle \mathbf{k}'|G|\mathbf{k}\rangle &= \left\langle \mathbf{k}' \left| \frac{1}{E - H_0} \right| \mathbf{k} \right\rangle \\ &= \frac{2m}{\hbar^2} \frac{1}{\mathbf{k}^2 - \mathbf{k}'^2}\, \delta(\mathbf{k} - \mathbf{k}').\end{aligned} \tag{14.14}$$

The coordinate representation of G can be found from its momentum representation, i.e. (14.14);

$$\begin{aligned}\langle \mathbf{r}'|G|\mathbf{r}''\rangle &= \int\int \langle \mathbf{r}'|\mathbf{k}'\rangle\ \langle \mathbf{k}'\,|G|\,\mathbf{k}''\rangle\, \langle \mathbf{k}''|\mathbf{r}''\rangle\, d^3k'\ d^3k'' \\ &= \frac{2m}{\hbar^2(2\pi)^3} \int e^{i\mathbf{k}'\cdot(\mathbf{r}'-\mathbf{r}'')} \frac{d^3k'}{\mathbf{k}^2 - \mathbf{k}'^2}.\end{aligned} \tag{14.15}$$

The last integral in (14.15) is singular at $k = k'$ and in order to have a well-defined G, we specify the path of integration by writing G as G_+ where

$$G_+ = \frac{1}{E - H_0 + i\epsilon}, \tag{14.16}$$

Let us define the Green function $G_+(\mathbf{r})$ by

$$\begin{aligned}G_+(\mathbf{r}) &= \frac{2m}{\hbar^2(2\pi)^3} \int \frac{e^{i\mathbf{q}\cdot\mathbf{r}}}{k^2 - q^2} d^3q \\ &= -\frac{2m}{\hbar^2(2\pi)^3} \int_0^\infty \int_0^\pi \int_0^{2\pi} \frac{e^{iqr\cos\theta}}{q^2 - k^2} q^2 dq \sin\theta\ d\theta d\phi \\ &= -\frac{2m}{2\hbar^2\pi^2 r} \int_0^\infty \frac{\sin qr}{(q^2 - k^2)} q dq = -\frac{2m}{\hbar^2 4\pi^2 r} \int_{-\infty}^\infty \frac{\kappa \sin\kappa}{\kappa^2 - (kr)^2} d\kappa \\ &= -\left(\frac{2m}{\hbar^2}\right)\left(\frac{1}{4\pi^2 r}\right) \\ &\times \left[\frac{1}{2i} \oint \frac{\kappa e^{i\kappa} d\kappa}{(\kappa - kr - i\epsilon)(\kappa + kr)} - \frac{1}{2i} \oint \frac{\kappa e^{-i\kappa}}{(\kappa - kr)(\kappa + kr + i\epsilon)}\right],\end{aligned} \tag{14.17}$$

where in the last square bracket we have replaced kr by $(kr + i\epsilon)$. Noting that kr is a positive quantity, and if we close the contour for the first exponential with a large semi-circle in the upper-half plane, then only the pole at $\kappa = kr + i\epsilon$ will contribute to the integral. Thus the first contour integral gives us $i\pi e^{ikr}$. For the second exponential we close the contour in the lower-half plane and then only the pole at $\kappa = -kr - i\epsilon$ will contribute, and the result will be $-i\pi e^{ikr}$. Therefore $G_+(r)$ which is the sum of these contributions assumes the simple form

$$G_+(r) = -\frac{m}{2\pi\hbar^2 r} e^{ikr}, \tag{14.18}$$

and for the matrix element $\langle \mathbf{r}'|G_+|\mathbf{r}''\rangle$ we get

$$\begin{aligned} \langle \mathbf{r}'|G_+|\mathbf{r}''\rangle &= \frac{2m}{\hbar^2 (2\pi)^3} \int e^{i\mathbf{k}'\cdot(\mathbf{r}'-\mathbf{r}'')} \frac{d^3k'}{\mathbf{k}^2 - \mathbf{k}'^2 + i\epsilon} \\ &= -\frac{m}{2\pi\hbar^2 |\mathbf{r}' - \mathbf{r}''|} e^{ik|\mathbf{r}'-\mathbf{r}''|}. \end{aligned} \tag{14.19}$$

With G_+, the scattered wave are outgoing waves from the center of the target. Writing $\psi(\mathbf{r})$ for $\langle \mathbf{r}|\psi\rangle$ and $e^{i\mathbf{k}\cdot\mathbf{r}}$ for $\langle \mathbf{r}|\mathbf{k}\rangle$ we find the total wave function from (14.10) to be

$$\psi(\mathbf{r}) = e^{i\mathbf{k}\cdot\mathbf{r}} - \frac{m}{2\pi\hbar^2} \int \frac{e^{ik|\mathbf{r}-\mathbf{r}'|}}{|\mathbf{r} - \mathbf{r}'|} V(\mathbf{r}')\, \psi(\mathbf{r}')\, d^3r'. \tag{14.20}$$

Next we observe that the asymptotic form of $\psi(\mathbf{r})$ obtained from (14.20) is

$$\psi(\mathbf{r}) \to e^{i\mathbf{k}\cdot\mathbf{r}} + f(\theta, \phi) \frac{e^{ikr}}{r}, \quad \text{as} \quad r \to \infty, \tag{14.21}$$

where

$$f(\theta, \phi) = -\frac{m}{2\pi\hbar^2} \int e^{i\mathbf{k}'\cdot\mathbf{r}'} V(\mathbf{r}')\, \psi(\mathbf{r}')\, d^3r', \tag{14.22}$$

and $\mathbf{k}' = k\hat{\mathbf{r}}$, $\hat{\mathbf{r}}$ being a unit vector along $\mathbf{r}$.

Now the incident flux I_0 is the probability current density of the incident beam, Eq. (4.259), in the z-direction. For the incident beam

$$\psi_{in}(\mathbf{r}) = e^{i\mathbf{k}\cdot\mathbf{r}} = e^{ikz}, \tag{14.23}$$

and therefore I_0 is given by

$$I_0 = \frac{\hbar k}{m}. \tag{14.24}$$

For the scattered particles we choose the scattered wave function as is given in (14.21), viz,

$$\psi_s(\mathbf{r}) = f(\theta, \phi) \frac{e^{ikr}}{r}, \tag{14.25}$$

and from the definition of the radial component of the current density $\mathbf{j}(\mathbf{r})$, Eq. (14.21), we obtain

$$j_r^{sc}(\mathbf{r}) = \frac{\hbar k}{m} \frac{1}{r^2} |f(\theta, \phi)|^2. \tag{14.26}$$

This is the number of particles at a distance r from the target passing through a unit surface area per unit time. Therefore the flux per unit solid angle is

$$I(\theta,\phi) = \frac{\hbar k}{m}|f(\theta,\phi)|^2. \tag{14.27}$$

By substituting from (14.24) and (14.27) in (14.1) we find the following expression for the differential cross section

$$\frac{d\sigma}{d\Omega} = |f(\theta,\phi)|^2. \tag{14.28}$$

Thus in order to determine the angular and the energy dependence of the differential cross section we need to solve the integral equation (14.20), for example by iteration, and then substitute for $\psi(\mathbf{r})$ in (14.22) to obtain the scattering amplitude $f(\theta,\phi)$ and consequently $\left(\frac{d\sigma}{d\Omega}\right)$.

When the potential V is a function of the radial coordinate r only, then we can find the contribution of different partial waves to the scattering amplitude. The decomposition of (14.20) in partial waves can be achieved in the following way:
Let us write $e^{i\mathbf{k}\cdot\mathbf{r}}$ and $\psi(\mathbf{r})$ as

$$e^{i\mathbf{k}\cdot\mathbf{r}} = e^{ikz} = \sum_{\ell=0}^{\infty} i^\ell(2\ell+1)j_\ell(kr)P_\ell(\cos\theta), \tag{14.29}$$

and

$$\psi_{\mathbf{k}}(\mathbf{r}) = \sum_{\ell=0}^{\infty} i^\ell(2\ell+1)\left(\frac{e^{i\delta_\ell}u_\ell(k,r)}{r}\right)P_\ell(\cos\theta), \tag{14.30}$$

where $u_\ell(k,r)$ is a real function. Let us expand $G_+(\mathbf{r}-\mathbf{r}')$ in terms of $P_\ell(\cos\Theta)$ where Θ is the angle between $\mathbf{r}$ and $\mathbf{r}'$,

$$\begin{aligned} G_+(\mathbf{r}-\mathbf{r}') &= -\left(\frac{2m}{\hbar^2}\right)\frac{1}{4\pi}\frac{\exp\left[ik\left|\mathbf{r}-\mathbf{r}'\right|\right]}{|\mathbf{r}-\mathbf{r}'|} \\ &= -\left(\frac{2m}{\hbar^2}\right)\frac{ik}{4\pi}\sum_{\ell}^{\infty}(2\ell+1)P_\ell(\cos\Theta)g_\ell^+(r,r'). \end{aligned} \tag{14.31}$$

Here

$$g_\ell^+(r,r') = \begin{cases} rr'j_\ell(kr')\,h_\ell(kr) & r>r' \\ rr'h_\ell(kr')\,j_\ell(kr) & r<r' \end{cases} \tag{14.32}$$

where $h_\ell(kr)$ is the Hankel function of the first kind [4]

$$h_\ell(kr) = j_\ell(kr) + in_\ell(kr) \to \frac{1}{kr}i^{-\ell-1}e^{ikr} \quad \text{as} \quad r\to\infty, \tag{14.33}$$

When we substitute $G_+(\mathbf{r}-\mathbf{r}')$ and $\psi(\mathbf{r}')$ in the right-hand side of Eq. (14.20) we encounter the following integral over the angular variables

$$J = \int\int P_{\ell'}(\cos\Theta)P_\ell(\cos\theta')\sin\theta' d\theta' d\phi'. \tag{14.34}$$

This integral can be found if we note that

$$\cos(\Theta) = \frac{\mathbf{r}\cdot\mathbf{r}'}{rr'} = \cos\theta\cos\theta' + \sin\theta\sin\theta'\cos(\phi-\phi'), \tag{14.35}$$

and therefore $P_{\ell'}(\cos\Theta)$ can be written as

$$\begin{aligned} P_{\ell'}(\cos\Theta) &= P_{\ell'}\left(\cos\theta\cos\theta' + \sin\theta\sin\theta'\cos(\phi-\phi')\right) \\ &= \frac{4\pi}{(2\ell'+1)}\sum_{m=-\ell'}^{\ell'} Y^*_{\ell',m}(\theta',\phi')\,Y_{\ell',m}(\theta,\phi), \end{aligned} \tag{14.36}$$

where

$$Y_{\ell',0}(\theta',\phi') = \sqrt{\frac{2\ell'+1}{4\pi}}P_{\ell'}(\cos\theta'). \tag{14.37}$$

We can simplify J in (14.34) using the orthogonality of the spherical harmonics $Y_{\ell,m}(\theta,\phi)$:

$$J = \frac{4\pi}{(2\ell'+1)}P_{\ell'}(\cos\theta)\delta_{\ell',\ell}. \tag{14.38}$$

Thus by substituting (14.29), (14.30) and (14.31) in (14.20) and performing the angular integration we obtain the following integral equation for the partial wave $u_\ell(k,r)$:

$$e^{i\delta_\ell}u_\ell(k,r) = rj_\ell(kr) - ike^{i\delta_\ell}\int_0^\infty g_\ell^+(r,r')\,v(r')\,u_\ell(k,r')\,dr', \tag{14.39}$$

where

$$v(r) = \frac{2m}{\hbar^2}V(r), \tag{14.40}$$

is the potential with the dimension of length^{-2}. The asymptotic form of (14.39) as $r\to\infty$ gives us two equations, one for the real part and the other for the imaginary part of (14.39)

$$\begin{aligned} u_\ell(k,r) \;&\to\; r\cos\delta_\ell\, j_\ell(kr) + krn_\ell(kr)\int_0^r j_\ell(kr')\,v(r')\,u_\ell(k,r')\,r'dr' \\ &+\; krj_\ell(kr)\int_r^\infty n_\ell(kr')\,v(r')\,u_\ell(k,r')\,r'dr', \end{aligned} \tag{14.41}$$

and

$$\sin\delta_\ell = -k\int_0^\infty j_\ell(kr')\,v(r')\,u_\ell(k,r')\,r'dr'. \tag{14.42}$$

From these equations, for a given ℓ, the partial wave phase shift, δ_ℓ, and the wave function $u_\ell(k,r)$ can be determined. We also note that when $V(\mathbf{r})$ is only a function of the radial distance r, $f(\theta,\phi)$ does not depend on ϕ.

The scattering amplitude $f(\theta)$, for central potentials, can also be expressed in terms of the phase shift δ_ℓ. Thus if we substitute the asymptotic form of

$e^{i\delta_\ell}u_\ell(k,r)$ from (14.39) and also expand $e^{i\mathbf{k}\cdot\mathbf{r}}$ as a sum over partial waves we obtain

$$\begin{aligned} f(\theta) &= -\sum_{\ell=0}^{\infty}(2\ell+1)P_\ell(\cos\theta)e^{i\delta_\ell}\int_0^\infty j_\ell\left(kr'\right)v\left(r'\right)u_\ell\left(k,r'\right)r'dr' \\ &= \frac{1}{k}\sum_{\ell=0}^{\infty}(2\ell+1)P_\ell(\cos\theta)e^{i\delta_\ell}\sin\delta_\ell. \end{aligned} \quad (14.43)$$

For $\theta=0$, since $P_\ell(1)=1$, $f(\theta=0)$, i.e. the forward scattering amplitude, simplifies

$$f(\theta=0)=\frac{1}{k}\sum_{\ell=0}^{\infty}(2\ell+1)e^{i\delta_\ell}\sin\delta_\ell. \quad (14.44)$$

Now if we calculate the total cross section σ_t, we find

$$\sigma_t=\int_0^\pi|f(\theta)|^2d\Omega=\frac{4\pi}{k^2}\sum_{\ell}^{\infty}(2\ell+1)\sin^2\delta_\ell. \quad (14.45)$$

By comparing (14.44) and (14.45) we obtain

$$\sigma_t=\frac{4\pi}{k}\mathrm{Im}f(0). \quad (14.46)$$

This important result which is called the optical theorem is a direct consequence of the conservation of probability [1]. Using the asymptotic form of $j_\ell(kr)$ and $n_\ell(kr)$ [4]

$$j_\ell(kr)\to\frac{1}{kr}\sin\left(kr-\frac{\ell\pi}{2}\right)\quad \text{as}\quad r\to\infty, \quad (14.47)$$

$$n_\ell(kr)\to-\frac{1}{kr}\cos\left(kr-\frac{\ell\pi}{2}\right)\quad \text{as}\quad r\to\infty, \quad (14.48)$$

in Eq. (14.41), we obtain the asymptotic form of $u_\ell(k,r)$ to be

$$u_\ell(k,r)\to e^{-i\delta_\ell}\sin\left(kr-\frac{\ell\pi}{2}+\delta_\ell\right)\quad \text{as}\quad r\to\infty. \quad (14.49)$$

We can also write (14.49) as

$$u_\ell(k,r)\to\frac{\exp\left(-\frac{i\ell}{2}\right)}{2i}\left[e^{ikr}-S_\ell(k)e^{-ikr}\right]\quad \text{as}\quad r\to\infty, \quad (14.50)$$

where, S_ℓ, the scattering matrix for the ℓ-th partial wave is defined by

$$S_\ell(k)=(-1)^\ell e^{-2i\delta_\ell}. \quad (14.51)$$

This relation shows that

$$S_\ell(-k)=S_\ell^{-1}(k), \quad (14.52)$$

or

$$\exp\left[2i\delta_\ell(-k)\right] = \exp\left[-2i\delta_\ell(k)\right], \tag{14.53}$$

from which it follows that

$$\delta_\ell(-k) = -\delta_\ell(k), \tag{14.54}$$

i.e. $\delta_\ell(k)$ is an odd function of k. Thus the partial wave scattering amplitude $f_\ell(k)$ can be written as

$$f_\ell(k) = |f_\ell(k)|e^{i\delta_\ell(k)}, \tag{14.55}$$

As an example of the solution of the wave equation for scattering let us consider the S-wave scattering of the two particles when the potential between them is given by the Eckart potential

$$V(r) = \frac{\hbar^2}{2m}v(r) = -\left(\frac{\hbar^2}{2m}\right)\frac{s\lambda\gamma^2 e^{-\gamma r}}{(1+\lambda e^{-\gamma r})^2}, \tag{14.56}$$

where s, λ and γ are constants. If we choose $s = 1$ then the wave function $u_0(r)$ satisfies the Schrödinger equation

$$u_0''(r) + k^2 u_0(r) = -\frac{2\lambda\gamma^2 e^{-\gamma r}}{(1+\lambda e^{-\gamma r})^2}\, u_0(r). \tag{14.57}$$

The general solution of this differential equation is given by

$$\begin{aligned} u_0(r) &= C\left\{\sin kr\left[1+\frac{\gamma^2}{\gamma^2+4k^2}\right]y(r) + \frac{2k\gamma}{\gamma^2+4k^2}y(r)\cos kr\right\} \\ &- D\left\{\cos kr\left[1+\frac{\gamma^2}{\gamma^2+4k^2}\right]y(r) - \frac{2k\gamma}{\gamma^2+4k^2}y(r)\sin kr\right\}, \end{aligned} \tag{14.58}$$

where

$$y(r) = -\frac{2\lambda e^{-\gamma r}}{(1+\lambda e^{-\lambda r})} \tag{14.59}$$

Now by requiring that $u_0(r=0) = 0$, then calculating the asymptotic form of $u_0(r)$ and comparing it with (14.49) we find $\cot\delta_0(k)$;

$$k\cot\delta_0(k) = \frac{(1-\lambda)\gamma}{4\lambda} + \frac{1+\lambda}{\lambda\gamma}k^2. \tag{14.60}$$

Thus the potential (14.56) gives us the effective range formula exactly (see Eq. (14.98)).

14.1 Determinantal Method in Potential Scattering

Let us consider the scattering experiment that we discussed earlier in this chapter and assume that the whole system (incident beam, the target and the scattered wave) is enclosed in a large sphere of radius R [6]–[8]. In the absence of the scattering potential the reduced form of the wave function $u_\ell^{(0)}(k,r)$ for the ℓ-th partial wave becomes

$$u_\ell^{(0)}(k,r) \sim \sin\left(k^{(0)}(\ell)r - \frac{\ell\pi}{2}\right), \tag{14.61}$$

where the (discrete) energy levels of the particle $E_n^{(0)}$ are related to $k_n^{(0)}(\ell)$ by the familiar relation

$$E_n^{(0)} = \frac{\hbar^2 k_n^{(0)\,2}(\ell)}{2m}. \tag{14.62}$$

We can determine $k_n^{(0)}(\ell)$ by imposing the boundary condition $u_\ell^{(0)}(r=R)=0$, to find

$$k_n^{(0)}(\ell)R - \frac{\pi\ell}{2} = n\pi. \tag{14.63}$$

In the presence of the potential $V(r)$ we write the asymptotic form of the wave function as in Eq. (14.49)

$$u_\ell(r) \sim e^{-i\delta_\ell}\sin\left(k_n(\ell)r - \frac{\ell\pi}{2} + \delta_\ell(E)\right), \tag{14.64}$$

where now $k_n(\ell)$ is found from $u_n(r=R)=0$ on $u_\ell(k,r)$, i.e.

$$k_n(\ell)R - \frac{\pi\ell}{2} + \delta_\ell(E) = n\pi, \tag{14.65}$$

with the corresponding energy $E_n = \frac{\hbar^2 k_n^2(\ell)}{2m}$. Thus the energy shift caused by the presence of the potential $V(r)$ is

$$\begin{aligned}\Delta E_n(\ell) &= E_n(\ell) - E_n^{(0)}(\ell) \approx \frac{dE_n(\ell)}{dk}\left(k_n(\ell) - k_n^{(0)}(\ell)\right)\\ &= -\frac{dE_n(\ell)}{dk}\frac{\delta_\ell(E)}{R}.\end{aligned} \tag{14.66}$$

The last term is found from Eqs. (14.63) and (14.65). Next we find $\frac{dE(\ell)}{dk}$ by noting that $dE(\ell)$ is the level spacing and is given by

$$\begin{aligned}dE(\ell) &= E_n^{(0)}(\ell) - E_{n-1}^{(0)}(\ell) = \frac{dE(\ell)}{dk}\left(k_n^{(0)}(\ell) - k_{n-1}^{(0)}(\ell)\right)\\ &= \frac{dE(\ell)}{dk}\left(\frac{\pi}{R}\right).\end{aligned} \tag{14.67}$$

By substituting for $\frac{dE(\ell)}{dk}$ from (14.67) in (14.66) we obtain the relation between the energy shift $\Delta E_n(\ell)$ and the phase shift $\delta_\ell(E)$:

$$\Delta E_n(\ell) = -\frac{1}{\pi}\delta_\ell(E)dE(\ell). \tag{14.68}$$

Thus as $R \to \infty$, $dE \to 0$ and the discrete spectrum approaches the continuum. For the rest of the present discussion we suppress the ℓ-dependence of the eigenvalues and write E_n for $E_n(\ell)$.

When we have a finite number of eigenvalues, we can find $E_n^{(0)}$ and E_n from the secular equations

$$\left|H_0 - E_n^{(0)}\right| = 0, \tag{14.69}$$

and

$$|H - E_n| = 0, \tag{14.70}$$

respectively. However for the problem that we are considering i.e. when there are infinite eigenvalues, the secular equations (14.69) and (14.70) do not exist. Rather than formulating the problem in terms of Eqs. (14.69) and (14.70) we consider the function $D(E)$ which is given by

$$\begin{aligned} D(E) &= \left|\frac{E-H}{E-H_0}\right| = \prod_k \left(\frac{E-E_k}{E-E_k^{(0)}}\right) \\ &= \prod_k \left(1 - \frac{\Delta E_k}{E-E_k^{(0)}}\right). \end{aligned} \tag{14.71}$$

This equation shows that if $D(E)$ is determined from the potential and the known eigenvalues $E_n^{(0)} = \frac{\hbar^2 k_n^{(0)2}}{2m}$ then the energies ΔE_k can be obtained. From these partial wave ΔE_k s, the phase shifts $\delta_\ell(E)$, the scattering amplitude $f(\theta)$ or the cross section σ_t can be calculated. To this end we write $D(E)$ as

$$D(E) = \left|1 - \frac{V}{E-H_0}\right| = |1 - G_0(E)V|, \tag{14.72}$$

where

$$G_0(E) = \frac{1}{E-H_0}. \tag{14.73}$$

Since we have assumed that V is a function of r only, we can consider the scattering for a fixed partial wave ℓ. Then the Hamiltonians H_0 and H are two operators both depending on r and ℓ;

$$H_0 = \frac{p_r^2}{2m} + \frac{\ell(\ell+1)}{2mr^2} = -\frac{\hbar^2}{2m}\frac{1}{r^2}\frac{d}{dr}\left(r^2\frac{d}{dr}\right) + \frac{\ell(\ell+1)}{2mr^2}, \tag{14.74}$$

and

$$H = H_0 + V(r). \tag{14.75}$$

In the following we write the ℓ-dependence explicitly when we are considering the phase shifts and the partial wave scattering amplitudes.

The function $D(E)$ has simple poles at the energies $E = E_k^{(0)}$, Eq. (14.71), and is equal to unity for $V = 0$, therefore we can write it in the form

$$D(E) = 1 + \sum_{E_j^{(0)}} \frac{r(j)}{E - E_j^{(0)}}. \tag{14.76}$$

By expanding $D(E)$ in powers of V, we find $r(k)$;

$$\begin{aligned} r(k) &= -\left\langle E_k^{(0)}|V|E_k^{(0)}\right\rangle \\ &+ \sum_{n=1}^{\infty} \frac{(-1)^{n+1}}{n!} \sum_{E_1^{(0)},\cdots E_n^{(0)}} \prod_{j=1}^{n} \left(\frac{1}{E_k^{(0)} - E_j^{(0)}}\right) \\ &\times \begin{vmatrix} \left\langle E_k^{(0)}|V|E_k^{(0)}\right\rangle & \left\langle E_k^{(0)}|V|E_1^{(0)}\right\rangle & \cdots & \left\langle E_k^{(0)}|V|E_n^{(0)}\right\rangle \\ \left\langle E_1^{(0)}|V|E_k^{(0)}\right\rangle & \left\langle E_1^{(0)}|V|E_1^{(0)}\right\rangle & \cdots & \left\langle E_1^{(0)}|V|E_n^{(0)}\right\rangle \\ \vdots & \vdots & \vdots & \vdots \\ \left\langle E_n^{(0)}|V|E_k^{(0)}\right\rangle & \left\langle E_n^{(0)}|V|E_1^{(0)}\right\rangle & \cdots & \left\langle E_n^{(0)}|V|E_n^{(0)}\right\rangle \end{vmatrix}. \end{aligned} \tag{14.77}$$

This series converges for all strengths of the potential $V(r)$ provided the zeroth and the first moment of the potential are finite (see Eqs. (14.4) and (14.5)). As can be seen from (14.71) at $E = E_k$, $D(E_k) = 0$, therefore

$$1 + \sum_{E_j^{(0)}} \frac{r(j)}{E_k - E_j^{(0)}} = 0. \tag{14.78}$$

Now as $dE \to 0$, the denominator in (14.76) becomes very small and we need to examine the behavior of $D(E)$ in this limit. For this we separate from the summation over $E_j^{(0)}$ the contribution from the levels $E_j^{(0)}$ lying within a range Δ of $E_k^{(0)}$. We choose Δ small enough so that $r\left(E_j^{(0)}\right)$ remains essentially constant in a range of 2Δ about $E_k^{(0)}$. Thus we have

$$\begin{aligned} \sum_{E_j^{(0)}} \frac{r\left(E_j^{(0)}\right) dE}{E_k - E_j^{(0)}} &\approx r\left(E_k^{(0)}\right) \sum_{|E_j^{(0)} - E_k^{(0)}| < \Delta} \frac{dE}{E_k - E_j^{(0)}} \\ &+ \sum_{|E_j^{(0)} - E_k^{(0)}| > \Delta} \frac{r\left(E_j^{(0)}\right) dE}{E_k - E_j^{(0)}}. \end{aligned} \tag{14.79}$$

Now within the range Δ of $E_k^{(0)}$ there are $2N$ levels, where $N = \frac{\Delta}{dE}$. As $dE \to 0$, $N \to \infty$ so that $\Delta = NdE$ remains finite. In this limit Eq. (14.79) can be written as

$$\sum_{E_j^{(0)}} \frac{r\left(E_j^{(0)}\right) dE}{E_k - E_j^{(0)}} \approx r\left(E_k^{(0)}\right) \lim_{N\to\infty} \lim_{dE\to 0} \sum_{n=-N}^{n=N} \frac{dE}{E_k - \left(E_k^{(0)} + n\, dE\right)} + \left[\int_0^{E_k^{(0)}-\Delta} + \int_{E_k^{(0)}+\Delta}^{\infty}\right] \frac{r\left(E'\right) dE'}{E_k - E'}, \tag{14.80}$$

where in the first term we have replaced $E_j^{(0)}$ by $E_k^{(0)} + ndE$ and the summation over $E_j^{(0)}$ by a summation over n. But in the second term since the relative change of the summand is of the order $\frac{1}{N}$, we have replaced the summation by integration. We also note that

$$\lim_{dE\to 0} \frac{E_k - E_k^{(0)}}{dE} = -\frac{1}{\pi}\delta_\ell(E). \tag{14.81}$$

Using this expression the summation over n in (14.80) can be carried out

$$\sum_{n=-\infty}^{\infty} \left[-\frac{1}{\pi}\delta_\ell(E_k) - n\right]^{-1} = -\pi \sum_{n=-\infty}^{\infty} \frac{1}{n\pi + \delta_\ell(E_k)} = -\pi \cot\delta_\ell(E_k). \tag{14.82}$$

In deriving this last relation we have used a result of the residue theorem, viz, [9]

$$\sum_{n=-\infty}^{\infty} g(n) = -\left[\sum_{\text{residues}} \pi\cot(\pi z) g(z)\right]_{\text{at the poles of } g(z)} \tag{14.83}$$

Therefore in the limit of $\Delta \to 0$, we have

$$\sum_{E_j^{(0)}} \frac{r(j)}{E_k - E_j^{(0)}} = -\pi r_\ell(E_k)\cot\delta_\ell(E_k) + \mathcal{P}\int_0^\infty \frac{r_\ell\left(E'\right)}{E_k - E'} dE', \tag{14.84}$$

where we have written the ℓ-dependence of $r(j)$ explicitly. Thus Eq. (14.78) in this limit takes the form

$$\pi r_\ell(E_k)\cot\delta_\ell(E_k) = 1 + \mathcal{P}\int_0^\infty \frac{r_\ell\left(E'\right)}{E_k - E'} dE'. \tag{14.85}$$

This relation together with (14.77) form the basic equations of scattering theory. Defining the wavenumber by $k = \sqrt{\frac{2mE_k}{\hbar}}$, we can write the partial wave scattering amplitude as

$$k f_\ell(E_k) = e^{i\delta_\ell(E_k)} \sin\delta_\ell(E_k). \tag{14.86}$$

This amplitude, $f_\ell(E_k)$, can be obtained from (14.85);

$$kf_\ell(E_k) = \frac{1}{\cot\delta_\ell(E_k) - i} = \frac{r_\ell(E_k)}{r_\ell(E_k)\cot\delta_\ell(E_k) - ir_\ell(E_k)}$$

$$= \pi \lim_{E\to E_k+i\epsilon}\left(\frac{r_\ell(E_k)}{1+\int_0^\infty \frac{r_\ell(E')dE'}{E-E'}}\right) = \lim_{E\to E_k+i\epsilon}\frac{\pi r_\ell(E_k)}{D(E)}. \tag{14.87}$$

From this point we omit the subscript k from E_k and write $E_k = E$. Using Eq. (14.85) we find an exact expression for $\tan\delta_\ell(E)$;

$$\tan\delta_\ell(E) = \frac{\pi r_\ell(E)}{1+\mathcal{P}\int_0^\infty \frac{r_\ell(E')dE'}{E-E'}}. \tag{14.88}$$

To the lowest order in $V(r)$ we keep only the first term of expansion of r_k, Eq. (14.77) and write it as

$$r_\ell(E) = -\frac{m}{\hbar^2 k}\langle E|V(r)|E\rangle, \tag{14.89}$$

or by introducing $v(r) = \frac{2m}{\hbar^2}V(r)$ we write

$$r_\ell(E) = -\langle E|v|E\rangle = \frac{-k}{\pi}\int_0^\infty r^2 j_\ell^2(kr)v(r)dr. \tag{14.90}$$

Substituting (14.89) and (14.90) in (14.88) we obtain

$$\tan\delta_\ell(E) = \frac{-k\int_0^\infty r^2 j_\ell^2(kr)\; v(r)dr}{1-\frac{\mathcal{P}}{\pi}\int_0^\infty \frac{k'dE'}{E-E'}\int_0^\infty j_\ell\left(k'r\right)^2 v(r)r^2dr}, \tag{14.91}$$

where $E = \frac{\hbar^2k^2}{2m}$ and $E' = \frac{\hbar^2k'^2}{2m}$.

For calculating the principal value integral in Eq. (14.91) we start with the following relation [10]:

$$\lim_{\epsilon\to o}\left\{\frac{2}{\pi}\int_0^\infty \frac{J_\mu^2(z)dz}{z^2-(\zeta+i\epsilon)^2}\right\} = J_\mu(\zeta)[iJ_\mu(\zeta) - N_\mu(\zeta)]. \tag{14.92}$$

Separating the real and imaginary parts of Eq. (14.92) and setting $\mu = \ell + \frac{1}{2}$ and $\zeta = kr$ we find

$$\frac{2\mathcal{P}}{\pi}\int_0^\infty \frac{z\; J_{\ell+\frac{1}{2}}^2(z)}{z^2-(kr)^2}dz = -J_{\ell+\frac{1}{2}}(kr)\; N_{\ell+\frac{1}{2}}(kr). \tag{14.93}$$

Now if we write this integral in terms of spherical Bessel functions $j_\ell(kr)$ and $n_\ell(kr)$ we have

$$\frac{\mathcal{P}}{\pi}\int_0^\infty \frac{k'dE'}{E'-E}j_\ell^2\left(k'r\right) = kj_\ell(kr)n_\ell(kr). \tag{14.94}$$

Thus to the first order in $r_\ell(E)$ we have a simple result for the phase shift [11]

$$\tan\delta_\ell(E) = \frac{-k\int_0^\infty r^2 j_\ell^2 v(r)dr}{1-k\int_0^\infty j_\ell(kr)\, n_\ell(kr)v(r)r^2dr}. \tag{14.95}$$

This relation is valid when the potential energy is small compared to the energy of the incident particle.

For low energy scattering we can find the effective range theory by observing that Eq. (14.85) can be written as

$$\pi r_\ell(E)\cot\delta_\ell(E) = 1 - \int_0^\infty \frac{r_\ell(E')}{E'}dE' + E\,\mathcal{P}\int_0^\infty \frac{r_\ell(E')\,dE'}{E'(E-E')}. \tag{14.96}$$

The last integral is nearly independent of energy for low energies, since most of the contribution to this integral comes from higher energies. For $E\to 0$, the behavior of $r_\ell(E)$ is that of a free particle wave function for the angular momentum ℓ. Since

$$j_\ell(kr) \to \frac{(kr)^\ell}{(2\ell+1)!!}, \tag{14.97}$$

as $k\to 0$, from (14.90) it follows that [4]

$$\lim_{k\to 0} r_\ell \to \mathcal{O}\left(k^{2\ell+1}\right), \tag{14.98}$$

and in this limit, i.e. $k\to 0$ the right-hand side of (14.96) can be expanded about $E\approx 0$ with the result that

$$k^{2\ell+1}\cot\delta_\ell(E) = A_\ell + B_\ell E, \tag{14.99}$$

A_ℓ and B_ℓ are energy independent constants. For S wave, $(\ell=0)$, we can write (14.99) as

$$k\cot\delta_0(E) = A_0 + B_0E = -\frac{1}{a} + \frac{1}{2}r_0k^2, \tag{14.100}$$

by replacing A_0 and B_0 by $A_0 = -\frac{1}{a}$ and $B_0 = \frac{m}{\hbar^2}r_0$. The two constants a and r_0 introduced in (14.100) have dimensions of length and are called the scattering length and the effective range respectively. The are important quantities in low energy nucleon-nucleon scattering, where the total cross section found from (14.45), with $\ell=0$ as the only significant term in the sum, becomes

$$\sigma_t = 4\pi a^2. \tag{14.101}$$

This result shows that for small energies of the incident particle, E, the scattering is isotropic and the cross section is independent of E.

14.2 Two Solvable Problems

In this section we will consider two important scattering problems where the cross section can be calculated analytically. The first is the scattering of a particle by a hard sphere, and the second is the Rutherford scattering.

Scattering by a Hard Sphere — An exactly solvable problem is the scattering of a particle from a hard sphere where the potential is very strong and repulsive

$$v(r) = \begin{cases} +\infty & \text{for } r \le a \\ 0 & \text{for } r > a \end{cases}. \tag{14.102}$$

From the integral equation satisfied by the wave function $u_\ell(k,r)$, it follows that this wave function is zero for $r \le a$, and for $r > a$, $u_\ell(k,r)$ is a combination of $rj_\ell(kr)$ and $rn_\ell(kr)$;

$$u_\ell(k,r) = rj_\ell(kr)\cos\delta_\ell(k) - rn_\ell(kr)\sin\delta_\ell(k). \tag{14.103}$$

This wave function must be zero inside as well as at the surface of the hard sphere. Therefore it must satisfy the boundary condition

$$u_\ell(k,a) = 0. \tag{14.104}$$

Thus from (14.103) and (14.104) it follows that

$$\tan\delta_\ell(k) = \frac{j_\ell(ka)}{n_\ell(ka)}, \tag{14.105}$$

and from this expression for $\tan\delta_\ell(k)$ we find the total cross section by substituting for $\sin\delta_\ell(k)$ in σ_t, Eq. (14.45)

$$\sigma_t = \frac{4\pi}{k^2}\sum_{\ell=0}^{\infty}(2\ell+1)\left(\frac{j_\ell^2(ka)}{j_\ell^2(ka)+n_\ell^2(ka)}\right). \tag{14.106}$$

For high energies when $ka \gg 1$, the contributions of the terms with $\ell > ka$ to the sum in (14.106) can be found from the expansion [4]

$$j_\ell(kr) \approx (ka)^{\ell+1}, \quad n_\ell(kr) \approx (ka)^{-\ell}, \quad \ell > ka \tag{14.107}$$

Let us note that terms with $\ell > ka$ correspond to the classical case where the particle pass by the sphere without hitting it ($\hbar\ell > mva$ where v is the speed of the particle). By substituting from (14.107) in (14.106), and observing that the terms with $\ell > ka$ are quite small and thus the sum is effectively is over a finite values ℓ, to a very good approximation we have

$$\sigma_t = \frac{4\pi}{k^2}\sum_{\ell=0}^{[ka]}(2\ell+1)\left(\frac{j_\ell^2(ka)}{j_\ell^2(ka)+n_\ell^2(ka)}\right), \tag{14.108}$$

where $[ka]$ is the integer closest to ka. When $\ell < ka$ we can approximate $j_\ell(kr)$ and $n_\ell(kr)$ by their asymptotic forms for large ka [4]

$$j_\ell(kr) = \sin\left(ka - \frac{\ell\pi}{2}\right), \quad n_\ell(kr) = -\cos\left(ka - \frac{\ell\pi}{2}\right). \tag{14.109}$$

With these approximations for $j_\ell(kr)$ and $j_\ell(kr)$, σ_t takes the simple form of

$$\sigma_t \approx \frac{4\pi}{k^2}\sum_{\ell=0}^{[ka]}(2\ell+1)\sin^2\left(ka - \frac{\ell\pi}{2}\right). \tag{14.110}$$

Now by expanding $\sin^2\left(ka - \frac{\ell\pi}{2}\right)$ we find

$$\sigma_t = \frac{4\pi}{k^2}\left\{\sin^2 ka\sum_{\ell=0}^{[ka]}(2\ell+1) + \cos 2ka\left(\sum_{\ell=\text{odd integers}}^{[ka]}(2\ell+1)\right)\right\}. \tag{14.111}$$

The two sums in (14.111) can be calculated in closed form with the result that

$$\sigma_t = \frac{4\pi}{k^2}\left\{-(ka+1)\sin^2 ka + \frac{1}{2}(ka+1)(ka+2)\right\}. \tag{14.112}$$

In the high energy limit only the term $\frac{1}{2}(ka)^2$ will be important, and in this limit we find

$$\sigma_t = 2\pi a^2. \tag{14.113}$$

At low energies the ratio in (14.108) becomes very small as $k \to 0$ except for $\ell = 0$. That is for S wave scattering from (14.105) we find

$$\tan\delta_0 = \frac{j_0(kr)}{n_0(kr)} = \tan(ka), \tag{14.114}$$

or

$$k\cot\delta_0 = k\cot(ka) = \frac{1}{a} - \frac{a}{3}k^2 + \mathcal{O}\left(k^4\right), \quad \text{as} \quad k \to 0. \tag{14.115}$$

Comparing this relation with (14.100) we have the effective range parameters for scattering by a hard sphere. In this limit the cross section becomes

$$\sigma_t = \frac{4\pi}{k^2}\tan^2(ka) \to 4\pi a^2 \quad \text{as} \quad k \to 0. \tag{14.116}$$

By calculating the terms in in (14.106) numerically we find that the total cross section at $k = 0$ is $4\pi a^2$, Eq. (14.116) and then it drops monotonically to the value of $2\pi a^2$ as $k \to \infty$, Eq. (14.113).

Rutherford Scattering — The scattering of two charged particles by Coulomb field is also exactly solvable. In this case the potential is long range and the conditions (14.4) and (14.5) are not satisfied. For the Coulomb potential the Schrödinger equation is separable in spherical polar coordinates as well as the parabolic coordinates. The latter coordinate system gives us the entire wave function and not the partial waves. Since we have to sum all partial waves to find the scattering amplitude, it is preferable to find the entire wave function and therefore we solve the wave equation in the parabolic coordinates. By choosing the z-axis to be the direction of the incident wave, the system composed of the projectile and the target is cylindrically symmetrical and the wave function is not dependent on the angle ϕ. The transformation from spherical polar to the parabolic coordinate is given by [4]

$$\xi = r + z = r(1 - \cos\theta), \quad \eta = r - z = r(1 + \cos\theta), \quad \phi = \phi. \tag{14.117}$$

In this coordinate system the Schrödinger equation with the Coulomb force

$$\pm \frac{Z_1 Z_2 e^2}{r} = \pm \frac{2 Z_1 Z_2 e^2}{\xi + \eta}, \tag{14.118}$$

is given by the partial differential equation

$$\frac{4}{\xi + \eta}\left[\frac{\partial}{\partial \xi}\left(\xi \frac{\partial \psi}{\partial \xi}\right) + \frac{\partial}{\partial \eta}\left(\eta \frac{\partial \psi}{\partial \eta}\right)\right] + \frac{1}{\xi\eta}\left(\frac{\partial^2 \psi}{\partial \phi^2}\right) \mp \frac{4 m Z_1 Z_2 e^2}{\hbar^2(\xi + \eta)}\psi + \frac{2mE}{\hbar^2}\psi = 0. \tag{14.119}$$

Since ψ does not depend on ϕ, we drop the term $\frac{1}{\xi\eta}\left(\frac{\partial^2 \psi}{\partial \phi^2}\right)$ in (14.119). Writing the solution as a product

$$\psi(\xi, \eta) = f_1(\xi) f_2(\eta), \tag{14.120}$$

and substituting it in (14.119), after separating the variables we obtain two ordinary differential equations

$$\frac{d}{d\xi}\left[\xi \frac{d f_1(\xi)}{d\xi}\right] + \left(\frac{mE}{2\hbar^2}\xi + \beta\right) f_1(\xi) = 0, \tag{14.121}$$

and

$$\frac{d}{d\eta}\left[\eta \frac{d f_2(\eta)}{d\eta}\right] + \left(\frac{mE}{2\hbar^2}\eta \mp \frac{m Z_1 Z_2 e^2}{\hbar^2} - \beta\right) f_2(\eta) = 0. \tag{14.122}$$

In these equations β is the separation constant. We want to solve these questions subject to the boundary conditions appropriate for scattering. First we note that the wave function for the incoming particle is

$$\psi_k(z) \sim e^{ikz} \quad \text{for} \quad -\infty < z < 0, \quad \text{as} \quad r \to \infty, \tag{14.123}$$

where $k = \sqrt{\frac{2mE}{\hbar^2}}$. In parabolic coordinates this condition translates to

$$\psi_k(z) \sim \exp\left[\frac{ik}{2}(\xi - \eta)\right], \quad \text{for all } \xi \text{ as } \eta \to \infty. \tag{14.124}$$

To satisfy these boundary conditions we write $f_1(\xi)$ and $f_2(\eta)$ as

$$f_1(\xi) = e^{\frac{ik\xi}{2}} \quad \text{for all } \xi, \tag{14.125}$$

and

$$f_2(\eta) \sim e^{\frac{-ik\eta}{2}} \quad \text{as} \quad \eta \to \infty. \tag{14.126}$$

Now (14.125) is a solution of (14.121) if we choose $\beta = -\frac{ik}{2}$. To simplify Eq. (14.122) further let us denote $\pm\frac{mZ_1Z_2e^2}{\hbar^2}$ by γ and introduce a new function of η which we denote by $w(\eta)$;

$$f_2(\eta) = e^{\frac{-ik\eta}{2}} w(\eta). \tag{14.127}$$

Then by substituting (14.127) in (14.122) we get the following equation for $w(\eta)$;

$$\eta \frac{d^2 w(\eta)}{d\,\eta^2} + (1 - ik\eta)\frac{dw(\eta)}{d\eta} - \gamma w(\eta) = 0. \tag{14.128}$$

The solution of this equation is the confluent hypergeometric function;

$$w(\eta) = {}_1F_1\left(-\frac{i\gamma}{k}, 1;\ ik\eta\right). \tag{14.129}$$

With this $w(\eta)$ we obtain the total wave function $\psi_k(\xi,\eta)$ to be

$$\psi_k(\xi,\eta) = {}_1F_1\left(-\frac{i\gamma}{k}, 1;\ ik\eta\right)\exp\left(\frac{ik(\xi-\eta)}{2}\right). \tag{14.130}$$

Since we are interested in the asymptotic form of $\psi_k(\xi,\eta)$ as $\eta \to \infty$, we find the asymptotic expression for ${}_1F_1\,(a,c;z)$ for large pure imaginary z;

$${}_1F_1\,(a,c;z) \to \frac{\Gamma(c)}{\Gamma(c-a)}(-z)^{-a} + \frac{\Gamma(c)}{\Gamma(a)}e^z z^{a-c}. \tag{14.131}$$

Using this expansion, we find the wave function for large η to be

$$\psi_k(\xi,\eta) \to \left[\frac{1}{\Gamma\left(1+\frac{i\gamma}{k}\right)}(-ik\eta)^{\frac{i\gamma}{k}} + \frac{1}{\Gamma\left(-\frac{i\gamma}{k}\right)}e^{ik\eta}(ik\eta)^{-\frac{i\gamma}{k}-1}\right]e^{i\frac{k}{2}(\xi-\eta)}. \tag{14.132}$$

With the help of the relation

$$\Gamma\left(1-\frac{i\gamma}{k}\right) = \frac{i\gamma}{k}\Gamma\left(-\frac{i\gamma}{k}\right). \tag{14.133}$$

we can write (14.132) as

$$\begin{aligned}\psi_k(\xi,\eta) \quad &\to \quad \frac{1}{\Gamma\left(1+\frac{i\gamma}{k}\right)}\exp\left\{i\left(k\frac{\xi-\eta}{2} + \frac{\gamma}{k}\ln(k\eta) - \frac{i\gamma\pi}{2k}\right)\right\} \\ &- \quad \frac{1}{\Gamma\left(1-\frac{i\gamma}{k}\right)}\left(\frac{\gamma}{k^2\eta}\right)\exp\left\{i\left(k\frac{\xi+\eta}{2} - \frac{\gamma}{k}\ln(k\eta) - \frac{i\gamma\pi}{2k}\right)\right\}\end{aligned} \tag{14.134}$$

Now we go back to the (r, z) coordinates and we write $\psi_k(r, z)$ as

$$\psi_k(\xi,\eta) \simeq \exp\left[i\left(kz + \frac{\gamma}{k}\ln[k(r-z)]\right)\right] - \frac{\Gamma\left[1+\frac{i\gamma}{k}\right]}{\Gamma\left[1-\frac{i\gamma}{k}\right]}\left(\frac{\gamma}{k^2}\right)\frac{\exp\left[ikr - \frac{i\gamma}{k}\ln[k(r-z)]\right]}{r-z}. \quad (14.135)$$

This expression for the wave function shows the incident wave which is the first term in (14.135) is distorted by a logarithmic term even at large distances. We have a similar distortion for the scattered wave which is the second term in (14.135). Writing the wave function in spherical polar coordinates with $z = r\cos\theta$ we obtain

$$\psi_k(r,\theta) \simeq \exp\left[i\left(kz + \frac{\gamma}{k}\ln[kr(1-\cos\theta)]\right)\right] + f(k,\theta)\frac{\exp\left[ikr - \frac{i\gamma}{k}\ln(kr)\right]}{r}, \quad (14.136)$$

where the scattering amplitude is

$$f(k,\theta) = -\coprod\frac{\Gamma\left[1+\frac{i\gamma}{k}\right]}{\Gamma\left[1-\frac{i\gamma}{k}\right]}\left(\frac{\gamma}{k^2}\right)\frac{\exp\left[-\frac{i\gamma}{k}\ln[k(1-\cos)]\right]}{1-\cos\theta}. \quad (14.137)$$

We can express the total scattering amplitude as a sum of partial wave amplitudes by expanding (14.137) in terms of $P_\ell(\cos\theta)$

$$f(k,\theta) = \sum_{\ell=0}^{\infty}(2\ell+1)\frac{1}{2ik}\left[\frac{\Gamma\left(\ell+1+\frac{i\gamma}{k}\right)}{\Gamma\left(\ell+1-\frac{i\gamma}{k}\right)} - 1\right]P_\ell(\cos\theta). \quad (14.138)$$

The important result found from $f(k,\theta)$ is that the exact differential cross section found for Coulomb (or Rutherford) scattering is the same as the classical cross section, i.e.

$$\frac{d\sigma}{d\Omega} = |f(k,\theta)|^2 = \frac{\gamma^2}{k^4\sin^4\left(\frac{\theta}{2}\right)} = \frac{\left(Z_1Z_2e^2\right)^2}{16E^2\sin^4\left(\frac{\theta}{2}\right)}. \quad (14.139)$$

14.3 Time-Dependent Scattering Theory

A different mathematical formulation of scattering theory which is closer to the Heisenberg idea of the time development of the motion of a particle is to consider the evolution of the state of the projectile in the course of time and its interaction with the target. In this formulation the incoming particle is characterized by its momentum $\mathbf{k}_{in}$ long before the scattering takes place, i.e. at $t \to -\infty$. After its interaction with the target, its asymptotic momentum is changed to $\mathbf{k}_{out}$. For elastic scattering the energy of the particle is conserved and $\mathbf{k}_{in}^2 = \mathbf{k}_{out}^2 = k^2$. Thus the projectile enters the interaction region sometime before $t = 0$ and leaves it sometime after $t = 0$. Working in momentum representation, we denote the time-independent basis set by $|\phi_{\mathbf{k}}\rangle$ where

$$H_0|\phi_{\mathbf{k}}\rangle = E(\mathbf{k})|\phi_{\mathbf{k}}\rangle, \quad E(\mathbf{k}) = \frac{\hbar^2\mathbf{k}^2}{2m}. \tag{14.140}$$

For simplicity in this section we set $\hbar = 2m = 1$. The time evolution of the system composed of the target and the projectile is governed by the Hamiltonian (14.7) and the time-dependent wave equation

$$i\frac{\partial}{\partial t}|\psi(t)\rangle = H|\psi(t)\rangle. \tag{14.141}$$

Since H does not depend explicitly on time the formal solution of (14.141) is given by

$$|\psi(t)\rangle = e^{-iHt}|\psi(0)\rangle. \tag{14.142}$$

Similarly if $|\phi(t)\rangle$ is a solution of the equation

$$i\frac{\partial}{\partial t}|\phi(t)\rangle = H_0|\phi(t)\rangle, \tag{14.143}$$

then

$$|\phi(t)\rangle = e^{-iH_0t}|\phi(0)\rangle. \tag{14.144}$$

Next we observe that in the limit of $t \to -\infty$, $H_0 = H$ and therefore we expect that the time development of the operator H_0 and H be the same. Consequently we assume that there exists a solution $|\phi_i(t)\rangle$ of Eq. (14.144) which, in the limit of $t \to -\infty$, equals a solution $|\psi_i(t)\rangle$ of the time-dependent equation (14.142) in the same limit

$$\lim_{t\to-\infty}|\psi_i(t)\rangle = \lim_{t\to-\infty}|\phi_i(t)\rangle. \tag{14.145}$$

Here the equality sign implies that $|\psi_i(t)\rangle$ converges strongly to $|\phi_i(t)\rangle$ in the indicated time limit [12]. Similarly using the same argument for the limit of $t \to +\infty$ we find

$$\lim_{t\to+\infty}|\psi_f(t)\rangle = \lim_{t\to+\infty}|\phi_f(t)\rangle. \tag{14.146}$$

Substituting for $|\psi_i(t)\rangle$ and $|\phi_i(t)\rangle$ from (14.142) and (14.144) we obtain the following result;

$$\lim_{t\to-\infty}\left\{e^{-iHt}|\psi_i(0)\rangle - e^{-iH_0t}|\phi_i(0)\rangle\right\} = 0. \tag{14.147}$$

Also in the limit of $t \to +\infty$ we have

$$\lim_{t\to+\infty}\left\{e^{-iHt}|\psi_f(0)\rangle - e^{-iH_0t}|\phi_f(0)\rangle\right\} = 0. \tag{14.148}$$

Now we multiply these two equations by e^{iHt} from the left and we find $|\psi_i(0)\rangle$ and $|\psi_f(0)\rangle$ in terms of $|\phi_i(0)\rangle$ and $|\phi_f(0)\rangle$;

$$|\psi_i(0)\rangle = \lim_{t\to-\infty} e^{iHt}e^{-iH_0t}|\phi_i(0)\rangle, \tag{14.149}$$

and

$$|\psi_f(0)\rangle = \lim_{t\to-\infty} e^{iHt}e^{-iH_0t}|\phi_f(0)\rangle. \tag{14.150}$$

By introducing the Møller operators $\Omega^\pm$ and these are defined by [13],[14]

$$\Omega^\pm = \lim_{t\to\mp} e^{iHt}e^{-iH_0t}, \tag{14.151}$$

we can write (14.149) and (14.150) as the following equations:

$$|\psi_i(0)\rangle = \Omega^+|\phi_i(0)\rangle, \tag{14.152}$$

and

$$|\psi_f(0)\rangle = \Omega^-|\phi_f(0)\rangle. \tag{14.153}$$

The essential property of the Møller operator which can be described as

$$H\Omega^\pm = \Omega^\pm H_0, \tag{14.154}$$

can be derived in the following way: We first observe that in the limit of $t \to -\infty$ not only Eq. (14.147) is true, but in addition the time derivative of (14.147) in this limit vanishes, i.e.

$$\begin{aligned} &\lim_{t\to-\infty}\left\{i\frac{\partial}{\partial t}\left[e^{-iHt}|\psi_i(0)\rangle - e^{-iH_0t}|\phi_i(0)\rangle\right]\right\} \\ = &\lim_{t\to-\infty}\left[e^{-iHt}H|\psi_i(0)\rangle - e^{-iH_0t}H_0|\phi_i(0)\rangle\right] = 0. \end{aligned} \tag{14.155}$$

Thus by multiplying (14.155) by e^{iHt} from the left we find

$$H|\psi_i(0)\rangle = \lim_{t\to-\infty}\left(e^{iHt}e^{-iH_0t}\right)H_0|\phi_i(0)\rangle. \tag{14.156}$$

The limit of the operator product inside the parenthesis as $t \to -\infty$ is Ω^+ according to Eq. (14.151). Thus

$$H\Omega^+|\phi_i(0)\rangle = \Omega^+H_0|\phi_i(0)\rangle. \tag{14.157}$$

Since (14.157) is true for an arbitrary state $|\phi_i(0)\rangle$, therefore we arrive at the result that

$$H\Omega^+ = \Omega^+H_0. \tag{14.158}$$

In a similar way we can verify that

$$H\Omega^- = \Omega^-H_0. \tag{14.159}$$

Noting that the initial state of the projectile is that of a free particle, we can characterize it by its momentum $\mathbf{k}$. For elastic scattering the magnitude of the momentum $\mathbf{k}$ remains unchanged in the course of scattering. Now from (14.152) and (14.153) we have

$$\left|\psi_{\mathbf{k}}^{\pm}\right\rangle = \Omega^{\pm}\left|\phi_{\mathbf{k}}\right\rangle, \tag{14.160}$$

that is we have used the subscript $\mathbf{k}$ to designate the state of the system. In the following discussion we assume that $|\phi_{\mathbf{k}}\rangle$ and $\left|\psi_{\mathbf{k}}^{\pm}\right\rangle$ are normalizable eigenstates of H_0 and H respectively, as was discussed in the previous section. Once we have reached the final result, we allow R, the radius of the sphere enclosing the system, go to infinity. It should be emphasize that this limiting process in no way will affect the results.

The kets defined by (14.160) are the eigenstates of H since

$$H\left|\psi_{\mathbf{k}}^{\pm}\right\rangle = H\Omega^{\pm}\left|\phi_{\mathbf{k}}\right\rangle = \Omega^{\pm}H_0\left|\phi_{\mathbf{k}}\right\rangle = E(\mathbf{k})\Omega^{\pm}\left|\phi_{\mathbf{k}}\right\rangle = E(\mathbf{k})\left|\psi_{\mathbf{k}}^{\pm}\right\rangle. \tag{14.161}$$

We can picture the particle entering the field of force with momentum $\mathbf{k}_{in}$ and after being scattered leaving the interaction region with momentum $\mathbf{k}_{out}$. Thus in the course of scattering the state of the particle has changed from $\left|\psi_{\mathbf{k}_{in}}^{\pm}\right\rangle$ to $\left|\psi_{\mathbf{k}_{out}}^{\pm}\right\rangle$, the former is the state at $t \to -\infty$ and the latter at $t \to +\infty$, Eqs. (14.152)-(14.153).

14.4 The Scattering Matrix

The scattering operator (or matrix) was first introduced by Heisenberg [15]. In this section we follow the method of Lehman Symanzik and Zimmermann (LSZ) to obtain the scattering matrix. We define the scattering matrix in terms of its matrix elements [17]

$$S_{\mathbf{k}_{out},\mathbf{k}_{in}} = \langle\psi_{\mathbf{k}_{out}}|\psi_{\mathbf{k}_{in}}\rangle. \tag{14.162}$$

This matrix element is the probability amplitude for a particle entering the interaction region with momentum $\mathbf{k}_{in}$ and emerging from this region with momentum $\mathbf{k}_{out}$. If we substitute for $\langle\psi_{\mathbf{k}_{out}}^{-}|$ and $|\psi_{\mathbf{k}_{in}}^{+}\rangle$ from (14.160) we obtain

$$S_{\mathbf{k}_{out},\mathbf{k}_{in}} = \left\langle\phi_{\mathbf{k}_{out}}\left|\left(\Omega^{-}\right)^{\dagger}\Omega^{+}\right|\phi_{\mathbf{k}_{in}}\right\rangle. \tag{14.163}$$

Noting that $|\phi_{\mathbf{k}_{out}}\rangle$ and $|\phi_{\mathbf{k}_{in}}\rangle$ are arbitrary kets belonging to a complete set of eigenvectors of H_0, we can define the S matrix by

$$S = \left(\Omega^{-}\right)^{\dagger}\Omega^{+}. \tag{14.164}$$

It is convenient to denote $\mathbf{k}_{in}$ and $\mathbf{k}_{out}$ states by just *in* and *out* respectively and write the matrix elements of S as

$$S_{out,in} = \langle\phi_{out}|S|\phi_{in}\rangle = \langle\psi_{out}^{-}|\psi_{in}^{+}\rangle. \tag{14.165}$$

In the following discussion we omit the superscript $+$ from Ω^+ and write

$$\Omega(t) \equiv \Omega^+(t) = e^{iHt}e^{-iH_0t}. \tag{14.166}$$

Using this Møller operator, from (14.160), (14.164) and (14.165) we get

$$\begin{aligned} S_{out,in} &= \lim_{t\to-\infty}\left\langle\psi^-_{out}|\Omega(t)|\phi_{in}\right\rangle = \lim_{t\to+\infty}\left\langle\psi^-_{out}|\Omega(t)|\phi_{in}\right\rangle \\ &- \left\{\lim_{t\to+\infty} - \lim_{t\to-\infty}\right\}\left\langle\psi^-_{out}|\Omega(t)|\phi_{in}\right\rangle. \end{aligned} \tag{14.167}$$

The first term in the right-hand side of (14.167) can be reduced to,

$$\lim_{t\to+\infty}\left\langle\psi^-_{out}|\Omega(t)|\phi_{in}\right\rangle = \left\langle\phi_{out}|\phi_{in}\right\rangle = \delta_{out,in}, \tag{14.168}$$

since $|\phi_{\mathbf{k}_{in}}\rangle$ and $|\phi_{\mathbf{k}_{out}}\rangle$ are a set of normalized eigenvalues of H_0. Now let us examine the second term in (14.167), which we will write as

$$\begin{aligned} &\left\{\lim_{t\to+\infty} - \lim_{t\to-\infty}\right\}\left\langle\psi^-_{out}|\Omega(t)|\phi_{in}\right\rangle = \int_{-\infty}^{+\infty}\left\langle\psi^-_{out}\left|\frac{d}{dt}\Omega(t)\right|\phi_{in}\right\rangle dt \\ &= \int_{-\infty}^{+\infty} i\left\langle\psi^-_{out}\left|e^{iHt}Ve^{-iH_0t}\right|\phi_{in}\right\rangle dt = 2\pi i\delta(E_{out}-E_{in})\left\langle\psi^-_{out}|V|\phi_{in}\right\rangle, \end{aligned} \tag{14.169}$$

where we have used the following relations:

$$e^{-iH_0t}|\phi_{in}\rangle = e^{-iE_{in}t}|\phi_{in}\rangle, \tag{14.170}$$

and

$$e^{iHt}|\psi^\pm_{out}\rangle = e^{iE_{out}t}|\psi^\pm_{out}\rangle. \tag{14.171}$$

Combining (14.167), (14.168) and (14.169) we find the matrix elements of $S_{out,in}$ expressed in terms of V;

$$S_{out,in} = \delta_{out,in} - 2\pi i\delta(E_{out}-E_{in})\left\langle\psi^-_{out}|V|\phi_{in}\right\rangle. \tag{14.172}$$

Alternatively we can write (14.167) as

$$S_{out,in} = \lim_{t\to+\infty}\left\langle\phi_{out}\left|\Omega^\dagger(t)\right|\psi^+_{in}\right\rangle, \tag{14.173}$$

and following the same steps as before we obtain

$$S_{out,in} = \delta_{out,in} - 2\pi i\delta(E_{out}-E_{in})\left\langle\phi_{out}|V|\psi^+_{in}\right\rangle. \tag{14.174}$$

Now if we compare (14.172) and (14.174) we find

$$T_{out,in} = \left\langle\psi^-_{out}|V|\phi_{in}\right\rangle = \left\langle\phi_{out}|V|\psi^+_{in}\right\rangle. \tag{14.175}$$

where Eq. (14.175) holds for the on-energy shell elements of the T matrix, i.e. elements of T for which $E_{out} = E_{in}$.

14.5 The Lippmann–Schwinger Equation

A very important method for the direct determination of the scattering amplitude is that of the Lippmann–Schwinger equation [16]. In order to derive this equation from the Møller operator, $\Omega(t)$, we start from Eq. (14.160) and solve it for $|\phi_{in}\rangle$,

$$|\phi_{in}\rangle = \lim_{t\to-\infty} \Omega^{-1}(t)|\psi_{in}^{+}\rangle. \tag{14.176}$$

Next we write (14.176) as

$$|\phi_{in}\rangle = \lim_{t\to-\infty} \Omega^{-1}(t)\theta(-t)|\psi_{in}^{+}\rangle, \tag{14.177}$$

where $\theta(t)$ is the step function. This last relation can be written as

$$\begin{aligned} |\phi_{in}\rangle &= -\left\{\lim_{t\to+\infty} - \lim_{t\to-\infty}\right\}\left(\Omega^{-1}(t)\theta(-t)\left|\psi_{in}^{+}\right\rangle\right) \\ &= -\int_{-\infty}^{+\infty} \frac{d}{dt}\left[\Omega^{-1}(t)\theta(-t)\right]\left|\psi_{in}^{+}\right\rangle dt. \end{aligned} \tag{14.178}$$

Using the relation

$$\frac{d}{dt}\theta(-t) = -\delta(t), \tag{14.179}$$

and the fact that $\theta(t)$ can be written as an integral

$$\theta(t) = \frac{1}{2\pi i}\int_{-\infty}^{\infty} \frac{e^{it\zeta}}{\zeta - i\epsilon}\, d\zeta, \tag{14.180}$$

we can write (14.178) as

$$\begin{aligned} |\phi_{in}\rangle &= -\int_{-\infty}^{+\infty}\left[-\delta(t) + \frac{1}{2\pi i}\int_{-\infty}^{+\infty}\frac{d\zeta}{\zeta - i\epsilon}e^{-i\zeta t}e^{-iH_0 t}(-iV)e^{-iHt}\right]\left|\psi_{in}^{+}\right\rangle dt \\ &= \left|\psi_{in}^{+}\right\rangle + \int_{-\infty}^{\infty}\frac{d\zeta}{\zeta - i\epsilon}\delta(\zeta - H_0 + E_{in})V\left|\psi_{in}^{+}\right\rangle \\ &= \left|\psi_{in}^{+}\right\rangle + \frac{1}{H_0 - E_{in} - i\epsilon}V\left|\psi_{in}^{+}\right\rangle. \end{aligned} \tag{14.181}$$

In deriving (14.181) we have interchanged the order of integration, an operation which is permissible for normalizable kets. We rearrange the terms in (14.181) to get the standard form of the integral equation for $\left|\psi_{in}^{+}\right\rangle$;

$$\left|\psi_{in}^{+}\right\rangle = |\phi_{in}\rangle + \frac{1}{E_{in} - H_0 + i\epsilon}V\left|\psi_{in}^{+}\right\rangle. \tag{14.182}$$

In a similar way we can obtain the equation for $\left|\psi_{in}^{-}\right\rangle$;

$$\left|\psi_{in}^{-}\right\rangle = |\phi_{in}\rangle + \frac{1}{E_{in} - H_0 - i\epsilon}V\left|\psi_{in}^{-}\right\rangle. \tag{14.183}$$

The ket $\left|\psi_{in}^{+}\right\rangle$ which is the solution of the integral equation (14.182) can be formally obtained in terms of $|\phi_{in}\rangle$. As it can be seen from (14.160)

$$\left|\psi_{in}^{+}\right\rangle = \lim_{t\to-\infty}\Omega(t)|\phi_{in}\rangle. \tag{14.184}$$

This equation can also be written as

$$\left|\psi_{in}^{+}\right\rangle = \lim_{t\to-\infty}\Omega(t)\theta(-t)|\phi_{in}\rangle = -\left\{\lim_{t\to+\infty} - \lim_{t\to-\infty}\right\}\Omega(t)\theta(-t)|\phi_{in}\rangle. \tag{14.185}$$

Again the right-hand side of (14.185) can be expressed as an integral

$$\left|\psi_{in}^{+}\right\rangle = -\int_{-\infty}^{+\infty}\frac{d}{dt}\left[\Omega(t)\theta(-t)\right]|\phi_{in}\rangle dt. \tag{14.186}$$

By expanding the integrand in (14.186) and substituting for $\frac{d\theta(-t)}{dt}$ and $\theta(-t)$ from (14.179) and (14.180) we get

$$\left|\psi_{in}^{+}\right\rangle = |\phi_{in}\rangle + \frac{1}{E_{in} - H + i\epsilon}V\left|\phi_{in}\right\rangle. \tag{14.187}$$

A similar equation, viz,

$$\left|\psi_{in}^{-}\right\rangle = |\phi_{in}\rangle + \frac{1}{E_{in} - H - i\epsilon}V\left|\phi_{in}^{+}\right\rangle, \tag{14.188}$$

can be derived for $\left|\psi_{in}^{-}\right\rangle$ [17],[18].

Integral Equation for the Transition Matrix — An important equation for calculating the scattering amplitude from the Fourier transform of the potential, is the transition matrix (or T matrix). The on-shell elements of the T-matrix, i.e. those elements subject to the condition $E_{out} = E_{in}$ are defined by Eq. (14.175). But for solving the integral equation for the T-matrix we need on-shell as well as off-shell matrix elements of T. To show this we write (14.182) as

$$\left|\psi_{\mathbf{k}_{in}}^{+}\right\rangle = |\phi_{\mathbf{k}_{in}}\rangle + \frac{1}{L^3}\sum_{\mathbf{q}}|\phi_{\mathbf{q}}\rangle\frac{1}{E_{in} - \frac{\hbar^2\mathbf{q}^2}{2m} + i\epsilon}\langle\phi_{\mathbf{q}}|V\left|\psi_{\mathbf{k}_{in}}^{+}\right\rangle, \tag{14.189}$$

where we have used a complete set of eigenvalues of H_0, viz, $|\phi_{\mathbf{q}}\rangle$, noting that the eigenvalue of H for $|\phi_{\mathbf{k}}\rangle$ is given by

$$E(\mathbf{q}) = \frac{\hbar^2\mathbf{q}^2}{2m}, \tag{14.190}$$

and that

$$\frac{1}{L^3}\sum_{\mathbf{q}}|\phi_{\mathbf{q}}\rangle\langle\phi_{\mathbf{q}}|, \tag{14.191}$$

is the unit operator. Now according to (3.233) we can replace the summation by integration over d^3q:

$$\frac{1}{L^3}\sum_{\mathbf{q}} |\phi_{\mathbf{q}}\rangle \langle\phi_{\mathbf{q}}| \to \frac{1}{(2\pi)^3}\int |\phi_{\mathbf{q}}\rangle \langle\phi_{\mathbf{q}}| d^3q, \tag{14.192}$$

is a unit operator. Next we multiply (14.189) from left by $\langle\phi_{\mathbf{k}'}| V$ to get

$$\langle\phi_{\mathbf{k}'}| V |\psi^+_{\mathbf{k}_{in}}\rangle = \langle\phi_{\mathbf{k}'}| V |\phi_{\mathbf{k}_{in}}\rangle + \frac{1}{(2\pi)^3}\int \langle\phi_{\mathbf{k}'}| V |\phi_{\mathbf{q}}\rangle \frac{d^3q}{\frac{\hbar^2}{2m}(\mathbf{k}^2_{in}-\mathbf{q}^2)+i\epsilon} \langle\phi_{\mathbf{q}}| V |\psi^+_{\mathbf{k}_{in}}\rangle, \tag{14.193}$$

where $E_{in} = \frac{\hbar^2\mathbf{k}^2_{in}}{2m}$. We can write this equation as

$$T(\mathbf{k}', \mathbf{k}_{in}) = V(\mathbf{k}', \mathbf{k}_{in}) + \frac{1}{(2\pi)^3}\int V(\mathbf{k}', \mathbf{q}) \frac{d^3q}{\frac{\hbar^2}{2m}(\mathbf{k}^2_{in}-\mathbf{q}^2)+i\epsilon} T(\mathbf{q}, \mathbf{k}_{in}). \tag{14.194}$$

In (14.194) the matrix elements of the potential, $V(\mathbf{k}', \mathbf{q})$, are the Fourier transforms of the potential $V(\mathbf{r})$;

$$V(\mathbf{k}', \mathbf{q}) = \int e^{-i(\mathbf{k}'-\mathbf{q})\cdot\mathbf{r}}\, V(\mathbf{r}) d^3r \tag{14.195}$$

Thus if $V(\mathbf{r})$ is known, whether it is a central potential or not, we find $V(\mathbf{k}', \mathbf{q})$ from (14.195) and substitute the result in (14.194) and solve the integral equation for $T(\mathbf{q}, \mathbf{k}_{in})$. Once the elements of this matrix are determined we set $q = k_{out}$ to find the on-shell elements of the transition matrix.

If we multiply (14.189) with $\langle\mathbf{r}|$ from left and note that

$$\langle\mathbf{r}|\psi^+_{\mathbf{k}_{in}}\rangle = \psi^+_{\mathbf{k}_{in}}(\mathbf{r}), \tag{14.196}$$

and

$$\langle\mathbf{r}|\phi_{\mathbf{k}_{in}}\rangle = e^{i\mathbf{k}\cdot\mathbf{r}}, \tag{14.197}$$

then we get an integral equation for $\psi_{\mathbf{k}_{in}}(\mathbf{r})$

$$\psi^+_{\mathbf{k}_{in}}(\mathbf{r}) = e^{i\mathbf{k}\cdot\mathbf{r}} + \int V(\mathbf{r}')\,\psi^+_{\mathbf{k}_{in}}(\mathbf{r}')\, d^3r' \int \frac{d^3q}{(2\pi)^3}\left[\frac{e^{-i\mathbf{q}\cdot(\mathbf{r}-\mathbf{r}')}}{\frac{\hbar^2}{2m}(\mathbf{k}^2_{in}-\mathbf{q}^2)+i\epsilon}\right]. \tag{14.198}$$

The $\mathbf{q}$ integration in (14.198) gives us the Green function $\langle\mathbf{r}'|G_+|\mathbf{r}''\rangle$, Eq. (14.19), so (14.198) can be written as

$$\psi^+_{\mathbf{k}_{in}}(\mathbf{r}) = e^{i\mathbf{k}\cdot\mathbf{r}} - \int \frac{m}{2\pi\hbar^2|\mathbf{r}-\mathbf{r}'|} e^{ik_{in}|\mathbf{r}-\mathbf{r}'|}\, V(\mathbf{r}')\psi^+_{\mathbf{k}_{in}}(\mathbf{r}')d^3r'. \tag{14.199}$$

For $r \to \infty$ the asymptotic form of $\psi^+_{\mathbf{k}_{in}}(\mathbf{r})$ is

$$\psi^+_{\mathbf{k}_{in}}(\mathbf{r})_{r\to\infty} \to e^{i\mathbf{k}\cdot\mathbf{r}} - \frac{m}{2\pi\hbar^2 r}\int e^{ik_{in}r}e^{i\mathbf{k}_{out}\cdot\mathbf{r}'}\, V(\mathbf{r}')\psi^+_{\mathbf{k}_{in}}(\mathbf{r}')d^3r', \tag{14.200}$$

with $\mathbf{k}_{out} = k_{in}\hat{\mathbf{r}}$. Comparing (14.200) with (14.22) we find that $f(\theta,\phi)$ is related to the on-shell matrix elements of $T(\mathbf{k}',\mathbf{k}_{in})$ by

$$f(\theta,\phi) = -\frac{m}{2\pi\hbar^2}T(\mathbf{k}_{out},\mathbf{k}_{in}). \tag{14.201}$$

In this way, by solving the integral equation for the T-matrix we find the total scattering amplitude $f(\theta,\phi)$.

Transition Matrix for Partial Waves — For central forces we can reduce the three-dimensional integral equation for $T(\mathbf{k}',\mathbf{k}_{in})$, Eq.(14.194), or for $\psi_{\mathbf{k}_{in}}(\mathbf{r})$, Eq. (14.200), to a set of one-dimensional linear integral equations each for a given partial wave ℓ. This can be achieved by either expanding $T(\mathbf{k}',\mathbf{k}_{in})$ or expanding $\psi_{\mathbf{k}_{in}}(\mathbf{r})$ in terms of the Legendre polynomial $P_\ell\left(\hat{\mathbf{k}}'\cdot\hat{\mathbf{k}}_{in}\right)$ ($\hat{\mathbf{k}}$ and $\hat{\mathbf{k}}_{in}$ are unit vectors). Here we consider the expansion $\psi_{\mathbf{k}_{in}}(\mathbf{r})$ in terms of $u_\ell(k,r)$, $(k_{in} = k)$ which we have already found, Eq. (14.30). In order to calculate the partial wave T-matrix, $t_\ell(k',k)$, we make use of the integral representation of $g^+_\ell(r,r')$, Eq. (14.32), which is

$$g^+_\ell(r,r') = \frac{2irr'}{\pi k}\int_0^\infty \frac{j_\ell(qr)j_\ell(kr')}{k^2-q^2+i\epsilon}dq. \tag{14.202}$$

By substituting this Green function in Eq. (14.39) we find

$$\begin{aligned} e^{i\delta_\ell(k)}u_\ell(k,r) &= rj_\ell(kr) + \frac{2}{\pi}\int_0^\infty \frac{rj_\ell(qr)}{k^2-q^2+i\epsilon}dq \\ &\times \int_0^\infty r'j_\ell(qr')\,v(r')\,e^{i\delta_\ell(k)}u_\ell(k,r')\,dr', \end{aligned} \tag{14.203}$$

Now we multiply the two sides of (14.203) by

$$rj_\ell(k'r)\,v(r)dr, \tag{14.204}$$

and integrate over r to find

$$t^+_\ell(k',k) = v_\ell(k',k) + \int_0^\infty \frac{v_\ell(k',q)\,t^+_\ell(q,k)}{k^2-q^2+i\epsilon}dq, \tag{14.205}$$

where

$$t^+_\ell(k',k) = \frac{2kk'}{\pi}\int_0^\infty r'j_\ell(kr')\,v(r')\,e^{i\delta_\ell(k)}u_\ell(kr')\,dr', \tag{14.206}$$

and

$$v(k',k) = \frac{2kk'}{\pi}\int_0^\infty r'j_\ell(kr')\,v(r')\,r'j_\ell(kr')\,dr'. \tag{14.207}$$

We can solve the integral equation (14.205) by iteration or by numerical methods, and find the complex quantity $t_\ell^+(k',k)$.

Once the integral equation (14.206) is solved, we obtain the partial wave phase shifts from the diagonal elements of $t_\ell^+(k',k)$, i.e.

$$\begin{aligned} t_\ell^+(k,k) &= -\frac{2k}{\pi}e^{i\delta_\ell(k)}\left(-k\int_0^\infty r' j_\ell(kr')\,v(r')\,u_\ell(k,r')\,dr'\right) \\ &= -\frac{2k}{\pi}e^{i\delta_\ell(k)}\sin\delta_\ell(k). \end{aligned} \tag{14.208}$$

Here we have replaced the quantity in the parenthesis by $\sin\delta_\ell(k)$ using Eq. (14.42).

Generalized Unitarity and the Optical Theorem — Let us write the general T^+ and T^- matrices in terms of their matrix elements, i.e.

$$T^+(\mathbf{q},\mathbf{k}) = \langle\phi_{\mathbf{q}}|V|\psi_{\mathbf{k}}^+\rangle, \tag{14.209}$$

and

$$T^-(\mathbf{q},\mathbf{k}) = \langle\psi_{\mathbf{q}}^-|V|\phi_{\mathbf{k}}\rangle. \tag{14.210}$$

As Eq. (14.175) shows when T^+ and T^- are defined for on-shell values $q^2 = k^2$, we have the relation

$$T^-(\mathbf{k}_{out},\mathbf{k}_{in}) = T^+(\mathbf{k}_{out},\mathbf{k}_{in}), \tag{14.211}$$

but otherwise they are not equal. We write $T^+(\mathbf{q},\mathbf{k})$ as

$$\begin{aligned} T^+(\mathbf{q},\mathbf{k}) &= \lim_{t\to-\infty}\langle\phi_{\mathbf{q}}|V\Omega(t)|\phi_{\mathbf{k}}\rangle \\ &= -\left\{\lim_{t\to+\infty} - \lim_{t\to-\infty}\right\}\langle\phi_{\mathbf{q}}|V\Omega(t)|\phi_{\mathbf{k}}\rangle + \lim_{t\to+\infty}\langle\phi_{\mathbf{q}}|V\Omega(t)|\phi_{\mathbf{k}}\rangle \\ &= -\int_{-\infty}^{+\infty}\left\langle\phi_{\mathbf{q}}\left|V\frac{d}{dt}\Omega(t)\right|\phi_{\mathbf{k}}\right\rangle dt + \langle\phi_{\mathbf{q}}|V|\psi_{\mathbf{k}}^-\rangle. \end{aligned} \tag{14.212}$$

The last term in (14.212) according to (14.209) is $(T^-)^*(\mathbf{k},\mathbf{q})$, and also

$$\frac{d\Omega(t)}{dt} = ie^{iHt}(H-H_0)e^{-iH_0t}, \tag{14.213}$$

therefore we can write (14.212) as

$$\begin{aligned} T^+(\mathbf{q},\mathbf{k}) - T^{-*}(\mathbf{k},\mathbf{q}) &= -\int_{-\infty}^{+\infty}\langle\phi_{\mathbf{q}}|Ve^{iHt}iVe^{-iH_0t}|\phi_{\mathbf{k}}\rangle\,dt \\ &= \frac{-i}{L^3}\int_{-\infty}^{+\infty}e^{-iE_kt}dt\sum_{\mathbf{p}}e^{iE_pt}\langle\phi_{\mathbf{q}}|V|\psi_{\mathbf{p}}^+\rangle\langle\psi_{\mathbf{p}}^+|V|\phi_{\mathbf{k}}\rangle. \end{aligned} \tag{14.214}$$

In the last integral we have inserted a complete set of states

$$\frac{1}{L^3}\sum_{\mathbf{p}} |\psi_{\mathbf{p}}^{+}\rangle\langle\psi_{\mathbf{p}}^{+}|, \tag{14.215}$$

which is a unit matrix. By carrying out the integration in (14.214) we obtain

$$T^{+}(\mathbf{q},\mathbf{k}) - T^{-*}(\mathbf{k},\mathbf{q}) = -\frac{2\pi i}{L^3}\sum_{\mathbf{p}} T^{+}(\mathbf{q},\mathbf{p})\delta(E_{\mathbf{p}} - E_{\mathbf{k}})T^{+*}(\mathbf{k},\mathbf{p}). \tag{14.216}$$

This important result is called generalized unitarity.

If we set $\mathbf{q} = \mathbf{k}$, then (14.216) reduces to

$$\begin{aligned} T^{+}(\mathbf{k},\mathbf{k}) - T^{-*}(\mathbf{k},\mathbf{k}) &= -2i \,\mathrm{Im}\, T(\mathbf{k},\mathbf{k}) \\ &= -\frac{2\pi i}{L^3}\sum_{\mathbf{p}} |T(\mathbf{k},\mathbf{p})|^2\delta(E_{\mathbf{p}} - E_{\mathbf{k}}), \end{aligned} \tag{14.217}$$

or simply

$$\mathrm{Im}\, T(\mathbf{k},\mathbf{k}) = -\frac{\pi}{L^3}\sum_{\mathbf{p}} |T(\mathbf{k},\mathbf{p})|^2\delta(E_{\mathbf{p}} - E_{\mathbf{k}}). \tag{14.218}$$

This is another way of expressing the optical theorem.

The Low Equation — A different equation for determining scattering matrix is the Low equation [19],[20]. When the potential does not support a bound state, this equation can be found by observing that

$$\begin{aligned} T^{+}(\mathbf{q},\mathbf{k}) &= \lim_{t\to-\infty} \langle\phi_{\mathbf{q}}|V\Omega(t)|\phi_{\mathbf{k}}\rangle \\ &= -\left\{\lim_{t\to+\infty} - \lim_{t\to-\infty}\right\}\langle\phi_{\mathbf{q}}|V\Omega(t)\theta(-t)|\phi_{\mathbf{k}}\rangle \\ &= -\int_{-\infty}^{+\infty}\frac{d}{dt}\langle\phi_{\mathbf{q}}|V\Omega(t)\theta(-t)|\phi_{\mathbf{k}}\rangle\, dt \\ &= \langle\phi_{\mathbf{q}}|V\Omega(t)|\phi_{\mathbf{k}}\rangle - \int_{-\infty}^{+\infty}\theta(-t)\langle\phi_{\mathbf{q}}|Ve^{iHt}iVe^{-iE_{\mathbf{k}}t}|\phi_{\mathbf{k}}\rangle dt. \end{aligned} \tag{14.219}$$

Now we insert a complete set of states (14.215) and use Eqs. (14.179) and (14.180) and change the order of integration in the last term of (14.219) to get

$$\begin{aligned} T^{+}(\mathbf{q},\mathbf{k}) &= V(\mathbf{q},\mathbf{k}) - \frac{1}{L^3}\sum_{\mathbf{p}}\int_{-\infty}^{+\infty}\frac{d\zeta}{\zeta - i\epsilon} \\ &\times\ \delta(\zeta - E_{\mathbf{p}} + E_{\mathbf{k}})\left\langle\phi_{\mathbf{q}}\left|V\right|\psi_{\mathbf{p}}^{+}\right\rangle\left\langle\psi_{\mathbf{p}}^{+}\left|V\right|\phi_{\mathbf{k}}\right\rangle. \end{aligned} \tag{14.220}$$

This relation can be simplified and after simplification we find the nonlinear Low equation;

$$T^{+}(\mathbf{q},\mathbf{k}) = V(\mathbf{q},\mathbf{k}) + \frac{1}{L^3}\sum_{\mathbf{p}}\frac{T^{+}(\mathbf{q},\mathbf{p})T^{+*}(\mathbf{k},\mathbf{p})}{E_{\mathbf{k}} - E_{\mathbf{p}} + i\epsilon}$$

$$\rightarrow \quad V(\mathbf{q},\mathbf{k}) + \frac{1}{(2\pi)^3}\int \frac{T^+(\mathbf{q},\mathbf{p})T^{+*}(\mathbf{k},\mathbf{p})}{E_{\mathbf{k}} - E_{\mathbf{p}} + i\epsilon}. \tag{14.221}$$

Just as we expanded the integral equation for the T-matrix, we can also expand the Low equation in terms of spherical harmonics with the result that for the ℓ-th partial wave we have

$$t_\ell^+(q,\ k) = v_\ell(q,\ k) + \int_0^\infty \frac{t_\ell^+(q,\ p)t_\ell^{+*}(k,p)}{k^2 - p^2 + i\epsilon}dp. \tag{14.222}$$

Symmetric and Antisymmetric Parts of the Transition Matrix — In our formulation of the scattering problem up to this point we have assumed that the potential is static and is local. The potential is static if it does not depend on the relative velocity nor on the angular momentum of the two interacting particles. The potential is local if it depends on the relative position of the two particles, $\mathbf{r}$, i.e. be of the form $V(\mathbf{r})$. But both of these conditions are not satisfied in all collisions. For instance in the case of nucleon-nucleon scattering the T-matrix found from a local static potential, $V(\mathbf{k},\mathbf{q})$, given by (14.195) does not fit the empirical data. For these cases one can generalize the concept of the potential and instead of using $V(\mathbf{r})$ works with a set of independent symmetric matrices $v_\ell(k,q)$ which are independent of each other for different ℓ s. The Hermiticity of the Hamiltonian (and thus of the interaction) restricts the matrix elements $v_\ell(k,q)$ to real symmetric matrices, with certain asymptotic behavior as k or q go to infinity [21]. For instance we can assume $v_0(k,q)$ for S wave scattering to be real and symmetric separable potential of the form

$$v_0(k,q) = -\frac{2\lambda\gamma kq}{\pi(1+\lambda)\sqrt{k^2+\frac{1}{4}\gamma^2}\sqrt{q^2+\frac{1}{4}\gamma^2}}, \tag{14.223}$$

where λ and γ are parameters. If we substitute this potential in (14.205), ($\ell = 0$), and solve for $t_0^+(q,k)$ and then set $q = k$ we obtain

$$t_0^+(k,k) = -\frac{2k^2}{\pi}\left[\frac{\gamma(1-\lambda)}{4\lambda} + \left(\frac{1+\lambda}{\gamma\lambda}\right)k^2 - ik\right]^{-1}. \tag{14.224}$$

The S wave phase shift found from (14.224) is

$$k\cot\delta_0 = \frac{\gamma(1-\lambda)}{4\lambda} + \frac{1+\lambda}{\lambda\gamma}k^2, \tag{14.225}$$

which is just the effective range formula, Eq. (14.100). If we compare (14.225) with (14.60) we notice that two different potentials (14.56) and (14.223) give identical S wave phase shifts for all energies.

Realizing that the phase shifts for all energies (and for all angular momenta) do not yield a unique interaction and that the potential is not an observable of the two-body system, we want to inquire about the possibility of replacing the most general form of of the potential matrix $v_\ell(q,k)$ by some other matrix $\sigma_\ell(q,k)$. This matrix, $\sigma_\ell(q,k)$, should generate the scattering result and at the same time contain experimental data. In order for $\sigma_\ell(q,k)$ to have the same information content as $v_\ell(q,k)$, our new interaction matrix must also be symmetric for each partial wave ℓ. One possible candidate for replacing $v_\ell(q,k)$ is the symmetric part of $t_\ell^+(q,k)e^{i\delta_\ell(k)}$ matrix [22]. The diagonal of this $\sigma_\ell(q,k)$ are given by $-\frac{2k}{\pi}\sin\delta_\ell(k)$ and are observables. Once the form of $\sigma_\ell(q,k)$ is assumed then we can use the properties of the T matrix to construct the antisymmetric part and thus the complete $t_\ell^+(q,k)$ matrix. For the sake of simplicity in the following formulation we assume that the interaction is not strong enough to allow for a bound state, but this approach can be extended to include a finite number of bound states as well [23]. Here we use units where $\hbar^2 = 2m = 1$.

Let us introduce a real matrix $\phi_\ell(q,k)$ which is related to $t_\ell^+(q,k)$ by

$$\phi_\ell(q,k) = e^{-i\delta_\ell(k)}t_\ell^+(q,k). \tag{14.226}$$

By substituting for $t_\ell^+(q,k)$ from (14.226) into Low's equation (14.221) we get

$$v_\ell(k,q) = \phi_\ell(k,q)\cos\delta_\ell(q) - \mathcal{P}\int_0^\infty \frac{\phi_\ell(k,p)\phi_\ell(q,p)}{q^2-p^2}dp. \tag{14.227}$$

Since $v_\ell(k,q) = v_\ell(q,k)$, by changing k and q in (14.227) and subtracting the resulting equation from (14.227) we find

$$\begin{aligned}\phi_\ell(k,q)\cos\delta_\ell(q) &= \phi_\ell(q,k)\cos\delta_\ell(k)\\ &- \mathcal{P}\int_0^\infty \phi_\ell(k,p)\phi_\ell(q,p)\left\{\frac{1}{k^2-p^2} - \frac{1}{q^2-p^2}\right\}dp.\end{aligned} \tag{14.228}$$

We find another equation satisfied by $\phi_\ell(k,q)$ from the partial wave T-matrix, Eqs. (14.205) and (14.226). This equation can be written as

$$\phi_\ell(k,q) = \int_0^\infty v_\ell(k,p)U_\ell(p,q)dp, \tag{14.229}$$

where

$$U_\ell(p,q) = \cos\delta_\ell(p)\delta(p-q) + \mathcal{P}\,\frac{\phi_\ell(p,q)}{q^2-p^2}. \tag{14.230}$$

Since the transpose of (14.229) is

$$\phi_\ell(q,k) = \int_0^\infty U_\ell^\dagger(k,p)v_\ell(p,q)dp, \tag{14.231}$$

therefore

$$\begin{aligned}\int_0^\infty U_\ell(q,j)\phi_\ell(q,k)dq &= \int_0^\infty U_\ell^\dagger(k,p)dp\int_0^\infty v_\ell(p,q)U_\ell(q,j)dq\\ &= \int_0^\infty U_\ell^\dagger(k,p)\phi_\ell(p,j)dp,\end{aligned} \tag{14.232}$$

where

$$U_\ell^\dagger(k,p) = \cos\delta_\ell(k)\delta(k-p) + \mathcal{P}\frac{\phi_\ell(p,k)}{k^2-p^2}. \tag{14.233}$$

By substituting (14.230) and (14.233) in (14.232) we find a different integral equation for $\phi_k(q,k)$;

$$\begin{aligned} &\phi_\ell(q,k)\cos\delta_\ell(q) - \phi_\ell(k,q)\cos\delta_\ell(k) \\ &= \mathcal{P}\int_0^\infty \phi_\ell(p,k)\phi_\ell(p,q)\left\{\frac{1}{k^2-p^2} - \frac{1}{q^2-p^2}\right\}dp. \end{aligned} \tag{14.234}$$

By adding and subtracting (14.228) and (14.234) we obtain the following relations

$$\begin{aligned} &\sigma_\ell(q,k)[\cos\delta_\ell(k) - \cos\delta_\ell(q)] \\ &= \mathcal{P}\int_0^\infty \left\{\frac{1}{k^2-p^2} - \frac{1}{q^2-p^2}\right\}\{\sigma_\ell(p,k)\alpha_\ell(q,p) - \alpha_\ell(p,k)\sigma_\ell(q,p)\}\,dp, \end{aligned} \tag{14.235}$$

and

$$\begin{aligned} &\alpha_\ell(q,k)[\cos\delta_\ell(k) + \cos\delta_\ell(q)] \\ &= \mathcal{P}\int_0^\infty \left\{\frac{1}{k^2-p^2} - \frac{1}{q^2-p^2}\right\}\{\sigma_\ell(k,p)\sigma_\ell(q,p) + \alpha_\ell(p,k)\alpha_\ell(p,q)\}\,dp. \end{aligned} \tag{14.236}$$

In these relations σ and α are the symmetric and antisymmetric parts of the ϕ matrix

$$\sigma_\ell(q,p) = \frac{1}{2}[\phi_\ell(q,p) + \phi_\ell(p,q)], \tag{14.237}$$

and

$$\alpha_\ell(q,p) = \frac{1}{2}[\phi_\ell(q,p) - \phi_\ell(p,q)]. \tag{14.238}$$

Equations (14.235) for $\sigma_\ell(q,k)$ and (14.236) for $\alpha_\ell(q,k)$ are both singular integral equations. To make them nonsingular we add the term

$$\mathcal{P}\int_0^\infty \alpha_\ell(q,k)\left[\frac{\sigma_\ell(q,q)}{q^2-p^2} - \frac{\sigma_\ell(k,k)}{k^2-p^2}\right]dp \equiv 0, \tag{14.239}$$

to (14.235) and the integral

$$\mathcal{P}\int_0^\infty \left\{\frac{1}{q^2-p^2} - \frac{1}{k^2-p^2}\right\}\sigma_\ell(k,k)\sigma_\ell(q,q)dp \equiv 0, \tag{14.240}$$

to (14.236).

The operator $U_\ell(p,q)$ defined by (14.230) can be decomposed into a symmetric and an antisymmetric part. Thus defining $A_\ell(k,p)$ and $S_\ell(p,q)$ by

$$A_\ell(k,p) = \mathcal{P}\,\frac{\sigma_\ell(p,k)}{k^2-p^2}, \tag{14.241}$$

and

$$S_\ell(p,q) = \delta(p-q)\cos\delta_\ell(q) + \frac{\alpha_\ell(p,q)}{q^2-p^2}, \tag{14.242}$$

we have

$$U_\ell(p,q) = A_\ell(p,q) + S_\ell(p,q). \tag{14.243}$$

These operators, in the absence of bound states, commute with each other;

$$\begin{aligned}
&\int_0^\infty \left[A_\ell(k,p)S_\ell(q,p) - S_\ell(k,p)A_\ell(p,q)\right]dp \\
= \;& \frac{1}{k^2-q^2}\left[\sigma_\ell(q,k)(\cos\delta_\ell(q) - \cos\delta_\ell(k))\right] \\
+ \;& \mathcal{P}\int_0^\infty \left\{\frac{1}{k^2-p^2} - \frac{1}{q^2-p^2}\right\}\left\{\sigma_\ell(p,k)\alpha_\ell(q,p) - \alpha_\ell(p,k)\sigma_\ell(q,p)\right\}dp = 0.
\end{aligned} \tag{14.244}$$

Similarly we can show that $U_\ell(p,q)$ is a unitary transformation, i.e.

$$\int_0^\infty \left[S_\ell(k,p)S_\ell(p,q) - A_\ell(k,p)A_\ell(p,q)\right]dp = \delta(k-q). \tag{14.245}$$

For the convergence of the integrals in Eq. (14.236), $\sigma(p,q)$ cannot become arbitrarily large when p or q tend to infinity or zero. Thus the integral over p in (14.236) converges provided that

$$\lim_{p\to\infty}\left[\frac{\sigma(p,q)}{p^{\frac{3}{2}+\epsilon}}\right] \to 0, \tag{14.246}$$

at the upper limit and

$$\lim_{p\to 0}\left[p^{\frac{1}{2}}\sigma(p,q)\right] \to \text{a constant}, \tag{14.247}$$

at the lower limit. In (14.246) ϵ is a small positive number.

An Example — As an example of determination of the t_ℓ matrix from its symmetric part, we consider the special case where $\phi_\ell(k,q)$ is separable, i.e. when it is expressible as

$$\phi_\ell(k,q) = \frac{N_\ell(k,q)}{|D_\ell(q)|} = \frac{f_\ell(k)f_\ell(q)}{|D_\ell(q)|}, \tag{14.248}$$

where f_ℓ, N_ℓ and $|D_\ell|$ are all real functions of their argument. Since N_ℓ is symmetric we have

$$N_\ell(k,q) = \sqrt{N_\ell(k,k)N_\ell(q,q)}, \quad \text{if} \quad N_\ell \ge 0, \tag{14.249}$$

and

$$N_\ell(k,q) = -\sqrt{N_\ell(k,k)N_\ell(q,q)}, \quad \text{if} \quad N_\ell \le 0. \tag{14.250}$$

The diagonal elements of $t_\ell^+(k,q)$ matrix in this case are given by

$$t_\ell^+(k,k) = e^{i\delta_\ell(k)}\frac{N_\ell(k,k)}{|D_\ell(k)|}. \tag{14.251}$$

Furthermore from (14.248)–(14.250) we can determine $\phi_\ell(k,q)$;

$$\phi_\ell(k,q) = \pm\sqrt{\phi_\ell(k,k)\phi_\ell(q,q)}\sqrt{\frac{|D_\ell(k)|}{|D_\ell(q)|}}. \tag{14.252}$$

Let us remember that the denominator $D_\ell(k)$ in (14.251) is an analytic function of the energy $E = k^2$. Next we write Eq. (14.251) as

$$e^{i\delta_\ell(E)}\phi_\ell(E) = t_\ell^+(E,E) = \frac{N_\ell(E,E)}{D_\ell(E)^+} = -2\frac{\sqrt{E}}{\pi}e^{i\delta_\ell(E)}\sin\delta_\ell(E), \tag{14.253}$$

where D_ℓ^+ is the denominator of $t_\ell(E,E)$. Noting that $N_\ell(E,E)$ is a real function of E, from (14.253) it follows that the complex function $D_\ell^+(E)$ satisfies the relation

$$D_\ell^+(E) = -\frac{\pi N_\ell(E,E)}{2\sqrt{E}}\left(\cot\delta_\ell(E) - i\right) \quad E = k^2 \geq 0. \tag{14.254}$$

Now by equating the imaginary parts of the two sides of (14.254) and replacing $N(E,E)$ from (14.253) we obtain

$$\operatorname{Im} D_\ell^+(E) = -e^{i\delta_\ell(E)}\sin\delta_\ell(E)\, D_\ell^+(E). \tag{14.255}$$

By applying Cauchy's theorem we find that $D_\ell^+(E)$ satisfies the Omnés–Mushkhelishvili equation, and that The solution of the equation for $D_\ell^+(E)$ is [1],[24],[25]

$$D_\ell^+(E) = \exp\left[-\frac{2}{\pi}\int_0^\infty \frac{\delta_\ell(E')\,dE'}{E'-E-i\epsilon}\right]. \tag{14.256}$$

Having found $D_\ell^+(E)$ we calculate $|D_\ell(q)|$

$$|D_\ell(q)| = \exp\left[-\frac{2\mathcal{P}}{\pi}\int_0^\infty \frac{\delta_\ell(p)\,p\,dp}{p^2-q^2}\right]. \tag{14.257}$$

Substituting for $|D_\ell(q)|$ and $\phi_\ell(q,q)$ in (14.252) we find $\phi_\ell(k,q)$, $\sigma_\ell(k,q)$ and $\alpha_\ell(k,q)$ to be

$$\begin{aligned}\phi_\ell(k,q) &= \pm\frac{2}{\pi}\sqrt{kq\sin\delta_\ell(k)\sin\delta_\ell(q)} \\ &\times \exp\left[-\frac{\mathcal{P}}{\pi}\int_0^\infty \frac{(k^2-q^2)\,\delta_\ell(p)\,p\,dp}{(p^2-q^2)(p^2-k^2)}\right],\end{aligned} \tag{14.258}$$

$$\sigma_\ell(k,q) = \pm\frac{2}{\pi}\sqrt{kq\sin\delta_\ell(k)\sin\delta_\ell(q)} \times \cosh\left[-\frac{\mathcal{P}}{\pi}\int_0^\infty \frac{(k^2-q^2)\,\delta_\ell(p)\,p\,dp}{(p^2-q^2)(p^2-k^2)}\right], \tag{14.259}$$

and

$$\alpha_\ell(k,q) = \pm\frac{2}{\pi}\sqrt{kq\sin\delta_\ell(k)\sin\delta_\ell(q)} \times \sinh\left[-\frac{\mathcal{P}}{\pi}\int_0^\infty \frac{(k^2-q^2)\,\delta_\ell(p)\,p\,dp}{(p^2-q^2)(p^2-k^2)}\right]. \tag{14.260}$$

These relations show that for a separable ϕ_ℓ-matrix defined by (14.248) one needs only to know the diagonal elements and then the off-diagonal elements are completely determined from these diagonal elements.

14.6 Analytical Properties of the Radial Wave Function

While we can solve the wave equation to find the phase shift $\delta_\ell(k)$ as a function of ℓ and k, it is important to find connections between the phase shift, bound state energies and the shape of the potential. In particular we want to answer the following questions:

(1) - If we know the phase shift $\delta_\ell(k)$ for a fixed partial wave and for all energies, $0 \le k < \infty$, can we find a unique potential? At a fixed energy, $E = \frac{\hbar^2 k^2}{2m}$, if the partial phase shifts, $\delta_\ell(k)$, are known for all ℓ, then is it possible to construct a local potential? We will consider the latter question later in this chapter.

(2) - Can we derive a relation between the phase shift at zero energy and the phase shift at infinite energy?

(3) - What is the relation between the scattering amplitude and the bound state energies?

For the sake of simplicity we consider the solution of the wave equation for zero angular momentum, but as we will see later, the results that we have found can be extended to other partial waves. Let us denote the regular solution of the Schrödinger equation for S wave by $u_0(k,r)$,

$$\frac{d^2u_0(k,r)}{d\,r^2} + k^2u_0(k,r) = v(r)u_0(k,r), \tag{14.261}$$

where $v(r) = \frac{2m}{\hbar^2}V(r)$. We assume that the potential $v(r)$ is of short range and that its first and second moment are finite (see Eqs. (14.5) and (14.6)).

If we solve the differential equation (14.261) with the boundary conditions

$$u_0(k,0)=0, \quad \text{and} \quad \left(\frac{du_0(k,r)}{dr}\right)_{r=0}=1, \tag{14.262}$$

then the wave function $u_0(k,r)$ is real and is dependent on k^2, i.e. $u_0(k,r)$ is an even function of k. Since the boundary conditions (14.262) do not depend on k^2 and the coefficient of $u_0(k,r)$ in (14.261) is an entire function of k^2, then the solution $\phi_0(k,r)$ is an entire function of k^2. In other words $u_0(k,r)$ is an analytic nonsingular function in the complex k-plane for all values of r (Poincaré theorem). To prove this result we first find the Green function for the differential operator $\frac{d^2}{dr^2}+k^2$ subject to the boundary condition (14.262)

$$G_k(r-r')=\begin{cases}\dfrac{\sin k(r-r')}{k} & r'<r\\ 0 \quad r>r'.\end{cases} \tag{14.263}$$

Using this Green function we write the solution of (14.261) as an integral equation

$$\begin{aligned}u_0(r) &= \frac{1}{k}\sin kr+\int_0^\infty G_k(r-r')\,v(r')\,u_0(k,r')\,dr'\\ &= \frac{1}{k}\sin kr+\int_0^r \frac{\sin k(r-r')}{k}v(r')\,u_0(k,r')\,dr'.\end{aligned} \tag{14.264}$$

This inhomogeneous integral equation with the upper limit r is a Volterra integral equation which can be solved by iteration. The iterative solution of (14.264) can be written as

$$u_0(k,r)=\sum_{n=0}^{\infty}u_0^{(n)}(k,r), \tag{14.265}$$

where

$$u_0^{(0)}(k,r)=\frac{1}{k}\sin kr, \tag{14.266}$$

and

$$u_0^{(n)}(k,r)=\int_0^r \frac{\sin k(r-r')}{k}v(r')\,u_0^{(n-1)}(k,r')\,dr'. \tag{14.267}$$

First we note that the kernel of the integral equation satisfies the inequality

$$\left|\frac{\sin k(r-r')}{k}\right|\le r. \tag{14.268}$$

This inequality allows us to find an upper bound to $\left|u_0^{(n)}(k,r)\right|$. Thus if we define $q(r)$ by

$$q(r)=\int_0^r v(r')\,r'dr', \tag{14.269}$$

then by successive iterations we have

$$\frac{1}{r}\left|u_0^{(1)}(k,r)\right| \le q(r), \tag{14.270}$$

$$\frac{1}{r}\left|u_0^{(2)}(k,r)\right| \le \int_0^r v(r')\,q(r')\,r'dr' = \int_0^{q(r)} q dq = \frac{1}{2!}q^2(r), \tag{14.271}$$

$$\frac{1}{r}\left|u_0^{(3)}(k,r)\right| \le \int_0^r v(r')\,\frac{1}{2!}q^2(r')\,r'dr' = \frac{1}{3!}q^3(r), \tag{14.272}$$

and so on. After the n-th iteration we obtain

$$\frac{1}{r}\left|u_0^{(n)}(k,r)\right| \le \frac{q^n(r)}{n!} \le \frac{M^n}{n!}, \tag{14.273}$$

where we have used Eq. (14.5) in writing the last term in (14.273). By substituting for for $u_0^{(n)}(k,r)$ in (14.265) we find an upper bound for $u_0(k,r)$;

$$|u_0(k,r)| \le r\sum_{n=0}^{\infty}\frac{M^n}{n!} = re^M. \tag{14.274}$$

The condition (14.268) is valid when the argument of the sine function is small. For larger arguments we can use the inequality

$$|\sin\zeta| \le 1, \quad \zeta = k\,(r-r'). \tag{14.275}$$

We can combine (14.268) and (14.275) and write an inequality which is valid for all values of ζ;

$$|\sin\zeta| \le B\frac{\zeta}{\zeta+1}, \quad \zeta \ge 0. \tag{14.276}$$

In this inequality B is a constant whose numerical value is immaterial. In order to extend these results to complex values of k, we replace ζ by z, and we note that $\sin z$ grows with increasing z as $\exp[|\mathrm{Im} z|]$. Thus instead of (14.276) we shall consider the inequality

$$|\sin z| \le B\frac{|z|}{|z|+1}\exp[|\mathrm{Im} z|]. \tag{14.277}$$

The convergence of the wave function for complex k can be proven in the same way as we proved the convergence for real nonnegative values of k. That is the inequality (14.277) guarantees the uniform convergence of (14.265) for arbitrary k, viz, after the n-th iteration we get

$$\left|u_n^{(0)}(k,r)\right| \le B\frac{r}{1+|kr|}\exp\left[|\mathrm{Im}\,(kr)|\right]\frac{M^n}{n!}. \tag{14.278}$$

Then the analyticity of $u_0(k,r)$ follows from the analyticity of each of the individual terms of $u_0^{(n)}(k,r)$. For the zero iteration the approximate solution

$$u_0^{(0)}(k,r) = \frac{\sin kr}{k}, \tag{14.279}$$

is an analytic function of k. Using this we can prove the analyticity of $u_0^{(1)}(k,r)$, $u_0^{(2)}(k,r), \cdots$ by induction. Thus we have shown that the solution of the Schrödinger equation is analytic in the open complex k-plane for all values of r [3].

14.7 The Jost Function

So far we have studied the analyticity of the solution of the wave equation when the initial conditions are independent k. Now let us first examine the $\ell = 0$, or the S-wave and introduce a different solution of the Schrödinger equation, $f_0(k,r)$, but with the boundary condition [26]–[28]

$$\lim_{r\to\infty} e^{ikr} f_0(k,r) = 1. \tag{14.280}$$

From this relation it is clear that the asymptotic form of $f_0(k,r)$ at infinity is

$$f_0(k,r) \to e^{-ikr}, \qquad r \to \infty. \tag{14.281}$$

We want to show that $f_0(k,r)$ is an analytic function of complex k for Im $k < 0$, and that it is continuous along the real axis (Im $k = 0$). Again we start with the integral equation for $f_0(k,r)$

$$f_0(k,r) = e^{-ikr} + \int_r^\infty \frac{\sin k\,(r'-r)}{k} v\,(r')\, f_0\,(k,r')\, dr'. \tag{14.282}$$

This function satisfies the radial Schrödinger equation for the S-wave;

$$\frac{d^2 f_0(k,r)}{d\,r^2} + k^2 f_0(k,r) = v(r) f_0(k,r), \tag{14.283}$$

provided that we can differentiate under the integral sign (this is permitted when the solution converges uniformly for all r).

We now introduce a function $g_0(k,r)$ which is related to $f_0(k,r)$ by

$$g_0(k,r) = f_0(k,r) e^{ikr}, \tag{14.284}$$

and substitute from this relation in Eq. (14.282) to find an integral equation for $g_0(k,r)$;

$$g_0(k,r) = 1 + \int_r^\infty D_k\,(r'-r)\, v\,(r')\, g_0\,(k,r')\, dr', \tag{14.285}$$

where

$$D_k(r'-r) = \frac{1}{k} e^{ikr} \sin k(r'-r) e^{-ikr'} = \frac{1}{2ik}\{1 - \exp[-2ik(r'-r)]\}, \tag{14.286}$$

Again we attempt to solve (14.285) by iteration and to this end we write

$$g_0(k,r) = \sum_{n=0}^{\infty} g_0^{(n)}(k,r), \tag{14.287}$$

where

$$g_0^{(0)}(k,r) = 1, \tag{14.288}$$

and

$$g_0^{(n)}(k,r) = \int_r^{\infty} D_k(r'-r)\, v(r')\, g_0^{(n-1)}(k,r')\, dr'. \tag{14.289}$$

Now we shall prove the convergence of the series (14.287) for Im $k < 0$. From the inequality

$$|D_k(r'-r)| \le r' - r \le r', \tag{14.290}$$

it follows that

$$g_0^{(1)}(k,r) \le \int_r^{\infty} |v(r')|\, r' dr', \tag{14.291}$$

or in general for the n-th iteration

$$\begin{aligned} \left|g_0^{(n)}(k,r)\right| &\le \frac{1}{n!}\left[\int_r^{\infty} |v(r')|\, r' dr'\right]^n \\ &\le \frac{1}{n!}\left[\int_0^{\infty} |v(r')|\, r' dr'\right]^n \le \frac{M^n}{n!}. \end{aligned} \tag{14.292}$$

The series (14.287) therefore converges as a consequence of the upper bound (14.292).

As we can see from the radial wave equation (14.261) and the boundary condition (14.281), in the region where $f_0(k,r)$ is analytic including the real axis, the Jost function satisfies the condition

$$f_0^*(-k^*,r) = f_0(k,r). \tag{14.293}$$

Now we want to see how the real solution of the Schrödinger equation, $u_0(k,r)$, is related to its complex solution $f_0(k,r)$. By calculating the Wronskian of $f_0(k,r)$ and $f_0(-k,r)$ we find

$$W[f_0(k,r),\ f_0(-k,r)] = 2ik. \tag{14.294}$$

This Wronskian is not zero, therefore $f_0(k,r)$ and $f_0(-k,r)$ are two independent solutions of (14.261). So any solution of (14.261) can be written as a linear combination of $f_0(k,r)$ and $f_0(-k,r)$. In particular we can write the real solution, $u_0(k,r)$, which we defined earlier, as

$$u_0(k,r) = \frac{1}{2ik}\left[f_0(k)f_0(-k,r) - f_0(-k)f_0(k,r)\right], \tag{14.295}$$

where the Jost function, $f_0(k)$, is related to $f_0(k,r)$ by

$$f_0(k) = f_0(k,0), \quad \text{and} \quad f_0(-k) = f_0(-k,0). \tag{14.296}$$

We can also define $f_0(k)$ by means of the Wronskian $W\left[f_0(k,r),\ u_0(k,r)\right]$. Thus with the help of (14.294), we have

$$W\left[f_0(k,r),\ u_0(k,r)\right] = f_0(k). \tag{14.297}$$

Finally we can write the asymptotic form of $u_0(k,r)$ as $r \to \infty$ in terms of the Jost function

$$u_0(k,r) \to -\frac{f_0(-k)}{2ik}\left\{e^{-ikr} - \frac{f_0(k)}{f_0(-k)}e^{ikr}\right\}, \quad r \to \infty. \tag{14.298}$$

By comparing (14.298) with (14.50) we obtain an important relation between the Jost function and the scattering matrix $S_0(k)$;

$$S_0(k) = \frac{f_0(k)}{f_0(-k)}. \tag{14.299}$$

Generalization to Higher Partial Waves — The results that we have found for the S wave, ($\ell = 0$), can be extended to higher partial waves by observing that the radial wave equation (14.261) will now have the additional singular term $\frac{\ell(\ell+1)}{r^2}$, or $u_\ell(k,r)$ satisfies the partial wave equation

$$\frac{d^2u_\ell(k,r)}{d\,r^2} + \left[k^2 - \frac{\ell(\ell+1)}{r^2} - v(r)\right]u_\ell(k,r) = 0. \tag{14.300}$$

The presence of this term, $\frac{\ell(\ell+1)}{r^2}$ changes the behavior of $u_\ell(k,r)$ at $r=0$. Here we also assume that the first and second moments of the potential, Eqs. (14.5) and (14.6), are finite. Very close to the origin the centrifugal energy term in (14.300) is much larger than $k^2 - v(r)$, and therefore we can ignore these two terms. In this limit (14.300) reduces to

$$\frac{d^2u_\ell(k,r)}{d\,r^2} - \frac{\ell(\ell+1)}{r^2}u_\ell(k,r) = 0. \tag{14.301}$$

The two independent solutions of (14.301) are

$$u_\ell(k,r) \to r^{\ell+1} \quad \text{as} \quad r \to 0, \tag{14.302}$$

$$u_\ell(k,r) \to r^{-\ell} \quad \text{as} \quad r \to 0, \tag{14.303}$$

and thus the regular solution of (14.300) satisfies the boundary condition (14.302) as $r \to 0$. We choose this boundary condition together with the boundary condition

$$\lim_{r\to\infty} f_\ell(k,r) \to e^{-ikr}, \tag{14.304}$$

to define the Jost solution for the ℓ-th partial wave. This solution satisfies a relation similar to (14.293),

$$f_\ell^*\left(-k^*, r\right) = f_\ell(k, r). \tag{14.305}$$

Now for the ℓ-th partial wave the real solution of the Schrödinger equation can be expressed as

$$u_\ell(k, r) = \frac{1}{2ik}\left[f_\ell(k) f_\ell(-k, r) - f_\ell(-k) f_\ell(k, r)\right], \tag{14.306}$$

where

$$f_\ell(k) = f_\ell(k, 0), \quad \text{and} \quad f_\ell(-k) = f_\ell(-k, 0). \tag{14.307}$$

Alternatively we can define $f_\ell(k)$ by the Wronskian

$$f_\ell(k) = W\left[f_\ell(k, r),\ u_\ell(k, r)\right]. \tag{14.308}$$

Equation (14.306) now shows us that the scattering matrix, $S_\ell(k)$, for the ℓ-th partial wave is

$$S_\ell(k) = (-1)^\ell \frac{f_\ell(k)}{f_\ell(-k)}. \tag{14.309}$$

14.8 Zeros of the Jost Function and Bound Sates

As we have seen in earlier chapters the stationary states of a system are associated with discrete energy states, and these states are related to the square integrable solutions of the Schödinger equation. We want to show that the zeros of the Jost function $f_\ell(k)$ in the lower half of the complex k-plane correspond to bound states. Denoting these zeros by k_n, $n = 1,\ 2\cdots$, we can write the eigenvalue equation for the bound states as

$$f_\ell(k_n) = 0. \tag{14.310}$$

By substituting $k = k_n$ in $u_\ell(k, r)$, noting that $f_\ell(k_n) = 0$, we find

$$u_\ell(k_n, r) = \frac{1}{\gamma_n} f_\ell(k_n, r), \tag{14.311}$$

where γ_n is a constant. This wave function vanishes at $r = 0$, and decreases exponentially for large r (remembering that Im $k < 0$). Therefore $u_\ell(k_n, r)$ is a square integrable function and describes a bounded motion, i.e. the particle cannot go far away from the center of force (see below). To relate these k_n s to the binding energies we need to show that all k_n^2 s are real and negative. Writing the differential equation for $u_\ell(k_n, r)$, Eq. (14.300), and for its complex conjugate, we find that the Wronskian so obtained satisfies the following equation

$$\frac{d}{dr} W\left[u_\ell(k_n, r),\ u_\ell^*(k_n, r)\right] = 2i \,\text{Im}\left(k_n^2\right) |u_\ell(k_n, r)|^2. \tag{14.312}$$

Since $u_\ell^*(k^*, r) = u_\ell(k, r)$ we have

$$u_\ell^*(k_n, r) = \frac{1}{\gamma_n^*} f_\ell(-k_n^*, r), \tag{14.313}$$

and this shows that when Im $k < 0$, then $u_\ell^*(k_n, r)$ decreases exponentially as $r \to \infty$. Using the fact that both $u_\ell(k_n, r)$ and $u_\ell^*(k_n, r)$ are zero at $r = 0$ and at $r = \infty$ we integrate (14.312) to obtain

$$\mathrm{Im}\left(k_n^2\right) \int_0^\infty |u_\ell(k_n, r)|^2 dr = 0. \tag{14.314}$$

From this equation it follows that

$$\mathrm{Im}\, k_n^2 = 0, \tag{14.315}$$

and

$$\int_0^\infty |u_\ell(k_n, r)|^2 dr \leq \infty. \tag{14.316}$$

Thus k_n^2 is real and $u_\ell(k_n, r)$ is a square integrable function of r. Now if k_n is real then $f_\ell(k_n)$ cannot be equal to zero since $f_\ell(-k_n)$ will also be zero according to (14.305), since the vanishing of both $f_\ell(k_n)$ and $f_\ell(-k_n)$ implies that $u_\ell(k_n, r)$ is zero everywhere, a result that follows from Eq. (14.308). Thus $f_\ell(k_n)$ can be zero only if $k_n^2 < 0$, i.e. the zeros of the Jost function $f_\ell(k)$ are in the lower half of k-plane on the imaginary axis

$$k_n = -i\kappa_n, \quad \kappa_n > 0. \tag{14.317}$$

We can also relate the bound state to the root of S matrix for $k = -i\kappa_n$ by writing the square integrability condition for the wave function $u_\ell(-i\kappa_n, r)$ for Eqs. (14.306) and (14.309). Thus for large r we must have

$$\int^\infty \left|(-1)^\ell e^{-\kappa_n r} - S_\ell(-i\kappa_n) e^{+\kappa_n r}\right|^2 dr < \infty, \quad \kappa_n > 0. \tag{14.318}$$

This condition can be satisfied provided that

$$S_\ell(-i\kappa_n) = 0, \quad \kappa_n > 0, \tag{14.319}$$

i.e. is the zeros of the $S_\ell(k)$ matrix located on the imaginary k-axis corresponding to the square integrable function describing the bound state. The symmetry property of the $S_\ell(k)$ matrix which we found earlier, Eq. (14.52), implies that $S_\ell(k)$ possesses simple poles at $k = i\kappa_n$. However $S_\ell(k)$ matrix also vanishes at the points where $f_\ell(-k)$ has poles [27]. These extra zeros of the $S_\ell(k)$ matrix do not represent bound states and they disappear whenever the Jost function is regular in the whole k-plane. For this to happen the potential must go faster to zero than any exponential

$$\int_0^\infty e^{\mu r} v(r) dr < \infty \quad \text{for any real positive } \mu. \tag{14.320}$$

Thus if the potential is identically zero for $r > R$, no matter how large R is, there will be no redundant poles of the $S_\ell(k)$ matrix.

For long range forces, e.g. the Coulomb force, we can find the bound state energies from the poles of the scattering amplitude, where this amplitude is given by (14.138). These poles are located at the points where

$$\Gamma\left(\ell+1-\frac{i\gamma}{k}\right)=\infty, \tag{14.321}$$

i.e. where

$$\ell+1-\frac{i\gamma}{k} = \text{ zero or negative integer.} \tag{14.322}$$

Calculating k from this condition and substituting it in $E=\frac{\hbar^2k^2}{2m}$ gives us the bound state energies for hydrogen like atoms when we set $Z'=1$.

14.9 Dispersion Relation

Since $f_\ell(k)$ ia an analytic function in the lower half k-plane, we can find a relation between Re $f_\ell(k)$ and Im $f_\ell(k)$. In Sec. 14.8 we proved that $f_\ell(k)$ is an entire function of k in the lower half of k-plane including the real axis, and on the real axis it satisfies the boundary condition (14.304). Noting that $f_\ell(k)-1$ approaches zero as $k\to\infty$ everywhere in the lower half of k-plane including the real axis, we can apply Cauchy's theorem and write for any value of k in the lower half plane

$$f_\ell(k)-1=\frac{1}{2\pi i}\oint_C\frac{f_\ell\left(k'\right)-1}{k'-k}dk'. \tag{14.323}$$

The contour C consists of a path on the real axis from $-k_c$ to $+k_c$, and a semi-circle of radius k_c in the lower half-plane where k lies within this closed contour. In the limit of k_c going to infinity, the contribution from the semi-circle vanishes and we have

$$f_\ell(k)=1-\frac{1}{2\pi i}\int_{-\infty}^{\infty}\frac{f_\ell\left(k'\right)-1}{k'-k}dk'. \tag{14.324}$$

If k approaches the real axis from below, then from (14.324) we obtain

$$f_\ell(k)=1-\frac{\mathcal{P}}{\pi i}\int_{-\infty}^{\infty}\frac{f_\ell\left(k'\right)-1}{k'-k}dk'. \tag{14.325}$$

Taking the real part of the two sides of (14.325) we find

$$\text{Re } f_\ell(k)=1-\frac{\mathcal{P}}{\pi}\int_{-\infty}^{\infty}\frac{\text{Im } f_\ell\left(k'\right)}{k'-k}dk'. \tag{14.326}$$

This is the dispersion relation connecting Re $f_\ell(k)$ to Im $f_\ell(k)$.

Now using the rule

$$\lim_{\epsilon\to 0^+}\frac{1}{x+i\epsilon}=\mathcal{P}\frac{1}{x}-i\pi\delta(x), \tag{14.327}$$

we can add i Im $f_\ell(k)$ to the two sides of Eq. (14.326) and obtain $f_\ell(k)$ as an integral over Im $f_\ell(k)$

$$f_\ell(k)=1-\frac{1}{\pi}\int_\infty^\infty\frac{\mathrm{Im}\, f_\ell\,(k')}{k'-k+i\epsilon}dk'. \tag{14.328}$$

To convert (14.328) into an integral equation for $f_\ell(k)$ we note that according (14.55) we can write

$$\mathrm{Im}\, f_\ell(k)=|f_\ell(k)|\sin\delta_\ell(k)=f_\ell(k)\sin\delta_\ell(k)e^{-i\delta_\ell(k)}. \tag{14.329}$$

Now if we Substitute for Im $f_\ell(k)$ in (14.328) we find the integral equation satisfied by $f_\ell(k)$;

$$f_\ell(k)=1-\frac{1}{\pi}\int_\infty^\infty\frac{f_\ell\,(k')\sin\delta_\ell\,(k')\,e^{-i\delta_\ell(k')}}{k'-k+i\epsilon}dk'. \tag{14.330}$$

Having obtained this integral equation for $f_\ell(k)$ we ask supposing that $\delta_\ell(k)$ is known for all k, then is it possible to solve this integral equation and find $f_\ell(k)$? The answer is, in general, no. The information about the phase shift $\delta_\ell(k)$ is not sufficient to determine $f_\ell(k)$, and we have to specify bound state energies. Let $E_n=-\frac{\hbar^2\kappa_n^2}{2m}$, ($n=1,\ 2\cdots N$) represent bound state energies, then these give us the location of the zeros of $f_\ell(k)$ in the lower half k-plane. Having this set of information enables us to find $f_\ell(k)$, from the solution of the singular integral equation known as Omnés-Mushkhelishvili type equation [1], [24], [25];

$$f_\ell(k)=\prod_{n=1}^N\left(1-\frac{\kappa_n^2}{k^2}\right)\exp\left[-\frac{1}{\pi}\int_{-\infty}^\infty\frac{\delta_\ell\,(k')}{k'-k+i\epsilon}\right]. \tag{14.331}$$

14.10 Central Local Potentials having Identical Phase Shifts and Bound States

Returning to the first question posed at the beginning of this section, we want to show by an explicit example that a central local potential cannot be uniquely determined by the phase shifts and bound state energies for a given ℓ state. For our example we choose the Eckart potential which we defined earlier and which is solvable for the S-wave, Eq. (14.56). We will consider the following two potentials:

(a) - We set $s = 6$ in (14.56) and choose $\lambda = 1$ so that the potential becomes

$$v(6,r) = -\frac{6\gamma^2 e^{-\gamma r}}{(1+e^{-\gamma r})^2} = -\frac{3\gamma^2}{2\cosh^2\left(\frac{1}{2}\gamma r\right)}. \tag{14.332}$$

The Jost solution $f_0^{(6)}(k,r)$ obtained from the wave equation is

$$f_0^{(6)}(k,r) = e^{ikr}\frac{4k^2 + 6i\mu(r)k + \gamma^2 - 3(\mu(r))^2}{(2k - i\gamma)(2k - 2i\gamma)}, \tag{14.333}$$

where $\mu(r)$ is given by

$$\mu(r) = \gamma\frac{e^{-\gamma r} - 1}{e^{-\gamma r} + 1}. \tag{14.334}$$

From the last two equations we find the Jost function to be

$$f_0^{(6)}(k) = \frac{4k^2 + \gamma^2}{(2k - i\gamma)(2k - 2i\gamma)} = \frac{2k + i\gamma}{2k - 2i\gamma}. \tag{14.335}$$

(b) - We now choose $s = 2$ and $\lambda_1 = 3$, and the potential becomes

$$v(2,r) = -\frac{6\gamma_1^2 e^{-\gamma_1 r}}{(1+e^{-\gamma_1 r})^2}. \tag{14.336}$$

The Jost solution for this potential is

$$f_0^{(2)}(k,r) = e^{ikr}\frac{2k + i\mu_1(r)}{(2k - i\gamma_1)}, \tag{14.337}$$

where

$$\mu_1(r) = \gamma_1\frac{\lambda_1 e^{-\gamma_1 r} - 1}{\lambda_1 e^{-\gamma_1 r} + 1}. \tag{14.338}$$

Again we find that the Jost function is given by

$$f_0^{(2)}(k) = \frac{2k + i\mu_1(0)}{2k - i\gamma_1} = \frac{2k + \frac{i}{2}\gamma_1}{2k - i\gamma_1}. \tag{14.339}$$

Therefore if we choose $\gamma_1 = 2\gamma$, $f_0^{(6)}(k)$ and $f_0^{(2)}(k)$ become identical. Thus the two potentials (14.332) and

$$v(2,r) = -\frac{24\gamma^2 e^{-2\gamma r}}{(1+3e^{-2\gamma r})^2} = -\frac{2\gamma^2}{\cosh^2[\gamma(r-a)]}, \quad a = \frac{\ln 3}{2\gamma} \tag{14.340}$$

are phase and bound state equivalent potentials [29]. The bound state for the two potentials are at the zero of $f_0(k)$, and this root is located at

$$k = -\frac{i\gamma}{2}, \tag{14.341}$$

and thus the bound state energy is

$$E_0 = -\frac{\hbar^2}{2m}\left(\frac{\gamma^2}{4}\right). \tag{14.342}$$

The phase shifts for both potentials are identical and satisfy the effective range formula exactly;

$$k\cot\delta_0(k) = -\frac{1}{3}\gamma + \frac{2}{3\gamma}k^2. \tag{14.343}$$

We observe that the force $\left(-\frac{\partial v(6,r)}{\partial r}\right)$ is attractive for all r, whereas $\left(-\frac{\partial v(2,r)}{\partial r}\right)$ is attractive for $r < a$ and is repulsive for $r > a$.

14.11 The Levinson Theorem

This theorem relates the difference between the scattering phase shifts at zero and at infinite energy to the number of bound states of the system. We use the analytic properties of the Jost function to prove this theorem. For a given partial wave, ℓ, from the definition of the scattering matrix in terms of the phase shift (14.51) we find the logarithmic derivative of $S_\ell(k)$ to be

$$\frac{1}{S_\ell(k)}\frac{dS_\ell(k)}{d\,k} = 2i\frac{d\delta_\ell(k)}{d\,k}. \tag{14.344}$$

Noting that $\delta_\ell(k)$ is an odd function of of k, $\delta(k) = -\delta(-k)$, we write the integral of the left-hand side of (14.344) as

$$\begin{aligned} I &= \int_{-\infty}^{\infty}\frac{1}{S_\ell(k)}\frac{dS_\ell(k)}{d\,k} = 4i\int_0^\infty\frac{d\delta_\ell(k)}{d\,k}dk \\ &= 4i\left\{\delta_\ell(\infty) - \delta_\ell(0)\right\} \end{aligned} \tag{14.345}$$

where we have regarded the phase shift as a continuous function of k in the range $0 \le k < \infty$. We can also write I in terms of the Jost function $f_\ell(k)$, Eq. (14.309)

$$\begin{aligned} I &= \int_{-\infty}^{\infty}\frac{d}{d\,k}\left\{\ln\, f_\ell(k) - \ln\, f_\ell(-k)\right\}dk \\ &= 2\int_{-\infty}^{\infty}\frac{d}{d\,k}\ln\, f_\ell(k)d\,k. \end{aligned} \tag{14.346}$$

The last integral can be evaluated by contour integration. We choose the contour to consist of the following parts: (a) a semi-circle C_1 of radius k_c in the lower half of k-plane and (b) a line extending from $k=-k_c-\varepsilon$ to $k=k_c+\varepsilon$, and (c) a semi-circle of radius ε centered at the origin. Thus

$$\frac{1}{2\pi i}\oint_C \frac{d}{d\,k}(\ln\ f_\ell(k))\,d\,k = \frac{1}{2\pi i}\int_{-k_c}^{k_c}\frac{d}{d\,k}(\ln\ f_\ell(k))\,d\,k$$
$$+\ \frac{1}{2\pi i}\int_{C_1}\frac{d}{d\,k}(\ln\ f_\ell(k))\,d\,k + \frac{1}{2\pi i}\int_{\varepsilon}\frac{d}{d\,k}(\ln\ f_\ell(k))\,d\,k. \quad (14.347)$$

In the limit of $k_c \to \infty$, the contour C will include all of the N_ℓ simple isolated zeros at the points $k_n(\ell) = -i\kappa_n(\ell)$, $n = 1, \cdots N_\ell$, and these are the poles of the integrand $\frac{d}{dk}\ln f_\ell(k)$. The first integral on the right-hand side of (14.347) is $(2\pi i)^{-1}\times\frac{I}{2}$. The second integral (along the semi-circle C_1) goes to zero as the radius of the semi-circle goes to infinity. The last integral along the semi-circle ε contributes an amount πi provided that $f_\ell(0)=0$, and this can happen only if $\ell=0$. Thus from Eqs. (14.345)–(14.347) we find

$$\delta_\ell(0)-\delta_\ell(\infty) = \begin{cases}\pi\left(N_0+\frac{1}{2}\right), & \ell=0, \quad f_0(0)=0\\ \pi N_\ell & \ell\neq 0\end{cases}. \quad (14.348)$$

If the potential has a finite range then $\delta_\ell(\infty)=0$, and $\delta_\ell(0)$ is determined by the number of bound states [30].

14.12 Number of Bound States for a Given Partial Wave

If the potential between two particles is attractive and is given by $V(r) = \frac{\hbar^2}{2m}v(r)$, then the reduced wave function, $u_\ell(r)$, can be chosen as a real function of r and is given by the solution of the Schödinger equation. If we replace $v(r)$ by $sv(r)$ where $0\le s\le 1$ and denote the number of bound states for the ℓ-th partial wave by $N_\ell(s)$, then $N_\ell(s)$ is an increasing function of s [31].

Consider the Schödinger equation for the potential $sv(r)$;

$$u_\ell''(r)+\left(k^2-\frac{\ell(\ell+1)}{r^2}\right)u_\ell(r) = -s|v(r)|u_\ell(r), \quad (14.349)$$

where $-s|v(r)|$ is the attractive potential, and where primes denote derivatives with respect to r. Differentiating (14.349) with respect to s we find

$$\frac{\partial u_\ell''(r)}{\partial\,s}+\frac{\partial k^2}{\partial\,s}u_\ell(r)+\left[k^2-\frac{\ell(\ell+1)}{r^2}+s|v(r)|\right]\frac{\partial u_\ell(r)}{\partial\,s} = -|v(r)|u_\ell(r). \quad (14.350)$$

By substituting for the terms in the square bracket in (14.350) from (14.349) and multiplying the result by $u_\ell(r)$ we get

$$\left(u_\ell(r)\frac{\partial u_\ell''(r)}{\partial\, s} - \frac{\partial u_\ell(r)}{\partial\, s}u_\ell''(r)\right) = -\left(\frac{\partial k^2}{\partial\, s}\right)u_\ell^2(r) - |v(r)|u_\ell^2(r)$$

$$= \frac{\partial}{\partial r}\left(u_\ell(r)\frac{\partial u_\ell'(r)}{\partial\, s} - \frac{\partial u_\ell(r)}{\partial\, s}u_\ell'(r)\right). \tag{14.351}$$

Now we integrate (14.351) from zero to infinity, and use the boundary conditions appropriate for the bound state, viz, $u_\ell(r)$ and $\frac{\partial u_\ell(r)}{\partial\, s}$ must vanish at $r = 0$ and at $r = \infty$. In this way we obtain

$$\frac{\partial k^2}{\partial s} = -\frac{\int_0^\infty |v(r)|u_\ell^2(r)dr}{\int_0^\infty u_\ell^2(r)dr} < 0. \tag{14.352}$$

This result shows that that the bound state energy is a decreasing function of s. As s decreases, the level with the least binding energy tends to zero, and by a further decrease of s this bound state disappears and $N_\ell(s)$ will be reduced by one. Hence the total number of bound states is equal to the number of zero energy bound states which we get as we vary s from 0 to 1. For this reason we want to determine the zero energy bound states for a fixed s.

The Schrödinger equation for $\mathbf{k} = 0$ is (see Eq. (14.20))

$$\psi(\mathbf{r}, 0) = -\frac{2m}{\hbar^2}s\int V\left(r'\right)\frac{1}{4\pi\left|\mathbf{r}-\mathbf{r}'\right|}\psi\left(\mathbf{r}', 0\right)d^3r'. \tag{14.353}$$

Since in this case we are considering bound states, there is no incoming wave and (14.20) becomes the homogeneous integral equation (14.353). We can decompose (14.353) in terms of partial waves exactly as we did for scattering problem. Thus we get

$$u_\ell(r, 0) = s\int_0^\infty \left|v\left(r'\right)\right| g_\ell\left(r, r'\right) u_\ell\left(r', 0\right) dr', \tag{14.354}$$

where

$$g\left(r, r'\right) = \frac{1}{2\ell+1}\begin{cases} \frac{r^{\ell+1}}{r'^{\,\ell}}, & r < r' \\ \frac{r'^{\,\ell+1}}{r^\ell}, & r > r' \end{cases}. \tag{14.355}$$

The kernel of the integral equation (14.354) is not symmetric, but we can make it symmetric by introducing a new kernel

$$\tilde{g}\left(r, r'\right) = \sqrt{|v(r)|\left|v\left(r'\right)\right|}\; g\left(r, r'\right), \tag{14.356}$$

and a new unknown

$$\tilde{u}_\ell(r, 0) = \sqrt{|v(r)|}\; u_\ell(r, 0), \tag{14.357}$$

and writing (14.354) for the n-th eigenvalue as

$$\tilde{u}_\ell^{(n)}(r, 0) = s_n\int_0^\infty \tilde{g}_\ell\left(r, r'\right)\tilde{u}_\ell^{(n)}\left(r', 0\right)dr'. \tag{14.358}$$

From this eigenvalue equation we can find the eigenfunctions $\tilde{u}_\ell^{(n)}(r,0)$ as well as the eigenvlaue s_n. A well-known result of the theory of integral equation states that if all the eigenvalues of (14.358) are known then $\tilde{g}(r,r')$ can be written as [32]

$$\tilde{g}(r,r') = \sum_{n=1}^{\infty} \frac{\tilde{u}_\ell^{(n)}(r,0)\tilde{u}_\ell^{(n)}(r',0)}{s_n}. \tag{14.359}$$

By setting $r = r'$ and integrating over r, then using the orthonormal properties of $\tilde{u}_\ell^{(n)}(r,0)$ s we obtain

$$\int_0^\infty \tilde{g}(r,r)\,dr = \sum_{n=1}^{\infty} \frac{1}{s_n} = \frac{1}{2\ell+1}\int_0^\infty r|v(r)|dr. \tag{14.360}$$

From the fact that $s_n < 1$, we find an inequality for the number of bound states N_ℓ;

$$N_\ell \le \sum_{n=1}^{N_\ell} \frac{1}{s_n} \le \sum_{n=1}^{\infty} \frac{1}{s_n} = \frac{1}{2\ell+1}\int_0^\infty r|v(r)|dr. \tag{14.361}$$

This result due to Bargmann [31] gives an upper bound for the number of bound states for a given partial wave.

For the two Eckart potentials $v(6,r)$ and $v(2,r)$, Eqs. (14.332) and (14.336), the integral

$$\int_0^\infty r|v(r)|dr, \tag{14.362}$$

can be evaluated exactly. The result for $v(2,r)$ is $2\ln 4$ and for $v(6,r)$ is $6\ln 2$. Thus N_1 is less than 1 and N_1 is about 1.38. We conclude that we can have a bound state for $\ell = 1$ for $v(6,r)$ but not for $v(2,r)$.

For bounds on the number of bound states for oscillating potentials see the work of Chadan and Grosse [34].

14.13 Analyticity of the S-Matrix and the Principle of Casuality

The analytic properties of the S-matrix that we have studied in preceding sections are closely associated with the principle of causality, i.e. to the fact that cause must precede the effect [35]. To demonstrate this connection we consider the scattering matrix for a potential having a finite range R and study the analytic properties of the S-matrix in the complex energy plane and for $\ell = 0$. For this case we can write the time-dependent wave function for $r > R$ as

$$\psi_0(E,r,t) = \frac{1}{r}\left\{e^{-ikr} - S_0(E)e^{ikr}\right\}\exp\left(-\frac{iEt}{\hbar}\right), \tag{14.363}$$

where E is related to k by the usual relation $E = \frac{\hbar^2 k^2}{2m}$. The first term in (14.363) is the incident wave and the second one is the scattered wave. Now we construct a wave packet which is localized both in space and in time

$$\psi_0(r,t) = \int_0^\infty A(E')\,\psi_0(E',r,t)\,dE', \tag{14.364}$$

where the coefficient $A(E')$ in the integrand determines the shape and the range of the localization. By substituting (14.364) in (14.363) we find that $\psi_0(r,t)$ can be written as the difference between two localized wave packets

$$\psi_0(r,t) = \psi_0^{inc}(r,t) - \psi_0^{sc}(r,t), \tag{14.365}$$

where

$$\psi_0^{inc}(r,t) = \int_0^\infty A(E')\,\frac{1}{r}\exp\left[-ik'r - \frac{i}{\hbar}E't\right]dE', \tag{14.366}$$

and

$$\psi_0^{sc}(r,t) = \int_0^\infty A(E')\,\frac{1}{r}S_0(E')\exp\left[-ik'r - \frac{i}{\hbar}E't\right]dE'. \tag{14.367}$$

Since the scattered wave is linearly related to the incident wave, (the linearity of the Schrödinger equation), there must be a linear relationship between the wave packets $\psi_0^{inc}(r,t)$ and $\psi_0^{sc}(r,t)$. To state this linearity assumption mathematically, we express $\psi_0^{sc}(r,t)$ at time t as a linear combination of the incoming wave at all previous times

$$\psi_0^{sc}(r,t) = \int_0^\infty \mathcal{K}(\tau)\,\psi_0^{inc}(r,\ t-\tau)d\tau. \tag{14.368}$$

Here the form of $\mathcal{K}(\tau)$ is determined by the properties of the system. If we multiply (14.368) by $\exp\left(\frac{i}{\hbar}Et\right)$ and integrate from $-\infty$ to ∞ we obtain

$$S_0(E)\,e^{2ikr} = \int_0^\infty \mathcal{K}(\tau)\exp\left(\frac{i}{\hbar}E\tau\right)d\tau. \tag{14.369}$$

Let us examine Eq. (14.369) when E is complex. This relation shows that the scattering matrix $S_0(E)$ is finite and single-valued in the upper half E-plane and does not go to infinity. For Im $E > 0$, the integrand in (14.369) contains the factor $\exp\left[-\frac{\mathrm{Im}\,(E)\tau}{\hbar}\right]$, $\tau > 0$ and since $\mathcal{K}(\tau)$ is finite for $0 \le \tau \le \infty$, the integral in (14.369) is convergent. The region of analyticity of $S_0(E)$ can be extended to the lower half plane by analytic continuation of (14.369). It should be emphasized that the analyticity of $S_0(E)$ follows only from the principle of causality and is not dependent on the nature of the interaction.

From the analytic properties of $S_0(E)$ in the E plane and the relation between E and k, we find that $S_0(k)$ is analytic in the first quadrant of the k-plane. Using the symmetry property of $S_0(k)$ matrix, Eq. (14.52) and also

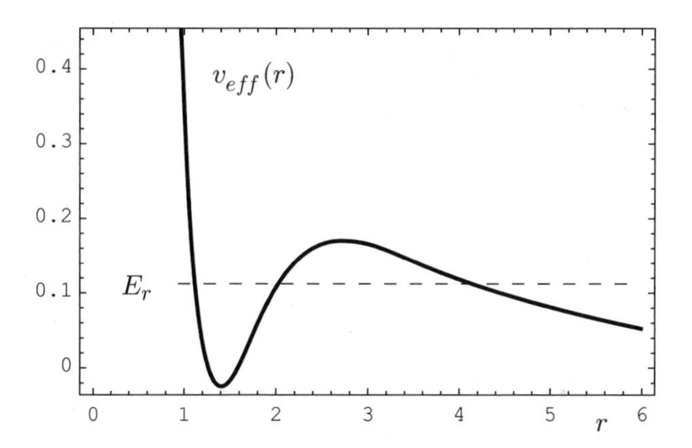

Figure 14.1: The effective potential, $v_{eff}(r)$, Eq. (14.372) plotted as a function r. The resonance energy E_r is positive and is below the maximum of the potential barrier.

$S_0^*(k^*) = S_0^{-1}(k)$, the analyticity region can be extended to the whole k plane except for the isolated poles.

For other partial waves similar analytic properties can be deduced from the principle of causality.

14.14 Resonance Scattering

In many scattering problems in atomic, nuclear and particle physics we encounter cases where the cross section for a given partial wave i.e.

$$\sigma_\ell(k) = 4\pi(2\ell+1)|f_\ell(k)|^2, \tag{14.370}$$

exhibits a sharp peak which we call a resonance. This happens when the potential is attractive close to the origin but becomes repulsive for large r. For $\ell = 0$, the potential itself must have both the attractive- and the repulsive part. However for $\ell \neq 0$ the potential $V(r)$ can be completely attractive, and the centripetal force $\frac{\ell(\ell+1)}{r^2}$ will provide the repulsive tail for the effective potential

$$v_{eff}(r) = v(r) + \frac{\ell(\ell+1)}{r^2}. \tag{14.371}$$

In Fig. 14.1 such an effective potential which is

$$v_{eff}(r) = v(6,r) + \frac{\ell(\ell+1)}{r^2}, \tag{14.372}$$

is plotted as a function of r, where $v(6,r)$ is given by Eckart potential, Eq. (14.340), and $\ell = 1$. In such a potential the particle can be trapped inside

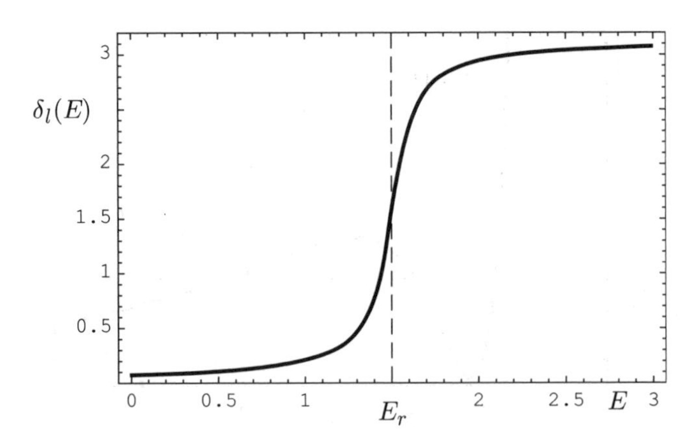

Figure 14.2: The partial wave phase shift shown as a function of the energy E, Eq. (14.377).

the well, when $E > 0$, say $E = E_r$, but such a trapped state has a finite lifetime, since the particle can escape to infinity by quantum tunneling [36]. If we calculate the scattering phase shift for this partial wave we find that δ_ℓ reaches the value of $\frac{\pi}{2}$ as the incident energy approaches the value $E = E_r$. In Fig. 14.2, the phase shift $\delta_\ell(E)$ is plotted versus the energy E. Here we observe the very rapid change of $\delta_{ell}(E)$ as E passes through E_r.

For $\ell > 0$ a resonance occurs when $\cot\delta_\ell(E)$ vanishes, and this happens when

$$\delta_\ell(E) = \left(n + \frac{1}{2}\right)\pi, \quad n \text{ an integer.} \tag{14.373}$$

Near the resonance we can expand $\cot\delta_\ell(E)$ in powers of E;

$$\cot\delta_\ell(E) = \cot\delta_\ell(E_r) - \frac{2}{\Gamma}(E - E_r) + \mathcal{O}\left((E - E_r)^2\right) \simeq -\frac{2}{\Gamma}(E - E_r), \tag{14.374}$$

where Γ is a constant and is given by

$$-\frac{2}{\Gamma} = \left[\frac{d}{dE}\cot\delta_\ell(E)\right]_{E=E_r}. \tag{14.375}$$

From this expanded form of $\cot\delta_\ell(E)$ we find the partial wave scattering amplitude near the resonance to be

$$\begin{aligned} f_\ell(E) &= \frac{1}{k\cot\delta_\ell(E) - i\,k} = \frac{1}{k}\frac{1}{\left[-\frac{2}{\Gamma}(E - E_r) - i\right]} \\ &= -\frac{\frac{1}{2}\Gamma}{k\left[(E - E_r) + \frac{i\Gamma}{2}\right]}. \end{aligned} \tag{14.376}$$

We also find that $\delta_\ell(E)$ near E_r to be of the form

$$\delta_\ell(E) = \frac{\pi}{2} + \tan^{-1}\left(\frac{E - E_r}{\frac{\Gamma}{2}}\right). \tag{14.377}$$

Substituting (14.376) in (14.370) we find the partial cross section to be

$$\sigma_\ell(E) = \frac{4\pi}{k^2}\frac{(2\ell+1)\left(\frac{\Gamma}{2}\right)^2}{(E-E_r)^2+\frac{1}{4}\Gamma^2}. \tag{14.378}$$

We note that as (14.374) shows the slope of $\delta_\ell(E)$ ar $E = E_r$ is equal to $\frac{2}{\Gamma}$. Thus the sharper the change in $\delta_\ell(E)$, the sharper is the resonance.

14.15 The Born Series

The simplest way of solving the integral equation (14.194) is by iterative-perturbation technique [4]. Using subscripts to indicate the order of the iteration we have

$$T_1^+(\mathbf{k}_{out}, \mathbf{k}_{in}) = V(\mathbf{k}_{out}, \mathbf{k}_{in}), \tag{14.379}$$

$$T_2^+(\mathbf{k}_{out}, \mathbf{k}_{in}) = V(\mathbf{k}_{out}, \mathbf{k}_{in}) + \frac{2m}{(2\pi)^3\hbar^2}\int \frac{V(\mathbf{k}_{out}, \mathbf{q})V(\mathbf{q}, \mathbf{k}_{in})}{k_{in}^2 - q^2 + i\epsilon}d^3q, \tag{14.380}$$

$$\begin{aligned} T_3^+(\mathbf{k}_{out}, \mathbf{k}_{in}) &= T_2^+(\mathbf{k}_{out}, \mathbf{k}_{in}) \\ &+ \left(\frac{2m}{\hbar^2}\right)^2\left(\frac{1}{(2\pi)^6}\right)\int \frac{V(\mathbf{k}_{out}, \mathbf{q})V(\mathbf{q}, \mathbf{p})V(\mathbf{p}, \mathbf{k}_{in})}{(k_{in}^2 - q^2 + i\epsilon)(k_{in}^2 - p^2 + i\epsilon)}d^3q d^3p, \end{aligned} \tag{14.381}$$

and similar equations for $T_4^+(\mathbf{k}_{out}, \mathbf{k}_{in})$, $T_5^+(\mathbf{k}_{out}, \mathbf{k}_{in})$ *etc.*

Once $T_n(\mathbf{k}_{out}, \mathbf{k}_{in})$ is determined the scattering amplitude to the n-th order of iteration can be obtained from (14.201);

$$f_{Bn}(\theta, \phi) = -\frac{m}{2\pi\hbar^2}T_n(\mathbf{k}_{out}, \mathbf{k}_{in}). \tag{14.382}$$

The rate of convergence of the series shown in (14.379)–(14.381), that is the Born series depends on the strength of the potential.

Let us consider the first term in the Born approximation f_{B1} which is given by

$$f_{B1}(\theta, \phi) = -\frac{m}{2\pi\hbar^2}V(\mathbf{k}_{out}, \mathbf{k}_{in}) = -\frac{m}{2\pi\hbar^2}\int e^{-i(\mathbf{k}_{in}-\mathbf{k}_{out})\cdot\mathbf{r}}\, V(\mathbf{r})d^3r. \tag{14.383}$$

We can write $f_{B1}(\theta, \phi)$ in terms of the momentum transfer vector $\mathbf{\Delta}$ which we define as

$$\mathbf{\Delta} = \mathbf{k}_{in} - \mathbf{k}_{out} = \frac{1}{\hbar}(\mathbf{p}_{in} - \mathbf{p}_{out}). \tag{14.384}$$

For elastic scattering $|\mathbf{k}_{in}| = |\mathbf{k}_{out}| = k$, therefore

$$\mathbf{\Delta}^2 = 2k^2 - 2k^2\cos\theta = 4k^2\sin^2\frac{\theta}{2}, \tag{14.385}$$

or

$$\Delta = 2k \sin \frac{\theta}{2}. \tag{14.386}$$

Thus we find a simple expression for $f_{B1}(\theta, \phi)$

$$f_{B1}(\theta, \phi) = -\frac{m}{2\hbar^2} \int e^{-i\mathbf{\Delta}\cdot\mathbf{r}}\, V(\mathbf{r}) d^3r \tag{14.387}$$

Now if the potential is central $V(\mathbf{r}) = V(r)$, then we can choose $\mathbf{\Delta}$ to be in the direction of the polar axis. By carrying out the integration over the angular coordinates of $\mathbf{r}$ we obtain

$$f_{B1}(k, \theta) = -\frac{2m}{\Delta\hbar^2} \int_0^\infty r \sin[\Delta(k,\theta) r]\, V(\mathbf{r}) dr. \tag{14.388}$$

This result shows that for central potentials f_{B1} is independent of the azimuthal angle ϕ.

We can expand $f_{B1}(k, \theta)$ in terms of $P_\ell(\cos\theta)$ noting that $\Delta^2 = 2k^2(1 - \cos\theta)$;

$$\begin{aligned} f_{B1}(k,\theta) &= -\frac{2m}{\hbar^2} \int_0^\infty r^2 \sum_{\ell=0}^\infty (2\ell+1) P_\ell(\cos\theta)\, j_\ell^2(kr) V(r) dr \\ &= \sum_{\ell=0}^\infty (2\ell+1) P_\ell(\cos\theta) \frac{\tan\delta_\ell^B}{k}, \end{aligned} \tag{14.389}$$

where

$$\tan\delta_\ell^B = -k \int_0^\infty j_\ell^2(kr) v(r) r^2 dr, \tag{14.390}$$

and where we have replaced $V(r)$ by $v(r)$, Eq. (14.40). We can obtain the same result from the integral equation for the partial wave function, (14.41), by noting that for a weak potential the solution of the Schrödinger equation can be approximated by

$$u_\ell(k, r) \approx r \cos\delta_\ell\, j_\ell(kr). \tag{14.391}$$

Now if we substitute (14.391) in (14.42) we find Eq. (14.390).

Convergence of the Born Series — Conditions for the convergence of the Born series can be obtained from the theory of the Fredholm integral equation of the second kind [32]. The inhomogeneous Fredholm equation is of the form

$$f(x) = g(x) + \lambda \int_a^b K(x,y) f(y) dy, \tag{14.392}$$

where $g(x)$ and $K(x, y)$ are both square integrable functions in the range $a \le x \le b$, and where λ is a parameter. The perturbative solution of (14.392) is given by Liouville-Neumann series [5]

$$f(x) = g(x) + \lambda g_1(x) + \lambda^2 g_2(x) + \cdots, \tag{14.393}$$

where

$$g_n(x) = \int_a^b [K(x,y)]^n g(y) dy, \tag{14.394}$$

provided that

$$|\lambda| \left[\int_a^b dx \int_a^b |K(x,y)|^2 dy\right] < 1. \tag{14.395}$$

When this condition is satisfied then the series (14.393) converges absolutely and uniformly in x in the interval $a \le x \le b$ [32].

Let us apply this result to the the integral equation (14.20) for the wave function. We note that this equation in its present form is not of Fredholm type. To make it a Fredholm equation we multiply it by $\sqrt{|V(\mathbf{r})|}$ and introduce $\Psi(\mathbf{r})$ and $\Phi(\mathbf{r})$ by

$$\Psi(\mathbf{r}) = \sqrt{|V(\mathbf{r})|}\psi(\mathbf{r}), \tag{14.396}$$

and

$$\Phi(\mathbf{r}) = \sqrt{|V(\mathbf{r})|} e^{i\mathbf{k}\cdot\mathbf{r}}. \tag{14.397}$$

Then Eq. (14.20) changes to

$$\Psi(\mathbf{r}) = \Phi(\mathbf{r}) + \int K(\mathbf{r},\mathbf{r}')\, \Psi(\mathbf{r}')\, d^3r. \tag{14.398}$$

In this equation both $\Phi(\mathbf{r})$ and $K(\mathbf{r},\mathbf{r}')$ are square integrable functions, and (14.398) is a Fredholm equation of the second kind. Therefore we can use the criterion for the convergence of the Born (or Liouville–Neumann) series which is

$$\left[\int |K(\mathbf{r},\mathbf{r}')|^2 d^3r d^3r'\right]^{\frac{1}{2}} < 1, \tag{14.399}$$

to find the condition that the potential, $V(\mathbf{r})$, has to satisfy. For real wave number $\mathbf{k}$, this condition can also be written as

$$\frac{1}{4\pi}\left[\int |V(\mathbf{r})|\,|V(\mathbf{r}')|\, \frac{1}{|\mathbf{r}-\mathbf{r}'|}\, d^3r\, d^3r'\right]^{\frac{1}{2}} < 1, \tag{14.400}$$

By defining $\mathcal{B}$ to be the maximum of the integral

$$\mathcal{B} = \max\, \frac{1}{4\pi}\int \frac{|V(\mathbf{r}')|}{|\mathbf{r}-\mathbf{r}'|}\, d^3r', \tag{14.401}$$

when $\mathbf{r}$ is varied, we find that the inequality (14.400) simplifies to

$$\mathcal{B} < 1, \quad \text{for all}\ \ \mathbf{r}. \tag{14.402}$$

For a central potential this inequality reduces to

$$\int_0^\infty r|V(r)|dr < 1. \tag{14.403}$$

But this is exactly Bergmann's condition derived in Sec. 14.12 for the nonexistence of a bound state. Thus we conclude that the Born series [37],[38] converges for all energies provided that $-|V(r)|$ does not support any bound state.

Second Born Approximation — As an example let us calculate the first and the second Born approximation for the Yukawa potential [4]

$$V(r) = V_0 \frac{e^{\mu r}}{\mu r}, \tag{14.404}$$

where V_0 and μ are constants. By substituting this potential in (14.383) and evaluating the integral we find

$$f_{B1}(k,\theta) = -\frac{mV_0}{2\mu\hbar^2\left(\mu^2+\Delta^2(k,\theta)\right)} = -\frac{mV_0}{2\mu\hbar^2\left(\mu^2+4k^2\sin^2\frac{\theta}{2}\right)}. \tag{14.405}$$

We find the second order of iteration from Eq. (14.380) and the Fourier transform of the potential (14.379);

$$\begin{aligned} & T_2^+(\mathbf{k}_{out},\mathbf{k}_{in}) = V(\mathbf{k}_{out},\mathbf{k}_{in}) \\ + \; & \frac{4mV_0^2}{\pi\mu^2}\int \frac{d^3q}{\left[\mu^2+(\mathbf{q}-\mathbf{k}_{in})^2\right]\left[\mu^2+(\mathbf{q}-\mathbf{k}_{out})^2\right]\left(k_{in}^2-q^2+i\epsilon\right)} \\ = \; & V(\Delta) + \frac{4mV_0^2}{\pi\mu^2\hbar^2}M(\Delta,k), \end{aligned} \tag{14.406}$$

where

$$\begin{aligned} M(\Delta,k) &= \frac{\pi^2}{\Delta\sqrt{\mu^4+4k^2\left(\mu^2+\Delta^2\right)}} \\ &\times \left\{\tan^{-1}\left[\frac{\mu\Delta}{\sqrt{\mu^4+4k^2\left(\mu^2+\Delta^2\right)}}\right]\right. \\ &+ \left.\frac{i}{2}\ln\left[\frac{\sqrt{\mu^4+4k^2\left(\mu^2+\Delta^2\right)}+2\Delta^2}{\sqrt{\mu^4+4k^2\left(\mu^2+\Delta^2\right)}-2\Delta^2}\right]\right\}. \end{aligned} \tag{14.407}$$

Substituting (14.407) in (14.201) gives us the scattering amplitude to the second order of the T-matrix.

For the scattering of a particle of charge Ze from a center of charge $Z'e$, i.e. Rutherford scattering we can derive the scattering amplitude by observing that the Coulomb potential can be written as

$$V_c = \lim_{\mu\to 0}\left[ZZ'e^2\mu\left(\frac{e^{-\mu r}}{\mu r}\right)\right]. \tag{14.408}$$

Thus by replacing $\frac{V_0}{\mu}$ by $ZZ'e^2$ in (14.405) and then taking the limit of $\mu \to 0$ we find the scattering amplitude $f_{B1}(\theta)$ to be

$$f_{B1}(\theta) = -\frac{mZZ'e^2}{2\hbar^2k^2\sin^2\frac{\theta}{2}} = -\frac{ZZ'e^2}{4E\sin^2\frac{\theta}{2}}, \tag{14.409}$$

and from $f_{B1}(\theta)$ we obtain the cross section

$$\frac{d\sigma}{d\Omega} = |f_{B1}(\theta)|^2 = \left(\frac{ZZ'e^2}{4E\sin^2\frac{\theta}{2}}\right)^2. \tag{14.410}$$

This result is identical with the classical cross section for scattering of a charged particle in the Coulomb field (Rutherford formula) [39]. It is also identical with the exact quantum mechanical result found by solving the Schrödinger equation in parabolic coordinates, Eq. (14.139) [40].

14.16 Impact Parameter Representation of the Scattering Amplitude

For high energy scattering a large number of phase shifts contribute to the scattering amplitude. Rather than calculating a large number of phase shifts and adding the partial wave scattering amplitudes, we use a different approach and replace the summation over partial waves by integration over the impact parameter [41]. Let us start with the integral equation for the T-matrix

$$T^+(\mathbf{p},\mathbf{k}) = V(\mathbf{p},\mathbf{k}) - \frac{1}{(2\pi)^3}\int \frac{V(\mathbf{p},\mathbf{q})T^+(\mathbf{q},\mathbf{k})}{q^2-k^2-i\epsilon}d^3q, \tag{14.411}$$

where we have set $\hbar^2 = 2m = 1$. We choose our coordinate system such that

$$\hat{\mathbf{k}} = (0,\ 0\ 1), \tag{14.412}$$

$$\hat{\mathbf{p}} = (\sin\theta\cos\phi,\ \sin\theta\sin\phi,\ \cos\theta), \tag{14.413}$$

and

$$\hat{\mathbf{q}} = (\sin\beta\cos\alpha,\ \sin\alpha\sin\beta,\ \cos\beta), \tag{14.414}$$

therefore

$$\cos\Theta = \hat{\mathbf{q}}\cdot\hat{\mathbf{p}} = \sin\theta\sin\beta\cos(\phi-\alpha) + \cos\theta\cos\beta. \tag{14.415}$$

Now we define

$$y = \sin\frac{\theta}{2}, \tag{14.416}$$

and introduce two new functions H^+ and B by the relations [42]

$$T^+(p,y,k) = -4\pi\int_0^\infty J_0(2pby)H^+(p,b,k)bdb, \tag{14.417}$$

$$V(p,y,k) = -4\pi\int_0^\infty J_0(2pby)B(p,b,k)bdb, \tag{14.418}$$

where $J_0(z)$ is the cylindrical Bessel function of order zero. By substituting (14.417) and (14.418) in (14.411) we have

$$\begin{aligned}
&\int_0^\infty J_0\left(2pb\sin\frac{\theta}{2}\right)[H^+(p,b,k)-B(p,b,k)]bdb \\
&= \frac{1}{2\pi^2}\int_0^\infty \frac{q^2dq}{q^2-k^2-i\epsilon}\int_0^\infty\int_0^\infty B(p,b',q)H^+(q,b'',k) \\
&\times \left\{\int J_0\left(2pb'\sin\frac{\Theta}{2}\right)J_0\left(2qb''\sin\frac{\beta}{2}\right)d\Omega_q\right\}b'\,db'\,b''\,db''. \quad (14.419)
\end{aligned}$$

Now let us consider the integral

$$I = \frac{1}{4\pi}\int J_0\left(2pb'\sin\frac{\Theta}{2}\right)J_0\left(2qb''\sin\frac{\beta}{2}\right)d\Omega_q. \quad (14.420)$$

We can simplify this integral by noting that [43]

$$\int_0^\pi J_0\left[\sqrt{z^2+Z^2-2zZ\cos\phi}\right]d\phi = \pi J_0(z)J_0(Z) \quad (14.421)$$

and

$$\int_0^{2\pi} J_0\left(2pb\sin\frac{\Theta}{2}\right)d\alpha = 2\pi J_0\left(2pb'\sin\frac{\theta}{2}\cos\frac{\beta}{2}\right)J_0\left(2pb'\sin\frac{\beta}{2}\cos\frac{\theta}{2}\right). \quad (14.422)$$

By substituting these two integrals in (14.420) we obtain

$$\begin{aligned}
I(p,y,q) &= \frac{1}{2}\int_{-1}^1 J_0\left(2qb''\sin\frac{\beta}{2}\right)J_0\left(2pb'\sin\frac{\theta}{2}\cos\frac{\beta}{2}\right) \\
&\times J_0\left(2pb'\sin\frac{\beta}{2}\cos\frac{\theta}{2}\right)d\cos\beta.
\end{aligned} \quad (14.423)$$

This last relation can also be written as

$$I(p,y,q) = \frac{1}{\pi}\int_0^\pi \frac{J_1(2\beta)}{\beta}d\phi, \quad (14.424)$$

where

$$\beta^2 = p^2b'^2+q^2b''^2-2pqb'b''\cos\frac{\theta}{2}\cos\phi. \quad (14.425)$$

Next if we define the propagator $G(p,q,b',b'';b)$ by

$$G(p,q,b',b'';b) = \frac{2p^2}{\pi}\int_0^\pi d\phi\int_0^\pi \sin\frac{\theta}{2}\cos\frac{\theta}{2}J_0\left(2pb\sin\frac{\theta}{2}\right)\frac{J_1[2\beta(\theta,\phi)]}{\beta(\theta,\phi)}d\theta, \quad (14.426)$$

and use

$$\int_0^\infty J_0\left(2pb\sin\frac{\theta}{2}\right) J_0\left(2pb\sin\frac{\theta'}{2}\right) b\,db = \frac{\delta\left(\sin\frac{\theta}{2} - \sin\frac{\theta'}{2}\right)}{4p^2\sin\frac{\theta}{2}}, \tag{14.427}$$

we can write $I(p, y, q)$ as

$$I(p, y, q) = \int_0^\infty G\left(p, q, b', b''; b\right) J_0(2pby) b\,db. \tag{14.428}$$

Finally by substituting for I in Eq. (14.419) and using (14.427) we find

$$\begin{aligned} H^+(p, b, k) &= B(p, b, k) + \frac{2}{\pi}\int_0^\infty \frac{q^2 dq}{q^2 - k^2 - i\epsilon} \\ \times &\int_0^\infty \int_0^\infty B\left(p, b', q\right) G\left(p, q, b', b''; b\right) H^+\left(q, b'', k\right) b'\,db'\,b''\,db''. \end{aligned} \tag{14.429}$$

This is the integral equation for $H^+(p, b, k)$ with G given by (14.429). In the high energy limit, Eq. (14.423) for I can be reduced to

$$\begin{aligned} I(p, y, q) &= \frac{1}{2}\int_{-1}^1 J_0\left(2qb''\sin\frac{\beta}{2}\right) J_0\left(2pb'\sin\frac{\theta}{2}\cos\frac{\beta}{2}\right) \\ &\times\ J_0\left(2pb'\sin\frac{\beta}{2}\cos\frac{\theta}{2}\right) d(\cos\theta) \\ &\to\ J_0\left(2pb'\sin\frac{\theta}{2}\right)\frac{\delta\left(b'' - \frac{p}{q}b\right)}{2qpb'} \quad \text{as} \quad p \text{ or } q \to \infty. \end{aligned} \tag{14.430}$$

Therefore in the high energy limit the kernel G becomes

$$G\left(q, p, b', b''; b\right) \approx \frac{1}{2pqb'b}\delta\left(b - b'\right)\delta\left(b'' - \frac{p}{q}b'\right). \tag{14.431}$$

Substituting for G from (14.431) in (14.429), we can simplify the integral equation (14.429)

$$H^+(p, b, k) = B(p, b, k) + \frac{1}{\pi}\int_0^\infty \frac{dq}{q^2 - k^2 - i\epsilon} B(p, b, q) H^+\left(q, \frac{p}{q}b, k\right). \tag{14.432}$$

Noting that

$$\frac{1}{q^2 - k^2 - i\epsilon} = \mathcal{P}\frac{1}{q^2 - k^2} + i\pi\delta\left(q^2 - p^2\right), \tag{14.433}$$

and in the high energy limit the contribution from the principal value integral is small compared with the contribution from the δ-function, we can evaluate the integral in (14.432) in this limit. Thus for the diagonal elements of H^+ we find [44]

$$H^+(k,b,k) = \frac{B(k,b,k)}{1-\frac{i}{2k}B(k,b,k)}. \tag{14.434}$$

While $B(k,b,q)$ can be determined from the inverse Bessel transform of (14.418) it is convenient to obtain it directly from the potential. For this we write $v(\mathbf{p},\mathbf{k})$ for a central force in the expanded form;

$$v(k,k\cos\theta) = 4\pi\sum_{\ell=0}^{\infty}(2\ell+1)P_\ell(\cos\theta)v_\ell(k,k), \tag{14.435}$$

where $v_\ell(k,k)$ is defined by (14.207)

$$v_\ell(k,k) = \frac{2k^2}{\pi}\int_0^\infty j_\ell^2(kr)v(r)r^2dr = k\int_0^\infty J_{\ell+\frac{1}{2}}^2(kr)v(r)rdr. \tag{14.436}$$

Next we replace $P_\ell(\cos\theta)$ in (14.435) by [4]

$$P_\ell\left(1-2y^2\right) = \int_0^\infty J_{2\ell+1}(x)J_0(xy)dx, \quad y=\sin\frac{\theta}{2}, \tag{14.437}$$

to get

$$v(k,y) = \frac{2\pi^2}{k^2}\int_0^\infty J_0(2kby)kdb\sum_{\ell=0}^{\infty}(2\ell+1)v_\ell(k,k)J_{2\ell+1}(2kb). \tag{14.438}$$

By comparing (14.418) with (14.438) we find $B(k,b,k)$,

$$B(k,b,k) = -\frac{\pi}{2kb}\sum_{\ell=0}^{\infty}(2\ell+1)v_\ell(k,k)J_{2\ell+1}(2kb). \tag{14.439}$$

In order to express $B(k,b,k)$ in terms of the potential $v(r)$ we substitute for $v_\ell(k,k)$ in (14.439),

$$v_\ell(k,k) = \frac{2k^2}{\pi}\int_0^\infty r^2 j_\ell^2(kr)v(r)dr, \tag{14.440}$$

and in addition we replace $j_\ell(kr)$ by $J_{\ell+\frac{1}{2}}(kr)$ in (14.440)

$$j_\ell^2(kr) = \frac{\pi}{2kr}J_{\ell+\frac{1}{2}}^2(kr), \tag{14.441}$$

to get

$$B(k,b,k) = -\frac{\pi}{b}\int_0^\infty v(r)rdr\left\{\sum_{\ell=0}^{\infty}(2\ell+1)J_{2\ell+1}(2kb)J_{\ell+\frac{1}{2}}^2(kr)\right\}. \tag{14.442}$$

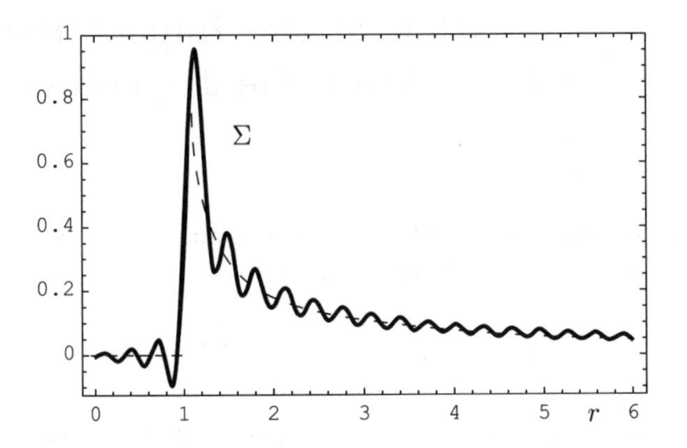

Figure 14.3: The sum in Eq. (14.443) is shown as a function of r, for $k = 10$ and $b = 1$. The dashed line in this figure is a plot of the right-hand side of Eq. (14.443).

For high energies, i.e. large k we have the asymptotic relation

$$\left\{\sum_{\ell=0}^{\infty}(2\ell+1)J_{2\ell+1}(2kb)J^2_{\ell+\frac{1}{2}}(kr)\right\} \to \frac{b\theta(r-b)}{\pi\sqrt{r^2-b^2}}, \tag{14.443}$$

where $\theta(r-b)$ is the step function. The validity of this approximation for large value of k is shown in Fig. 14.3. Substituting for the sum in (14.442) we find a simple expression for $B(k,b,k)$;

$$B(k,b,k) = -\int_b^{\infty}\frac{rv(r)}{\sqrt{r^2-b^2}} = -\int_0^{\infty} v\left(\sqrt{z^2+b^2}\right)dz. \tag{14.444}$$

The diagonal elements of the T^+-matrix in high energy limit are obtained by substituting (14.444) in (14.434) and (14.417). To simplify the resulting T-matrix let us define the impact parameter phase shift $\delta(b)$ by

$$\tan\delta(b) = -\frac{1}{2k}\int_0^{\infty} v\left(\sqrt{z^2+b^2}\right)dz, \tag{14.445}$$

then $H^+(k,b,k)$, Eq. (14.434), becomes

$$H^+(k,b,k) = -ik\left[e^{2i\delta(b)}-1\right]. \tag{14.446}$$

Having obtained $H^+(k,b,k)$ from (14.417) we find the energy- and the angular-dependence of the diagonal elements of the T-matrix

$$T^+\left(k,\sin\frac{\theta}{2}\right) = 4\pi ik\int_0^{\infty} J_0\left(2kb\sin\frac{\theta}{2}\right)\left[e^{2i\delta(b)}-1\right]bdb. \tag{14.447}$$

Finally we get the scattering amplitude from Eq. (14.201). Writing the factors m and $\hbar$ explicitly we have

$$f(\theta,k) = -ik\int_0^{\infty} J_0\left(2kb\sin\frac{\theta}{2}\right)\left[e^{2i\delta(b)}-1\right]bdb. \tag{14.448}$$

14.17 Determination of the Impact Parameter Phase Shift from the Differential Cross Section

For elastic scattering, we can find the scattering amplitude from $\sigma(\theta)$ when f is a function of θ only. We write Eq. (14.28) as

$$f(\theta) = |f(\theta)|e^{i\nu(\theta)} = \sqrt{\frac{d\sigma(\theta)}{d\Omega}}e^{i\nu(\theta)}, \tag{14.449}$$

where $\nu(\theta)$ which is the phase of f to be determined. From the unitarity condition, Eq. (14.216), written in terms of f, i.e.

$$\frac{1}{2i}\left[f(\mathbf{k}_{out}, \mathbf{k}_{in}) - f^*(\mathbf{k}_{in}, \mathbf{k}_{out})\right] = \frac{k}{4\pi}\int f^*(\mathbf{p}, \mathbf{k}_{out})f(\mathbf{p}, \mathbf{k}_{in})d\Omega_{\mathbf{p}} \tag{14.450}$$

and Eq. (14.449) we obtain an integral equation for $\nu(\theta)$ [45]

$$\sqrt{\frac{d\sigma(\theta)}{d\Omega}}\sin\nu(\theta) = \frac{k}{4\pi}\int\sqrt{\frac{d\sigma\left(\theta'\right)}{d\Omega'}\frac{d\sigma\left(\theta''\right)}{d\Omega''}}\cos\left[\nu\left(\theta'\right) - \nu\left(\theta''\right)\right]d\Omega_{\mathbf{p}}, \tag{14.451}$$

where $\cos\theta = \hat{\mathbf{k}}_{out}\cdot\hat{\mathbf{k}}_{in}$, $\cos\theta' = \hat{\mathbf{p}}\cdot\hat{\mathbf{k}}_{in}$ and $\cos\theta'' = \hat{\mathbf{p}}\cdot\hat{\mathbf{k}}_{out}$ and again we have set $2m = \hbar = 1$. The integral in (14.451) is over all possible directions of the vector $\mathbf{p}$. From Eqs. (14.448) and (14.449) it follows that

$$\sqrt{\frac{d\sigma(\theta)}{d\Omega}}\sin\nu(\theta) = k\int_0^\infty J_0\left(2kb\sin\frac{\theta}{2}\right)[1 - 2\cos(2\delta(k,b))]bdb. \tag{14.452}$$

Next we change the variable from the scattering angle θ to the momentum transfer $\Delta = 2k\sin\frac{\theta}{2}$,

$$\sqrt{\frac{d\sigma}{d\Omega}(\Delta)}\sin\nu(\Delta) = k\int_0^\infty J_0\left(b\Delta\right)[1 - 2\cos(2\delta(k,b))]bdb. \tag{14.453}$$

For high energies Δ can be regarded as a variable changing from zero to infinity. In this limit we can invert Eq. (14.453) and find $\delta(k,b)$ in terms of $\frac{d\sigma}{d\Omega}(\Delta)$

$$2\sin^2\delta(k,b) = \frac{1}{k}\int_0^\infty J_0\left(b\Delta\right)\sqrt{\frac{d\sigma}{d\Omega}(\Delta)}\sin\nu(\Delta)\Delta d\Delta, \tag{14.454}$$

where we have used the Fourier–Bessel integral

$$\int_0^\infty J_n(b\Delta)b\,db\int_0^\infty J_n\left(b'\Delta\right)F\left(b'\right)b'\,db' = F(b). \tag{14.455}$$

Thus from the measurement of $\frac{d\sigma(\theta)}{d\Omega}$ for high energy scattering we obtain $\delta(k,b)$.

Now we want to see how we can determine the interaction potential $v(r)$ if $\delta(k,b)$ is known. The simplest way is to start with Eq. (14.445),

$$\tan\delta(k,b) = -\frac{1}{2k}\int_0^\infty v\left(\sqrt{b^2+z^2}\right)dz = -\frac{1}{2k}\int_b^\infty \frac{v(r)dr}{\sqrt{r^2-b^2}}. \tag{14.456}$$

Here the left side of (14.456) is assumed to be known and we want to find $v(r)$. Thus (14.456) is an integral equation of Abel type for $v(r)$. The solution of (14.456) is given by [46]

$$v(r) = \frac{4k}{\pi r}\frac{d}{dr}\int_0^\infty \tan\delta\left[k, \sqrt{r^2+x^2}\right]dx. \tag{14.457}$$

The potential found from (14.457) is not static but is energy-dependent. An alternative method is to use the Born approximation for the partial wave phase shift, Eq. (14.390), to determine the potential $v(r)$. Thus using the sum

$$\sum_{\ell=0}^\infty (-1)^\ell(2\ell+1)j_\ell^2(kr) = j_0(2kr), \tag{14.458}$$

we have

$$\begin{aligned}\sum_{\ell=0}^\infty (-1)^\ell(2\ell+1)\tan\delta_\ell^B &= -k\int_0^\infty v(r)r^2dr\sum_{\ell=0}^\infty(-1)^\ell(2\ell+1)j_\ell^2(kr)\\ &= -\int_0^\infty v(r)r^2 j_0(2kr)dr.\end{aligned} \tag{14.459}$$

Now (14.459) is an integral equation for $v(r)$ which can be solved by the inverse Fourier–Bessel transform;

$$v(r) = -\frac{8}{\pi r}\int_0^\infty\left\{\sum_{\ell=0}^\infty(-1)^\ell(2\ell+1)\tan\delta_\ell^B(k)\right\}\sin(2kr)dk. \tag{14.460}$$

We can also write (14.460) in terms of the phase shift $\delta(k,b)$ by replacing the sum by an integral. To this end we change ℓ to ρ where $\rho = kb = \ell + \frac{1}{2}$;

$$v(r) = -\frac{16}{\pi r}\int_0^\infty \sin(2kr)dk\int_0^\infty \rho\sin(\pi\rho)\tan\delta_\ell^B(\rho,k)d\rho. \tag{14.461}$$

A better way of inverting (14.456) is to use the sum

$$\sum_{\ell=0}^\infty(-1)^\ell\ell(\ell+1)(2\ell+1)j_\ell^2(kr) = -kj_1(2kr), \tag{14.462}$$

and again replace the summation by integration to obtain

$$v(r) = \frac{16}{\pi r}\int_0^\infty j_1(2kr)dk\int_0^\infty\rho\left(\rho^2-\frac{1}{4}\right)\sin(\pi\rho)\tan\delta_\ell^B(\rho,k)d\rho. \tag{14.463}$$

We note that in (14.463) the phase shifts for higher partial waves are weighed more in the integral over ρ. Now the phase shifts for higher partial waves are determined more accurately in the Born approximation, and therefore we expect that $v(r)$ obtained from (14.463) to be closer to the actual potential than the one found from (14.460).

For a detailed account of the inversion of the empirical data to determine the potential the reader is referred to the excellent monograph by Chadan and Sabatier [47].

14.18 Elastic Scattering of Identical Particles

In this section we consider the scattering of two identical particles when (a) - the target and projectile are bosons, and (b) - the two particles are fermions. For scattering of two particles, since we can only observe the particles far from the scattering region, we have situations shown in Fig. 14.4 (a) and (b), where detectors D_1 and D_2 detect the scattered particles. In the case of identical particles we do not know which detector registers which particle. As we will see we have different cross sections depending whether the particles are fermions or bosons.

Exchange Effects in the Case of Bosons — If $\mathbf{r}_1$ and $\mathbf{r}_2$ denote the coordinates of the two particles and $\mathbf{r}$ is defined by $\mathbf{r} = \mathbf{r}_1 - \mathbf{r}_2$, then the asymptotic form of the wave function according to Eq. (14.21) is

$$\psi(\mathbf{r})_{r\to\infty} \to e^{ikr\cos\theta} + \frac{f(\theta)}{r}e^{ikr}. \tag{14.464}$$

For the case of two bosons we have to symmetrize this wave function so that by interchanging $\mathbf{r}_1$ and $\mathbf{r}_2$, $\psi(\mathbf{r}_1, \mathbf{r}_2)$ should remain unchanged. Now in this permutation of $\mathbf{r}_1$ and $\mathbf{r}_2$, $\mathbf{r}$ will change sign and becomes $-\mathbf{r}$. This vector in polar coordinates has the coordinates by $(r,\ \pi - \theta)$. Therefore the asymptotic form symmetric wave function $\psi_s(\mathbf{r})$ becomes

$$\begin{aligned}\lim_{r\to\infty}\psi_s(\mathbf{r}) &= \lim_{r\to\infty}\left[\psi(\mathbf{r}) + \psi(-\mathbf{r})\right] \\ &\to \left[e^{ikr\cos\theta} + e^{-ikr\cos\theta} + \frac{f(\theta) + f(\pi-\theta)}{r}e^{ikr}\right].\end{aligned} \tag{14.465}$$

The first two terms in (14.465) represent incoming particles one along the positive and the other along the negative z-axis. The scattered wave $\frac{e^{ikr}}{r}$ has the coefficient $f(\theta) + f(\pi - \theta)$, and thus the cross section will be

$$\frac{d\sigma_s}{d\Omega} = |f(\theta) + f(\pi - \theta)|^2. \tag{14.466}$$

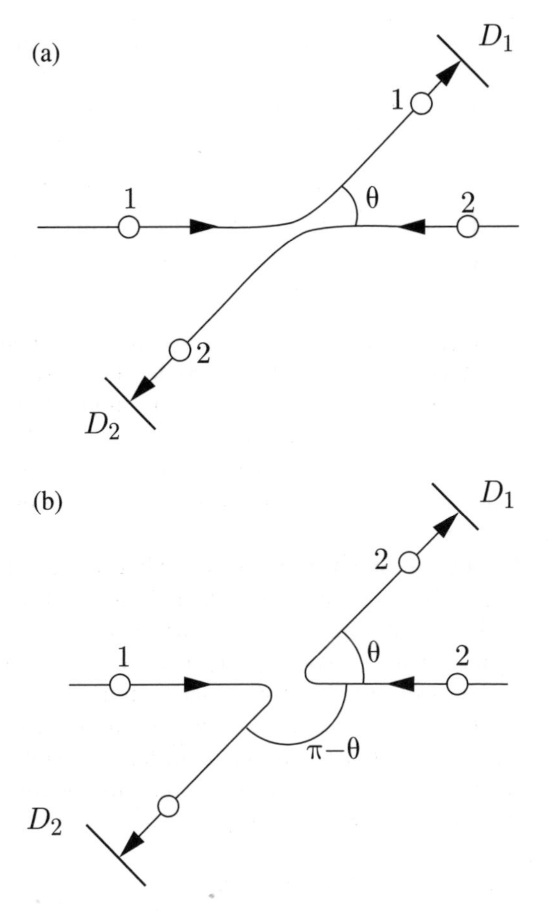

Figure 14.4: Scattering of two particles in their center of mass system. (a) - Particle 1 is detected in D_1 and 2 in the detector D_2. (b) - Particle 2 is detected in the detector D_1 and particle 1 is detected in D_2. For two identical particles (a) and (b) are indistinguishable.

For the case of two distinguishable particles the differential cross section for observing one of the two particles at θ, would be the sum of the two cross sections

$$\frac{d\sigma}{d\Omega} = |f(\theta)|^2 + |f(\pi - \theta)|^2. \tag{14.467}$$

The difference between the two cross sections (14.466) and (14.467) which is given by

$$2\mathrm{Re}\left\{f^*(\theta)f(\pi - \theta)\right\}, \tag{14.468}$$

is the effect of exchange symmetry and it arises from the symmetrization of the wave function.

If we consider the scattering of two charged particle by Coulomb force, e.g. for $\alpha - \alpha$ scattering, from $f(\theta)$ for Rutherford scattering (14.409) and (14.466), we obtain

$$\frac{d\sigma_s}{d\Omega} = \left(\frac{Z^2e^2}{2mv^2}\right)^2 \left\{\frac{1}{\sin^4\frac{\theta}{4}} + \frac{1}{\cos^4\frac{\theta}{4}} + \frac{2\cos\left[\left(\frac{Z^2e^2}{\hbar v}\right)\ln\tan^2\frac{\theta}{2}\right]}{\sin^2\frac{\theta}{2}\cos^2\frac{\theta}{2}}\right\}, \quad (14.469)$$

where v is the relative speed of the two particles and m is their reduced mass, $m = \frac{1}{2}m_\alpha$, where m_α is the mass of each α particle. This scattering of identical bosons is called Mott scattering [48]. The last term in (14.469) is the exchange effect, and is a purely quantum mechanical effect. This can be seen by considering the limit of the last term in (14.469) as $\hbar \to 0$. In this limit $\frac{Z^2e^2}{\hbar v} \to \infty$, and the last term oscillates rapidly, and if we average this term over a small angle it will vanish.

Exchange Effect for Scattering of Identical Fermions — If two spin $\frac{1}{2}$ particles scatter off each other, then the total spin of the system is either zero (for singlet state) and one (for triplet state). Now the total wave function which is a product of a spin part and a space part must be antisymmetric, therefore for singlet spin state the coordinate wave function must be symmetric. As we have seen before for a symmetric coordinate wave function the total cross section $\frac{d\sigma_s}{d\Omega}$ is given by (14.466). On the other hand for the triplet state the spins are parallel and thus the coordinate wave function must be antisymmetric, and thus the asymptotic form of $\psi_t(\mathbf{r})$ becomes

$$\begin{aligned}\lim_{r\to\infty}\psi_t(\mathbf{r}) &= \lim_{r\to\infty}\left[\psi(\mathbf{r}) - \psi(-\mathbf{r})\right] \\ &\to \left[e^{ikr\cos\theta} - e^{-ikr\cos\theta} + \frac{f(\theta) - f(\pi-\theta)}{r}e^{ikr}\right]. \quad (14.470)\end{aligned}$$

From this asymptotic wave function we get the cross section $\frac{d\sigma_t}{d\Omega}$ where

$$\frac{d\sigma_t}{d\Omega} = |f(\theta) - f(\pi-\theta)|^2. \quad (14.471)$$

For the Coulomb scattering of two fermions in singlet state, the wave function being symmetric, we get the same cross section as the one we found for two bosons, that is (14.469). On the other hand for triplet state from (14.471) we have

$$\frac{d\sigma_t}{d\Omega} = \left(\frac{Z^2e^2}{2mv^2}\right)^2 \left\{\frac{1}{\sin^4\frac{\theta}{4}} + \frac{1}{\cos^4\frac{\theta}{4}} - \frac{2\cos\left[\left(\frac{Z^2e^2}{\hbar v}\right)\ln\tan^2\frac{\theta}{2}\right]}{\sin^2\frac{\theta}{2}\cos^2\frac{\theta}{2}}\right\}. \quad (14.472)$$

When both the particles in the target and in the beam are unpolarized then we must find the average of the singlet and triplet cross sections. Noting that there is one spin state for singlet and three for triplet, therefore this average is

$$\begin{aligned}\frac{d\sigma}{d\Omega} &= \frac{1}{4}\frac{d\sigma_s}{d\Omega} + \frac{3}{4}\frac{d\sigma_t}{d\Omega} \\ &= \left(\frac{Z^2e^2}{2mv^2}\right)^2 \left\{ \frac{1}{\sin^4\frac{\theta}{4}} + \frac{1}{\cos^4\frac{\theta}{4}} - \frac{\cos\left[\left(\frac{Z^2e^2}{\hbar v}\right)\ln\tan^2\frac{\theta}{2}\right]}{\sin^2\frac{\theta}{2}\cos^2\frac{\theta}{2}} \right\}.\end{aligned} \tag{14.473}$$

Here we have assumed that each spin state has the same probability.

14.19 Transition Probability

The transition between the initial state $|in\rangle = |i\rangle$ and the final state $|out\rangle = |f\rangle$ is given by Eq. (14.175) which we write as

$$S_{fi} = \delta_{fi} - 2\pi i\delta(E_f - E_i)T_{fi}. \tag{14.474}$$

From this expression we can find the transition probability per unit time which is physically an important observable quantity. We observe that the transition takes place between different states, i.e. $i \neq f$, and thus the term δ_{fi} will not contribute to S_{fi}, but as long as $E_f = E_i$, and the energy is conserved, the second term will contribute. We also note that in the case of transition into continuum, the energy eigenstates can be highly degenerate.

Let us first define the transition amplitude $S_{fi}(\mathcal{T})$ for a long but finite time, which we will denote by $\mathcal{T}$,

$$S_{fi}(\mathcal{T}) = -T_{fi}\left(\frac{i}{\hbar}\right)\int_{-\frac{1}{2}\mathcal{T}}^{\frac{1}{2}\mathcal{T}} e^{\frac{i}{\hbar}(E_i-E_f)t}dt, \tag{14.475}$$

so that

$$S_{fi} = \lim_{\mathcal{T}\to\infty} S_{fi}(\mathcal{T}). \tag{14.476}$$

Since $S_{fi}(\mathcal{T})$ is the transition amplitude over a period $\mathcal{T}$ of observation, the transition probability is $|S_{fi}(\mathcal{T})|^2$ over the same period, and the transition probability per unit time is

$$w_{i\to f} = \lim_{\mathcal{T}\to\infty}\frac{|S_{fi}(\mathcal{T})|^2}{\mathcal{T}}. \tag{14.477}$$

Now for the limit of large $\mathcal{T}$ we have

$$\lim_{\mathcal{T}\to\infty}\frac{\sin^2\left[\frac{1}{2\hbar}(E_i - E_f)\mathcal{T}\right]}{\left(\frac{1}{2\hbar}\right)^2\mathcal{T}(E_i - E_f)^2} \to 2\pi\hbar\delta(E_i - E_f). \tag{14.478}$$

By substituting (14.478) in (14.477) we find a simple expression for $w_{i\to f}$

$$w_{i\to f} = \frac{2\pi}{\hbar}\delta(E_i - E_f)|T_{fi}|^2, \quad i \neq f. \tag{14.479}$$

The singularity seen in (14.479) originates from the assumption that transition is into a sharp state in continuum. In fact we want to calculate the transition from the initial state i into a group of states centered about $E_i = E_f$. If we denote the density of final states in the neighborhood of E_f by $\rho_f(E)$, then the transition probability into this group of states can be expressed in a simple form:

$$w_{i\to f} = \frac{2\pi}{\hbar}\int_{E_f-\frac{1}{2}dE}^{E_f+\frac{1}{2}dE} \delta(E_i - E_f)|T_{fi}|^2\rho_f(E)dE = \frac{2\pi}{\hbar}\rho_f(E)|T_{fi}|^2. \tag{14.480}$$

This relation is the well-known Fermi's golden rule.

14.20 Transition Probabilities for Forced Harmonic Oscillator

Under the action of a time-dependent force an initial stationary state may pass to another stationary state. Here we want to determine the probability of this transition using the Heisenberg equations of motion. A very simple but instructive example of using the S- matrix discussed earlier in this chapter is to calculate the transition probabilities for an oscillator coupled to a force which is only a function of time [49]–[51]. For instance if we have a particle with charge q bound by a harmonic force and placed in an electric field $E_x(t)$ parallel to the x-axis then its total Hamiltonian is

$$H = \frac{p^2}{2m} + \frac{1}{2}m\omega^2x^2 - xF(t), \tag{14.481}$$

where the force $F(t) = qE_x$ is only a function of time. We assume that $F(t)$ goes to zero as $t \to \pm\infty$ and that its time Fourier transform

$$f(\omega) = \int_{-\infty}^{\infty} F(t)e^{i\omega t}dt, \tag{14.482}$$

is well-defined. We now write H in terms of creation and annihilation operators, Eqs. (5.35) and (5.36);

$$H = \omega a^\dagger a - \frac{1}{\sqrt{2}\alpha}\left(a^\dagger + a\right)F(t), \tag{14.483}$$

where $\alpha = \sqrt{m\omega}$ and we have set $\hbar = 1$. The Heisenberg equation for $a(t)$ is

$$i\frac{da(t)}{dt} = [a(t),\ H] \tag{14.484}$$

or by calculating the commutator we can write it as

$$\left(-i\frac{d}{dt}+\omega\right)a(t)=\frac{1}{\sqrt{2}\alpha}F(t). \tag{14.485}$$

Next let us define the *in*- and *out*- operators from the Heisenberg equation (14.485) by a limiting procedure. We have assumed that $F(t)\to 0$ as $t\to\pm\infty$, therefore asymptotically $a(t)\to a(0)e^{-i\omega t}$ and $a^\dagger(t)\to a^\dagger(0)e^{i\omega t}$. With these asymptotic conditions we define a_{out} and a_{in} by

$$\lim_{t\to+\infty}\left[\langle out|a(t)|in\rangle e^{i\omega t}\right]=\langle out|a_{out}|in\rangle, \tag{14.486}$$

and

$$\lim_{t\to-\infty}\left[\langle out|a(t)|in\rangle e^{i\omega t}\right]=\langle out|a_{in}|in\rangle. \tag{14.487}$$

Similarly for $a_{out}^\dagger$ we have

$$\lim_{t\to+\infty}\left[\langle out\left|a^\dagger(t)\right|in\rangle e^{-i\omega t}\right]=\left\langle out\left|a^\dagger_{out}\right|in\right\rangle, \tag{14.488}$$

and the corresponding relation for $a^\dagger(t)$ for the limit of $t\to-\infty$. From the properties of the creation and annihilation operators it follows that

$$a^\dagger_{out}|n,\ out\rangle=\sqrt{n+1}\,|n+1,\ out\rangle, \tag{14.489}$$

$$\left(a^\dagger_{out}\right)^n|0,\ out\rangle=\sqrt{n!}\,|n,\ out\rangle, \tag{14.490}$$

$$a_{out}|n,\ out\rangle=\sqrt{n}\,|n-1,\ out\rangle, \tag{14.491}$$

and

$$a_{out}|0,\ out\rangle=0, \tag{14.492}$$

with a similar set of equations for *in*-states.

Next we define the unitary S-matrix by

$$S=\sum_j|j,\ in\rangle\langle j,\ out| \tag{14.493}$$

with its inverse which is

$$S^{-1}=\sum_k|k,\ out\rangle\langle k,\ in|. \tag{14.494}$$

We can easily verify that

$$S^\dagger S=SS^\dagger=\sum_k|k,\ out\rangle\langle k,\ out|=1, \tag{14.495}$$

where 1 is the unit matrix. The matrix elements of S can be determined from (14.494);

$$\begin{aligned} S_{k,j} &= \langle k,\ out|S|j,\ out\rangle = \sum_i \langle k,\ out|i,\ in\rangle\langle i,\ out|j,\ out\rangle \\ &= \sum_i \delta_{ij}\langle k,\ out|i,\ in\rangle = \langle k,\ out|j,\ in\rangle. \end{aligned} \tag{14.496}$$

From this S-matrix we find the transition probability from the initial state k to the final state j;

$$P_{k\to j} = |\langle j,\ out|k,\ in\rangle|^2. \tag{14.497}$$

In order to calculate $P_{k\to j}$ we start by evaluating the matrix element $S_{j,0}$;

$$S_{j,0} = \langle j,\ out|0,\ in\rangle. \tag{14.498}$$

With the help of Eqs. (14.486) and (14.491) we write $S_{j,0}$ as

$$\begin{aligned} S_{j,0} &= \frac{1}{\sqrt{j}}\langle j-1,\ out|a_{out}|0,\ in\rangle \\ &= \frac{1}{\sqrt{j}}\lim_{t\to+\infty}\left[\langle j-1,\ out|a(t)|0,\ in\rangle e^{i\omega t}\right]. \end{aligned} \tag{14.499}$$

This matrix element can also be written as an integral over $a(t)$, i.e.

$$\begin{aligned} S_{j,0} &= \sqrt{j}\int_{-\infty}^{+\infty}\frac{d}{dt}\left[\langle j-1,\ out|a(t)|0,\ in\rangle e^{i\omega t}\right]dt \\ &+ \sqrt{j}\lim_{t\to-\infty}\left[\langle j-1,\ out|a(t)|0,\ in\rangle e^{i\omega t}\right]. \end{aligned} \tag{14.500}$$

Since

$$\lim_{t\to-\infty} a(t)|0,\ in\rangle = a_{in}|0,\ in\rangle = 0, \tag{14.501}$$

the last term in (14.500) will not contribute to $S_{j,0}$. By taking the derivative of the integrand in (14.500) and substituting for $\frac{da(t)}{dt}$ from the Heisenberg equation (14.485) we obtain

$$S_{j,0} = \frac{i}{\sqrt{2j}\alpha}\int_{-\infty}^{+\infty} e^{i\omega t}\langle j-1,\ out|F(t)|0,\ in\rangle dt. \tag{14.502}$$

The force $F(t)$ is a c-number, and can be taken out of the expression for the expectation value, then the integral in (14.502) gives us the Fourier transform of $F(t)$;

$$S_{j,0} = \frac{i}{\sqrt{2j}\alpha}f(\omega)\langle j-1,\ out|0,\ in\rangle. \tag{14.503}$$

By comparing (14.503) with (14.498) we observe that $S_{j,0}$ can be expressed in terms of $S_{j-1,0}$;

$$S_{j,0} = \frac{i}{\sqrt{2j}\alpha}f(\omega)S_{j-1,0}. \tag{14.504}$$

Now repeating this process $j-1$ times we arrive at

$$S_{j,0} = [if(\omega)]^j \frac{1}{\sqrt{j!}}(\sqrt{2}\alpha)^{-j} S_{0,0}. \tag{14.505}$$

Thus the probability for $j \to 0$ transition becomes

$$P_{j\to 0} = |S_{j,0}|^2 = \frac{1}{(2\alpha^2)^j} \frac{1}{j!} |f(\omega)|^{2j}\, |S_{0,0}|^2. \tag{14.506}$$

Let us introduce the dimensionless quantity ξ by

$$\xi = \frac{1}{2\alpha^2}|f(\omega)|^2, \tag{14.507}$$

and observe that the sum of probabilities for transition from $j = 0, 1, 2\cdots$ to the ground state 0 must be equal to unity;

$$\sum_{j=0}^{\infty} P_{j\to 0} = |S_{0,0}|^2 \sum_{j=0}^{\infty} \frac{\xi^j}{j!} = |S_{0,0}|^2 e^{\xi} = 1. \tag{14.508}$$

Using this argument we find that $|S_{0,0}|^2 = e^{-\xi}$. By substituting this result in (14.506) we obtain $P_{j\to 0}$ in terms of the known quantity ξ;

$$P_{j\to 0} = \left(\frac{\xi^j}{j!}\right) e^{-\xi}. \tag{14.509}$$

As (14.509) shows the transition probability between an excited level j and the ground state 0 has the form of a Poisson distribution.

Having found the transition from a higher level to the ground state we now proceed to find the transition probability from one excited level j to a lower level k, $(j \geq k)$. In this case we start with

$$\begin{aligned} S_{j,k} &= \langle j,\ out|k,\ in\rangle = \frac{1}{\sqrt{k}}\left\langle j,\ out\left|a_{in}^{\dagger}\right| k-1,\ in\right\rangle \\ &= \frac{1}{\sqrt{k}} \lim_{t\to-\infty}\left(\left\langle j,\ out\left|a^{\dagger}(t)\right| k-1,\ in\right\rangle e^{-i\omega t}\right). \end{aligned} \tag{14.510}$$

Again we write (14.510) as an integral over $a^{\dagger}(t)$

$$\begin{aligned} S_{j,k} &= -\frac{1}{\sqrt{k}} \int_{-\infty}^{+\infty} e^{-i\omega t} \left\langle j,\ out\left|\left[\frac{da^{\dagger}(t)}{dt} - i\omega a^{\dagger}(t)\right]\right| k-1,\ in\right\rangle dt \\ &+ \frac{1}{\sqrt{k}} \lim_{t\to+\infty}\left(\left\langle j,\ out\left|a^{\dagger}(t)\right| k-1,\ in\right\rangle e^{-i\omega t}\right), \end{aligned} \tag{14.511}$$

and we can simplify and write it as

$$S_{j,k} = \frac{i}{\sqrt{2k}\alpha} \int_{-\infty}^{+\infty} e^{-i\omega t} F^*(t) dt\, \langle j,\ out|k-1,\ in\rangle$$

$$
\begin{aligned}
&+ \sqrt{\frac{j}{k}}\,\langle j-1,\ out|k-1,\ in\rangle \\
&= \frac{i}{\sqrt{2k\alpha}} f^*(\omega) S_{j,k-1} + \sqrt{\frac{j}{k}} S_{j-1,k-1}.
\end{aligned} \tag{14.512}
$$

This relation is a recurrence relation connecting $S_{j,k}$ to both $S_{j,k-1}$ and $S_{j-1,k-1}$. By iterating the recurrence relation (14.512) we arrive at the result [49]

$$
S_{j,k} = \sum_{l=0}^{k} \frac{k!}{(k-l)!l!} [if^*(t)]^l \frac{1}{\sqrt{2}\alpha} \left[\frac{j!}{k!(j-k+l)!}\right]^{\frac{1}{2}} S_{j-k+l,0}. \tag{14.513}
$$

We have already determined $S_{j-k+l,0}$, Eq. (14.505), with $j \to j-k-l$. If we substitute from that equation in (14.513) we find

$$
S_{j,k} = \frac{\sqrt{k!j!}}{(\sqrt{2}\alpha)^{j-k}} i^{j-k} (f(\omega))^{j-k} e^{-\frac{\xi}{2}} \sum_{l=0}^{k} \frac{(-\xi)^l}{(j-k+l)!(k-l)!l!}. \tag{14.514}
$$

Thus the probability for $j \to k$ transition is

$$
P_{j\to k} = |S_{j,k}|^2 = k!j!\xi^{j-k} e^{-\xi} \left[\sum_{l=0}^{k} \frac{(-\xi)^l}{(j-k+l)!(k-l)!l!}\right]^2. \tag{14.515}
$$

The sum in (14.515) can be written in terms of Laguerre polynomials [43];

$$
P_{j\to k} = \frac{k!}{j!} \xi^{j-k} e^{-\xi} \left[L_k^{j-k}(\xi)\right]^2, \quad j \ge k. \tag{14.516}
$$

For $j < k$ we simply interchange the integers j and k in (14.516);

$$
P_{j\to k} = \frac{j!}{k!} \xi^{k-j} e^{-\xi} \left[L_j^{k-j}(\xi)\right]^2, \quad j < k. \tag{14.517}
$$

From this expression we can determine the mean-energy transform ΔE_k;

$$
\begin{aligned}
\Delta E_k &= \omega \sum_{j=0}^{\infty} (j-k) P_{j\to k} \\
&= \omega \sum_{j=k}^{\infty} (j-k) P_{j\to k} + \omega \sum_{j=0}^{k-1} (k-j) P_{k\to j}.
\end{aligned} \tag{14.518}
$$

Using the properties of the Laguerre polynomials we can write the last factor in (14.516) as

$$
\left[L_k^{j-k}(\xi)\right]^2 = \frac{j!}{k!} \sum_{l=0}^{k} \frac{\xi^{2l} L_{k-l}^{j-k+2l}(2\xi)}{(j-k+l)!l!}. \tag{14.519}
$$

Substituting for $P_{j\to k}$ and $P_{k\to j}$ in (14.518) we obtain

$$\begin{aligned}\frac{\Delta E_k}{\omega} &= \sum_{n=0}^{\infty} n e^{-\xi}\xi^n \sum_{l=0}^{k} \frac{\xi^{2l} L_{k-l}^{n+2l}(2\xi)}{(n+l)!l!} \\ &- \sum_{n=1}^{k} n e^{-\xi}\xi^n \sum_{l=0}^{k-n} \frac{\xi^{2l} L_{k-n-l}^{n+2l}(2\xi)}{(n+l)!l!}. \end{aligned} \tag{14.520}$$

Next we set $n = j - k$ in (14.516) and $n = k - j$ in (14.520), and also change the indices

$$l \to l + n, \quad n \to -n. \tag{14.521}$$

These changes allow us to combine the second sum in (14.520) with the first with the result that

$$\frac{\Delta E_k}{\omega} = e^{-\xi} \sum_{l=0}^{k} \sum_{n=-l}^{\infty} \frac{n\xi^{n+2l} L_{k-l}^{n+2l}(2\xi)}{(n+l)!l!}. \tag{14.522}$$

One more change, viz, $n \to n + l$ gives us

$$\frac{\Delta E_k}{\omega} = e^{-\xi} \sum_{l=0}^{k} \sum_{n=0}^{\infty} \frac{(n-l)\xi^{n+l} L_{k-l}^{n+l}(2\xi)}{n!l!}. \tag{14.523}$$

The Laguerre polynomials can be obtained from the Rodrigues representation [43]

$$L_n^{\beta} = \frac{e^{\xi}\xi^{-\beta}}{n!} \frac{d^n}{d\xi^n} \left(e^{-\xi}\xi^{n+\beta}\right), \tag{14.524}$$

and if we use this expression in (14.522) we find

$$\frac{\Delta E_k}{\omega} = e^{\xi} \sum_{l=0}^{k} \sum_{n=0}^{\infty} \frac{(n-l)}{n!l!(k-l)!2^{n+l}} \frac{d^{k-l}}{du^{k-l}} \left(e^{-u}u^{k+n}\right), \tag{14.525}$$

where $u = 2\xi$. We observe that from the sum

$$\sum_{n=0}^{\infty} \frac{(n-l)\left(\frac{u}{2}\right)^n}{n!} = e^{\frac{u}{2}} \left(\frac{u}{2} - l\right), \tag{14.526}$$

and (14.522) we obtain

$$\begin{aligned}\frac{\Delta E_k}{\omega} e^{-\xi} &= \sum_{l=0}^{k} \frac{2^{-l-1}}{l!(k-l)!} \frac{d^{k-l}}{du^{k-l}} \left(e^{-\frac{u}{2}} u^{k+1}\right) \\ &- \sum_{l=0}^{k} \frac{2^{-l}}{(l-1)!(k-l)!} \frac{d^{k-l}}{du^{k-l}} \left(e^{-\frac{u}{2}} u^{k}\right). \end{aligned} \tag{14.527}$$

Making use of the binomial expansion and a shift of the index in the second term we find

$$\begin{aligned} e^{-\xi}\frac{\Delta E_k}{\omega} &= \frac{1}{k!}\left(1+\frac{d}{d\xi}\right)^k\left(e^{-\xi}\xi^{k+1}\right) \\ &- \frac{1}{(k-1)!}\left(1+\frac{d}{d\xi}\right)^{k-1}\left(e^{-\xi}\xi^k\right). \end{aligned} \tag{14.528}$$

Noting that

$$\left(1+\frac{d}{d\xi}\right)\left(e^{-\xi}x^k\right) = ke^{-\xi}\xi^{k-1}, \tag{14.529}$$

equation (14.523) can be reduced to

$$\frac{\Delta E_k}{\omega} = \xi = \frac{1}{2m\omega}|f(\omega)|^2. \tag{14.530}$$

Classical Expression for the Transfer of Energy — The result that we have found for $\frac{\Delta E_k}{\omega}$ is identical to the classical expression for the energy transfer and is independent of the Planck constant $\hbar$. Consider the classical forced oscillator with the natural frequency ω

$$\ddot{x}(t) + \omega^2 x(t) = \frac{1}{m}F(t), \tag{14.531}$$

with $F(t) \to 0$ as $t \to \pm\infty$. Let us assume that initially the oscillator is at rest with zero velocity $x(-\infty) = \dot{x}(-\infty) = 0$, and thus its total energy is zero. In order to integrate (14.531) we introduce $z(t)$ by

$$z(t) = \dot{x}(t) - i\omega x(t). \tag{14.532}$$

Using the variable $z(t)$ we write (14.531) as

$$\dot{z}(t) + i\omega z(t) = \frac{1}{m}F(t). \tag{14.533}$$

For integrating this first order differential equation we multiply it by $e^{i\omega t}$ and integrate it

$$\int_{-\infty}^{t}\left(\dot{z}(t) + i\omega z(t)\right)e^{i\omega t}dt = \frac{1}{m}\int_{-\infty}^{t} F(t)e^{i\omega t}dt. \tag{14.534}$$

By integrating the left-hand side of (14.534) by parts we find

$$\lim_{t\to+\infty}\left(z(t)e^{i\omega t}\right) = \frac{1}{m}\int_{-\infty}^{t} F(t)e^{i\omega t}dt = \frac{1}{m}f(\omega). \tag{14.535}$$

The total energy of the oscillator as $t \to +\infty$ becomes

$$E(t\to+\infty) = \frac{m}{2}\left(\dot{x}^2 + \omega^2x^2\right) = \frac{m}{2}|z(t\to+\infty)|^2. \tag{14.536}$$

Noting that the oscillator initially has zero energy, therefore the energy transfer is

$$\Delta E = E(t \to +\infty) - E(t \to -\infty) = \frac{m}{2}|z(t \to +\infty)|^2 = \frac{1}{2m}|f(\omega)|^2, \quad (14.537)$$

which is the same as the quantum mechanical result (14.530).

Bibliography

[1] For a detailed discussion of the general theory of scattering see M.L. Goldberger and K.M. Watson, *Collision Theory*, (John Wily & Sons, New York, 1964).

[2] N.F. Mott and N.S.W. Massey, *The Theory of Atomic Collisions*, (Oxford University Press, Oxford, 1965).

[3] A.G. Sitenko, *Scattering Theory*, (Springer-Verlag, Berlin 1991), Chapter 7.

[4] P.M. Morse and H. Feshbach, *Methods of Theoretical Physics*, Part II, (McGraw-Hill, New York, 1953).

[5] See for example P.J. Collins, *Differential and Integral Equations*, (Oxford University Press, 2006), Chapter 8.

[6] M. Baker, Determinantal approach to meson-nucleon scattering, Ann. Phys. 271 (1958).

[7] E.J. Heller, T.N. Rescigno and W.P. Reinhardt, Extraction of scattering information from Fredholm determinants calculated in an L^2 basis: A Chebyschev discretization of continuum, Phys. Rev. A 8, 2946 (1973).

[8] B.O. Kerbikov, Scattering theory in a system with confinement, Theo. Math. Phys. 65, 1225 (1985).

[9] For a simple account of the summation of series by the calculus of residues see E.G. Phillips, *Functions of a Comples Variable with Applications*, (Oliver and Boyd, Edinburgh, 1961) Sec. 50.

[10] G.N. Watson, *Theory of Bessel Functions*, (Cambridge University Press, New York, 1958), p. 429.

[11] D.S. Falk, Equivalence of the Brysk approximation and the deteminantal method, Phys. Rev. 129, 2340 (1963).

[12] For a discussion of this point see A.Z. Capri, *Nonrelativistic Quantum Mechanics*, (World Scientific, Singapore, 2002), p. 406.

[13] C. Møller, General properties of the characteristic matrix in the theory of elementary particles, I, Kgl. Danske Videnske Vid. Sels. Mat.-Phys. Medd. 23, 1 (1945).

[14] C. Møller, General properties of the characteristic matrix in the theory of elementary particles, II, Kgl. Danske Videnske Vid. Sels. Mat.-Phys. Medd. 24, 19 (1946).

[15] W. Heisenberg, Die beobachtbaren Gröβ in der Theorie der Elementarteilechen, II, Z. Physik, 120, 673 (1943).

[16] B.A. Lippmann and J. Schwinger, Variational principles for scattering process. I, Phys. Rev. 79, 469 (1950).

[17] H. Lehman, K. Symanzik and W. Zimmermann, On the formulation of quantized field theory, Nuovo Cimento, 2, 425 (1955).

[18] R. Rockmore, A reduction formalism for potential scattering, Ann. Phys. 20, 375 (1962).

[19] F.E. Low, boson-fermion scattering in the Heisenberg representation, Phys. Rev. 97, 1392 (1955).

[20] H. Fukutome, Low's scattering equation and S-matrix, Prog. Theo. Phys. 17, 383 (1957).

[21] A concise account of the theory and application of separable potentials in nuclear physics is given in G.E. Brown and A.D. Jackson, *The Nucleon-Nucleon Interaction*, (North-Holland, Amsterdam, 1976) Chapter V.

[22] M. Baranger, R. Giraud, S.K. Mukhopadhyay and P. Sauer, Off-energy-shell continuation of the nucleon-nucleon transition matrix, Nucl. Phys. A138, 1 (1969).

[23] W. van Dijk and M. Razavy, Constuction of the T-matrix from its symmetric part, Nucl. Phys. A 159, 161 (1970).

[24] R. Omnès, On the solution of certain singular integral equations of qantum field theory, Nuovo Cimento, 8, 316 (1958).

[25] N.I. Mushkhelishvili, *Singular Integral Equations*, (Noordhoff, Groningen, 1955).

[26] R. Jost, On the false zero eigenvalues of the S-matrix, Helv. Phys. Acta, 20, 256 (1947).

[27] D. ter Haar, On the redundant zeros in the theory of the Heisenberg matrix, Physica, 12, 501 (1946).

[28] R. Jost and W. Kohn, Construction of a potential from a phase shift, Phys. Rev. 87, 977 (1952).

[29] V. Bargmann, On the connection between phase shifts and scattering potentials, Rev. Mod. Phys. 21, 488 (1949).

[30] N. Levinson, On the uniqueness of the potential in a Schrödinger equation for a given asymptotic phase, Danske Videnskab, Mat.-Phys. Medd. 26, 9 (1949).

[31] V. Bargmann, On the number of bound states in a central field of force, Proc. N. A. S. U.S., 38, 961 (1952).

[32] See, for example, D. Porter and D.G. Stirling, *Integral Equations*, (Cambridge University Press, New York), Chapter 4.

[33] V. Bargmann, Remarks on the determination of a central field of force from the elastic scattering phase shift, Phys. Rev. 75, 301 (1949).

[34] K. Chadan and H. Grosse, New bounds on the number of bound states, J. Phys. A 16, 955 (1983).

[35] H.M. Nussenzweig, *Causality and Dispersion Relation*, (Academic Press, 1972).

[36] M. Razavy, *Quantum Theory of Tunneling*, (World Scientific, Singapore, 2003).

[37] M. Born, Quantenmechanik der Stoβvorgänge, Z. Physik, 38, 803 (1926).

[38] See for example C.J. Joachain, *Quantum Collision Theory*, (North-Holland, Amsterdam, 1975), Chapter 8.

[39] L.D. Landau and E.M. Lifshitz, *Mechanics*, (Pergamon Press, Oxford, 1960), p. 53.

[40] L.D. Landau and E.M. Lifshitz, *Quantum Mechanics, Nonrelativistic Theory*, (Pergamon Press, Oxford, 1958), p. 419.

[41] For the original and a very interesting discussion of this method see, R. Glauber in *Lectures in Theoretical Physics*, Vol. I. (Intersciece Publishers, New York, 1959).

[42] M. Razavy, Impact parameter representation of the scattering amplitude, Can. J. Phys. 49, 1885 (1971).

[43] See, for instance, I.S. Gradshetyn and I.M. Ryzhik, *Tables of Integrals, Series and Products*, Fourth Edition, (Academic Press, New York, 1965), p. 1037.

[44] R. Blakenbecler and M.L. Goldberger, Behavior of scattering amplitudes at high energies, bound states, and resonances, Phys. Rev. 126, 766 (1962).

[45] T.Y. Wu and T. Ohmura, *Quantum Theory of Scattering*, (Prentice-Hall, Englewood Cliffs, 1962).

[46] E.T. Whittaker and G.N. Watson, *A Course of Modern Analysis*, (Cambridge University Press, 1948), p. 229.

[47] K. Chadan and P.C. Sabatier, *Inverse Problems in Quantum Scattering Theory*, Second Edition, (Springer-Verlag, Berlin, 1989).

[48] N.F. Mott, Collision between two electrons, Proc. Roy. Soc. 126, 259 (1930).

[49] L.M. Scarfone, Transition probabilities for the forced quantum oscillator, Am. J. Phys. 32, 158 (1964).

[50] D.M. Gilby and F.O. Goodman, Quantum theory of the forced harmonic oscillator, Am. J. Phys. 34, 143 (1966).

[51] R.W. Fuller, S.M. Harris and E.L. Slaggie, S-matrix solution for the forced harmonic oscillator, Am. J. Phys. 31, 431 (1963).

Chapter 15

Quantum Diffraction

How a wave train representing an incident particle propagates in space and time and how sharply the wave front can be defined for such a wave? These questions arise naturally in studying a number of phenomena including the problem of the decay of an unstable state through tunneling [1],[2]. This problem which is related to the energy-time uncertainty relation, mathematically resembles the problem of Fraunhofer diffraction in optics. For this reason it is generally referred to as diffraction in time [3],[4]. In the first part of this chapter we study the simple and exactly solvable case of a beam of particles which is incident on a completely absorbing shutter and this shutter opens suddenly. We want to determine the profile of the probability density after the shutter opens.

In the second part we consider the time-independent aspect of scattering of an incident wave, again by a completely absorbing sphere. In nearly all scattering processes diffraction patterns are present and in particular when the scatterer is a composite system which may be regarded as a homogeneous structureless body then the target can be approximated by an absorptive (imaginary) potential. This pattern show up in the angular distribution of the particles undergoing the diffraction scattering of Fraunhofer type where there is a pronounced maximum in the forward direction when the scattering angle θ satisfies the inequality

$$\theta \le \frac{\lambda}{R}. \tag{15.1}$$

In this relation λ is the de Broglie wavelength of the particle, $\lambda = \frac{2\pi}{k}$, and R is the radius of the absorptive target. Quantum diffraction of this type has been observed in the scattering of hadrons by atomic nuclei [5]. In this case if the energy of the projectile (hadron) is large enough to satisfy the inequality $\lambda \ll R$, then the nuclei become strongly absorptive. The scattering caused by the strongly absorptive nucleus for short wavelengths of the incident particles is analogous to the diffraction of light by a black disk in optics.

15.1 Diffraction in Time

In this section we study the problem of propagation of a beam of particles each of unit mass $m = 1$ and with momentum k ($\hbar = 1$) coming from $x \to -\infty$ and impinging on a completely absorbing shutter located at $x = 0$. The shutter can be partially or perfectly reflecting or completely absorbing, but the nature of shutter will not affect the state at the other side of the shutter. The initial wave function for the incident beam is given by

$$\phi(x) = \begin{cases} e^{ikx} & x \le 0 \\ 0 & x > 0 \end{cases}. \tag{15.2}$$

Let us consider the time evolution of this wave function if at $t = 0$ we remove the shutter. Thus for $t > 0$ we solve the time-dependent Schrödinger equation

$$i\frac{\partial\psi(x,t)}{\partial t} = -\frac{1}{2}\frac{\partial^2\psi(x,t)}{\partial x^2}, \tag{15.3}$$

with the initial condition

$$\psi(x,0) = e^{ikx}\theta(-x), \tag{15.4}$$

assuming for the moment that the shutter is completely absorbing. This can be done with the help of the Green function

$$G(x-x',t) = \frac{1}{\sqrt{2\pi t}} e^{\frac{-i\pi}{4}} \exp\left[\frac{i(x-x')^2}{2t}\right]. \tag{15.5}$$

This Green function is the solution of the inhomogeneous wave function

$$i\frac{\partial G}{\partial t} + \frac{1}{2}\frac{\partial^2 G(x,t)}{\partial x^2} = -\delta(x-x')\,\delta(t-t'). \tag{15.6}$$

By the standard technique of solving an inhomogeneous differential equation with the aid of the Green function from Eqs. (15.3), (15.4) and (15.6) we find that $\psi(x,t)$ is given by

$$\begin{aligned} \psi(x,t) &= 2M(x,k,t) = \int_{-\infty}^{\infty} G(x-x',t)\, e^{ikx'}\theta(-x')\, dx' \\ &= \frac{1}{\sqrt{2}} e^{-\frac{i\pi}{4}} \exp\left(ikx - \frac{i}{2}k^2 t\right) \int_{-\infty}^{t} \exp\left(\frac{i\pi}{2}u^2\right) du, \end{aligned} \tag{15.7}$$

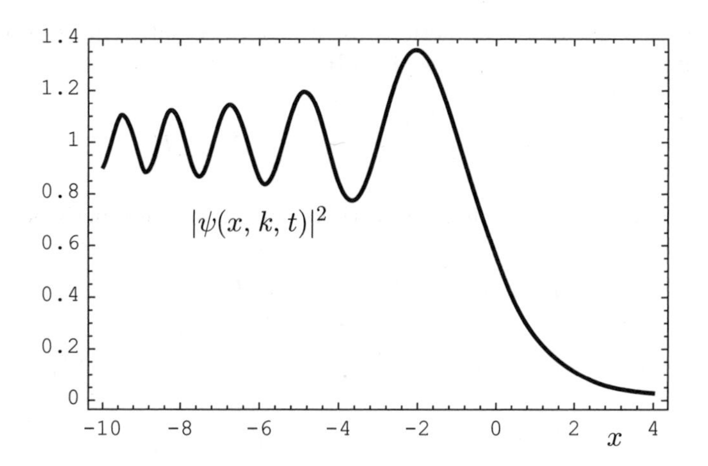

Figure 15.1: The probability density $|\psi(x,k,t)|^2$ for fixed time, $t=2$, plotted as a function of x showing the diffraction pattern to the right of the shutter.

where $M(x,k,t)$ is the Moshinsky function [4] and u is defined by

$$u = \frac{1}{\sqrt{\pi t}}\left(x' - x\right) + k\sqrt{\frac{t}{\pi}}. \tag{15.8}$$

The function $M(x,k,t)$ can also be written in terms of the error function, erfc,

$$M(x,k,t) = \exp\left[i\left(kx - \frac{k^2 t}{2}\right)\right] \operatorname{erfc}\left[(1-i)\left(\frac{x-kt}{\sqrt{4t}}\right)\right]. \tag{15.9}$$

If the shutter is not a perfect absorber but it partially reflects the wave, then $\psi(x,k,t)$ in (15.3) must be replaced by

$$\psi(x,k,t) = \begin{cases} 2\left\{e^{ikx} + R(k)e^{-ikx}\right\} & x \le 0 \\ 0 \quad x > 0 \end{cases} \tag{15.10}$$

where $R(k)$ is the reflection amplitude for the surface at $x=0$. In this case the time-dependent wave function becomes

$$\psi(x,k,t) = M(x,k,t) + R(k)M(x,-k,t). \tag{15.11}$$

For simplicity we choose the special case of $R(k) = 1$. Now if we close the shutter at $t=T$ we want to determine the spread in energy associated with the wave function

$$\psi(x,k,t) = M(x,k,t) + M(x,-k,t), \tag{15.12}$$

defined in the interval $0 \le x \le \infty$. The time-dependent probability density associated with $\psi(x,k,t)$ is plotted at a fixed time in Fig. 15.1. Now we consider a complete set of eigenfunctions for $0 \le x \le \infty$ satisfying the boundary condition $\phi(k', x) = 0$. This set is given by

$$\phi(E', x) = \sqrt{\frac{2}{\pi k'}} \sin(k'x), \quad E' = \frac{1}{2}k'^{\,2}, \tag{15.13}$$

and the members of this set form an orthonormal group of functions:

$$\int_0^\infty \phi(E, x)\phi(E', x)\, dx = \delta(E - E'). \tag{15.14}$$

We expand the wave function $\psi(x, t)$ in terms of the set $\{\phi(k', x)\}$

$$\int_0^\infty \psi(x, k, t)\phi(k', x)\, dx = B(k, k', t), \tag{15.15}$$

where $B(k, k', t)$ measures the overlap between $\psi(x, k, t)$ and $\phi(k', x)$.

$$\begin{aligned} B(k, k', t) &= \sqrt{\frac{2k'}{\pi}} \left[\frac{-1}{k^2 - k'^{\,2}} M_0(k, t) \right. \\ &+ \left. \frac{1}{2k'(k - k')} M_0(k', t) - \frac{1}{2k'(k + k')} M_0(-k', t) \right], \end{aligned} \tag{15.16}$$

with $M_0(k, t)$ defined by

$$M_0(k, t) = M(x = 0, k, t) = e^{-i\frac{k^2 t}{2}} \operatorname{erfc}\left[-(1 - i)\frac{k\sqrt{t}}{2} \right]. \tag{15.17}$$

Now suppose that we open the shuttered at $t = 0$ and close it at $t = T$, then we want to find the spread energy of the state (15.12) which is defined in the interval $0 \le x \le \infty$. We find this energy spread by calculating $P(E, E', T) = |B(k, k', T)|^2$;

$$\begin{aligned} P(E, E', t) &= \frac{2k'}{\pi} \left(\frac{1}{k^2 - k'^{\,2}} \right)^2 |[M_0(k, T) + M_0(-k, T)] \\ &- [M_0(k', T) + M_0(-k', T)]|^2. \end{aligned} \tag{15.18}$$

Since

$$\operatorname{erfc}(z) + \operatorname{erfc}(-z) = \frac{2}{\sqrt{\pi}} \int_{-\infty}^{\infty} e^{-u^2} du = 2, \tag{15.19}$$

therefore from (15.17) we have

$$M_0(k, T) + M_0(-k, T) = 2 \exp\left(-\frac{ik^2 T}{2} \right). \tag{15.20}$$

By substituting for $M_0(k,T)$ and $M_0(-k,T)$ from (15.20) in (15.18) we simplify $P(k,k',T)$ and now we write it in terms of E and E';

$$P(E,E',t) = \frac{8}{\pi}\sqrt{2E'}\frac{1}{(E-E')^2}\sin^2\left[\frac{(E-E')T}{2}\right]. \tag{15.21}$$

This is yet another version of the time-energy uncertainty relation that we have discussed earlier.

15.2 High Energy Scattering from an Absorptive Target

The high energy scattering of a beam of particles from a composite system can be described by a complex (or an optical) potential

$$v(\mathbf{r}) = v_R(\mathbf{r}) + iv_I(\mathbf{r}), \tag{15.22}$$

where the imaginary part of the potential is negative definite function of $\mathbf{r}$

$$v_I(\mathbf{r}) \le 0 \quad \text{for all } \mathbf{r}. \tag{15.23}$$

Due to the presence of $v_I(\mathbf{r})$ which is an absorptive potential, the conservation of current and hence the conservation of the number of scattered particles is violated since some of the incoming particles are absorbed by the target.

A simple way of formulating the high energy scattering of particles is to use the impact parameter approximation discussed in the previous chapter. In this approach we find that the scattering amplitude when the interaction $v(\mathbf{r})$ is complex, Eq. (14.445) changes to

$$\tan[\delta(b)+i\eta(b)] = -\frac{1}{2k}\int_0^\infty \left\{v_R\left(\sqrt{z^2+b^2}\right) + iv_I\left(\sqrt{z^2+b^2}\right)\right\}dz. \tag{15.24}$$

That is the phase shift also becomes a complex function of b and k. Noting that for high energies the phases are small we can replace $\tan[\delta(b)+i\eta(b)]$ by $\delta(b)+i\eta(b)$ in (15.24). Substituting for these phase in Eq. (14.447) for $f(\theta,k)$ we obtain

$$f(\theta,k) = -ik\int_0^\infty J_0\left(2kb\sin\frac{\theta}{2}\right)\left[e^{-2i\delta(b)-2\eta(b)} - 1\right] b\,db. \tag{15.25}$$

A simple model of high energy scattering illustrates the phenomenon of diffraction in quantum theory [6],[7]. Suppose that the potential is confined to a sphere of radius a which and has a large negative imaginary part which makes it absorptive.

$$V(r) = \begin{cases} -i\infty & r<a \\ 0, & r>a \end{cases}. \tag{15.26}$$

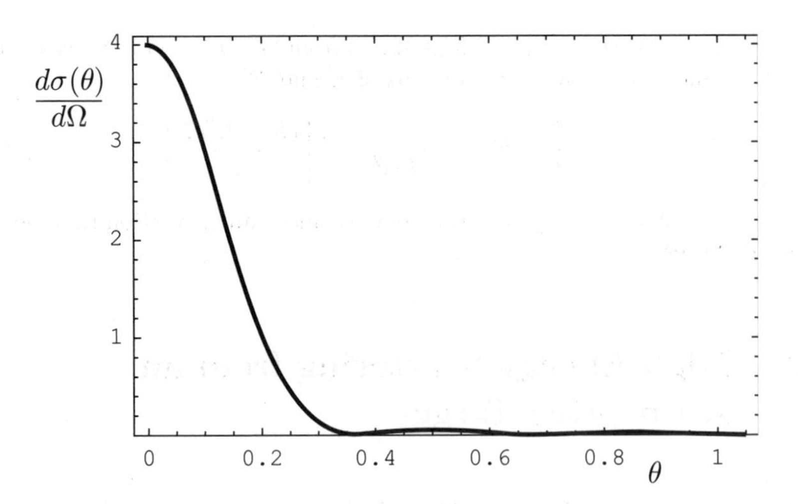

Figure 15.2: The angular dependence of the differential cross section is shown for scattering from an absorptive sphere.

We also assume that the incident particles have high energies so that Eqs. (14.445) and (14.448) are valid for the description of scattering. From Eqs. (15.26) and (14.445) we obtain

$$e^{2i\delta(b)} = \begin{cases} 0 & b < a \\ 1, & b > a \end{cases}. \tag{15.27}$$

Using (15.27) we find the scattering amplitude analytically

$$f(\theta) = -ik\int_0^a J_0\left(2kb\sin\frac{\theta}{2}\right) bdb = ia\frac{J_1\left(2ka\sin\frac{\theta}{2}\right)}{2\sin\frac{\theta}{2}}. \tag{15.28}$$

The diffraction cross section found from (15.28) is

$$\begin{aligned}\frac{d\sigma(\theta)}{d\Omega} &= |f(\theta)|^2 = \left|-ik\int_0^a J_0\left(2kb\sin\frac{\theta}{2}\right) bdb\right|^2 \\ &= \left(ka^2\right)^2 \frac{J_1^2\left(2ka\sin\frac{\theta}{2}\right)}{\left(2ka\sin\frac{\theta}{2}\right)^2},\end{aligned} \tag{15.29}$$

and this has the form of a well-known result of Fraunhofer diffraction scattering by a sphere (see Fig. 15.2. By integrating (15.29) we find that the total cross section is

$$\begin{aligned}\sigma_{total} &= \frac{4\pi}{k}\mathrm{Im} f(\theta = 0) = 2\int\left[1 - \mathrm{Re}\left(e^{2i\delta(b)}\right)\right] 2\pi bdb \\ &= 2\pi a^2.\end{aligned} \tag{15.30}$$

Bibliography

[1] M. Razavy, *Quantum Theory of Tunneling*, (World Scientific, Singapore, 2003), p. 61.

[2] W. van Dijk and Y. Nogami, Novel expressions for the wave function of a decaying quantum state, Phys. Rev. Lett. 83, 2867 (1999).

[3] M. Moshinsky, Diffraction in time, Phys. Rev. 88, 625 (1952).

[4] M. Moshinsky, Diffraction in time and the time-energy uncertainty relation, Am. J. Phys. 44, 1037 (1976).

[5] H.A. Bethe, A continuum theory of compound nucleus, Phys. Rev. 57, 1125 (1940).

[6] R.J. Glauber in *Lectures in Theoretical Physics*, Vol. 1, edited by W.E. Brittin and L.G. Dunham, (Interscience Publishers, 1959).

[7] W.E. Frahn, Quantum Diffraction, Rivista Nuovo Cimento, 7, 499 (1977).

Chapter 16

Motion of a Charged Particle in Electromagnetic Field and Topological Quantum Effects for Neutral Particles

Among the fundamental forces of nature that we encounter in quantum domain none is better known and understood than the electromagnetic interaction. The Maxwell equations and the Lorentz force law give us a complete picture of the classical electrodynamics. However as we will see there are some features in quantum description which do not have classical analogues. We will have a detailed discussion of these features later.

We want to set up the problem of interaction of charged particles with an external electromagnetic field, not by postulating the minimal coupling rule as is usually done, but follow a method originally due to Feynman in which two of the Maxwell's equations, together with the Lorentz force are derived from the Heisenberg equations of motion [1].

Let us consider a single particle of mass m and charge e with position $\mathbf{r} = (x_1, x_2, x_3)$ obeying the Heisenberg equation

$$m\frac{d\dot{x}_k}{dt} = F_k\left(\mathbf{r}, \dot{\mathbf{r}}, t\right), \tag{16.1}$$

where F_k is the k-th component of the force. Among the components of velocity and position we have the canonical commutation relations which we write as

$$[x_j,\ x_k] = 0, \tag{16.2}$$

and

$$m\,[x_j,\ \dot{x}_k] = i\hbar\delta_{jk}. \tag{16.3}$$

Now from (16.1) and (16.3) it follows that

$$[x_j,\ F_k] + m\,[\dot{x}_j,\ \dot{x}_k] = 0. \tag{16.4}$$

According to the rules governing these operators, the Jacobi identity must be satisfied (see Eq. (3.130));

$$[x_l,\ [\dot{x}_j,\ \dot{x}_k]\,] + [\dot{x}_j,\ [\dot{x}_k,\ x_l]\,] + [\dot{x}_k,\ [x_l,\ \dot{x}_j]\,] = 0. \tag{16.5}$$

In this relation if we put $m[x_l,\ \dot{x}_j] = i\hbar\delta_{lj}$, and for $[\dot{x}_j,\ \dot{x}_k]$ we substitute from (16.4) we get

$$[x_l, [x_j,\ F_k]\,] = 0. \tag{16.6}$$

In addition Eq. (16.4) implies that

$$[x_j,\ F_k] = -[x_k,\ F_j]. \tag{16.7}$$

This property of the commutator $[x_i,\ F_k]$ allows us to express it in terms of ε_{jkl}, the completely antisymmetric tensor (Levi–Civita symbol) defined by (1.32) and (1.33), with all other components being equal to zero. Thus we have

$$[x_j,\ F_k] = -\frac{ie\hbar}{m}\sum_l \varepsilon_{jkl}B_l, \tag{16.8}$$

where e is the charge of the electron, c is the velocity of light and $\mathbf{B}$ is a function of $\mathbf{r}$ and t (all equations are given in cgs units). This vector, $\mathbf{B}$, cannot be a function of $\dot{\mathbf{r}}$ since then Eq. (16.6) cannot be satisfied. At this point we want to inquire about the most general form of F_k which satisfies (16.8) and can be dependent on $\mathbf{r}$ and $\dot{\mathbf{r}}$. Clearly $\mathbf{F}$ can be a function of $\mathbf{r}$ and t, but can also depend linearly on $\dot{\mathbf{r}}$. Solving Eq. (16.8) for F_j we find the general solution to be

$$F_j = e\left[E_j + \frac{1}{2c}\sum_{kl}\varepsilon_{jkl}\left(\dot{x}_k B_l + B_l\dot{x}_k\right)\right], \tag{16.9}$$

where E_k depends on $\mathbf{r}$ and t and not $\dot{\mathbf{r}}$. We have also made use of the Weyl's rule or symmetrization rule (Secs. 3.5 and 3.6) to write the operator $\dot{x}_k B_\ell$ is symmetrized form.

From Eqs. (16.4) and (16.8) it follows that

$$B_l = -\frac{im^2c}{2\hbar e}\sum_{j,k}\varepsilon_{jkl}\,[\dot{x}_j,\ \dot{x}_k]\,. \tag{16.10}$$

If we calculate the commutator $[x_j,\, B_l]$ we find

$$\begin{aligned}[x_j,\, B_l] &= -\frac{im^2c}{2\hbar c}\sum_{j,k}\varepsilon_{jkl}\,[x_i,\, [\dot{x}_j,\, \dot{x}_k]\,] \\ &= -\frac{im^2c}{2\hbar c}\sum_{j,k}\varepsilon_{jkl}\,\{[\,[x_i,\, \dot{x}_j]\,,\, \dot{x}_k] + [\dot{x}_j,\, [x_i,\, \dot{x}_k]\,]\} = 0,\end{aligned} \tag{16.11}$$

where we have used the Jacobi identity to simplify (16.11. This result shows that B_l cannot be a function of $\dot{\mathbf{r}}$. Again by writing the Jacobi identity for the components of $\dot{\mathbf{r}}$ as

$$\sum_{j,k}\varepsilon_{jkl}\,[\dot{x}_l,\, [\dot{x}_j,\, \dot{x}_k]\,] = 0, \tag{16.12}$$

and combining it with (16.10) we obtain

$$\sum_l [\dot{x}_l,\, B_l] = 0. \tag{16.13}$$

Thus Eq. (16.13) with

$$m\dot{x}_l = -i\hbar\frac{\partial}{\partial x_l}, \tag{16.14}$$

leads to the Maxwell equation;

$$\sum_l \frac{\partial B_l}{\partial x_l} = \nabla\cdot\mathbf{B} = 0. \tag{16.15}$$

Next we take the time derivative of $\mathbf{B}$, Eq. (16.10), to find

$$\begin{aligned}\frac{d\mathbf{B}}{dt} &= \frac{\partial\mathbf{B}}{\partial t} + \mathcal{O}_S\left(\dot{\mathbf{r}}\cdot\nabla\mathbf{B}\right) \\ &= -\frac{im^2c}{\hbar e}\left(\ddot{\mathbf{r}}\wedge\dot{\mathbf{r}} - \dot{\mathbf{r}}\wedge\ddot{\mathbf{r}}\right) = -\frac{imc}{2\hbar e}\left(\mathbf{F}\wedge\dot{\mathbf{r}} + \dot{\mathbf{r}}\wedge\mathbf{F}\right) \\ &= -\frac{imc}{\hbar}\left\{(\mathbf{E}\wedge\dot{\mathbf{r}} - \dot{\mathbf{r}}\wedge\mathbf{E}) + \mathcal{O}_S\left(\dot{\mathbf{r}}\wedge\mathbf{B}\right)\wedge\dot{\mathbf{r}} - \dot{\mathbf{r}}\wedge\mathcal{O}_S\left(\dot{\mathbf{r}}\wedge\mathbf{B}\right)\right\},\end{aligned} \tag{16.16}$$

where $\mathcal{O}_S$ denotes that the operators are symmetrized Sec. 3.5,

$$\left[\mathcal{O}_S\left(\dot{\mathbf{r}}\wedge\nabla\mathbf{B}\right)\right]_j = \frac{1}{2}\sum_k\left(\dot{x}_k\frac{\partial B_j}{\partial x_k} + \frac{\partial B_j}{\partial x_k}\dot{x}_k\right). \tag{16.17}$$

Thus the l-th component of $\frac{d\mathbf{B}}{dt}$ obtained by differentiating B_l, Eq. (16.16) is

$$\frac{dB_l}{dt} = -\frac{im^2c}{\hbar c}\sum_{j,k}\varepsilon_{jkl}\,[\dot{x}_j,\, \ddot{x}_k]\,. \tag{16.18}$$

We also note that $\mathbf{E}$ (like $\mathbf{B}$) cannot depend on $\dot{\mathbf{r}}$, since

$$\begin{aligned}
[x_l,\, E_i] &= \frac{1}{e}[x_l,\, F_i] - \frac{1}{c}\sum_{j,k}\varepsilon_{ijk}\mathcal{O}_S\left([x_l,\, \dot{x}_j]\, B_l + \dot{x}_j\, [x_l,\, B_l]\right) \\
&= -\frac{i\hbar}{mc}\sum_k \varepsilon_{lik}B_k - \frac{i\hbar}{mc}\sum_k \varepsilon_{ijk}\delta_{lj}B_k = 0. \qquad (16.19)
\end{aligned}$$

Returning to Eq. (16.16) we observe that the expression in the first parenthesis of (16.16) is simply the curl of $\mathbf{E}$;

$$-\frac{im}{2\hbar}\left(\mathbf{E}\wedge\dot{\mathbf{r}} - \dot{\mathbf{r}}\wedge\mathbf{E}\right) = -\nabla\wedge\mathbf{E}. \qquad (16.20)$$

For the other two terms we write the i-th component as

$$\begin{aligned}
&\left[\mathcal{O}_S\left(\dot{\mathbf{r}}\wedge\mathbf{B}\right)\wedge\dot{\mathbf{r}} - \dot{\mathbf{r}}\wedge\mathcal{O}_S\left(\dot{\mathbf{r}}\wedge\mathbf{B}\right)\right]_i = \sum_{jk}\varepsilon_{ijk}\left[r_j,\, \left(\mathcal{O}\left(\dot{\mathbf{r}}\wedge\mathbf{B}\right)\right)_k\right] \\
&= \frac{1}{2}\sum_{jk}\sum_{ln}\varepsilon_{ijk}\varepsilon_{kln}\left[\dot{x}_j,\, \dot{x}_l B_n + B_n\dot{x}_l\right] \\
&= \frac{1}{2}\sum_j\left[\dot{x}_j,\, \dot{x}_i B_j + B_j\dot{x}_i - \dot{x}_j B_j - B_i\dot{x}_j\right] \\
&= \frac{1}{2}\sum_j\left\{[\dot{x}_j,\, \dot{x}_i]\, B_j + B_j\, [\dot{x}_j,\, \dot{x}_i] - \dot{x}_j\, [\dot{x}_j,\, B_i] - [\dot{x}_j,\, B_i]\, \dot{x}_j\right\}.
\end{aligned} \qquad (16.21)$$

We can simplify this last equation by observing that with the help of (16.5)

$$\sum_j\left\{[\dot{x}_j,\, \dot{x}_i]\, B_j + B_j\, [\dot{x}_j,\, \dot{x}_i]\right\} \sim \sum_{jk}\varepsilon_{ijk}\left(B_k B_j + B_j B_k\right) = 0, \qquad (16.22)$$

so that the first two terms of (16.21) do not contribute. Now by examining the last two terms of (16.21) we find

$$-\frac{1}{2}\sum_j\left\{\dot{x}_j\, [\dot{x}_j,\, B_i] + [\dot{x}_j,\, B_i]\, \dot{x}_j\right\} = \frac{i\hbar}{2m}\sum_j\left(\dot{x}_j\frac{\partial B_j}{\partial x_j} + \frac{\partial B_j}{\partial x_j}\dot{x}_j\right). \qquad (16.23)$$

By substituting (16.20)–(16.23) in (16.16) we obtain the Faraday law:

$$\frac{1}{c}\frac{\partial\mathbf{B}}{\partial t} = -\nabla\wedge\mathbf{E}. \qquad (16.24)$$

Having found two of Maxwell's equations (16.15) and (16.20) we observe that the other two of the Maxwell equations (in cgs units)

$$\nabla\cdot\mathbf{E} = 4\pi\rho, \qquad (16.25)$$

and

$$-\frac{1}{c}\frac{\partial \mathbf{E}}{\partial t} + \nabla \wedge \mathbf{B} = \frac{4\pi}{c}\mathbf{j}, \tag{16.26}$$

define the charge and the current density respectively [1].

To conclude our discussion, it is worthwhile to make the following observations:

(1) - Whether the two Maxwell equations (16.25) and (16.26) can be used to define ρ and $\mathbf{j}$ has been the subject of discussion by some authors [2]–[5]. It is well-known that the Maxwell equations are invariant under Lorentz transformation, whereas Newton's second law is invariant under Galilean transformation. We can ask how it is possible to obtain a truly relativistic set of Maxwell's equation from the non-relativistic law of motion. The fact is that the two equations that we have found Eqs. (16.15) and (16.24) are indeed compatible with the Galilean invariance, whereas the other two equations (16.25) and (16.26) are not. One can show that a Galilean invariant theory of electromagnetism requires keeping (16.25) but replacing (16.26) by [5],[6]

$$\nabla \wedge \mathbf{B} = \frac{4\pi}{c}\mathbf{j}. \tag{16.27}$$

(2) - While Feynman derived the Maxwell equations starting with the Heisenberg equations of motion (16.1), one can find the same result using the Poisson formulation of classical dynamics [3],[4]. Classically it can be shown that any acceleration-independent generalized force for which the equation of motion can be derived from a Lagrangian must be of the form given by (16.9) [3].

If we write Eqs. (16.1) and (16.9) as vector equations

$$m\frac{d\dot{\mathbf{r}}}{dt} = m[\,[\mathbf{r},\, H],\, H] = e\mathbf{E} + \frac{e}{2c}[\mathbf{v} \wedge \mathbf{B} - \mathbf{B} \wedge \mathbf{v}]. \tag{16.28}$$

The Hamiltonian which gives us this equation of motion is

$$H = \frac{1}{2m}\left(\mathbf{p} - \frac{e}{c}\mathbf{A}\right)^2. \tag{16.29}$$

The Heisenberg's equation of motion is local since the force on the electron only depends on electric and magnetic fields at the position of the electron. However since $\mathbf{r}$ and H in (16.28) do not commute with each other, we can have nonlocal effects such as Aharonov–Bohm effect which we will discuss next.

16.1 The Aharonov–Bohm Effect

The Heisenberg equation (16.28) shows that the observables for the motion of a charged particle in an electric- or magnetic field or both are dependent on

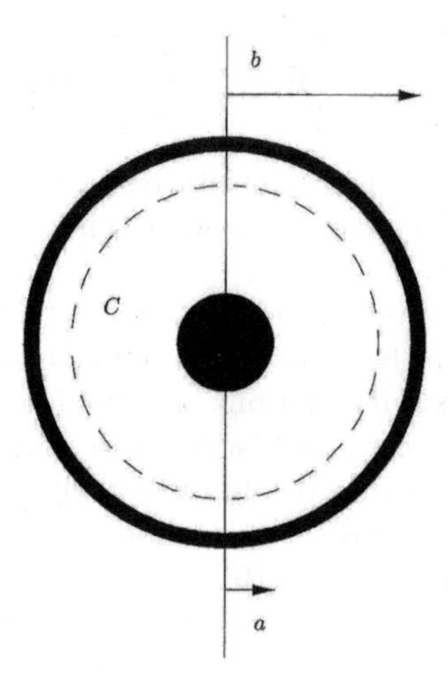

Figure 16.1: The magnetic field **B** is confined inside a cylinder of radius $\rho = a$ and the electron can move in the space between the two cylinders, $a < \rho < b$. While the electron does not feel the magnetic force yet its energy eigenvalues depend on the flux Φ of this field.

these fields and not on the electromagnetic potentials **A** and ϕ. But if we try to calculate the average force acting on the particle or calculate the expectation values of **r** or **v**, then from (16.28) we have

$$m\frac{d}{dt}\langle\psi|\mathbf{v}|\psi\rangle = e\langle\psi|\mathbf{E}|\psi\rangle + \frac{e}{2c}\langle\psi[\mathbf{v}\wedge\mathbf{B} - \mathbf{B}\wedge\mathbf{v}]\psi\rangle. \tag{16.30}$$

This expression shows that the wave function of the problem $|\psi\rangle$ which is dependent on the potentials **A** and ϕ through the Hamiltonian (16.29) will have an effect on the average force felt by the particle [9],[10]. Thus the motion of the charged particle is affected by the presence of the electromagnetic field in the regions where the particle does not enter. This is counterintuitive particularly when we accept the Heisenberg equation (16.28) as the equation of motion of the charged particle, since the force appears to be local. The Aharonov–Bohm effect can be observed in bound states as well as scattering processes. First let us consider the exactly solvable example which shows the Aharonov–Bohm effect in bound state problems.

Aharonov–Bohm Effect in Bound States — An electron of charge e is constrained to move between two concentric cylinders of radii a and b shown in Fig. 16.1. An external magnetic flux Φ goes up along the axis of the inner cylinder and returns uniformly along the surface of the outer cylinder i.e. **B** is nonzero for $\rho < a$. Thus the electron is not exposed to the external magnetic field. The problem has cylindrical symmetry, therefore we use cylindrical coordinates (ρ, ϕ, z) to solve the problem. For $a < \rho < b$ the magnetic potential **A** satisfies two conditions:

(a) - $\mathbf{B} = \nabla\wedge\mathbf{A} = 0$.

(b) - For any closed path C that encircles the inner cylinder according to the Stokes theorem we have

$$\oint_C \mathbf{A} \cdot d\mathbf{l} = \int_S \nabla \wedge \mathbf{A} \cdot d\mathbf{S} = \int_S \mathbf{B} \cdot d\mathbf{S} = \Phi, \tag{16.31}$$

where Φ is the flux through the inner cylinder. The components of $\mathbf{A}$ satisfying these conditions are given by

$$A_\rho = A_z = 0, \quad A_\phi = \frac{\Phi}{2\pi\rho}. \tag{16.32}$$

In the space between the cylinders, the Hamiltonian for the motion of the electron according to (16.29) is

$$H = -\frac{\hbar^2}{2m}\nabla^2 + \frac{i\hbar e}{2mc}\left[2\mathbf{A} \cdot \nabla + (\nabla \cdot \mathbf{A})\right] + \frac{e^2}{2mc^2}\mathbf{A} \cdot \mathbf{A}. \tag{16.33}$$

This Hamiltonian expressed in cylindrical coordinates (ρ, ϕ, z) is of the form

$$\begin{aligned} H &= -\frac{\hbar^2}{2m}\left(\frac{\partial^2}{\partial\rho^2} + \frac{1}{\rho}\frac{\partial}{\partial\rho} + \frac{1}{\rho^2}\frac{\partial^2}{\partial\phi^2} + \frac{\partial^2}{\partial z^2}\right) \\ &+ \frac{i\hbar e\Phi}{2m\rho^2 c}\frac{\partial}{\partial\phi} + \frac{e^2\Phi^2}{8\pi^2 mc^2\rho^2}. \end{aligned} \tag{16.34}$$

From this Hamiltonian we find the solution to the time-dependent Schrödinger equation

$$H\psi_{n,m,k_z}(\rho, \phi, z) = E(n, m, k_z)\psi_{n,m,k_z}(\rho, \phi, z), \tag{16.35}$$

is separable and can be written as

$$\psi_{n,m,k_z}(\rho, \phi, z) = R_n(\rho)e^{im\phi}e^{ik_z z}, \tag{16.36}$$

where m is an integer and $\hbar k_z$ is the momentum of the electron in the z direction. Substituting (16.34) and (16.36) in (16.35) we find that $R_n(\rho)$ is the solution of the differential equation

$$\frac{d^2 R_n}{d\,\rho^2} + \frac{1}{\rho}\frac{dR_n}{d\rho} + \left(\beta^2 - \frac{\mu^2}{\rho^2}\right)R_n = 0, \tag{16.37}$$

where β and μ are defined by

$$\beta_n = \left[\frac{2mE(n, m, k_z)}{\hbar^2} - k_z^2\right]^{\frac{1}{2}}, \quad \text{and} \quad \mu = m - \frac{e\Phi}{c\hbar}. \tag{16.38}$$

The solution of (16.37) is a combination of the Bessel function J_μ and the Neumann function N_μ [15]

$$R_n(\rho) = A_n J_\mu\left(\frac{\rho}{\beta_n}\right) + B_n N_\mu\left(\frac{\rho}{\beta_n}\right). \tag{16.39}$$

Now we impose the condition that the wave function must be zero at the surfaces of the inner and outer cylinders. In this way we find the eigenvalue equation

$$J_\mu\left(\frac{b}{\beta_n}\right)N_\mu\left(\frac{a}{\beta_n}\right)-J_\mu\left(\frac{a}{\beta_n}\right)N_\mu\left(\frac{b}{\beta_n}\right)=0. \tag{16.40}$$

Thus for a given set of Φ, m and k_z we can solve the eigenvalue equation (16.40) for $\beta_n(\mu, a, b)$, and then from (16.38) find the discrete set of eigenvalues $E(n, m, k_z)$. These energies will depend on Φ through μ which appears in (16.38).

If the concentric cylinders are of finite length L and they are covered at the top and bottom, and the wave function must vanish at $z = 0$ and $z = L$, then $k_z = \frac{j\pi}{L}$, $j = 0, 1, 2\cdots$ and β_n becomes a function of the quantum number j.

A simpler case is the one where the electron is constrained to move on a circle of radius $\mathcal{R}$, where the Hamiltonian H, Eq. (16.34), reduces to [11]–[14]

$$H' = -\frac{1}{2m\mathcal{R}^2}\left(L_z+\frac{e\Phi}{2\pi c}\right)^2. \tag{16.41}$$

In this relation

$$L_z = -i\hbar\frac{\partial}{\partial\phi}, \tag{16.42}$$

where L_z is the z component of the angular momentum operator. The eigenvalues of H' are

$$-\frac{\hbar^2\mu^2}{2m\mathcal{R}^2} = -\frac{\hbar^2}{2m\mathcal{R}^2}\left(m'+\frac{e\Phi}{2\pi\hbar c}\right)^2 = -\frac{\hbar^2}{2m\mathcal{R}^2}\left(m'+\frac{\Phi}{\Phi_0}\right)^2, \tag{16.43}$$

where m' is an integer and Φ_0 is the quantum of flux $\Phi_0 = \frac{2\pi\hbar c}{e}$. For the present case we do not need to use the single-valuedness of the wave function to show that m' is an integer. Since $\hbar\mu$ is the eigenvalue of the operator $L_z + \frac{e\Phi}{2\pi c}$, if we denote its eigenkets by $|\mu\rangle$ then we have

$$\left(L_z+\frac{e\Phi}{2\pi c}\right)|\mu\rangle = \hbar\mu|\mu\rangle. \tag{16.44}$$

Now if

$$|\mu\rangle = a_1|\mu_1\rangle + a_2|\mu_2\rangle, \tag{16.45}$$

is a superposition of two eigenvectors of $L_z + \frac{e\Phi}{2\pi c}$, then rotation by 2π about the z axis must yield the same $|\mu\rangle$ with a possible phase factor,

$$\hat{R}(2\pi)|\mu\rangle = e^{2\pi i\mu_1}a_1|\mu_1\rangle + e^{2\pi i\mu_2}a_2|\mu_2\rangle. \tag{16.46}$$

This state must be the same as (16.45) multiplied by a phase factor, therefore

$$e^{2\pi i(\mu_2-\mu_1)} = 1, \tag{16.47}$$

or $\mu_2 - \mu_1 = m'$ must be an integer. The fact that $m' + \frac{\Phi}{\Phi_0}$ in (16.43) is observable and m' is an integer implies that the eigenvalues of H' are dependent on the flux modulo Φ_0 where

$$\Phi_0 = \frac{2\pi\hbar c}{e} = \oint \mathbf{A} \cdot d\mathbf{l}. \tag{16.48}$$

According to Yang, in quantum mechanics, the force

$$-e\left[\mathbf{E} + \frac{1}{2c}(\mathbf{v} \wedge \mathbf{B} - \mathbf{B} \wedge \mathbf{v})\right], \tag{16.49}$$

underdescribes electromagnetism, since the magnetic potential, $\mathbf{A}$, has observable effects. On the other hand

$$\oint \mathbf{A} \cdot d\mathbf{l} = \Phi_0 \tag{16.50}$$

overdescribes it because the result of Aharonov–Bohm effect depends on Φ modulo Φ_0. Only the quantity

$$\exp\left[-\left(\frac{ie}{\hbar c}\right)\int \mathbf{A} \cdot d\mathbf{l}\right], \tag{16.51}$$

gives a complete description of the physics [7],[8].

Aharonov–Bohm Effect in Scattering — Let us consider a two-slit experiment where a beam of electrons from a source S reach the point O on the screen by going through the first or the second slit. The superposition of these two waves will produce an interference pattern. Just behind the slits we have a shielded homogeneous magnetic field coming out of the plane of the diagram. The magnetic field is cylindrically symmetric and therefore we formulate the problem by writing the wave equation in cylindrical coordinates. To simplify the calculation we consider the limit where the diameter of the solenoid tends to zero while the total flux remains finite. The wave equation for the Hamiltonian (16.34) can be written as

$$\begin{aligned} H\psi &= -\frac{\hbar^2}{2m}\left(\nabla - \frac{ie}{\hbar c}\mathbf{A}\right)^2 \psi \\ &= -\frac{\hbar^2}{2m}\left[\frac{\partial^2}{\partial\rho^2} + \frac{1}{\rho}\frac{\partial}{\partial\rho} + \frac{1}{\rho^2}\left(\frac{\partial}{\partial\phi} + i\alpha\right)^2 + \frac{\partial^2}{\partial z^2}\right]\psi = E\psi, \end{aligned} \tag{16.52}$$

where

$$\alpha = \frac{\Phi}{\Phi_0} = -\frac{e\Phi}{2\pi\hbar c}. \tag{16.53}$$

For the sake of simplicity we consider a weak magnetic field such that $0 \le \alpha < 1$, but the same method can be modified and used when $\alpha > 1$ [18]. Again we write the wave function as the product of three terms:

$$\psi(\rho, \phi, z) = \psi_{m'}(\rho) e^{im'\phi} e^{ik_z z}. \tag{16.54}$$

By substituting (16.54) in (16.52) we find

$$\left[\frac{d^2}{d\,\rho^2}+\frac{1}{\rho}\frac{d}{d\rho}-\frac{(m'+\alpha)^2}{\rho^2}+k^2\right]\psi_{m'}(\rho)=0, \tag{16.55}$$

where

$$(k^2+k_z^2)=\frac{2mE}{\hbar^2}, \tag{16.56}$$

and

$$m'=0,\ \pm 1,\ \pm 2\cdots. \tag{16.57}$$

The regular solution of (16.55), i.e. the solution which is finite at $\rho = 0$ is the Bessel function of order $m' + \alpha$, therefore the solution of (16.52) is

$$\psi(\rho,\phi,z)=\sum_{m'=-\infty}^{\infty} e^{im'\phi}e^{ik_z z}J_{|m'+\alpha|}(k\rho). \tag{16.58}$$

Assuming that the wave enters from $+x$ direction and moves towards $-x$ direction, the scattering takes place at $x = 0$. An idealized arrangement for verifying the Aharonov–Bohm effect is shown in Fig. 16.2. Since the coordinate z does not play a role in this scattering, we will suppress the z-dependence and consider ψ to be a function of ρ and ϕ.

The incident wave which can be written as

$$\psi_{inc}=e^{-ikx}e^{-i\alpha\phi},\quad x\ge 0. \tag{16.59}$$

is single valued ($\phi = 0$ is a line along the direction of the incoming wave). The reason that we have distorted e^{-ikx} by a factor $e^{-i\alpha\phi}$ is because in the present case the scattering potential does not vanish as $\rho \to \infty$. This is a consequence of the fact that the interaction is between the beam of electrons and the field of infinitely long solenoid.

The total wave function $\psi(\rho, \phi, \alpha)$, Eq. (16.58), which reduces to (16.59) for $\phi = 0$ (or for $x \ge 0$) is given by [9]

$$\psi(\rho,\phi,\alpha)=\sum_{m'=-\infty}^{\infty}(-i)^{|m'+\alpha|}J_{|m'+\alpha|}(k\rho)e^{im'\phi}. \tag{16.60}$$

Now we want to find the asymptotic form of $\psi(\rho, \phi, \alpha)$ when $\rho \to \infty$. For this we first carry out the summation over m'. The wave function (16.60) can be written as the sum of three terms:

$$\begin{aligned}\psi(\rho,\phi,\alpha) &= \sum_{m'=1}^{\infty}(-i)^{|m'+\alpha|}J_{|m'+\alpha|}(k\rho)e^{im'\phi}\\ &+ \sum_{m'=1}^{\infty}(-i)^{|m'-\alpha|}J_{|m'-\alpha|}(k\rho)e^{-im'\phi}+(-i)^{|\alpha|}J_{|\alpha|}(k\rho).\end{aligned} \tag{16.61}$$

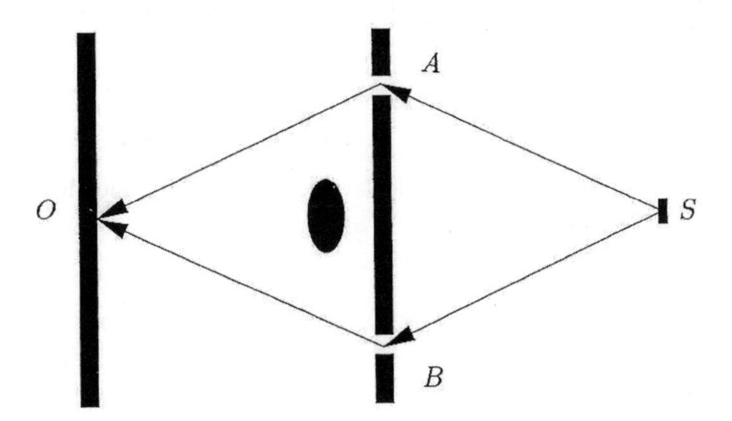

Figure 16.2: Schematic double-slit experiment to verify the Aharonov–Bohm effect. The electron beam coming from $x > 0$ passes through two slits and produces interference pattern on the screen to the left. The solenoid is located just behind the screen as is shown in the figure. The interference pattern formed on the screen is dependent on the magnetic flux through the shielded cylinder.

Let us denote the first sum by $\psi_1(\rho, \phi, \alpha)$, then from (16.61) it is clear that the second sum $\psi_2(\rho, \phi)$ can be found from $\psi_1(\rho, \phi)$;

$$\psi_2(\rho, \phi, \alpha) = \psi_1(\rho, -\phi, -\alpha). \tag{16.62}$$

The asymptotic expansion of the first term $\psi_1(\rho, \phi, \alpha)$, in powers of $(k\rho)^{-\frac{1}{2}}$ can be calculated, and the leading term in this expansion is [9]

$$\psi_1(\rho, \phi, \alpha) \to \frac{\sqrt{-i}}{2\sqrt{2\pi}}\left[(-1)^{\alpha}\frac{e^{ik\rho}}{\sqrt{k\rho}}\frac{1+e^{i\phi}}{1+\cos\phi} + i\frac{e^{-ik\rho}}{\sqrt{k\rho}}\frac{1-e^{i\phi}}{1-\cos\phi}\right]. \tag{16.63}$$

Once the two parts $\psi_1(\rho, \phi, \alpha)$ and $\psi_2(\rho, \phi, \alpha)$ are added together, then in the limit of $\rho \to \infty$ we have

$$\begin{aligned}\psi_1(\rho, \phi, \alpha) + \psi_2(\rho, \phi, \alpha) \quad &\to \quad \frac{\sqrt{-i}}{\sqrt{2\pi}}\left[i\frac{e^{-ik\rho}}{\sqrt{k\rho}} + \frac{e^{ik\rho}}{\sqrt{k\rho}}\frac{\cos\left(\pi\alpha - \frac{1}{2}\phi\right)}{\cos\frac{\phi}{2}}\right] \\ &+ \quad e^{-i(k\rho\cos\phi + \alpha\phi)}.\end{aligned} \tag{16.64}$$

The last term in (16.61) has the asymptotic behavior of the Bessel function [15]

$$(-i)^{|\alpha|}J_{|\alpha|}(k\rho) \to (-i)^{|\alpha|}\left(\frac{2}{\pi k\rho}\right)^{\frac{1}{2}}\cos\left(k\rho - \frac{1}{4}\pi - \frac{1}{2}|\alpha|\pi\right). \tag{16.65}$$

By using this result and collecting all the terms we find the asymptotic form of $\psi(\rho, \phi, \alpha)$ for large ρ to be

$$\psi(\rho, \phi, \alpha) \to e^{-i(\alpha\phi + k\rho\cos\phi)} + \frac{e^{ik\rho}}{\sqrt{2\pi i k\rho}} \sin(\pi\alpha) \frac{e^{-i\frac{\phi}{2}}}{\cos\frac{\phi}{2}}. \tag{16.66}$$

The first term in (16.66) is the incident wave, Eq. (16.59), and the second one is the scattered wave.

If we write $\psi(\rho, \phi, \alpha)$ as [16]

$$\psi(\rho, \phi, \alpha) \to \psi_{inc}(\rho, \phi, \alpha) + \frac{f(\phi)}{\sqrt{\rho}} e^{ik\rho} \quad \text{as} \quad \rho \to \infty, \tag{16.67}$$

where $f(\phi)$ is the scattering amplitude, then by comparing (16.66) and (16.67) we find $f(\phi)$

$$f(\phi) = \frac{e^{-i\frac{\phi}{2}}}{\sqrt{2ik\pi}} \left(\frac{\sin(\pi\alpha)}{\cos\frac{\phi}{2}} \right). \tag{16.68}$$

The differential cross section for this scattering found from (16.68) is

$$\frac{d\sigma(\phi)}{d\Omega} = |f(\phi)|^2 = \frac{1}{2\pi k} \left[\frac{\sin^2(\pi\alpha)}{\cos^2\left(\frac{\phi}{2}\right)} \right]. \tag{16.69}$$

Equation (16.66) clearly shows that on the line $\phi = \pi$ this asymptotic formula breaks down. In the exact solution of (16.60) which can be found directly, the second term which is multi-valued around $\phi = \pi$, combines with the multi-valued first term and the result is a single-valued wave function.

When $\alpha = n + \frac{1}{2}$, then the summation in (16.60) can be done exactly. Using the integral representation of the Bessel function [17]

$$J_{m'+\frac{1}{2}}(z) = \frac{2\left(\frac{z}{2}\right)^{m'+\frac{1}{2}}}{\sqrt{\pi}\Gamma(m'+1)} \int_0^1 \left(1 - t^2\right)^{m'} \cos(zt)dt, \tag{16.70}$$

we can sum over m' in (16.60) and obtain the simple result [9]

$$\psi(\rho, \phi, n + \frac{1}{2}) = \sqrt{\frac{i}{2}} e^{-i\left(\frac{\phi}{2} + k\rho\cos\phi\right)} \int_0^{\sqrt{k\rho(1+\cos\phi)}} \exp\left(iz^2\right) dz. \tag{16.71}$$

This wave function which is independent of n can also be written in terms of the Fresnel integrals S and C [15]

$$\begin{aligned} \psi(\rho, \phi, n + \frac{1}{2}) &= \sqrt{\frac{\pi i}{4}} e^{-i\left(\frac{\phi}{2} + k\rho\cos\phi\right)} \\ &\times \left[C\left(\sqrt{\frac{2}{\pi}k\rho(1+\cos\phi)}\right) + iS\left(\sqrt{\frac{2}{\pi}k\rho(1+\cos\phi)}\right) \right]. \end{aligned} \tag{16.72}$$

As (16.72) shows $\psi\left(\rho,\phi,n+\frac{1}{2}\right)$ is single-valued and vanishes on the line $\phi=\pi$.

A simpler way of finding the scattering amplitude is by the method of partial wave expansion which we discussed in detail in Chapter 14. Noting that the regular solution of the Schrödinger equation (16.52) is given by Eq. (16.58) and expressing the Bessel function in terms of the Hankel functions $H^{(1)}_{|m'+\alpha|}(k\rho)$ and $H^{(2)}_{|m'+\alpha|}(k\rho)$, we have

$$\psi(\rho,\phi,\alpha)=\frac{1}{2}\sum_{m'=-\infty}^{\infty}a_{m'}e^{im'\phi}\left[H^{(1)}_{|m'+\alpha|}(k\rho)+H^{(2)}_{|m'+\alpha|}(k\rho)\right]. \tag{16.73}$$

The asymptotic forms of the Hankel functions as $k\rho\to\infty$ are given by [15]

$$H^{(1)}_{|m+\alpha|}(k\rho)\to\sqrt{\frac{2}{\pi k\rho}}\exp\left[i\left(k\rho-\frac{(m+\alpha)\pi}{2}-\frac{\pi}{4}\right)\right]\left(1+\mathcal{O}\left(\frac{1}{k\rho}\right)\right), \tag{16.74}$$

and

$$H^{(2)}_{|m+\alpha|}(k\rho)\to\sqrt{\frac{2}{\pi k\rho}}\exp\left[-i\left(k\rho-\frac{(m+\alpha)\pi}{2}-\frac{\pi}{4}\right)\right]\left(1+\mathcal{O}\left(\frac{1}{k\rho}\right)\right). \tag{16.75}$$

Using these asymptotic forms we can express the total wave function $\psi(\rho,\phi,\alpha)$ as $k\rho\to\infty$ by

$$\begin{aligned}\psi(\rho,\phi,\alpha)&=\frac{1}{2}\sum_{m'=-\infty}^{\infty}a_{m'}e^{im'\phi}\\ &\times\left\{\exp\left[i\left(k\rho-|m'+\alpha|\frac{\pi}{2}-\frac{\pi}{4}\right)\right]+\exp\left[-i\left(k\rho-|m'+\alpha|\frac{\pi}{2}-\frac{\pi}{4}\right)\right]\right\}\\ &\times\left(1+\mathcal{O}\left(\frac{1}{k\rho}\right)\right).\end{aligned} \tag{16.76}$$

To find the scattering amplitude we write the asymptotic form of $\psi(\rho,\phi,\alpha)$ as

$$\begin{aligned}\psi(\rho,\phi,\alpha)&=\psi_{inc}+\psi_s=e^{-ik\rho\cos\phi-i\alpha\phi}+\psi_s(\rho,\phi,\alpha)\\ &\to e^{-ik\rho\cos\phi-i\alpha\phi}+\frac{e^{ik\rho}f(\phi)}{\sqrt{\rho}}e^{-i\alpha\phi}\left(1+\mathcal{O}\left(\frac{1}{k\rho}\right)\right),\end{aligned} \tag{16.77}$$

where for the incident wave we have added an extra factor $e^{-i\alpha\phi}$ to account for the distortion of the incoming wave caused by the long range effect of an infinite solenoid. Expanding the incident wave $e^{-ik\rho\cos\phi}$ in terms of the Bessel function we have

$$\begin{aligned}e^{-ik\rho\cos\phi}&=\sum_{n=-\infty}^{\infty}(-i)^n e^{i\,n\phi}J_n(k\rho)\\ &\to\sum_{n=-\infty}^{\infty}(-i)^n e^{i\,n\phi}\left(\frac{2}{\pi k\rho}\right)\cos\left(k\rho-\frac{n\pi}{2}-\frac{\pi}{2}\right).\end{aligned} \tag{16.78}$$

By substituting (16.78) in (16.77) and comparing the coefficients of different terms with those of (16.76) we obtain

$$a_{m'} = \exp\left[-\frac{i\pi}{2}\,|m'+\alpha|\right]. \tag{16.79}$$

We also find

$$f(\phi) = \frac{1}{\sqrt{2\pi\, ik}} \sum_{m=-\infty}^{\infty} e^{im(\phi-\pi)}\left(e^{2i\delta_m}-1\right), \tag{16.80}$$

where δ_m is the partial wave phase shift which depends on α

$$\delta_m = -\frac{\pi}{2}|m+\alpha| + \frac{\pi}{2}|m|. \tag{16.81}$$

In the Aharonov–Bohm formulation of this scattering problem the sum over m was carried out before taking the limit of $k\rho \to \infty$, whereas in the partial wave expansion this order was reversed. The result found by the partial wave decomposition for $f(\phi)$ is

$$f(\phi) = \frac{1}{\sqrt{2\pi ik}}\left[-2\pi\delta(\phi-\pi)(1-\cos\pi\alpha) - i\sin\pi\alpha\, e^{\frac{-i\phi}{2}}\cos\frac{\phi}{2}\right]. \tag{16.82}$$

But the cross section obtained from the two methods are identical [18].

For experimental confirmation of the Aharonov–Bohm effect see [19] and [20].

16.2 Time-Dependent Interaction

The nonrelativistic motion of a charged particle in an external electromagnetic field where the scalar potential is zero and the vector potential is given by

$$\mathbf{A} = \frac{1}{2}B(t)\mathbf{k}\wedge\mathbf{r}, \tag{16.83}$$

can be reduced to the equation of motion for a harmonic oscillator for complex q and p. Here $B(t)\mathbf{k}$ is the time-dependent magnetic field which is assumed to be along the z-axis and $\mathbf{r}$ is the position of the particle.

The Hamiltonian for the motion of the charged particle is

$$\begin{aligned} H &= \frac{1}{2m}\left(\mathbf{p}-\frac{e}{c}\mathbf{A}\right)^2 = \frac{1}{2m}\left(p_x^2+p_y^2+p_z^2\right) \\ &+ \frac{e^2B^2(t)}{8mc^2}\left(x^2+y^2\right) + \frac{eB(t)}{2mc}(yp_x - xp_y). \end{aligned} \tag{16.84}$$

In this case z is an ignorable coordinate, i.e. it does not appear in the Hamiltonian. Therefore we only need to consider the motion in the xy plane. Then this

Hamiltonian for the two-dimensional motion can also be written in cylindrical polar coordinates:

$$H = \frac{1}{2m}\left[p_r^2 + \frac{p_\theta^2}{r^2}\right] + \frac{e^2 B^2(t)}{8mc^2} r^2 - \frac{eB(t)}{2mc} p_\theta, \tag{16.85}$$

where p_r is the radial momentum and p_θ is the angular momentum of the particle.

A transformation of x, y, p_x and p_y of the form

$$\begin{cases} Q(t) = (x+iy)\exp\{\frac{ie}{2mc}\int^t B(t')dt'\} \\ P(t) = \frac{c}{e}(p_x + ip_y)\exp\{\frac{ie}{2mc}\int^t B(t')dt'\}, \end{cases} \tag{16.86}$$

together with the Hamiltonian (16.84) give us the equations of motion for $P(t)$ and $Q(t)$;

$$\dot{Q} = \frac{e}{mc} P, \tag{16.87}$$

$$\dot{P} = -\frac{e}{4mc} B^2(t) Q. \tag{16.88}$$

By eliminating P between (16.87) and (16.88) we obtain the equation of motion for the time-dependent harmonic oscillator:

$$\ddot{Q} + \Omega^2(t) Q = 0, \tag{16.89}$$

with

$$\Omega^2(t) = \frac{e^2}{4m^2c^2} B^2(t), \tag{16.90}$$

This transformation is not canonical since the Poisson bracket $\{Q, P\}$ is zero.

16.3 Harmonic Oscillator with Time-Dependent Frequency

We studied the classical formulation of this problem earlier in Sec. 1.12. Here we present a different formulation of the same problem which is based on the expansion of the operator $I(p(t), q(t), t)$ in terms of of the basis set $T_{m,n}$ s. Let the operator $I(p(t), q(t), t)$ be a first integral of the equation of motion

$$\ddot{q}(t) + \Omega^2(t) q(t) = 0, \tag{16.91}$$

then by the definition of a first integral

$$\frac{dI}{dt} = \frac{\partial I}{\partial t} + i[H(t),\ I], \tag{16.92}$$

where

$$H(t) = \frac{1}{2}p^2 + \frac{1}{2}\Omega^2(t)q^2. \tag{16.93}$$

Now we write $H(t)$ and expand $I(p(t), q(t), t)$ in terms of the basis set $T_{m,n}(t)$;

$$H(t) = \frac{1}{2}T_{2,0} + \frac{1}{2}\Omega^2(t)T_{0,2}, \tag{16.94}$$

and

$$I = \sum_{m,n} I_{m,n}(t)T_{m,n}(t). \tag{16.95}$$

By substituting (16.94) and (16.95) in (16.92) and equating the coefficients of $T_{m,n}$ we find a set of linear differential-difference equation;

$$\frac{dI_{m,n}}{dt} + (n+1)I_{m-1,n+1} - (m+1)\Omega^2(t)I_{m+1,n-1} = 0. \tag{16.96}$$

Next we use the method of generating function to solve (16.96) [21]. Let us assume that we know the initial value $I_{m,n}(t=0)$ of $I_{m,n}$. Let us define $G(x, y, t)$ by

$$G(x, y, t) = \sum_{m,n} I_{m,n}x^m y^n, \tag{16.97}$$

then by multiplying (16.96) by $x^m y^n$ and summing over m and n we obtain a first order partial differential equation;

$$\frac{\partial G}{\partial t} + x\frac{\partial G}{\partial y} - \Omega^2(t)y\frac{\partial G}{\partial x} = 0. \tag{16.98}$$

The characteristic equation for (16.98) is [22]

$$dt = -\frac{dx}{\Omega^2(t)y} = \frac{dy}{x}. \tag{16.99}$$

This equation can be written as a matrix equation;

$$\frac{d}{dt}\begin{bmatrix} y \\ x \end{bmatrix} = \begin{bmatrix} 0 & 1 \\ -\Omega^2(t) & 0 \end{bmatrix}\begin{bmatrix} y \\ x \end{bmatrix}. \tag{16.100}$$

The general solution of (16.100) can be expressed in terms of the initial operators q_0 and p_0

$$\begin{cases} x \equiv p(t) = g_1(t)p_0 + f_1(t)q_0 \\ \\ y \equiv q(t) = g(t)p_0 + f(t)q_0 \end{cases}. \tag{16.101}$$

By substituting (16.101) in (16.100) we find the matrix differential equation

$$\frac{d}{dt}\begin{bmatrix} g(t) & f(t) \\ g_1(t) & f_1(t) \end{bmatrix} = \begin{bmatrix} 0 & 1 \\ -\Omega^2(t) & 0 \end{bmatrix}\begin{bmatrix} g(t) & f(t) \\ g_1(t) & f_1(t) \end{bmatrix}. \tag{16.102}$$

This matrix equation shows that $g(t)$ and $f(t)$ are independent solutions of the the classical motion of the harmonic oscillator with time-dependent frequency (see the next section).

16.4 Heisenberg's Equations for Harmonic Oscillator with Time-Dependent Frequency

Let us suppose that the classical equation of motion for the particle is given by (16.89) and we want to solve it using the Heisenberg method, i.e. take (16.89) and write it as an operator equation

$$\ddot{q} + \Omega^2(t)q = 0. \tag{16.103}$$

Equation (16.103) together with the canonical commutation relation $[p, q] = -i$ must be solved to find the eigenvalues (we have set $m = 1$ and $\hbar = 1$). Note that here the momentum p is different from P defined in (16.86), and is given by $p = \dot{q}$, which is the mechanical momentum of the particle.
Since (16.103) is linear in q, the general solution of this operator equation is

$$q(t) = f(t)q_0 + g(t)p_0, \tag{16.104}$$

where $f(t)$ and $g(t)$ are functions of t satisfying the differential equations

$$\ddot{f}(t) + \Omega^2(t)f(t) = 0, \quad \ddot{g}(t) + \Omega^2(t)g(t) = 0, \tag{16.105}$$

and q_0 and p_0 are the initial coordinate and momentum operators with the commutation relation $[q_0, p_0] = i$. The initial conditions for f and g are

$$f(0) = \dot{g}(0) = 1, \quad \dot{f}(0) = g(0) = 0, \tag{16.106}$$

and the Wronskian of these two scalar functions $f(t)$ and $g(t)$ is

$$f(t)\dot{g}(t) - g(t)\dot{f}(t) = 1. \tag{16.107}$$

These together with the equation for $p(t)$,

$$p(t) = \dot{q}(t) = \dot{f}(t)q_0 + \dot{g}(t)p_0, \tag{16.108}$$

imply the constancy of the commutation relation

$$[q(t), p(t)] = [q_0, p_0] = i. \tag{16.109}$$

The Hamiltonian which generates the equation of motion for this system is

$$H(t) = \frac{1}{2}p^2(t) + \frac{1}{2}\Omega^2(t)q^2(t), \tag{16.110}$$

and this Hamiltonian is not a constant of motion. If we substitute for $q(t)$ and $p(t)$ we find

$$\begin{aligned} H(t) &= \frac{1}{2}\left[\left(\dot{f}^2 + \Omega^2(t)f^2\right)q_0^2 + \left(\dot{g}^2 + \Omega^2(t)g^2\right)p_0^2\right. \\ &+ \left.\left(\dot{f}\dot{g} + \Omega^2(t)fg\right)(q_0p_0 + p_0q_0)\right]. \end{aligned} \tag{16.111}$$

Next we find the expectation value of $H(t)$ between the states of the oscillator at time t, i.e. we define $|n\rangle$ by

$$H(t=0)|n\rangle = E_n|n\rangle = \Omega_0\left(n+\frac{1}{2}\right)|n\rangle, \tag{16.112}$$

where in this relation we have denoted $\Omega(t=0)$ by Ω_0. Using any of these states, $|n\rangle$, we first calculate the expectation values of q_0^2, p_0^2, and $(p_0q_0+q_0p_0)$;

$$\langle n\left|q_0^2\right|n\rangle = \frac{1}{\Omega_0}\left(n+\frac{1}{2}\right), \tag{16.113}$$

$$\langle n\left|p_0^2\right|n\rangle = \left\langle n\left|-\frac{\partial^2}{\partial q_0^2}\right|n\right\rangle = \Omega_0\left(n+\frac{1}{2}\right), \tag{16.114}$$

and

$$\langle n\left|p_0q_0+q_0p_0\right|n\rangle = 0. \tag{16.115}$$

Substituting for these terms in the expectation value of $H(t)$, we find

$$\langle n\left|H(t)\right|n\rangle = \frac{1}{2}\Omega_0\left(n+\frac{1}{2}\right)\left[\left(\dot{g}^2+\Omega^2(t)g^2\right)+\frac{1}{\Omega_0^2}\left(\dot{f}^2+\Omega^2(t)f^2\right)\right]. \tag{16.116}$$

We want to show that the time-dependence of (16.116) can be written in terms of a single function of time $\rho(t)$ where

$$\rho(t) = \left(f^2(t)+\Omega_0^2g^2(t)\right)^{\frac{1}{2}}. \tag{16.117}$$

By differentiating $\rho(t)$ twice and substituting from Eqs. (16.105) and (16.107) we find that $\rho(t)$ which is a dimensionless quantity is the solution of the nonlinear equation

$$\ddot{\rho}(t)+\Omega^2(t)\rho(t)-\frac{\Omega_0^2}{\rho^3(t)} = 0, \tag{16.118}$$

and that the initial conditions for this differential equation are

$$\rho(0)=1, \quad \dot{\rho}(0)=0. \tag{16.119}$$

In order to express the expectation value of the Hamiltonian in terms of $\rho(t)$ we use Eq. (16.117) to get

$$\left(\rho(t)\dot{\rho}(t)\right)^2+\Omega_0^2 = \left(f^2+\Omega_0^2g^2\right)\left(\dot{f}^2+\Omega_0^2\dot{g}^2\right) = \rho^2(t)\left(\dot{f}^2+\Omega_0^2\dot{g}^2\right). \tag{16.120}$$

Thus

$$\frac{1}{\rho^2(t)}\left[\left(\rho(t)\dot{\rho}(t)\right)^2+\Omega_0^2\right]+\Omega^2(t)\rho^2(t) = \left(\dot{f}^2+\Omega^2(t)f^2\right)+\Omega_0^2\left(\dot{g}^2+\Omega^2(t)g^2\right). \tag{16.121}$$

By comparing (16.116) and (16.121) we find the desired result, i.e. the diagonal elements of the Hamiltonian [23]

$$\langle n|H(t)|n\rangle = \frac{1}{2\Omega_0}\left(n+\frac{1}{2}\right)\left[\dot{\rho}^2(t)+\frac{\Omega_0^2}{\rho^2(t)}+\Omega^2(t)\rho^2(t)\right]. \tag{16.122}$$

There is another way of finding the energy eigenvalues of $H(t)$. In this method we find a first integral of motion, $I(p,q,t)$, which is also a conserved quantity for a harmonic oscillator with time-dependent frequency [24]–[27]. Since we want $I(p,q,t)$ to be a constant of motion, therefore its total time derivative must be zero

$$\frac{dI}{dt} = \frac{\partial I}{\partial t} + \frac{1}{i}[I,H] = 0, \quad \hbar = 1. \tag{16.123}$$

We assume that this first integral is a homogeneous quadratic function of p and q and in addition we want it to be Hermitian. Let us write this operator as

$$I(t) = \frac{1}{2}\left[\alpha(t)q^2 + \beta(t)p^2 + \gamma(t)(pq+qp)\right], \tag{16.124}$$

and note that $I(p,q,t)$ is a self-adjoint operator

$$I(p,q,t) = I^\dagger(p,q,t). \tag{16.125}$$

We also write the Hamiltonian in a slightly different form

$$H(t) = \frac{1}{2m}\left[p^2 + m^2\Omega^2(t)q^2\right]. \tag{16.126}$$

This is similar to $H(t)$ defined by (16.110) with the mass m introduced in its definition. From (16.123), (16.124) and (16.126) it follows that

$$\begin{aligned} &\frac{1}{2}\left[\left(\dot{\alpha}(t) - 2m\Omega^2(t)\gamma(t)\right)q^2 + \left(\dot{\beta}(t) + \frac{2}{m}\gamma(t)\right)p^2\right. \\ &+ \left.\left(\dot{\gamma}(t) + \frac{1}{m}\alpha(t) - m\Omega^2(t)\beta(t)\right)(pq+qp)\right] = 0. \end{aligned} \tag{16.127}$$

Equation (16.127) must be satisfied by all matrix elements of I, therefore the following relations must be true:

$$\dot{\alpha}(t) = 2m\Omega^2(t)\gamma(t), \tag{16.128}$$

$$\dot{\beta}(t) = -\frac{2}{m}\gamma(t), \tag{16.129}$$

and

$$\dot{\gamma}(t) = m\Omega^2(t)\beta(t) - \frac{1}{m}\alpha(t). \tag{16.130}$$

Now let us introduce a function $\sigma(t)$ by the relation

$$\sigma^2(t) = \beta(t), \tag{16.131}$$

then Eq. (16.129) can be written as

$$\gamma(t) = -m\sigma(t)\dot{\sigma}(t). \tag{16.132}$$

By substituting (16.131) and (16.132) in (16.130) we find $\alpha(t)$

$$\alpha(t) = m^2\left[\left(\dot{\sigma}^2(t) + \sigma(t)\ddot{\sigma}(t)\right) + \Omega^2(t)\sigma^2(t)\right]. \tag{16.133}$$

The two functions $\gamma(t)$ and $\alpha(t)$ found in (16.132) and (16.133) must satisfy Eq. (16.128). Thus we get a differential equation for $\sigma(t)$;

$$\sigma(t)\frac{d}{dt}\left[m^2\left(\ddot{\sigma}(t) + \Omega^2(t)\sigma(t)\right)\right] + 3m^2\left[\dot{\sigma}(t)\left(\ddot{\sigma}(t) + \Omega^2(t)\sigma(t)\right)\right] = 0. \tag{16.134}$$

This equation can be integrated to yield

$$m^2\ddot{\sigma}(t) + m^2\Omega^2(t)\sigma(t) - \frac{c}{\sigma^3(t)} = 0, \tag{16.135}$$

where c is the constant of integration and is real. We now make a scale change by writing

$$\sigma(t) = \frac{c^{\frac{1}{4}}\rho(t)}{\sqrt{m\Omega_0}}, \tag{16.136}$$

and by substituting this expression in (16.135) we find Eq. (16.118). Written in terms of $\rho(t)$, the invariant operator $I(p,q,t)$ becomes

$$I(p,q,t) = \frac{1}{2}\left\{\frac{1}{\rho^2(t)}q^2 + \left[\rho(t)p - m\dot{\rho}(t)q\right]^2\right\}. \tag{16.137}$$

Having found the explicit form of $I(p,q,t)$ we can use it to determine the eigenvalues of $H(t)$ by the method of factorization of Sec. 8.4.

Here we use a representation in which $I(p,q,t)$, and not $H(p,q,t)$, is diagonal. Let $|n\rangle$ be a state such that

$$I(p,q,t)|n\rangle = n|n\rangle \tag{16.138}$$

and

$$\langle n'|I(p,q,t)|n\rangle = \delta_{n,n'}. \tag{16.139}$$

Since $I(p,q,t)$ depends on time, we will show that the eigenvalue n is time-independent. We first find the partial derivative of (16.138) with respect to t;

$$\frac{\partial I}{\partial t}|n\rangle + I\frac{\partial}{\partial t}|n\rangle = \frac{\partial n}{\partial t}|n\rangle + n\frac{\partial}{\partial t}|n\rangle. \tag{16.140}$$

At the same time from Eq. (16.123) we have

$$i\frac{\partial I}{\partial t}|n\rangle + I(p,q,t)H|n\rangle - nH|n\rangle = 0. \tag{16.141}$$

Also we find that the scalar product of (16.141) with $\langle n'|$ is

$$i\left\langle n'\left|\frac{\partial I}{\partial t}\right|n\right\rangle + (n'-n)\langle n'|H|n\rangle = 0, \tag{16.142}$$

and therefore

$$\left\langle n\left|\frac{\partial I}{\partial t}\right|n\right\rangle = 0. \tag{16.143}$$

Now we observe that from the scalar product of (16.140) with $\langle n|$ it follows that

$$\frac{\partial n}{\partial t} = 0. \tag{16.144}$$

Thus the eigenvalues of $I(p,q,t)$ do not depend on time.

For the calculating the matrix elements of $I(p,q,t)$ we introduce the lowering and the raising operators, a and $a^\dagger$, by

$$a = \frac{1}{\sqrt{2}}\left[\frac{1}{\rho(t)}q + i\left(\rho(t)p - m\dot{\rho}(t)q\right)\right], \tag{16.145}$$

and

$$a^\dagger = \frac{1}{\sqrt{2}}\left[\frac{1}{\rho(t)}q - i\left(\rho(t)p - m\dot{\rho}(t)q\right)\right]. \tag{16.146}$$

We note that these operators satisfy the canonical commutation relation

$$\left[a, a^\dagger\right] = 1, \tag{16.147}$$

and that

$$aa^\dagger = I(p,q,t) + \frac{1}{2}. \tag{16.148}$$

The ground state of the system is defined by

$$a|0\rangle = 0, \tag{16.149}$$

and the other states can be obtained from the ground state by applying the raising (or creation operator),

$$|n+1\rangle = \frac{1}{\sqrt{n+1}}a^\dagger|n\rangle, \quad n = 0, 1, \cdots. \tag{16.150}$$

Since n does not depend on time from (16.150) we have

$$\frac{\partial}{\partial t}|n+1\rangle = \frac{1}{\sqrt{n+1}}\frac{\partial}{\partial t}a^\dagger|n\rangle, \quad n = 0, 1, \cdots. \tag{16.151}$$

Also Eq. (16.149) can be used to obtain the ground state wave function

$$\psi_0(q,t) = \frac{e^{i\delta_0(t)}}{\left(\pi\rho^2(t)\right)^{\frac{1}{4}}}\exp\left\{-\frac{q^2}{2\rho^2(t)}\left(1 - im\dot{\rho}(t)\rho(t)\right)\right\}, \tag{16.152}$$

and from this result with the aid of (16.146) and (16.150) we can construct the excited state wave functions.

We can use the definition of $a^\dagger$ and a, Eqs. (16.145) and (16.146) and write the diagonal elements of the Hamiltonian as

$$\begin{aligned}\langle n|H|n\rangle &= \frac{m}{2}\left(\dot{\rho}^2(t)+\Omega^2(t)\rho^2(t)+\frac{1}{m^2\rho^2(t)}\right)\langle n\left|aa^\dagger+a^\dagger a\right|n\rangle \\ &= \frac{m}{4}\left(n+\frac{1}{2}\right)\left(\dot{\rho}^2(t)+\Omega^2(t)\rho^2(t)+\frac{1}{m^2\rho^2(t)}\right), \qquad (16.153)\end{aligned}$$

which has the same form as Eq. (16.122).

In addition using this method enables us to calculate the off-diagonal elements of H as well [27];

$$\begin{aligned}\langle n'|H|n\rangle &= \frac{1}{4}\left\{\left[m\left(\dot{\rho}^2(t)-\rho(t)\ddot{\rho}(t)\right)-2i\left(\frac{\dot{\rho}(t)}{\rho(t)}\right)\right][n(n-1)]^{\frac{1}{2}}\delta_{n'+2,n}\right. \\ &+ \left.\left[m\left(\dot{\rho}^2(t)-\rho(t)\ddot{\rho}(t)\right)+2i\left(\frac{\dot{\rho}(t)}{\rho(t)}\right)\right][(n+1)(n+2)]^{\frac{1}{2}}\delta_{n',n+2}\right\}, \quad n'\neq n.\end{aligned} \qquad (16.154)$$

When $\Omega(t)$ is constant and is equal to Ω_0, then $\rho(t) = \rho(0) = 1$ and (16.153) and (16.154) reduce to a single equation

$$\langle n'|H|n\rangle = \Omega_0\left(n+\frac{1}{2}\right)\delta_{n',n}, \qquad (16.155)$$

which is the diagonal matrix elements for the Hamiltonian of a simple harmonic oscillator with frequency Ω_0. This problem can also be solved by the method of integration of the operator differential equation of Sec. 16.3 which we applied to solve the problem of simple harmonic oscillator. For this we expand the operator conjugate to the Hamiltonian operator, $\Theta(p,q,t)$, in terms of the basis set of $T_{m,n}(p,q)$ operators,

$$\Theta(p,q,t) = \sum_{m,n}\alpha_{m,n}(t)T_{m,n}(p,q). \qquad (16.156)$$

In Sec. 16.3 we showed how we can calculate the time-dependent coefficients $\alpha_{m,n}(t)$. These coefficients are given in terms of $\sigma(t)$ as follows [28]

$$\alpha_{0,\,0} = \frac{\sqrt{c}\,\Omega_0}{\sigma^2(t)}\tan^{-1}\left[-\frac{\sigma^2(t)}{\sqrt{c}\,\Omega_0}\frac{d}{dt}\ln\sigma(t)\right], \qquad (16.157)$$

$$\begin{aligned}\alpha_{-1-2m,\,1+2m} &= \frac{(-1)^m}{(1+2m)}\left(\frac{\sqrt{c}\,\Omega_0}{\sigma^2(t)}\right)^{2+2m} \\ &\times\ {}_2F_1\left[-\frac{1}{2}-m,-m,\frac{1}{2},-\frac{\sigma^4(t)}{c\,\Omega_0^2}\left(\frac{d}{dt}\ln\sigma(t)\right)^2\right],\end{aligned} \qquad (16.158)$$

and

$$\alpha_{-2-2m,\,2+2m} = (-1)^m \left(\frac{d}{dt}\ln\sigma(t)\right)\left(\frac{\sqrt{c}\,\Omega_0}{\sigma^2(t)}\right)^{2+2m} \times {}_2F_1\left[-\frac{1}{2}-m,-m,\frac{3}{2},-\frac{\sigma^4(t)}{c\,\Omega_0^2}\left(\frac{d}{dt}\ln\sigma(t)\right)^2\right], \tag{16.159}$$

where $m = 0,\ 1,\ 2\cdots$ and ${}_2F_1$ is the hypergeometric function. For the special case of $\Omega(t) = \text{constant}$, only the coefficients $\alpha_{-1-2m,\,1+2m}$ are nonzero and are given by $\frac{(-1)^m}{(2m+1)}$ which is the same as (??).

16.5 Neutron Interferometry

The close mathematical analogy between the Schrödinger equation describing the low energy (or thermal) neutrons on one hand, and the propagation of light waves as is formulated by Maxwell's equation on the other, suggests that one can find quantum analogues for most of the phenomena occurring for the light waves. This is true although photons are bosons whereas neutrons are fermions.

Neutrons are electrically neutral particles each having a mass of $m_n = 1.67\times10^{-27}$ Kg. and a magnetic moment of $\mu_n = -1.913\left(\frac{e\hbar}{2m_nc}\right)$. In this section we show how neutron interferometry can be used to test the spinor nature of neutrons. Later we will show how by splitting a beam of neutrons into two parts and then recombining the two beams we can detect the presence of the classical gravitational potential.

As is shown in Fig. 16.3, a neutron beam at S is split into two beams. The lower beam passes through a magnetic field which causes a precession of the magnetic moment of the neutron and thus a change in the phase of the wave function. When the two beams are combined at S, the intensity will depend on this change of phase as will be shown below (see Eq. (16.166)). First let us consider the passage of a neutron through a uniform magnetic field. The precession of the magnetic moment after a very short time dt is given by

$$d\phi = \omega_L dt = \frac{\mu_n B}{\hbar}dt. \tag{16.160}$$

Integrating this expression, assuming a constant speed v_n for the neutrons we get

$$\phi = \frac{\mu_n}{\hbar}\int Bdt = \frac{\mu_n}{\hbar v_n}\int_{path}\mathbf{B}\cdot d\mathbf{l}. \tag{16.161}$$

If the neutron beam is unpolarized then the neutrons reaching the observation point O through the path SAO will have no change in phase and the wave

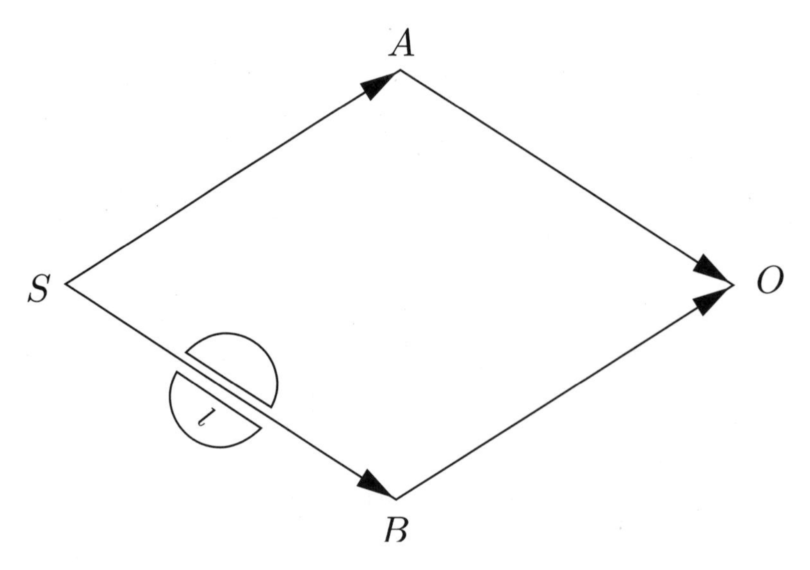

Figure 16.3: The neutron beam at S is split into two beams, and then these beams are recombined at O. The lower beam passes through a magnetic field, and this results in the interference pattern at O.

function is

$$\psi_{SAO} = \psi_A = \frac{1}{\sqrt{2}}\begin{bmatrix} 1 \\ 1 \end{bmatrix}. \tag{16.162}$$

On the other hand the beam passing through the lower arm SBO interacts with the magnetic field over a length l in its path and this causes a change in the phase of the wave function

$$\psi_{SBO} = \psi_B = \frac{1}{\sqrt{2}}\begin{bmatrix} e^{\frac{i\phi}{2}} \\ e^{\frac{-i\phi}{2}} \end{bmatrix}. \tag{16.163}$$

This is found from the action of the operator T_ϕ, Eq. (9.166) on ψ_A where ϕ is given by (16.161). When the two beams recombine at O, the total wave function is the sum of ψ_A and ψ_B;

$$\psi_{total} = \psi_A + \psi_B = \frac{1}{\sqrt{2}}\begin{bmatrix} 1 + e^{\frac{i\phi}{2}} \\ 1 + e^{\frac{-i\phi}{2}} \end{bmatrix}, \tag{16.164}$$

and this gives us the following probability density

$$|\psi_{total}|^2 = \frac{1}{2}\left(\left|1 + e^{\frac{i\phi}{2}}\right|^2 + \left|1 + e^{\frac{-i\phi}{2}}\right|^2\right) = 2\left(1 + \cos\frac{\phi}{2}\right). \tag{16.165}$$

Denoting the intensity of the beam by $I(\phi)$ and the intensity in the absence of the magnetic field by $I(0)$, we have

$$\frac{I(\phi)}{I(0)} = \frac{|\psi_{total}|^2(\phi)}{|\psi_{total}|^2(0)} = \frac{1}{2}\left(1 + \cos\frac{\phi}{2}\right). \tag{16.166}$$

If we keep the length of the field l fixed and vary B, then ϕ and consequently $I(\phi)$ will change. But $I(\phi)$ has maxima at $\phi = 0,\ \ 4\pi \cdots$, i.e. we get the same intensity, not by a rotation by 2π, but by a rotation through 4π for spin $\frac{1}{2}$ particles.

16.6 Gravity-Induced Quantum Interference

Another example where the potential has an observable effect in quantum mechanics is the gravity-induced quantum interference. For the formulation of this problem let us first consider the motion of a particle in a constant field of gravity.

For the motion of a falling body the classical equation of motion is

$$m_n \ddot{\mathbf{r}} = -m_n \nabla V_g = -m_n g \hat{\mathbf{k}}. \tag{16.167}$$

In this case the motion does not depend on the mass m_n since it drops out of the two side of the equation. This is a consequence of the equality of the gravitational and inertial masses. But in quantum mechanics the analogue of (16.167) is the Schrödinger equation

$$\left(-\frac{\hbar^2}{2m_n}\nabla^2 + m_n V_g\right)\psi = i\hbar\frac{\partial \psi}{\partial t}, \tag{16.168}$$

and this shows that the wave function is dependent on the combination $\hbar/m$ [40]. Now let us study a remarkable phenomena where the motion of a particle such as a neutron shows observable quantum phase differences [41].

Let us consider an arrangement shown in Fig. 16.4. The upper and the lower arms $SABO$ and $SCDO$ have the same path lengths for the neutrons. The experiment is done by splitting a beam of neutrons at the source point S into two beams, and allowing them to recombine at the point O in Fig. 16.4.

If the particles are thermal neutrons, the size of the wave packet for each particle will be much smaller than the macroscopic dimensions of the two arms, and thus we can consider a classical trajectory for the motion, i.e. we can use the WKB approximation. Thus the wave function for a neutron passing through the upper arm is

$$\psi_1(x_O) = \psi(x_S)\exp\left(-\frac{i}{\hbar}\int_{x_S}^{x_O} p_u(x)dx\right), \tag{16.169}$$

and for a particle reaching the point O through lower arm is

$$\psi_2(x_O) = \psi(x_S)\exp\left(-\frac{i}{\hbar}\int_{x_S}^{x_O} p_l(x)dx\right), \tag{16.170}$$

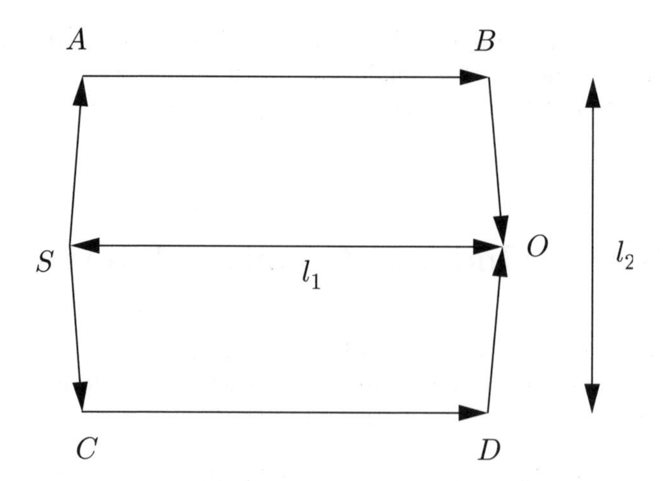

Figure 16.4: At the point S a beam of neutrons is split into two beams one travelling along $SABO$ and the other along $SCDO$. At the point O the two beams are recombined. By rotating the loop about the SO axis one observes an interference pattern when the intensity at O is plotted as a function of the angle of rotation of the loop.

where in these relations $p_u(x)$ and $p_l(x)$ are defined by

$$p_u(x) = \sqrt{2m_n[E - V_u(x)]}, \tag{16.171}$$

and

$$p_l(x) = \sqrt{2m_n[E - V_l(x)]}, \tag{16.172}$$

where m_n is the mass of a single neutron. Since the potential energy $V_{u,l}(x)$ is small relative to the total energy of the neutrons, we have

$$p_u(x) \approx \sqrt{2m_nE}\left(1 - \frac{V_u(x)}{2E} + \cdots\right), \tag{16.173}$$

and

$$p_l(x) \approx \sqrt{2m_nE}\left(1 - \frac{V_l(x)}{2E} + \cdots\right). \tag{16.174}$$

When the two arms lie in a horizontal plane, V_u is the same as V_l, and since the potentials are the same and the path lengths are identical we have coherent waves arriving at S. By rotating the loop about the horizontal axis through an angle δ, the potential V_u becomes larger than V_l by an amount $m_ngl_2 \sin\delta$, where l_2 is the length $AC = BD$ of the arms. Thus from Eqs. (16.169),(16.170) and (16.173) and (16.174) it follows that the wavepacket arriving at O via the upper arm suffers a phase change

$$\exp\left(-\frac{im_ngl_2T\sin\delta}{\hbar}\right), \tag{16.175}$$

relative to that of the wavepacket reaching O by the lower arm. Here T is the time that takes for the wavepacket to move from A to B (or from C to D).

We can write the phase difference between the neutrons going through the path $SABO$ and those going by $SCDO$ as

$$\Delta\phi = \phi_{SABO} - \phi_{SCDO} = -\frac{m_n^2 g l_1 l_2 \lambda \sin\delta}{\hbar^2}, \quad \lambda = \frac{2\pi\hbar}{m_n v}, \tag{16.176}$$

where we have replaced the time T by

$$T = \frac{l_1}{v_{\text{wavepacket}}} = \frac{l_1 2\pi\lambda m}{\hbar}. \tag{16.177}$$

That is we have expressed time in terms of the length l_1 of the arm and $\bar{\lambda}$, the de Broglie wavelength of the neutron. With the thermal neutrons with the wavelength $\bar{\lambda} = 8.92$ Å and the surface area of the loop $l_1 l_2 = 10$ cm^2 we get $\Delta\phi = 55.6$ radians. By rotating the loop about the SO axis we find that because of the interference between the two beams we have a pattern similar to the interference pattern in optics, exhibiting a series of maxima and minima. Thus we expect to see $\frac{55.6}{2\pi} \approx 9$ oscillations [41], [42]. Experimentally this pattern with these oscillations have been observed [43]. That this phenomenon is of purely quantum mechanical origin can be seen from the limit of $\hbar \to 0$, where the pattern disappears. This experiment also shows that in quantum mechanics gravity is not purely a geometric factor since the result depends on $(m_n/\hbar)^2$.

16.7 Quantum Beats in Waveguides with Time-Dependent Boundaries

Another nonclassical feature of quantum particles, that is those obeying the Schrödinger equation, is their interference, as we observed in connection with the Aharonov–Bohm effect and also in connection with the observability of the gravitational potential. An interesting and important example of the interference occurs when two waves, with frequencies close to each other interfere. The classical example of two tuning forks, or two weakly coupled harmonic oscillators have their analogues in quantum mechanics [29].

Let us consider a beam of particles, e.g. cold neutrons, split in two parts, the two parts are passed through identical waveguides and then brought back together at the position of the detector where they can interfere constructively. If the boundaries of one of the two waveguides starts moving while the other remains stationary, the wave arriving at the detector will have different frequencies, and the absolute value of the total wave function at the terminal point oscillates in time and shows the presence of interference and quantum beats. There are two factors responsible for this change of phase:

(a) - As the walls of the waveguide expands, the eigenvalues of the Hamiltonian will change, and so the time-dependent factor

$$\exp\left(\frac{-i\int E(t)dt}{\hbar}\right), \tag{16.178}$$

in the Schrödinger equation causes a "dynamical" phase shift [30],[31].

(b) - In addition the center of each eigenfunction along the direction of motion will change, (the center still remains at the middle of the waveguide), thus the eigenfunctions will acquire additional phase, this time "geometrical phase".

Let us consider a waveguide formed from surfaces at $x = -a$, $x = a$, $y = -L(t)$ and $y = L(t)$, where a is a constant and $L(t)$ is a smooth function of time. A monoenergetic beam of identical particles which is incident from the negative z-axis, travels parallel to the z-axis and emerges in the direction of the positive z-axis. Denoting the mass of each particle by m_n, and setting $\hbar = 1$, we can find the wave function from the solution of the time-dependent Schrödinger equation

$$i\frac{\partial\psi}{\partial t} = -\frac{1}{2m_n}\nabla^2\psi, \tag{16.179}$$

but now with the time-dependent boundary conditions:

$$\psi(x = \pm a,\ y,\ z,\ t) = 0, \tag{16.180}$$

and

$$\psi(x,\ y = \pm L(t),\ z,\ t) = 0. \tag{16.181}$$

The symmetry of the boundary condition shows that there is no mixing between even and odd parity states. Therefore we will just consider the solutions of Eq. (16.179) for even parity states;

$$\psi(x,\ y,\ z,\ t) = \psi(-x,\ y,\ z,\ t), \tag{16.182}$$

and

$$\psi(x,\ y,\ z,\ t) = \psi(x,\ -y,\ z,\ t). \tag{16.183}$$

A wave traveling in this waveguide in the direction of positive z axis and satisfying the symmetry conditions (16.182) and (16.183) is given by

$$\begin{aligned}\psi(x,\ y,\ z,\ t) &= \sum_{n=0}^{\infty}\sum_{m=0}^{\infty}\cos\left[\left(n+\frac{1}{2}\right)\frac{\pi y}{L(t)}\right]\cos\left[\left(m+\frac{1}{2}\right)\frac{\pi x}{a}\right] \\ &\times \exp\left[-ik_3(nm)z\right]\phi_{nm}(k_3,t),\end{aligned} \tag{16.184}$$

where $k_3(mn)$ s are wave numbers for the propagating modes.

To find the time-dependent coefficients $\phi_{nm}(k_3,t)$ we substitute (16.184) in (16.179), multiply the result by

$$\cos\left[\left(j+\frac{1}{2}\right)\frac{\pi y}{L(t)}\right]\cos\left[\left(m+\frac{1}{2}\right)\frac{\pi x}{a}\right], \tag{16.185}$$

and integrate over x from $-a$ to a and over y from $-L(t)$ to $L(t)$. The result shows that $\phi_{nm}(k_3,t)$ satisfies the following coupled ordinary differential equations

$$\begin{aligned} i\frac{d\phi_{jm}(t)}{dt} &+ \left(\frac{i}{2\pi}\right)\left(\frac{d}{dt}\ln L(t)\right)\sum_{n=0}^{\infty} A(j,n)\phi_{nm}(t) \\ &+ \frac{1}{2m_n}\left\{\left(j+\frac{1}{2}\right)^2\frac{\pi^2}{L^2(t)} + k_m^2 + k_3^2\right\}\phi_{jm}(t) = 0, \end{aligned} \tag{16.186}$$

where $A(j,n)$ is a dimensionless matrix with elements

$$\begin{cases} A(j,n) = -A(n,j) = \dfrac{(-1)^{j-n+1}\pi(2j+1)(2n+1)}{(j-n)(j+n+1)} & j \neq n \\ A(n,n) = \pi \end{cases}, \tag{16.187}$$

and

$$k_m^2 = \frac{\left(m+\frac{1}{2}\right)^2\pi^2}{a^2}. \tag{16.188}$$

To simplify Eq. (16.186) further we first suppress the index m in the equation for ϕ_{nm}. Then we change $\phi_j(t)$ to $\theta_j(t)$, and study the solution for a fixed m, $\phi_j(t) \equiv \phi_{jm}(t)$ to $\theta_j(t)$ where the latter is defined by

$$\phi_j(t) = \frac{1}{\sqrt{L(t)}}\exp\left[\frac{i}{2m_n}\left(k_3^2 + k_m^2\right)t\right]\theta_j(t). \tag{16.189}$$

With this change we get a simpler equation for $\theta_j(t)$;

$$i\frac{d\theta_j(t)}{dt} + \left(\frac{1}{2m_n}\right)\left[\frac{\left(j+\frac{1}{2}\right)^2\pi^2}{L^2(t)}\right]\theta_j(t) + \left(\frac{i}{2\pi}\right)\frac{d}{dt}\ln L(t)\sum_{n\neq j} A(j,n)\theta_n(t) = 0. \tag{16.190}$$

Now we calculate

$$\theta_j^*(t)\frac{d\theta_j(t)}{dt} + \theta_j(t)\frac{d\theta_j^*(t)}{dt}, \tag{16.191}$$

using (16.190) and its complex conjugate we find a simple form for the conservation of flux

$$\frac{d}{dt}\sum_{j=0}^{\infty}|\theta_j(t)|^2 = 0, \tag{16.192}$$

Next we want to determine $\theta_j(t)$ by solving (16.190) and for integrating this equation we need to know the initial conditions. These conditions depend on the way that $L(t)$ behaves as $t \to \infty$. Here we only consider those cases where $L(-\infty)$ tends to a finite nonzero value and in the same limit of $t \to \infty$, $(d\ln L(t)/dt)$ goes to zero. Thus for large negative values of t we write

$$\begin{aligned}\psi_{m,j,k_3}(x,\ y,\ z,\ t) &= \cos\left[\left(j+\frac{1}{2}\right)\frac{\pi y}{L(t)}\right]\cos\left[\left(m+\frac{1}{2}\right)\frac{\pi x}{a}\right]\\ &\times\ \exp(-ik_3 z)\exp[-iE(-\infty)t].\end{aligned} \tag{16.193}$$

By substituting (16.193) in (16.179) we find $E(-\infty)$

$$E(-\infty) = \frac{1}{2m_n}\left\{k_3^2 + \left(j+\frac{1}{2}\right)^2\frac{\pi^2}{L^2(-\infty)} + \left(m+\frac{1}{2}\right)^2\frac{\pi^2}{a^2}\right\}. \tag{16.194}$$

For propagating modes k_3^2 must be positive, therefore $E(-\infty)$ must satisfy the inequality

$$2m_n E(-\infty) - \left\{\left(j+\frac{1}{2}\right)^2\frac{\pi^2}{L^2(-\infty)} + \left(m+\frac{1}{2}\right)^2\frac{\pi^2}{a^2}\right\} > 0. \tag{16.195}$$

Obviously when $E(-\infty)$ is finite, only a finite number of modes can propagate. In order to have any propagating mode at all the energy $E(-\infty)$ has to satisfy the inequality

$$E(-\infty) > \frac{\pi^2}{8m_n}\left(\frac{1}{L^2(-\infty)} + \frac{1}{a^2}\right). \tag{16.196}$$

For finite $L(-\infty)$ we observe that there will be j modes for which $\theta(-\infty)$ s are not zero. That is as $t \to \infty$ we have

$$\begin{cases}\theta_j(-\infty) = \frac{1}{\sqrt{J}}, & \text{for}\quad j = 0,\ 1,\ 2\cdots J-1\\ \theta_j(-\infty) = 0, & \text{for}\quad j = J,\ J+1,\cdots\end{cases}. \tag{16.197}$$

Quantum Beats — Suppose that a beam of identical monoenergetic particles, (e.g. ultracold neutrons with wavelength of the order 10^3 Å and average speed of 5 m/s) is split up into two parts. These two beams pass through two rectangular waveguides with identical geometries and then they are brought together at the position of the detector by identical passages as is shown in Fig. 16.5. In this case we have constructive interference between the two waves. Now if the two boundaries at $y = \pm L(t)$ of one of the waveguides start moving smoothly then the wave function of the waves arriving at the detector from the two waveguides will have the same amplitude but different phases:

$$\psi_S(y,\ z,\ t) = \sum_{j=0}^{J-1}\exp(-ik_3 z)\cos\left[\frac{\left(j+\frac{1}{2}\right)\pi y}{L(-\infty)}\right]\exp[-iE(-\infty)t], \tag{16.198}$$

and

$$\psi_M(y,\ z,\ t) = \sum_{n=0}^{N-1}\exp(-ik_3 z)\cos\left[\frac{\left(n+\frac{1}{2}\right)\pi y}{L(t)}\right]\theta_n(t), \tag{16.199}$$

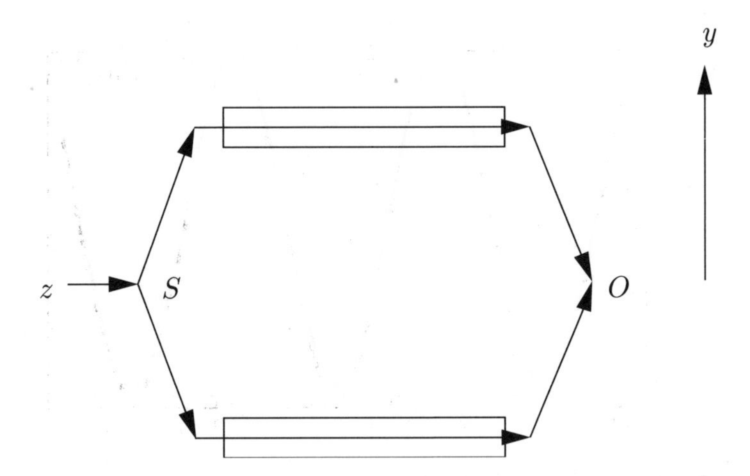

Figure 16.5: Sketch of an experiment demonstrating the change of phase of a wave passing through a waveguide with moving boundaries. Here the walls of the upper waveguide is moving in the y direction and their displacements are given by $\pm L(t)$. The lower waveguide has fixed walls.

for stationary and moving boundary waveguides respectively. For simplicity in these equations we have suppressed the x dependence and the index m. Having found the waves arriving by two different paths, we can determine the total wave function at the detector;

$$\psi(y,\ z,\ t) = \psi_M(y,\ z,\ t) + \psi_S(y,\ z,\ t). \tag{16.200}$$

The z component of the velocity of the particle is given by $\hbar k_3/m_n$, where k_3 is related to the initial energy $E(-\infty)$ by Eq. (16.194). If l denotes the length of the waveguide and T denotes the traversal time of the particles, then $T = \frac{lm_n}{\hbar k_3}$. Let ν be a frequency such that for an integer N, $L\left(-\frac{N}{\nu}\right) - L(-\infty)$ and $L\left(\frac{N}{\nu}\right) - L(-\infty)$ are small quantities, now if $2N\nu T < 1$ then the particles passing through the waveguide with moving walls will experience the full force due to the complete transverse motion of these walls. For example we can choose

$$L(t) = L_1(t) = \pi\left[1 + \frac{\sin^2 \nu\pi t}{\cosh 2\nu t}\right], \tag{16.201}$$

or

$$L(t) = L_2(t) = \pi\left[1 + \sin^2 \nu\pi t\right], \quad 0 \le t \le \frac{N}{\nu}, \tag{16.202}$$

then $L\left(\pm\frac{N}{\nu}\right) - L(-\infty)$ is zero. To find the partial wave projection of $\psi(y,\ z,\ t)$ we multiply (16.200) by $\cos\left[\left(j + \frac{1}{2}\right)\pi y/L(-\infty)\right]$ and integrate from $-L(-\infty)$ to $L(-\infty)$. If $L(-\infty) = L(\infty)$, then we have the simple result

$$\psi(j,\ z,\ t) = \exp(-ik_3 z)L(-\infty)\left\{e^{-iE(-\infty)t} + \theta_j(t)\right\}. \tag{16.203}$$

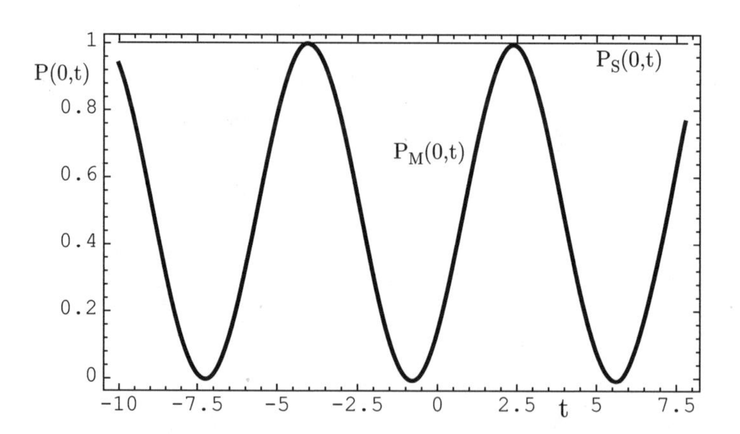

Figure 16.6: The probabilities obtained at the point O Fig. 16.5 from the waves arriving from the two identical waveguides: (a) - $P_S(0,t)$ is the probability when both are stationary and (b) - $P_M(0,t)$ is the probability when one waveguide is stationary and the other has moving walls with $L(t) = L_1(t)$.

From these partial waves we can calculate the probability of finding particles at the position of the detector.

For simplicity let us choose $E(-\infty)$ so that only one propagating mode, $j = m = 0$, is allowed, i.e. $E(-\infty)$ can have any value between

$$\frac{\pi^2}{8m_n}\left[\frac{1}{L^2(-\infty)} + \frac{1}{a^2}\right], \tag{16.204}$$

and

$$\frac{9\pi^2}{8m_n}\left[\frac{1}{L^2(-\infty)} + \frac{1}{a^2}\right]. \tag{16.205}$$

Then by integrating (16.190) and substituting for θ_j in (16.203) we can calculate $\psi(j,\ z,\ t))$. Finally $P_M(0,t)$ can be found from normalized $\psi(j,\ z,\ t))$;

$$P_M(0,\ t) = \frac{1}{4}|\psi(j,\ z,\ t)|^2 == \frac{1}{4}\left|\exp(-ik_3 z)L(-\infty)\left\{e^{-iE(-\infty)t} + \theta_j(t)\right\}\right|^2. \tag{16.206}$$

The probability $P_M(0,\ t)$ shows a typical interference pattern, indicating a change in the phase of the wave caused by the motion of the walls. This probability is shown in Fig. 16.6 when the wall boundaries are $\pm L(t) = \pm L_1(t)$, Eq. (16.201).

Forces Generated by the Motion of the Boundaries — The change of the phase of the wave can be attributed to the action of a rather complicated time-dependent force. Let us inquire if it is possible to find a potential which simulates the effect of the motion of the boundaries [32]. For this we examine Eq. (16.190) for $\theta_j(t)$. This equation simplifies if the two time-dependent coefficients $\frac{d}{dt}[\ln L(t)]$ and $L^{-2}(t)$ are proportional to each other, that is

$$\frac{d}{dt}\ln L(t) = \frac{\omega L_0^2}{2L^2(t)}, \tag{16.207}$$

where $\omega L_0^2/2$ is the constant of proportionality. By integrating this equation we find

$$L^2(t) = L_0^2(\omega t + \delta), \tag{16.208}$$

with δ being the constant of integration.

If we change the variable t to τ, where τ defined by

$$\tau = \frac{1}{\omega}\ln\left(\frac{\omega t + \delta}{\delta}\right), \tag{16.209}$$

we obtain the following set of linear differential equations for $\theta_j(\tau)$,

$$\begin{aligned} i\frac{d\theta_j(\tau)}{d\tau} &+ \left(\frac{\pi^2}{2m_n L_0^2}\right)\left(j+\frac{1}{2}\right)^2 \theta_j(t) \\ &+ \frac{i\omega}{4\pi}\sum_{n\neq j} A(j,n)\,\theta_n(t). \end{aligned} \tag{16.210}$$

This is a linear homogeneous differential equation with constant coefficients. We use the trial solution

$$\theta_j(t) = \theta_j(0)e^{i\Omega\tau}, \tag{16.211}$$

to find the following eigenvalue equation for Ω;

$$\det\left[\left\{\Omega - \left(j+\frac{1}{2}\right)^2 \frac{\pi^2}{2m_n L_0^2}\right\} - \frac{i\omega}{4\pi}A(j,n)(1-\delta_{nj})\right] = 0. \tag{16.212}$$

Since the matrix

$$\left[-\frac{i\omega}{4\pi}A(j,n)(1-\delta_{nj})\right], \tag{16.213}$$

is Hermitian, the eigenvalues, Ω_α are real. Thus if the walls are moving according to (16.208), the eigenvalues change from $\left(j+\frac{1}{2}\right)^2 \frac{\pi^2}{2m_n L_0^2}$ to Ω_α. The change in the eigenvalues can be attributed to a potential $U(y)$ acting on particles trapped within the interval $-L_0 \le y \le L_0$.

In order to find $U(y)$ we observe that the matrix elements of U obtained with the unperturbed wave function $\left(\cos\frac{(n+\frac{1}{2})\pi y}{L_0}\right)$ must be the same as the off-diagonal elements in (16.212), i.e.

$$-\frac{i\omega}{4\pi}A(j,n)(1-\delta_{nj}) = \int_{L_0}^{L_0}\cos\left[\frac{\left(j+\frac{1}{2}\right)\pi y}{L_0}\right]U(y)\cos\left[\frac{\left(n+\frac{1}{2}\right)\pi y}{L_0}\right]dy, \tag{16.214}$$

where $A(j, n)$ is given by (16.187). Solving this integral equation for $U(y)$ we find that $U(y)$ is given by the Hermitian operator

$$U(y) = -\frac{i\omega}{2L_0}\left(2y\frac{\partial}{\partial y} + 1\right) = \frac{\omega}{2L_0}(yp_y + p_y y). \tag{16.215}$$

A more general form of the motion of the walls, viz,

$$L^2(t) = L_0^2\left(\alpha t^2 + \omega t + \delta\right), \tag{16.216}$$

also gives us a solvable form of differential equation for $\theta_j(t)$ [33].

16.8 Spin Magnetic Moment

A classical particle with charge e and mass m having an angular momentum $\mathbf{L}$ has a magnetic moment

$$\boldsymbol{\mu}_L = \frac{e}{2mc}\mathbf{L}. \tag{16.217}$$

This relation is also true in quantum mechanics except that since L_i s do not commute, $(\boldsymbol{\mu}_L)_i$ s also have nonvanishing commutators. Now for a particle with spin $\mathbf{S}$ there is an additional contribution to the magnetic moment which we denote by $\boldsymbol{\mu}_S$

$$\boldsymbol{\mu}_S = \frac{ge}{2mc}\mathbf{S}, \tag{16.218}$$

where g is called the Landé g factor. For an electron e is negative and

$$g - 2 = \frac{\alpha}{2\pi} + \mathcal{O}\left(\alpha^2\right), \tag{16.219}$$

where α is the fine structure constant

$$\alpha = \frac{e^2}{\hbar c} \approx \frac{1}{137.04}. \tag{16.220}$$

In this relation for g the terms on the right-hand side of (16.219) are corrections coming from relativistic quantum electrodynamics. These corrections will be ignored in our non-relativistic discussion. Thus for an electron

$$\boldsymbol{\mu}_{total} = \boldsymbol{\mu}_L + \boldsymbol{\mu}_S = \frac{-|e|}{2mc}(\mathbf{L} + \mathbf{S}), \tag{16.221}$$

where m is the mass of the electron. For protons and neutrons we have

$$\boldsymbol{\mu}_S(proton) = \frac{5.59|e|}{2m_p c}\mathbf{S}, \tag{16.222}$$

and

$$\boldsymbol{\mu}_S(neutron) = \frac{-3.83|e|}{2m_n c}\mathbf{S}. \tag{16.223}$$

In these relations m_p and m_n denote the masses of proton and neutron respectively.

If we place an electron with magnetic moment $\boldsymbol{\mu}_S$ in a magnetic field $\mathbf{B}(\mathbf{r}, t)$ then the Hamiltonian will have an additional term which expresses the interaction between $\mathbf{S}$ and $\mathbf{B}$

$$H_S = -\boldsymbol{\mu}_S \cdot \mathbf{B}(\mathbf{r}, t). \tag{16.224}$$

By adding this additional term to the Hamiltonian of a charged particle in a Coulomb field (e.g. the problem of hydrogen atom) we get

$$H = \frac{1}{2m}\left(\mathbf{p} - \frac{e}{c}\mathbf{A}\right)^2 - \frac{e^2}{r} - \frac{e}{mc}\mathbf{S}\cdot\mathbf{B}. \tag{16.225}$$

Assuming a weak magnetic field, we can expand H in (16.225) in powers of $\mathbf{B}$ and retain only the terms linear in $\mathbf{B}$

$$\begin{aligned} H &= H_0 - \frac{e}{2mc}(\mathbf{p}\cdot\mathbf{A} + \mathbf{A}\cdot\mathbf{p}) - \frac{e}{mc}\mathbf{S}\cdot\mathbf{B} \\ &= H_0 - \frac{e\mathbf{B}}{2mc}\cdot(\mathbf{L} + 2\mathbf{S}) = H_0 - \boldsymbol{\mu}_{total}\cdot\mathbf{B} \\ &= H_0 + H', \end{aligned} \tag{16.226}$$

where H_0 is the Hamiltonian for the electron in the Coulomb field

$$H_0 = \frac{1}{2m}\mathbf{p}^2 - \frac{e^2}{r}. \tag{16.227}$$

If we choose the magnetic field to be along the positive z axis then H' simplifies;

$$H' = H'_L + H'_S = -\frac{eB}{2mc}(L_z + 2S_z). \tag{16.228}$$

In the absence of spin the energy levels will split due to the contribution of H'_L, i.e.

$$(\Delta E)_{no\ spin} = \langle n, \ell, \bar{m}\,|H'_L|\,n, \ell, \bar{m}\rangle = -\frac{e\hbar}{2mc}\bar{m}B = -\mu_B\bar{m}B. \tag{16.229}$$

Here we have denoted the eigenvalues of L_z by $\bar{m}$ and where $\mu(B)$ is the Bohr magneton,

$$\mu_B = \left|\frac{e\hbar}{2mc}\right|. \tag{16.230}$$

This perturbation removes the $(2\ell+1)$-fold degeneracy of the eigenvalues which is common to all central force problems. Thus the spectral lines of an atom placed in a magnetic field will split and this splitting is called the Zeeman effect. Such a splitting is shown in Fig. 16.7.

There is a further splitting of the levels caused by the spin of the electron. Again to the first order perturbation theory from (16.228) we have

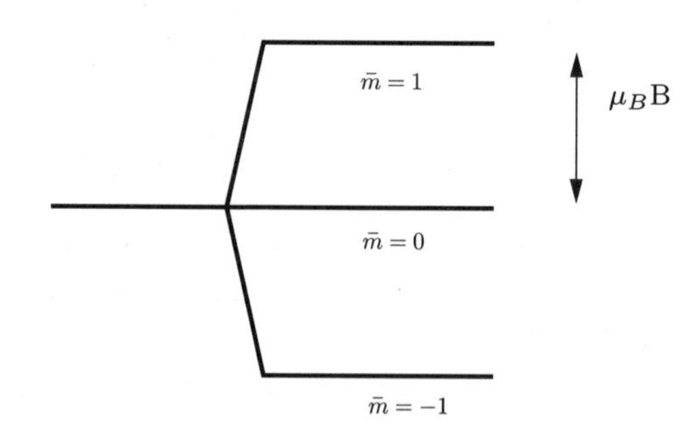

Figure 16.7: Removal of the degeneracy of the P state of the hydrogen atom in the presence of the magnetic field when the spin is neglected (Zeeman effect).

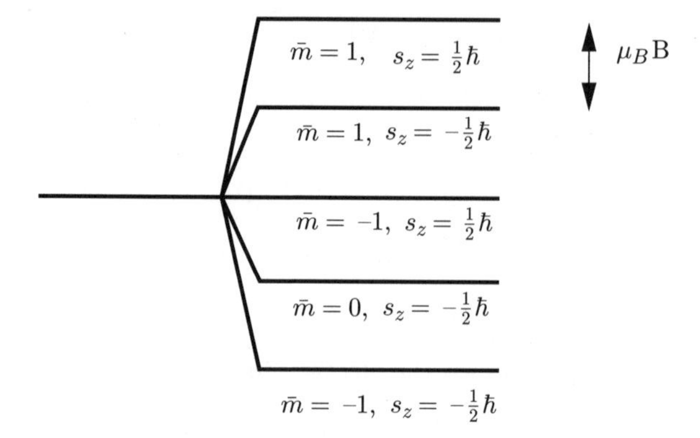

Figure 16.8: Same as in Fig. 16.7 but the coupling of the magnetic field to the spin is not neglected (anomalous Zeeman effect).

$$\Delta E = \langle n, \ell, \bar{m}, s_z \left| H' \right| n, \ell, \bar{m}, s_z \rangle = \mu_B B \left(\bar{m} + \frac{2s_z}{\hbar} \right), \tag{16.231}$$

where s_z is the eigenvalue of S_z. We note here that each $\bar{m}$ level is split into two levels, depending on the direction of the electron spin. In our discussion we have used the first order perturbation theory to calculate the shift in the energy levels caused by an external magnetic field. This splitting, shown in Fig. 16.8, is called anomalous Zeeman effect. For an exact formulation of the problem we can write the Schrödinger equation for the motion of a spinning electron in a magnetic field. In this situation the wave function has two components $\psi_\uparrow$ and $\psi_\downarrow$ depending on whether the spin is up or is down. In this case the Schrödinger equation becomes a matrix equation

$$\begin{aligned} i\hbar \frac{\partial}{\partial t} \begin{bmatrix} \psi_\uparrow \\ \psi_\downarrow \end{bmatrix} &= \left[\frac{1}{2m} \left(-i\hbar\nabla - \frac{e}{c}\mathbf{A}(\mathbf{r}, t) \right)^2 \cdot 1 \right] \begin{bmatrix} \psi_\uparrow \\ \psi_\downarrow \end{bmatrix} \\ &- \left(\frac{ge\hbar}{4mc} \boldsymbol{\sigma} \cdot \mathbf{B}(\mathbf{r}, t) \right) \begin{bmatrix} \psi_\uparrow \\ \psi_\downarrow \end{bmatrix} + V(\mathbf{r}, t) \cdot 1 \begin{bmatrix} \psi_\uparrow \\ \psi_\downarrow \end{bmatrix}. \end{aligned} \tag{16.232}$$

In this equation $V(\mathbf{r},t)$ is the potential energy and 1 is the unit 2×2 matrix. The coupling between $\psi_\uparrow$ and $\psi_\downarrow$ is provided by the term

$$\boldsymbol{\sigma} \cdot \mathbf{B} = \begin{bmatrix} B_z & B_x - iB_y \\ B_x + iB_y & B_z \end{bmatrix}, \tag{16.233}$$

in (16.232). Equation (16.232) is known as the Pauli equation [34].

16.9 Stern–Gerlach Experiment

The Stern–Gerlach experiment may be regarded as one of the most important experiments in the development of quantum mechanics [35]. In this experiment a beam of hydrogen atoms is sent through an inhomogeneous magnetic field, as is shown in Fig. 16.9. The atoms are in the ground state, i.e. the electrons are in singlet S state with zero orbital angular momentum. However because of the spin magnetic moment, the beam is split into two parts of equal intensity. This splitting shows that all electrons have a magnetic moment with the same absolute value, but with two possible orientations, parallel and antiparallel to the magnetic field. Denoting the spin magnetic moment by $\boldsymbol{\mu}_S$, the force felt by the electron because of the inhomogeneous magnetic field is

$$\mathbf{F} = -\nabla(-\boldsymbol{\mu}_S \cdot \mathbf{B}) == (\boldsymbol{\mu}_S \cdot \nabla)\mathbf{B}, \tag{16.234}$$

For the magnetic field in the Stern–Gerlach experiment we consider a field with two components $B_x(\mathbf{r})$ and $B_z(\mathbf{r})$, i.e.

$$\mathbf{B}(\mathbf{r}) = \mathbf{i}B_x(\mathbf{r}) + \mathbf{k}B_z(\mathbf{r}). \tag{16.235}$$

The magnetic field must satisfy the Maxwell equation $\nabla \cdot \mathbf{B} = 0$ and this condition is satisfied if $\mathbf{B}(\mathbf{r})$ is derived from the vector potential

$$\mathbf{A}(\mathbf{r}) = \mathbf{j}B_0(y)a(\mathbf{r}), \tag{16.236}$$

since then

$$\mathbf{B}(\mathbf{r}) = \nabla \wedge \mathbf{A}(\mathbf{r}) = \left(-\mathbf{i}\,\frac{\partial a}{\partial z} + \mathbf{k}\,\frac{\partial a}{\partial x}\right) B_0(y), \tag{16.237}$$

and $\nabla \cdot \mathbf{B} = 0$. Now the Hamiltonian for a neutral particle interacting with the magnetic field $\mathbf{B}$ is

$$H = \frac{1}{2m}\mathbf{p}^2 + \mu\,\boldsymbol{\sigma} \cdot \mathbf{B}(\mathbf{r}), \tag{16.238}$$

where $\boldsymbol{\mu} = -\mu\,\boldsymbol{\sigma}$ is the magnetic moment of the atom, (Eq. 16.224). Since the spin is coupled to the magnetic field, $\boldsymbol{\sigma}$ will be changing as a function of time.

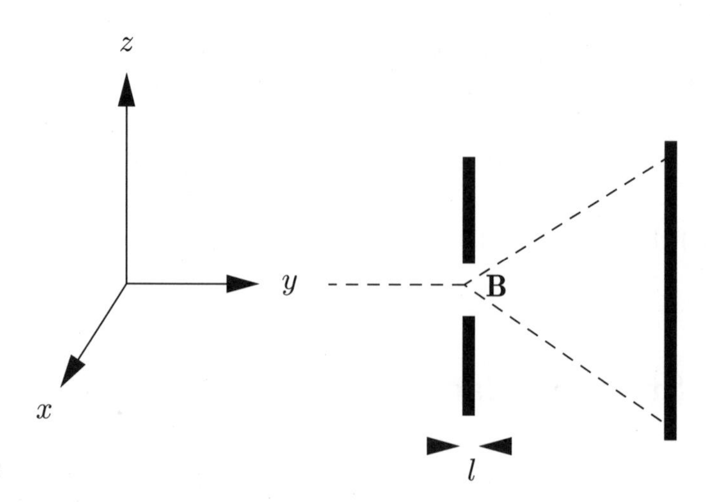

Figure 16.9: A beam of spin $\frac{1}{2}$ particles passing through an inhomogeneous magnetic field $\mathbf{B}(\mathbf{r})$ is split into two components.

From the Hamiltonian operator we find the Heisenberg equations of motion for the three vectors $\mathbf{r}$, $\mathbf{p}$ and $\boldsymbol{\sigma}$

$$\frac{dr_j}{dt} = \frac{i}{\hbar}[H,\, r_j] = \frac{1}{m}p_j, \tag{16.239}$$

$$\frac{dp_j}{dt} = \frac{i}{\hbar}[H,\, p_j] = -\mu\sum_k \sigma_k \frac{\partial B_k}{\partial r_j}, \tag{16.240}$$

$$\frac{d\sigma_j}{dt} = \frac{i}{\hbar}[H,\, \sigma_j] = \mu[\boldsymbol{\sigma}\wedge\mathbf{B}(\mathbf{r})]_j. \tag{16.241}$$

To simplify these equations further, we choose a symmetric field where $a(\mathbf{r})$ and $\mathbf{B}$ are given by

$$a(\mathbf{r}) = xz, \quad \text{and} \quad \mathbf{B}(\mathbf{r}) = B_0(y)\,(-\mathbf{i}x + \mathbf{k}z)\,. \tag{16.242}$$

With this simple form of $\mathbf{B}(\mathbf{r})$ we find nine equations for position, momentum and the spin of the particle;

$$\dot{x} = \frac{1}{m}p_x, \quad \dot{y} = \frac{1}{m}p_y \quad \dot{z} = \frac{1}{m}p_z, \tag{16.243}$$

$$\dot{p}_x = \mu B_0(y)\sigma_x, \quad \dot{p}_y = (-\mu\sigma_x x + \mu\sigma_z z)\frac{\partial B_0(y)}{\partial y}, \quad \dot{p}_z = -\mu B_0(y)\sigma_z, \tag{16.244}$$

$$\dot{\sigma}_x = -\frac{\mu}{\hbar}B_0(y)\sigma_y z, \quad \dot{\sigma}_y = \frac{\mu}{\hbar}B_0(y)(\sigma_x z + \sigma_z x), \quad \dot{\sigma}_z = \frac{\mu}{\hbar}B_0(y)\sigma_y x. \tag{16.245}$$

The complete integration of these coupled equations is difficult. To simplify the problem, let us assume that the field B_0 is nonzero only over a small interval $0 \le y \le l$ and write

$$B_0(y) = b\Delta(y), \tag{16.246}$$

where $\Delta(y)$ is a sharply peaked function of unit amplitude. Here we assume that the field $B_0(y)$ is small while its gradient is large enough to cause significant deflection.

Now we write the equations of motion in the frame attached to and moving with the atom, taking the velocity in the y direction to be constant [36]. In this frame we have

$$\frac{d}{dt} = \frac{\partial}{\partial t} + v\frac{\partial}{\partial y}. \tag{16.247}$$

Writing Eqs. (16.245) in this frame we have

$$v\frac{\partial}{\partial y}\sigma_x(y) = -\frac{\mu}{\hbar}B_0(y)\sigma_y(y)\ z, \tag{16.248}$$

$$v\frac{\partial}{\partial y}\sigma_y(y) = \frac{\mu}{\hbar}B_0(y)[\sigma_x(y)\ z + \sigma_z(y)\ x], \tag{16.249}$$

$$v\frac{\partial}{\partial y}\sigma_z(y) = \frac{\mu}{\hbar}B_0(y)\sigma_y(y)\ x. \tag{16.250}$$

Since $B_0(y)$ is sharply peaked around $y = 0$, $\Delta(y)$ simulates a delta function and for this choice we can integrate Eqs. (16.248)–(16.250). Expressing the final result in terms of $t = \frac{y}{v}$, we have for $t > 0$;

$$\sigma_x(t) = \sigma_x(0) - \frac{\mu b}{v\hbar}\sigma_y(0)\ z(0), \tag{16.251}$$

$$\sigma_y(t) = \sigma_y(0) + \frac{\mu b}{v\hbar}\left(\sigma_x(0)\ z(0) + \sigma_z(0)\ x(0)\right), \tag{16.252}$$

$$\sigma_z(t) = \sigma_z(0) + \frac{\mu b}{v\hbar}\sigma_y(0)\ x(0). \tag{16.253}$$

Having found $\boldsymbol{\sigma}(t)$, we now proceed to determine $\mathbf{p}(t)$. For this we substitute Eqs. (16.251)–(16.253) in (16.244) to obtain

$$p_x(t) = p_x(0) + \frac{\mu b}{v}\left[\sigma_x(0) - \frac{\mu b}{\hbar v}\sigma_y(0)\ z(0)\right], \tag{16.254}$$

and

$$p_z(t) = p_z(0) - \frac{\mu b}{v}\left[\sigma_z(0) - \frac{\mu b}{\hbar v}\sigma_y(0)\ x(0)\right], \tag{16.255}$$

all for $t > 0$. The momentum in the y direction in this approximation remains constant. Finally from (16.243), (16.254) and (16.255) we find $x(t)$ and $z(t)$

$$x(t) = x(0) + \frac{p_x(0)t}{m} + \frac{\mu bt}{mv}\left[\sigma_x(0) - \frac{\mu b}{v\hbar}\sigma_y(0)\ z(0)\right], \tag{16.256}$$

$$z(t) = z(0) + \frac{p_z(0)t}{m} - \frac{\mu bt}{mv}\left[\sigma_z(0) + \frac{\mu b}{v\hbar}\sigma_y(0)\, x(0)\right]. \tag{16.257}$$

If $\frac{\mu b}{v} \ll 1$, then we can keep the first order contribution in $\frac{\mu b}{v}$ and write the solution of the Heisenberg's equations as

$$x(t) = x(0) + \frac{p_x(0)t}{m} + \frac{\mu bt}{mv}\sigma_x(0), \tag{16.258}$$

$$z(t) = z(0) + \frac{p_z(0)t}{m} - \frac{\mu bt}{mv}\sigma_z(0). \tag{16.259}$$

Now we calculate the expectation value of these equations with the spinor $|a\rangle = \begin{bmatrix}\alpha \\ \beta\end{bmatrix}$,

$$\langle x(t)\rangle = \langle x(0)\rangle + \frac{\langle p_x(0)\rangle t}{m} + \frac{\mu bt}{mv}\langle a|\sigma_x(0)|a\rangle, \tag{16.260}$$

$$\langle z(t)\rangle = \langle z(0)\rangle + \frac{\langle p_z(0)\rangle t}{m} - \frac{\mu bt}{mv}\langle a|\sigma_z(0)|a\rangle. \tag{16.261}$$

Equation (16.261) shows that for a spin up particle $|a\rangle = \begin{bmatrix}1 \\ 0\end{bmatrix}$ will be deflected in the $-z$ direction by a distance $\left(-\frac{\mu bt}{mv}\right)$. However $\langle\uparrow|\sigma_x(0)|\uparrow\rangle = 0$, and therefore there is no deflection in the x direction. It should be noted that here the average x deflection is zero, but a spin up particle will be deflected along the x axis with equal probabilities for $-x$ and $+x$ directions [36].

For other accounts of this important experiment the reader is referred to [37]–[39].

16.10 Precession of Spin Magnetic Moment in a Constant Magnetic Field

The coupling between the magnetic moment of the particle the electromagnetic field causes the spin to precess, and this precession of spin is of fundamental importance in the applied aspects of quantum theory. The Hamiltonian for the spinning particle in a magnetic field is

$$H_S = -\frac{g}{2}\frac{e}{mc}\mathbf{S}\cdot\mathbf{B} = -\frac{g}{4}\frac{e\hbar}{mc}\boldsymbol{\sigma}\cdot\mathbf{B}. \tag{16.262}$$

From this Hamiltonian we obtain the Heisenberg equation of motion for the i-th component of the spin

$$
\begin{aligned}
i\hbar\frac{dS_i(t)}{dt} &= [S_i(t),\ H_S] = \sum_j -\left(\frac{ge}{2mc}\right)[S_i(t),\ S_j(t)]\,B_j \\
&= -\frac{ige\hbar}{2mc}\sum_{j,k}\varepsilon_{ijk}S_kB_j.
\end{aligned} \tag{16.263}
$$

We can write (16.263) as a vector equation

$$
\frac{d\mathbf{S}(t)}{dt} = \frac{ge}{2mc}\mathbf{S}(t)\wedge\mathbf{B} = \boldsymbol{\mu}_S(t)\wedge\mathbf{B}. \tag{16.264}
$$

For a magnetic field which is time-independent we can integrate the equation of motion (16.264). Assuming that the magnetic field is in the direction of the positive z axis, then Eq. (16.264) simplifies to

$$
\frac{dS_x(t)}{dt} = \frac{geB}{2mc}S_y, \tag{16.265}
$$

$$
\frac{dS_y(t)}{dt} = -\frac{geB}{2mc}S_x, \tag{16.266}
$$

and

$$
\frac{dS_z(t)}{dt} = 0. \tag{16.267}
$$

These are linear and coupled equations which can be solved easily to yield

$$
S_x(t) = S_x(0)\cos\omega t + S_y(0)\sin\omega t, \tag{16.268}
$$

$$
S_y(t) = -S_x(0)\sin\omega t + S_y(0)\cos\omega t, \tag{16.269}
$$

and

$$
S_z(t) = S_z(0), \tag{16.270}
$$

where

$$
\omega = \frac{geB}{2mc}. \tag{16.271}
$$

The time-dependence of the ket $|\psi(t)\rangle$ can be found from Eq. (4.11):

$$
|\psi(t)\rangle = \exp\left(-\frac{iH_St}{\hbar}\right)|\psi(0)\rangle = \exp\left(\frac{i\omega tS_z}{\hbar}\right)|\psi(0)\rangle. \tag{16.272}
$$

16.11 Spin Resonance

We have seen how the magnetic moment of a particle precesses around the direction of a static magnetic field. Now if in addition to this field we apply a time-dependent magnetic field at right angle to the original static magnetic field, then we can have a resonance condition that we will discuss now.

As it can be seen from Fig. 16.10 the magnetic moment of the particle denoted by $\boldsymbol{\mu}$ precesses around $\mathbf{B}_0$. The alternating $\mathbf{B}'(t)$ in general will have little effect on the precession of $\boldsymbol{\mu}$ since the torque that it generates will average out to zero. However if the magnetic field rotates with the angular frequency

$$\omega = \omega_{precession} = \frac{g\mu B}{\hbar}, \tag{16.273}$$

i.e. if it stays in phase with the frequency of the precession, the torque will act continuously and it will have a dramatic effect on the rotational motion of the particle. As we can see from Eq. (16.231)

$$(\Delta E)_S = \frac{g\mu s_z B}{\hbar} = \pm\frac{g\mu B}{2}, \tag{16.274}$$

where we have replaced μ_B by μ. Therefore the splitting between the parallel and antiparallel configurations of spin is given by

$$\Delta E = (\Delta E)_+ - (\Delta E)_- = g\mu B. \tag{16.275}$$

Thus electromagnetic radiation of angular frequency

$$\omega = \frac{\Delta E}{\hbar} = \frac{g\mu B}{\hbar}, \tag{16.276}$$

can be absorbed by this system. To find the condition for resonance in the setup shown in Fig. 16.10 we solve the equation of motion for the spinning particle.

Let us write $\mathbf{B}_0$ and $\mathbf{B}'$ in terms of their components

$$\mathbf{B}_0 = (0,\ 0,\ B_0), \tag{16.277}$$

and

$$\mathbf{B}' = (B'\cos\omega t,\ -B'\sin\omega t,\ 0)\,. \tag{16.278}$$

The spin Hamiltonian (16.262) for this case is

$$\begin{aligned} H_S &= -\frac{g\mu}{2}\mathbf{B}\cdot\sigma = -\frac{g\mu}{2}\left(B'_x\sigma_x + B'_y\sigma_y + B'_z\sigma_z\right) + B_0\sigma_z \\ &= -\frac{g\mu}{2}\begin{bmatrix} B_0 & B'e^{i\omega t} \\ B'e^{-i\omega t} & -B_0 \end{bmatrix}. \end{aligned} \tag{16.279}$$

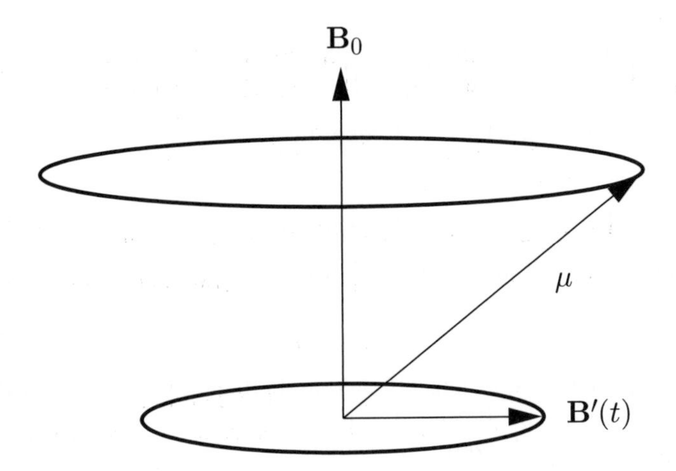

Figure 16.10: Precession of the magnetic moment of a particle around the static field $\mathbf{B}_0$ and an additional small rotating magnetic field.

The time-development of the components of the spin can be found from the Schrödinger equation

$$i\hbar\frac{d}{dt}\begin{bmatrix}\alpha_+(t)\\ \alpha_-(t)\end{bmatrix} = H_S\begin{bmatrix}\alpha_+(t)\\ \alpha_-(t)\end{bmatrix}$$
$$= \frac{-\hbar}{2}\begin{bmatrix}\omega_0 & \omega' e^{i\omega t}\\ \omega' e^{-i\omega t} & -\omega_0\end{bmatrix}\begin{bmatrix}\alpha_+(t)\\ \alpha_-(t)\end{bmatrix}, \tag{16.280}$$

where ω_0 and ω' are defined by

$$\omega_0 = \frac{g\mu B_0}{\hbar} \quad \text{and} \quad \omega' = \frac{g\mu B'}{\hbar}, \tag{16.281}$$

and ω is the frequency of the external field which varies sinusoidally. The matrix equation (16.280) gives us two coupled first order differential equations for α_+ and α_-;

$$\begin{cases} -2i\frac{d\alpha_+(t)}{dt} = \omega_0\alpha_+(t) + \omega' e^{i\omega t}\alpha_-(t) \\ -2i\frac{d\alpha_-(t)}{dt} = -\omega_0\alpha_-(t) + \omega' e^{-i\omega t}\alpha_+(t) \end{cases}. \tag{16.282}$$

For the initial conditions we assume that the spinor is pointing up, i.e.

$$\begin{bmatrix}\alpha_+(0)\\ \alpha_-(0)\end{bmatrix} = \begin{bmatrix}1\\ 0\end{bmatrix}. \tag{16.283}$$

We are looking for periodic solutions for the spin components, i.e. we want to have

$$\alpha_+(t) = A_+ e^{i\omega_+ t} \quad \text{and} \quad \alpha_-(t) = A_- e^{i\omega_- t}. \tag{16.284}$$

In these relations A_+, A_-, ω_+ and ω_- are constants to be determined. Substituting (16.284) in (16.282) we obtain the following equations

$$2\omega_+ A_+ = \omega_0 A_+ + \omega' A_- \exp[i(\omega + \omega_- - \omega_+)t], \tag{16.285}$$

and

$$2\omega_- A_- = -\omega_0 A_- + \omega' A_+ \exp[-i(\omega + \omega_- - \omega_+)t]. \tag{16.286}$$

These equations show that for A_+ and A_- to be constants we must have

$$\omega = \omega_+ - \omega_-. \tag{16.287}$$

Using (16.287) the two equations (16.285) and (16.286) become coupled homogeneous linear equations in A_+ and A_-. For a nontrivial solution the determinant of the coefficients must be zero;

$$\begin{vmatrix} 2\omega_+ - \omega_0 & -\omega' \\ -\omega' & 2\omega_- + \omega_0 \end{vmatrix} = 0. \tag{16.288}$$

If we eliminate ω_+ between this equation and Eq. (16.287) we find a quadratic equation for the unknown frequency ω_-. Solving this equation we obtain ω_- and ω_+:

$$\omega_- = -\frac{\omega}{2} \pm \Delta\omega = -\frac{\omega}{2} \pm \frac{1}{2}\sqrt{(\omega - \omega_0)^2 + \omega'^{\,2}}, \tag{16.289}$$

and

$$\omega_+ = \frac{\omega}{2} \pm \Delta\omega. \tag{16.290}$$

From (16.284) and (16.289) we obtain the general solution for $\alpha_-(t)$;

$$\alpha_-(t) = A\exp\left(-\frac{i\omega t}{2} + i\Delta\omega t\right) + B\exp\left(-\frac{i\omega t}{2} - i\Delta\omega t\right), \tag{16.291}$$

where A and B are constants. By imposing the initial conditions (16.283) on $\alpha_-(t)$ and $\alpha_+(t)$ we finally find $\alpha_-(t)$ to be

$$\alpha_-(t) = \frac{i\omega'}{2\Delta\omega}\sin(\Delta\omega t)\exp\left(-\frac{i\omega t}{2}\right). \tag{16.292}$$

Thus the probability for the spin flip from the initial condition $|\alpha_-(0)|^2 = 0$ to $|\alpha_-(t)|^2$ is given by [3]

$$\mathcal{P}_\downarrow(t) = |\alpha_-(t)|^2 = \left[\frac{\omega'^{\,2}}{\omega'^{\,2} + (\omega - \omega_0)^2}\right]\sin^2(\Delta\omega t). \tag{16.293}$$

For a very sharp resonance, ω' must be very small compared to ω_0 and this means that B' must be small. An examination of Eq. (16.293) shows that $\mathcal{P}_\downarrow(t)$ reaches its maximum when $\Delta\omega\, t = \frac{\pi}{2}$, however this maximum probability is very small. When the condition of resonance is reached, i.e. when $\omega = \omega_0$, then $\mathcal{P}_{\downarrow}\left(\frac{\pi}{\Delta\omega}\right) = 1$ and the spin has flipped with certainty. Now let us consider the case of the free or unpaired protons which is of great interest in the nuclear magnetic resonance (NMR). The parameters for having a resonance in this case when the magnetic field is 1 $Tesla$, are

$$f' \approx 40\ \text{MHz}, \quad \lambda = 7\ \text{m} \quad \text{and} \quad \Delta E = 2\times 10^{-7}\ \text{eV}, \tag{16.294}$$

Here f' is the resonant frequency, λ is the wavelength and ΔE is the photon energy. The value of f' shows that the time-dependent magnetic field must have a frequency in the range of radio waves.

16.12 A Simple Model of Atomic Clock

Another interesting application of spin resonance condition is in the operation of an atomic clock. The principle used in making of an atomic clock is a tunable cavity (or cavities) which can be tuned for a maximum microwave amplitude.

There are many different types of atomic clocks some more accurate than others [44]–[46]. A simple model of such a clock consists of two resonant cavities aligned along the y axis and separated from each other as is shown in Fig. 16.11. In this model a beam, consisting of particles with their spin along the positive z axis, enters the first cavity in the direction of the y axis and emerges on the right hand side of the second cavity [34],[47]. Inside each cavity there is the same radio frequency field $B_1 \cos\omega t$ in the x direction. In addition there is a uniform static field B_0 along the positive z axis, and this field acts inside the cavities as well as in the space between them. We want to calculate the probability of a particle entering the first cavity with spin up to emerge from the second cavity with spin down. The spin wave function of the system satisfies the Schrödinger equation

$$i\hbar\frac{d}{dt}|\psi(t)\rangle = -\frac{e\hbar}{2mc}\left[B_0\sigma_z + B'(t)\sigma_x\right]|\psi(t)\rangle. \tag{16.295}$$

Now we introduce a new wave function $|\psi_1(t)\rangle$ by the relation

$$|\psi(t)\rangle = \exp\left(\frac{i\omega_0 t\sigma_z}{2}\right)|\psi_1(t)\rangle, \tag{16.296}$$

where $\omega_0 = \frac{eB_0}{mc}$. Substituting for $|\psi(t)\rangle$ from (16.296) in (16.295) we find

$$\begin{aligned} i\frac{d|\psi_1(t)\rangle}{dt} &= -\frac{eB'(t)}{2mc}\exp\left(\frac{-i\omega_0 t\sigma_z}{2}\right)\sigma_x \exp\left(\frac{i\omega_0 t\sigma_z}{2}\right)|\psi_1(t)\rangle \\ &= -\frac{eB'(t)}{2mc}\sigma_x \exp\left(i\omega_0 t\sigma_z\right)|\psi_1(t)\rangle. \end{aligned} \tag{16.297}$$

We assume that at the time $t = 0$ a spinor enters the first cavity and

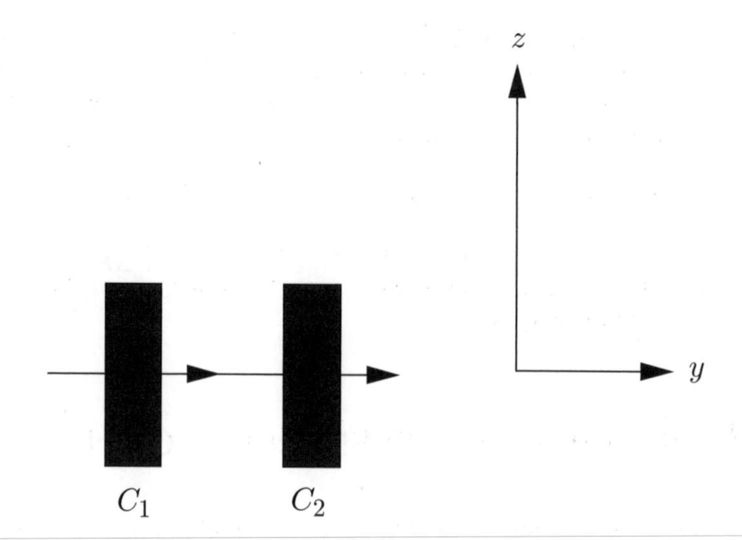

Figure 16.11: The cavities shown in this figure form a part of an atomic clock. Particles with spin up enter the first cavity in the direction of the arrow and emerge from the second cavity with their spins aligned in either positive or negative z axis.

therefore the initial condition for the differential equation (16.297) is

$$|\psi(0)\rangle = |\psi_1(0)\rangle = |\mathbf{k}\uparrow\rangle, \tag{16.298}$$

where $\mathbf{k}$ is a unit vector in the positive z axis. With the initial condition (16.298) we can integrate Eq. (16.297). Because of the smallness of $B'(t)$ we only need the solution of $|\psi_-(t)\rangle$ to the first order in $B'(t)$;

$$|\psi_1(t)\rangle = |\psi_1(0)\rangle + \frac{ie}{2mc}\int_0^t B'(t')\,\sigma_x \exp\left(i\omega_0 t'\sigma_z\right)|\psi_1(0)\rangle. \tag{16.299}$$

At a later time $t = t_2$ when the spinor leaves the second cavity the component of its wave function with spin down will be

$$\langle\mathbf{k}\downarrow|\psi_1(t_2)\rangle = \frac{ie}{2mc}\int_0^{t_2} B'(t)e^{i\omega_0 t}dt. \tag{16.300}$$

Let us denote the time that the particle leaves the first cavity by τ and the time when it emerges from the second cavity by $t_2 = T + \tau$, then the integral in (16.300) can be written as the sum of two terms

$$\int_0^{t_2} B'(t)e^{i\omega_0 t}dt = \int_0^{\tau} B'\cos(\omega t)e^{i\omega_0 t}dt + \int_T^{\tau+T} B'\cos(\omega t)e^{i\omega_0 t}dt. \tag{16.301}$$

The main contribution to these integrals come from the neighborhood of $\omega \approx \omega_0$, thus we can ignore the contributions of the terms of the type $e^{i(\omega_0+\omega)t}$ in the

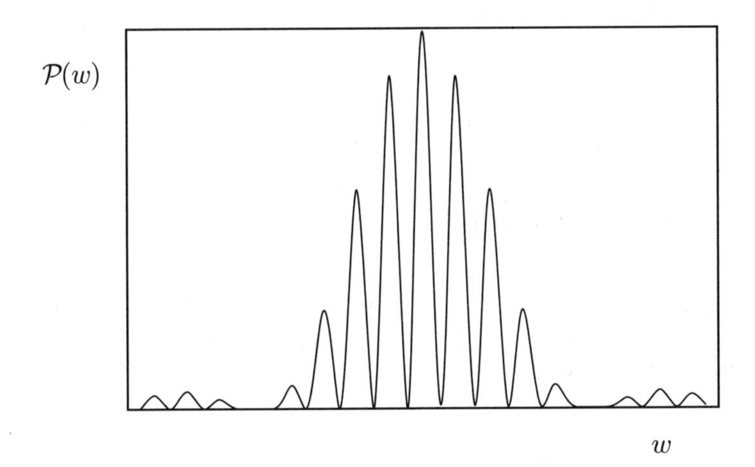

Figure 16.12: The probability of spin flip for the particles emerging from the second cavity of Fig. 16.11. This probability is plotted as a function of the frequency of the applied r.f. field.

integrated. Using this approximation we find the transition amplitude to be

$$\langle \mathbf{k} \downarrow |\psi_1(\tau + T)\rangle = \frac{eB'}{4mc}\left(\frac{e^{i(\omega_0-\omega)\tau} - 1}{\omega_0 - \omega}\right)\left(1 + e^{i(\omega_0-\omega)T}\right). \tag{16.302}$$

The amplitude in (16.302) consists of two parts, one which is proportional to $e^{i(\omega_0-\omega)T}$ is the amplitude for the spin flip in the second cavity. The other term shows the amplitude for spin flip in the first cavity. From Eq. (16.302) we find the probability for the spin flip to be

$$\mathcal{P}(\omega) = \left(\frac{eB'}{2mc}\right)^2 \left(\frac{\sin\frac{(\omega_0-\omega)\tau}{2}}{\frac{\omega_0-\omega}{2}}\right)^2 \left(\frac{1+\cos(\omega_0-\omega)T}{2}\right). \tag{16.303}$$

Let us examine this equation for constant values of τ and T, assuming that τ is much smaller than T. We observe that $\mathcal{P}$ consists of two ω-dependent factors, the first is the enveloping curve and has the form of a single-slit diffraction pattern, while the second term $[1 + \cos(\omega_0 - \omega)T]$ represents a rapidly oscillating function of ω. This rapidly oscillating term has a maximum at $\omega = \omega_0$ where the probability for spin flip is also maximum. The dependence of $\mathcal{P}(\omega)$ on ω is displayed in Fig. 16.12. By observing the rapid oscillations of $\mathcal{P}(\omega)$ one can accurately tune ω to ω_0. The spin of the particle in the two cavities feels the r.f. (radio frequency) field, and the difference in the phase of this r.f. field which the spin feels give rise to the interference pattern.

For accurate atomic clock the spin of an electron in caesium atom is used instead of the molecular beam. The applied constant field in this case is the internal field of the electron and

$$\omega_0 = \frac{1}{\hbar}(E_{\text{spin up}} - E_{\text{spin down}}), \tag{16.304}$$

is a constant and thus we have a very accurate measurement of the frequency and of the time.

16.13 Berry's Phase

In our formulation of the adiabatic approximation of Sec. 11.6 we found a phase $\gamma_n(t)$ which was defined by Eq. (11.162) and it appeared through the dependence of H on time

$$\frac{d\gamma_n(t)}{dt} = i \int \psi_n^*(\mathbf{r}, t) \frac{\partial}{\partial t} \psi_n(\mathbf{r}, t) d^3 r. \tag{16.305}$$

There we argued that in most applications we can ignore this phase. But for nonholonomic systems, viz the systems that do not return to their original state when transported adiabatically around a closed path, the phase $\gamma_n(t)$, cannot be ignored. Now let us consider the case where the time-dependence of the Hamiltonian arises from its dependence on a set of k parameters $R_1(t), \cdots R_k(t)$. These parameters can be considered as a vector $\mathbf{R}$ with k components

$$\mathbf{R}(t) = (R_1(t),\ R_2(t) \cdots R_k(t)). \tag{16.306}$$

We assume that $\mathbf{R}(t)$ varies slowly with time and that it satisfies the condition for the validity of the adiabatic approximation, viz,

$$\left| \frac{1}{R_j(t)} \frac{dR_j(t)}{dt} \right| \ll \left| \frac{1}{E_n(t)} \frac{dE_n(t)}{dt} \right|, \tag{16.307}$$

where E_n s are the instantaneous eigenvalues of $H(t)$, Eq. (11.143). We also assume that $\mathbf{R}$ is a periodic function of time with period T;

$$\mathbf{R}(0) = \mathbf{R}(T). \tag{16.308}$$

Since the Hamiltonian depends on time through $\mathbf{R}(t)$, the time-dependence of the wave function also will be determined by its dependence on this vector. Thus we can write (16.305) as

$$\frac{d\gamma_n(t)}{dt} = i \int \sum_{j=1}^{k} \left\{ \psi_n^*(\mathbf{r}, R_j(t)) \frac{d}{dR_j} \psi_n(\mathbf{r}, R_j(t)) \frac{dR_j(t)}{dt} \right\} d^3 r. \tag{16.309}$$

This equation can also be written in the ket notation as

$$\frac{d\gamma_n(t)}{dt} = i \left\langle n, \mathbf{R} \right| \nabla_{\mathbf{R}} \left| n, \mathbf{R} \right\rangle \cdot \frac{d\mathbf{R}(t)}{dt}. \tag{16.310}$$

According to Berry, this phase, $\gamma_n(t)$, is observable for a nonholonomic system, when the time evolution brings the parameter $\mathbf{R}(t)$ back to the starting point $\mathbf{R}(0)$ defined by (16.308) [48]. We can write the phase for a complete cycle $\gamma_n(T)$ as

$$\begin{aligned} \dot{\gamma}_n(T) &= \alpha(t) = i\int_0^T \langle n, \mathbf{R}|\nabla_{\mathbf{R}}|n, \mathbf{R}\rangle \cdot \frac{d\mathbf{R}(t)}{dt} \\ &= i\oint_C \langle n, \mathbf{R}|\nabla_{\mathbf{R}}|n, \mathbf{R}\rangle \cdot d\mathbf{R}. \end{aligned} \tag{16.311}$$

The line integral in (16.311) is around a closed curve C in the parameter space.

Let us define a vector $\mathbf{A}_n(\mathbf{R})$ by

$$\mathbf{A}_n(\mathbf{R}) = i\,\langle n, \mathbf{R}|\nabla_{\mathbf{R}}|n, \mathbf{R}\rangle\,, \tag{16.312}$$

so that

$$\gamma_n(T) = \oint \mathbf{A}_n(\mathbf{R}) \cdot d\mathbf{R}. \tag{16.313}$$

This $\mathbf{A}_n(\mathbf{R})$ is similar to the vector potential in electrodynamics, but in a k-dimensional space. If we redefine the phase of the eigenstate by

$$|n, \mathbf{R}\rangle \to \exp[i\varphi_n(\mathbf{R})]\;|n, \mathbf{R}\rangle, \tag{16.314}$$

then we have

$$\mathbf{A}_n(\mathbf{R}) \to \mathbf{A}_n(\mathbf{R}) - \nabla_{\mathbf{R}}\varphi_n(\mathbf{R}), \tag{16.315}$$

which is analogous to the gauge transformation. Now we want to show that the observable Berry's phase does not depend on the choice of gauge $\varphi_n(\mathbf{R})$. Consider the generalized form of the Stokes theorem for a vector with k components, and let us apply it to the right-hand side of (16.313)

$$\begin{aligned} \gamma_n(T) &= \int_0^T \mathbf{A}_n \cdot d\mathbf{R} = \int_S (\nabla_{\mathbf{R}} \wedge \mathbf{A}_n(\mathbf{R})) \cdot d\mathbf{S} \\ &\to \int_s \nabla_{\mathbf{R}} \wedge (\mathbf{A}_n(\mathbf{R}) - \nabla_{\mathbf{R}}\varphi_n(\mathbf{R})) \cdot d\mathbf{S} \\ &= \int_S \nabla_{\mathbf{R}} \wedge \mathbf{A}_n \cdot d\mathbf{S}, \end{aligned} \tag{16.316}$$

where $\mathbf{S}$ is a smooth surface that is bounded by a simple closed curve in the parameter space. Therefore $\gamma_n(T)$ which is the geometric phase is just the flux of the vector field

$$\mathbf{V}_n(\mathbf{R}) = \nabla_{\mathbf{R}} \wedge \mathbf{A}_n, \tag{16.317}$$

through the surface s and does not depend on $\varphi_n(\mathbf{R})$.

As in the case of electromagnetic field $\mathbf{V}_n(\mathbf{R})$ will have nontrivial structure in the presence of a source. This will be the case if two or more of the fast eigenvalues $E_n(\mathbf{R})$ become degenerate for some values of $\mathbf{R}$. To show this connection between the degeneracy and the presence of the Berry's phase we first introduce the notations

$$\nabla_{\mathbf{R}}|n,\mathbf{R}\rangle = |\nabla_{\mathbf{R}}(n,\mathbf{R})\rangle, \tag{16.318}$$

and

$$\nabla_{\mathbf{R}}\langle n,\mathbf{R}| = \langle\nabla_{\mathbf{R}}(n,\mathbf{R})|, \tag{16.319}$$

where for the sake of simplicity we have suppressed the time dependence of $\mathbf{R}(t)$. Now we express $\mathbf{V}_n(\mathbf{R})$ which is a real vector as

$$\begin{aligned}\mathbf{V}_n(\mathbf{R}) &= i\nabla_{\mathbf{R}}\wedge\langle n,\mathbf{R}|\nabla_{\mathbf{R}}|n,\mathbf{R}\rangle \\ &= i\left\langle\nabla_{\mathbf{R}}(n,\mathbf{R})|\wedge\nabla_{\mathbf{R}}(n,\mathbf{R})\right\rangle \\ &= -\mathrm{Im}\left\langle\nabla_{\mathbf{R}}(n,\mathbf{R})|\wedge\nabla_{\mathbf{R}}(n,\mathbf{R})\right\rangle.\end{aligned} \tag{16.320}$$

By introducing a complete a set of states we write (16.320) as

$$\mathbf{V}_n(\mathbf{R}) = -\mathrm{Im}\sum_k\left\{\langle\nabla_{\mathbf{R}}(n,\mathbf{R})|k,\mathbf{R}\rangle\wedge\langle k,\mathbf{R}|\nabla_{\mathbf{R}}(n,\mathbf{R})\rangle\right\}, \tag{16.321}$$

We note that the normalization of the wave function

$$\langle n,\mathbf{R}|n,\mathbf{R}\rangle = 1, \tag{16.322}$$

has the gradient

$$\langle n,\mathbf{R}|\nabla_{\mathbf{R}}(n,\mathbf{R})\rangle + \langle\nabla_{\mathbf{R}}(n,\mathbf{R})|n,\mathbf{R}\rangle = 2\mathrm{Re}\,\langle n,\mathbf{R}|\nabla_{\mathbf{R}}(n,\mathbf{R})\rangle = 0. \tag{16.323}$$

Thus the term with $k = n$ in the intermediate states is real and does not contribute to the sum.

The time-dependent eigenvalue equation (11.143) for this case takes the form

$$H(\mathbf{R})|n,\mathbf{R}\rangle = E_n(\mathbf{R})|n,\mathbf{R}\rangle. \tag{16.324}$$

By taking the gradient of this expression we find

$$\begin{aligned}(\nabla_{\mathbf{R}})H(\mathbf{R})|n,\mathbf{R}\rangle &+ H(\mathbf{R})\nabla_{\mathbf{R}}|n,\mathbf{R}\rangle = (\nabla_{\mathbf{R}}(E_n(\mathbf{R}))|n,\mathbf{R}\rangle \\ &+ E_n(\mathbf{R})\nabla_{\mathbf{R}}|n,\mathbf{R}\rangle.\end{aligned} \tag{16.325}$$

Now we multiply (16.325) by $\langle k,\mathbf{R}|$ with $k\neq n$ and use (16.318) (16.319) to get

$$\langle k,\mathbf{R}|\nabla_{\mathbf{R}}H(\mathbf{R})|n,\mathbf{R}\rangle = (E_n(\mathbf{R}) - E_k(\mathbf{R}))\langle k,\mathbf{R}|\nabla_{\mathbf{R}}(n,\mathbf{R})\rangle. \tag{16.326}$$

If we substitute this relation in Eq. (16.321) we find the following expression for $\mathbf{V}_n(\mathbf{R})$;

$$\mathbf{V}_n(\mathbf{R}) = -\mathrm{Im}\sum_{k\neq n}\frac{\langle n,\mathbf{R}|\nabla_{\mathbf{R}}H(\mathbf{R})|k,\mathbf{R}\rangle\wedge\langle k,\mathbf{R}|\nabla_{\mathbf{R}}H(\mathbf{R})|n,\mathbf{R}\rangle}{(E_k(\mathbf{R}) - E_n(\mathbf{R}))^2}. \tag{16.327}$$

Now if for a given time t

$$E_k(\mathbf{R}(t)) - E_n(\mathbf{R}(t)) = 0, \tag{16.328}$$

then $\mathbf{V}_n(\mathbf{R})$ diverges and this means that a field source exists.

Berry's Phase for a Particle with Spin in a Slowly Varying Magnetic Field — A simple system which exhibits the geometrical phase in its motion is that of a spinning particle in a time-dependent magnetic field $\mathbf{B}(t)$ [48]–[50]. The Hamiltonian for the system is given by Eq. (16.224) which we write as

$$H_S(\mathbf{B}) = -\mu \mathbf{S} \cdot \mathbf{B}. \tag{16.329}$$

The eigenfunctions and the eigenvalues for a fixed value of the magnetic field are

$$[H_S(\mathbf{B}) - E_{\bar{m}}(\mathbf{B})] \, |\bar{m}\mathbf{B}\rangle = 0, \tag{16.330}$$

where

$$E_{\bar{m}}(\mathbf{B}) = -\mu \bar{m} B, \quad \bar{m} = \ell,\ \ell - 1 \cdots - \ell, \qquad \ell \text{ an integer.} \tag{16.331}$$

For the time-dependent magnetic field the wave function is the solution of the Schrödinger equation

$$\left[i\frac{\partial}{\partial t} - H(\mathbf{B}(t))\right] |\bar{m}, \mathbf{B}(t)\rangle = 0, \quad \hbar = 1, \tag{16.332}$$

and this equation is subject to the initial condition

$$|\bar{m}, \mathbf{B}(0)\rangle = |\bar{m}, \mathbf{B}\rangle. \tag{16.333}$$

The time-dependence of the wave function comes only through its dependence on $\mathbf{B}(t)$, therefore

$$\frac{d}{dt}|\bar{m}, \mathbf{B}(t)\rangle = \frac{d\mathbf{B}(t)}{dt} \cdot |\nabla_{\mathbf{B}}(\bar{m}, \mathbf{B}(t))\rangle. \tag{16.334}$$

As we discussed earlier the Berry's phase is given by $\gamma_{\bar{m}}(T) = \gamma_{\bar{m}}(C)$ where C is the closed curve

$$\gamma_{\bar{m}}(T) = \gamma_{\bar{m}}(C) = \oint_C \langle \bar{m}, \mathbf{B}(t) | \nabla_{\mathbf{B}}(\bar{m}, \mathbf{B}(t))\rangle \cdot d\mathbf{B}, \tag{16.335}$$

and the flux $\mathbf{V}_{\bar{m}}(\mathbf{B})$ according to (16.327) has the form

$$\mathbf{V}_{\bar{m}}(\mathbf{B}) = -\mathrm{Im} \sum_{\bar{n} \neq \bar{m}} \frac{\langle \bar{m}, \mathbf{B} | \nabla_{\mathbf{B}} H(\mathbf{B}) | \bar{n}, \mathbf{B}\rangle \wedge \langle \bar{n}, \mathbf{B} | \nabla_{\mathbf{B}} H(\mathbf{B}) | \bar{m}, \mathbf{B}\rangle}{(E_{\bar{m}}(\mathbf{B}) - E_{\bar{n}}(\mathbf{B}))^2}. \tag{16.336}$$

We can simplify this expression by substituting from Eqs. (16.329)–(16.331) in it;

$$\mathbf{V}_{\bar{m}}(\mathbf{B}) = -\frac{1}{B^2} \mathrm{Im} \sum_{\bar{n} \neq \bar{m}} \frac{\langle \bar{m}, \mathbf{B} | \mathbf{S} | \bar{n}, \mathbf{B}\rangle \wedge \langle \bar{n}, \mathbf{B} | \mathbf{S} | \bar{m}, \mathbf{B}\rangle}{(\bar{m} - \bar{n})^2}, \tag{16.337}$$

where $\bar{m}$ and $\bar{n}$ are the eigenvalues of $\frac{1}{B}(\mathbf{S}\cdot\mathbf{B})$. The nonzero matrix elements of $\frac{1}{B}(\mathbf{S}\cdot\mathbf{B})$ are $\bar{n}=\bar{m}\pm 1$, therefore $(\bar{m}-\bar{n})^2=1$. To the sum in (16.337) we add the term $\bar{m}=\bar{n}$ since $\langle\bar{m}|\mathbf{S}|\bar{m}\rangle\wedge\langle\bar{m}|\mathbf{S}|\bar{m}\rangle$ is zero. Hence we can write $\mathbf{V}_{\bar{m}}(\mathbf{B})$ as

$$\begin{aligned}\mathbf{V}_{\bar{m}}(\mathbf{B}) &= -\frac{1}{B^2}\mathrm{Im}\,\langle\bar{m},\mathbf{B}|\mathbf{S}\wedge\mathbf{S}|\bar{m},\mathbf{B}\rangle \\ &= -\frac{1}{B^2}\langle\bar{m},\mathbf{B}|\mathbf{S}|\bar{m},\mathbf{B}\rangle. \end{aligned} \tag{16.338}$$

Here we have used $\mathbf{S}\wedge\mathbf{S}=i\mathbf{S}$, $(\hbar=1)$, Eq. (9.111). If we rotate the axes so that the z axis is in the direction of $\mathbf{B}$, then $\mathbf{V}_{\bar{m}}(\mathbf{B})$ will also be in the z direction. Then the expectation value shown in (16.338) can be easily evaluated and the result is

$$\mathbf{V}_{\bar{m}}(\mathbf{B}) = -\frac{\bar{m}\mathbf{B}}{B^3}. \tag{16.339}$$

For the special case of spin $\frac{1}{2}$ particle the Hamiltonian $H_S(\mathbf{B})$ takes the simple form

$$\begin{aligned}H_S(\mathbf{B}) &= -\mu\mathbf{S}\cdot\mathbf{B} = -\frac{\mu}{2}\boldsymbol{\sigma}\cdot\mathbf{B} \\ &= -\frac{\mu}{2}\begin{bmatrix} B_z & B_x-iB_y \\ B_x+iB_y & -B_z\end{bmatrix}. \end{aligned} \tag{16.340}$$

By diagonalizing this Hamiltonian we find that its eigenvalues are given by

$$E_{\pm} = \pm\frac{\mu}{2}\left[B_x^2+B_y^2+B_x^2\right]^{\frac{1}{2}} = \pm\frac{\mu}{2}|\mathbf{B}|. \tag{16.341}$$

Thus we have a degeneracy when $\mathbf{B}=0$. The Berry phase found from this flux is

$$\gamma_{\bar{m}}(T) = -\bar{m}\int_s \frac{\mathbf{B}\cdot ds}{B^3} = -\bar{m}\Omega(C), \tag{16.342}$$

where $\Omega(C)$ is the solid angle subtended by the closed curve C as seen from the origin of the vector $\mathbf{B}$. This result also indicates the geometrical nature of the Berry's phase, since $\gamma_{\bar{m}}(T)$ does not depend on the magnitudes of $\mathbf{S}$ and $\mathbf{B}$, but only on the eigenvalue $\bar{m}$.

By changing the closed curve C we can change the phase $\exp[i\gamma_{\bar{m}}(T)]$, and this phase can be made observable by superimposing two beams, one passed through the magnetic field and the other not. This superposition produces an observable interference pattern of the form $I_{\bar{m}}=\frac{1}{2}(1+\cos\gamma_{\bar{m}}(T))$. For the choice of a curve C to be a circle around the cone of angle ϑ, we find the intensity pattern to be

$$I_{\bar{m}}(\vartheta) = \cos^2\left[\bar{m}\pi\left(1-\cos\frac{\vartheta}{2}\right)\right]. \tag{16.343}$$

The state of a spinor aligned with the $\mathbf{R}$ direction which is specified by the angles ϑ and φ is

$$|\uparrow,\mathbf{R}\rangle = \begin{bmatrix}\cos\frac{\vartheta}{2} \\ e^{i\varphi}\sin\frac{\vartheta}{2}\end{bmatrix} \tag{16.344}$$

as is given by (9.162). The flux $\mathbf{V}_n(\mathbf{R})$, Eq. (16.320) in this case is

$$\mathbf{V}_{\uparrow}(\mathbf{R}) = \nabla_{\mathbf{R}} \wedge \mathbf{A}_{\uparrow}(\mathbf{R}), \tag{16.345}$$

where

$$\mathbf{A}_{\uparrow}(\mathbf{R}) = i\langle\uparrow, \mathbf{R}|\nabla_{\mathbf{R}}|\uparrow, \mathbf{R}\rangle = -\hat{\varphi}\frac{1}{R\sin\vartheta}\sin^2\frac{\vartheta}{2}. \tag{16.346}$$

By substituting (16.346) in (16.345) we find the flux

$$\mathbf{V}_{\uparrow}(\mathbf{R}) = \nabla_{\mathbf{R}} \wedge \mathbf{A}_{\uparrow}(\mathbf{R}) = \frac{\hat{\mathbf{R}}}{R\sin\vartheta}\frac{\partial}{\partial\vartheta}(\sin\vartheta A_{\varphi}) = -\frac{\hat{\mathbf{R}}}{2R^2}. \tag{16.347}$$

The flux can be viewed as the flux produced by a magnetic monopole of "charge" $-\frac{1}{2}$ located at the origin. We can calculate the Berry phase from $\mathbf{A}_{\uparrow}(\mathbf{R})$;

$$\gamma_{\uparrow}(C) = \oint \mathbf{A}_{\uparrow}(\mathbf{R}) \cdot d\mathbf{R} = 2\pi R\sin\vartheta A_{\varphi} = \pm\pi(1 - \cos\vartheta). \tag{16.348}$$

In this case the solid angle swept out by this trajectory is

$$\Omega_{\uparrow}(C) = \int_0^{\vartheta} \sin\vartheta' d\vartheta' \int_0^{2\pi} d\varphi = 2\pi(1 - \cos\vartheta), \tag{16.349}$$

i.e. the Berry's phase in this instance is

$$\gamma_{\uparrow}(C) = \pm\frac{1}{2}\Omega_{\uparrow}(C). \tag{16.350}$$

Connection with the Aharonov–Bohm Effect — Let us consider a magnetic field with a simple flux line of magnitude Φ

$$\Phi = \oint \mathbf{A}(\mathbf{R}) \cdot d\mathbf{R}, \tag{16.351}$$

and a quantum mechanical system consisting of a particle with charge e confined to a box located at $\mathbf{R}$ and not penetrated by the flux line (Fig. 16.13). In the absence of the flux, the Hamiltonian of the system depends on $\mathbf{r}$, and its conjugate momentum $\mathbf{p} = -i\hbar\nabla_{\mathbf{r}}$

$$H = H\left(-i\hbar\nabla_{\mathbf{r}},\ \mathbf{r} - \mathbf{R}\right). \tag{16.352}$$

The eigenfunction is $\psi_n(\mathbf{r} - \mathbf{R})$ and the eigenvalues E_n s are independent of $\mathbf{R}$. The state $|n, \mathbf{R}\rangle$ are solutions of

$$H\left[\left(-i\hbar\nabla_{\mathbf{r}} - \frac{e}{c}\mathbf{A}(\mathbf{r})\right),\ (\mathbf{r} - \mathbf{R})\right]|n, \mathbf{R}\rangle = E_n|n, \mathbf{R}\rangle. \tag{16.353}$$

This equation has an exact solution and the resulting wave function is

$$\langle\mathbf{r}|n, \mathbf{R}\rangle = \exp\left[\frac{ie}{\hbar c}\int_{\mathbf{R}}^{\mathbf{r}} \mathbf{A}\left(\mathbf{r}'\right) \cdot d\mathbf{r}'\right]\psi_n(\mathbf{r} - \mathbf{R}). \tag{16.354}$$

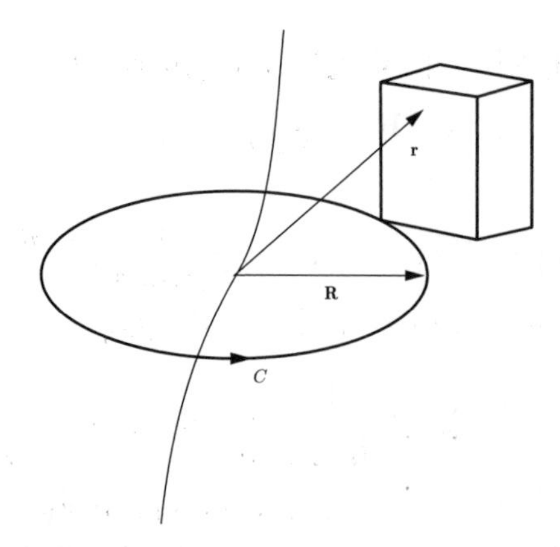

Figure 16.13: A charged particle is confined in a box. When the box is transported round a flux line the wave function of the particle will acquire a geometrical phase.

These wave functions are single valued functions of $\mathbf{r}$. Now if the box is transported around a curve C which encloses the flux line, then we have

$$\begin{aligned} & \langle n,\mathbf{R}|\nabla_{\mathbf{R}}|n,\mathbf{R}\rangle \\ = & \int \psi_n^*(\mathbf{r}-\mathbf{R})\left\{-\frac{ie}{\hbar c}\mathbf{A}(\mathbf{R})\psi_n(\mathbf{r}-\mathbf{R})+\nabla_{\mathbf{R}}\psi_n(\mathbf{r}-\mathbf{R})\right\}d^3r \\ = & \frac{-ie}{\hbar c}\mathbf{A}(\mathbf{R}). \end{aligned} \tag{16.355}$$

Thus according to Eq. (16.311) the Berry phase for this case is

$$\gamma_n(C)=\frac{e}{\hbar c}\oint_C \mathbf{A}(\mathbf{R})\cdot d\mathbf{R}=\frac{e}{\hbar c}\Phi, \tag{16.356}$$

and is independent of the quantum number n. In this example the analogy between Im $\langle n,\mathbf{R}|\nabla_{\mathbf{R}}|n,\mathbf{R}\rangle$ and the magnetic potential becomes evident.

Bibliography

[1] F.J. Dyson, Feynman's proof of the Maxwell equations, Am. J. Phys. 58, 209 (1990).

[2] J.L. Anderson, Comment on "Feynman's proof of the Maxwell equations", Am. J. Phys. 59, 86 (1991).

[3] R.J. Hughes, On Feynman's proof of the Maxwell equations, Am. J. Phys. 60, 301 (1992).

[4] J.F. Carnena, L.A. Ibort, G. Marmo and A. Stern, The Feynman problem and the inverse problem for Poissón dynamics, Phys. Rep. 263, 153 (1995).

[5] N. Dombey, Comment on "Feynman's proof of the Maxwell equations," by F.J. Dyson, Am. J. Phys. 59, 85 (1991).

[6] M. Le Bellac and J-M. Levy-Lebelond, Galilean electromagnetism, Nuovo Cimento, 14 B, 217 (1973).

[7] N. Hokkyo, Feynman's proof of Maxwell equations and Yang's unification of electromagnetic and gravitational Aharonov–Bohm effects, Am. J. Phys. 72, 345 (2004).

[8] C.N. Yang in *Proceedings of the International Symposium on Foundations of Quantum Mechanics in the Light of New Technology*, Edited by S. Kamafuchi, (Physical Society of Japan, Tokyo, 1983), p. 5.

[9] Y. Aharonov and D. Bohm, Significance of electromagnetic potentials in the quantum theory, Phys. Rev. 115, 485 (1959).

[10] Y. Aharonov and D. Bohm, Further considerations on electromagnetic potentials in quantum theory, Phys. Rev. 123, 1511 (1961).

[11] V.F. Weisskopf, in *Lectures in Theoretical Physics*, Vol. III, Edited by W.E. Brittin, B.W. Downs and J. Downs, (Intersience Publishers, New York, 1961), p. 54.

[12] I.S. Gradshetyn and I.M. Ryzhik, *Tables of Integrals, Series and Products*, Fourth Edition, (Academic Press, New York, 1965) p. 962.

[13] S. Olariu and I. Iovitzu Popescu, The quantum effects of electromagnetic fluxes, Rev. Mod. Phys. 57, 339 (1985).

[14] M. Peshkin and A. Tonomura, *The Aharonov–Bohm Effect*, Lecture Notes in Physics, (Springer-Verlag, 1989), p. 340.

[15] I.S. Gradshteyn and I.M. Ryzhik, *Tables of Integrals, Series and Products*, (Academic Press, 1965), p. 951.

[16] For a simple account of quantum-mechanical scattering in two dimensions for short-range potentials the reader is referred to a paper by R. Lapidus, Quantum-mechanical scattering in two dimensions, Am. J. Phys. 50, 45 (1982).

[17] M. Abramowitz and I.A. Stegun, *Handbook of Mathematical Functions*, (Dover Publications, New York, 1970), p. 360.

[18] C.R. Hagen, Aharonov–Bohm scattering amplitude, Phys. Rev. D 41, 2015 (1990).

[19] A. Tonomura, O Noboyuki, T. Matsuda, T. Kawasaki, J. Endo, S. Yano and H. Yamada, Evidence for Aharonov–Bohm effect with magnetic field completely shielded from electron wave, Phys. Rev. Lett. 56, 792 (1986).

[20] H. Batelaan and A. Tonomura, The Aharonov–Bohm effects: Variations on subtle theme, Phys. Today 62, 2 (2009).

[21] R.E. Bellman and K.L. Cooke, *Differential-Difference Equations*, (Academic Press, New York, 1963).

[22] See for instance, C.R. Chester, *Techniques in Partial Differential Equations*, (McGraw-Hill, New York, 1971), Chapter 8.

[23] M. Razavy, Time-dependent harmonic oscillator in a box, Phys. Rev. 44, 2384 (1991).

[24] H.R. Lewis, Jr. Classical and quantum systems with time-dependent harmonic-oscillator-type Hamiltonians, Phys. Rev. Lett. 18, 510 (1967).

[25] H.R. Lewis, Jr. Class of invariants for classical and quantum time-dependent harmonic oscillators, J. Math. Phys. 9, 1976 (1968).

[26] H.R. Lewis, Jr. Motion of a time-dependent harmonic oscillator and of a charged particle in a class of time-dependent axially symmetric electromagnetic field, Phys. Rev. 172, 1313 (1968).

[27] H.R. Lewis, Jr. and W.B. Riesenfeld, Quantum theory in time-dependent electromagnetic field J. Math. Phys. 10, 1458 (1969).

[28] G. Landolfi, Weyl-ordered series for the angle variable of the time-dependent oscillator, J. Phys. A 41, 1 (2008).

[29] P.H.E. Meijer and T. Tanaka, Quantum mechanics of beats between weakly coupled oscillators, Am. J. Phys. 31, 161 (1963).

[30] J.-M. Levy-Lablond, A geometrical quantum phase effect, Phys. Lett. A125, 441 (1987).

[31] D.M. Greenberger, A new non-local effect in quantum mechanics, Physica B 151, 374 (1988).

[32] M. Razavy, Quantum beats in waveguides with time-dependent boundaries, Phys. Rev. A 48, 3486 (1993).

[33] M. Razavy, Second quantization in a box with a moving wall, Lett. Nuovo Cimento, 37, 449 (1983).

[34] G. Baym, *Lectures on Quantum Mechanics*, (W.A. Benjamin, Reading, 1969).

[35] W. Gerlach and O. Stern, Das magnetische moment des silberatoms, Z. Phys. 9, 353 (1922).

[36] M.O. Scully, W.E. Lamb, Jr. and A. Barut, On the theory of the Stern–Gerlach apparatus, Found. Phys. 17, 575 (1987).

[37] G. Potel, F. Barranco, S. Cruz-Barrios and J. Gómez-Cmacho, Quantum mechanical description of Stern–Gerlach experiment, Phys. Rev. A 71, 052106 (2005).

[38] E.B. Manoukian and A. Rotjanakusol, Quantum dynamics of Stern–Gerlach (S-G) effect, Eur. J. Phys. 25, 253 (2003).

[39] J. Porter, R.F. Pettifer and D.R. Leadly, Direct demonstration of the transverse Stern–Gerlach effect, Am. J. Phys. 71, 1103 (2003).

[40] For the question of the role of inertial mass and of gravitational mass in quantum mechanics see D.M. Greenberger, The role of equivalence in quantum mechanics, Ann. Phys. 47, 116 (1968).

[41] J.J. Sakurai, *Modern Quantum Mechanics*, Revised Edition, (Addison-Wesley, Reading, 1994), p. 126.

[42] R.W. Robinett, *Quamtum Mechanics, Classical Results, Modern Systems, and Visualised Examples*, (Oxford University Press, 1997), p. 472.

[43] R.A. Calella, A.W. Overhauser and S.A. Werner, Observation of gravitationally induced quantum interference, Phys. Rev. Lett. 34, 1472 (1975).

[44] F.G. Major, *The Quantum Beat: Principle and Applications of Atomic Clocks* (Springer, New York, 2007).

[45] N.F. Ramsey, History of early atomic clocks, Metrologia, 42, S1 (2005).

[46] J. Vanier and C. Audoin, The classical caesium beam frequency standard: fifty years later, Metrologia, 42, S31 (2005).

[47] N.F. Ramsey, A molecular beam resonance method with separate oscillating fields, Phys. Rev. 78, 695 (1950).

[48] M.V. Berry, Quantal phase factors accompanying adiabatic changes, Proc. Roy. Soc. A 392, 45 (1984).

[49] K. Gottfried and T-M Yan, *Quantum Mechanics: Fundamentals*, (Springer, New York, 2003).

[50] B.R. Holstein, *Topics in Advanced Quantum Mechanics*, (Addison-Wesley, Reading, 1991), p. 245.

Chapter 17

Quantum Many-Body Problem

In this chapter we want to study the physical properties of a system with N particles, where these particles interact through two-body forces. It is well-known that in classical mechanics, there are no known exactly solvable problems for a system of N interacting particles. Even the general case of the three-body problem cannot be solved exactly. Similarly in quantum mechanics there is hardly an interesting problem with realistic interaction which is solvable. Therefore we have to seek approximate solutions which preserves the essential features of the whole problem or, alternatively we should be able to demonstrate the validity of the approximation. An outstanding feature of the latter case is the Hartree self-consistent field [1], in which every electron in an atom has its own wave function and eigenvalue.

We start this chapter with the study of the important problem of the two-electron atom which is a three-body problem. Using perturbation theory and also variational method we find the ground state energy for helium-like atoms. Then we study the Hartree–Fock method when the number of particles, N, is finite and not very large, e.g. for atoms or atomic nuclei [2],[3]. Then we use the technique of second quantization to study the problems of many-bosons and many-fermion.

17.1 Ground State of Two-Electron Atom

We will now consider "helium-like" atom where two electrons move in the Coulomb field of the nucleus with a charge $-Ze$. The spectra of an atom with two electrons was first obtained by Heisenberg in 1926, and was considered as a great success of the new quantum theory [9]. Among the examples of such a system we have the helium atom and singly ionized lithium, Li^+, and doubly ionized beryllium Be^{++} atoms. All these atomic problems can be described by the Hamiltonian

$$H(1,2) = H_0(1,2) + V(\mathbf{r}_1 - \mathbf{r}_2). \tag{17.1}$$

where

$$H_0(1,2) = -\frac{\hbar^2}{2m}\left(\nabla_1^2 + \nabla_2^2\right) - Ze^2\left(\frac{1}{r_1} + \frac{1}{r_2}\right), \tag{17.2}$$

is the Hamiltonian for the two electrons in the Coulomb field of the nucleus and

$$V(\mathbf{r}_1 - \mathbf{r}_2) = \frac{e^2}{|\mathbf{r}_1 - \mathbf{r}_2|}, \tag{17.3}$$

is the Coulomb potential between the two electrons. The ground state energy of $H_0(1,2)$ is twice that of the ground state of a hydrogen like atom (see Eq. (9.252))

$$E_0 = -\frac{Z^2e^2}{a_0}, \tag{17.4}$$

where a_0 is the Bohr radius, $a_0 = \frac{\hbar^2}{me^2}$, and m is the mass of the electron. The ground state wave function for H_0 is

$$\psi_0(\mathbf{r}_1, \mathbf{r}_2) = \phi_{1s}(\mathbf{r}_1)\phi_{1s}(\mathbf{r}_2) \to \frac{1}{\pi}\left(\frac{Z}{a_0}\right)^2 \exp\left[-\frac{Z}{a_0}(r_1 + r_2)\right]. \tag{17.5}$$

Here ϕ_{1s} denotes the ground state of the hydrogen-like atom, Eq. (9.207), and 1 stands for the principle quantum number. We observe that in (17.5) the space part of the wave function is symmetric, therefore the spin part must be antisymmetric, or the total spin must be zero, i.e. $n = 1$. This state has a lower energy than the one where the spins of the two particles are parallel. From the first order perturbation theory we have

$$E_0^P = E_0 + \langle\phi_0|V|\psi_0\rangle, \tag{17.6}$$

where

$$\langle\phi_0|V|\psi_0\rangle = \int \phi_{1s}^2(r_1)\frac{e^2}{|\mathbf{r}_1 - \mathbf{r}_2|}\phi_{1s}^2(r_2)d^3r_1d^3r_2. \tag{17.7}$$

To evaluate (17.7) we expand $|\mathbf{r}_1 - \mathbf{r}_2|^{-1}$ as

$$\frac{1}{|\mathbf{r}_1 - \mathbf{r}_2|} = \begin{cases} \frac{4\pi}{r_1} \sum_{\ell,m} \frac{1}{(2\ell+1)} \left(\frac{r_2}{r_1}\right)^\ell Y^*_{\ell,m}(\theta_1,\phi_1) Y_{\ell,m}(\theta_2,\phi_2) & if \;\; r_1 > r_2 \\ \frac{4\pi}{r_2} \sum_{\ell,m} \frac{1}{(2\ell+1)} \left(\frac{r_1}{r_2}\right)^\ell Y^*_{\ell,m}(\theta_1,\phi_1) Y_{\ell,m}(\theta_2,\phi_2) & if \;\; r_2 > r_1. \end{cases} \tag{17.8}$$

In this expression θ_1, ϕ_1 and θ_2, ϕ_2 are the polar angles of the radius vectors $\mathbf{r}_1$ and $\mathbf{r}_2$ respectively. Since apart from $|\mathbf{r}_1 - \mathbf{r}_2|^{-1}$ there are no other terms in the integrand of (17.7) which depend on θ s and ϕ s, therefore after integration over angular variables only the term with $\ell = m = 0$ will survive and (17.7) becomes

$$\begin{aligned} \langle \psi_0 | V | \psi_0 \rangle &= \frac{4e^2}{\pi} \left(\frac{Z}{a_0}\right)^6 \\ &\times \int_0^\infty e^{-\frac{2Zr_1}{a_0}} \left[\frac{1}{r_1} \int_0^{r_1} e^{-\frac{2Zr_2}{a_0}} r_2^2 dr_2 + \int_{r_1}^\infty e^{-\frac{2Zr_2}{a_0}} r_2 dr_2\right] r_1^2 dr_1 \\ &= \frac{5Ze^2}{8a_0}. \end{aligned} \tag{17.9}$$

By substituting (17.4) and (17.9) in (17.6) we find the result of the first order perturbation theory to be

$$E_0^P = -\frac{Ze^2}{a_0}\left(Z - \frac{5}{8}\right). \tag{17.10}$$

A more accurate result can be found by using the variational method outlined in Chapter 12. For the trial wave function we choose the same wave function as (17.5) but now we replace Z by the variational parameter ζ,

$$\psi_0(r_1, r_2) = \frac{1}{\pi}\left(\frac{\zeta}{a_0}\right)^3 \exp\left[-\frac{\zeta(r_1 + r_2)}{a_0}\right]. \tag{17.11}$$

By substituting (17.11) in the Hamiltonian (17.1) using (17.2) and (17.3) and following the same steps that led us to (17.10) we obtain

$$E_0^V(\zeta) = \frac{e^2}{a_0}\left[\zeta^2 - \left(2Z - \frac{5}{8}\right)\zeta\right], \tag{17.12}$$

a result which reduces to (17.10) when we choose $\zeta = Z$. But now we minimize $E(\zeta)$ with respect to ζ, i.e. we set

$$\frac{dE_0^V(\zeta)}{d\zeta} = 0, \tag{17.13}$$

and we obtain the only root of this equation which is

$$\zeta = \zeta_0 = Z - \frac{5}{16}. \tag{17.14}$$

If we replace ζ by this value in (17.12) we find the ground state energy of the system to be

$$E_0^V = -\left(\frac{e^2}{a_0}\right)\left(Z^2 - \frac{5}{8}Z + \frac{25}{256}\right). \tag{17.15}$$

The number $\zeta_0 = Z - \frac{5}{16}$ is called the effective nuclear charge for the two electron atom.

In order to compare these quantum mechanical results with the experimental measurements we calculate the ionization energy, E_{ion}, which is the energy required to remove one electron from the atom. This is equal to the difference of the energy $-\frac{Z^2e^2}{2a_0}$ of the remaining electrons in the field of the charge $-Ze$ and E_0. Here we can either use $E_0 = E_0^P$, Eq. (17.10) or $E_0 = E_0^V$, given by (17.15). If we use E_0^P from the variational method we get

$$E_{ion}^P = -E_0^P - \frac{Z^2e^2}{a_0} = \frac{Ze^2}{2a_0}\left(Z - \frac{5}{4}\right). \tag{17.16}$$

On the other hand using our variational calculation we find E_{ion}^V to be

$$E_{ion}^V = -E_0^V - \frac{Z^2e^2}{a_0} = \frac{e^2}{2a_0}\left(Z^2 - \frac{5}{4}Z + \frac{25}{128}\right). \tag{17.17}$$

Correlation Effects — In our variational calculation leading to Eq. (17.17) we did not take into account the fact the motion of one of the electrons affects the other electron and therefore the simple wave function given by (17.5) cannot give an accurate description of the system. In general we can have two types of correlation, the angular and the radial. The former which is harder to use in variational calculation reflects the fact that the electrons are more likely to be found on opposite sides of the nucleus rather than the same side. The radial correlation, on the other hand, implies that if one electron comes nearer to the nucleus, the other is forced to move out. Here we consider only the radial correlation. Assuming that the two electrons shield each other differently, we write the effective nuclear charge as seen by the electron nearer to the nucleus by $(1+\nu)\zeta$ and the one further away by $(1-\nu)\zeta$. For zero total spin (spin singlet) the spatial variational wave function has to be symmetric, and as we argued it has to depend on the two parameters ν and ζ.

$$\begin{aligned}\psi_0(r_1, r_2) &= N\left\{\exp\left[-(1+\nu)\zeta\left(\frac{r_1}{a}\right) - (1-\nu)\zeta\left(\frac{r_2}{a}\right)\right]\right. \\ &+ \left.\exp\left[-(1-\nu)\zeta\left(\frac{r_1}{a}\right) - (1+\nu)\zeta\left(\frac{r_2}{a}\right)\right]\right\},\end{aligned} \tag{17.18}$$

where N is the normalization constant. Using this wave function in place of (17.11) we find that the minimum value of energy occurs when

$$\zeta = \frac{16Z - 4\delta - \delta^2 + 16Z\delta^3 - 5\delta^3}{8\left(2 - \delta + \delta^4\right)}, \tag{17.19}$$

where $\delta = 1 - \nu^2$. The ground state energy found from this value of ζ is

$$E = -\left(\frac{e^2}{a_0}\right)\left[\frac{\left(16Z - 4\delta - \delta^2 + 16Z\delta^3 - 5\delta^3\right)^2}{128\left(1+\delta^3\right)\left(2-\delta+\delta^4\right)}\right]. \tag{17.20}$$

The last column of TABLE XIV shows results of the variational calculation when the radial correlation effect is included.

TABLE XIV: Ionization energies for helium-like atoms calculated from first order perturbation and by simple variational method and these are compared with the observed values. All energies are in the unit of $\frac{e^2}{a_0}$.

Atom	Experimental value	E^P_{ion}	E^V_{ion}	Radial correlation
He	0.903	0.75	0.85	0.875
Li^+	2.7798	2.62	2.72	2.75
Be^{++}	5.6560	5.50	5.60	5.625
C^{+++}	14.4070	14.25	14.35	

17.2 Hartree and Hartree–Fock Approximations

In the last section we showed how the ground state energy of a two-electron atom can be obtained by perturbation and or by variational methods. For larger atoms, we assume a model where an electron feels the Coulomb force of the nucleus plus some average field due to the distribution of all other electrons. Now we want to determine the average field and the wave function for an electron moving in the combined fields of the nucleus and this average field [1]–[4]. We observe that the average force felt by the j-th electron depends on the states of other electrons, and this in turn is dependent on the average field in which these electrons are moving.

The simplest N-body wave function for the system is the product of one electron wave functions

$$\psi(1,\ 2,\cdots,N) = \phi(\mathbf{r}_1,s_1)\phi(\mathbf{r}_2,s_2)\cdots\phi(\mathbf{r}_N,s_N). \tag{17.21}$$

We can allow for the exclusion principle by imposing the condition that the single particle wave functions be orthogonal to each other and we will suppress the spin quantum numbers, $s_1, s_2 \cdots s_N$. In this way we guarantee that no two electrons will be in the same state. If we know the wave function $\phi_j(\mathbf{r})$, then the average charge density at a point $\mathbf{r}'$ due to the j-th electron will be $e|\phi_j(\mathbf{r}')|^2$.

This charge density produces a field at the point $\mathbf{r}$ with the potential energy

$$\int \frac{e^2}{|\mathbf{r}-\mathbf{r}'|}|\phi_j(\mathbf{r}')|^2 d^3r'. \tag{17.22}$$

Now for any other electron, say k-th, the average potential energy is the sum of two terms

$$V_k(\mathbf{r}) = \int \frac{e^2}{|\mathbf{r}-\mathbf{r}'|}\sum_{j\neq k}|\phi_j(\mathbf{r}')|^2 d^3r' - \frac{Ze^2}{r}. \tag{17.23}$$

where the first term on the right-hand side arises from all of the other electrons and the last term is the Coulomb potential of the nucleus. Thus the wave equation for the motion of the k-th particle is given by

$$\left[-\frac{\nabla^2}{2m} + V_k\right]\phi_k(\mathbf{r}) = \mathcal{E}_k\phi_k(\mathbf{r}), \quad \hbar = 1, \tag{17.24}$$

where $\mathcal{E}_k$ is the energy associated with the motion of the k-th particle.

If we start with the N-particle Hamiltonian and find its expectation value with the wave function $\psi(1,\ 2,\cdots,N)$, Eq. (17.21), we find

$$\begin{aligned}\langle\psi|H\psi\rangle &= -\sum_k\int\left[\frac{|\nabla\phi_k(\mathbf{r})|^2}{2m} + \frac{Ze^2}{r}|\phi_k(\mathbf{r})|^2\right]d^3r' \\ &+ \frac{1}{2}\sum_{j,k}\int\int\frac{e^2}{|\mathbf{r}-\mathbf{r}'|}|\phi_j(\mathbf{r}')|^2|\phi_k(\mathbf{r})|^2 d^3r d^3r'.\end{aligned} \tag{17.25}$$

Now by minimizing (17.25) with respect to $\phi_k(\mathbf{r})$ subject to the condition that $\phi_k(\mathbf{r})$ s must remain orthogonal to each other, we obtain the nonlinear equation for $\phi_k(\mathbf{r})$;

$$\left(-\frac{\nabla^2}{2m} - \frac{Ze^2}{r}\right)\phi_k(\mathbf{r}) + \int\frac{e^2}{|\mathbf{r}-\mathbf{r}'|}\sum_j\phi_j(\mathbf{r}')\left[\phi_j^*(\mathbf{r}')\phi_k(\mathbf{r})\right]d^3r = \mathcal{E}_k\phi_k(\mathbf{r})d^3r'. \tag{17.26}$$

Here the parameter $\mathcal{E}_k$ is introduced as a Lagrange multiplier for the subsidiary condition that the wave function is normalized,

$$\langle\psi|\psi\rangle = 1. \tag{17.27}$$

Thus the Hartree equation, (17.24), can be found from a variational principle. The nonlinear integro-differential equation (17.26) can be solved by iteration. The iterative scheme is designed in such a way that the resulting equations in any order of iteration are linear differential equations. For this linear set we write (17.26) as

$$\begin{aligned}\left(-\frac{\nabla^2}{2m} - \frac{Ze^2}{r}\right)\phi_k^{(\alpha+1)}(\mathbf{r}) &+ \int\frac{e^2}{|\mathbf{r}-\mathbf{r}'|}\sum_j\left|\phi_j^{(\alpha)}(\mathbf{r}')\right|^2\phi_k^{(\alpha+1)}(\mathbf{r})\,d^3r' \\ &= \mathcal{E}_k^{(\alpha+1)}\phi_k^{(\alpha+1)}(\mathbf{r}), \quad \alpha = 0,\ 1,\cdots.\end{aligned} \tag{17.28}$$

By starting with a set of trial wave functions $\phi_k^{(0)}(\mathbf{r}), j = 1,\ 2\cdots N$, we calculate $\phi_k^{(1)}(\mathbf{r}), j = 1,\ 2\cdots N$ and $\mathcal{E}_k^{(1)}$ and continue this iteration until the desired accuracy is reached. From Eq. (17.26) it follows that

$$\begin{aligned}\mathcal{E}_k &= \int \phi_k^*(\mathbf{r})\left(-\frac{\nabla^2}{2m} - \frac{Ze^2}{r}\right)\phi_k(\mathbf{r})d^3r \\ &+ \sum_{k\neq j}\int \phi_k^*(\mathbf{r})\frac{e^2}{|\mathbf{r}-\mathbf{r}'|}\sum_j \phi_j(\mathbf{r}')\,\phi_j^*(\mathbf{r}')\,\phi_k(\mathbf{r})d^3rd^3r'. \end{aligned} \tag{17.29}$$

Now the total energy of all the particles in the system is

$$E = \sum_{k=1}^{N}\mathcal{E}_k - \frac{1}{2}\sum_{k,j\neq k}\int \phi_k^*(\mathbf{r})\phi_j(\mathbf{r}')\frac{e^2}{|\mathbf{r}-\mathbf{r}'|}\phi_j^*(\mathbf{r}')\,\phi_k(\mathbf{r})d^3rd^3r', \tag{17.30}$$

since in the sum $\sum_{k=1}^{N}\mathcal{E}_k$, the electrostatic interaction is taken into account twice.

The Hartree–Fock Approximation — In the Hartree approximation the symmetric wave function (17.21) is used. In order to include the exchange effect we replace the symmetric wave function with the antisymmetrized wave function of the type given by Slater determinant (6.107) [2]–[4]

$$\psi_A(\mathbf{r}_1,\cdots,\mathbf{r}_N) = \frac{1}{\sqrt{N!}}\begin{vmatrix}\phi_1(\mathbf{r}_1) & \phi_1(\mathbf{r}_2) & \cdots & \phi_1(\mathbf{r}_N)\\ \phi_2(\mathbf{r}_1) & \phi_2(\mathbf{r}_2) & \cdots & \phi_2(\mathbf{r}_N)\\ \cdots & \cdots & \cdots & \cdots \\ \phi_N(\mathbf{r}_1) & \phi_N(\mathbf{r}_2) & \cdots & \phi_N(\mathbf{r}_N)\end{vmatrix}. \tag{17.31}$$

By calculating $\langle\psi|H\psi\rangle$ using (17.31) we find that in this case the expectation value of the Hamiltonian is

$$\begin{aligned}\langle\psi|H\psi\rangle &= -\sum_k\int\left[\frac{|\nabla\phi_k(\mathbf{r})|^2}{2m} + \frac{Ze^2}{r}|\phi_k(\mathbf{r})|^2\right]d^3r' \\ &+ \frac{1}{2}\sum_{j,k}\int\int\frac{e^2}{|\mathbf{r}-\mathbf{r}'|}|\phi_j(\mathbf{r}')|^2|\phi_k(\mathbf{r})|^2d^3rd^3r' \\ &- \frac{1}{2}\sum_{j,k}\delta_{s_j,s_k}\int\int\frac{e^2}{|\mathbf{r}-\mathbf{r}'|}\phi_j^*(\mathbf{r})\,\phi_j(\mathbf{r}')\,\phi_k^*(\mathbf{r}')\,\phi_k(\mathbf{r})\,d^3rd^3r'.\end{aligned} \tag{17.32}$$

The term δ_{s_j,s_k} in the last term of this equation means that the sum over j should have the same spin as the sum over k. Also note that the term $j = k$ does not appear in the right-hand side. The nonlinear wave equation derived by minimizing the functional (17.32) subject to the orthogonality of $\phi_k(\mathbf{r})$ s is

$$\begin{aligned}&\left(-\frac{\nabla^2}{2m} - \frac{Ze^2}{r}\right)\phi_k(\mathbf{r}) + \int\frac{e^2}{|\mathbf{r}-\mathbf{r}'|}\sum_j\phi_j^*(\mathbf{r}') \\ &\times\ \left[\phi_k(\mathbf{r})\phi_j(\mathbf{r}') - \delta_{s_j,s_k}\phi_k(\mathbf{r}')\phi_j(\mathbf{r})\right]\,d^3r' = \mathcal{E}_k\phi_k(\mathbf{r}).\end{aligned} \tag{17.33}$$

The sum of the two terms in the bracket in (17.33) is the amplitude that the electrons with the same spin and in the states j and k can be found at $\mathbf{r}$ and $\mathbf{r}'$.

Hartree–Fock Method Applied to the Helium Atom — As an example of the application of the Hartree–Fock method let us calculate the energy levels of the ground and the first excited states of the helium atom. For the ground state the two electrons have opposite spins, so, as we see from Eq. (17.32) there is no exchange term. Thus for the ground state of the electron in the $1S$ state we write (17.32) as

$$\mathcal{E}_{1S}\phi_{1S}(r) = \left[-\frac{\hbar^2}{2m}\nabla^2 - \frac{2e^2}{r} + e^2\int\frac{|\phi_{1S}(r')|^2}{|\mathbf{r}-\mathbf{r}'|}d^3r'\right]\phi_{1S}(r). \tag{17.34}$$

This nonlinear equation can be solved by numerical iteration.

The more interesting case of the first excited state of helium atom is the one where one electron is in $1S$ state and the other is in $2S$ state. In this case the interaction consists of three potentials, two direct and one exchange

$$V_{1S}(r) = e^2\int[\phi_{1S}(r')]^2\frac{d^3r'}{|\mathbf{r}-\mathbf{r}'|}, \tag{17.35}$$

$$V_{2S}(r) = e^2\int[\phi_{2S}(r')]^2\frac{d^3r'}{|\mathbf{r}-\mathbf{r}'|}, \tag{17.36}$$

and

$$V^{ex}(r) = e^2\int\phi_{2S}(r')\,\phi_{1S}(r')\frac{d^3r'}{|\mathbf{r}-\mathbf{r}'|}. \tag{17.37}$$

Considering the spin states of the system we observe that we can have spin singlet (or opposite spins) for the two electrons or spin triplet (when spin are parallel). In the first case of spin singlet we have parahelium and for this atom the exchange term is positive. For this situation the Hartree–Fock equations are:

$$\mathcal{E}_{1S}\phi_{1S}(r) = \left[-\frac{\hbar^2}{2m}\nabla^2 - \frac{2e^2}{r} + V_{2S}(r)\right]\phi_{1S}(r) + V^{ex}(r)\phi_{2S}(r), \tag{17.38}$$

$$\mathcal{E}_{2S}\phi_{1S}(r) = \left[-\frac{\hbar^2}{2m}\nabla^2 - \frac{2e^2}{r} + V_{1S}(r)\right]\phi_{2S}(r) + V^{ex}(r)\phi_{1S}(r). \tag{17.39}$$

For the spin triplet state or orthohelium the exchange term is negative and the corresponding equations become

$$\mathcal{E}_{1S}\phi_{1S}(r) = \left[-\frac{\hbar^2}{2m}\nabla^2 - \frac{2e^2}{r} + V_{2S}(r)\right]\phi_{1S}(r) - V^{ex}(r)\phi_{2S}(r), \tag{17.40}$$

$$\mathcal{E}_{2S}\phi_{1S}(r) = \left[-\frac{\hbar^2}{2m}\nabla^2 - \frac{2e^2}{r} + V_{1S}(r)\right]\phi_{2S}(r) - V^{ex}(r)\phi_{1S}(r). \tag{17.41}$$

To find the energies of the ortho- and para-states we can use perturbation theory and calculate the average value of the Hamiltonian in these states. Thus for the energy of the parahelium we have

$$E_S = \langle \psi_S | H | \psi_S \rangle = \mathcal{E}_{1S} + \mathcal{E}_{2S} + Q + A. \tag{17.42}$$

In the same way for the energy of the orthohelium we get

$$E_A = \langle \psi_A | H | \psi_A \rangle = \mathcal{E}_{1S} + \mathcal{E}_{2S} + Q - A. \tag{17.43}$$

In these expressions Q and A represent the following integrals:

$$Q = \int |\phi_{1S}(\mathbf{r})|^2 \frac{e^2}{|\mathbf{r} - \mathbf{r}'|} |\phi_{2S}(\mathbf{r}')|^2 d^3r d^3r', \tag{17.44}$$

and

$$A = \int \phi_{1S}(\mathbf{r})\phi_{2S}(\mathbf{r}') \frac{e^2}{|\mathbf{r} - \mathbf{r}'|} \phi_{1S}(\mathbf{r}')\phi_{2S}(\mathbf{r}) d^3r d^3r'. \tag{17.45}$$

From these results it follows that $E_A < E_S$, or the orthohelium has the lower energy. We note that this triplet state cannot decay to the ground state by emitting a photon because of the conservation of total spin. As we can see from Eq. (17.28), for helium and other heavier atoms, we can find an effective potential

$$\int \frac{e^2}{|\mathbf{r} - \mathbf{r}'|} \sum_j \left| \phi_j^{(\alpha)}(\mathbf{r}') \right|^2 \phi_k^{(\alpha+1)}(\mathbf{r})\, d^3r' \tag{17.46}$$

and the eigenvalue equation (17.28) and in this way we can calculate the total energy as the sum of one-electron eigenvalues. If we take this effective potential to be spherically symmetric, then we can write the wave function in terms of a radial function and spherical harmonics with quantum numbers ℓ and m, and just like any central force problem with a principal quantum number n. The valence electrons, i.e. those in the outermost shell, determine the physical and chemical properties of the elements. The results of the Hartree–Fock approximation for valence energy levels of some elements are tabulated in TABLE XV.

TABLE XV: Valence energy levels calculated by the Hartree–Fock method (in electron-volts). For comparison the ionization potentials for these elements are also shown.

	n	ℓ	Valence Energy Levels (eV)	Ionization Potential (eV)
Na	3	0	−4.96	5.14
Mg	3	0	−6.89	7.64
Al	3	1	−5.71	5.98
Si	3	1	−7.59	8.15
P	3	1	−9.54	10.48
S	3	1	−11.60	10.36
Cl	3	1	−13.78	13.01
Ar	3	1	−16.08	15.75

Hartree–Fock Method for Calculating Nuclear Energy Levels — This method can also be used to find the energies of the nucleons in the nucleus. In this case, unlike the problem of electrons in the atom, there is no central potential, however we can assume that each nucleon moves in an average potential created by other nucleons. This single particle potential should be determined in a way to best reproduce the empirical data. Thus we write the Hamiltonian for the motion of A nucleons in a nucleus as

$$H = \sum_{i=1}^{A} -\frac{\hbar^2}{2m}\nabla_i^2 + \frac{1}{2}\sum_{i,j=1}^{A} V(\mathbf{r}_i, \mathbf{r}_j). \tag{17.47}$$

For this Hamiltonian, just as in the atomic Hartree–Fock problem we have a set of wave functions $\psi_k(\mathbf{r})$ satisfying the nonlinear integro-differential equation

$$\begin{aligned} -\frac{\nabla^2}{2m}\psi_k(\mathbf{r}) \quad &+ \quad \sum_{j=1}^{A} V(\mathbf{r}-\mathbf{r}')\,\psi_j^*(\mathbf{r}')\left[\psi_j(\mathbf{r}')\,\psi_k(\mathbf{r}) - \psi_k(\mathbf{r}')\,\psi_j(\mathbf{r})\right] d^3r' \\ &= \quad \mathcal{E}_k\psi_k(\mathbf{r}). \end{aligned} \tag{17.48}$$

Now we define the local Hartree potential as

$$U(\mathbf{r}) = \int V(\mathbf{r}-\mathbf{r}')\sum_{j=1}^{A}\left|\psi_j(\mathbf{r}')\right|^2 d^3r', \tag{17.49}$$

and the much weaker nonlocal or exchange potential which is

$$U_{ex}(\mathbf{r},\mathbf{r}') = -V(\mathbf{r}-\mathbf{r}')\sum_{j=1}^{A}\psi_j^*(\mathbf{r}')\,\psi_j(\mathbf{r}). \tag{17.50}$$

Noting that the force felt by a nucleon in a nucleus goes to zero for distances larger than the radius of the nucleus, a realistic choice of $U(\mathbf{r}_i)$ must have a short range. However in calculating bound state energies, for the sake of simplicity, we choose a harmonic oscillator potential as the first approximation for $U(\mathbf{r}_i)$. If we ignore the nonlocal interaction, then we obtain the eigenvalue equation

$$\left[-\frac{\hbar^2}{2m}\nabla^2 + U^{(0)}(\mathbf{r})\right]\psi_i^{(0)}(\mathbf{r}) = \mathcal{E}_i^{(0)}\psi_i^{(0)}(\mathbf{r}), \tag{17.51}$$

which can be solved exactly if $U^{(0)}(\mathbf{r})$ is a slvable potential. Next using $\psi_i^{(0)}(\mathbf{r})$ we can include not only the contribution of the residual interaction $U(\mathbf{r})-U^{(0)}(\mathbf{r})$ but also the contribution of the exchange interaction. Adding this potential to $U^{(0)}(\mathbf{r})$ we find a new wave equation

$$\begin{aligned} & -\frac{\hbar^2}{2m}\nabla^2 + U^{(1)}(\mathbf{r})\psi_i^{(1)}(\mathbf{r}) \\ + & \frac{1}{2}\sum_{i,j} U_{ex}^{(0)}(\mathbf{r},\mathbf{r}')\,\psi_i^{(1)}(\mathbf{r}')\,d^3r' = \mathcal{E}_i^{(1)}\psi_i^{(1)}(\mathbf{r}). \end{aligned} \tag{17.52}$$

We continue this iteration process until a stable solution is found. In general if $|U^{(0)}(\mathbf{r})| \gg |U^{(1)}(\mathbf{r})|$, only a few iterations are needed to find an accurate solution. We can include the exchange part of the Hartree–Fock formulation as we did for the problem of electrons in atoms.

Testing the Accuracy of the Hartree–Fock Approximation — We can test the accuracy of the nuclear Hartree–Fock approximation in a simple model. This model consists of two identical harmonic oscillators interacting through harmonic forces. Setting the mass of each oscillator equal to unity and taking $\hbar = 1$, the Hamiltonian of the system can be written as [5]–[7]

$$\begin{aligned} H &= \frac{1}{2}\left(\mathbf{p}_1^2+\mathbf{r}_1^2\right) + \frac{1}{2}\left(\mathbf{p}_2^2+\mathbf{r}_2^2\right) + \kappa\left[\frac{1}{\sqrt{2}}\left(\mathbf{r}_1-\mathbf{r}_2\right)\right]^2 \\ &= \frac{1}{2}\left(\mathbf{P}^2+\mathbf{R}^2\right) + \frac{1}{2}\left[\mathbf{p}^2 + (2\kappa+1)\mathbf{r}^2\right], \end{aligned} \tag{17.53}$$

where $\mathbf{r}_1$ and $\mathbf{r}_2$ denote the coordinates of the two particles and $\mathbf{p}_1$ and $\mathbf{p}_2$ are their momenta. The center of mass and the relative coordinates introduced here are defined by

$$\mathbf{R} = \frac{1}{\sqrt{2}}\left(\mathbf{r}_1+\mathbf{r}_2\right), \quad \mathbf{r} = \frac{1}{\sqrt{2}}\left(\mathbf{r}_1-\mathbf{r}_2\right), \tag{17.54}$$

with $\mathbf{P}$ and $\mathbf{p}$ are their conjugate momenta. The coupling constant for the interaction between the two oscillators is denoted by κ. For the ground we find the wave function and the energy of the Hamiltonian (17.53) to be

$$\psi = \frac{1}{\pi^{\frac{3}{2}}}(2\kappa+1)^{\frac{3}{8}}\exp\left[-\frac{1}{2}\mathbf{R}^2\right]\exp\left[-\frac{1}{2}\sqrt{2\kappa+1}\,\mathbf{r}^2\right], \tag{17.55}$$

and

$$E = \left(\frac{3}{2}\right)\left[1+\sqrt{2\kappa+1}\right]. \tag{17.56}$$

Now let us apply the Hartree–Fock approximation to this problem. The ground state wave function is symmetric and is given by

$$\psi'(\mathbf{r}_1, \mathbf{r}_2) = \phi(\mathbf{r}_1)\phi(\mathbf{r}_2), \tag{17.57}$$

where $\phi(\mathbf{r}_1)$ is the solution of the integro-differential equation

$$\begin{aligned}\mathcal{E}\phi(\mathbf{r}_1) &= \frac{1}{2}\left(\mathbf{p}_1^2+\mathbf{r}_1^2\right)\phi(\mathbf{r}_1) + \left[\int \kappa\phi^*(\mathbf{r}_2)\left\{\frac{1}{\sqrt{2}}(\mathbf{r}_1-\mathbf{r}_2)\right\}^2\phi(\mathbf{r}_2)d^3r_2\right]\phi(\mathbf{r}_1)\\ &= \frac{1}{2}\left(\mathbf{p}_1^2+(\kappa+1)\mathbf{r}_1^2\right)\phi(\mathbf{r}_1) + \frac{\kappa}{2}\left[\int \phi^*(\mathbf{r}_2)\mathbf{r}_2^2\phi(\mathbf{r}_2)d^3r_2\right]\phi(\mathbf{r}_1),\end{aligned} \tag{17.58}$$

where we have simplified the result noting that due to parity symmetry, the term $\mathbf{r}_1\cdot\mathbf{r}_2$ in the integrand will not contribute to the integral.

The exact solution of $\phi(\mathbf{r}_1)$ can be found by solving (17.58)

$$\phi(\mathbf{r}_1) = \pi^{-\frac{3}{4}}(\kappa+1)^{\frac{3}{8}}\exp\left[-\frac{1}{2}\sqrt{\kappa+1}\,\mathbf{r}_1^2\right], \tag{17.59}$$

where the eigenvalue $\mathcal{E}$ is given by

$$\mathcal{E} = \frac{3}{2}\sqrt{\kappa+1}\left(\frac{3\kappa+2}{2\kappa+2}\right). \tag{17.60}$$

By rewriting the Hamiltonian as

$$H = \frac{1}{2}\left[\mathbf{p}_1^2+(\kappa+1)\mathbf{r}_1^2\right] + \frac{1}{2}\left[\mathbf{p}_2^2+(\kappa+1)\mathbf{r}_2^2 - \kappa\mathbf{r}_1\cdot\mathbf{r}_2\right], \tag{17.61}$$

we calculate the expectation value with $\phi(\mathbf{r}_1)$ given by (17.59) to find the approximate value of the energy

$$E' = \langle\phi(\mathbf{r}_1)\phi(\mathbf{r}_2)|H|\phi(\mathbf{r}_1)\phi(\mathbf{r}_2)\rangle = 3\sqrt{1+\kappa}. \tag{17.62}$$

Comparing E and E', Eqs. (17.56) and (17.62) we observe that for small κ,

$$E \approx E' \approx 3 + \frac{3}{2}\kappa. \tag{17.63}$$

Even for strong coupling between the two oscillators $\kappa \approx 1$ the approximate energy is 0.96 of the exact value. In addition in this simple model we can determine the overlap between the exact and the approximate wave functions and this gives us a measure of the accuracy of the approximate wave function. Thus from Eqs. (17.55) and (17.57) we find $\langle\psi|\psi'\rangle$ to be

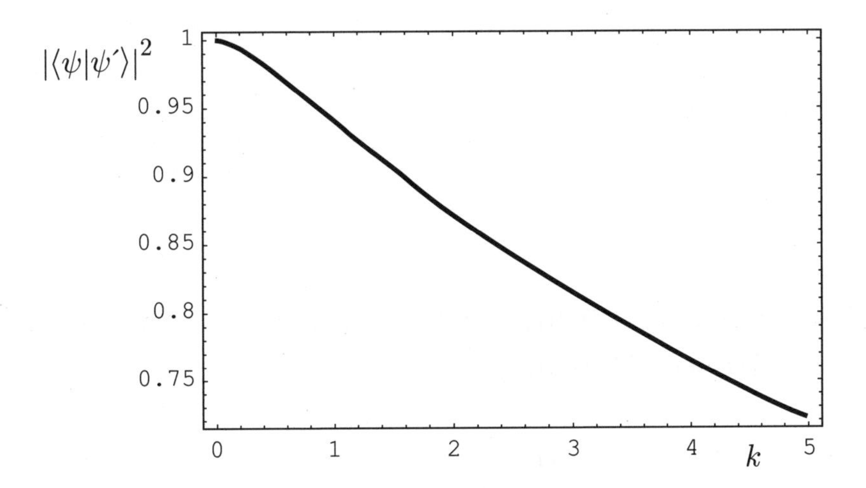

Figure 17.1: The square of the overlap between the exact and Hartree–Fock wave functions is shown as a function of the strength of the coupling constant.

$$\langle\psi|\psi'\rangle = \frac{(\kappa+1)^{\frac{3}{4}}(2\kappa+1)^{\frac{3}{8}}}{\left[\frac{1}{2}\left\{\sqrt{\kappa+1}+\sqrt{2\kappa+1}\right\}\right]^{\frac{3}{2}}\left[\frac{1}{2}\left(1+\sqrt{\kappa+1}\right)\right]^{\frac{3}{2}}}. \tag{17.64}$$

The square of $\langle\psi|\psi'\rangle$ as a function of κ is plotted in Fig. 17.1. As this figure shows for a wide range of the strength of the coupling constant the approximate wave function is accurate, viz, $\langle\psi|\psi'\rangle \approx 1$

For other solvable examples of the Hartree–Fock method see [8]–[10].

17.3 Second Quantization

Our approach to the approximate solution of the many-body problem will be based on the method of second quantization. This method has the advantage that it incorporates the statistics of the particles, i.e. whether they are bosons or fermions. The basis set that we will use here is the vector space composed of the abstract state vectors of the form

$$|n_{k_1}, n_{k_2}\cdots n_{k_\infty}\rangle, \tag{17.65}$$

and this space is usually referred to as Fock space. In this space k denotes the set of quantum numbers, e.g. energy or momentum of the particle, n_{k_1} is the number of particles in the eigenstate k_1, n_{k_2} particles in the eigenstate k_2 and so on.

The basis set is assumed to be complete and orthogonal, i.e. these states satisfy the following orthogonality and completeness conditions

$$\left\langle n'_{k_1}, n'_{k_2}\cdots n'_{k_\infty}\middle| n_{k_1}, n_{k_2}\cdots n_{k_\infty}\right\rangle = \delta_{n'_{k_1} n_{k_1}}\delta_{n'_{k_2} n_{k_2}}\cdots\delta_{n'_{k_\infty} n_{k_\infty}}, \tag{17.66}$$

and

$$\sum_{n_{k_1},n_{k_2}\cdots n_{k_\infty}} |n_{k_1}, n_{k_2} \cdots n_{k_\infty}\rangle\langle n_{k_1}, n_{k_2} \cdots n_{k_\infty}| = 1, \tag{17.67}$$

where the sum (17.67) is over all possible occupation numbers $n_{k_1}, n_{k_2} \cdots n_{k_\infty}$, and 1 is the unit matrix or operator.

17.4 Second-Quantized Formulation of the Many-Boson Problem

We now introduce the set of creation and and annihilation operators, $\left\{a_{k_i}^\dagger(t)\right\}$, and $\{a_{k_i}(t)\}$ for bosons, and by the following relations (see for example Eqs. (9.131) and (9.133) of Chapter 9)

$$a_{k_i}^\dagger\, |n_{k_1}, n_{k_2} \cdots, n_{k_i}, \cdots n_{k_\infty}\rangle = \sqrt{n_{k_i}+1}\, |n_{k_1}, n_{k_2} \cdots n_{k_i}+1, \cdots n_{k_\infty}\rangle, \tag{17.68}$$

and

$$a_{k_i}\, |n_{k_1}, n_{k_2} \cdots, n_{k_i}, \cdots n_{k_\infty}\rangle = \sqrt{n_{k_i}}\, |n_{k_1}, n_{k_2} \cdots n_{k_i}-1, \cdots n_{k_\infty}\rangle. \tag{17.69}$$

Using these two relations we define the number operator for the state k_i

$$\hat{n}_{k_i} = a_{k_i}^\dagger a_{k_i}, \tag{17.70}$$

with the property that

$$\begin{aligned} a_{k_i}^\dagger a_{k_i}\, |n_{k_1}, n_{k_2} \cdots, n_{k_i}, \cdots n_{k_\infty}\rangle &= n_{k_i} |n_{k_1}, \cdots n_{k_i} \cdots n_{k_\infty}\rangle, \\ n_{k_i} &= 0,\ 1, \cdots \infty. \end{aligned} \tag{17.71}$$

The fact that n_{k_i} is greater or equal to zero follows from the scaler product

$$\begin{aligned} n_{k_i} &= \left\langle n_{k_1}, n_{k_2} \cdots, n_{k_i}, \cdots n_{k_\infty} \left| a_{k_i}^\dagger a_{k_i} \right| n_{k_1}, n_{k_2} \cdots, n_{k_i}, \cdots n_{k_\infty} \right\rangle \\ &= \sum \left\langle n_{k_1}, n_{k_2} \cdots, n_{k_i}, \cdots n_{k_\infty} \left| a_{k_i}^\dagger \right| n_{k_1'}, n_{k_2'} \cdots, n_{k_i'}, \cdots n_{k_\infty'} \right\rangle \\ &\times \left\langle n_{k_1'}, n_{k_2'} \cdots, n_{k_i'}, \cdots n_{k_\infty'} \left| a_{k_i} \right| n_{k_1}, n_{k_2} \cdots, n_{k_i}, \cdots n_{k_\infty} \right\rangle \\ &= \sum \left| \left\langle n_{k_1'}, n_{k_2'} \cdots, n_{k_i'}, \cdots n_{k_\infty'} \left| a_{k_i} \right| n_{k_1}, n_{k_2} \cdots, n_{k_i}, \cdots n_{k_\infty} \right\rangle \right|^2 \\ &\geq 0, \end{aligned} \tag{17.72}$$

where the summation is over all integers $n_{k_1}', n_{k_2'}, \cdots n_{k_\infty'}$. From the commutation relations

$$\left[a_{k_i}, a_{k_j}^\dagger\right] = \delta_{k_i k_j}, \quad \left[a_{k_i}^\dagger, a_{k_i}^\dagger\right] = 0, \quad \left[a_{k_i}, a_{k_j}\right] = 0, \tag{17.73}$$

we get

$$\left[a^\dagger_{k_i}a_{k_i}, a_{k_j}\right] = -a_{k_i}\delta_{k_ik_j}, \tag{17.74}$$

and using this we can easily show that the operator a_{k_i} acting on an eigenstate $|n_{k_1}, n_{k_2}\cdots n_{k_\infty}\rangle$ produces a new eigenstate but now with the eigenvalue $(n_{k_i}-1)$

$$\begin{aligned} & a^\dagger_{k_i}a_{k_i}\left(a_{k_i}|n_{k_1}, n_{k_2}\cdots, n_{k_i}, \cdots n_{k_\infty}\rangle\right) \\ = \; & a_{k_i}\left(a^\dagger_{k_i}a_{k_i}\right)|n_{k_1}, \cdots n_{k_i}\cdots n_{k_\infty}\rangle, \\ + \; & \left[a^\dagger_{k_i}a_{k_i},\, a_{k_i}\right]|n_{k_1}, \cdots n_{k_i}\cdots n_{k_\infty}\rangle \\ = \; & (n_{k_i}-1)\left(a_{k_i}|n_{k_1}, \cdots n_{k_i}\cdots n_{k_\infty}\rangle\right). \end{aligned} \tag{17.75}$$

In exactly the same way we can show that from the commutation relation

$$\left[a^\dagger_{k_i}a_{k_i}, a^\dagger_{k_j}\right] = a^\dagger_{k_i}\delta_{k_ik_j}, \tag{17.76}$$

we can construct an eigenstate

$$\left(a^\dagger_{k_i}|n_{k_1}, n_{k_2}\cdots, n_{k_i}, \cdots n_{k_\infty}\rangle\right), \tag{17.77}$$

for which the eigenvalue is $n_{k_i}+1$.

A state like (17.65) is related to the vacuum state

$$|0,\ 0, \cdots 0,\ 0\rangle \tag{17.78}$$

by the action of the creation operators

$$\begin{aligned} |n_{k_1}, n_{k_2}\cdots, n_{k_\infty}\rangle = \; & \left[n_{k_1}!, n_{k_2}!\cdots, n_{k_\infty}!\right]^{-\frac{1}{2}} \\ \times \; & \left[\left(a^\dagger_{k_1}\right)^{n_{k_1}}\left(a^\dagger_{k_2}\right)^{n_{k_2}}\cdots\left(a^\dagger_{k_\infty}\right)^{n_{k_\infty}}\right]|0, \cdots, 0\rangle. \end{aligned} \tag{17.79}$$

In fact if we choose (17.79) as the definition of the state $|n_{k_1}, n_{k_2}\cdots, n_{k_\infty}\rangle$, then using the commutation relations we can verify Eqs. (17.68) and (17.69).

Now we want to see how we can express quantum mechanical operators in Fock space. In the many-body systems that we will be studying we assume that the operators involve the coordinates of one or two particles. Let us first consider an operator which acts only on a single particle.

A typical example is provided by the kinetic energy operator

$$\hat{T} = \sum_n \frac{\hat{\mathbf{p}}_n^2}{2m}, \tag{17.80}$$

where $\hat{\mathbf{p}}_n$ is the momentum operator acting on the n-th particle. Let $\hat{\mathcal{O}}_1$ represents a one-body operator which we assume is diagonal in the basis $|k\rangle$

$$\hat{\mathcal{O}}_1 = \sum_i \mathcal{O}_{k_i}|k_i\rangle\langle k_i|, \tag{17.81}$$

where

$$\mathcal{O}_{k_i} = \left\langle k_i \left| \hat{\mathcal{O}}_1 \right| k_i \right\rangle. \tag{17.82}$$

In this representation we have

$$\begin{aligned}
&\left\langle n'_{k_1}, n'_{k_2} \cdots, n'_{k_i}, \cdots n'_{k_\infty} \left| \hat{\mathcal{O}}_1 \right| n_{k_1}, n_{k_2} \cdots, n_{k_i}, \cdots n_{k_\infty} \right\rangle \\
&= \sum_i \mathcal{O}_{k_i} n_{k_i} \left\langle n'_{k_1}, n'_{k_2} \cdots, n'_{k_i}, \cdots n'_{k_\infty} | n_{k_1}, n_{k_2} \cdots, n_{k_i}, \cdots n_{k_\infty} \right\rangle \\
&= \left\langle n'_{k_1}, n'_{k_2} \cdots, n'_{k_i}, \cdots n'_{k_\infty} \left| \sum_i \mathcal{O}_{k_i} a^\dagger_{k_i} a_{k_i} \right| n_{k_1}, n_{k_2} \cdots, n_{k_i}, \cdots n_{k_\infty} \right\rangle.
\end{aligned} \tag{17.83}$$

This relation is true for any set of states, therefore it must be true for the second quantized form of $\mathcal{O}_1$;

$$\mathcal{O}_1 = \sum_{k=0}^{\infty} \mathcal{O}_k \hat{n}_k = \sum_{k=0}^{\infty} \left\langle k \left| \hat{\mathcal{O}}_1 \right| k \right\rangle a^\dagger_k a_k. \tag{17.84}$$

The one-body operator $\hat{\mathcal{O}}_1$ in (17.84) which is written in the diagonal representation shows that this operator can be written as to the product of the eigenvalue of one-body operator in the state $|k\rangle$ times the number of the particles in that state.

We can also write (17.84) in a representation where $\hat{\mathcal{O}}_1$ is not diagonal. In this representation $\mathcal{O}_1$ is

$$\mathcal{O}_1 = \sum_{k=0, j=0}^{\infty} \left\langle k \left| \hat{\mathcal{O}}_1 \right| j \right\rangle a^\dagger_k a_j. \tag{17.85}$$

This is a straightforward generalization of (17.84) and reduces to it when $\hat{\mathcal{O}}_1$ is diagonal.

Next let us consider the operator $\hat{\mathcal{O}}_2$, or the two-body operator in the Fock space. The operator $\hat{\mathcal{O}}_2$ which is acting on states in the two-particle space has the form

$$\hat{\mathcal{O}}_2 = \sum_{i,j} |k_i k_j\rangle\langle k_i k_j|\mathcal{O}_2|\ell_i \ell_j\rangle\langle \ell_i \ell_j|. \tag{17.86}$$

Using (17.86) we find that the action of $\hat{\mathcal{O}}_2$ on the state of the many-body system is expressible as

$$\begin{aligned}
&\hat{\mathcal{O}}_2 |k_1, \cdots k_i \cdots k_j \cdots k_\infty\rangle \\
&= \sum_{\ell_i \ell_j} \langle \ell_i \ell_j | \mathcal{O}_2 | k_i k_j\rangle \, |k_1, \cdots \ell_i \cdots \ell_j \cdots k_\infty\rangle.
\end{aligned} \tag{17.87}$$

We also observe that the matrix elements of $\hat{\mathcal{O}}_2$ do not depend on which pair of particles it is acting upon, i.e. the matrix elements $\langle \ell_i \ell_j|\mathcal{O}_2|k_i k_j\rangle$ are the same for any pair of the particles as long as the same quantum numbers k and ℓ are involved. Summing $\hat{\mathcal{O}}_2(i,j)$ over all pairs $i \neq j$ we have

$$\sum_{i>j} \hat{\mathcal{O}}_2(i,j)|k_1,\cdots k_i\cdots k_j\cdots k_\infty\rangle$$
$$= \frac{1}{2}\sum_{i\neq j}\sum_{\ell_i,\ell_j} \langle \ell_i\ell_j|\mathcal{O}_2|k_ik_j\rangle\,|k_1,\cdots \ell_i\cdots \ell_j\cdots k_\infty\rangle. \tag{17.88}$$

As in the case of $\hat{\mathcal{O}}_1$, we will show that the second quantized form of $\hat{\mathcal{O}}_2$ is

$$\mathcal{O}_2 = \frac{1}{2}\sum_{k,\ell,m,n} \langle k,\ell\,|\mathcal{O}_2|\,m,n\rangle\, a_k^\dagger a_\ell^\dagger a_m a_n. \tag{17.89}$$

To show this we calculate the following commutator

$$\begin{aligned}\left[\hat{\mathcal{O}}_2,\, a_{k_i}^\dagger\right] &= \frac{1}{2}\sum_{k,\ell,m,n} \langle k,\ell\,|\mathcal{O}_2|\,m,n\rangle\, a_k^\dagger a_\ell^\dagger \left[a_m a_n,\, a_{k_i}^\dagger\right]\\ &= \sum_{k,\ell,n} \langle k,\ell\,|\mathcal{O}_2|\,k_i,n\rangle\, a_k^\dagger a_\ell^\dagger a_n\\ &= \sum_{\ell_i,\ell_{j'},k_{j'}} \langle \ell_i\ell_{j'}\,|\mathcal{O}_2|\,k_i,k_{j'}\rangle\, a_{\ell_i}^\dagger a_{\ell_{j'}}^\dagger a_{k_i}.\end{aligned} \tag{17.90}$$

In arriving at Eq. (17.90) we have used the symmetry of the matrix elements of $\mathcal{O}_2$

$$\langle k\ell|\mathcal{O}_2|mn\rangle = \langle mn|\mathcal{O}_2|k\ell\rangle. \tag{17.91}$$

Next we note that

$$\hat{\mathcal{O}}_2|k_1,\cdots k_i\cdots k_j\cdots k_\infty\rangle = \hat{\mathcal{O}}_2\left(a_{k_1}^\dagger, a_{k_2}^\dagger\cdots a_{k_\infty}^\dagger\right)|0\rangle$$
$$= \sum_i\left(a_{k_1}^\dagger\cdots\left[\hat{\mathcal{O}}_2, a_{k_i}^\dagger\right]\cdots a_{k_\infty}^\dagger|0\rangle\right)$$
$$= \sum_i\sum_{\ell_i,\ell_{j'},k_{j'}} \langle \ell_i\ell_{j'}\,|\mathcal{O}_2|\,k_ik_{j'}\rangle\left(a_{k_1}^\dagger\cdots a_{\ell_i}^\dagger a_{\ell_{j'}}^\dagger, a_{k_{j'}} a_{k_{i+1}}^\dagger\cdots a_{k_\infty}^\dagger|0\rangle\right). \tag{17.92}$$

We can reduce Eq. (17.92) by noting that for any function $\mathcal{F}(\ell_{i'},k_{i'})$

$$\sum_{\ell_{i'},k_{i'}} \mathcal{F}(\ell_{i'},k_{i'})\left[a_{\ell_{i'}}^\dagger a_{k_{i'}}^\dagger,\; a_{k_i}\right] = \sum_{\ell_{i'}}\mathcal{F}(\ell_{i'},k_i)a_{\ell_{i'}}^\dagger, \tag{17.93}$$

with the result

$$\hat{\mathcal{O}}_2|k_1,\cdots k_i\cdots k_j\cdots k_\infty\rangle =$$
$$= \sum_i\sum_{i<j}\sum_{\ell_i,\ell_j} \langle \ell_i\ell_j\,|\mathcal{O}_2|\,k_ik_j\rangle\left(a_{k_1}^\dagger\cdots a_{\ell_i}^\dagger\cdots a_{\ell_j}^\dagger\cdots a_{k_\infty}^\dagger|0\rangle\right), \tag{17.94}$$

and this is the same as (17.88) for an arbitrary many-body state

$$|k_1 \cdots \ell_i \cdots \ell_j \cdots k_\infty\rangle. \tag{17.95}$$

Having found the operator forms of $\hat{\mathcal{O}}_1$ and $\hat{\mathcal{O}}_2$ we can write the Hamiltonian for a many-boson system as

$$\hat{H} = \sum_{k\ell} \langle k\,|T|\,\ell\rangle\, a_k^\dagger a_\ell + \frac{1}{2}\sum_{k\ell mn} \langle k\ell|V|mn\rangle a_k^\dagger a_\ell^\dagger a_m a_n. \tag{17.96}$$

17.5 Many-Fermion Problem

Fermions are particles which obey the Pauli exclusion principle (Sec. 6.6). According to this principle no two particles can occupy the same quantum state.

Let us consider a state $|n\rangle$ in which all fermions are in the lowest level of the system with spin up. There are two such states: the state with one particle is shown by $|1\rangle$, and the state with no particle is denoted by $|0\rangle$. Now we introduce the creation and annihilation operators for fermions for the lowest level, $b_0^\dagger$ and b_0, by the following relations:

$$b_0|0\rangle = 0, \quad b_0|1\rangle = |0\rangle, \tag{17.97}$$

$$b_0^\dagger|0\rangle = |1\rangle, \quad b_0^\dagger|1\rangle = 0. \tag{17.98}$$

Matrix representations of b_0 and $b_0^\dagger$ can be found from (17.97) and (17.98), and these have the same forms as σ_+ and σ_- which we introduced earlier, Eqs. (8.196) and (8.197). We note that $b_0^\dagger|1\rangle = 0$ means that two fermions cannot occupy the same state and $b_0|0\rangle = 0$ means that we cannot remove a particle from vacuum state. From Eqs. (17.97) and (17.98) it follows that

$$b_0 b_0^\dagger + b_0^\dagger b_0 \equiv \left[b_0,\, b_0^\dagger\right]_+ = 1, \tag{17.99}$$

where $\left[b_0,\, b_0^\dagger\right]_+ = 1$ is the anti-commutator. In addition, from these definitions it follows that

$$b_0^2 = 0, \quad \left(b_0^\dagger\right)^2 = 0. \tag{17.100}$$

These relations imply that it is impossible to remove two fermions from the same state and it is also impossible to have two fermions in the same level. The general form of the anti-commutation relations for other quantum levels of the system are:

$$\left[b_k,\, b_j^\dagger\right]_+ = \delta_{jk}, \quad [b_k,\, b_j]_+ = \left[b_k^\dagger,\, b_j^\dagger\right]_+ = 0. \tag{17.101}$$

As in the case of bosons for a system with n_0 particles in the ground state, n_1 particles in the first excited state, etc. we write the state of the system as

$$|n_0, n_1 \cdots\rangle. \tag{17.102}$$

This state can be obtained from the vacuum $|0\rangle = |0,\ 0,\ 0\cdots\rangle$ by the action of the operator $b_k^\dagger$;

$$|n_0, n_1, n_2\cdots\rangle = \cdots \left(b_2^\dagger\right)^{n_2}\left(b_1^\dagger\right)^{n_1}\left(b_0^\dagger\right)^{n_0}|0\rangle, \tag{17.103}$$

with no factorials since $n_k! = 1$ for $n_k = 0$ or $n_k = 1$. For fermions just like bosons we define the number operator in the state k by (see Eq. (9.127))

$$\hat{N}_k = b_k^\dagger b_k. \tag{17.104}$$

Using this operator we write the one-body operator $\hat{\mathcal{O}}_1$ in the second quantized form as

$$\hat{\mathcal{O}}_1 = \sum_k \langle k|Q|k\rangle \hat{N}_k = \sum_k \langle k|\mathcal{O}_1|k\rangle\, b_k^\dagger b_k, \tag{17.105}$$

or we write it as

$$\hat{\mathcal{O}}_1 = \sum_{k,j} \langle k|\mathcal{O}_1|j\rangle\, b_k^\dagger b_j. \tag{17.106}$$

To find the second quantized representation of $\hat{\mathcal{O}}_2$ we follow exactly the same procedure as the one we used for interacting bosons, but now we replace the commutation relations by anti-commutation relations. The final result for the Hamiltonian of an interacting system of fermions is similar to (17.96), viz,

$$\hat{H} = \sum_{k\ell} \langle k\,|T|\,\ell\rangle\, b_k^\dagger b_\ell + \frac{1}{2}\sum_{k\ell mn} \langle k\ell|V|mn\rangle b_k^\dagger b_\ell^\dagger b_m b_n. \tag{17.107}$$

The time evolution of $b_k^\dagger$ and b_k can be found from the Heisenberg equation. If we consider a system of noninteracting fermions with the Hamiltonian $H = \sum_k E_k b_k^\dagger b_k$, then

$$\begin{aligned} i\hbar\frac{db_n}{dt} &= [b_n,\ H] = \left[b_n,\ \sum_k E_k b_k^\dagger b_k\right] \\ &= \sum_k E_k\left(b_n b_k^\dagger b_k - b_k^\dagger b_k b_n\right) = \sum_k E_k\left[\left(\delta_{nk} - b_k^\dagger b_n\right) b_k - b_k^\dagger b_k b_n\right] \\ &= E_n b_n + \sum_k E_k\left(-b_k^\dagger b_n b_k - b_k^\dagger b_k b_n\right) = E_n b_n. \end{aligned} \tag{17.108}$$

Let us remind ourselves that fermion operators do not satisfy the conditions of skew symmetry, Leibnitz property and Jacobi identity, Eqs. (3.127), (3.129),

(3.130), that the boson operators satisfy. Thus we do not have any correspondence between the Poisson bracket and the anticommutators for fermion operators. In the case of fermions, those operators that are linear in b_k and $b_k^\dagger$ cannot be measured classically whereas operators bilinear in creation and annihilation operators have classical limits and are measurable.

Noninteracting Fermions — As an introduction to the physics of many-body we want to calculate some of the physical properties of a simple system of fermions. For this we consider a system of spin $\frac{1}{2}$ fermions enclosed in a large cubical box with sides of length L. At the end of the calculation we will take the limit of $L \to \infty$. In such a uniform infinite medium, the physical properties must remain invariant under spatial transformation. Therefore we choose single particle wave functions satisfying the periodic boundary condition and we write them as

$$\phi_{\mathbf{p}}(\mathbf{r}) = \frac{1}{L^{\frac{3}{2}}} e^{i\mathbf{p}\cdot\mathbf{r}}, \quad \hbar = 1, \tag{17.109}$$

where

$$p_x = \frac{2\pi n_x}{L}, \quad n_x = 0, \ \pm 1, \ \pm 2 \cdots, \tag{17.110}$$

and similar forms for p_y and p_z. Let us denote the creation operator for a particle of momentum $\mathbf{p}$ and spin s by $b_{\mathbf{p}s}^\dagger$ and the annihilation operator by $b_{\mathbf{p}s}$ The amplitude at the point $\mathbf{r}$ for finding the particle created by the action of $b_{\mathbf{p}s}^\dagger$ is $L^{\frac{-3}{2}} e^{i\mathbf{p}\cdot\mathbf{r}}$. Thus by superposition of momentum states we obtain the amplitude for adding a particle of spin s to the point $\mathbf{r}$ to be

$$\psi_s^\dagger(\mathbf{r}) = \sum_{\mathbf{p}} \frac{1}{L^{\frac{3}{2}}} e^{-i\mathbf{p}\cdot\mathbf{r}} b_{\mathbf{p}s}^\dagger. \tag{17.111}$$

The Hermitian adjoint of $\psi_s^\dagger(\mathbf{r})$ i.e.

$$\psi_s(\mathbf{r}) = \sum_{\mathbf{p}} \frac{1}{L^{\frac{3}{2}}} e^{i\mathbf{p}\cdot\mathbf{r}} b_{\mathbf{p}s}. \tag{17.112}$$

is the amplitude for removing a particle of spin s from the point $\mathbf{r}$. The operators $\psi_s^\dagger(\mathbf{r})$ and $\psi_s(\mathbf{r})$ defined by (17.111) and (17.112) satisfy the anti-commutation relations

$$\psi_s(\mathbf{r})^\dagger \psi_{s'}^\dagger(\mathbf{r}') + \psi_{s'}^\dagger(\mathbf{r}')\, \psi_s^\dagger(\mathbf{r}) = 0, \tag{17.113}$$

$$\psi_s(\mathbf{r}) \psi_{s'}(\mathbf{r}') + \psi_{s'}(\mathbf{r}')\, \psi_s(\mathbf{r}) = 0, \tag{17.114}$$

and

$$\begin{aligned} \psi_s(\mathbf{r})\psi_{s'}^\dagger(\mathbf{r}') &+ \psi_{s'}^\dagger(\mathbf{r}')\, \psi_s(\mathbf{r}) \\ &= \sum_{\mathbf{p}} \sum_{\mathbf{p}'} \frac{1}{L^3} e^{i\mathbf{p}\cdot\mathbf{r}} e^{-i\mathbf{p}'\cdot\mathbf{r}'} \left(b_{\mathbf{p}s} b_{\mathbf{p}'s'\dagger} + b_{\mathbf{p}'s'\dagger} b_{\mathbf{p}} s \right) \\ &= \sum_{\mathbf{p}} \sum_{\mathbf{p}'} \frac{1}{L^3} e^{i\mathbf{p}\cdot\mathbf{r} - i\mathbf{p}'\cdot\mathbf{r}'} \delta_{\mathbf{p}\mathbf{p}'} \delta_{ss'}, \end{aligned} \tag{17.115}$$

where in arriving at (17.113)–(17.115) we have used the anti-commutation relation (17.101) for fermions. By replacing the summation by integration, i.e. taking the limit of $L \to \infty$, we obtain

$$\psi_s(\mathbf{r})\psi^\dagger_{s'}(\mathbf{r}') + \psi^\dagger_{s'}(\mathbf{r}')\psi_s(\mathbf{r}) = \delta(\mathbf{r}-\mathbf{r}')\,\delta_{ss'}, \quad \hbar = 1. \tag{17.116}$$

From the operators $\psi^\dagger_s(\mathbf{r})$ and $\psi_s(\mathbf{r})$ we can define the density operator by

$$\rho(\mathbf{r}) = \sum_s \rho_s(\mathbf{r}) = \sum_s \psi^\dagger_s(\mathbf{r})\psi_s(\mathbf{r}). \tag{17.117}$$

Now if we integrate $\rho(\mathbf{r})$ over the large box of volume L^3 we find the number operator $\hat{N}$

$$\hat{N} = \int \rho(\mathbf{r}) d^3r. \tag{17.118}$$

In order to relate $\bar{N}$ to the creation and annihilation operators we substitute for $\psi^\dagger_s(\mathbf{r})$ and $\psi_s(\mathbf{r})$ from (17.111) and (17.112) in (17.117) and (17.118) to obtain

$$\hat{N} = \frac{1}{L^3}\sum_s\sum_{\mathbf{p}\mathbf{p}'} b^\dagger_{\mathbf{p}s} b_{\mathbf{p}'s}\int e^{i(\mathbf{p}'-\mathbf{p})\cdot\mathbf{r}} d^3r. \tag{17.119}$$

Since

$$\int e^{i(\mathbf{p}'-\mathbf{p})\cdot\mathbf{r}} d^3r = L^3\delta_{\mathbf{p}\mathbf{p}'}, \tag{17.120}$$

we find

$$\hat{N} = \sum_{\mathbf{p}} b^\dagger_{\mathbf{p}s} b_{\mathbf{p}s}, \tag{17.121}$$

and this is the sum of N_k s defined by (17.104).

Now let us consider a gas of noninteracting spin $\frac{1}{2}$ particles. For such a system the ground state $|\Phi_0\rangle$ is a state where different momentum states are filled up to some maximum momentum p_F, called the Fermi momentum. Thus we find the expectation value of the number operator of the state $|\Phi_0\rangle$ to be

$$\begin{aligned} n_{\mathbf{p}} &= \left\langle \Phi_0 \left| b^\dagger_{\mathbf{p}\uparrow} b_{\mathbf{p}\uparrow} + b^\dagger_{\mathbf{p}\downarrow} b_{\mathbf{p}\downarrow} \right| \Phi_0 \right\rangle \\ &= \begin{cases} 2 & \text{for } |\mathbf{p}| \le p_F \\ 0 & \text{for } |\mathbf{p}| \ge p_F \end{cases}. \end{aligned} \tag{17.122}$$

The total number of the particles in this case is

$$N = \sum_{\mathbf{p}} n_{\mathbf{p}} = \sum_{|\mathbf{p}|\le p_F} 2, \tag{17.123}$$

where the sum is over all states within the sphere $|\mathbf{p}| \le p_F$ (Fermi sphere). Replacing the summation (17.123) by integration we get

$$2L^3\int_0^{p_F}\frac{d^3p}{(2\pi)^3} = \frac{p_F^3}{3\pi^2}L^3. \tag{17.124}$$

If n denotes the average particle density $n = \frac{N}{L^3}$, then from (17.124) we have

$$p_F^3 = 3\pi^2 n. \tag{17.125}$$

We can also calculate the expectation value of the density operator $\rho(\mathbf{r})$, Eq. (17.117), with the ground state $|\Phi_0\rangle$

$$\begin{aligned}\langle\Phi_0|\rho(\mathbf{r})|\Phi_0\rangle &= \frac{1}{L^3}\sum_{\mathbf{pp}'s}\exp\left[-i\left(\mathbf{p}-\mathbf{p}'\right)\cdot\mathbf{r}\right]\left\langle\Phi_0\left|b^\dagger_{\mathbf{p}s}b_{\mathbf{p}'s}\right|\Phi_0\right\rangle \\ &= \frac{1}{L^3}\sum_{\mathbf{p}} n_{\mathbf{p}} = n.\end{aligned} \tag{17.126}$$

i.e. the density of such a fermion gas is uniform and is equal to n. The energy of the system is found by noting that

$$\begin{aligned}E_0 &= \sum_{\mathbf{p}s}\frac{p^2}{2m}\left\langle\Phi_0\left|b^\dagger_{\mathbf{p}s}b_{\mathbf{p}s}\right|\Phi_0\right\rangle = \sum_{\mathbf{p}s}\frac{p^2}{2m}n_{\mathbf{p}} \\ &\to 2L^3\int_0^{p_F}\frac{p^2}{2m}\frac{d^3p}{(2\pi)^3} = \frac{3}{5}\left(\frac{p_F^2}{2m}\right)N.\end{aligned} \tag{17.127}$$

A weakly interacting many-fermion system has a ground state $|\Phi_0'\rangle$ which differs slightly from $|\Phi_0\rangle$ defined by (17.122). Because of the weak interaction between the fermions the original state $|\Phi_0\rangle$ changes and some of the occupied levels inside the Fermi sphere ($|\mathbf{p}| < p_F$) will be emptied and these particles fill the same number of levels with energies greater than $E_F = \frac{p_F^2}{2m}$. Thus we can characterize the state of this many-fermion system by indicating which states above the Fermi level are occupied and which states with energies lower than E_F are unoccupied. These unoccupied states are called "holes" and such a description of the states of fermions is referred to as "hole representation".

Starting with the ground state $|\Phi_0\rangle$ with the energy E_0, Eq. (17.127), we observe that any excitation of the system corresponds to a creation of a pair of particles: a particle with the energy $E_k > E_F$ and a "hole" with the energy $E_k < E_F$. Other excited states are generated by the creation of a number of pairs of particles and holes. On the other hand the transition from a higher to a lower energy state of the system is possible by the mechanism of annihilation of pairs. In order to describe the creation or annihilation of holes we introduce two new operators $\beta_{k,s}$ and $\beta^\dagger_{k,s}$ for $E_k < E_F$. If the state k has a momentum $\mathbf{p}_k$ and the state $-k$ is associated with momentum $-\mathbf{p}_k$, then the hole creation and annihilation operators are related to $b_{k,s}$ and $b_{-k,-s}$ by

$$\beta^\dagger_{k,s} = b_{-k,-s}, \quad \text{and} \quad \beta_{k,s} = b^\dagger_{-k,-s}, \quad E_k \le E_F. \tag{17.128}$$

These hole operators also satisfy the anti-commutation relations

$$\begin{aligned}\left[\beta^\dagger_{k,s},\, \beta_{k,s'}\right]_+ &= \delta_{kn}\delta_{ss'}, \quad E_k \le E_F \\ \left[\beta_{k,s},\, \beta_{k,s'}\right]_+ &= \left[\beta^\dagger_{k,s},\, \beta^\dagger_{k,s'}\right]_+ = 0.\end{aligned} \tag{17.129}$$

In this representation the ground state is determined by the conditions

$$\begin{cases} b_{k,s}|\Phi_0\rangle = 0 & \text{if } E_k > E_F \\ \beta_{k,s}|\Phi_0\rangle = 0 & \text{if } E_k < E_F \end{cases}. \tag{17.130}$$

We note that in the hole representation the particles and holes are always created and annihilated in pairs [11]–[13].

17.6 Pair Correlations Between Fermions

The state of a single particle in a many fermion system influences the state of all of the other particles in the system, i.e. there is a tendency for the particles with the same quantum numbers to avoid each other as is implied by the exclusion principle. This tendency which is the cause of correlation between the particles is present whether the particles are interacting with each other or not. Thus we do have correlations between noninteracting fermions which is the result of exchange effects. In principle we can have correlations between any number of interacting or noninteracting particles, however we confine our attention to the simplest case and the lowest order, viz, correlations between a pair of noninteracting fermions.

Pair Correlation Function for Fermions — The density operator for a particle having spin s and the diagonal element $\rho_s(\mathbf{r})$, Eq. (17.117), is given by

$$\hat{\rho}_s(\mathbf{r},\mathbf{r}') = \psi_s^\dagger(\mathbf{r})\psi_s(\mathbf{r}'). \tag{17.131}$$

This operator has the symmetry property

$$\hat{\rho}_s(\mathbf{r},\mathbf{r}') = \hat{\rho}_s^\dagger(\mathbf{r}',\mathbf{r}) \tag{17.132}$$

In order to determine the ground state expectation values of $\hat{\rho}(\mathbf{r},\mathbf{r}')$ we employ the expansion of $\psi_s^\dagger(\mathbf{r})$ and $\psi_s(\mathbf{r})$ (17.111) and (17.112). Anticipating that at the end we want to take the limit as L tends to infinity, we write $b_s^\dagger(\mathbf{p})$ and $b_s(\mathbf{p})$ for $b_{\mathbf{p}s}^\dagger$ and $b_{\mathbf{p}s}$ respectively. And in this limit we replace the summation by integration. In this way we obtain

$$\begin{aligned} \rho_s(\mathbf{r},\mathbf{r}') &= \langle\Phi_0|\hat{\rho}_s(\mathbf{r},\mathbf{r}')|\Phi_0\rangle \\ &= \frac{1}{(2\pi)^6}\int \exp(-i\mathbf{p}\cdot\mathbf{r}+i\mathbf{p}'\cdot\mathbf{r}')\left\langle\Phi_0\left|b_s^\dagger(\mathbf{p})b_s(\mathbf{p}')\right|\Phi_0\right\rangle. \end{aligned} \tag{17.133}$$

We can evaluate the matrix element $\left\langle\Phi_0\left|b_s^\dagger(\mathbf{p})b_s(\mathbf{p}')\right|\Phi_0\right\rangle$ in (17.133) by noting that this matrix element is nonzero only if $\mathbf{p}=\mathbf{p}'$ and at the same time $|\mathbf{p}|\le p_F$.

Thus we have

$$\begin{aligned}\rho_s(\mathbf{r},\mathbf{r}') &= \frac{1}{(2\pi)^3}\int \exp\left(-i\mathbf{p}\cdot(\mathbf{r}-\mathbf{r}')\right)\theta(p_F-|\mathbf{p}|)d^3p \\ &= \frac{p_F^3}{3\pi^2}\frac{(\sin\zeta-\zeta\cos\zeta)}{\zeta^3} = \left(\frac{3n}{2}\right)\left(\frac{j_1(\zeta)}{\zeta}\right),\end{aligned} \tag{17.134}$$

where $j_1(\zeta)$ is the spherical Bessel function, $\zeta = p_F|\mathbf{r}-\mathbf{r}'|$, and n, the average particle density is defined by Eq. (17.125). By setting $\mathbf{r}=\mathbf{r}'$ (17.134) reduces to (17.126);

$$\langle\Phi_0|\rho(\mathbf{r})|\Phi_0\rangle = 2\lim_{\zeta\to 0}\left[\left(\frac{3n}{2}\right)\frac{j_1(\zeta)}{\zeta}\right] = n. \tag{17.135}$$

Now let us calculate the probability amplitude of finding a particle with spin s at the point $\mathbf{r}$ (inside the Fermi sphere) when another particle with spin s' is at the point $\mathbf{r}'$. In the case of a homogeneous free gas of fermions, this pair-correlation function which here is denoted by $\mathcal{C}_{ss'}$ is a function of the relative distance between two particles $|\mathbf{r}-\mathbf{r}'|$. We normalize this function of $|\mathbf{r}-\mathbf{r}'|$ by requiring that when $s\neq s'$

$$\mathcal{C}_{ss'}(|\mathbf{r}-\mathbf{r}'|) = 1, \tag{17.136}$$

i.e. the particles with opposite spin can approach each other closely, and $\mathcal{C}_{ss'}$ is independent of $|\mathbf{r}-\mathbf{r}'|$. However if the spin of the two particles are equal, $s=s'$, then from the exclusion principle we expect that

$$\mathcal{C}_{ss'}(|\mathbf{r}-\mathbf{r}'|)\to 0 \quad \text{as} \quad |\mathbf{r}-\mathbf{r}'|\to 0, \tag{17.137}$$

i.e. the probability of finding two particles having the same set of quantum numbers at the same point is zero. In addition if the two particles are far apart, then their motions are uncorrelated and

$$\mathcal{C}_{ss'}(|\mathbf{r}-\mathbf{r}'|)\to 1 \quad \text{as} \quad |\mathbf{r}-\mathbf{r}'|\to \infty. \tag{17.138}$$

Subject to these conditions we want to determine the spatial dependence of $\mathcal{C}_{ss'}$. Now the field operators $\psi_s(\mathbf{r})$ and $\psi_{s'}(\mathbf{r}')$ annihilate two particles with spins s and s' and at the positions $\mathbf{r}$ and $\mathbf{r}'$. If the ground state $|\Phi_0^N\rangle$ consists of N particles, then the transition amplitude for $|\Phi_0^N\rangle$ changing into an $(N-2)$ particle state $\left|\Phi_\beta^{N-2}\right\rangle$ is given by

$$\left\langle\Phi_\beta^{N-2}\left|\psi_{s'}(\mathbf{r}')\psi_s(\mathbf{r})\right|\Phi_0^N\right\rangle. \tag{17.139}$$

By summing over all possible final states we find the total probability for the annihilation of the two particles to be;

$$\sum_\beta\left|\left\langle\Phi_\beta^{N-2}\left|\psi_{s'}(\mathbf{r}')\psi_s(\mathbf{r})\right|\Phi_0^N\right\rangle\right|^2$$

$$= \left\langle \Phi_0^N \left| \psi_s^\dagger(\mathbf{r}') \psi_{s'}^\dagger(\mathbf{r}) \sum_\beta \left|\Phi_\beta^{N-2}\right\rangle \left\langle \Phi_\beta^{N-2}\right| \psi_{s'}(\mathbf{r}') \psi_s(\mathbf{r}) \right| \Phi_0^N \right\rangle$$

$$= \left\langle \Phi_0^N \left| \psi_s^\dagger(\mathbf{r}') \psi_{s'}^\dagger(\mathbf{r}) \psi_{s'}(\mathbf{r}') \psi_s(\mathbf{r}) \right| \Phi_0^N \right\rangle. \quad (17.140)$$

Another way of formulating this problem is to remove a particle with spin s at $\mathbf{r}$ from the system, leaving $N-1$ particles in the state $\left|\Phi_s^{N-1}\right\rangle = \psi_s(\mathbf{r})\left|\Phi_0^N\right\rangle$. The density distribution of particles with spin s' in this new state is

$$\left\langle \Phi_s^{N-1} \left| \psi_{s'}^\dagger(\mathbf{r}') \psi_{s'}^\dagger(\mathbf{r}') \right| \Phi_s^{N-1} \right\rangle$$

$$= \left\langle \Phi_0^N \left| \psi_s^\dagger(\mathbf{r}') \psi_{s'}^\dagger(\mathbf{r}) \psi_{s'}(\mathbf{r}') \psi_s(\mathbf{r}) \right| \Phi_0^N \right\rangle$$

$$= \left(\frac{n}{2}\right)^2 \mathcal{C}_{ss'}(|\mathbf{r}-\mathbf{r}'|) \quad (17.141)$$

and by definition this is the pair-correlation function $\mathcal{C}_{ss'}(|\mathbf{r}-\mathbf{r}'|)$. The multiplicative constant $\left(\frac{n}{2}\right)^2$ guarantees the correct normalization for $\mathcal{C}_{ss'}$. We can use either (17.140) or (17.141) to determine the pair correlation function. Here we start with (17.140) and substitute for $\psi_s^\dagger(\mathbf{r})$ and $\psi_s(\mathbf{r})$ using Eqs. (17.111) and (17.112) to obtain the following expression for $\left(\frac{n}{2}\right)^2 \mathcal{C}_{ss'}(|\mathbf{r}-\mathbf{r}'|)$

$$\left(\frac{n}{2}\right)^2 \mathcal{C}_{ss'}(|\mathbf{r}-\mathbf{r}'|) = \int \left\langle \Phi_0^N \left| b_s^\dagger(\mathbf{p}) b_{s'}^\dagger(\mathbf{q}) b_{s'}(\mathbf{q}') b_s(\mathbf{p}') \right| \Phi_0^N \right\rangle$$

$$\times \left[e^{-i(\mathbf{p}-\mathbf{p}')\cdot\mathbf{r}} e^{i(\mathbf{q}-\mathbf{q}')\cdot\mathbf{r}}\right] \frac{d^3p}{(2\pi)^3} \frac{d^3q}{(2\pi)^3} \frac{d^3p'}{(2\pi)^3} \frac{d^3q'}{(2\pi)^3}. \quad (17.142)$$

Now the anticommutation relations between the fermion operators are

$$\left[b_s^\dagger(\mathbf{p}),\, b_{s'}^\dagger(\mathbf{p}')\right]_+ = \left[b_s(\mathbf{p}),\, b_{s'}(\mathbf{p}')\right]_+ = 0, \quad (17.143)$$

and

$$\left[b_s^\dagger(\mathbf{p}),\, b_{s'}(\mathbf{p}')\right]_+ = (2\pi)^3 \delta_{ss'} \delta(\mathbf{p}-\mathbf{p}'). \quad (17.144)$$

The factor $(2\pi)^3$ on the right-hand side of Eq. (17.144) comes from the expression $\frac{d^3p}{(2\pi)^3}$ which is the volume element in momentum space. Using these operators we calculate the matrix element

$$K(\mathbf{p},\mathbf{q},\mathbf{p}',\mathbf{q}') = \left\langle \Phi_0^N \left| b_s^\dagger(\mathbf{p}) b_{s'}^\dagger(\mathbf{q}) b_{s'}(\mathbf{q}') b_s(\mathbf{p}') \right| \Phi_0^N \right\rangle. \quad (17.145)$$

For calculating K we observe that for $s = s'$, the operators $b_s(\mathbf{q}')\, b_s(\mathbf{q}')$ and $b_s^\dagger(\mathbf{p})\, b_s^\dagger(\mathbf{p})$ are zero, and we conclude that K vanishes unless $\mathbf{p} \neq \mathbf{p}'$ and $\mathbf{q} \neq \mathbf{q}'$. Thus we consider the terms with $\mathbf{p} = \mathbf{q}'$ and $\mathbf{q} = \mathbf{p}'$ in our calculation.

From these results and from the commutator (17.144) for $\mathbf{q}$ and $\mathbf{p}'$ we find

$$K(\mathbf{p},\mathbf{q},\mathbf{p}',\mathbf{q}') = -(2\pi)^3 \delta_{ss'} \delta(\mathbf{p}'-\mathbf{q}) \left\langle \Phi_0^N \left| b_s^\dagger(\mathbf{p})\, b_{s'}(\mathbf{q}') \right| \Phi_0^N \right\rangle$$

$$
\begin{aligned}
&+ \left\langle \Phi_0^N \left| b_s^\dagger(\mathbf{p})\, b_s(\mathbf{p}')\, b_{s'}^\dagger(\mathbf{q})\, b_{s'}(\mathbf{q}') \right| \Phi_0^N \right\rangle \\
&= (2\pi)^6 \delta(\mathbf{p}'-\mathbf{p})\, \delta(\mathbf{q}'-\mathbf{q}) \left\langle \Phi_0^N \left| b_s^\dagger(\mathbf{p})\, b_s(\mathbf{p}') \right| \Phi_0^N \right\rangle \\
&\times \left\langle \Phi_0^N \left| b_{s'}^\dagger(\mathbf{q})\, b_{s'}(\mathbf{q}') \right| \Phi_0^N \right\rangle \\
&- (2\pi)^6 \delta_{ss'} \delta(\mathbf{q}'-\mathbf{p})\, \delta(\mathbf{p}'-\mathbf{q}) \left\langle \Phi_0^N \left| b_s^\dagger(\mathbf{p})\, b_s(\mathbf{p}) \right| \Phi_0^N \right\rangle .
\end{aligned}
\tag{17.146}
$$

By substituting for the matrix element $K(\mathbf{p}, \mathbf{q}, \mathbf{p}', \mathbf{q}')$, Eq. (17.146), in (17.142), we note that in the latter equation we get two contributions, one from the first term of $K(\mathbf{p}, \mathbf{q}, \mathbf{p}', \mathbf{q}')$ and the other from the second term. The first term makes the exponentials in (17.142) equal to one and the second term makes them equal to $\exp[-i(\mathbf{p}+\mathbf{q})\cdot(\mathbf{r}-\mathbf{r}')]$. Thus

$$
\begin{aligned}
\left(\frac{n}{2}\right)^2 \mathcal{C}_{ss'}(|\mathbf{r}-\mathbf{r}'|) &= \int \theta(p_F - |\mathbf{p}|)\frac{d^3p}{(2\pi)^3} \int \theta(p_F - |\mathbf{q}|)\frac{d^3q}{(2\pi)^3} \\
-\delta_{ss'} \int e^{-i\mathbf{p}\cdot(\mathbf{r}-\mathbf{r}')}\, \theta(p_F - |\mathbf{p}|)\frac{d^3p}{(2\pi)^3} &\int e^{-i\mathbf{q}\cdot(\mathbf{r}-\mathbf{r}')}\, \theta(p_F - |\mathbf{q}|)\frac{d^3q}{(2\pi)^3} .
\end{aligned}
\tag{17.147}
$$

By carrying out the integrations over $\mathbf{p}$ and $\mathbf{q}$ we find $C_{ss'}$ in terms of $\rho_s(\mathbf{r}, \mathbf{r}')$, Eq. (17.134);

$$
\left(\frac{n}{2}\right)^2 \mathcal{C}_{ss'}(|\mathbf{r}-\mathbf{r}'|) = \left(\frac{n}{2}\right)^2 \delta_{ss'} |\rho_s(\mathbf{r}, \mathbf{r}')|^2. \tag{17.148}
$$

If we combine (17.148) with the value of $\mathcal{C}_{ss'}$ when $s \neq s'$, Eq. (17.136), we obtain the general form of $\mathcal{C}_{ss'}$;

$$
\mathcal{C}_{ss'}(|\mathbf{r}-\mathbf{r}'|) = 1 - \delta_{ss'} \left(\frac{3j_1(\zeta)}{\zeta}\right)^2 , \tag{17.149}
$$

where $\zeta = p_F |\mathbf{r}-\mathbf{r}'|$.

As this equation shows $\mathcal{C}_{ss'}(|\mathbf{r}-\mathbf{r}'|)$ oscillates between zero and one and asymptotically approaches unity as $|\mathbf{r}-\mathbf{r}'| \to \infty$, i.e. Eq. (17.137) is satisfied. $\mathcal{C}_{ss'}$ also goes to zero as $|\mathbf{r}-\mathbf{r}'| \to 0$. The latter result shows that for relative distances shorter than the inverse Fermi momentum, the probability of finding two fermions with parallel spins is reduced since we have

$$
\mathcal{C}_{ss'} \le 1; \quad \text{for} \quad 0 \le |\mathbf{r}-\mathbf{r}'| \le \frac{\pi}{4p_F}. \tag{17.150}
$$

A plot of $\mathcal{C}_{ss}(\zeta)$ versus *zeta* is shown in Fig. 17.2. The part of this graph between $\zeta = 3$ and $\zeta = 7$ is oscillatory, but the oscillations have small amplitudes.

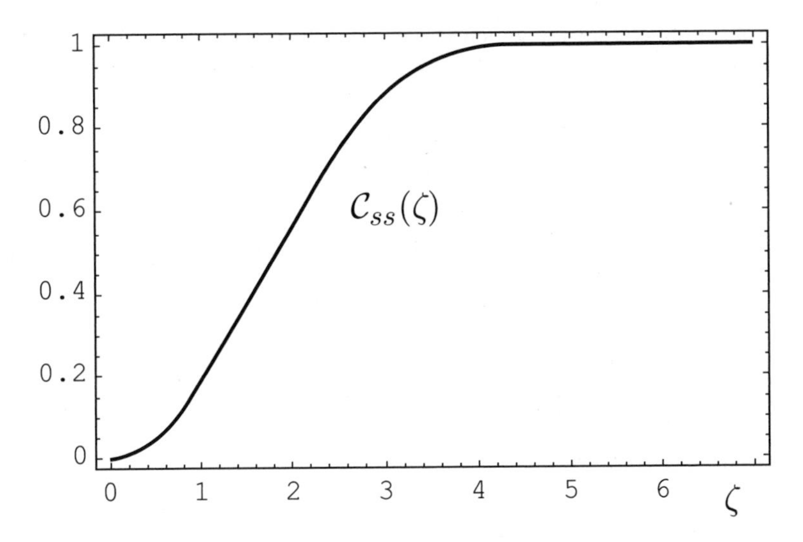

Figure 17.2: Correlation function $\mathcal{C}_{ss}(\zeta)$ for two identical fermions plotted as a function of dimensionless number $\zeta = k_F\,|\mathbf{r} - \mathbf{r}'|$, $\mathbf{r} - \mathbf{r}'$ being their relative distance.

17.7 Uncertainty Relations for a Many-Fermion System

For a system of noninteracting fermions we found that the Fermi momentum is related to the number of particles per unit volume, Eq. (17.125), or

$$p_F^3 V = 3\pi^2 N. \tag{17.151}$$

If we denote the average interparticle spacing by $\bar{d}$, then $V = \frac{4\pi}{3}\bar{d}^3$, and thus we get

$$p_F\bar{d} = \left(\frac{9\pi}{4}\right)^{\frac{1}{3}} \approx 1.919 N^{\frac{1}{3}}\hbar, \tag{17.152}$$

where we have written the Planck constant in the equation. The average momentum for the system can be found from

$$\bar{p^2} = \frac{\int_0^{p_F} p^2 d^3p}{\int_0^{p_F} d^3p} = \frac{3}{5}p_F^2. \tag{17.153}$$

From Eqs. (17.152) and (17.153) we have

$$\bar{p}\bar{d} = \sqrt{\frac{5}{3}}\left(\frac{9\pi}{4}\right)^{\frac{1}{3}} \approx 2.4775 N^{\frac{1}{3}}\hbar. \tag{17.154}$$

This is a special form of the uncertainty relation for the noninteracting fermions.

Now let us consider the same problem for a system of interacting fermions.

A many fermion system for which the exact solution is known for any number of identical particles each of unit mass is described by the Hamiltonian

$$H = \sum_{j=1}^{N} \frac{1}{2}\mathbf{p}_j^2 + \sum_{j}\sum_{i<j} \frac{1}{2}\omega^2 (\mathbf{r}_i - \mathbf{r}_j)^2 . \tag{17.155}$$

Let us denote the total momentum by $\mathbf{P} = \sum_i^N \mathbf{p}_i$ and the position of the center of mass by $\mathbf{R} = \frac{1}{N}\sum_{i=1}^N \mathbf{r}_i$, then using the identity

$$\sum_{j=1}^{N}\sum_{i<j}^{N} (\mathbf{r}_i - \mathbf{r}_j)^2 = N\sum_{i=1}^{N} \mathbf{r}_i^2 - N^2\mathbf{R}^2, \tag{17.156}$$

we can write the Hamiltonian of the system as

$$H = H_0 + \frac{1}{2N}\mathbf{P}^2, \tag{17.157}$$

where

$$H_0 = \sum_{i=1}^{N} \left(\frac{1}{2}\mathbf{p}_i^2 + \frac{1}{2}N\omega^2\mathbf{r}_i^2\right) - \left(\frac{1}{2N}\mathbf{P}^2 + \frac{1}{2}N^2\omega^2\mathbf{R}^2\right). \tag{17.158}$$

This relation shows that we have a system of uncoupled oscillators moving in a common central field. As is well-known in the theory of nuclear shell model there is a difficulty related to the motion of the center of the potential well. We note that the potential is an average over the motion of the particles and as such its center has no fundamental physical significance, but coincides with the center of mass. If we fix the center of the potential to be at the origin, then the center of mass oscillates about the origin and there is energy associated with this motion, but this energy is fictitious since in reality it is the center of mass that is fixed [15].

The Hamiltonian H_0, Eq. (17.158), consists of two commuting parts, and this Hamiltonian can be diagonalized without any difficulty with the result that

$$E(n_k) = \omega\sqrt{N}\left[\sum_{k=1}^{N}\left(n_k + \frac{1}{2}\mathcal{D}\right) - \left(\mathcal{N} + \frac{1}{2}\mathcal{D}\right)\right], \quad \hbar = 1. \tag{17.159}$$

In this relation $\mathcal{D}$ denotes the dimension of space, and the quantum number $\mathcal{N}$ is fixed by the choice of the quantum number n_k. For a system of interacting bosons, the ground state corresponds to the state where all n_k s are zero and $\mathcal{N}$ is also zero and therefore the energy of the ground state for a $\mathcal{D}$ dimensional boson system is

$$E_G^{\mathcal{D}}(bosons) = \frac{1}{2}\mathcal{D}\omega\sqrt{N}(N-1). \tag{17.160}$$

In the case of a system of interacting fermions in one dimension the ground state can be found by setting $n_k = k - 1$, with $k = 1,\ 2\cdots N$ and with $\mathcal{N} = 0$. Thus we have

$$E_G^{\mathcal{D}}(fermions) = \frac{1}{2}\omega\sqrt{N}\left(N^2 - 1\right), \quad (\mathcal{D} = 1). \tag{17.161}$$

For a three-dimensional system of fermions we have to account for the accidental degeneracy of the one-particle state. In a three-dimensional harmonic oscillator the total degeneracy at the level n is $\frac{1}{2}(n+1)(n+2)$. Thus the total number of particles filling up the levels up to K is

$$\sum_{n=0}^{K} \frac{1}{2}(n+1)(n+2) = \frac{(K+1)(K+2)(K+2)}{6}. \tag{17.162}$$

Now if the level K is filled but $K+1$ is not (K is the last one-particle oscillator level to be filled), then the total number of particles N lies between the two integers

$$\frac{1}{6}(K+1)(K+2)(K+3) \leq N < \frac{1}{6}(K+2)(K+3)(K+4). \tag{17.163}$$

The ground state energy for this three-dimensional system is

$$\begin{aligned} & E_G^{\mathcal{D}=3}(fermions) \\ = \ & \omega\sqrt{N}\left[\left(K + \frac{5}{2}\right)N - \frac{1}{24}(K+1)(K+2)(K+3)(K+4) - \frac{3}{2}\right]. \end{aligned} \tag{17.164}$$

In the limit of large N from Eq. (17.163) it follows that $K \approx (6N)^{\frac{1}{3}}$, and in this limit $E_G^{\mathcal{D}=3}$ becomes

$$E_G^{\mathcal{D}=3}(fermions) \geq \alpha\omega N^{\frac{4}{3}}\left(N - \frac{1}{2}\right)^{\frac{1}{2}}, \tag{17.165}$$

where

$$\alpha = \left(\frac{3^4}{2^5}\right)^{\frac{1}{3}} \approx 1.36. \tag{17.166}$$

Having found the energy of this N fermion system, we consider two useful quantities: the mean squared momentum $\bar{p}$ which is defined by

$$\bar{p}^2 = \frac{1}{N}\left\langle \sum_i \mathbf{p}_i^2 \right\rangle, \tag{17.167}$$

and the mean two particle distance $\bar{d}$ defined by

$$\frac{1}{2}N(N-1)\bar{d}^2 = \left\langle \sum_j \sum_{i<j} (\mathbf{r}_i - \mathbf{r}_j)^2 \right\rangle. \tag{17.168}$$

Now according Raleigh-Ritz variational principle we have

$$E_G^{\mathcal{D}}(N) \le \langle H \rangle = \frac{1}{2} N \bar{p}^2 + \frac{1}{4}\omega^2 N(N-1)\bar{d}^2, \tag{17.169}$$

where $E_G^{\mathcal{D}}(N)$ is the ground state energy. For either $\mathcal{D} = 1$ or $\mathcal{D} = 3$, $E_G^{\mathcal{D}}(N)$ is a linear function of ω, therefore the inequality (17.169) is a quadratic function of ω. In order to preserve these inequalies the quadratic form should have no real roots, i.e. the discriminant must be negative. Thus for $\mathcal{D} = 1$ we have

$$\Delta = N\left(N^2-1\right)^2 - 2N^2(N-1)\left(\bar{d}\bar{p}\right)^2 \le 0, \tag{17.170}$$

and for $\mathcal{D} = 3$ we get

$$\Delta = \alpha^2 N^{\frac{8}{3}} N - \frac{1}{2} N^2 (N-1)\left(\bar{d}\bar{p}\right)^2 \le 0. \tag{17.171}$$

From these inequalities we deduce that

$$\bar{p}\bar{d} \ge \frac{1}{\sqrt{2}} N\hbar, \quad \mathcal{D} = 1, \tag{17.172}$$

and

$$\bar{p}\bar{d} \ge \sqrt{2}\alpha N^{\frac{1}{3}}\hbar, \quad \mathcal{D} = 3. \tag{17.173}$$

Here we have explicitly written the Planck's constant $\hbar$. The first inequality shows that a system of N one-dimensional fermions the average length is larger that N times the average de Broglie length $\frac{\hbar}{\bar{p}}$. The second equality shows that one cannot put more than one fermion in a phase space cell of the volume of the order $\hbar^3$. This result supports the standard interpretation of the Pauli exclusion principle [14].

17.8 Pair Correlation Function for Noninteracting Bosons

Consider an ideal system of bosons, i.e. a collection of identical particles with spin zero or integer spin enclosed in a cubic box of volume L^3 satisfying the periodic boundary conditions Eq. (17.110). For this system we define the correlation function $\mathcal{C}\left(|\mathbf{r}-\mathbf{r}|\right)$ exactly the same way as we did for fermions, Eq. (17.141). We denote the state of the system by $|\Phi\rangle$, where

$$|\Phi\rangle = |n_{\mathbf{p}_0},\, n_{\mathbf{p}_1},\, n_{\mathbf{p}_2}\cdots\rangle, \tag{17.174}$$

and observe that the density of this state is given by

$$\left\langle \Phi|\psi^\dagger(\mathbf{r})\psi(\mathbf{r})|\Phi\right\rangle = \frac{1}{L^3}\sum_{\mathbf{p}} n_{\mathbf{p}} = n. \tag{17.175}$$

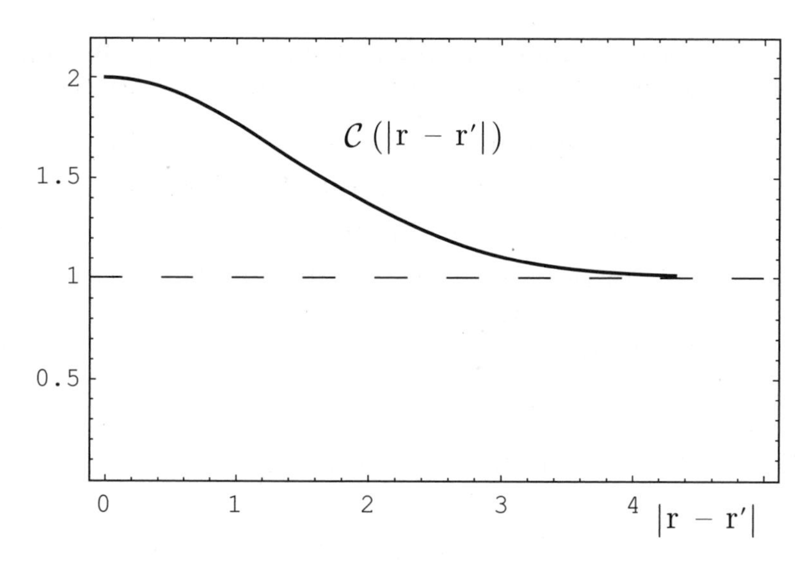

Figure 17.3: Correlation function $\mathcal{C}\left(|\mathbf{r}-\mathbf{r}'|\right)$ for two identical bosons as a function of their separation $|\mathbf{r}-\mathbf{r}'|$.

Writing $\mathcal{C}\left(|\mathbf{r}-\mathbf{r}|\right)$ in terms of boson operators $\psi^\dagger(\mathbf{r})$ and $\psi(\mathbf{r})$, we have

$$\left(\frac{n}{2}\right)^2 \mathcal{C}\left(|\mathbf{r}-\mathbf{r}|\right) = \left\langle\Phi|\psi^\dagger(\mathbf{r})\psi^\dagger\left(\mathbf{r}'\right)\psi\left(\mathbf{r}'\right)\psi\left(\mathbf{r}\right)|\Phi\right\rangle, \tag{17.176}$$

where $\psi^\dagger(\mathbf{r})$ and $\psi^\dagger(\mathbf{r})$ are defined by relations similar to those of $\psi_s^\dagger(\mathbf{r})$ and $\psi_s^\dagger(\mathbf{r})$, Eqs. (17.111) and (17.112);

$$\psi^\dagger\left(\mathbf{r}\right) = \frac{1}{L^{\frac{3}{2}}}\sum_{\mathbf{p}} e^{-i\mathbf{p}\cdot\mathbf{r}} a^\dagger(\mathbf{p}), \tag{17.177}$$

and

$$\psi\left(\mathbf{r}\right) = \frac{1}{L^{\frac{3}{2}}}\sum_{\mathbf{p}} e^{i\mathbf{p}\cdot\mathbf{r}} a(\mathbf{p}). \tag{17.178}$$

Here the creation and annihilation operators satisfy the commutation relation

$$\left[a^\dagger(\mathbf{p}),\ a^\dagger\left(\mathbf{p}'\right)\right] = \left[a(\mathbf{p}),\ a\left(\mathbf{p}'\right)\right] = 0, \tag{17.179}$$

and

$$\left[a^\dagger(\mathbf{p}),\ a\left(\mathbf{p}'\right)\right] = (2\pi)^3\delta\left(\mathbf{p}-\mathbf{p}'\right). \tag{17.180}$$

By substituting (17.179) and (17.180) in (17.176) we find the expectation value of the products of four operators.

Let us consider a typical term like $\left\langle\Phi\left|a^\dagger(\mathbf{p})a^\dagger(\mathbf{q})a\left(\mathbf{q}'\right)a\left(\mathbf{p}'\right)\right|\Phi\right\rangle$. This term does not vanish only if $\mathbf{p} = \mathbf{p}'$, $\mathbf{q} = \mathbf{q}'$ or if $\mathbf{p} = \mathbf{q}'$, $\mathbf{q} = \mathbf{p}'$. But if $\mathbf{p} = \mathbf{q}'$ these are not distinct cases. Now we write $\mathbf{p}'$ and $\mathbf{q}'$ in terms of $\mathbf{p}$ and $\mathbf{q}$ to get

$$\begin{aligned}
& \left\langle \Phi \left| a^\dagger(\mathbf{p}) a^\dagger(\mathbf{q}) a(\mathbf{q}') \, a(\mathbf{p}') \right| \Phi \right\rangle \\
= & \ (1 - \delta_{\mathbf{pq}}) \left\{ \delta_{\mathbf{pp}'} \delta_{\mathbf{qq}'} \left\langle \Phi \left| a^\dagger(\mathbf{p}) a^\dagger(\mathbf{q}) a(\mathbf{q}) \, a(\mathbf{p}) \right| \Phi \right\rangle \right. \\
+ & \left. \ \delta_{\mathbf{pp}'} \delta_{\mathbf{qq}'} \left\langle \Phi \left| a^\dagger(\mathbf{p}) a^\dagger(\mathbf{q}) a(\mathbf{p}) \, a(\mathbf{q}) \right| \Phi \right\rangle \right\} \\
+ & \ \delta_{\mathbf{pq}} \delta_{\mathbf{pp}'} \delta_{\mathbf{qq}'} \left\langle \Phi \left| a^\dagger(\mathbf{p}) a^\dagger(\mathbf{p}) a(\mathbf{p}') \, a(\mathbf{p}) \right| \Phi \right\rangle \\
= & \ (1 - \delta_{\mathbf{pq}}) \left(\delta_{\mathbf{pp}'} \delta_{\mathbf{qq}'} + \delta_{\mathbf{pq}'} \delta_{\mathbf{qp}'} \right) n_{\mathbf{p}} n_{\mathbf{q}} \\
+ & \ \delta_{\mathbf{pq}} \delta_{\mathbf{pp}'} \delta_{\mathbf{qq}'} n_{\mathbf{p}} (n_{\mathbf{p}} - 1).
\end{aligned} \tag{17.181}$$

Thus from (17.176)–(17.178) we obtain

$$\begin{aligned}
\left(\frac{n}{2}\right)^2 \mathcal{C}\left(|\mathbf{r} - \mathbf{r}|\right) &= \left\langle \Phi \right| \psi^\dagger(\mathbf{r}) \psi^\dagger(\mathbf{r}') \, \psi(\mathbf{r}') \, \psi(\mathbf{r}) \left| \Phi \right\rangle \\
&= n^2 + \left| \frac{1}{L^3} \sum_{\mathbf{p}} n_{\mathbf{p}} \, e^{-i\mathbf{p}\cdot(\mathbf{r}-\mathbf{r}')} \right|^2 - \frac{1}{L^6} \sum_{\mathbf{p}} n_{\mathbf{p}} (n_{\mathbf{p}} + 1).
\end{aligned} \tag{17.182}$$

If we compare this pair correlation function with the corresponding correlation function for fermions, Eq. (17.149), we find that they differ from each other in two ways:

(a) - The sign of the second term for fermions is negative, indicating an effective repulsion from the exchange symmetry of fermions, whereas in the case of bosons, as Eq. (17.182) shows the exchange symmetry is positive and attractive.

(b) - There is an additional term in (17.182) for the boson pair correlation function, but this term is smaller than the other two by a factor of $\frac{1}{L^3}$.

In the limit of $L \to \infty$ we can write this equation for $\mathcal{C}$ as

$$\left(\frac{n}{2}\right)^2 \mathcal{C}\left(|\mathbf{r} - \mathbf{r}|\right) = n^2 + \left| \frac{1}{(2\pi)^3} \int n(\mathbf{p}) \, e^{-i\mathbf{p}\cdot(\mathbf{r}-\mathbf{r}')} \right|^2 d^3p. \tag{17.183}$$

Now let us calculate the pair correlation function for a beam of noninteracting bosons, e.g. a photon beam. We choose $n(\mathbf{p})$ to be a smoothly varying function of $\mathbf{p}$. For instance a Gaussian function

$$n(\mathbf{p}) = \alpha e^{-\frac{\beta}{2}(\mathbf{p}-\mathbf{p}_0)^2}, \tag{17.184}$$

represents a beam of particles of momentum centered about $\mathbf{p}_0$. By substituting $n(\mathbf{p})$ in (17.183) we find

$$\left(\frac{n}{2}\right)^2 \mathcal{C}\left(|\mathbf{r} - \mathbf{r}|\right) = n^2 \left(1 + e^{-\frac{1}{\beta}|\mathbf{r}-\mathbf{r}'|^2} \right), \tag{17.185}$$

where in the parenthesis the second term shows the effect of exchange. This result shows that the probability for two bosons to be close together is increased due to the exchange effect (see Fig. 17.3), and that this probability for two bosons to be next to each other is twice the value of the probability for the two to be far apart [16],[17] .

17.9 Bogoliubov Transformation for a Many-Boson System

The many boson system described by the second quantized Hamiltonian (17.96) is solvable only for a very few special cases. For instance for a Hamiltonian containing only pairing-forces of the form

$$H_R = \sum_{\mathbf{p}} \frac{\mathbf{p}^2}{2m} a_{\mathbf{p}}^\dagger a_{\mathbf{p}} + \frac{\lambda}{2} \sum_{\mathbf{pp'}} \theta(p)\theta(p')\, a_{\mathbf{p}}^\dagger a_{-\mathbf{p}}^\dagger a_{-\mathbf{p'}} a_{\mathbf{p'}}, \tag{17.186}$$

where $\theta(p)$ is the step function

$$\theta(p) = \begin{cases} 1 & \text{for } 0 \le p \le P \\ 0 & \text{for } p > P \end{cases}, \tag{17.187}$$

can be diagonalized exactly [18].

A simpler model, that of Bassichis and Foldy which is described by the Hamiltonian [19]

$$\begin{aligned} H_{BF} &= a_+^\dagger a_+ + a_-^\dagger a_- + g\left[a_0^\dagger a_0 \left(a_+^\dagger a_+ + a_-^\dagger a_-\right)\right. \\ &+ \left. a_0^{\dagger 2} a_+ a_- + a_0^2 a_+^\dagger a_-^\dagger\right] + F g a_+^\dagger a_-^\dagger a_+ a_-, \end{aligned} \tag{17.188}$$

can be obtained from (17.96) for N particles occupying $+$, $-$ and 0 momentum states [19]. The exact solution for this model will be considered later.

An approximate way of obtaining the energy eigenvalues of this Hamiltonian is by the method of Bogoliubov transformation. This method is of considerable interest in the theory of superfluidity of ^{4}He. The helium four atoms are bosons with zero spin and at very low temperature they exhibit the properties of a superfluid. The interaction between two ^{4}He atoms i and j, $V(i,j)$, is weak

$$V(i,j) = V(|\mathbf{r}_i - \mathbf{r}_j|), \tag{17.189}$$

and this weakness of the potential is essential in developing the following approximate method for the solution of the problem. For the complete set of states we choose plane waves $L^{-3/2}e^{i\mathbf{p}\cdot\mathbf{r}}$ with periodic boundary conditions with

$$p_x = \frac{2n_x\pi}{L}, \quad p_y = \frac{2n_y\pi}{L}, \quad p_z = \frac{2n_z\pi}{L}, \tag{17.190}$$

where each n_x, n_y and n_z can take positive or negative integers or zero.

In order to evaluate the potential matrix element in (17.96) using plane waves we write

$$\langle k\ell|V|mn\rangle = \frac{1}{L^6}\int\int e^{i\mathbf{p}_k\cdot\mathbf{r}}e^{i\mathbf{p}_\ell\cdot\mathbf{r}'}V\left(|\mathbf{r}-\mathbf{r}'|\right)e^{-i\mathbf{p}_m\cdot\mathbf{r}}e^{-i\mathbf{p}_n\cdot\mathbf{r}'}d^3rd^3r'. \quad (17.191)$$

By changing the variables in (17.191) from $\mathbf{r}$ and $\mathbf{r}'$ to $\boldsymbol{\rho} = \mathbf{r}-\mathbf{r}'$ and $\mathbf{R} = \frac{1}{2}(\mathbf{r}+\mathbf{r}')$ and carrying out the integration over $\mathbf{R}$ we find

$$\langle k\ell|V|mn\rangle = \frac{1}{L^3}\delta_{\mathbf{p}_k+\mathbf{p}_\ell-\mathbf{p}_m-\mathbf{p}_n}\tilde{V}(|\mathbf{p}_\ell-\mathbf{p}_n|), \quad (17.192)$$

where $\tilde{V}(|\mathbf{p}_\ell-\mathbf{p}_n|)$ is the Fourier transform of $V(\rho)$

$$\begin{aligned}\tilde{V}(|\mathbf{p}_\ell-\mathbf{p}_n|) &= \int V(\boldsymbol{\rho})e^{i(\mathbf{p}_\ell-\mathbf{p}_n)\cdot\boldsymbol{\rho}}d^3\rho \\ &= \frac{4\pi}{|\mathbf{p}_\ell-\mathbf{p}_n|}\int V(\rho)\rho\,\sin(|\mathbf{p}_\ell-\mathbf{p}_n|\rho)d\rho. \end{aligned} \quad (17.193)$$

The sum in (17.96) over $\mathbf{p}_k, \mathbf{p}_\ell, \mathbf{p}_m$ and $\mathbf{p}_n$ shows that the total momentum of the two particles is conserved, as indicated by the Kronecker δ in (17.192) and that the annihilation of the pair of bosons with momenta $\mathbf{p}_m$ and $\mathbf{p}_n$ is accompanied by the creation of a pair with momenta $\mathbf{p}_k$ and $\mathbf{p}_\ell$. For the first term in (17.96) we note that using plane waves we have

$$\sum_{\mathbf{p}_k,\mathbf{p}_\ell}\langle k|T|\ell\rangle a_k^\dagger a_k \to \sum_k \frac{\mathbf{p}_k^2}{2m}a_k^\dagger a_k. \quad (17.194)$$

Thus we can write (17.96) as

$$\hat{H} = \sum_k \frac{\mathbf{p}_k^2}{2m}a_k^\dagger a_k + \frac{1}{2L^3}\left(\sum_{\mathbf{p}_k+\mathbf{p}_\ell=\mathbf{p}_m+\mathbf{p}_n}\tilde{V}(|\mathbf{p}_\ell-\mathbf{p}_n|)a_k^\dagger a_\ell^\dagger a_m a_n\right). \quad (17.195)$$

The Hamiltonian $\hat{H}$ commutes with the number operator $\hat{N} = \sum_k a_k^\dagger a_k$, a result which can be verified using the commutation relations (17.73). When $V = 0$, the ground state of the system corresponds to the condensation of all particles in the state of lowest energy, i.e. $\mathbf{p} = 0$. But when V is a weakly repulsive force as in the case of liquid helium, ^{4}He, most of the particles will be in the lowest state. If the total number of particles in the system is N, then the number of atoms in the zero momentum state n_0 is very close to N.

If we denote the eigenvalue of the operators $a_0^\dagger a_0$ by n_0, then the eigenvalue of $a_0 a_0^\dagger$ is given by n_0+1. Now let us arrange the contributions to the potential energy in (17.195) when most of the particles are in the ground state. Expanding the sum in (17.195) in terms of the number of factors of a_0 and $a_0^\dagger$ we have typical terms like

$$
\begin{aligned}
& E_0 = a_0^\dagger a_0^\dagger a_0 a_0 \\
& V_1 = a_0^\dagger a_0^\dagger a_{\mathbf{k}} a_{-\mathbf{k}} \quad V_2 = a_{\mathbf{k}}^\dagger a_{-\mathbf{k}}^\dagger a_0 a_0 \quad V_3 = a_0^\dagger a_{\mathbf{k}}^\dagger a_{\mathbf{k}} a_0 \quad V_4 = a_{\mathbf{k}}^\dagger a_0^\dagger a_{\mathbf{k}} a_0 \\
& V_5 = a_{\mathbf{k}}^\dagger a_{\mathbf{j}}^\dagger a_{\mathbf{k}+\mathbf{j}} a_0 \quad V_6 = a_0^\dagger a_{\mathbf{k}+\mathbf{j}}^\dagger a_{\mathbf{k}} a_{\mathbf{j}} \\
& V_7 = a_{\mathbf{k}}^\dagger a_{\mathbf{j}}^\dagger a_{\mathbf{n}} a_{\mathbf{k}+\mathbf{j}-\mathbf{n}},
\end{aligned}
\tag{17.196}
$$

where $\mathbf{k}$, $\mathbf{j}$, and $\mathbf{n}$ denote the momenta $\mathbf{p}_k$, $\mathbf{p}_j$ and $\mathbf{p}_n$ all different from zero. The first term E_0 is proportional to n_0^2, and the other contributions from V_1 to V_4 are all of the order n_0. Of the remaining terms V_5 and V_6 are of the order $\sqrt{n_0}$. Now in the limit of large L^3, the contributions from $a_0^\dagger a_0$ and $a_0 a_0^\dagger$ to the potential energy will be proportional to $\frac{n_0}{L^3}$ and $\frac{n_0+1}{L^3}$ respectively. Thus

$$
\frac{a_0 a_0^\dagger}{L^3} - \frac{a_0^\dagger a_0}{L^3} \approx \left(\frac{1}{L^3}\right)_{L\to\infty} \to 0. \tag{17.197}
$$

This result shows that in the limit of large L, a_0 and $a_0^\dagger$ commute with each other and thus can be regarded as c-numbers.

Following Bogoliubov we introduce a new set of operators $A_{\mathbf{k}}$ and $A_{\mathbf{k}}^\dagger$ by the following relations [20],[21]

$$
A_{\mathbf{k}} = a_0^\dagger \frac{1}{\sqrt{n_0}} a_{\mathbf{k}}, \quad A_{\mathbf{k}}^\dagger = a_{\mathbf{k}}^\dagger \frac{1}{\sqrt{n_0}} a_0, \tag{17.198}
$$

and write the Hamiltonian $\hat{H}$ in terms of $A_{\mathbf{k}}$ and $A_{\mathbf{k}}^\dagger$;

$$
\begin{aligned}
\hat{H} &= \frac{N^2}{2L^3}\tilde{V}(0) + \sum_{\mathbf{k}\neq 0} \frac{\hbar^2 k^2}{2m} A_{\mathbf{k}}^\dagger A_{\mathbf{k}} \\
&+ \frac{n_0}{2L^3} \sum_{\mathbf{k}\neq 0} \tilde{V}(\mathbf{k}) \left[A_{\mathbf{k}}^\dagger A_{-\mathbf{k}}^\dagger + A_{\mathbf{k}} A_{-\mathbf{k}} + 2A_{\mathbf{k}}^\dagger A_{\mathbf{k}}\right] + \hat{H}',
\end{aligned}
\tag{17.199}
$$

where as we have indicated the sum is over states with nonzero momenta, and the Fourier transform of the potential $\tilde{V}(\mathbf{k})$ is an even function of $\mathbf{k}$. From the definitions of $A_{\mathbf{k}}^\dagger$ and $A_{\mathbf{k}}$ it follows that

$$
\sum_{\mathbf{k}\neq 0} A_{\mathbf{k}}^\dagger A_{\mathbf{k}} = \sum_{\mathbf{k}\neq 0} a_{\mathbf{k}}^\dagger a_{\mathbf{k}} = N - n_0, \tag{17.200}
$$

and thus the operators $A_{\mathbf{k}}^\dagger$ and $A_{\mathbf{k}}$ are small (of the order of N^{-1} [24]). The operator $\hat{H}'$ which includes the products of three or four $A_{\mathbf{k}}^\dagger$ and $A_{\mathbf{k}}$ operators can be neglected compared with the other terms in (17.199). Using this approximation we can write $\hat{H}$ as

$$
\hat{H} = \frac{N^2 \tilde{V}(0)}{2L^3} + \frac{1}{2}\sum_{\mathbf{k}\neq 0} \hat{H}_{\mathbf{k}}, \tag{17.201}
$$

where

$$\hat{H}_{\mathbf{k}} = \left(\frac{\hbar^2\mathbf{k}^2}{m} + \frac{V(\mathbf{k})n_0}{L^3}\right) A^\dagger_{\mathbf{k}} A_{\mathbf{k}} + \frac{\tilde{V}(\mathbf{k})n_0}{L^3}\left(A^\dagger_{\mathbf{k}} A^\dagger_{-\mathbf{k}} + A_{\mathbf{k}} A_{-\mathbf{k}}\right). \quad (17.202)$$

We observe that $\hat{H}_{\mathbf{k}}$ is quadratic in $A^\dagger_{\mathbf{k}}$, $A_{\mathbf{k}}$, $A^\dagger_{-\mathbf{k}}$ and $A_{-\mathbf{k}}$, and this allows us to diagonalize this Hamiltonian with the help of the canonical transformation

$$\alpha_1(\mathbf{k}) = A_{\mathbf{k}} \cosh\phi(\mathbf{k}) + A^\dagger_{-\mathbf{k}} \sinh\phi(\mathbf{k}), \quad (17.203)$$

$$\alpha_2(\mathbf{k}) = A^\dagger_{\mathbf{k}} \sinh\phi(\mathbf{k}) + A_{-\mathbf{k}} \cosh\phi(\mathbf{k}). \quad (17.204)$$

The coefficients of the transformations (17.203) and (17.204) are real and are spherically symmetric. That these transformations are canonical can be verified by calculating the commutators:

$$\left[\alpha_1^\dagger(\mathbf{k}),\ \alpha_1(\mathbf{k}')\right] = \delta_{\mathbf{k},\mathbf{k}'}, \quad (17.205)$$

and

$$[\alpha_1(\mathbf{k}),\ \alpha_1(\mathbf{k}')] = \left[\alpha_1^\dagger(\mathbf{k}),\ \alpha_1^\dagger(\mathbf{k}')\right] = 0. \quad (17.206)$$

From Eqs (17.203) and (17.204) it follows that for $\phi(\mathbf{k})$ to be an even function of $\mathbf{k}$ then $\alpha_2(\mathbf{k}) = \alpha_1(\mathbf{k})$. Substituting from (17.203) and (17.204) in (17.202) and setting the coefficients of $\alpha_2(\mathbf{k})\alpha_1(\mathbf{k})$ and $\alpha_2^\dagger(\mathbf{k})\alpha_1^\dagger(\mathbf{k})$ equal to zero we obtain

$$\sinh\phi(\mathbf{k}) = \frac{D(\mathbf{k})}{[1 - D^2(\mathbf{k})]^{\frac{1}{2}}}, \quad \cosh\phi(\mathbf{k}) = \frac{1}{[1 - D^2(\mathbf{k})]^{\frac{1}{2}}}, \quad (17.207)$$

where

$$D(\mathbf{k}) = \frac{L^3}{n_0\tilde{V}(\mathbf{k})}\left[\frac{\hbar^2\mathbf{k}^2}{2m} + \frac{n_0\tilde{V}(\mathbf{k})}{L^3} - E(\mathbf{k})\right], \quad (17.208)$$

and $E(\mathbf{k})$ is given by the following relation

$$E(\mathbf{k}) = \left[\left(\frac{\hbar^2\mathbf{k}^2}{2m}\right)^2 + \frac{\hbar^2\mathbf{k}^2 n_0\tilde{V}(\mathbf{k})}{mL^3}\right]^{\frac{1}{2}}. \quad (17.209)$$

The diagonal form of $\hat{H}$ then becomes

$$\hat{H} = \frac{n^2\tilde{V}(0)}{2L^3} + \sum_{\mathbf{k}\neq 0} E_0(\mathbf{k}) + \sum_{\mathbf{k}\neq 0} E(\mathbf{k})\alpha_1^\dagger(\mathbf{k})\alpha_1(\mathbf{k}), \quad (17.210)$$

where

$$E_0(\mathbf{k}) = \frac{D^2(\mathbf{k})E(\mathbf{k})}{D^2(\mathbf{k}) - 1}. \quad (17.211)$$

Thus for low-lying excited states of helium atoms the Hamiltonian $\hat{H}$ represents a collection of noninteracting "quasi-particles" each with the energy $E(\mathbf{k})$ and this energy is given by

$$E(\mathbf{k}) = \left[\left(\frac{\hbar^2\mathbf{k}^2}{2m}\right)^2 + \frac{\hbar^2\mathbf{k}^2 N\tilde{V}(\mathbf{k})}{mL^3}\right]^{\frac{1}{2}}, \tag{17.212}$$

where we have replaced n_0 in (17.209) by N. When $|\mathbf{k}|$ is very small we can expand (17.212) in powers of $|\mathbf{k}|$ and keep the linear term in $|\mathbf{k}|$;

$$E(\mathbf{k}) = \left[\frac{N\tilde{V}(0)}{mL^3}\right]^{\frac{1}{2}} \hbar|\mathbf{k}| + \cdots. \tag{17.213}$$

We note that for the stability of the ground state energy, $E_0(\mathbf{k})$ must be real, therefore as (17.211) shows $E(\mathbf{k})$ is real too and this implies that

$$\tilde{V}(0) = \int V(\rho)d^3\rho > 0. \tag{17.214}$$

From this result we conclude that for the stability of the ground state at low $|\mathbf{k}|$ the interaction energy between ^{4}He atoms must be positive or the interaction force on the whole must be repulsive.

While (17.213) gives us the low momentum behavior of $E(\mathbf{k})$, for large $|\mathbf{k}|$ from (17.209) we get

$$E(\mathbf{k}) \approx \frac{\hbar^2\mathbf{k}^2}{2m} + \frac{N\tilde{V}(\mathbf{k})}{L^3}. \tag{17.215}$$

The velocity with which these quasi-particles at very low temperatures move is the group velocity which is given by

$$v_s = \frac{1}{\hbar}\left(\frac{\partial E(\mathbf{k})}{\partial k}\right)_{\mathbf{k}=0} = \left(\frac{N\tilde{V}(0)}{mL^3}\right)^{\frac{1}{2}}. \tag{17.216}$$

Thus whereas the quasi-particle energy at low momenta is linear in $\mathbf{k}$

$$E(\mathbf{k}) \approx v_s|\mathbf{k}|, \tag{17.217}$$

this energy becomes a quadratic function of $\mathbf{k}$ for large $\mathbf{k}$. In this limit the energy is just the kinetic energy of bosons. The dependence of $E(\mathbf{k})$ on the momentum $\hbar k$ for a repulsive force, $\tilde{V}(0) > 0$, is plotted in Fig. 17.4.

From the energy $E(\mathbf{k})$ we can define a critical velocity v_{cr} by the relation

$$\text{minimum of } \frac{E(\mathbf{k})}{\hbar k} \equiv v_{cr} > 0. \tag{17.218}$$

Now according to Landau's theory, the state of superfluidity can occur only if the quasi-particles are moving with velocities less that the critical velocity v_{cr}

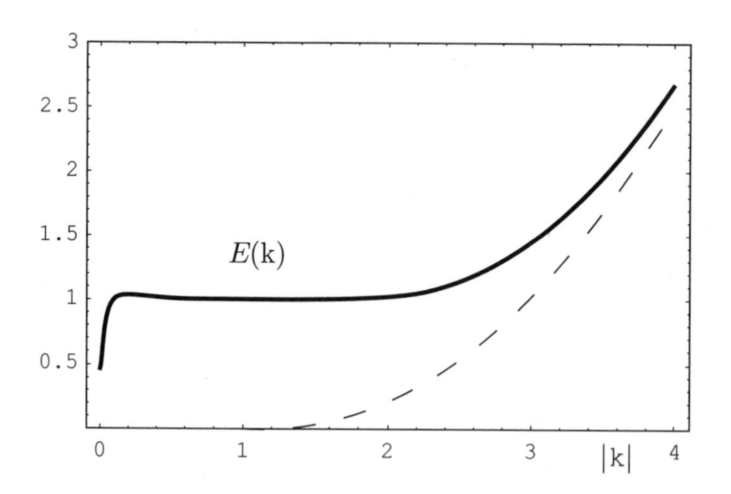

Figure 17.4: The energy spectrum of a quasi-particle in a superfluid. The dashed line shows the kinetic energy of the particles (e.g. ^{4}He).

which in this case is nonzero [22].

For the liquid helium at very low temperatures the critical speed (which is the speed of the first sound) is about $v_s = 238$ m/s. In addition to the linear dependence near $\mathbf{k} = 0$ as shown in Fig. 17.4, $E(\mathbf{k})$, has a minimum in the middle near k_0. The excitations near this dip are called rotons [12].

Accuracy of the Bogoliubov Approximation for Many-Boson Problem — To test the accuracy of the Bogoliubov transformation for bosons, Bassichis and Foldy have solved the model given by (17.188). In addition to the Hamiltonian $\hat{H}_{BF}$, this model possesses two other constants of motion

$$\hat{N} = a_+^\dagger a_+ + a_-^\dagger a_- + a_0^\dagger a_0, \tag{17.219}$$

and

$$\hat{\mathcal{D}} = a_+^\dagger a_+ - a_-^\dagger a_-. \tag{17.220}$$

The state of the system $|\Psi\rangle$ will depend on the constants N, $\mathcal{D}$, the eigenvalues of $\hat{N}$ and $\hat{\mathcal{D}}$ and also on the number of the particles n, which is the eigenvalue of $a_-^\dagger a_-$. For simplicity we set $\mathcal{D} = 0$, and for this special case we solve the eigenvalue equation for $\hat{H}_{BF}$

$$\hat{H}_{BF}|\Psi\rangle = E|\Psi\rangle. \tag{17.221}$$

Here $|\Psi\rangle$ depends on N, n and $\mathcal{D}$ and also on the coefficient of expansion C_n;

$$|\Psi\rangle = \sum_{n=0}^{\frac{1}{2}N} C_n|N, \mathcal{D} = 0, n\rangle, \tag{17.222}$$

i.e. the state $|\Psi\rangle$ depends on the total number of particles and the number of particles in $(+)$ or $(-)$ states n, and these are equal since we have chosen $\mathcal{D} = 0$.

By substituting (17.222) in (17.221) we find that C_n s are the roots of the linear difference equation

$$\begin{aligned} & g\left[(N-2n+2)(N-2n+1)n^2\right]^{\frac{1}{2}} C_{n-1} \\ + & \left[2n+g(N-2n)2n-Fgn^2-E\right] C_n \\ + & g(n+1)\left[(N-2n-1)(N-2n)\right]^{\frac{1}{2}} C_{n+1}=0. \end{aligned} \tag{17.223}$$

Our choice of $\mathcal{D}=0$ means that the maximum number of particles either in (+) or in (−) levels is $\frac{1}{2}N$. Therefore

$$C_{-1}=C_{\frac{1}{2}N+1}=0, \tag{17.224}$$

and in addition we have the normalization conditions

$$\sum_{n=0}^{\frac{1}{2}N} C_n^2=1. \tag{17.225}$$

All of the eigenvalues E_n are obtained by diagonalizing the tri-diagonal matrix formed from the coefficients of C_n s in (17.223).

Let us now apply the Bogoliubov approximation to this solvable model. For this we replace both $a_0^\dagger$ and a_0 in (17.188) by $\sqrt{n_0}$ and keep the terms proportional to n_0 and n;

$$\hat{H}_{BF} \approx a_+^\dagger a_+ + a_-^\dagger a_- + g n_0\left(a_+^\dagger a_+ + a_-^\dagger a_- + a_+ a_- + a_+^\dagger a_-^\dagger\right). \tag{17.226}$$

We note that in this reduced Hamiltonian the term $Fga_+^\dagger a_+ a_-^\dagger a_-$ does not appear and thus the approximation can be valid only for a certain range of values of F. Also as a result of this approximation N is no longer a constant of motion, but $\hat{\mathcal{D}}$ still is. Since (17.226) is quadratic in a s and $a^\dagger$ s, we can diagonalize it by the transformation

$$a_+ = \cosh\phi\,\alpha_+ - \sinh\phi\,\alpha_-^\dagger, \quad a_- = \cosh\phi\,\alpha_- - \sinh\phi\,\alpha_+^\dagger, \tag{17.227}$$

$$a_+^\dagger = \cosh\phi\,\alpha_+^\dagger - \sinh\phi\,\alpha_-, \quad a_-^\dagger = \cosh\phi\,\alpha_-^\dagger - \sinh\phi\,\alpha_+, \tag{17.228}$$

where

$$\cosh\phi = \left[\frac{1+gn_0+E_0}{2E_0}\right]^{\frac{1}{2}}, \quad \sinh\phi = \left[\frac{1+gn_0-E_0}{2E_0}\right]^{\frac{1}{2}}, \tag{17.229}$$

$$E_0=(1+2gn_0)^{\frac{1}{2}}, \tag{17.230}$$

and

$$\cosh\phi\sinh\phi = \frac{gn_0}{2E_0}. \tag{17.231}$$

With these substitution $\hat{H}_{BF}$ becomes

$$\hat{H}_{BF} = E_0 - 1 - gn_0 + E_0\left(\alpha_+^\dagger \alpha_+ + \alpha_-^\dagger \alpha_-\right). \tag{17.232}$$

From (17.232) it is clear that the ground state energy of this Hamiltonian is

$$E_G = E_0 - 1 - gn_0 = (1 + 2gn_0)^{\frac{1}{2}} - 1 - gn_0. \tag{17.233}$$

In order to determine n_0, the number of particles in the ground state, we note that

$$\begin{aligned} \left\langle \Psi \left| a_+^\dagger a_+ + a_-^\dagger a_- + a_0^\dagger a_0 \right| \Psi \right\rangle &= n_0 + \left\langle \Psi \left| a_+^\dagger a_+ + a_-^\dagger a_- \right| \Psi \right\rangle \\ &= n_0 + 2\sinh^2\phi = N. \end{aligned} \tag{17.234}$$

By substituting for $\sinh\phi$ from (17.229) in (17.234) we find that n_0 is related to N by

$$N = n_0 - 1 + \frac{1 + gn_0}{\sqrt{1 + 2gn_0}}. \tag{17.235}$$

In TABLES XVI and XVII we have compared the results obtained from the solution of the exact eigenvalue equation with those found from the Bogoliubov approximation for a range of g and N values [23]. The probability amplitudes for finding the particles in the ground and in the first excited states, $|C_0|$ and $|C_1|$ are also given in the following two tables:

TABLE XVI: The energies of the ground and the first excited states given as a function of N ($F = 0,\ g = 0.01$).

N	E_0(exact)	$\lvert C_0\rvert$	E_0(Bog)	E_1(exact)	$\lvert C_1\rvert$	E_1(Bog)
20	−0.0162	0.9966	−0.0168	2.32	0.083	2.35
100	−0.2668	0.9634	−0.268	3.19	0.258	3.20
200	−0.7625	0.9242	−0.764	3.70	0.353	3.71

TABLE XVII: The energies of the ground and the first excited states calculated as a function of the coupling constant g ($F = 0$, $N = 64$).

g	E_0(exact)	$\lvert C_0\rvert$	E_0(Bog)	E_1(exact)	$\lvert C_1\rvert$	E_1(Bog)
0.01	−0.1291	0.9792	−0.1300	2.875	0.199	2.89
0.1	−3.653	0.8195	−3.685	3.807	0.467	3.745
1	−51.923	0.5828	−53.642	−25.94	0.457	30.93
5	−278.09	0.4928	−295.68	−193.45	0.402	−245.05

The results shown in these two tables indicate that the Bogoliubov approximation works well for a range of N and g values particularly for larger N and smaller g. Not only the ground state is accurately determined by this approximation, but for the first excited state we have also acceptable results provided that $g \ll 1$. When the coupling Fg is not zero the approximation breaks down as F approaches F_T, where F_T is a number which depends on N and g. For instance when $N = 64$ and $g = 0.01$, then for $F = 6.25634, C_0 = 0.9781$ whereas for $F = 6.25635$, C_0 becomes very small, $C_0 \sim 10^{-6}$.

17.10 Scattering of Two Quasi-Particles

In the previous sections of this chapter we mentioned that an exact solution of a many-boson or a many-fermion for realistic type of forces cannot be found. But as a first approximation the interacting many-body system may be replaced by a system of non-interacting quasi-particles.

For the many-boson systems the quasi-particles have one attribute, that is, their energy $E(\mathbf{k})$. We write the Hamiltonian in the second quantized form as

$$H = \sum_{\mathbf{k}} (\varepsilon(\mathbf{k}) - \mu) a_{\mathbf{k}}^\dagger a_{\mathbf{k}} + \frac{1}{2L^3} \sum_{\mathbf{jlmn}} \delta_{\mathbf{j+l,m+n}} \tilde{V}(\mathbf{j} - \mathbf{m}) a_{\mathbf{j}}^\dagger a_{\mathbf{l}}^\dagger a_{\mathbf{m}} a_{\mathbf{n}}, \tag{17.236}$$

where $\varepsilon_{\mathbf{k}} = \frac{\mathbf{p}_k^2}{2m}$ and μ is the chemical potential. In principle we can diagonalize this Hamiltonian and write it as

$$H = \sum_{\mathbf{k}} \varepsilon_{\mathbf{k}} N_{\mathbf{k}} + \sum_{\mathbf{kj}} \varepsilon_{\mathbf{kj}} N_{\mathbf{k}} N_{\mathbf{j}} + \cdots, \tag{17.237}$$

where $N_{\mathbf{k}}$ is the number of quantum particles in the state $\mathbf{k}$. Evidently $\varepsilon_{\mathbf{k}}$'s are the exact single particle energies, but the system is not completely specified in terms of $\varepsilon_{\mathbf{k}}$'s alone. For a complete dynamical description of the system the quantities $\varepsilon_{\mathbf{jk}}, \varepsilon_{\mathbf{ijk}}$, etc. are needed. In other words the usual description of quasi-particles as a system of non-interacting excitations of the bosons is approximate in nature. But we can write the exact Hamiltonian (17.236) as the Hamiltonian for quasi-particles plus residual terms representing the interaction between these quasi particles. As far as the potential $V(\mathbf{r})$ is concerned, we assume that it is composed of an attractive tail and a strong short range repulsive part.

The creation and annihilation for the bosons and for quasi-particles are given by

$$\left[a_{\mathbf{k}},\, a_{\mathbf{j}}^{\dagger}\right] = \delta_{\mathbf{k},\mathbf{j}}, \tag{17.238}$$

and

$$\left[\beta_{\mathbf{k}},\, \beta_{\mathbf{j}}^{\dagger}\right] = \delta_{\mathbf{k},\mathbf{j}}. \tag{17.239}$$

To the lowest order, we write $a_{\mathbf{k}}$ as a linear combination of $\beta_{\mathbf{k}}$ and $\beta_{-\mathbf{k}}^{\dagger}$ for all values of $\mathbf{k}$ including $k = 0$ momentum state [24],

$$\begin{cases} a_{\mathbf{k}} = h(k)\beta_{\mathbf{k}} + g(k)\beta_{-\mathbf{k}}^{\dagger} \\ a_{-\mathbf{k}}^{\dagger} = h(k)\beta_{-\mathbf{k}}^{\dagger} + g(k)\beta_{\mathbf{k}} \end{cases}, \tag{17.240}$$

where $h(k)$ and $g(k)$ are real functions of $\mathbf{k}$ and

$$h(k) = h(-k), \quad \text{and} \quad g(k) = g(-k). \tag{17.241}$$

Also from (17.238) and (17.239) we have the following relation between $h(k)$ and $g(k)$

$$h^2(k) - g^2(k) = 1. \tag{17.242}$$

Denoting the vacuum state of quasi-particles by $|0\rangle$ we find the number of particles in the state $\mathbf{k}$ to be

$$n_k = \left\langle 0 \left| a_{\mathbf{k}}^{\dagger} a_{\mathbf{k}} \right| 0 \right\rangle = g^2(k). \tag{17.243}$$

Thus the total number of the particles, N, is

$$N = \sum_k n_k^2 = \sum_k g^2(k), \tag{17.244}$$

and this last relation fixes the chemical potential μ. Next we substitute for $a_{\mathbf{k}}$ and $a_{-\mathbf{k}}^{\dagger}$ from (17.240) in (17.236) and arrange all the terms in normal order and then we equate the coefficient of the term $\beta_{\mathbf{k}}^{\dagger}\beta_{-\mathbf{k}}^{\dagger} + \beta_{\mathbf{k}}\beta_{-\mathbf{k}}$ to zero. Thus we obtain the following relation

$$\left[\varepsilon_{\mathbf{k}} - \mu + \mathcal{A}(\mathbf{k})\right] h(k)g(k) + \frac{1}{2}\Delta(\mathbf{k})\left[h^2(k) + g^2(k)\right] = 0, \tag{17.245}$$

where

$$\mathcal{A}(\mathbf{k}) = \frac{1}{L^3}\sum_{\mathbf{j}}\left[\tilde{V}(0) + \tilde{V}(\mathbf{k}-\mathbf{j})\right]g^2(j) = \mathcal{A}(-\mathbf{k}), \tag{17.246}$$

and

$$\Delta(\mathbf{k}) = \frac{1}{L^3}\sum_{\mathbf{j}}\tilde{V}(\mathbf{k}-\mathbf{j})h(j)g(j) = \Delta(-\mathbf{k}). \tag{17.247}$$

We can write the Hamiltonian in terms of the quasi-particle operators as

$$H = \mathcal{E} + \sum_{\mathbf{k}} E_{\mathbf{k}}\beta_{\mathbf{k}}^{\dagger}\beta_{-\mathbf{k}} + H_I, \tag{17.248}$$

where

$$\mathcal{E} = \sum_{\mathbf{k}}\left\{\left[\varepsilon_{\mathbf{k}} - \mu + \frac{1}{2}\mathcal{A}(\mathbf{k})\right]g^2(k) + \frac{1}{2}\Delta(\mathbf{k})h(k)g(k)\right\}, \tag{17.249}$$

and

$$E_{\mathbf{k}} = \left[\varepsilon_{\mathbf{k}} - \mu + \mathcal{A}(\mathbf{k})\right]\left[h^2(k) + g^2(k) + 2\Delta(\mathbf{k})h(k)g(k)\right], \tag{17.250}$$

or

$$E_{\mathbf{k}} = \left\{\left[\varepsilon_{\mathbf{k}} - \mu + \mathcal{A}(\mathbf{k})\right]^2 - \Delta^2(\mathbf{k})\right\}^{\frac{1}{2}}. \tag{17.251}$$

This last relation is found by substituting (17.245) and (17.242) in (17.249).

The residual Hamiltonian, H_I, when expressed in terms of $\beta_{\mathbf{k}}$ and $\beta_{\mathbf{k}}^{\dagger}$ contains quartic terms of the form

$$\begin{aligned} H_I &= \frac{1}{2L^3}\sum_{\mathbf{jlmn}}\delta_{\mathbf{j}+\mathbf{l},\mathbf{m}+\mathbf{n}}\left[F_1\beta_{\mathbf{j}}^{\dagger}\beta_{-\mathbf{l}}\beta_{\mathbf{m}}\beta_{\mathbf{n}} + F_2\beta_{\mathbf{j}}^{\dagger}\beta_{\mathbf{l}}^{\dagger}\beta_{\mathbf{m}}\beta_{\mathbf{n}}\right. \\ &+ \left. F_3\beta_{\mathbf{j}}^{\dagger}\beta_{\mathbf{l}}^{\dagger}\beta_{-\mathbf{n}}^{\dagger}\beta_{\mathbf{m}} + F_4\beta_{\mathbf{j}}^{\dagger}\beta_{\mathbf{l}}^{\dagger}\beta_{-\mathbf{m}}^{\dagger}\beta_{-\mathbf{n}}^{\dagger} + F_5\beta_{-\mathbf{j}}^{\dagger}\beta_{-\mathbf{l}}^{\dagger}\beta_{\mathbf{m}}\beta_{\mathbf{n}}\right]. \end{aligned} \tag{17.252}$$

In this residual Hamiltonian H_I, $F_1 \cdots F_5$ are all dependent on the variables $\mathbf{j}$, $\mathbf{l}$, $\mathbf{m}$ and $\mathbf{n}$ through the functions $\tilde{V}$, g and h. For example F_2 is given by

$$\begin{aligned} F_2 &= 2\left[\tilde{V}(\mathbf{j}-\mathbf{m}) + \tilde{V}(\mathbf{j}+\mathbf{l})\right]h(j)g(l)h(m)g(n) \\ &+ \tilde{V}(\mathbf{j}-\mathbf{m})[h(j)h(l)h(m)h(n) + g(j)g(l)g(m)g(n)]. \end{aligned} \tag{17.253}$$

The Scattering Matrix — To determine the scattering matrix let us find the Heisenberg equation of motion for $\beta_{\mathbf{q}}^{\dagger}$. In what follows we set $\hbar = 1$.

$$\frac{d\beta_{\mathbf{q}}^{\dagger}(t)}{dt} = i\left[H,\ \beta_{\mathbf{q}}^{\dagger}(t)\right] = iE_{\mathbf{q}}\beta_{\mathbf{q}}^{\dagger}(\mathbf{q}) + iI\left[\mathbf{q},\ \beta_{\mathbf{q}}(t)\right], \tag{17.254}$$

where

$$I\left[\mathbf{q},\ \beta_{\mathbf{q}}(t)\right] = \frac{1}{L^3}\sum_{\mathbf{jlm}} \delta_{\mathbf{j}+\mathbf{q},\mathbf{m}+\mathbf{l}}\tilde{V}\left(\mathbf{q}-\mathbf{l}\right) \times \mathcal{O}_N\left[h(q)a^\dagger_{\mathbf{l}}a^\dagger_{\mathbf{m}}a_{\mathbf{j}} + g(q)a^\dagger_{-\mathbf{j}}a_{-\mathbf{l}}a_{-\mathbf{m}}\right]. \quad (17.255)$$

In this relation the symbol $\mathcal{O}_N$ denotes the normal ordering of operators, i.e. after substituting for $a_{\mathbf{k}}$ and $a_{-\mathbf{k}^\dagger}$ from (17.240), we rearrange the operators so that the creation operators are to the left of annihilation operators.

Now let $\mathbf{p}$ and $\mathbf{q}$ be the initial and $\mathbf{m}$ and $\mathbf{n}$ be the final momenta of the two quasi-particles. As we have seen before the scattering matrix can be written as

$$S_{\mathbf{m},\mathbf{n};\ \mathbf{p},\mathbf{q}} = \langle \mathbf{m},\mathbf{n}\ out|\mathbf{q},\mathbf{p}\ in\rangle = \left\langle \mathbf{m},\mathbf{n}\ out\left|\beta^{\dagger\ in}_{\mathbf{q}}\right|\mathbf{p}\ out\right\rangle = \lim_{t\to-\infty}\left\langle \mathbf{m},\mathbf{n}\ out\left|\beta^{\dagger\ in}_{q}(t)\right|\mathbf{p}\ out\right\rangle e^{-iE_{\mathbf{q}}t}. \quad (17.256)$$

Using the method of LSZ discussed in Chapter 14 we find the matrix elements of the S matrix to be [26],[27]

$$(S-1)_{\mathbf{m},\mathbf{n};\ \mathbf{p},\mathbf{q}} = -\int_{-\infty}^{\infty}\left\langle \mathbf{m},\mathbf{n}\ out\left|\frac{d\beta^\dagger_{\mathbf{q}}(t)}{dt} - iE_{\mathbf{q}}\beta^\dagger_{\mathbf{q}}(t)\right|\mathbf{p}\ out\right\rangle e^{-iE_{\mathbf{q}}t}dt. \quad (17.257)$$

Substituting for $\frac{d\beta^\dagger_{\mathbf{q}}(t)}{dt}$ from (17.254) in(17.257) we get

$$(S-1)_{\mathbf{m},\mathbf{n};\ \mathbf{p},\mathbf{q}} = -i\int_{-\infty}^{\infty}\langle \mathbf{m},\mathbf{n}\ out\left|I[\mathbf{q},\ \beta_{\mathbf{q}}(t)]\right|\mathbf{p}\ out\rangle e^{-iE_{\mathbf{q}}t}dt. \quad (17.258)$$

To the lowest order the contribution to the scattering matrix comes from $I\left[\mathbf{q},\beta^{out}_{\mathbf{q}}(t)\right]$ in (17.258). This functional is found by iterating (17.254) once, i.e. since

$$\frac{d\beta^{\dagger\ out}_{\mathbf{q}}(t)}{dt} = iE_{\mathbf{q}}\beta^{\dagger\ out}_{\mathbf{q}}(t), \quad (17.259)$$

we have

$$\frac{d\beta^\dagger_{\mathbf{q}}(t)}{dt} \approx iE_{\mathbf{q}}\beta^\dagger_{\mathbf{q}}(t) + iI\left[\mathbf{q},\ \beta^{\dagger\ out}_{\mathbf{q}}(t)\right]. \quad (17.260)$$

Using this approximate form we can calculate the matrix element in (17.258) and after carrying out the integration we get

$$(S-1)_{\mathbf{m},\mathbf{n};\ \mathbf{p},\mathbf{q}} = -2\pi i\delta(E_{\mathbf{m}} + E_{\mathbf{n}} - E_{\mathbf{q}} - E_{\mathbf{p}})\tilde{V}(\mathbf{m},\mathbf{n};\ \mathbf{p},\mathbf{q}), \quad (17.261)$$

where

$$\tilde{V}(\mathbf{m},\mathbf{n};\ \mathbf{p},\mathbf{q}) = \frac{1}{L^3\delta_{\mathbf{m}+\mathbf{n},\mathbf{p}+\mathbf{q}}\left\{\tilde{V}_1 + \tilde{V}_2 + \tilde{V}_3\right\}}. (17.262)$$

The interaction terms $\tilde{V}_1$, $\tilde{V}_2$ and $\tilde{V}_3$ are given by

$$\tilde{V}_1 = \tilde{V}(\mathbf{q}-\mathbf{n})[h(q)h(m) + g(p)g(n)][h(p)h(n) + g(p)g(n)], \quad (17.263)$$

$$\tilde{V}_2 = \tilde{V}(\mathbf{q}-\mathbf{n})[h(q)h(n)+g(q)g(n)][h(p)h(m)+g(p)g(m)], \qquad (17.264)$$

and

$$\begin{aligned}\tilde{V}_3 &= \tilde{V}(\mathbf{q}+\mathbf{p})\left\{g(p)h(q)[g(m)h(n)+h(m)g(n)]\right.\\ &+ \left. h(p)g(q)[h(m)g(n)+g(m)h(n)]\right\}. \end{aligned} \qquad (17.265)$$

Thus the $\tilde{V}$ matrix is just the Born term for the elastic scattering of two quasi-particles.

The G-Matrix — The S matrix that we studied in the previous section is expressible in powers of the potential $\tilde{V}$ of which (17.262) is the first term. If the two quasi-particles interact strongly with each other, e.g. at short distances the repulsive force can be very large, then we can consider the scattering of two quasi-particles in the average field of force produced by the presence of other quasi-particles. For this we consider a system composed of two quasi-particles and assume that the Hamiltonian (17.236) can be written as

$$H = E + \sum_{\mathbf{k}} \omega_{\mathbf{k}} \alpha_{\mathbf{k}}^{\dagger} \alpha_{\mathbf{k}}, \qquad (17.266)$$

where in writing (17.266) we have omitted higher order terms such as

$$\sum_{\mathbf{ij}} \varepsilon_{\mathbf{ij}} \alpha_{\mathbf{i}}^{\dagger} \alpha_{\mathbf{j}}^{\dagger} \alpha_{\mathbf{i}} \alpha_{\mathbf{j}} + \cdots. \qquad (17.267)$$

Thus as before, the quasi-particles are characterized by their energy-momentum relation $\omega_{\mathbf{k}}$. We write the total Hamiltonian H as the sum of two terms

$$H = H_0 + H_I, \qquad (17.268)$$

where

$$H_0 = \mathcal{E} + \sum_{\mathbf{k}} E_{\mathbf{k}} \beta_{\mathbf{k}}^{\dagger} \beta_{\mathbf{k}}. \qquad (17.269)$$

Here the quantities $\mathcal{E}$, $E_{\mathbf{k}}$ and H_I are given by Eqs. (17.248) , (17.252) and (17.252). Since E and $\mathcal{E}$ are c-numbers, they can be absorbed in H and H_0 respectively. For a state of momentum $\mathbf{k}$ the difference measures the energy shift caused by the interaction, but $E_{\mathbf{k}}$ is known and at this stage $\omega_{\mathbf{k}}$ is unknown.

Consider the expansion of the product of two annihilation operator $\alpha_{\mathbf{p}}\alpha_{\mathbf{q}}$ in terms of $\beta_{\mathbf{p}}\beta_{\mathbf{q}}$;

$$\alpha_{\mathbf{p}}\alpha_{\mathbf{q}} = \beta_{\mathbf{p}}\beta_{\mathbf{q}} + \sum_{\mathbf{u},\mathbf{v}\neq\mathbf{p},\mathbf{q}} K(\mathbf{p},\mathbf{q};\mathbf{u},\mathbf{v})\, \beta_{\mathbf{u}}\beta_{\mathbf{v}} + \cdots. \qquad (17.270)$$

Noting that both H and H_0 are diagonal operators, the former in $\alpha_{\mathbf{k}}$'s and the latter in $\beta_{\mathbf{k}}$'s we find the following commutation relations

$$[H,\ \alpha_{\mathbf{p}}\alpha_{\mathbf{q}}] = -(\omega_{\mathbf{p}} + \omega_{\mathbf{q}})\alpha_{\mathbf{p}}\alpha_{\mathbf{q}}, \qquad (17.271)$$

and

$$[H_0,\ \beta_{\mathbf{u}}\beta_{\mathbf{v}}] = -(E_{\mathbf{u}} + E_{\mathbf{v}})\beta_{\mathbf{u}}\beta_{\mathbf{v}}. \tag{17.272}$$

If we substitute (17.270) in (17.271) we get

$$[H_I,\ \alpha_{\mathbf{p}}\alpha_{\mathbf{q}}] = [H - H_0,\ \alpha_{\mathbf{p}}\alpha_{\mathbf{q}}] = (E_{\mathbf{p}} + E_{\mathbf{q}} - \omega_{\mathbf{p}} - \omega_{\mathbf{q}})\beta_{\mathbf{p}}\beta_{\mathbf{q}}$$
$$+ \sum_{\mathbf{u},\mathbf{v}\neq\mathbf{p},\mathbf{q}} K(\mathbf{u},\mathbf{v};\mathbf{p},\mathbf{q})(E_{\mathbf{u}} + E_{\mathbf{v}} - \omega_{\mathbf{p}} - \omega_{\mathbf{q}})\beta_{\mathbf{u}}\beta_{\mathbf{v}}. \tag{17.273}$$

We can find the kernel $K(\mathbf{u},\mathbf{v};\mathbf{p},\mathbf{q})$ by taking the matrix element of (17.273) between the states $\langle 0|$ and $\beta_{\mathbf{x}}\beta_{\mathbf{y}}|0\rangle$ where $\mathbf{x}$ and $\mathbf{y}$ are assumed to be different from $\mathbf{p}$ and $\mathbf{q}$;

$$K(\mathbf{u},\mathbf{v};\mathbf{p},\mathbf{q}) = \frac{\left\langle 0\left|[H_I,\ \alpha_{\mathbf{p}}\alpha_{\mathbf{q}}]\,\beta_{\mathbf{u}}^\dagger\beta_{\mathbf{v}}^\dagger\right|0\right\rangle}{(E_{\mathbf{u}} + E_{\mathbf{v}} - \omega_{\mathbf{p}} - \omega_{\mathbf{q}})}. \tag{17.274}$$

Substituting (17.274) in (17.270) we have the operator equation

$$\alpha_{\mathbf{p}}\alpha_{\mathbf{q}} = \beta_{\mathbf{p}}\beta_{\mathbf{q}} + \frac{1}{2}\sum_{\mathbf{u},\mathbf{v}\neq\mathbf{p},\mathbf{q}} \frac{\left\langle 0\left|[H_I,\ \alpha_{\mathbf{p}}\alpha_{\mathbf{q}}]\,\beta_{\mathbf{u}}^\dagger\beta_{\mathbf{v}}^\dagger\right|0\right\rangle}{(E_{\mathbf{u}} + E_{\mathbf{v}} - \omega_{\mathbf{p}} - \omega_{\mathbf{q}})}\beta_{\mathbf{u}}\beta_{\mathbf{v}}. \tag{17.275}$$

Let us note that we sum over all values of $\mathbf{u}$ and $\mathbf{v}$ not equal to $\mathbf{p}$ and $\mathbf{q}$ and thus we count each pair twice and to compensate for this we have the factor $\frac{1}{2}$ in front of the sum.

Now we define the matrices $\tilde{V}$ and G by their matrix elements

$$\left\langle \mathbf{m},\mathbf{n}\left|\tilde{V}\right|\mathbf{p},\mathbf{q}\right\rangle = -\left\langle 0\left|[H_I,\ \beta_{\mathbf{p}}\beta_{\mathbf{q}}]\,\beta_{\mathbf{m}}^\dagger\beta_{\mathbf{n}}^\dagger\right|0\right\rangle, \tag{17.276}$$

and

$$\left\langle \mathbf{m},\mathbf{n}\left|G\right|\mathbf{p},\mathbf{q}\right\rangle = -\left\langle 0\left|[H_I,\ \alpha_{\mathbf{p}}\alpha_{\mathbf{q}}]\,\beta_{\mathbf{p}}^\dagger\beta_{\mathbf{q}}^\dagger\right|0\right\rangle. \tag{17.277}$$

From these relations and (17.275) we find the following integral equation for the G matrix

$$\begin{aligned}\left\langle \mathbf{m},\mathbf{n}\left|G\right|\mathbf{p},\mathbf{q}\right\rangle &= \left\langle \mathbf{m},\mathbf{n}\left|\tilde{V}\right|\mathbf{p},\mathbf{q}\right\rangle \\ &+ \frac{1}{2}\sum_{\mathbf{u},\mathbf{v}\neq\mathbf{p},\mathbf{q}} \frac{\left\langle \mathbf{m},\mathbf{n}\left|\tilde{V}\right|\mathbf{u},\mathbf{v}\right\rangle\left\langle \mathbf{u},\mathbf{v}\left|G\right|\mathbf{p},\mathbf{q}\right\rangle}{(\omega_{\mathbf{p}} + \omega_{\mathbf{q}} - E_{\mathbf{u}} - E_{\mathbf{v}})}\beta_{\mathbf{u}}\beta_{\mathbf{v}}.\end{aligned} \tag{17.278}$$

The integral equation (17.278) relates the G-matrix to the effective interaction between quasi-particles $\tilde{V}$. By substituting the explicit form of H_I, Eq. (17.252), in (17.276) we find the matrix elements of $\tilde{V}$ which is the same as the one given in (17.262).

Next we want to determine the energies $\omega_{\mathbf{p}}$ and $\omega_{\mathbf{q}}$ in (17.278). For this we first calculate

$$[\alpha_{\mathbf{p}}\alpha_{\mathbf{p}},\ H] = 2\omega_{\mathbf{p}}\alpha_{\mathbf{p}}\alpha_{\mathbf{p}} = \left[\alpha_{\mathbf{p}}\alpha_{\mathbf{p}},\ \sum_{\mathbf{k}} E_{\mathbf{k}}\beta_{\mathbf{k}}^\dagger\beta_{\mathbf{k}} + H_I\right]. \tag{17.279}$$

Then we find the matrix elements of (17.279) between the state $\langle 0|$ and $|\beta^\dagger_{\mathbf{p}}\beta^\dagger_{\mathbf{p}}|0\rangle$, i.e.

$$\begin{aligned} 2\omega_{\mathbf{p}}\left\langle 0\left|\alpha_{\mathbf{p}}\alpha_{\mathbf{p}}\beta^\dagger_{\mathbf{p}}\beta^\dagger_{\mathbf{p}}\right|0\right\rangle &= \left\langle 0\left|\left[\alpha_{\mathbf{p}}\alpha_{\mathbf{p}},\ \sum_{\mathbf{k}} E_{\mathbf{k}}\beta^\dagger_{\mathbf{k}}\beta_{\mathbf{k}}\right]\beta^\dagger_{\mathbf{p}}\beta_{\mathbf{p}}\right|0\right\rangle \\ &- \left\langle 0\left|\left[H_I,\ \alpha^\dagger_{\mathbf{p}}\alpha_{\mathbf{p}}\right]\beta^\dagger_{\mathbf{p}}\beta^\dagger_{\mathbf{p}}\right|0\right\rangle. \end{aligned} \tag{17.280}$$

Now from Eq. (17.275) we obtain

$$\left\langle 0\left|\alpha_{\mathbf{p}}\alpha_{\mathbf{q}}\beta^\dagger_{\mathbf{p}}\beta^\dagger_{\mathbf{q}}\right|0\right\rangle = 2, \tag{17.281}$$

and

$$\left\langle 0\left|\left[\alpha_{\mathbf{p}}\alpha_{\mathbf{p}},\ \sum_{\mathbf{k}} E_{\mathbf{k}}\beta^\dagger_{\mathbf{k}}\beta_{\mathbf{k}}\right]\beta^\dagger_{\mathbf{p}}\beta_{\mathbf{p}}\right|0\right\rangle = 4E_{\mathbf{p}}. \tag{17.282}$$

If we substitute these results in (17.280) we find that $\omega_{\mathbf{p}}$ is related to the diagonal elements of G;

$$\omega_{\mathbf{p}} = E_{\mathbf{p}} + \frac{1}{4}\langle \mathbf{p},\mathbf{p}|G|\mathbf{p},\mathbf{p}\rangle. \tag{17.283}$$

17.11 Bogoliubov Transformation for Fermions Interacting through Pairing Forces

For a system of N interacting fermions the second quantized Hamiltonian is given by (17.107) where each of the quantum numbers, say k denotes the momentum $\mathbf{p}_k$ as well as the spin s. Now we write the spin of particles explicitly and define the number operator as

$$\hat{N} = \sum_{k,s} b^\dagger_{k,s} b_{k,s}. \tag{17.284}$$

This number operator commutes with the Hamiltonian and is a constant of motion. We choose N, the number of particles in the system, to be very large and at the same time allow the volume tend to infinity, $L^3 \to \infty$, in such a way that the average particle density $n = \frac{N}{L^3}$ stays finite. In this limit we assume that the wave function is a plane wave, Eq. (17.112), and just as we have seen for a system of bosons, Eq. (17.195), the corresponding Hamiltonian for fermions can be written as

$$\begin{aligned} \hat{H} &= \sum_{k,s} \frac{\mathbf{p}_k^2}{2m} b^\dagger_{k,\,s} b_{k,\,s} \\ &- \frac{1}{2L^3} \sum_{\mathbf{p}_k+\mathbf{p}_\ell=\mathbf{p}_m+\mathbf{p}_n} \tilde{V}(|\mathbf{p}_\ell - \mathbf{p}_n|) b^\dagger_{k,\,s} b^\dagger_{\ell,\,s'} b_{m,\,s''} b_{n,\,s'''}, \end{aligned} \tag{17.285}$$

where

$$\tilde{V}(|\mathbf{p}_k - \mathbf{p}_n|) = \frac{-4\pi}{|\mathbf{p}_k - \mathbf{p}_n|} \int_0^\infty V(\rho)\rho \sin(|\mathbf{p}_k - \mathbf{p}_n|\rho)d\rho. \tag{17.286}$$

Note that for an attractive potential $V(\rho) < 0$ and therefore $\tilde{V}(|\mathbf{p}_k - \mathbf{p}_n|)$ will be a positive function. For the approximate diagonalization of (17.285) we want to keep the number of particles, N, fixed. For this we introduce the chemical potential μ by adding a term $-\mu \sum_k b_{k,\,s}^\dagger b_{k,\,s}$ to the above Hamiltonian. We also assume that the interaction between fermions depends only on their relative distance between the particles i.e. $|\mathbf{r}_i - \mathbf{r}_j|$. With these assumptions (17.285) can be written as

$$\hat{H} = \sum_{k,s} \left(\frac{\mathbf{p}_k^2}{2m} - \mu\right) b_{k\,s}^\dagger b_{k\,s} - \frac{1}{2L^3} \sum_{\mathbf{p}_k+\mathbf{p}_\ell=\mathbf{p}_m+\mathbf{p}_n} \tilde{V}(|\mathbf{p}_\ell - \mathbf{p}_n|) b_{k\,s_1}^\dagger b_{\ell\,s_2}^\dagger b_{m\,s_2} b_{n\,s_1}, \tag{17.287}$$

To simplify (17.287) further we note that when $\tilde{V}(|\mathbf{p}_\ell - \mathbf{p}_n|)$ has a range shorter than the average wavelength of the relative motion of a pair of fermions. Here we assume that the interaction between fermions with parallel spins is weak since they do not come close to each other and thus in this approximation, the interaction will be between fermions with opposite spins, i.e. in (17.287) we set $s_2 = -s_1$. At the same time we isolate those terms in the interaction for which the relative momentum is conserved.

$$\mathbf{p}_k + \mathbf{p}_l = \mathbf{p}_m + \mathbf{p}_n \neq 0. \tag{17.288}$$

Since the contribution from $s_1 = \frac{1}{2}$ is the same as that of $s_1 = -\frac{1}{2}$, we can write (17.287) as

$$\hat{H} = \hat{H}_0 + \hat{H}', \tag{17.289}$$

where

$$\hat{H}_0 = 2\sum_k e(\mathbf{p}_k) b_{k,\,\frac{1}{2}}^\dagger b_{k,\,\frac{1}{2}} - \frac{1}{L^3} \sum_{k,n} \tilde{V}(|\mathbf{p}_k - \mathbf{p}_n|) b_{k,\,\frac{1}{2}}^\dagger b_{-n,\,-\frac{1}{2}}^\dagger b_{-k,\,-\frac{1}{2}} b_{n,\,\frac{1}{2}}, \tag{17.290}$$

and

$$e(\mathbf{p}_k) = \frac{\mathbf{p}_k^2}{2m} - \mu, \tag{17.291}$$

where in (17.290) k and $-k$ refer to the momenta $\mathbf{p}_k$ and $-\mathbf{p}_k$ respectively. The Hamiltonian $\hat{H}_0$, Eq. (17.290), in which fermions with opposite momenta appear in the potential is called pairing Hamiltonian. The operator $\hat{H}'$ contains all of the other terms in the potential which are not included in $\hat{H}_0$.

We can diagonalize $\hat{H}_0$ by a Bogoliubov transformation by introducing quasi-particle operators $B_{k,\,0}$, $B_{k,\,1}$, $B_{k,\,0}^\dagger$ and $B_{k,\,1}^\dagger$:

$$b_{k,\,\frac{1}{2}} = u(k) B_{k,\,0} + v(k) B_{k,\,1}^\dagger, \tag{17.292}$$

and

$$b_{-k,\,-\frac{1}{2}} = u(k)B_{k,\,1} - v(k)B^{\dagger}_{k,\,0}, \tag{17.293}$$

where $u(k)$ and $v(k)$ are real even functions of k which will be determined later. For the transformation (17.292),(17.293) to be canonical when B_{k0} and B_{k1} satisfy the anti-commutation relations,

$$\left[B^{\dagger}_{k,\,0},\, B_{k,\,0}\right]_{+} = \left[B^{\dagger}_{k,\,1},\, B_{k,\,1}\right]_{+} = 1, \tag{17.294}$$

$$\left[B_{k,\,0},\, B_{k,\,1}\right]_{+} = \left[B^{\dagger}_{k,\,0},\, B^{\dagger}_{k,\,1}\right]_{+} = 0, \tag{17.295}$$

we find that $u(k)$ and $v(k)$ must satisfy the condition

$$u^2(k) + v^2(k) = 1. \tag{17.296}$$

By replacing the fermion creation and annihilation operators

$$b_{k,\,\frac{1}{2}},\;\; b^{\dagger}_{-k,\,-\frac{1}{2}}\;\cdots, \tag{17.297}$$

by

$$B_{k,\,0},\;\; B^{\dagger}_{-k,\,-1}\cdots, \tag{17.298}$$

rearranging the Hamiltonian $\hat{H}_0$ so that the creation operators are to the left of annihilation operators, we obtain

$$\hat{H}_0 = E_0(\mathbf{p}_k) + \hat{H}^{(0)}_0 + \hat{H}^{(1)}_0 + \hat{H}^{(2)}_0, \tag{17.299}$$

where $E_0(\mathbf{p}_k)$ is a c-number multiplied by the unit operator.

$$E_0(\mathbf{p}_k) = 2\sum_k e(\mathbf{p}_k)v^2(k) - \frac{1}{L^3}\sum_{k,n}\tilde{V}(|\mathbf{p}_k - \mathbf{p}_n|)u(n)v(n)u(k)v(k). \tag{17.300}$$

The operator $\hat{H}^{(0)}_0$ is diagonal in the number representation

$$\begin{aligned}\hat{H}^{(0)}_0 &= \sum_k \Big\{e(\mathbf{p}_k)\left(u(k)^2 - v(k)^2\right) \\ &+ \frac{2u(k)v(k)}{L^3}\sum_n \tilde{V}(|\mathbf{p}_k - \mathbf{p}_n|)u(n)v(n)\Big\} \\ &\times \left(B^{\dagger}_{k,0}B_{k,0} + B^{\dagger}_{k,1}B_{k,1}\right),\end{aligned} \tag{17.301}$$

and $\hat{H}^{(1)}_0$ has the non-diagonal part, but this operator is also quadratic in B_k s

$$\begin{aligned}\hat{H}^{(1)}_0 &= \sum_k\left\{2e(\mathbf{p}_k)u(k)v(k) - \frac{u^2(k) - v^2(k)}{L^3}\sum_n \tilde{V}(|\mathbf{p}_k - \mathbf{p}_n|)u(n)v(n)\right\} \\ &\times \left(B^{\dagger}_{k,0}B^{\dagger}_{k,1} + B_{k,1}B_{k,0}\right).\end{aligned} \tag{17.302}$$

Finally the operator $\hat{H}_0^{(2)}$ contains products of four of the quasi-fermion operators, and requires more elaborate techniques for its inclusion in this formulation.

As we mentioned earlier $u(k)$ and $v(k)$ are arbitrary except for the constraint (17.296). Now we choose $u(k)$ and $v(k)$ so that the coefficient of the non-diagonal operators in $\hat{H}_0^{(1)}$ vanishes, i.e.

$$2e(\mathbf{p}_k)u(k)v(k) - \frac{u^2(k) - v^2(k)}{L^3}\sum_n \tilde{V}(|\mathbf{p}_k - \mathbf{p}_n|)u(n)v(n) = 0. \tag{17.303}$$

This condition is the same as the one found by requiring $E_0(k)$, Eq. (17.300), be a minimum. Rather than working with $u(k)$ and $v(k)$, we introduce a function $\Delta(k)$;

$$\Delta(k) = \frac{1}{L^3}\sum_n u(n)v(n)\tilde{V}(|\mathbf{p}_k - \mathbf{p}_n|), \tag{17.304}$$

and using Eqs. (17.296) and (17.303) we can write $u(k)$ and $v(k)$ in terms of $\Delta(k)$;

$$u^2(k) = \frac{1}{2}\left[1 + \frac{e(\mathbf{p}_k)}{\sqrt{\Delta^2(k) + e^2(\mathbf{p}_k)}}\right], \tag{17.305}$$

and

$$v^2(k) = \frac{1}{2}\left[1 - \frac{e(\mathbf{p}_k)}{\sqrt{\Delta^2(k) + e^2(\mathbf{p}_k)}}\right]. \tag{17.306}$$

By substituting (17.305) and (17.306) in (17.304) we find that $\Delta(k)$ satisfies the nonlinear integral equation (called BCS gap equation) [25]

$$\Delta(k) = \frac{1}{2L^3}\sum_n \frac{\tilde{V}(|\mathbf{p}_k - \mathbf{p}_n|)\Delta(n)}{\sqrt{\Delta^2(n) + e^2(\mathbf{p}_n)}}. \tag{17.307}$$

In terms of $\Delta(k)$ the Hamiltonian $\hat{H}_0^{(0)}$ can be written as

$$\hat{H}_0^{(0)} = \sum_k \sqrt{\Delta^2(k) + e^2(\mathbf{p}_k)}\left(B_{k,0}^\dagger B_{k,0} + B_{k,1}^\dagger B_{k,1}\right), \tag{17.308}$$

and from this relation it is evident that the energy of excitation of a quasi-fermion is

$$E_0(\mathbf{p}_k) = \sqrt{\Delta^2(k) + e^2(\mathbf{p}_k)}. \tag{17.309}$$

Let us consider the elementary excitation of a fermion with momentum $\mathbf{p}_k$. There are two kinds of excitation: one defined by $B_{k,0}^\dagger B_{k,0}$ and the other by $B_{k,1}^\dagger B_{k,1}$;

$$B_{k,0}^\dagger B_{k,0}|n_{k_0}, n_{k_1}\rangle = n_{k_0}|n_{k_0}, n_{k_1}\rangle, \tag{17.310}$$

and

$$B_{k,1}^\dagger B_{k,1}|n_{k_0}, n_{k_1}\rangle = n_{k_1}|n_{k_0}, n_{k_1}\rangle. \tag{17.311}$$

Now the difference between the energy of excitation of a particle and a quasi-particle is given by

$$E(\mathbf{p}_k) - e(\mathbf{p}_k) = \sqrt{e^2(\mathbf{p}_k) + \Delta^2(k)} - e(\mathbf{p}_k). \tag{17.312}$$

The trivial solution of Eq. (17.307) is obtained if we set $\Delta(\mathbf{k}) = 0$ for all $\mathbf{k}$, and this implies $u(k)v(k) = 0$, a result which follows from Eq. (17.304). In this case we consider the following possibilities:

$$\begin{cases} u(k) = 1, \quad v(k) = 0 \quad \text{if} \;\; e(\mathbf{p}_k) = \dfrac{\mathbf{p}_k^2}{2m} - \mu > 0 \\ u(k) = 0, \quad v(k) = 1 \quad \text{if} \;\; e(\mathbf{p}_k) = \dfrac{\mathbf{p}_k^2}{2m} - \mu < 0 \end{cases}. \tag{17.313}$$

To interpret this result we find the inverse of the canonical transformations (17.292) and (17.293), viz,

$$B_{k,0} = u(k)b_{k,\frac{1}{2}} - v(k)b^{\dagger}_{-k,-\frac{1}{2}}, \tag{17.314}$$

and

$$B_{k,1} = u(k)b_{-k,-\frac{1}{2}} + v(k)b^{\dagger}_{k,\frac{1}{2}}, \tag{17.315}$$

When Eq. (17.313) is satisfied, we observe that for $|\mathbf{p}| > p_F$, or outside the Fermi sphere $u(k) = 1$, $v(k) = 0$ and $e(\mathbf{p}_k) > 0$, and both of the operators $B_{k,0} = b_{k,\frac{1}{2}}$ and $B_{k,1} = b^{\dagger}_{k,\frac{1}{2}}$ annihilate fermions (or spin $\frac{1}{2}$ particles) with quantum numbers $(\mathbf{p}_k, \frac{1}{2})$ and $(-\mathbf{p}_k, -\frac{1}{2})$ respectively. Inside the Fermi sphere $u(k) = 0, v(k) = 1$ and $e(\mathbf{p}_k) < 0$, and the operators $B_{k,0} = -b_{-k,-\frac{1}{2}}$ and $B_{k,1} = -b^{\dagger}_{k,\frac{1}{2}}$ annihilate holes (or creates fermions) in the states $(-\mathbf{p}_k, -\frac{1}{2})$ and $(\mathbf{p}_k, \frac{1}{2})$. These results are for the trivial solution of the gap equation, i.e. for $\Delta(k) = 0$.

Now let us suppose that the force $\tilde{V}(\mathbf{p}_k - \mathbf{p}_n)$ is attractive and strong so that we have the inequality

$$\frac{1}{2L^3}\sum_n \frac{\tilde{V}(|\mathbf{p}_k - \mathbf{p}_n|)}{|e(\mathbf{p}_n)|} > 1. \tag{17.316}$$

To find the approximate solution of the gap equation we first write the chemical potential μ as

$$\mu = \frac{\hbar^2 k_0^2}{2m}, \tag{17.317}$$

and this defines the momentum $\hbar k_0$. Then we consider the special case where $\tilde{V}(|\mathbf{p}_k - \mathbf{p}_n|)$ is a constant, $\tilde{V}(\mathbf{p}_k - \mathbf{p}_n) = \tilde{V}$ when $|\mathbf{p}_k|$ and $|\mathbf{p}_n|$ lie within $\hbar(k_0 - q)$ and $\hbar(k_0 + q)$. For this special case Eq. (17.307) for Δ becomes

$$\frac{\tilde{V}}{2L^3}\sum_{\hbar(k_0-q)\le|\mathbf{p}_k|\le\hbar(k_0+q)} \frac{1}{\sqrt{e^2(\mathbf{p}_n) + \Delta^2}} = 1. \tag{17.318}$$

Since we have assumed a strong potential, Δ is larger than the spacing between the levels of $e(\mathbf{p}_k)$ and we replace the summation in (17.318) by integration

$$\sum_{\mathbf{p}_k} \to \frac{L^3}{8\pi^3}\int d^3k. \tag{17.319}$$

In addition we can approximate d^3k by

$$d^3k = 4\pi k_0^2 dk, \tag{17.320}$$

and we also use the approximate form of $e(\mathbf{p}_k)$;

$$e(\mathbf{p}_k) \approx \frac{\hbar k_0(p_k - \hbar k_0)}{m}, \tag{17.321}$$

in (17.318). By substituting these approximate terms Eq. (17.318), this equation becomes

$$\frac{\tilde{V}k_0^2}{4\pi^2}\int_{-q}^{q}\frac{dk}{\left[\Delta^2 + \left(\frac{\hbar^2 k_0 k}{m}\right)^2\right]^{\frac{1}{2}}} = 1. \tag{17.322}$$

By carrying out the integration in (17.322) and then solving the resulting expression for Δ we obtain

$$\Delta = \left(\frac{2\hbar^2 k_0 q}{m}\right)\frac{\exp\left(-\frac{D}{\tilde{V}}\right)}{1-\exp\left(-\frac{D}{\tilde{V}}\right)}, \tag{17.323}$$

where D is defined by

$$D = \frac{2\pi^2\hbar^2}{mk_0}. \tag{17.324}$$

We note that the dependence of Δ on $\tilde{V}$ shows that this solution cannot be found by the perturbation theory, since this amounts to an expansion of $e^{-\frac{D}{\tilde{V}}}$ in powers of $\tilde{V}$ around $\tilde{V} = 0$.

The ground state energy $E_0(\mathbf{p}_k)$ can be expressed in terms of $\Delta(k)$ and $e(\mathbf{p}_k)$ by substituting for $u(k)$ and $v(k)$ in (17.300) and after simplifying the result we get

$$E_0(\mathbf{p}_k) = \sum_{\mathbf{p}_k}\frac{e(\mathbf{p}_k)\left[\sqrt{e(\mathbf{p}_k)^2+\Delta^2(k)} - e(\mathbf{p}_k)\right] - \frac{1}{2}\Delta^2(k)}{\sqrt{e(\mathbf{p}_k)^2+\Delta^2(k)}}. \tag{17.325}$$

The result of the exact solution for the function $u^2(k)$ and $v^2(k)$ are shown by solid curves in Fig. 17.5, when $\Delta(k) \neq 0$. The dashed lines show these solutions for the trivial case of $\Delta(k) = 0$, Eq. (17.313).
For the trivial solution of this equation if $\Delta(k) = 0$, then $E_0(\mathbf{p}_k) = 0$, however

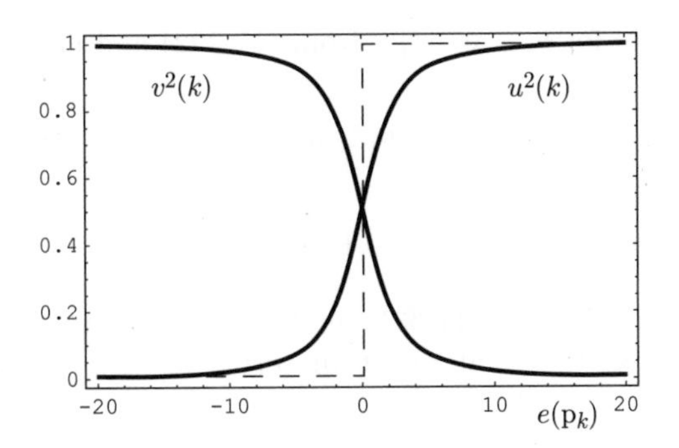

Figure 17.5: Plots of $u^2(k)$ and $v^2(k)$ as functions of the energy $e(\mathbf{p}_k)$. For $\Delta = 0$, $u^2(k)$ and $v^2(k)$ are shown by dashed lines.

if $\Delta(k) \neq 0$, $E_0(\mathbf{p}_k)$ is negative. Thus the non-trivial solution has a lower energy and is the more favorable state of the system. For $\Delta \neq 0$ the energy of a quasi-particle is given by (17.309) which in terms of k_0 can be written as

$$E(\mathbf{p}_k) = \left[\left\{\frac{\hbar^2}{2m}\left(k^2 - k_0^2\right)\right\}^2 + \Delta^2(k)\right]^{\frac{1}{2}}. \tag{17.326}$$

This energy, $E(\mathbf{p}_k)$ for large $|\mathbf{p}_k| = \hbar k$ tends to $\frac{\hbar^2 k^2}{2m}$, but as $\mathbf{p}_k$ tends to the Fermi momentum , $\mathbf{p}_F$, i.e. as $k \to k_0$, $E(\mathbf{p}_k)$ approaches a finite value

$$\lim E(\mathbf{p}_k) \to \Delta(k_0) \quad \text{as} \quad |\mathbf{p}_k| \to \hbar k_0. \tag{17.327}$$

We observe that the difference in the energy of the ground state and the first excited state is $\Delta(k_0)$, and when $\Delta(k_0) \neq 0$ then there is an energy gap in the excitation of this system of fermions. Since the creation and annihilation of quasi-particles are superpositions of fermion and hole states, therefore excitations with $\Delta \neq 0$ correspond to a collective motion of the system.

The only unknown parameter which is left in the expressions for $E_0(\mathbf{p}_k)$ and $E(\mathbf{p}_k)$ is the wave number k_0 which is related to the chemical potential μ by $\mu = \frac{\hbar^2 k_0^2}{2m}$. We can find k_0 from the number of fermions per unit volume, $\frac{N}{L^3}$. If N denotes the total number of particles in the system, and $|\Phi_0\rangle$ is the ground state wave function corresponding to the absence of quasi-particles, then

$$N = \left\langle \Phi_0 \left| \hat{N} \right| \Phi_0 \right\rangle = \left\langle \Phi_0 \left| \sum_{k,s} b_{k,s}^\dagger b_{k,s} \right| \Phi_0 \right\rangle. \tag{17.328}$$

Therefore the action of the operators $B_{k,0}$ and $B_{k,1}$ on $|\Phi_0\rangle$ yield the following results

$$B_{k,0}|\Phi_0\rangle = 0, \tag{17.329}$$

and

$$B_{k,1}|\Phi_0\rangle = 0. \tag{17.330}$$

Now if we substitute for $b_{k,s}$ and $b^\dagger_{k,s}$ from (17.292) and (17.293) and sum over s we find

$$\begin{aligned}\hat{N} &= \sum_k \Big[2v^2(k) + \left(u^2(k) - v^2(k)\right)\left(B^\dagger_{k,0}B_{k,0} + B^\dagger_{k,1}B_{k,1}\right) \\ &+ 2u(k)v(k)\left(B^\dagger_{k,0}B^\dagger_{k,1} + B_{k,0}B_{k,1}\right)\Big]. \end{aligned} \tag{17.331}$$

Next we substitute $\hat{N}$ from (17.331) in (17.328) and then use (17.329) and (17.330) and express the final result in terms of $\Delta(k)$, Eqs. (17.305) and (17.306);

$$N = 2\sum_k v^2(k) = \sum_k \left[1 - \frac{k^2 - k_0^2}{\sqrt{\left(k^2 - k_0^2\right)^2 + \left(\frac{2m}{\hbar^2}\Delta(k)\right)^2}}\right]. \tag{17.332}$$

Finally by replacing the summation by integration as indicated in Eq. (17.319) we get

$$\frac{N}{L^3} = \frac{1}{2\pi^2}\int_0^\infty \left[1 - \frac{k^2 - k_0^2}{\sqrt{\left(k^2 - k_0^2\right)^2 + \left(\frac{2m}{\hbar^2}\Delta(k)\right)^2}}\right] k^2 dk. \tag{17.333}$$

From this equation we can determine k_0^2 in terms of the density of fermions $\frac{N}{L^3}$.

17.12 Damped Harmonic Oscillator

One of the exactly solvable systems with an infinite degrees of freedom is that of the motion of a particle linearly coupled to a heat bath. The heat bath can consist of a collection of oscillators or a vector or a scalar field. This coupling results in the damping motion of the particle which is of great interest in the quantum theory of radiation and the theory of the natural line width [28],[29]. There are a number of models of this type for which the Heisenberg equations of motion are exactly solvable. In this section we will study a model originally discussed by Unruh and Zurek [30] and was later solved by Harris [31] using Heisenberg's equations of motion. Other problems such as the motion of a harmonically bound radiating electron has the same mathematical structure and can be solved in the same way [32].

The Harris model is given by the Lagrangian

$$L_q = \int_L^L \mathcal{L}\left(Q, \dot{Q}, \frac{\partial y}{\partial t}, \frac{\partial y}{\partial x}\right) dx, \tag{17.334}$$

where the Lagrangian density $\mathcal{L}$ is

$$\mathcal{L}\left(Q, \dot{Q}, \frac{\partial y}{\partial t}, \frac{\partial y}{\partial x}\right) = \frac{1}{2}\left[\left(\frac{\partial y}{\partial t}\right)^2 - \left(\frac{\partial y}{\partial x}\right)^2\right]$$

$$+ \quad \frac{1}{2}\delta(x)\left[\dot{Q}^2 - \omega_0^2 Q^2 - 2\epsilon Q\left(\frac{\partial y}{\partial t}\right)\right]. \tag{17.335}$$

In writing this Lagrangian we have assumed that a particle of unit mass is coupled to a string of length $2L$ and that the string is fixed at the two ends, i.e.

$$y(-L,t) = y(L,t) = 0. \tag{17.336}$$

Next we expand $y(x,t)$ in the interval $-L < x < L$ as a Fourier series

$$y(x,t) = \sum_k \frac{1}{\sqrt{L}} Q_k(t)\cos(kx), \tag{17.337}$$

where to satisfy the boundary conditions k must take the values

$$k = \left(n + \frac{1}{2}\right)\frac{\pi}{L}, \quad n = 0,\ 1,\ 2\cdots. \tag{17.338}$$

Terms involving $\sin(kx)$ do not couple to the motion of the of the oscillator and therefore are not included in the sum. By substituting for $y(x,t)$ from (17.337) in (17.335) and (17.334) we find the Lagrangian to be

$$L_q = \frac{1}{2}\left(\dot{Q}^2 - \omega_0^2 Q^2\right) + \frac{1}{2}\sum_k\left(\dot{q}_k^2 - k^2 q_k^2\right) - \frac{\epsilon}{\sqrt{L}}\sum_k \dot{q}_k Q. \tag{17.339}$$

From this Lagrangian we find the Hamiltonian for this coupled system

$$H = \frac{1}{2}\left(P^2 + \omega_0^2 Q^2\right) + \frac{1}{2}\sum_k\left[\left(p_k + \frac{\epsilon Q}{\sqrt{L}}\right)^2 + k^2 q_k^2\right], \tag{17.340}$$

where P, Q, p_k and q_k are operators satisfying the canonical commutation relations

$$[Q,\ P] = [q_k,\ p_k] = i\hbar. \tag{17.341}$$

From these relations with the help of the Heisenberg equation of motion

$$i\hbar\dot{Q} = [Q,\ H], \quad i\hbar\dot{P} = [P,\ H], \tag{17.342}$$

$$i\hbar\dot{q}_k = [q_k,\ H], \quad i\hbar\dot{q}_k = [q_k,\ H], \tag{17.343}$$

we find the operator differential equations for q_k and Q:

$$\ddot{Q} + \omega_0^2 Q = -\frac{\epsilon}{\sqrt{L}}\sum_k \dot{q}_k, \tag{17.344}$$

and

$$\ddot{q}_k + k^2 q_k = \frac{\epsilon}{\sqrt{L}}\dot{Q}. \tag{17.345}$$

These coupled equations are linear in the operators Q and q_k and can be solved analytically. The solution of Eq. (17.343) with the initial conditions $q_k(0)$ and $p_k(0)$ is given by

$$q_k(t) = \frac{\epsilon}{\sqrt{L}} \int_0^t \cos\left[k\left(t-t'\right)\right] Q\left(t'\right) dt' + \left(q_k(0)\cos(kt) + p_k(0)\frac{\sin(kt)}{k}\right). \tag{17.346}$$

Summing (17.346) over all k s and noting that

$$\frac{1}{L}\sum_k \cos\left[k\left(t-t'\right)\right] = \delta\left(t-t'\right), \tag{17.347}$$

we obtain

$$\frac{\epsilon}{\sqrt{L}}\sum_k q_k(t) = \lambda Q(t) + \frac{\epsilon}{\sqrt{L}}\sum_k \left(q_k(0)\cos(kt) + p_k(0)\frac{\sin(kt)}{k}\right), \tag{17.348}$$

where $\lambda = \frac{1}{2}\epsilon^2$. Next we try to eliminate all q_k s from the equation (17.344) and for this we differentiate (17.348) and we substitute for $\sum_k \dot{q}_k(t)$. In this way we find

$$\ddot{Q} + \lambda\dot{Q}(t) + \omega_0^2 Q = \frac{\epsilon}{\sqrt{L}}\sum_k \left[q_k(0)k\sin(kt) - p_k(0)\cos(kt)\right]. \tag{17.349}$$

As the commutation relation

$$\left[q_k(0), p_k(0)\right] = i\hbar, \tag{17.350}$$

shows we cannot set the right-hand side of (17.349) equal to zero, therefore (17.349) will be an inhomogeneous differential equation for $Q(t)$. Let us write the general solution of the operator equation (17.349) as

$$Q(t) = Q(0)f_1(t) + P(0)f_2(t) + \frac{\epsilon}{\sqrt{L}}\sum_k \left[q_k(0)f_3(k,t) - p_k(0)f_4(k,t)\right], \tag{17.351}$$

and then the equation for the momentum operator $P(t)$ can be obtained by differentiating $Q(t)$;

$$P(t) = \dot{Q}(t). \tag{17.352}$$

The time-dependent functions $f_1(t)$, $f_2(t)$, $f_3(k,t)$ and $f_4(k,t)$ will be determined later.

If we assume that at $t = 0$ all of the oscillators q_k are in the ground state, then we can find the average of $Q(t)$ and $P(t)$ over the ground state of the scalar field. If $|0\rangle$ denotes this ground state we have

$$\langle 0|q_k(0)|0\rangle = \langle 0|p_k|0\rangle = 0, \tag{17.353}$$

then by averaging over the state $|0\rangle$ we find

$$\frac{d}{dt}\langle 0|P(t)|0\rangle + \lambda\langle 0|P(t)|0\rangle + \omega_0^2\langle 0|Q(t)|0\rangle = 0, \tag{17.354}$$

and

$$\frac{d}{dt}\langle 0|Q(t)|0\rangle = \langle 0|P(t)|0\rangle. \tag{17.355}$$

From Eq. (17.351) it follows that

$$\langle 0|Q(t)|0\rangle = \langle 0|Q(0)|0\rangle f_1(t) + \langle 0|P(0)|0\rangle f_2(t), \tag{17.356}$$

and by substituting this in Eqs. (17.354) and (17.355) we find that both $f_1(t)$ and $f_2(t)$ are solutions of the differential equation

$$\ddot{f}_i + \lambda\dot{f}_i + \omega_0^2 f_i = 0, \quad i = 1,\ 2, \tag{17.357}$$

but with the boundary conditions

$$f_1(0) = 1, \quad \dot{f}_1(0) = 0, \tag{17.358}$$

and

$$f_2(0) = 0, \quad \dot{f}_2(0) = 1. \tag{17.359}$$

For small coupling $\lambda < \omega_0$ the solution of (17.357) with the boundary conditions (17.358) and (17.359) are damped sinusoidal motions:

$$f_1(t) = e^{-\frac{\lambda t}{2}}\left[\cos(\omega t) + \frac{\lambda}{2\omega}\sin(\omega t)\right], \tag{17.360}$$

and

$$f_2(t) = \frac{1}{\omega}e^{-\frac{\lambda t}{2}}\sin(\omega t), \tag{17.361}$$

with the shifted frequency ω;

$$\omega = \left(\omega_0^2 - \frac{\lambda^2}{4}\right)^{\frac{1}{2}}. \tag{17.362}$$

The other two functions $f_3(k,t)$ and $f_4(k,t)$ can be determined by substituting (17.351) in (17.349) and matching the coefficients of $q_k(0)$ and $p_k(0)$ on the two sides. In this way we find that these two functions are the special solutions of the inhomogeneous differential equations

$$\ddot{f}_3(k,t) + \lambda\dot{f}_3(k,t) + \omega_0^2 f_3(k,t) = k\sin(kt), \tag{17.363}$$

and

$$\ddot{f}_4(k,t) + \lambda\dot{f}_4(k,t) + \omega_0^2 f_4(k,t) = \cos(kt). \tag{17.364}$$

The solutions of these equations are given by

$$\begin{aligned} f_3(k,t) &= \frac{k^2 e^{s_+ t}}{(s_+ - s_-)(s_+^2 + k^2)} + \frac{ike^{ikt}}{2(k+is_+)(k+is_-)} \\ &+ \frac{k^2 e^{s_- t}}{(s_- - s_+)(s_-^2 + k^2)} - \frac{ike^{-ikt}}{2(k-is_-)(k-is_+)}, \end{aligned} \tag{17.365}$$

and

$$f_4(k,t) = \frac{s_+ e^{s_+ t}}{(s_+ - s_-)(s_+^2 + k^2)} + \frac{e^{ikt}}{2(k + is_+)(k + is_-)}$$
$$+ \frac{s_- e^{s_- t}}{(s_- - s_+)(s_-^2 + k^2)} + \frac{e^{-ikt}}{2(k - is_-)(k - is_+)}, \quad (17.366)$$

where $s_\pm = \frac{\lambda}{2} \pm i\omega$ are the roots of the quadratic equations

$$s^2 + \lambda s + \omega_0^2 = 0. \quad (17.367)$$

From Eqs. (17.355) and (17.356) we can determine the commutator of the averages of $Q(t)$ and $P(t)$;

$$[\langle 0|Q(t)|0\rangle, \langle 0|P(t)|0\rangle] = [\langle 0|Q(0)|0\rangle, \langle 0|P(0)|0\rangle]$$
$$\times \left(f_1(t)\dot{f}_2(t) - \dot{f}_1(t)f_2(t)\right) = [\langle 0|Q(0)|0\rangle, \langle 0|P(0)|0\rangle]\, e^{-\lambda t}. \quad (17.368)$$

This relation shows that when the motion of the particle is averaged over $q_k(0)$ and $p_k(0)$, the commutator $[\langle 0|Q(t)|0\rangle, \langle 0|P(t)|0\rangle]$ decreases exponentially in time. The reason for the time-dependence of the commutation relation (17.365) is the fact that we have not included the contribution of the commutator $[q_k(0), p_k(0)] = i\hbar$ in our calculation. If we calculate $[Q(t), P(t)]$ using the exact relations (17.351) and (17.352) we obtain

$$[\langle 0|Q(t)|0\rangle, \langle 0|P(t)|0\rangle] = [\langle 0|Q(0)|0\rangle, \langle 0|P(0)|0\rangle]$$
$$\times \left\{f_1(t)\dot{f}_2(t) - \dot{f}_1(t)f_2(t)\right\} - \frac{\epsilon^2}{L}\sum_k [q_k(0), p_k(0)]$$
$$\times \left\{f_3(k,t)\dot{f}_4(k,t) - \dot{f}_3(k,t)f_4(k,t)\right\}, \quad (17.369)$$

Using $f_3(k,t)$ and $f_4(k,t)$ we calculate the coefficient of $[q_k(0), p_k(0)]$ in (17.369)

$$\frac{\epsilon^2}{L}\sum_k \left\{f_3(k,t)\dot{f}_4(k,t) - \dot{f}_3(k,t)f_4(k,t)\right\} = -\left(1 - e^{-\lambda t}\right). \quad (17.370)$$

By substituting (17.370) in (17.369) we find

$$[Q(t), P(t)] = i\hbar. \quad (17.371)$$

i.e. the exact commutator remains a constant of motion.

Next let us calculate the expectation values of $Q^2(t)$ and $P^2(t)$ over the ground state of the string. These are given by

$$\langle 0|Q^2(t)|0\rangle = \langle 0|Q^2(0)|0\rangle f_1^2(t) + \langle 0|P^2(0)|0\rangle f_2^2(t)$$
$$+ \langle 0|Q(0)P(0) + P(0)Q(0)|0\rangle f_1(t)f_2(t) + \frac{\lambda}{2\pi}g_1(t), \quad (17.372)$$

$$\begin{aligned}\left\langle 0\left|P^2(t)\right|0\right\rangle &= \left\langle 0\left|Q^2(0)\right|0\right\rangle \dot{f}_1^2(t) + \left\langle 0\left|P^2(0)\right|0\right\rangle \dot{f}_2^2(t) \\ &+ \langle 0|Q(0)P(0)+P(0)Q(0)|0\rangle \dot{f}_1(t)\dot{f}_2(t) + \frac{\lambda}{2\pi} g_2(t),\end{aligned} \tag{17.373}$$

and

$$\langle 0|Q(t)P(t)+P(t)Q(t)|0\rangle = \frac{d}{dt}\left\langle 0\left|Q^2(t)\right|\right\rangle. \tag{17.374}$$

where

$$g_1(t) = \hbar\int_0^\infty k\left[\frac{1}{k^2}f_3^2(k,t) + f_4^2(k,t)\right]dk, \tag{17.375}$$

and

$$g_2(t) = \hbar\int_0^\infty k\left[\frac{1}{k^2}\dot{f}_3^2(k,t) + \dot{f}_4^2(k,t)\right]dk. \tag{17.376}$$

The last two equations are found by taking the limit of $L \to \infty$ and in this limit replace the summation over k by integration, i.e.

$$\sum_k \to \frac{L}{2}\int_{-\infty}^{\infty} dk. \tag{17.377}$$

As we can see from Eq. (17.376) $g_2(t)$ is logarithmically divergent and thus $\left\langle 0\left|P^2\right|0\right\rangle$ is infinite, and this is a defect of the model. We can remedy this defect by making ϵ a function of k in such a way that the high frequency oscillators decouple from the motion of the particle. This has been done in van Kampen's model [32] and in Ullersma's model [33]. For a detailed discussion of this and other models the reader is referred to the book "*Classical and Quantum Dissipative Systems*" and references therein [34].

The Schrödinger–Langevin Equation — We have found the equations of motion for damped harmonic oscillator, (17.354),(17.355), in the Heisenberg picture. What is the form of the Schrödinger equation for this damped motion? To find the answer to this question we first simplify the notation and define the operators $\bar{P}$ and $\bar{Q}$ by

$$\bar{P}(t) = \langle 0|P(t)|0\rangle, \quad \text{and} \quad \bar{Q}(t) = \langle 0|O(t)|0\rangle. \tag{17.378}$$

Then the operator equations (17.354),(17.355) become

$$\frac{d}{dt}\bar{P}(t) + \lambda\bar{P}(t) + \omega_0^2\bar{Q}(t) = 0, \tag{17.379}$$

and

$$\frac{d}{dt}\bar{Q}(t) = \bar{P}(t). \tag{17.380}$$

These equations are given in the Heisenberg picture where the operators are time-dependent. To find the corresponding Schrödinger equation we will use the well-known result of the connection between the Heisenberg picture and the Schrödinger picture, (see Eq. (4.12)); and

$$\langle \bar{P} \rangle = \langle \psi(\bar{Q},t)|\bar{P}|\psi(\bar{Q},t)\rangle, \tag{17.381}$$

where $\psi(\bar{Q},t)$ is the time-dependent wave function and $\bar{P}$ is the momentum operator. Noting that in the Schrödinger picture $\bar{P} = -i\hbar\frac{\partial}{\partial \bar{Q}}$, we have

$$\begin{aligned}\langle \bar{P} \rangle &= \langle \psi(\bar{Q},t)|\bar{P}\psi(\bar{Q},t)\rangle = \langle \bar{P}\psi(\bar{Q},t)|\psi(\bar{Q},t)\rangle \\ &= \frac{\hbar}{2i}\int\left[\frac{\partial\psi(\bar{Q},t)}{\partial\bar{Q}}\psi^*(\bar{Q},t) - \frac{\partial\psi^*(\bar{Q},t)}{\partial\bar{Q}}\psi(\bar{Q},t)\right]d\bar{Q}.\end{aligned} \tag{17.382}$$

Now we ask wether it is possible to write $\langle \bar{P} \rangle$ which is present in the Heisenberg equation, as a potential term in the Schrödinger equation. Assuming the existence of such a potential term $V_L(\bar{Q},t)$ we can write the time-dependent Schrödinger equation as

$$i\hbar\frac{\partial\psi}{\partial t} = -\frac{\hbar^2}{2m}\frac{\partial^2\psi}{\partial\bar{Q}^2} + \frac{1}{2}\bar{Q}^2\psi + V_L(\bar{Q},t)\psi = 0. \tag{17.383}$$

Now if we differentiate (17.382) with respect to time we have

$$\frac{d\left\langle \bar{P}(t)\right\rangle}{dt} = \left\langle \frac{\partial\psi}{\partial t}\bar{P}\psi\right\rangle + \left\langle \psi\bar{P}\frac{\partial\psi}{\partial t}\right\rangle. \tag{17.384}$$

By substituting from Eq. (17.383) in (17.384) we obtain

$$\frac{d\left\langle \bar{P}(t)\right\rangle}{dt} = -\omega_0^2\left\langle \bar{Q}(t)\right\rangle - \left\langle \frac{\partial V_L(\bar{Q},t)}{\partial\bar{Q}}\right\rangle. \tag{17.385}$$

Comparing Eq. (17.385) with the expectation value of the operator equation (17.379) we find that

$$\left\langle \frac{\partial V_L(\bar{Q},t)}{\partial\bar{Q}}\right\rangle = \lambda\langle \bar{P}(t)\rangle. \tag{17.386}$$

Writing this solution in the expanded form and substituting for $\langle \bar{P} \rangle$ from (17.382) we get

$$\begin{aligned}&\int \psi^*(\bar{Q},t)\frac{\partial V_L(\bar{Q},t)}{\partial\bar{Q}}\psi(\bar{Q},t)d\bar{Q} \\ &= \frac{\lambda\hbar}{2i}\int\left[\frac{\partial\psi(\bar{Q},t)}{\partial\bar{Q}}\psi^*(\bar{Q},t) - \frac{\partial\psi^*(\bar{Q},t)}{\partial\bar{Q}}\psi(\bar{Q},t)\right]d\bar{Q}.\end{aligned} \tag{17.387}$$

This equation must be true for any acceptable wave function $\psi(\bar{Q},t)$, therefore we conclude that $V_L(\bar{Q},t)$ is of the form

$$V_L(\bar{Q},t) = \frac{\lambda\hbar}{2i}\ln\left[\frac{\psi(\bar{Q},t)}{\psi^*(\bar{Q},t)}\right] + W(t), \tag{17.388}$$

where $W(t)$ is a function of time. This function can be determined by requiring that the expectation value of the energy $\langle\psi(\bar{Q},t)\,|E(t)|\,\psi(\bar{Q},t)\rangle$ be equal to the sum of the expectation values of the kinetic energy, the potential energy $\frac{1}{2}\omega_0^2\bar{Q}^2$, and the potential $V_L(\bar{Q},t)$ [35]. The essential feature of this equation is its nonlinearity. Thus while the whole system satisfies the linear Schrödinger equation, for a subsystem, i.e. the central oscillator, the wave equation is nonlinear. This is an important point in the discussion of decoherence in the measurement process.

Removal of High Frequency Divergence for Many-Body Problems — In most of the solvable models the Hamiltonian is a quadratic function of p_k s and q_k s as in the case of Harris's model or Ullersma's model [34]. Any one of these Hamiltonians can be diagonalized and cast in the form of the Hamiltonian of an infinitely many non-interacting harmonic oscillators. This is achieved by a series of canonical transformations [32]. The final result is of the general form

$$H = \frac{1}{2}\sum_{k=1}^{\infty}\left(p_k^2 + \omega_k^2 q_k^2\right), \tag{17.389}$$

where ω_k is determined by the natural frequencies of the oscillators in the system and the coupling between them. If $\omega_k \to 0$ as $k \to \infty$ then the energy of the system will be finite. Otherwise the energy eigenvalues will be infinite. An example of the latter case is provided by considering the zero-point energy of the wave in a string which is given by

$$E^{(0)} = \sum_{k=1}^{\infty}\frac{1}{2}\hbar\omega_k. \tag{17.390}$$

If the string is of length L and is fixed at both ends then $\omega_k = \frac{k\pi c}{L}$ and therefore

$$E^{(0)} = \sum_{k=1}^{\infty}\left(\frac{\hbar\pi c}{2L}\right)k \to \infty. \tag{17.391}$$

In most of the problems this zero-point energy is omitted since the energy differences are observables of interest.

We noted earlier that at least for the harmonic oscillator the commutation relation cannot be uniquely determined from the equations of motion and the Hamiltonian. One can ask whether this lack of uniqueness of the commutator may be used to eliminate some of the divergencies occurring in the quantum theory of systems with infinite degrees of freedom. This question has been studied by Schweber for a simple quantized scalar field theory [36].

Let us write the general form of the commutation relation (5.34) with $\hbar$ appearing explicitly;

$$\left([p_k,\ q_k] + i\hbar\right)^2 = -\hbar^2 w_k^2. \tag{17.392}$$

In this commutator w_k is given by

$$w_k^{(n)} = (-1)^n\left[\left(\frac{2E_k^{(0)}(\omega_k)}{\hbar\omega_k}\right) - 1\right], \tag{17.393}$$

where $E_k^{(0)}(\omega_k)$ is the zero point energy of the k-th oscillator. We can choose $E_k^{(0)}(\omega_k)$ as we like, as long as it is greater than zero and less than $\frac{1}{2}\hbar\omega_k$. Now the energy eigenvalues of the Hamiltonian (17.389) are given by

$$\sum_{k=1}^{\infty} E_k^{(n)} = \sum_{k=1}^{\infty}\left(n\hbar\omega_k + E_k^{(0)}(\omega_k)\right). \tag{17.394}$$

The eigenvalues for finite number of excitations remain finite provided that for a given oscillator $E_k^{(0)}(\omega_k)$ goes to zero as $\omega_k \to \infty$. For instance we can choose $E_k^{(0)}(\omega_k)$ to be

$$E_k^{(0)}(\omega_k) = \frac{1}{2}\hbar\Omega_k \exp\left(-\frac{\omega_k^2}{\Omega_k^2}\right), \tag{17.395}$$

where Ω_k is a constant with the dimension of angular frequency. For this particular choice of $E_k^{(0)}(\omega_k)$ the result of calculation of the matrix elements of q_k, can be found analytically, e.g.

$$\langle 0_k|q_k|1_k\rangle = \langle 1_k|q_k|0_k\rangle = \sqrt{\frac{E_k^{(0)}(\omega_k)}{\omega_k^2}} = \sqrt{\frac{\hbar\Omega_k}{2\omega_k^2}}\exp\left(-\frac{\omega_k^2}{2\Omega_k^2}\right). \tag{17.396}$$

This result shows a cut-off for high frequencies and hence the removal of some of the divergent quantities from the problem.
The problem with this idea is that the relativistic invariance prevents such a cut-off in quantized field theory [36].

Bibliography

[1] D.R. Hartree, Wave mechanics of an atom with non-Coulomb central field, I. Theory and methods, II. Some results and discussion. Proc. Camb. Phil. Soc. 24, 89 (1928).

[2] V. Fock, Näherungsmethode zur Lösung des quantenmechanischen Mehrkörper problems, Z. Phys. 61, 126 (1930).

[3] J.C. Slater, A simplification of the Hartree–Fock method, Phys. Rev. 81, 385 (1951).

[4] For an excellent and detailed account of the Hartree–Fock method see J.C. Slater, *Quantum Theory of Atomic Structure*, Vol. II, (McGraw-Hill, New York, 1960), Chap. 17.

[5] M. Moshinsky, How good is the Hartree–Fock approximation, Am. J. Phys. 35, 52 (1968).

[6] M. Moshinsky, How good is the Hartree–Fock approximation, Errata, Am. J. Phys. 36, 763 (1968).

[7] M. Moshinsky, How good is the Hartree–Fock approximation, II. The case of closed shells, Am. J. Phys. 58, 456 (1970).

[8] M. de Llano, A solvable self-consistent nuclear model, Am. J. Phys. 41, 484 (1973).

[9] Y. Nogami, M. Vallières and W. van Dijk, Hartree–Fock approximation for the one-dimensional "helium atom".

[10] L.L. Foldy, An intersting exactly soluble one-dimensional Hartree–Fock problem, Am. J. Phys. 44, 1192 (1976).

[11] N.H. March, W.H. Young and S. Sampanthar, *The Many-Body Problem in Quantum Mechanics*, (Cambridge University Press, London, 1967), Chapter 3.

[12] A.L. Fetter and J.D. Walecka, *Quantum Theory of Many-Particle Systems*, (McGraw-Hill, New York, 1971), Chapter 1.

[13] D.S. Koltun and J.M. Eisenberg, *Quantum Mechanics of Many Degrees of Freedom*, (John Wiley & Sons, New York, 1971), Chapter 1.

[14] J.-M. Lévy-Leblond, Generalized uncertainty relations for many-fermion system, Phys. Lett. 26 A, 540 (1968).

[15] See for instance, M.A. Preston, *Physics of the nucleous*, (Addison-Wesley, Reading, 1962).

[16] W. Greiner, *Quantum Mechanics, Special Chapters*, (Springer-Verlag, Berlin), p. 218.

[17] G. Baym, *Lectures on Quantum Mechanics*, (W.A. Benjamin, Reading, 1969), p. 427.

[18] R.W. Richardson, Exactly solvable many-boson models, J. Math. Phys. 9, 1327 (1968).

[19] W.H. Bassichis and L.L. Foldy, Analysis of the Bogoliubov method applied to a simple boson model, Phys. Rev. A 133, 935 (1964).

[20] N.N. Bogoliubov, On the theory of superfluidity, J. Phys. USSR, 11, 23 (1947), translated in D. Pines *The Many-Body Problem*, (W.A. Benjamin, New York, 1961) p. 292.

[21] N.N. Bogoliubov and N.N. Bogoliubov Jr. *Introduction to Quantum Statistical Mechanics*, translated by V.P. Gupta, (World Scientific, Singapore, 1982).

[22] A.S. Davidov, *Quantum Mechanics*, Second Edition, translated by D. ter Haar, (Pergamon Press, Oxford 1965), p. 356.

[23] M. Razavy and E.S. Krebes, Theory of Bruekner and Sawada applied to a many-boson model, Can. J. Phys. 51, 292 (1973).

[24] W.A. Evans and Y. Imry, On the pairing theory of the Bose superfluid, Nuovo Cimento, 63 B, 155, (1960).

[25] J. Bardeen, L.N. Cooper and J.P. Schrieffer, Theory of superconductivity, Phys. Rev. 108, 1175 (1957).

[26] M. Razavy and M. Binder, Scattering of two quasi-particles, Can. J. Phys. 49, 501 (1971).

[27] H. Lehman, K. Symanzik and W. Zimmermann, On the formulation of the quantized field theory Nuovo Cimento, 2, 425 (1955).

[28] W.H. Louisell, *Quantum Properties of Radiation*, (John Wiley & Sons, New York, 1073).

[29] V. Weisskopf and E. Wigner, Berechnung der natürlichen Linenbreite auf Grund der Diracschen Lichttheorie, Z. Physik, 63, 54 (1930).

[30] W.G. Unruh and W.H. Zurek, Reduction of a wavepacket in quantum Brownian motion, Phys. Rev. D 40, 1071 (1989).

[31] E.G. Harris, Quantum theory of the damped harmonic oscillator, Phys. Rev. 42 A, 3685 (1990).

[32] N.G. van Kampen, Contribution to the quantum theory of light, Dans. mat. fys. Medd, 26, 77 (1951).

[33] Ullersma, An exactly solvable model for the Brownian motion, I. Derivation of the Langevin equation, Physica 32, 27 (1966).

[34] M. Razavy, *Classical and Quantum Dissipative Systems*, (Imperial College Press, 2005).

[35] M.D. Kostin, On the Schrödinger–Langevin equation, J. Chem. Phys. 57, 3589 (1972).

[36] S. Schweber, A note on commutators in quantized field theory, Phys. Rev. 78, 613 (1950).

Chapter 18

Quantum Theory of Free Electromagnetic Field

The free electromagnetic field can be described by the electric and magnetic fields $\mathbf{E}$ and $\mathbf{B}$ satisfying Maxwell's equations (16.24)–(16.26) but with no charge or current present $\rho = 0$ and $\mathbf{J} = 0$. Alternatively we can write these fields in terms of the electromagnetic potentials $\mathbf{A}$ and ϕ, related to $\mathbf{E}$ and $\mathbf{B}$ by

$$\mathbf{E} = -\frac{1}{c}\frac{\partial \mathbf{A}}{\partial t} - \nabla\phi, \tag{18.1}$$

and

$$\mathbf{B} = \nabla \wedge \mathbf{A}. \tag{18.2}$$

Since there are no sources present, we choose the Coulomb gauge for ϕ and $\mathbf{A}$;

$$\phi = 0, \quad \nabla \cdot \mathbf{A} = 0. \tag{18.3}$$

From Eqs. (16.24)–(16.26) and (18.1)–(18.3) we find that $\mathbf{A}$ is the solution of the wave equation

$$\nabla^2\mathbf{A} - \frac{1}{c^2}\frac{\partial^2 \mathbf{A}}{\partial t^2} = 0. \tag{18.4}$$

We can write the classical Hamiltonian for this field in terms of $\mathbf{E}$ and $\mathbf{B}$ or in terms of $\mathbf{A}$:

$$H = \frac{1}{8\pi}\int \left(\mathbf{E}^2 + \mathbf{B}^2\right) d^3r = \frac{1}{8\pi}\int \left\{\frac{1}{c^2}\left|\frac{\partial \mathbf{A}}{\partial t}\right|^2 + |\nabla \wedge \mathbf{A}|^2\right\} d^3r. \tag{18.5}$$

To quantize this Hamiltonian we want to express the field in terms of a set of discrete variables. To this end we assume a large cubical box of volume L^3 and impose periodic boundary conditions on the magnetic potential $\mathbf{A}$

$$\mathbf{A}(0, y, z, t) = \mathbf{A}(L, y, z, t), \tag{18.6}$$

$$\mathbf{A}(x, 0, z, t) = \mathbf{A}(x, L, z, t), \tag{18.7}$$

$$\mathbf{A}(x, y, 0, t) = \mathbf{A}(x, y, L, t), \tag{18.8}$$

Next we write $\mathbf{A}$ as a Fourier series;

$$\mathbf{A}(\mathbf{r}, t) = \sum_{\mathbf{k}, k_z > 0} \sum_{\sigma=1}^{2} \left(\frac{2\pi\hbar c^2}{L^3 \omega_k}\right)^{\frac{1}{2}} \mathbf{u}_{\mathbf{k},\sigma} \left\{a_{\mathbf{k}\sigma}(t) e^{i\mathbf{k}\cdot\mathbf{r}} + a^*_{\mathbf{k}\sigma}(t) e^{-i\mathbf{k}\cdot\mathbf{r}}\right\}. \tag{18.9}$$

In this expression the factor $\left(\frac{2\pi\hbar c^2}{L^3\omega_k}\right)^{\frac{1}{2}}$ is the normalization coefficient and $\mathbf{u}_{\mathbf{k},1}$ and $\mathbf{u}_{\mathbf{k},2}$ are two orthogonal vectors denoting the polarization of $\mathbf{A}$. Now in order to satisfy Eq. (18.3) we need to impose the condition that

$$\mathbf{u}_{\mathbf{k},1} \cdot \mathbf{k} = \mathbf{u}_{\mathbf{k},2} \cdot \mathbf{k} = 0. \tag{18.10}$$

In addition to satisfy the boundary conditions (18.6)–(18.8) the components of the vector $\mathbf{k}$ must be of the form

$$(n_x, n_y, n_z)\frac{2\pi}{L}, \tag{18.11}$$

where n_x, n_y and n_z are integers. We note that $\mathbf{A}$ is a real vector and since both $e^{i\mathbf{k}\cdot\mathbf{r}}$ and $e^{-i\mathbf{k}\cdot\mathbf{r}}$ are included in each term of (18.9) we have restricted the summation over $\mathbf{k}$ to the one-half of the $\mathbf{k}$ space, and therefore the condition $k_z > 0$. If we substitute (18.9) in (18.4) we find that the Fourier components $a_{\mathbf{k},\sigma}(t)$ satisfy the equations of motion

$$\frac{d^2 a_{\mathbf{k},\sigma}(t)}{d\,t^2} + \omega_k^2 a_{\mathbf{k},\sigma}(t) = 0, \tag{18.12}$$

where $\omega_k = ck$ is the angular frequency. We can write the general solution of (18.12) in terms of complex exponentials

$$a_{\mathbf{k},\sigma}(t) = a^{(1)}_{\mathbf{k},\sigma}(0) e^{-i\omega_k t} + a^{(2)}_{\mathbf{k},\sigma}(0) e^{i\omega_k t}. \tag{18.13}$$

To get rid of the condition $k_z > 0$ we extend the definition of $a_{\mathbf{k},\sigma}(t)$ to include positive as well as negative k_z s

$$\begin{cases} a_{\mathbf{k},\sigma}(0) = a^{(1)}_{\mathbf{k},\sigma}(0) & \text{for} \quad k_z > 0 \\ a_{\mathbf{k},\sigma}(0) = a^{(1)}_{-\mathbf{k},\sigma}(0) & \text{for} \quad k_z < 0 \end{cases}, \tag{18.14}$$

and write $\mathbf{A}(\mathbf{r}, t)$ as a sum over all $\mathbf{k}$ s

$$\mathbf{A}(\mathbf{r}, t) = \sum_{\mathbf{k},\sigma} \left(\frac{2\pi\hbar c^2}{L^3 \omega_k}\right)^{\frac{1}{2}} \mathbf{u}_{\mathbf{k},\sigma} \left\{a_{\mathbf{k}\sigma}(t) e^{i\mathbf{k}\cdot\mathbf{r}} + a^*_{\mathbf{k}\sigma}(t) e^{-i\mathbf{k}\cdot\mathbf{r}}\right\}, \tag{18.15}$$

where

$$a_{\mathbf{k},\sigma}(t) = a_{\mathbf{k},\sigma}(0)e^{-i\omega_k t}, \tag{18.16}$$

Next we substitute the last expression that we have found for $\mathbf{A}(\mathbf{r},t)$ in the Hamiltonian (18.5) and carry out the integration over the volume using the integral

$$\int_{L^3} e^{i(\mathbf{k}-\mathbf{k}')\cdot\mathbf{r}} d^3r = L^3\delta_{\mathbf{k},\mathbf{k}'}, \tag{18.17}$$

and we get a simple form for the Hamiltonian operator

$$H = \frac{1}{2}\sum_{\mathbf{k},\sigma}\hbar\omega_k\left(a_{\mathbf{k},\sigma}a^*_{\mathbf{k},\sigma} + a^*_{\mathbf{k},\sigma}a_{\mathbf{k},\sigma}\right). \tag{18.18}$$

In obtaining (18.18) we have used the orthogonality condition of $\mathbf{u}_{\mathbf{k}\sigma}$;

$$\mathbf{u}_{\mathbf{k}\sigma}\cdot\mathbf{u}_{\mathbf{k},\sigma'} = \delta_{\sigma,\sigma'}. \tag{18.19}$$

The Energy Associated with the Electromagnetic Field — The quantum mechanical Hamiltonian correspondig to (18.18) is found by taking $a_{\mathbf{k},\sigma}$ to be the annihilation operator for a photon of momentum $\hbar\mathbf{k}$ and frequency ω_k. Its adjoint which is $a^\dagger_{\mathbf{k},\sigma}$ replaces $a^*_{\mathbf{k},\sigma}$ in (18.18). These two operators satisfy the canonical commutation relations

$$\left[a_{\mathbf{k},\sigma},\, a^\dagger_{\mathbf{k}',\sigma'}\right] = \delta_{\mathbf{k},\mathbf{k}'}\delta_{\sigma,\sigma'}, \tag{18.20}$$

$$\left[a_{\mathbf{k},\sigma},\, a_{\mathbf{k}',\sigma'}\right] = \left[a^\dagger_{\mathbf{k},\sigma},\, a^\dagger_{\mathbf{k}',\sigma'}\right] = 0. \tag{18.21}$$

Rearranging the second term in (18.18) using the commutation relation we find the quantized Hamiltonian for the free electromagnetic field to be

$$H = \sum_{\mathbf{k},\sigma}\hbar\omega_k\left(a^\dagger_{\mathbf{k},\sigma}a_{\mathbf{k},\sigma} + \frac{1}{2}\right). \tag{18.22}$$

The Momentum of Electromagnetic Field — The momentum of the electromagnetic field can be found from the classical expression for the Poynting vector $\mathbf{S}$ [1]

$$\mathbf{P} = \int_{L^3}\frac{\mathbf{E}\wedge\mathbf{B}}{4\pi c}d^3r = \frac{1}{c^2}\mathbf{S}. \tag{18.23}$$

By substituting for $\mathbf{E}$ and $\mathbf{B}$ in terms of the magnetic potential $\mathbf{A}$, as is given in Eqs. (18.1) and (18.2) and by integrating over the volume L^3 we obtain

$$\mathbf{P} = \sum_{\mathbf{k},\sigma}\hbar\mathbf{k}a^\dagger_{\mathbf{k},\sigma}a_{\mathbf{k},\sigma}. \tag{18.24}$$

We can interpret the results of Eqs. (18.22) and (18.24) in the following way: In the representation where $a^\dagger_{\mathbf{k},\sigma}a_{\mathbf{k}}$ is diagonal, i.e.

$$a^\dagger_{\mathbf{k},\sigma}a_{\mathbf{k},\sigma}|n_{\mathbf{k},\sigma}\rangle = n_{\mathbf{k},\sigma}|n_{\mathbf{k},\sigma}\rangle, \tag{18.25}$$

the electromagnetic field consists of $n_{\mathbf{k},\sigma}$ photons each with the energy $\hbar\omega_k$, momentum $\hbar\mathbf{k}$ and polarization σ. The operators $a^\dagger_{\mathbf{k},\sigma}$ and $a_{\mathbf{k},\sigma}$ are the creation and annihilation operators for photons with these properties.

In quantum field theory the equations of motion are in the form of wave equations, and they are obtained by the variation of the Hamiltonian density operator. From these two, i.e. the wave equation and the Hamiltonian density, we can derive the commutation relation between the wave field and the corresponding momentum density. This can be done in a way similar to the Wigner's derivation of the commutator for the harmonic oscillator. Here again we observe that the commutator cannot be obtained uniquely. But if we impose the requirement of the Lorentz invariance, then the resulting commutator found in this way will depend on a parameter, and by varying this parameter nothing interesting will result [2].

Uncertainty Principle for the Quantized Electromagnetic Field — In the classical electromagnetic theory the momentum density of the field is given by

$$\frac{d\mathbf{p}}{dV} = \frac{1}{4\pi c}(\mathbf{E}\wedge\mathbf{B}), \tag{18.26}$$

For a small volume δV, the x component of momentum is [1]

$$\frac{\Delta p_x}{\delta V} \sim \frac{1}{4\pi c}(\mathbf{E}\wedge\mathbf{B})_x = \frac{1}{4\pi c}(E_yB_z - E_zB_y). \tag{18.27}$$

Now assuming that the uncertainty in Δp_x will be of the same order as the uncertainties in the values of the field components, we have

$$\Delta p_x \Delta x \sim \frac{\delta V}{4\pi c}\Delta E_y \Delta B_z \Delta x \geq \frac{\hbar}{2}. \tag{18.28}$$

Writing Eq. (18.28) as

$$\Delta E_y \Delta B_z \geq 2\pi c\hbar \left(\frac{1}{\Delta x \delta V}\right). \tag{18.29}$$

we have the products of the uncertainties in E_y and B_z fields or E_z and B_y. This result and similar results for $\Delta E_x \Delta B_y$ and $\Delta E_z \Delta B_x$ indicate that by trying to measure $\mathbf{E}$ and $\mathbf{B}$ fields on increasingly smaller distances, the values of these fields become increasingly uncertain.

Now the energy per unit volume of the electromagnetic field is equal to $\frac{1}{8\pi}\left(\mathbf{E}^2 + \mathbf{B}^2\right)$, therefore the energy contained in this field fluctuates widely at short distances [3].

18.1 Coherent State of the Radiation Field

Coherent state of a field describes a quantized oscillating field which is closest to a classical wave. Since the classical electromagnetic fields $\mathbf{E}$ and $\mathbf{H}$ can be characterized by their amplitudes as well as their phases, we need to find quantum states where the expectation values of these field exhibit these characteristics [4],[5]. From Eq. (18.9) for $\mathbf{A}$ we can determine $\mathbf{E}$ for the radiation field where the scalar potential ϕ is zero. For simplicity we consider a single mode for the field

$$\mathbf{E} = -\frac{1}{c}\frac{\partial \mathbf{A}}{\partial t} = -i\left(\frac{2\pi\hbar\omega}{L^3}\right)^{\frac{1}{2}} \mathbf{u}\left[ae^{i\mathbf{k}\cdot\mathbf{r}} - a^\dagger e^{-i\mathbf{k}\cdot\mathbf{r}}\right], \tag{18.30}$$

where L^3 is the volume of the space under consideration. Here we have dropped the subscripts $\mathbf{k}$ and σ from $\mathbf{u}$, and a and $a^\dagger$ stand for $a(t)$ and $a^\dagger(t)$. The summation over these indices can easily be reintroduced. Supposing that there are n photons in the field, then from $\langle n|a|n\rangle = \langle n\left|a^\dagger\right|n\rangle = 0$ it follows that

$$\langle n|\mathbf{E}|n\rangle = 0. \tag{18.31}$$

But in the photon number description $|n\rangle$, the expectation value of the $\mathbf{E}^2$ is not zero

$$\langle n\left|\mathbf{E}^2\right|n\rangle = \frac{4\pi\hbar\omega}{L^3}\left(n + \frac{1}{2}\right). \tag{18.32}$$

Equation (18.31) shows that when n photons are present in the field and they have random phases, then the average of $\mathbf{E}$ over the field is zero.

The "uncertainties" of the electric and magnetic fields for a state in which n photons are present can be found from the expectation values of $\mathbf{E}$, $\mathbf{E}^2$, $\mathbf{H}$, and $\mathbf{H}^2$. These are

$$\begin{aligned} \Delta E &= \left\{\langle n\left|\mathbf{E}^2\right|n\rangle - \langle n|\mathbf{E}|n\rangle^2\right\}^{\frac{1}{2}} \\ &= \left(\frac{4\pi\hbar\omega}{L^3}\right)^{\frac{1}{2}}\sqrt{\left(n+\frac{1}{2}\right)}, \end{aligned} \tag{18.33}$$

$$\begin{aligned} \Delta H &= \left\{\langle n\left|\mathbf{H}^2\right|n\rangle - \langle n|\mathbf{H}|n\rangle^2\right\}^{\frac{1}{2}} \\ &= \left(\frac{\hbar\omega c^2}{4\pi L^3}\right)^{\frac{1}{2}}\sqrt{\left(n+\frac{1}{2}\right)}, \end{aligned} \tag{18.34}$$

and

$$\Delta E\Delta H = \frac{\hbar\omega c}{L^3}\left(n + \frac{1}{2}\right). \tag{18.35}$$

Now to introduce the coherent states we first need to define the phase operator. The phase operator, $\widehat{\varphi}$, is given by [6]–[8]

$$e^{i\widehat{\varphi}} = \left(\hat{N} + 1\right)^{-\frac{1}{2}} a, \tag{18.36}$$

and

$$e^{-i\widehat{\varphi}} = a^\dagger\left(\hat{N} + 1\right)^{-\frac{1}{2}}, \tag{18.37}$$

where $\hat{N}$ is the number operator $\hat{N}|n\rangle = n|n\rangle$. From the relation $aa^\dagger = \hat{N}+1$ it follows that

$$e^{\widehat{i\varphi}}e^{-\widehat{i\varphi}} = 1. \tag{18.38}$$

It is important to note that the order of operations in Eqs. (18.36) and (18.37) are important, since e.g. $e^{\widehat{i\varphi}}$ as defined by (18.36) is not a unitary operator, but it is "one-sided" unitarity [8]. To show this result let us calculate $e^{\widehat{i\varphi}}|n\rangle$ and $e^{-\widehat{i\varphi}}|n\rangle$,

$$e^{\widehat{i\varphi}}|n\rangle = \left(\hat{N}+1\right)^{-\frac{1}{2}} a|n\rangle = \left(\hat{N}+1\right)^{-\frac{1}{2}} \sqrt{n}\,|n-1\rangle = |n-1\rangle, \tag{18.39}$$

and

$$e^{-\widehat{i\varphi}}|n\rangle = a^\dagger \left(\hat{N}+1\right)^{-\frac{1}{2}} |n\rangle = a^\dagger (n+1)^{-\frac{1}{2}} |n\rangle = |n+1\rangle. \tag{18.40}$$

Thus the only nonvanishing matrix elements of $e^{\widehat{i\varphi}}$ and $e^{-\widehat{i\varphi}}$ are

$$\left\langle n-1 \left| e^{\widehat{i\varphi}} \right| n \right\rangle = 1, \tag{18.41}$$

$$\left\langle n+1 \left| e^{-\widehat{i\varphi}} \right| n \right\rangle = 1, \tag{18.42}$$

and from these it follows that [8]

$$\left\langle k \left| e^{\widehat{i\varphi}} e^{-\widehat{i\varphi}} \right| n \right\rangle = \delta_{kn}, \tag{18.43}$$

and

$$\left\langle k \left| e^{-\widehat{i\varphi}} e^{\widehat{i\varphi}} \right| n \right\rangle = \delta_{kn} - \delta_{k0}\delta_{n0}. \tag{18.44}$$

As we have seen in Sec. 4.4 it is possible to introduce two Hermitian operators $\sin\widehat{\varphi}$ and $\cos\widehat{\varphi}$ by

$$\cos\widehat{\varphi} = \frac{1}{2}\left[e^{\widehat{i\varphi}} + e^{-\widehat{i\varphi}}\right], \tag{18.45}$$

and

$$\sin\widehat{\varphi} = \frac{1}{2i}\left[e^{\widehat{i\varphi}} - e^{-\widehat{i\varphi}}\right], \tag{18.46}$$

and these have nonzero matrix elements

$$\langle n-1\,|\cos\widehat{\varphi}|\,n\rangle = \langle n\,|\cos\widehat{\varphi}|\,n-1\rangle = \frac{1}{2}, \tag{18.47}$$

and

$$\langle n-1\,|\sin\widehat{\varphi}|\,n\rangle = -\langle n\,|\sin\widehat{\varphi}|\,n-1\rangle = \frac{1}{2i}. \tag{18.48}$$

Again for $\cos\widehat{\varphi}$ and $\sin\widehat{\varphi}$ we find the uncertainties

$$\Delta\cos\widehat{\varphi} = \Delta\sin\widehat{\varphi} = \frac{1}{\sqrt{2}}, \tag{18.49}$$

$$\Delta \cos\widehat{\varphi}\, \Delta \sin\widehat{\varphi} = \frac{1}{2}. \tag{18.50}$$

Next we define the phase state $|\varphi\rangle$ by [9],[10]

$$|\varphi\rangle = \lim_{s\to\infty} \frac{1}{\sqrt{s+1}} \sum_{n=0}^{s} e^{in\varphi}|n\rangle. \tag{18.51}$$

From the orthonormal properties of the number state $|n\rangle$ it is easy to show that $|\phi\rangle$ is normalized

$$\langle\varphi|\varphi\rangle = 1. \tag{18.52}$$

Let us now consider the state $\cos\hat{\varphi}|\varphi\rangle$

$$\begin{aligned} \cos\hat{\varphi}|\varphi\rangle &= \frac{1}{2}\lim_{s\to\infty}\frac{1}{\sqrt{s+1}}\left\{\sum_{n=0}^{s} e^{in\varphi}e^{i\hat{\varphi}}|n\rangle + \sum_{n=0}^{s} e^{in\varphi}e^{-i\hat{\varphi}}|n\rangle\right\} \\ &= \frac{1}{2}\lim_{s\to\infty}\frac{1}{\sqrt{s+1}}\left\{\sum_{n=0}^{s} e^{in\varphi}|n-1\rangle + \sum_{n=0}^{s} e^{in\varphi}|n+1\rangle\right\}, \end{aligned} \tag{18.53}$$

where we have used (18.39) and (18.40) to get the last term. We can simplify (18.53) by writing it as

$$\begin{aligned} \cos\hat{\varphi}|\varphi\rangle &= \frac{1}{2}\lim_{s\to\infty}\frac{1}{\sqrt{s+1}}\left\{e^{i\varphi}\sum_{k=0}^{s-1} e^{ik\varphi}|k\rangle + e^{-i\varphi}\sum_{k=1}^{s+1} e^{ik\varphi}|k\rangle\right\} \\ = \quad & \cos\varphi|\varphi\rangle + \frac{1}{2}\lim_{s\to\infty}\frac{1}{\sqrt{s+1}}\left\{e^{is\varphi}|s+1\rangle - e^{i(s+1)\varphi}|s\rangle - e^{-i\varphi}|0\rangle\right\}, \end{aligned} \tag{18.54}$$

From this relation we conclude that $|\varphi\rangle$ is not an exact eigenfunction of $\cos\widehat{\varphi}$, but asymptotically may be regarded as its eigenfunction with the eigenvalue $\cos\varphi$. The diagonal elements of $\cos\widehat{\varphi}$ and $\sin\widehat{\varphi}$ are given by

$$\langle\varphi\,|\cos\widehat{\varphi}|\,\varphi\rangle = \cos\varphi\left[1 - \lim_{s\to\infty}\frac{1}{s+1}\right] \to \cos\varphi, \tag{18.55}$$

and

$$\langle\varphi\,|\sin\widehat{\varphi}|\,\varphi\rangle = \sin\varphi\left[1 - \lim_{s\to\infty}\frac{1}{s+1}\right] \to \sin\varphi, \tag{18.56}$$

respectively.

Now using Eq. (18.51) let us calculate the probability of finding n photons in a state of well-defined phase $|\varphi\rangle$:

$$\mathcal{P}_n = |\langle n|\varphi\rangle|^2 = \lim_{s\to\infty}\frac{1}{s+1}. \tag{18.57}$$

This result shows there is an equal but very small probability of any number of photons to be in the field, and that the magnitudes of **E** and **H** are completely undetermined when the phase is exactly known.

Coherent States of Photons — In quantum theory the coherent state is defined as a state whose dynamics is as close to the dynamics of the classical motion as possible. As we will see below, the coherent state of a harmonic oscillator can be described by the motion of a Gaussian wave packet where the center moves exactly as the classical harmonic oscillator.

The eigenstate of the annihilation operator for a photon having a wave number **k** and polarization σ is defined as the coherent state $|\alpha_{\mathbf{k},\sigma}\rangle$ for the quantized electromagnetic field,

$$a_{\mathbf{k},\sigma}|\alpha_{\mathbf{k},\sigma}\rangle = \alpha_{\mathbf{k},\sigma}|\alpha_{\mathbf{k},\sigma}\rangle. \tag{18.58}$$

Since $a_{\mathbf{k},\sigma}$ is not a Hermitian operator $\alpha_{\mathbf{k},\sigma}$, in general, is a complex number. Again we will consider a single mode of the field and we will omit the subscripts **k** and σ. From the definition of the cohrrent state it follows that

$$\langle n|a|\alpha\rangle = \alpha\langle n|\alpha\rangle, \tag{18.59}$$

or

$$\langle n|\alpha\rangle = \frac{\alpha}{\sqrt{n}}\langle n-1|\alpha\rangle = \frac{\alpha^n}{\sqrt{n!}}\langle 0|\alpha\rangle. \tag{18.60}$$

Introducing the complete set of states

$$\sum_{n=0}^{\infty}|n\rangle\langle n| = 1, \tag{18.61}$$

where 1 is the unit operator, we have

$$|\alpha\rangle = \sum_{n=0}^{\infty}|n\rangle\langle n|\alpha\rangle = \langle 0|\alpha\rangle\sum_{n=0}^{\infty}\frac{\alpha^n}{\sqrt{n!}}|n\rangle. \tag{18.62}$$

Now we normalize the eigenket $|\alpha\rangle$ using (18.62)

$$\begin{aligned}\langle\alpha|\alpha\rangle &= \langle\alpha|0\rangle\langle 0|\alpha\rangle\sum_{n=0}^{\infty}\frac{\alpha^{*\,n}\alpha^n}{n!}\\ &= \langle\alpha|0\rangle\langle 0|\alpha\rangle\exp\left(|\alpha|^2\right) = 1.\end{aligned} \tag{18.63}$$

Thus we have

$$\langle\alpha|0\rangle = \langle 0|\alpha\rangle = \exp\left(-\frac{1}{2}|\alpha|^2\right). \tag{18.64}$$

By substituting (18.64) in (18.62) we find the normalized coherent state $|\alpha\rangle$ to be

$$|\alpha\rangle = \exp\left(-\frac{1}{2}|\alpha|^2\right)\sum_{n=0}^{\infty}\frac{\alpha^n}{\sqrt{n!}}|n\rangle. \tag{18.65}$$

From (18.65) we find a useful expression which is the scalar product of two coherent states $\langle\alpha'|$ and $|\alpha\rangle$;

$$\begin{aligned}\langle\alpha'|\alpha\rangle &= \exp\left(-\frac{1}{2}|\alpha|^2 - \frac{1}{2}|\alpha'|\right)\sum_{n=0}^{\infty}\sum_{j=0}^{\infty}\frac{\alpha^n\alpha'^j}{\sqrt{n!}\sqrt{j!}}\langle n|j\rangle \\ &= \exp\left(-\frac{1}{2}|\alpha-\alpha'|^2\right). \end{aligned} \tag{18.66}$$

We can write Eq. (18.65) as

$$|\alpha\rangle = \exp\left(-\frac{1}{2}|\alpha|^2\right)e^{\alpha a^\dagger}|0\rangle, \tag{18.67}$$

where we have used the fact that (see Eq. (16.150))

$$|n\rangle = \frac{\left(a^\dagger\right)^n}{\sqrt{n!}}|0\rangle. \tag{18.68}$$

From the expression of $|\alpha\rangle$ in terms of $|n\rangle$, Eq. (18.65), we can calculate the wave packet associated with the motion of a simple harmonic oscillator. The scalar product $\langle x|n\rangle = \psi_n(x)$ is the wave function in the coordinate space when the quantum number is n, which is given by Eq. (8.92). Thus the wave packet associated with the time-dependent coherent state is

$$\begin{aligned}\langle x|\alpha(t)\rangle &= \psi_\alpha(x,t) = \left(\frac{\beta}{\pi}\right)^{\frac{1}{4}}\exp\left(-\frac{1}{2}|\alpha(t)|^2\right)\sum_{n=0}^{\infty} \\ &\times \exp\left(-\frac{1}{2}\beta^2x^2\right)\frac{\alpha^n(t)}{\sqrt{2^n}}\frac{1}{n!}H_n(\beta x). \end{aligned} \tag{18.69}$$

We can simplify this expression by noting that: (a)

$$\sum_{n=0}^{\infty}\frac{\xi^k}{k!}H_k(\beta x) = \exp\left[-\xi^2 + 2\beta\xi x\right]. \tag{18.70}$$

(b) That the time-dependence of $\alpha(t)$ is the same as $a(t)$, Eq. (8.96),

$$\alpha(t) = \sqrt{\frac{\beta}{2}}Ae^{-i\omega t}, \quad \beta = \frac{m\omega}{\hbar}, \tag{18.71}$$

where A is the amplitude of the oscillation. We have also introduced the factor $\sqrt{\frac{\beta}{2}}$ to make $\alpha(t)$ a dimensionless function of time and the resulting wave finction $\langle x|\alpha\rangle$ normalized. Now if we substitute (18.70) and (18.71) in (18.69), and calculate the square of $|\langle x|\alpha(t)\rangle|$, we get a time-dependent Gaussian wave packet whose center moves according to laws of classical mechanics;

$$|\psi_\alpha(x,t)|^2 = \langle x|\alpha(t)\rangle|^2 = \frac{\beta}{\sqrt{\pi}}\exp\left[-\beta^2(x - A\cos\omega t)^2\right]. \tag{18.72}$$

Returning to our coherent state formulation of the photon field, from Eq. (18.58) it follows that for a single mode of the field

$$\langle\alpha|a|\alpha\rangle = \alpha, \tag{18.73}$$

and also

$$\left\langle\alpha\left|a^\dagger\right|\alpha\right\rangle = \alpha^*. \tag{18.74}$$

Similarly straightforward calculations show that

$$\left\langle\alpha\left|\hat{N}\right|\alpha\right\rangle = \left\langle\alpha\left|a^\dagger a\right|\alpha\right\rangle = |\alpha|^2, \tag{18.75}$$

$$\left\langle\alpha\left|\hat{N}\right|\alpha\right\rangle + 1 = \left\langle\alpha\left|aa^\dagger\right|\alpha\right\rangle = |\alpha|^2 + 1, \tag{18.76}$$

$$\left\langle\alpha\left|\hat{N}^2\right|\alpha\right\rangle = \left\langle\alpha\left|a^\dagger a a^\dagger a\right|\alpha\right\rangle = |\alpha|^4 + |\alpha|^2, \tag{18.77}$$

$$\left\langle\alpha\left|a^2\right|\alpha\right\rangle = \alpha^2, \tag{18.78}$$

and

$$\left\langle\alpha\left|a^{\dagger\,2}\right|\alpha\right\rangle = \alpha^{*\,2}. \tag{18.79}$$

To find the uncertainty in the number of photons in the state $|\alpha\rangle$, from Eqs. (18.75)–(18.77) we calculate ΔN;

$$\begin{aligned}\Delta N &= \left\langle\alpha\left|\left(\hat{N} - \left\langle\alpha\left|\hat{N}\right|\alpha\right\rangle\right)^2\right|\alpha\right\rangle^{\frac{1}{2}} = \left(\left\langle\alpha\left|\hat{N}^2\right|\alpha\right\rangle - \left\langle\alpha\left|\hat{N}\right|\alpha\right\rangle^2\right)^{\frac{1}{2}} \\ &= \left(|\alpha|^4 + |\alpha|^2 - |\alpha|^4\right)^{\frac{1}{2}} = |\alpha| = \left\langle\alpha\left|\hat{N}\right|\alpha\right\rangle^{\frac{1}{2}}.\end{aligned} \tag{18.80}$$

The relative uncertainty calculated from (18.80) is inversely proportional to $|\alpha|$;

$$\frac{\Delta N}{\left\langle\alpha\left|\hat{N}\right|\alpha\right\rangle} = \frac{1}{\left\langle\alpha\left|\hat{N}\right|\alpha\right\rangle^{\frac{1}{2}}} = \frac{1}{|\alpha|}. \tag{18.81}$$

This uncertainty becomes very small when there are a large number of photons in the same mode of the field. Now in the limit of large number of photons the expectation value of **E** (or **H**) behaves like a classical field.

We note that α is a complex number and can be written as

$$\alpha = |\alpha|e^{i\theta}. \tag{18.82}$$

Thus using $\langle\alpha|a|\alpha\rangle$ and $\langle\alpha|a^\dagger|\alpha\rangle$ as given by Eqs. (18.73) and (18.74) and calculating the expectation value of **E**, Eq. (18.30), we find

$$\langle\alpha|\mathbf{E}|\alpha\rangle = -2\sqrt{\frac{2\pi\hbar\omega}{L^3}}\mathbf{u}\,|\alpha|\,\sin(\mathbf{k}\cdot\mathbf{r} - \omega t + \theta), \tag{18.83}$$

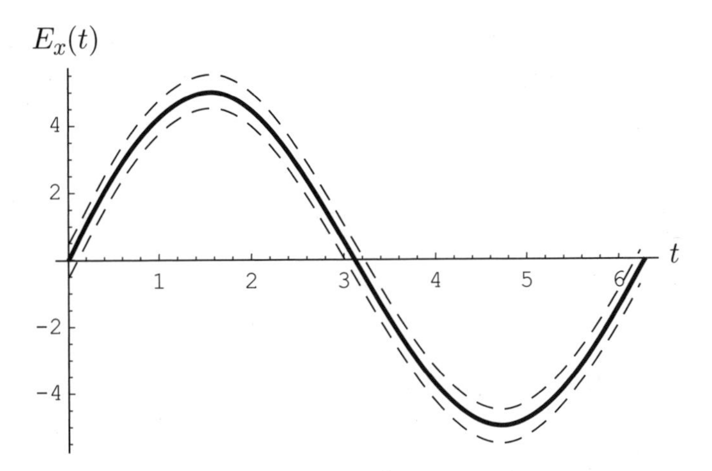

Figure 18.1: The electric field of a single-mode coherent state, Eq. (18.83), with $\left\langle\alpha\left|\hat{N}\right|\alpha\right\rangle = 25$ shown here as a solid line. The uncertainty in the electric field is a band with constant width, ΔE, Eq. (18.85) shown by dashed lines.

which is very similar to its classical counterpart.

We can also calculate the probability of finding n photons in a given coherent state $|\alpha\rangle$. This is obtained from the amplitude $\langle n|\rangle$;

$$\mathcal{P} = |\langle n|\alpha\rangle|^2 = \frac{|\alpha|^{2n}}{n!} e^{-|\alpha|^2}, \tag{18.84}$$

Finally the uncertainty in the electric field when the field is given in the coherent state representation can be found from

$$\Delta E = \left[\left\langle\alpha\left|\mathbf{E}^2\right|\alpha\right\rangle - \langle\alpha|\mathbf{E}|\alpha\rangle^2\right]^{\frac{1}{2}} = \left(\frac{2\pi\hbar\omega}{L^3}\right)^{\frac{1}{2}}, \tag{18.85}$$

As expected ΔE in the classical limit of $\hbar \to 0$, goes to zero. Now ΔE is independent of $\left\langle\alpha\left|\hat{N}\right|\alpha\right\rangle$, while $\langle\alpha|\mathbf{E}|\alpha\rangle$, Eq. (18.83), depends on $|\alpha| = \left\langle\alpha\left|\hat{N}\right|\alpha\right\rangle^{-1}$, therefore as the number of photons increases the uncertainty in $\langle\alpha|\mathbf{E}|\alpha\rangle$ decreases and the electric (magnetic) field behaves more and more like a classical wave (see Fig. 18.1).

18.2 Casimir Force

We begin this section by solving a simple problem of the zero-point energy density of a one-dimensional wave (e.g. a longitudinal wave in an elastic rod).

A One-Dimensional Model Exhibiting the Casimir Force — Let us first consider a model which exhibits Casimir effect [11]. The characteristics (angular) frequencies for a one-dimensional resonator of length L are

$$\omega_n(L) = \left(\frac{c\pi}{L}\right) n, \qquad n = 1,\ 2\ \cdots, \tag{18.86}$$

where c is the velocity of light. The zero-point energy density which is

$$u(L) = \frac{\pi}{L}\sum_{n=1}^{\infty} \hbar\omega_n, \tag{18.87}$$

becomes infinite. However the difference between the zero-point energy densities of a resonator of length L and a resonator of infinite length

$$\Delta u(L) = u_L - \lim_{L'\to\infty} \Delta u\left(L'\right), \tag{18.88}$$

remains finite. To show this we use a cut-off factor $\exp\left(-\alpha\frac{\omega_n(L)}{\omega_1(L)}\right)$ and write

$$\begin{aligned}
\Delta u(L) &= \pi\hbar \lim_{\alpha\to 0}\left[\frac{1}{L}\sum_{n=1}^{\infty}\omega_n(L)e^{-\alpha\frac{\omega_n(L)}{\omega_1(L)}} - \lim_{L'\to\infty}\frac{1}{L'}\sum_{n=1}^{\infty}\omega_n\left(L'\right)e^{-\alpha\frac{\omega_n(L')}{\omega_1(L')}}\right] \\
&= \frac{\pi\hbar c}{2L^2}\lim_{\alpha\to 0}\left\{\sum_{n=1}^{\infty} ne^{-\alpha n} - \int_0^{\infty} ne^{-\alpha n}dn\right\}.
\end{aligned} \tag{18.89}$$

The quantity in the curley bracket can be written as an infinite series involving Benoullis's numbers B_j (see below)

$$\begin{aligned}
\Delta u(L) &= -\frac{\pi\hbar}{2L^2}\lim_{\alpha\to 0}\left(\frac{d}{d\alpha}\sum_{j=1}^{\infty}\frac{B_j}{j!}\alpha^{j-1}\right) \\
&= -\frac{B_2}{8}\frac{2\pi\hbar c}{L^2} = -\frac{\pi}{24}\frac{\hbar c}{L^2},
\end{aligned} \tag{18.90}$$

where in the last step the value of $B_2 = \frac{1}{6}$ has been used. Thus the presence of the two boundaries at $x = 0$ and $x = L$ changes the zero point energy of this simple system by an amount

$$\Delta\mathcal{E} = L\Delta u(L) = -\frac{\pi}{24}\frac{\hbar c}{L}. \tag{18.91}$$

In this case the force which is attractive and is directed inside the cavity is given by

$$\mathcal{F} = -\frac{d}{dL}(\Delta\mathcal{E}) = -\frac{\pi\hbar c}{24L^2}. \tag{18.92}$$

18.3 Casimir Force Between Parallel Conductors

The last term in the Hamiltonian (18.22) represents the zero point energy of the field in vacuum and is a divergent quantity

$$H = \frac{\hbar}{2} \sum_{\mathbf{k},\sigma} \omega_k \to \infty. \tag{18.93}$$

While it can be argued that the absolute value of energy is of no significance, and can be subtracted from the result, this argument is not always true. According to Casimir the zero point energy exists and has observable effects [12],[13]. We have already studied the quantization of the electromagnetic field in a cubical box of volume L^3. Now we want to see if there is a change in the energy if we put a conducting plane at $x = R$, keeping R finite while letting L tend to infinity. Let us denote the energy in the volume L^3 by $\mathcal{E}_L$. When the conducting plane is placed at $x = R$ we have the energy $\mathcal{E}_R$ for the volume between $x = 0$ and $x = R$ and the energy $\mathcal{E}_{L-R}$ for the volume limited to $R \le x \le L$. Each of these zero point energies are divergent. As $L \to \infty$ we have

$$\mathcal{E}_L = 2\frac{\hbar}{2} \sum_{\mathbf{k}} ck \to \hbar c \frac{L^3}{(2\pi)^3} \int\int\int \sqrt{k_x^2 + k_y^2 + k_z^2}\, dk_x dk_y dk_z, \tag{18.94}$$

$$\mathcal{E}_{L-R} = 2\frac{\hbar}{2} c \frac{L^2}{(2\pi)^3} (L-R) \int\int\int \sqrt{k_x^2 + k_y^2 + k_z^2}\, dk_x dk_y dk_z, \tag{18.95}$$

and

$$\mathcal{E}_R = 2\frac{\hbar}{2} c \frac{L^2}{(2\pi)^2} \sum_{n=-\infty}^{\infty} \int\int\int \sqrt{\left(\frac{2\pi n}{R}\right)^2 + k_y^2 + k_z^2}\, dk_y dk_z, \tag{18.96}$$

where we have accounted for the two degrees of polarization by the factor 2 in these equations and in the last relation we have kept R large but finite. While each of these energies are infinite the difference

$$\Delta\mathcal{E} = \mathcal{E}_R + \mathcal{E}_{L-R} - \mathcal{E}_L, \tag{18.97}$$

is finite. We write this difference $\Delta\mathcal{E}$ as

$$\Delta\mathcal{E} = \frac{\hbar c L^2}{(2\pi)^2} \int\int \times \left\{ \sum_{n=-\infty}^{\infty} \sqrt{\left(\frac{2\pi n}{R}\right)^2 + k_y^2 + k_z^2} - \frac{R}{2\pi} \int_{-\infty}^{\infty} \sqrt{k_x^2 + k_y^2 + k_z^2}\, dk_x \right\} dk_x dk_y. \tag{18.98}$$

Next we write the integral over k_y and k_z in polar coordinates with

$$dk_y dk_z = 2\pi k_\perp dk_\perp, \tag{18.99}$$

and

$$k_\perp^2 = k_y^2 + k_z^2. \tag{18.100}$$

With these changes we obtain

$$\frac{\Delta\mathcal{E}}{L^2} = \frac{\hbar c}{2\pi}\int_0^\infty \left\{\sum_{n=1}^{+\infty}\sqrt{k_\perp^2 + \frac{n^2\pi^2}{R^2}} - \int_0^\infty \sqrt{k_\perp^2 + k_x^2}\, dk_x\right\} k_\perp dk_\perp. \tag{18.101}$$

To simplify the result further we replace $k_\perp$ by $z = \frac{k_\perp R}{\pi}$ and thus we find the energy per unit area $\Delta\mathcal{E}/L^2$ to be

$$\frac{\Delta\mathcal{E}}{L^2} = \frac{\hbar c}{2\pi}\left(\frac{\pi}{R}\right)^3 \left\{\frac{1}{2}E(0) + \sum_{n=1}^{\infty} E(n) - \int_0^\infty E(n)\, dn\right\}, \tag{18.102}$$

where

$$E(n) = \int_0^\infty \sqrt{z^2+n^2}\, z\, dz. \tag{18.103}$$

Since this integral is divergent we introduce a cut-off Λ to make it finite and at the end of calculation we let Λ go to infinity. Thus we write

$$E(n,\Lambda) = \int_0^\Lambda \sqrt{z^2+n^2}\, z\, dz = \frac{1}{3}\left[\left(\Lambda^2+n^2\right)^{\frac{3}{2}} - n^3\right]. \tag{18.104}$$

Now that we have a convergent sum and a convergent integral we can evaluate the quantity given in the bracket in (18.102). We note that from trapezoidal approximation

$$\int_a^b f(x)\, dx \approx F_N(a,b) = h\left(\frac{1}{2}f(a) + \sum_{n=1}^{N-1} f(a+nh) + \frac{1}{2}\right), \tag{18.105}$$

where $h = \frac{(b-a)}{N}$, we have the Euler-Maclaurin formula

$$\int_a^b \left[f(x) - F_N(a,b)\right]\, dx = -\frac{B_2}{2!}h^2\left[f'(x)\right]_a^b - \frac{B_4}{4!}h^4\left[f'''(x)\right]_a^b + \cdots. \tag{18.106}$$

Here B_n's are Bernoulli's numbers [14]

$$B_0 = 1, \quad B_2 = \frac{1}{6}, \quad B_4 = -\frac{1}{30}\cdots. \tag{18.107}$$

Using this formula we calculate the terms in bracket Eq. (18.102)

$$\begin{aligned}\frac{1}{2}E(0,\Lambda) \quad &+ \quad \sum_{n=1}^{+\infty} E(n,\Lambda) - \int_0^{+\infty} E(n,\Lambda)\, d\,n \\ &= \quad -\frac{1}{6\cdot 2!}\left[\frac{dE(n,\Lambda)}{dn}\right]_{n=0} + \frac{1}{30\cdot 4!}\left[\frac{d^3E(n,\Lambda)}{dn^3}\right]_{n=0} + \cdots,\end{aligned} \tag{18.108}$$

with all higher derivatives of $E(n,\Lambda)$ being zero independent of the value of Λ. From Eqs. (18.104) and (18.108) we calculate the derivatives of $E(n,\Lambda)$;

$$E'(0,\Lambda)=0, \quad E'''(0,\Lambda)=-6. \tag{18.109}$$

Finally we find $\frac{\Delta\mathcal{E}}{L^2}$ to be

$$\frac{\Delta\mathcal{E}}{L^2}=-\frac{\pi^2}{720}\frac{\hbar c}{R^3}, \tag{18.110}$$

for the energy per unit surface area between the two conducting planes. The force between the two plates is found by differentiating (18.110) with respect to R and this gives us

$$\frac{\Delta\mathcal{F}}{L^2}=-\frac{\pi^2}{240}\frac{\hbar c}{R^4}, \tag{18.111}$$

a result which was first derived by Casimir [12].

For the experimental evidence of the existence of this force see [15] and [16].

18.4 Casimir Force in a Cavity with Conducting Walls

An interesting and solvable case where we can find the Casimir force exactly is in a cavity with conducting walls bounded by $0\le x\le L_1, 0\le y\le L_2$ and $0\le z\le L_3$. This problem has been solved by Lukosz using the method outlined earlier for a one-dimensional cavity. In this way one finds the shift in the energy density $\Delta u(L_1,L_2,L_3)$ to be

$$\begin{aligned}\Delta u(L_1,L_2,L_3) &= -\frac{\hbar c}{16\pi^2}\left\{\sum_{m_1=-\infty}^{\infty\,\prime}\sum_{m_2=-\infty}^{\infty\,\prime}\right.\\ &\times \sum_{m_3=-\infty}^{\infty\,\prime}\frac{1}{\left(m_1^2L_1^2+m_2^2L_2^2+m_3^2L_3^2\right)^2}\\ &- \left.\frac{\pi^3}{3}\frac{1}{L_1L_2L_3}\left(\frac{1}{L_1}+\frac{1}{L_2}+\frac{1}{L_3}\right)\right\},\end{aligned} \tag{18.112}$$

where $\sum_{m_i=-\infty}^{\infty\,\prime}$ indicates summation over all integers except $m_i=0$. From the shift in the energy density we find the shift in the zero point energy to be

$$\Delta\mathcal{E}=L_1L_2L_3\Delta u(L_1,L_2,L_3), \tag{18.113}$$

and thus the force per unit area is

$$\mathcal{F}_i(L_1,L_2,L_3)=-\frac{\partial}{\partial L_i}\left[L_i\Delta u(L_1,L_2,L_3)\right]. \tag{18.114}$$

Now let us consider the following special cases of this problem:

(1) - If $R = L_1 \ll L_2 = L_3 = L$, then we have

$$\Delta u(R) = -\frac{\hbar c \pi^2}{720 R^4} \tag{18.115}$$

where we have used the summation formula

$$\zeta(4) = \sum_{m=1}^{\infty} \frac{1}{m^4} = \frac{\pi^4}{90} \tag{18.116}$$

From this result we find that the force per unit area is

$$-\frac{\partial}{\partial R}(R\Delta u(R)) = -\frac{\hbar c \pi^2}{240 R^4} \tag{18.117}$$

which is the same as (18.111).

(2) - When $L_3 \gg L_1 = L_2 = R$, then the cavity becomes a long cylinder along the z-axis with square cross section. Here we make use of the Epstein ζ function [11]

$$\sum_{m_1=-\infty}^{+\infty}{}' \sum_{m_2=-\infty}^{+\infty}{}' \frac{1}{(m_1^2 + m_2^2)^2} = 4G\,\zeta(2), \tag{18.118}$$

where G is a numerical constant

$$G = \sum_{n=0}^{\infty} \frac{(-1)^n}{(2n+1)^2} = 0.91596, \tag{18.119}$$

and

$$\zeta(2) = \sum_{n=1}^{\infty} \frac{1}{n^2} = \frac{\pi^2}{6}. \tag{18.120}$$

Thus for $\Delta u(R, R)$ we obtain

$$\Delta u(R, R) = -\frac{\hbar G}{24 R^4} \tag{18.121}$$

In this case again we find that the energy density is negative.

(3) - Finally for a cubical cavity with $L_1 = L_2 = L_3 = R$ the energy density assumes the simple form

$$\Delta u(R, R, R) = -\frac{2\pi \hbar c}{32 R^4}\left[-1 + \frac{1}{\pi^3} \sum_{m_1=-\infty}^{+\infty}{}' \sum_{m_2=-\infty}^{+\infty}{}' \sum_{m_3=-\infty}^{+\infty}{}' \frac{1}{(m_1^2 + m_2^2 + m_3^2)^2}\right]. \tag{18.122}$$

The Epstein ζ function which is the sum over m_1, m_2 and m_3 in this equation has the numerical value of

$$\sum_{m_1=-\infty}^{+\infty}{}' \sum_{m_2=-\infty}^{+\infty}{}' \sum_{m_3=-\infty}^{+\infty}{}' \frac{1}{(m_1^2 + m_2^2 + m_3^2)^2} = 16.5323. \tag{18.123}$$

Thus the energy density for a cubical cavity is

$$\Delta\mathcal{E} = R^3 \Delta u(R) \approx 0.0916 \left(\frac{\hbar c}{R}\right), \tag{18.124}$$

and this leads to a repulsive force, i.e. the zero-point energy fluctuations of the electromagnetic field tends to expand the cube [11].

Bibliography

[1] J.D. Jackson, *Classical Electrodynamics*, (John Wiley & Sons, New York, 1998).

[2] S. Schweber, A note on commutators in quantized field theory, Phys. Rev. 78, 613 (1950).

[3] R.W. Robinett, *Quamtum Mechanics, Classical Results, Modern Systems, and Visualised Examples*, (Oxford University Press, 1997), p. 481.

[4] R. Glauber, Coherent and incoherent states of the radiation field, Phys. Rev. 131, 2766 (1963).

[5] W. Vogel, D-G. Welsch and S. Wallentowitz, *Quantum Optics, An Introduction*, Second Edition, (Wiley-VCH, Berlin, 2001), Chpter 3.

[6] L. Susskind and J. Glowgower, Quantum mechanical phase and time operator, Physics 4, 49 (1964).

[7] P. Carruthers and M.M. Nieto, Coherent states and number-phase uncertainty relation, Phys. Rev. Lett. 14, 387 (1965).

[8] P. Carruthers and M.M. Nieto, Phase and angle variables in quantum mechanics, Rev. Mod. Phys. 40, 411 (1968).

[9] R. Louden, *The Quantum Theory of Light*, (Oxford University Press, Oxford, 1973).

[10] D.T. Pegg and S.M. Barnett, Unitary phase operator in quantum mechanics, Europhys. Lett. 6, 483 (1988).

[11] W. Lukosz, Electromagnetic zero-point energy and radiation pressure for a rectangular cavity, Physica, 56, 109 (1971).

[12] H.B.G. Casimir, On the attraction between two perfectly conducting plates, Proc. K. Ned. Akad. Wet, 51, 393 (1948).

[13] E.G. Harris, *A Pedestrian Approach to Quantum Field Theory*, (Wiley-Interscience, New York, 1972), Chapter 10.

[14] M. Abramovitz and I.A. Stegun, *Handbook of Mathematical Functions*, (Dover Publications. New York, 1965), p. 806.

[15] M.J. Sparnaay, Measurements of attractive forces between flat plates, Physica, 24, 751 (1958).

[16] I.I. Abrikosova and B.V. Deriagin, Direct measurement of the molecular attraction of solid bodies, Sov. Phys. JETP, 3, 819 (1957).

Chapter 19

Interaction of Radiation with Matter

In this chapter we consider the interaction of an atom with quantized electromagnetic field. For a collection of charged particles and atoms interacting with each other and with an external field we can follow a similar method, but here for the sake of simplicity we consider a single atom. For such a system the Hamiltonian is the sum of the Hamiltonians for the particle, the radiation field, and the interaction between these two. Thus we write

$$\Delta H = H_{atom} + H_{rad} + H_{int}, \tag{19.1}$$

where

$$\Delta H_{atom} = \frac{1}{2m}\mathbf{p}^2 + V(\mathbf{r}), \tag{19.2}$$

$$H_{rad} = \frac{1}{8\pi}\int \left(\mathbf{E}^2 + \mathbf{H}^2\right) d^3r, \tag{19.3}$$

and

$$H_{int} = -\frac{e}{mc}\mathbf{p}\cdot\mathbf{A}(\mathbf{r}) + \frac{e^2}{2mc^2}\mathbf{A}^2(\mathbf{r}). \tag{19.4}$$

The interaction Hamiltonian is the same found by coupling the magnetic potential $\mathbf{A}$ to the momentum of the particle that we have seen in Chapter 16. (Note that here e denotes the charge of electron, and m its mass). For a weak external field the second term in (19.4) is small compared to the first term. This is usually the case when the fields described by $\mathbf{A}$ are small compared to the atomic field of $\frac{e}{a_0^2}$, where a_0 is the Bohr's radius. Keeping just the first term in (19.4) and substituting for $\mathbf{A}$ from (18.9) we find the perturbation H';

$$H' = -\frac{e}{mc}\mathbf{p}\cdot\mathbf{A}(\mathbf{r}) = -\frac{e}{mc}\sum_{\mathbf{k},\sigma}\left(\frac{2\pi\hbar c^2}{\Omega\omega_k}\right)^{\frac{1}{2}}\mathbf{p}\cdot\mathbf{u}_{\mathbf{k},\sigma}\left[a_{\mathbf{k},\sigma}e^{i\mathbf{k}\cdot\mathbf{r}} + a^\dagger_{\mathbf{k},\sigma}e^{-i\mathbf{k}\cdot\mathbf{r}}\right]. \tag{19.5}$$

Thus in this approximation we have the simplified Hamiltonian

$$H = H_{atom} + H_{rad} + H' = H_0 + H'. \tag{19.6}$$

Now the eigenstates of the unperturbed Hamiltonian, H_0, is of the form

$$|\, atom + rad\,\rangle = |\, i\rangle|\cdots n_{\mathbf{k},\sigma}\cdots\rangle, \tag{19.7}$$

where i denotes the initial quantum numbers of the atom and $n_{\mathbf{k},\sigma}$ is the number of photons with the momentum $\hbar\mathbf{k}$ and polarization σ at $t = 0$. The perturbation H' contains one creation and one annihilation operator, only transitions in which the number of photons changes by $+1$ or -1 will be induced. Had we included the term with $\mathbf{A}^2$ in the perturbation, then either two photons would be emitted or two absorbed, or one would be emitted and one absorbed [1],[2].

Emission of Light — We assume that initially the atom is in the state $|\, i\rangle$ and the perturbation H' causes it to decay to a state $|\, f\rangle$ with the emission of a photon of momentum $\hbar\mathbf{k}$ and polarization σ. Thus the initial state of the system, $|I\rangle$, is

$$|I\rangle = |\, i\rangle_{atom}|\cdots n_{\mathbf{k},\sigma},\cdots\rangle_{rad}, \tag{19.8}$$

and finally the system will be in the state $|F\rangle$ where

$$|F\rangle = |f\rangle_{atom}|\cdots n_{\mathbf{k},\sigma}+1,\cdots\rangle_{rad}. \tag{19.9}$$

The matrix element of the perturbation H' between $\langle F|$ and $|I\rangle$ is

$$\langle F\,|H'|\,I\rangle = -\frac{e}{mc}\sqrt{n_{\mathbf{k},\sigma}+1}\sum_{\mathbf{k},\sigma}\left(\frac{2\pi\hbar c^2}{L^3\omega_k}\right)^{\frac{1}{2}}\left\langle f\,\left|\mathbf{p}\cdot\mathbf{u}_{\mathbf{k},\sigma}e^{-i\mathbf{k}\cdot\mathbf{r}}\right|\,i\right\rangle_{atom}, \tag{19.10}$$

where we have used

$$a^\dagger_{\mathbf{k},\sigma}\,|\cdots n_{\mathbf{k},\sigma}\cdots\rangle = \sqrt{n_{\mathbf{k},\sigma}+1}\,|\cdots n_{\mathbf{k},\sigma}+1\cdots\rangle. \tag{19.11}$$

From Eq. (11.180) we can calculate the transition rate between the initial and final states of the atom for emission

$$\begin{aligned}\Gamma_{i\to f} &= \frac{2\pi}{\hbar}\left|\langle F\,|H'|\,I\rangle\right|^2\delta(E_F - E_I)\\ &= \frac{2\pi}{\hbar}\left(\frac{e}{mc}\right)^2\left(\frac{2\pi\hbar c^2}{L^3\omega_k}\right)(n_{\mathbf{k},\sigma}+1)\left|\left\langle f\,\left|\mathbf{p}\cdot\mathbf{u}_{\mathbf{k},\sigma}\,e^{-i\mathbf{k}\cdot\mathbf{r}}\right|\,i\right\rangle\right|^2\\ &\times\ \delta(E_f - E_i + \hbar\omega),\end{aligned} \tag{19.12}$$

where E_i and E_f refer to the initial and final energies of the atom which, by the law of conservation of energy is,

$$E_F - E_I = E_f - E_i + \hbar\omega. \tag{19.13}$$

We observe that even when $n_{\mathbf{k},\sigma} = 0$, the transition rate $\Gamma_{i\to f}$ is not zero. This special case, viz, $n_{\mathbf{k},\sigma} = 0$ is called "spontaneous" emission of light. Otherwise we call it "stimulated" emission.

Let us first consider the case of spontaneous emission. For this we write $\Gamma_{i\to f}$ as

$$\Gamma_{i\to f} = \frac{4\pi^2 e^2}{m^2 L^3}\sum_{\mathbf{k},\sigma}\frac{1}{\omega_k}\left|\langle f\left|\mathbf{p}\cdot\mathbf{u}_{\mathbf{k},\sigma}\, e^{-i\mathbf{k}\cdot\mathbf{r}}\right|i\rangle\right|^2\delta(E_f - E_i + \hbar\omega). \tag{19.14}$$

Now we sum over the polarization vectors of the photons. These are shown by the two unit vectors $\mathbf{u}_{\mathbf{k},1}$ and $\mathbf{u}_{\mathbf{k},2}$. We choose the plane formed by the two vectors $\mathbf{k}$ and $\mathbf{p}$ as the xy-plane, and take $\mathbf{u}_{\mathbf{k},1}$ to be along the z axis. Then in the xy-plane $\mathbf{u}_{\mathbf{k},2}$ will be perpendicular to $\mathbf{k}$ (see Eq. (18.10)). If θ denotes the angle between $\mathbf{p}$ and $\mathbf{k}$, we have

$$\sum_{\sigma=1,2}\left|\langle f\left|\mathbf{p}\cdot\mathbf{u}_{\mathbf{k},\sigma}\, e^{-i\mathbf{k}\cdot\mathbf{r}}\right|i\rangle\right|^2 = \left|\langle f\left|\mathbf{p}\, e^{-i\mathbf{k}\cdot\mathbf{r}}\right|i\rangle\right|^2\sin^2\theta. \tag{19.15}$$

Next we replace the summation over $\mathbf{k}$ by integration

$$\sum_{\mathbf{k}} \to \frac{L^3}{(2\pi)^3}\int d^3k, \quad \text{as} \quad \Omega\to\infty, \tag{19.16}$$

and also expand the exponential in (19.15)

$$e^{-i\mathbf{k}\cdot\mathbf{r}} = 1 - i\mathbf{k}\cdot\mathbf{r} + \frac{1}{2}(\mathbf{k}\cdot\mathbf{r})^2 + \cdots. \tag{19.17}$$

This expansion converges rapidly for atomic decays since the eigenket $|i\rangle$ corresponds to a bound state and has a range of the order of Bohr radius, $a_0 = \frac{\hbar^2}{me^2} = 5\times 10^{-11}cm$. Thus $\mathbf{r}$ is of the order of magnitude $\mathbf{k}\cdot\mathbf{r} \approx ka_0 \approx 0.5\times 10^{-3}$ for a visible light of wavelength $\lambda = 6\times 10^{-9}m$. By retaining just the first term of expansion in (19.17) we have the so called " electric dipole" approximation as it will be explained later. The next order term which is the matrix element of $(\mathbf{k}\cdot\mathbf{r})$ $(\mathbf{p}\cdot\mathbf{u}_{\mathbf{k},\sigma})$ can be written as

$$\begin{aligned}(\mathbf{k}\cdot\mathbf{r})(\mathbf{p}\cdot\mathbf{u}_{\mathbf{k},\sigma}) &= \frac{1}{2}(\mathbf{u}_{\mathbf{k},\sigma}\cdot\mathbf{p}\,\mathbf{k}\cdot\mathbf{r} + \mathbf{u}_{\mathbf{k},\sigma}\cdot\mathbf{r}\,\mathbf{p}\cdot\mathbf{k}) \\ &+ \frac{1}{2}(\mathbf{u}_{\mathbf{k},\sigma}\cdot\mathbf{p}\,\mathbf{k}\cdot\mathbf{r} - \mathbf{u}_{\mathbf{k},\sigma}\cdot\mathbf{r}\,\mathbf{p}\cdot\mathbf{k}) \\ &= \frac{1}{2}(\mathbf{u}_{\mathbf{k},\sigma}\cdot\mathbf{p}\,\mathbf{k}\cdot\mathbf{r} + \mathbf{u}_{\mathbf{k},\sigma}\cdot\mathbf{r}\,\mathbf{p}\cdot\mathbf{k}) \\ &+ \frac{1}{2}(\mathbf{k}\wedge\mathbf{r})\cdot(\mathbf{r}\wedge\mathbf{p}),\end{aligned} \tag{19.18}$$

The first term is called the electric quadrupole and the second which is related to $\mathbf{L}\cdot\mathbf{B}$ is called the magnetic dipole [3]. These are $\frac{Ze^2}{\hbar c} \approx \frac{Z}{137}$ times smaller than the electric dipole for an atom with the nuclear charge Z. Thus for small Z we need to consider the electric dipole radiation only.

In the electric dipole approximation, (19.14) reduces to

$$\Gamma_{i\to f} = \frac{e^2}{2\pi m^2}\int |\langle f|\mathbf{p}|i\rangle|^2 \sin^2\theta\, \delta(E_j - E_k + \hbar\omega_k)\frac{d^3k}{\omega_k}. \tag{19.19}$$

In order to carry out the integration, we choose spherical coordinates for the k-space, taking k_z to be in the direction of $\langle f|\mathbf{p}|i\rangle$. The element of volume in this space is

$$d^3k = k^2 dk\, \sin\theta d\theta d\phi \to \sin\theta d\theta \frac{\omega_k^2 d\omega_k}{c^3} d\phi. \tag{19.20}$$

In the k-space (or ω_k-space) we can carry out the integration over θ, ϕ and ω_k with the result that

$$\Gamma_{i\to f} = \frac{4e^2}{3m^2c^3\hbar}\omega_{if}\, |\langle f|\mathbf{p}|i\rangle|^2, \tag{19.21}$$

where $\Gamma_{i\to f}$ has the dimension of $(time)^{-1}$ and

$$\omega_{if} = \frac{E_i - E_f}{\hbar}. \tag{19.22}$$

Now we will show the how in Eq. (19.21) the matrix element $e\langle f|\mathbf{p}|i\rangle$ is related to the expectation value of the dipole moment. Noting that $\mathbf{p}$ is the mechanical momentum of the electron we have

$$\begin{aligned}\langle f|\mathbf{p}|i\rangle &= \left\langle f\left| m\frac{d\mathbf{r}}{dt}\right| i\right\rangle = -\frac{im}{\hbar}\langle f|\mathbf{r}H - H\mathbf{r}|i\rangle \\ &= \frac{im}{\hbar}(E_i - E_f)\langle f|\mathbf{r}|i\rangle = im\,\omega_{if}\langle f|\mathbf{r}|i\rangle.\end{aligned} \tag{19.23}$$

Substituting for $\langle f|\mathbf{p}|i\rangle$ from (19.23) in (19.21) we obtain

$$\Gamma_{i\to f} = \frac{4\omega_{if}^3}{3\hbar c^3}\, |\langle f|e\mathbf{r}|i\rangle|^2. \tag{19.24}$$

Here $-e\mathbf{r}$ is the dipole moment of the radiating electron and as (19.24) shows the transition rate is proportional to the square of the matrix element of the dipole moment.

Assuming that the electron is harmonically bound, i.e.

$$\frac{d^2\mathbf{r}}{dt^2} = -\omega_{if}^2\mathbf{r}, \tag{19.25}$$

we can substitute for $\mathbf{r}$ from (19.25) in (19.24) to get

$$\hbar\omega_{if}\Gamma_{i\to f} = \frac{4e^2}{3c^3}\left|\left\langle f\left|\frac{d^2r}{dt^2}\right|i\right\rangle\right|^2. \tag{19.26}$$

This relation is similar to the classical formula of Larmor, Eq. (2.1), which states that the total power radiated by an accelerated (non-relativistic) electron is [4]

$$P = \frac{d\mathcal{E}}{dt} = \frac{2e^2}{3c^3}\left|\frac{d^2\mathbf{r}}{d\,t^2}\right|^2 . \tag{19.27}$$

Absorption and Stimulated Emission of Light — Since for the absorption of light we must have photons present in the field we take $n_{\mathbf{k},\sigma} \neq 0$ in the initial state, $|\cdots n_{\mathbf{k},\sigma}, \cdots\rangle_{rad}$, of Eq. (19.8). The annihilation operator in H', Eq. (19.5), removes a photon from the field, therefore the final state is given by

$$|\,F\rangle = |f\rangle_{atom}|\cdots n_{\mathbf{k},\sigma} - 1\cdots\rangle_{rad}. \tag{19.28}$$

Following the method that we used earlier to calculate the transition rate for the spontaneous emission, when applied to the present problem gives us

$$\Gamma_{i\to f} = \frac{4\pi^2 e^2}{m^2 L^3 \omega_k} n_{\mathbf{k},\sigma} \left|\langle f\,|\mathbf{p}\cdot\mathbf{u}_{\mathbf{k},\sigma}\, e^{+i\mathbf{k}\cdot\mathbf{r}}|\,i\rangle\right|^2 \delta(E_i - E_f + \hbar\omega). \tag{19.29}$$

For this case we can determine the cross section for absorption of a photon by an atom by observing that the flux of photons of momentum $\hbar\mathbf{k}$ and polarization σ is

$$\Phi = \frac{n_{\mathbf{k},\sigma} c}{L^3}. \tag{19.30}$$

By dividing (19.29) by the flux Φ we obtain the cross section for this process:

$$\sigma_{i\to f}(\mathbf{k},\sigma) = \frac{4\pi^2 e^2}{m^2 c\,\omega_k} \left|\langle f\,|\mathbf{p}\cdot\mathbf{u}_{\mathbf{k},\sigma}\, e^{+i\mathbf{k}\cdot\mathbf{r}}|\,i\rangle\right|^2 \delta(E_i - E_f + \hbar\omega). \tag{19.31}$$

This relation should be integrated over a narrow frequency range of $\hbar\omega_k = E_f - E_i$.

By a similar argument we find that the rate of a downward transition induced by the incident beam from initial state $|f\,\rangle$ to the final state $|i\,\rangle$ is given by

$$\Gamma_{f\to i}^{ind\;em} = \Gamma_{i\to f}^{abs}. \tag{19.32}$$

As we will see later neither in the emitted nor in the absorbed light the spectral lines are infinitely sharp as it seems to be indicated by the δ-functions in (19.14) and (19.31).

19.1 Theory of Natural Line Width

The emitted light from an atom is a wave train of finite duration. A Fourier analysis of the finite wave train shows that the spectral line has a finite width.

This means that the delta function appearing in (19.14) and (19.29) is a result of the approximation used in the calculation. Exactly solvable models for emission and absorption like the Wigner–Weisskopf model exhibit a decay for the initial state which is approximately exponential [5],[6].

Consider a simple case where there is no photon in the initial state and only one photon of momentum $\hbar\mathbf{k}$ in the final state. Thus the initial and final states are:

$$| I\rangle = | 0 \rangle_{photon}|i \rangle_{atom} = | 0; i\rangle. \tag{19.33}$$

$$| F\rangle = | 1_{\mathbf{k},\sigma}\rangle_{photon}|f \rangle_{atom} = | \mathbf{k},\sigma; f\rangle. \tag{19.34}$$

The time evolution of the state of this system is given by

$$i\hbar\frac{\partial}{\partial t}|\psi(t)\rangle = (H_0 + H'(t))\,|\psi(t)\rangle, \tag{19.35}$$

where $H'(t)$ is given by (19.5). We write the ket $|\psi(t)\rangle$ in terms of the time-dependent coefficients $C_{i0}(t)$ and $C_{f\mathbf{k},\sigma}(t)$, i.e.

$$|\psi(t)\rangle = C_{i0}(t)|\, 0; i\rangle + \sum_{\mathbf{k},\sigma} C_{f\mathbf{k},\sigma}(t)|\,\mathbf{k},\sigma; f\rangle. \tag{19.36}$$

Substituting (19.36) in (19.35) and equating the coefficients of $|\, 0; i\rangle$ and $|\mathbf{k},\sigma; f\rangle$ we find

$$i\hbar\frac{dC_{i0}(t)}{dt} = E_i C_{i0} + \sum_{\mathbf{k},\sigma}\langle 0; i\,|H'|\,\mathbf{k},\sigma; f\rangle\, C_{f\mathbf{k},\sigma}(t), \tag{19.37}$$

and

$$i\hbar\frac{dC_{f\mathbf{k},\sigma}(t)}{dt} = (E_f + \hbar\omega)\, C_{f\mathbf{k},\sigma} - \langle\mathbf{k},\sigma; f\,|H'|\,0; i\rangle\, C_{i0}(t). \tag{19.38}$$

The initial conditions for these differential equations are:

$$C_{i0}(0) = 1, \quad C_{f\mathbf{k},\sigma}(0) = 0. \tag{19.39}$$

To find the solutions of the coupled equations (19.37) and (19.38) we write the Fourier series for $C_{i0}(t)$ and $C_{f\mathbf{k},\sigma}(t)$

$$C_{i0}(t) = \sum_{\omega} p(\omega) C_0(\omega)\exp[-i(E_i + \hbar\omega)t], \tag{19.40}$$

$$C_{f\mathbf{k},\sigma}(t) = \sum_{\omega} p(\omega) C_{\mathbf{k},\sigma}(\omega)\exp[-i(E_i + \hbar\omega)t], \tag{19.41}$$

where $p(\omega)$ is given by

$$p(\omega) = C_0^*(\omega)C_{i0}(t=0) + \sum_{\mathbf{k},\sigma} C_{f\mathbf{k},\sigma}^*(\omega)C_{\mathbf{k},\sigma}(t=0) = C_0^*(\omega). \tag{19.42}$$

Now we substitute (19.40) and (19.41) in (19.37) and (19.38) to find a coupled algebraic equations:

$$\hbar\omega C_0(\omega) = \sum_{\mathbf{k},\sigma} \langle 0; i\,|H'|\,\mathbf{k},\sigma; f\rangle\, C_{\mathbf{k},\sigma}(\omega), \tag{19.43}$$

$$\hbar\omega C_{\mathbf{k},\sigma}(\omega) = (E_f + \hbar\omega_k - E_i)C_{\mathbf{k},\sigma}(\omega) - \langle \mathbf{k},\sigma; f\,|H'|\,0; i\rangle\, C_0(\omega). \tag{19.44}$$

By eliminating $C_{\mathbf{k},\sigma}(\omega)$ between (19.44) and (19.43) we find the eigenvalue equation for ω;

$$\begin{aligned}\hbar\omega &= -\sum_{\mathbf{k},\sigma} \frac{\left|\langle \mathbf{k},\sigma; f\,|H'|\,0; i\rangle\right|^2}{E_f - E_i + \hbar(\omega_k - \omega)} \\ &\to -\frac{2\Omega}{(2\pi)^3}\int \frac{\left|\langle \mathbf{k},\sigma; f\,|H'|\,0; i\rangle\right|^2}{E_f - E_i + \hbar(\omega_k - \omega)} d^3k \\ &= -g\hbar \int \frac{\left|\langle f|\mathbf{p}|i|\rangle\right|^2}{E_f - E_i + \hbar(\omega_k - \omega)}\omega_k d\omega_k, \end{aligned} \tag{19.45}$$

where g is a constant;

$$g = \frac{4e^2}{3\pi c^3 m^2}. \tag{19.46}$$

We normalize $C_0(\omega)$ and $C_{\mathbf{k},\sigma}(\omega)$ by requiring that

$$C_0^*(\omega)C_0\left(\omega'\right) + \sum_{\mathbf{k},\sigma} C_{\mathbf{k},\sigma}^*(\omega)C_{\mathbf{k},\sigma}\left(\omega'\right) = \delta_{\omega,\omega'}. \tag{19.47}$$

Setting $\omega = \omega'$ in (19.47) and substituting for $C_{\mathbf{k},\sigma}(\omega)$ from (19.44) we find $C_0(\omega)$

$$C_0(\omega) = \left\{1 + g\hbar \int \frac{|\langle f|\mathbf{p}|i\rangle^2\omega_k}{E_f - E_i + \hbar\omega_k - \hbar\omega}\, d\omega_k\right\}^{-\frac{1}{2}}. \tag{19.48}$$

Now if we define $\Delta(\omega)$ by

$$\Delta(\omega) = \omega + \left(\frac{E_f - E_i}{\hbar}\right) + g\int_0^\infty \frac{|\langle f|\mathbf{p}|i\rangle|^2\omega_k d\omega_k}{E_f - E_i + \hbar(\omega_k - \omega)}, \tag{19.49}$$

then

$$C_0(\omega) = \left[\frac{d\Delta(\omega)}{d\omega}\right]^{-\frac{1}{2}}. \tag{19.50}$$

From Eq. (19.44) we get $C_{\mathbf{k},\sigma}(\omega)$;

$$C_{\mathbf{k},\sigma}(\omega) = -\left(\frac{|\,\langle 0; i\,|H'|\,\mathbf{k},\sigma; f\rangle}{E_f - E_i + \hbar\omega_k - \hbar\omega}\right)\left[\frac{d\Delta(\omega)}{d\omega}\right]^{-\frac{1}{2}}. \tag{19.51}$$

We also obtain $C_{i0}(t)$ by substituting $C_0(\omega)$ from (19.50) in (19.40);

$$C_{i0}(t) = \sum_{\omega_j} |C_0(\omega_j)|^2 e^{-i\omega_j t} = \sum_{\omega_j} \frac{e^{-i\omega_j t}}{\left(\frac{d\Delta(\omega)}{d\omega}\right)_{\omega_j}}, \tag{19.52}$$

where ω_j s are the roots of $\Delta(z)$;

$$\Delta(z) = z + \left(\frac{E_f - E_i}{\hbar}\right) + g \int_0^\infty \frac{|\langle f|\mathbf{p}|i\rangle|^2 \omega_k d\omega_k}{E_f - E_i + \hbar(\omega_k - z)} = 0. \tag{19.53}$$

Equation (19.52) has a form that enables us to express $C_{i0}(t)$ as a contour integral

$$C_{i0}(t) = \frac{1}{2\pi i} \exp\left(-\frac{iE_f t}{\hbar}\right) \oint_C \frac{e^{-izt}}{\Delta(z)}, \tag{19.54}$$

where the contour contains all roots ω_j of $\Delta(z) = 0$. The residues at the poles of $z = \omega_j$ of the integrand in (19.52) are

$$2\pi i \frac{e^{-i\omega_j t}}{\left(\frac{d\Delta(\omega)}{d\omega}\right)_{\omega_j}}, \tag{19.55}$$

and these are the same as those given by (19.52), Thus we can use (19.54) to determine $C_{i0}(t)$.

In order to evaluate $C_{i0}(t)$ and $C_{f\mathbf{k},\sigma}(t)$ by the method of contour integration we first define $\Delta^I(z)$ and $\Delta^I(z)$ by the analytic continuation of $\Delta(z)$

$$\Delta(z) = \begin{cases} \Delta^I(z) & \text{for} \quad 0 < \arg z < 2\pi \\ \Delta^{II}(z) & \text{for} \quad -2\pi < \arg z < 0 \end{cases}, \tag{19.56}$$

then we find that as $z \to x - i\epsilon$ the two parts are connected by

$$\Delta^{II}(z) = \Delta^I(z) + \frac{2ig\pi}{\hbar} z\, |\langle f|\mathbf{p}|i\rangle|^2. \tag{19.57}$$

For the sake of simplicity we assume that $\Delta^{II}(z)$ has only one root [7],[8]

$$\Delta^{II}(z_0) = 0, \quad \text{for} \quad z_0 = \frac{1}{\hbar}\left(E_r - E_f - \frac{i\hbar}{2}\Gamma\right), \tag{19.58}$$

where E_r and Γ are constants to be determined. By substituting (19.58) in (19.53) and equating the real and imaginary parts of $\Delta^{II}(z_0)$ we find two equations for E_r and Γ;

$$\begin{aligned} E_r - E_i \quad &+ \quad g\hbar \int_0^\infty \frac{(\hbar\omega_k + E_f - E_r)|\langle f|\mathbf{p}|i\rangle|^2 \omega_k}{(\hbar\omega_k + E_f - E_r)^2 + \frac{1}{4}\hbar^2\Gamma^2} d\omega_k \\ &+ \quad \frac{g\hbar\Gamma}{2} |\langle f|\mathbf{p}|i\rangle|^2_{z_0} = 0, \end{aligned} \tag{19.59}$$

and

$$
\begin{aligned}
&\frac{\hbar}{2}\Gamma\left\{1+g\hbar\int_0^\infty \frac{|\langle f|\mathbf{p}|i\rangle|^2\omega_k}{(\hbar\omega_k+E_f-E_r)^2+\frac{1}{4}\hbar^2\Gamma^2}d\omega_k\right\}\\
&= \frac{\pi g}{2\hbar}(E_r-E_f)\,|\langle f|\mathbf{p}|i\rangle|^2_{z_0},
\end{aligned}
\tag{19.60}
$$

These two equations can be solved for Γ and the shift in the energy of the initial state E_r. An approximate solution can be found by expanding Eqs. (19.59) and (19.60) by assuming that the coupling constant g is small. Thus expanding these equations in powers of g and retaining terms proportional to g we obtain

$$
E_r - E_i \approx -g\hbar\,\mathcal{P}\int_0^\infty \frac{|\langle f|\mathbf{p}|i\rangle|^2\omega_k}{(\hbar\omega_k+E_f-E_r)}d\omega_k, \tag{19.61}
$$

and

$$
\Gamma \approx \frac{\pi g}{\hbar^2}(E_f-E_i)|\langle f|\mathbf{p}|i\rangle|^2_{z_0} \approx \frac{4e^2}{3m^2c^3\hbar}\omega_{if}|\langle f|\mathbf{p}|i\rangle|^2_{z_0}, \tag{19.62}
$$

where $\omega_{rf}\approx\omega_{if}$ is the frequency of the emitted photon. The second equation is the same as $\Gamma_{i\to f}$, Eq. (19.21), found from the first order perturbation theory.

Once the root of $\Delta^{II}(z)$ in the second Riemann sheet is determined, we can calculate $C_{i0}(t)$ and $C_{f\mathbf{k},\sigma}$ by contour integration. Thus we write $C_{i0}(t)$ as

$$
C_{i0}(t) = \frac{1}{2\pi i}\exp\left(-i\frac{E_f}{\hbar}\right)\left\{\int_{\mathcal{C}_1}+\int_{\mathcal{C}_2}\right\}\frac{e^{-izt}}{\Delta(z)}, \tag{19.63}
$$

where the contour $\mathcal{C}_1$ is composed of the following parts:

(a) - A straight line in the upper half z plane in the first Riemann sheet extending from $R+i\varepsilon$ to $-\varepsilon+i\varepsilon$ where it bends, crosses the x axis and joins the line in the lower half of z plane extending from $-\varepsilon+i\varepsilon$ to $-R-i\varepsilon$. A semi-circle of radius R in the second Riemann sheet completes the contour $\mathcal{C}_1$.

(b) - The contour $\mathcal{C}_2$ consists of two straight lines, one in the first sheet extending from $-\infty+i\varepsilon$ to $-\delta+i\varepsilon$ and in the second sheet from $-\delta-i\varepsilon$ to $-\infty-i\varepsilon$. These two lines are joined at $x=-\delta$. The quantities δ and ε are small positive numbers whereas R is a very large positive quantity. The denominator $\Delta(z)$ in (19.63) is either $\Delta^I(z)$ or $\Delta^{II}(z)$ depending on whether the contour is on the first or second Riemann sheet. Evaluating the integrals in (19.63) we find

$$
\begin{aligned}
C_{i0}(t) &= \frac{\exp\left(-i\frac{E_r t}{\hbar}-\frac{1}{2}\Gamma t\right)}{\left(\frac{d\Delta^{II}(z)}{dz}\right)_{z_0}}\\
&- g\int_{-\infty}^0 \exp\left(-i\frac{E_f t}{\hbar}\right)\frac{|\langle f|\mathbf{p}|i\rangle|^2 x\,e^{-ixt}}{\Delta^I(x)\Delta^{II}(x)}dx.
\end{aligned}
\tag{19.64}
$$

with a similar relation for $C_{f\mathbf{k},\sigma}(t)$.

A simple approximate form of $C_{i0}(t)$ can be obtained by noting that in the limit of small g or e^2 we have

$$\left(\frac{d\Delta^{II}(z)}{dz}\right)_{z_0} \approx 1 + \mathcal{O}(g), \tag{19.65}$$

and thus we get

$$C_{i0}(t) \approx \exp\left[-i\frac{E_r t}{\hbar} - \frac{\Gamma t}{2}\right] + \mathcal{O}(g). \tag{19.66}$$

Now by substituting (19.66) in (19.38) and integrating the inhomogeneous differential equation with the boundary condition (19.39) we obtain the approximate form of $C_{f\mathbf{k},\sigma}(t)$;

$$C_{f\mathbf{k},\sigma}(t) \approx -\langle \mathbf{k}\sigma; f\,|H'|\,0; i\rangle \left(\frac{e^{\frac{-iE_r t}{\hbar}}\, e^{-\frac{\Gamma t}{2}} - e^{\frac{-i}{\hbar}(E_f + \hbar\omega_k)t}}{E_f - E_r + \hbar\omega_k + \frac{i}{2}\hbar\Gamma}\right) + \mathcal{O}(g). \tag{19.67}$$

The exponential decay of the probability of the atom to be in the state i, i.e. $|C_{i0}(t)|^2$ is a result of the approximations that we have made. In general these decays are nonexponential [7]–[9].

Calculation of the Lifetime of a Decaying State — In order to have an idea about the lifetime of an unstable atomic system let us calculate this lifetime, Γ^{-1} for the $2P_0$ state of the hydrogen atom when it decays to the ground state $1S_0$ [10]. The wave functions for these states are given by (see Eqs. (9.221) and (9.222))

$$\psi_{100}(r,\theta,\phi) = \frac{1}{\sqrt{\pi a_0^3}} \exp\left(-\frac{r}{a_0}\right), \tag{19.68}$$

and

$$\psi_{210}(r,\theta,\phi) = \frac{\sqrt{2}}{8\sqrt{\pi a_0^3}} \left(\frac{r}{a_0}\right) \exp\left(-\frac{r}{2a_0}\right) \cos\theta. \tag{19.69}$$

The frequency of the emitted photon is given by

$$\hbar\omega_{if} = \hbar\omega_{1S,2P} = E_{1S} - E_{2P} = \frac{me^4}{2\hbar^2}\left(1 - \frac{1}{4}\right) = \frac{3e^2}{8a_0}. \tag{19.70}$$

The angular integration leaves the z component of $\mathbf{r}$ as the only nonzero part of the matrix element. Thus Eq. (19.29) (or (19.62)) reduces to

$$\Gamma = \frac{4}{3}\left(\frac{3}{8}\right)^3 \left(\frac{c}{a_0^3}\right)\left(\frac{e^2}{\hbar c}\right)^4 \left|\langle\psi_{210}(r,\theta)|r\cos\theta|\psi_{100}\rangle\right|^2, \tag{19.71}$$

The matrix element in (19.71) can be evaluated easily

$$\begin{aligned} &\langle\psi_{210}(r,\theta)|r\cos\theta|\psi_{100}(r)\rangle \\ &= \frac{\sqrt{2}}{8\pi a_0^3}\int\left(\frac{r}{a_0}\right) e^{\frac{r}{2a_0}}\cos\theta(r\cos\theta)e^{-\frac{r}{a_0}r^2}\,dr\sin\theta d\theta d\phi \\ &= 4\sqrt{2}\left(\frac{2}{3}\right)^5 a_0. \end{aligned} \tag{19.72}$$

Substituting this result in (19.71) we obtain

$$\Gamma = \frac{4}{3}\left(\frac{3}{8}\right)^3\left(\frac{e^2}{\hbar c}\right)^4 2^5 \left(\frac{2}{3}\right)^{10}\left(\frac{c}{a_0}\right). \tag{19.73}$$

Thus the lifetime for this unstable state is

$$\frac{1}{\Gamma} = \left(\frac{3}{2}\right)^8\left(\frac{\hbar c}{e^2}\right)^4\left(\frac{c}{a_0}\right) \approx 1.6\times 10^{-8}s. \tag{19.74}$$

19.2 The Lamb Shift

As we seen earlier, Eq. (19.61), the coupling of electron to the electromagnetic field causes a shift of the spectral line. This shift is observable e.g. the energy difference in the $2S$ and $2P_{\frac{1}{2}}$ levels of the hydrogen atom which, as we have seen must be degenerate. These levels are also degenerate in the Dirac's relativistic theory of H atom. The beautiful experiment of Lamb and Retherford showed that there is a small energy difference of 1057 megacycle between the $2S$ and $2P_{\frac{1}{2}}$ states [11]. A non-relativistic account of the Lamb shift is given by Bethe, which we will discuss in this section [12].

Let us consider an electron in the state $|n\rangle$ with energy E_n. Again we assume that the interaction between the electromagnetic field and the electron is given by (19.5) and again we ignore the term which is quadratic in $\mathbf{A}$. By the coupling to the field $\mathbf{A}(\mathbf{r})$, the electron can emit spontaneously a photon by making transition to the state $|j\rangle$ and then absorbing the same photon and coming back to the state $|n\rangle$. The second order contribution to the energy E_n caused by this emission and absorption is (see Eq. (11.24))

$$\Delta E_n^{(2)} = \sum_j \sum_{\mathbf{k},\sigma} \frac{\langle n;0\,|H'|\,\mathbf{k},\sigma;j\rangle\,\langle \mathbf{k},\sigma;j\,|H'|\,n;0\rangle}{E_n - E_j - \hbar\omega_k}. \tag{19.75}$$

Here $|i;0\rangle$ denotes the initial state with electron in in the state $|n\rangle$ when there is no photon and $|\mathbf{k},\sigma;j\rangle$ is the intermediate state with the electron in state $|j\rangle$ when a single photon of momentum $\hbar\mathbf{k}$ and polarization σ is present. Thus the energy of this intermediate state is given by $E_j + \hbar\omega_k$. The matrix elements of H' in the dipole approximation are given by Eq. (19.10),

$$\langle n;0\,|H'|\,\mathbf{k},\sigma;j\rangle = -\frac{e}{mc}\left(\frac{2\pi\hbar c^2}{\Omega\omega_k}\right)^{\frac{1}{2}}\langle n|\mathbf{p}\cdot\mathbf{u}_{\mathbf{k},\sigma}|j\rangle. \tag{19.76}$$

By integrating over the angular variable θ and ϕ as we did earlier, Eq. (19.20), and summing over σ we find

$$\int\sum_\sigma |\langle n|\mathbf{p}\cdot\mathbf{u}_{\mathbf{k},\sigma}|j\rangle|^2\, 2\pi\sin\theta d\theta = \frac{8\pi}{3}|\langle n|\mathbf{p}|j\rangle|^2. \tag{19.77}$$

Substituting (19.76) and (19.77) in (19.75) we find

$$\Delta E_n^{(2)} = \frac{2e^2\hbar}{3\pi m^2 c^3}\int_0^\infty \omega_k d\omega_k \sum_j \frac{|\langle n|\mathbf{p}|j\rangle|^2}{E_n - E_j - \hbar\omega_k}. \tag{19.78}$$

We notice that the integral in (19.78) is divergent, i.e. there is an infinite shift in the energy of electron. Bethe observed that if we make a similar calculation for a free electron, again the result is infinite. Consider the emission and absorption of a photon by a free electron where in the dipole approximation we find the energy shift to be

$$\Delta E_{\mathbf{p}}^{(2)} = -\frac{2e^2}{3\pi m^2 c^3}\sum_{\mathbf{q}} |\langle \mathbf{q}|\mathbf{p}|\mathbf{p}\rangle|^2 \int_0^\infty d\omega_k, \tag{19.79}$$

and this shift can be written as

$$\Delta E_{\mathbf{p}}^{(2)} = -\frac{2e^2}{3\pi m^2 c^3}\left|\langle \mathbf{q}\left|\mathbf{p}^2\right|\mathbf{p}\rangle\right| \int_0^\infty d\omega_k = -\frac{2e^2}{3\pi m^2 c^3}\mathbf{p}^2 \int_0^\infty d\omega_k. \tag{19.80}$$

The energy shift is proportional to $\mathbf{p}^2$ and thus it can be combined with the (zero order) kinetic energy and the sum can be written as

$$\begin{aligned} E_{\mathbf{p}} \approx E_{\mathbf{p}} + \Delta E_{\mathbf{p}}^{(2)} &= \frac{1}{2}\mathbf{p}^2\left(\frac{1}{m_0} - \frac{4e^2}{3\pi m^2 c^3}\int_0^\infty d\omega_k\right) \\ &= \frac{1}{2m}\mathbf{p}^2. \end{aligned} \tag{19.81}$$

We can interpret this result in the following way: The mass m_0 in the expression of the kinetic energy of the electron,

$$E_{\mathbf{p}}^{(0)} = \frac{\mathbf{p}^2}{2m_0}, \tag{19.82}$$

refers to the "bare" mass of the electron, i.e. the mass when there is no coupling to the electromagnetic field. But this is a fictitious quantity since there is no way that we can turn off the interaction. So what is measured as the mass of the electron is m which is given by (19.81) and we know that this is a finite quantity. This change of mass from m_0 to m is called "mass renormalization".

When the electron is bound then the expectation value of $\mathbf{p}^2$ in the eigenstate $|\mathbf{p}\rangle$ should be replaced by

$$\Delta E_n^{(2)} = -\frac{2e^2\hbar}{3\pi m^2 c^3}\sum_j \int_0^\infty \frac{\left|\langle j\left|\mathbf{p}\right| n\rangle\right|^2}{\hbar\omega_k}\omega_k d\omega_k. \tag{19.83}$$

We can attribute the divergence of Eq. (19.78) to the infinite change that we have found in the observed mass of the electron. For example if we write the Hamiltonian of the hydrogen atom as

$$H = \frac{1}{2m_0}\mathbf{p}^2 - \frac{e^2}{r} - \frac{e}{m_0 c}\mathbf{p}\cdot\mathbf{A}, \tag{19.84}$$

then we can substitute for $\frac{\mathbf{p}^2}{2m_0}$ from (19.81) and write the Hamiltonian as

$$H = \frac{1}{2m}\mathbf{p}^2 - \frac{e^2}{r} + \left\{ -\frac{e}{m_0 c}\mathbf{p}\cdot\mathbf{A} + \frac{2\mathbf{p}^2 e^2}{3\pi\, m^2 c^3}\int d\omega_k \right\}. \tag{19.85}$$

The last term in the curly bracket is the effective interaction of an electron of renormalized mass m with the radiation field. In this formulation we have included the electromagnetic interaction (19.80) in H and also in m as defined by (19.81). In order to avoid counting this interaction twice, we subtract $\Delta E_{\mathbf{p}}$, Eq. (19.81), from ΔE_n, Eq. (19.78) to obtain the shift in the energy level $\delta E_n^{(2)}$;

$$\delta E_n^{(2)} = \frac{2e^2\hbar}{3\pi m^2 c^3}\int_0^\infty \left(\sum_j \frac{|\langle j|\mathbf{p}|n\rangle|^2}{E_n - E_j - \hbar\omega_k} + \frac{\langle n\,|\mathbf{p}^2|n|\rangle}{\hbar\omega_k} \right)\omega_k d\omega_k, \tag{19.86}$$

where in getting the last term of this relation we have used the completeness relation

$$\langle n\,|\mathbf{p}^2|\,n\rangle = \sum_j |\langle j|\mathbf{p}|n\rangle|^2. \tag{19.87}$$

Writing (19.86) as

$$\delta E_n^{(2)} = \frac{2e^2}{3\pi m^2 c^3}\sum_j |\langle j|\mathbf{p}|n\rangle|^2 \int_0^\infty \frac{(E_n - E_j)}{E_n - E_j - \hbar\omega_k} d\omega_k, \tag{19.88}$$

we observe that the last integral over ω_k is still divergent, but now logarithmically. In the relativistic formulation of the problem $\delta E_n^{(2)}$ turns out to be a finite number. To get a finite result for the non-relativistic formulation, Bethe argued that the upper limit of the integral in (19.88) should be $\frac{mc^2}{\hbar}$, which corresponds to a photon energy equal to the rest mass energy of the electron,

$$\hbar\omega_k = mc^2. \tag{19.89}$$

Using this upper limit cut-off we find $\delta E_n^{(2)}$ to be

$$\delta E_n^{(2)} = \frac{2e^2}{3\pi\hbar\, m^2 c^3}\sum_j |\langle j|\mathbf{p}|n\rangle|^2 (E_n - E_j)\ln\left|\frac{mc^2}{E_j - E_n}\right|, \tag{19.90}$$

where we have neglected $(E_j - E_n)$ compared to mc^2. To evaluate the right-hand side of (19.90) we first define the average excitation energy by

$$\{\ln|E_j - E_n|\}_{ave} = \frac{\sum_j |\langle j|\mathbf{p}|n\rangle|^2 (E_j - E_n)\ln|E_j - E_n|}{\sum_j |\langle j|\mathbf{p}|n\rangle|^2 (E_j - E_n)}, \tag{19.91}$$

and then we write $\delta E_n^{(2)}$ as

$$\delta E_n^{(2)} = \frac{2e^2}{3\pi\hbar\, m^2 c^3}\left(\ln\left|\frac{mc^2}{E_j - E_n}\right|\right)_{ave}\sum_j |\langle j|\mathbf{p}|n\rangle|^2 (E_j - E_n). \tag{19.92}$$

The sum over j in this relation can be calculated by noting that;

$$\begin{aligned}\sum_j |\langle j|\mathbf{p}|n\rangle|^2(E_j - E_n) &= \langle n|\mathbf{p}(H_0 - E_n)\cdot\mathbf{p}|n\rangle \\ &= -\frac{1}{2}\langle n|\,[\mathbf{p},\ [\mathbf{p},\ H_0]\,]\,|n\rangle,\end{aligned} \tag{19.93}$$

where H_0 is the hydrogen atom Hamiltonian. Using the commutator

$$H_0\mathbf{p} - \mathbf{p}H_0 = i\hbar\nabla V = i\hbar\nabla\left(-\frac{e^2}{r}\right), \tag{19.94}$$

we calculate the right-hand side of (19.93)

$$\frac{\hbar^2}{2}\left\langle n\left|\nabla^2\left(-\frac{e^2}{r}\right)\right|n\right\rangle = -\frac{e^2\hbar^2}{2}\int|\psi_n(\mathbf{r})|^2\frac{1}{r}d^3r = 2\pi e^2\hbar^2|\psi_n(0)|^2. \tag{19.95}$$

By substituting (19.95) in (19.92) we find the final form for $\delta E_n^{(2)}$

$$\delta E_n^{(2)} = \frac{4e^4\hbar}{3m^2c^3}\,|\psi_n(0)|^2\left(\ln\left|\frac{mc^2}{E_j - E_n}\right|\right)_{ave}. \tag{19.96}$$

The wave function $\psi_n(0)$ vanishes for states with $\ell \neq 0$. For S states we have

$$|\psi_n(0)|^2 = \frac{1}{\pi}\left(\frac{1}{na_0}\right)^3, \tag{19.97}$$

where $a_0 = \frac{\hbar^2}{me^2}$ is the Bohr radius.

From the numerical calculation of $(E_j - E_n)_{ave}$, Bethe found an approximate value of 1040 megacycle for the $2S$ wave in good agreement with the experimental result of 1057 megacycle.

Let us note that the two assumptions used in Bethe's approximate calculation are not compatible with each other. In the first place he had assumed the validity of the dipole approximation, $\exp(-i\mathbf{k}\cdot\mathbf{r}) \approx 1$, Eqs. (19.17) and (19.76). The second approximation in Bethe's approach was the high frequency cut-off introduced by assuming that the maximum photon energy is equal to the rest mass of the electron, $mc^2 = \hbar\omega = \hbar kc$ and this was used to derive (19.90). Now the electron is localized within the Bohr radius, a_0, therefore the assumption that $\mathbf{k}\cdot\mathbf{r}$ is much less than one is violated since

$$\max\ (\mathbf{k}\cdot\mathbf{r}) = k_{max}a_0 = \frac{mc}{\hbar}\left(\frac{\hbar^2}{me^2}\right)\frac{\hbar c}{e^2} \approx 137. \tag{19.98}$$

The correct relativistic calculation with the inclusion of the vacuum polarization effect gives us an accurate value of the Lamb shift which agrees very well with the empirical result [13].

19.3 Heisenberg's Equations for Interaction of an Atom with Radiation

In our formulation of the problem of interaction of an atom with electromagnetic field we discussed the mechanism for the transition from an initial state to a final state with the emission or absorption of a single photon. Now we want to study a model where an atom with $2M+1$ quantum levels interacts with the electromagnetic field, and that this interaction is accompanied with the emission or absorption of a number of photons. Thus the two-state atom considered earlier will corresponds to the special case where $M = \frac{1}{2}$ and emission or absorption of a single photon. In the present problem we assume that photons of unit energy ($\hbar = \omega = 1$) are being absorbed or emitted by the atom, therefore the Hamiltonian for the field is simply given by the number operator $a^\dagger a$. The atomic system in this model has $2M+1$ equally spaced energy levels with the level spacing ϵ. We assume that the Hamiltonian for the atom alone is given by ϵM_z, where M_z is the z component of the angular momentum operator. In addition we choose the coupling between the atom and the quantized electromagnetic field, in the dipole approximation to be of the form

$$\frac{\lambda}{2}\left(M_+ a + M_- a^\dagger\right), \tag{19.99}$$

where M_+ and M_- are raising and lowering operators given by (9.53). Thus the total Hamiltonian for the atom, the radiation field and the interaction between them is

$$H = a^\dagger a + \epsilon M_z + \frac{\lambda}{2}\left(M_+ a + M_- a^\dagger\right), \tag{19.100}$$

where $M_\pm$ and M_z satisfy the commutation relation (9.56).

There are two constants of motion associated with this Hamiltonian:

$$\mathbf{M}^2 = \frac{1}{2}(M_+M_- + M_-M_+) + M_z^2, \tag{19.101}$$

and

$$\mathcal{C} = a^\dagger a + M_z. \tag{19.102}$$

These two operators commute with the total Hamiltonian (19.100) and also commute with each other,

$$\left[\mathbf{M}^2,\, \mathcal{C}\right] = 0. \tag{19.103}$$

In the absence of interaction, i.e. when $\lambda = 0$, the eigenstates of the Hamiltonian for the atom are given by

$$M_z|m\rangle = m|m\rangle, \qquad m = -j,\ -j+1 \cdots j-1,\ j, \tag{19.104}$$

and for the field, one has the eigenstates $|n\rangle$, and the eigenvalue equation for the number operator

$$a^\dagger a|n\rangle = n|n\rangle, \qquad n = 0,\ 1,\ 2\cdots. \tag{19.105}$$

We can express the eigenstates of the total Hamiltonian in terms of states with definite m and n;

$$|E,\gamma\rangle = \sum_{m,n} \delta_{\gamma,m+n} B(m,n)|m\rangle|n\rangle, \tag{19.106}$$

where $|m\rangle|n\rangle$ denotes the state m of the atom and n of the radiation field. In Eq. (19.106), $\gamma = m+n$ is the eigenvalue of the operator $\mathcal{C}$, and $B(m,n)$ is the coefficient of expansion. The standard method of solving this problem for the energy $\mathcal{E}$ of the atom at a given time t can be formulated in terms of the initial wave function $|\psi(0)\rangle$. Thus expanding $|\psi(0)\rangle$ in terms of the states $|E,\gamma\rangle$ we have

$$|\psi(0)\rangle = \sum_{E,\gamma} \alpha(E,\gamma)|E,\gamma\rangle. \tag{19.107}$$

The time development of $|\psi(0)\rangle$ can be easily found since $|E,\gamma\rangle$ is an eigenstate of H,

$$|\psi(t)\rangle = e^{-iHt}|\psi(0)\rangle = \sum_{E,\gamma} \alpha(E,\gamma)e^{-iEt}|E,\gamma\rangle. \tag{19.108}$$

Using this time-dependent wave function, we calculate the expectation value of ϵM_z which is the energy of atom, $\mathcal{E}(t)$:

$$\begin{aligned} \mathcal{E}(t) &= \epsilon\langle\psi(t)|M_z|\psi(t)\rangle = \epsilon \sum_{E,E',\gamma} \alpha^*\left(E',\gamma\right)\alpha(E,\gamma) \\ &\times \exp\left[i\left(E'-E\right)t\right]\left\langle E',\gamma|M_z|E,\gamma\right\rangle. \end{aligned} \tag{19.109}$$

Now in the Heisenberg picture we first determine the operator $M_z(t)$ where

$$M_z(t) = e^{iHt}M_z(0)e^{-iHt}, \tag{19.110}$$

and then calculate the matrix element

$$\mathcal{E}(t) = \epsilon\left\langle\psi(0)|M_z(t)|\psi(0)\right\rangle. \tag{19.111}$$

This relation shows that we can obtain $\mathcal{E}(t)$ if we have the Heisenberg equation for $M_z(t)$. It is convenient to define a parameter β

$$\beta = \epsilon - 1 \tag{19.112}$$

which measures detuning from the resonance. Furthermore we introduce three operators $\mathcal{A}$, $\mathcal{B}$ and $\mathcal{D}$ by

$$\mathcal{A} = H - \mathcal{C}, \tag{19.113}$$

$$\mathcal{B} = \frac{\lambda}{2}\left[M_+a - M_-a^\dagger\right], \tag{19.114}$$

and

$$\mathcal{D} = \frac{\lambda}{2}\left[M_+a + M_-a^\dagger\right] = \mathcal{A} - \beta M_z. \tag{19.115}$$

The operator $\mathcal{C}$ is a constant of motion, and so is $\mathcal{A}$. From the definition of M_+ and M_- we have

$$M_+M_- = \mathbf{M}^2 - M_z^2 + M_z \quad \text{and} \quad M_-M_+ = \mathbf{M}^2 - M_z^2 - M_z. \tag{19.116}$$

From these operators and the Heisenberg equation $i\dot{F} = [F,\ H]$ we find the following equations for the time derivatives of M_z and $\mathcal{B}$;

$$i\frac{dM_z(t)}{dt} = \mathcal{B}, \tag{19.117}$$

$$i\frac{d\mathcal{B}(t)}{dt} = \hat{\Omega}^2 M_z(t) - \frac{\lambda^2}{2}\left[3M_z^2 - \mathbf{M}^2\right] - \beta\mathcal{A}. \tag{19.118}$$

In the last equation $\hat{\Omega}$ which is a frequency operator and is a constant of motion is defined by

$$\hat{\Omega}^2 = \beta^2 + \lambda^2\left(\mathcal{C} + \frac{1}{2}\right), \tag{19.119}$$

By differentiating (19.117) with respect to t and substituting for $\frac{d\mathcal{B}(t)}{dt}$ from (19.118) we obtain

$$\frac{d^2M_z(t)}{d\,t^2} + \hat{\Omega}^2 M_z(t) = \frac{\lambda^2}{2}\left(3M_z^2(t) - \mathbf{M}^2\right) + \beta\mathcal{A}, \tag{19.120}$$

In this equation $\hat{\Omega}^2$, $\mathbf{M}^2$ and $\mathcal{A}$ are all constant operators. The operators $\hat{\Omega}^2$ and $\mathbf{M}^2$ commute with $M_z(t)$ and $\mathcal{B}(t)$ for all times, whereas the operator $\mathcal{A} = H - \mathcal{C}$ does not. Equation (19.120) is a nonlinear differential equation for M_z. This equation can be solved analytically for some special cases [14].

Two-Level Atom — When $j = \frac{1}{2}$, then $\mathbf{M}^2 = \frac{3}{4}$ and $M_z^2 = \frac{1}{4}$ for all times. Then Eq. (19.120) reduces to

$$\frac{d^2M_z(t)}{d\,t^2} + \hat{\Omega}^2 M_z(t) = \beta\mathcal{A}. \tag{19.121}$$

This is a linear differential equation for the operator M_z for which the general solution is a sinusoidal function of time

$$M_z(t) = \left[M_z(0) - \frac{\beta}{\hat{\Omega}^2}\mathcal{A}\right]\cos\left(\hat{\Omega}t\right) - \frac{i\mathcal{B}(0)}{\hat{\Omega}}\sin\left(\hat{\Omega}t\right) + \frac{\beta}{\hat{\Omega}^2}\mathcal{A}. \tag{19.122}$$

Let us consider the time-dependence of $M_z(t)$ when the two level system interacts with coherent radiation. For simplicity we discuss the case of exact resonance by setting the detuning parameter β given by (19.112) equal to zero. Then Eq. (19.122) reduces to

$$M_z(t) = M_z(0)\cos\hat{\Omega}t - \frac{i\mathcal{B}}{\hat{\Omega}}\sin\hat{\Omega}t, \tag{19.123}$$

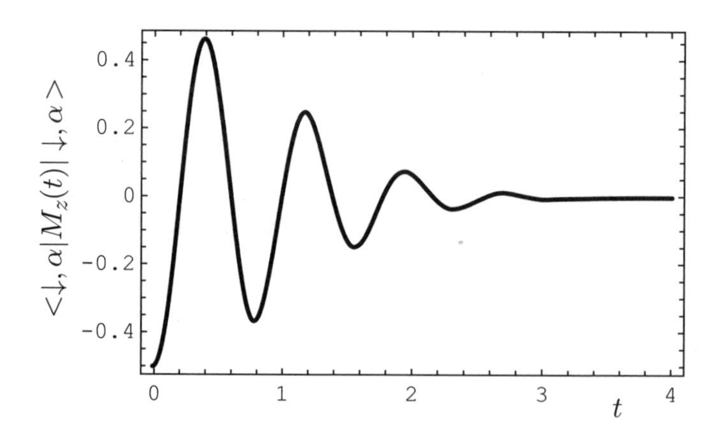

Figure 19.1: Expectation value of M_z for a two-level atom using coherent state wave function plotted as a function of time. Here the collapse of the initial matrix element is shown. This result has been found from Eq. (19.125) for $\bar{N} = 16$ and under the resonance condition $\beta = 0$.

Here $\hat{\Omega}$ for $M_z(0) = -\frac{1}{2}$ is given by

$$\begin{aligned} \hat{\Omega}^2 &= \lambda^2\left(\mathcal{C}+\frac{1}{2}\right) \\ &= \lambda^2\left(a^\dagger a + M_z + \frac{1}{2}\right) = \lambda^2 j, \end{aligned} \tag{19.124}$$

where j is the eigenvlaue of $a^\dagger a$. In a two level system $M_z(0) = \pm\frac{1}{2}$, therefore we can use the notations of up $\uparrow$ and down $\downarrow$ of spin matrices to denote the state of the atom. For the initial condition we choose the atomic state to be $M_0 = |\downarrow\rangle$, and the coherent field $|\alpha\rangle$ given by (18.67). The mean photon number in the initial state according to (18.75) is $\bar{N} = \langle\alpha\left|a^\dagger a\right|\alpha\rangle = |\alpha|^2$. By imposing these initial conditions on the solution of Eq. (19.123), the expectation value of the $M_z(t)$ becomes (see Eqs. (18.65) and (19.124)) [14],[15]

$$\langle\downarrow,\alpha|M_z(t)|\downarrow,\alpha\rangle = -\frac{1}{2}e^{-\bar{N}}\sum_{j=0}^{\infty}\frac{(\bar{N})^j}{j!}\cos\left(\lambda t\sqrt{j}\right). \tag{19.125}$$

The time-dependence of this matrix element is shown for short times in Fig. 19.1 and for longer times in Fig. 19.2. This calculation is done with the coupling constant $\lambda = 1$ with an average number of photons $\bar{N} = 16$. The first figure shows that the matrix element $\langle\downarrow,\alpha|M_z(t)|\downarrow,\alpha\rangle$ oscillates initially and then collapses. For much longer times there is a revival of the matrix elements and this revival happens periodically but each time with a slightly smaller amplitude. This phenomenon of collapse and revival can be seen in Fig. 19.2. A detailed discussion of this topic with analytical approximation to the sum in (19.125) can be found in the work of Eberly and collaborators [16].

Multilevel Atom Interacting with Radiation — As we have seen

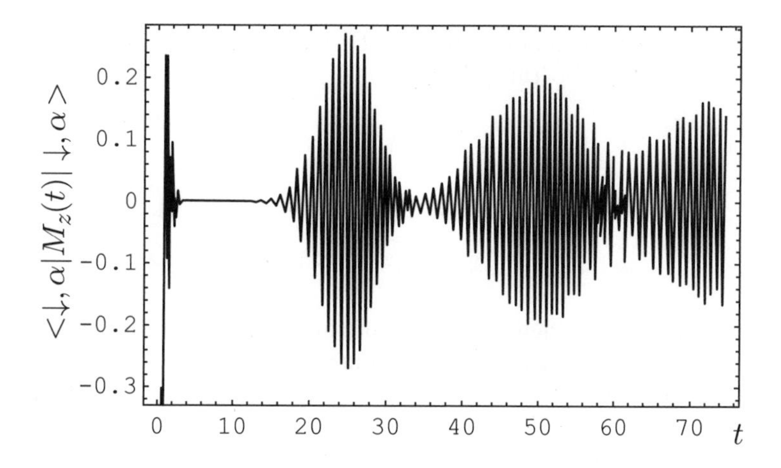

Figure 19.2: Plot of the matrix elements $\langle\downarrow,\alpha|M_z(t)|\downarrow,\alpha\rangle$ for larger t showing the revival of this quantity after its initial collapse.

above, the solution for a two-level atom or molecule for the case of resonance, $\beta = 0$, can easily be found. For $m > 2$, the problem is more complicated. We first derive the Heisenberg equations for $M_z(t)$ in general case and then discuss the three-level problem. Again our starting point is Eq. (19.120). We differentiate this equation twice to get a fourth order differential equation for $M_z(t)$ [14]

$$\begin{aligned}\frac{d^4 M_z}{d\,t^4} &+ \hat{\Omega}^2 \frac{d^2 M_z}{d\,t^2} = \frac{3}{2}\lambda^2 \frac{d^2 M_z}{d\,t^2} \\ &= \frac{3}{2}\lambda^2 \frac{d}{dt}\left(M_z \frac{dM_z}{dt} + \frac{dM_z}{dt} M_z\right) \\ &= \frac{3}{2}\lambda^2 \left\{\left[M_z,\ \frac{d^2 M_z}{d\,t^2}\right]_+ - 2\mathcal{B}^2\right\},\end{aligned} \tag{19.126}$$

where $[\ ,\ ]_+$ denotes the anticommutator. Here we have used (19.117) to simplify the result. We also note that the anticommutator in this relation can be written in terms of M_z with the help of Eq. (19.120). From Eqs. (19.114) and (19.115) we obtain

$$\mathcal{B}^2 = \mathcal{D}^2 - \frac{\lambda^2}{2}\left(M_+ M_- a a^\dagger + M_- M_+ a^\dagger a\right). \tag{19.127}$$

Using (19.116) and $\mathcal{D} = \mathcal{A} - \beta M_z$, this equation takes the form

$$\mathcal{B}^2 = \hat{\Omega}^2 M_z^2 - \beta\,[A,\ M_z]_+ + \lambda^2 M_z\left(\mathbf{M}^2 - M_z^2 - \frac{1}{2}\right) - \Gamma, \tag{19.128}$$

where Γ is a constant operator

$$\Gamma = \lambda^2\left(\mathcal{C} + \frac{1}{2}\right)\mathbf{M}^2 - \mathcal{A}^2. \tag{19.129}$$

By substituting the anticommutator and $\mathcal{B}^2$ in Eq. (19.126) we find the differential equation for the operator $M_z(t)$;

$$\begin{aligned} \frac{d^4 M_z}{d\,t^4} &+ 5\hat{\Omega}^2 \frac{d^2 M_z}{d\,t^2} + 4\hat{\Omega}^4 M_z \\ &- \frac{3}{2}\lambda^2 \left\{\lambda^2 M_z \left(5M_z^2 - 3\mathbf{M}^2 + 1\right) + 3\beta\left[\mathcal{A},\, M_z\right]_+\right\} \\ &= \left\{4\beta\hat{\Omega}^2\mathcal{A} + \lambda^2\left(\hat{\Omega}^2 - 3\beta^2\right)\mathbf{M}^2 - 3\lambda^2\mathcal{A}^2\right\}, \end{aligned} \tag{19.130}$$

where the terms on the right-hand side of this equation add up to a constant operator. This equation has a cubic nonlinearity, M_z^3, and at first sight seems more complicated than the original equation Eq. (19.120). However, in its present form it is suitable for solving the problem of radiation from three-level atom or molecule. The reason being that when the quantum number m takes the values $(-1,\ 0,\ 1)$ corresponding to the three levels, then the operator M_z satisfies the equation $M_z^3 = M_z$. Thus Eq. (19.130) is in fact a linear differential equation for M_z. The same technique of differentiation and elimination can be used when we have systems with more than three levels. Again the three-level atom in resonance with the field has a simpler solution than the non-resonant case.

Setting $M_z^2 = M_z$, $\mathbf{M}^2 = 0$ and $\beta = 0$ in (19.130) reduces this equation to an inhomogeneous linear fourth-order differential equation for M_z;

$$\begin{aligned} \frac{d^4 M_z}{d\,t^4} &+ 5\hat{\Omega}^2 \frac{d^2 M_z}{d\,t^2} + 4\hat{\Omega}^4 M_z \\ &= \lambda^2\left(2\hat{\Omega}^2 - 3\mathcal{A}^2\right), \end{aligned} \tag{19.131}$$

where $\hat{\Omega}^2$ is given by (19.119) with $\beta = 0$. Now introducing a new operator N_z by

$$N_z(t) = M_z(t) - \frac{\lambda^2}{4\hat{\Omega}^4}\left(2\hat{\Omega}^2 - 3\mathcal{A}^2\right), \tag{19.132}$$

and replacing $M_z(t)$ by $N_z(t)$, we find that $N_z(t)$ satisfies a homogeneous differential equation:

$$\frac{d^4 N_z}{d\,t^4} + 5\hat{\Omega}^2 \frac{d^2 N_z}{d\,t^2} + 4\hat{\Omega}^4 N_z = 0. \tag{19.133}$$

We can find the characteristic frequencies of $N_z(t)$ by writing

$$N_z(t) = N_z(0)e^{i\omega t}, \tag{19.134}$$

where upon substitution of (19.134) in (19.133) we obtain a quartic equation for ω

$$\omega^4 - 5\hat{\Omega}^2\omega^2 + 4\hat{\Omega}^2 = 0. \tag{19.135}$$

This equation has four roots:

$$\omega = \pm\hat{\Omega}, \quad \text{and} \quad \omega = \pm 2\hat{\Omega}. \tag{19.136}$$

From this result we conclude that the time-dependence of $M_z(t)$ may be expressed in terms of harmonic functions of ωt with ω given by (19.136) where the coefficients are linear combinations of the values of $M_z(0)$, $\left(\frac{dM_z}{dt}\right)_{t=0}$, $\left(\frac{d^2M_z}{d\,t^2}\right)_{t=0}$ and $\left(\frac{d^3M_z}{d\,t^3}\right)_{t=0}$. These coefficients are found from Eq. (19.131) and from Eq. (19.120) and its derivative [14].

The Jaynes–Cummings Model — The model that we just studied above is a variant of the Jaynes–Cummings model which we will discuss now. Setting $\hbar = 1$ and choosing the energy of the photon $\hbar\omega$ to be the unit of energy, we write the Hamiltonian as [17],[18],

$$\begin{aligned} H &= H_{field} + H_{atom} + H_{int} \\ &= \epsilon\sigma_z + a^\dagger a + \frac{\lambda}{2}\left(\sigma_+ a + \sigma_- a^\dagger\right). \end{aligned} \tag{19.137}$$

We will divide this Hamiltonian into two parts H_1 and H_2;

$$\begin{aligned} H &= H_1 + H_2 = \left(a^\dagger a + \sigma_z\right) \\ &+ \left[\beta\sigma_z + \frac{\lambda}{2}\left(\sigma_+ a + \sigma_- a^\dagger\right)\right]. \end{aligned} \tag{19.138}$$

In this relations $\sigma_\pm = \sigma_x \pm \sigma_y$ where σ_x, σ_y and σ_z are Pauli matrices introduced in Sec. 9.4, and $\beta = \epsilon - 1$ is the detuning parameter introduced earlier, Eq. (19.112). As is evident from (19.137) the atom in this model has two levels, spin up $|\uparrow\rangle$, and spin down, $|\downarrow\rangle$. The two parts of H_1 and H_2 commute with each other

$$[H_1,\ H_2] = 0, \tag{19.139}$$

and this result enables us to write the time evolution operator $U(t)$ as a product

$$U(t) = \exp(-iHt) = \exp(-iH_1t)\exp(-iH_2t). \tag{19.140}$$

We can easily diagonalize the first part, i.e. $U_1(t)$ where

$$U_1(t) = \exp(-iH_1t) = \exp\left(-ia^\dagger at\right)\begin{bmatrix} e^{\frac{-it}{2}} & 0 \\ 0 & e^{\frac{it}{2}} \end{bmatrix}. \tag{19.141}$$

For the second part, $\exp(-iH_2t)$, we expand the operator $U_2(t)$;

$$\begin{aligned} U_2(t) &= \exp(-iH_2t) = \sum_{n=0}^{\infty}\frac{(-it)^n}{n!}(H_2)^n \\ &= \sum_{n=0}^{\infty}\frac{(-it)^n}{n!}\begin{bmatrix} \frac{1}{2}\beta & \frac{\lambda a}{2} \\ \frac{\lambda a^\dagger}{2} & -\frac{1}{2}\beta \end{bmatrix}^n. \end{aligned} \tag{19.142}$$

Let j be an integer, then

$$\begin{bmatrix} \frac{1}{2}\beta & \frac{\lambda a}{2} \\ \frac{\lambda a^\dagger}{2} & -\frac{1}{2}\beta \end{bmatrix}^{2j} = \begin{bmatrix} \Lambda^j & 0 \\ 0 & \varphi^j \end{bmatrix}, \tag{19.143}$$

where

$$\varphi = \frac{\lambda^2}{4} a^\dagger a + \frac{1}{4}\beta^2, \tag{19.144}$$

and

$$\Lambda = \varphi + \frac{\lambda^2}{4}, \tag{19.145}$$

Also from (19.143) we obtain

$$\begin{bmatrix} \frac{1}{2}\beta & \frac{\lambda}{2} a \\ \frac{\lambda}{2} a^\dagger & -\frac{1}{2}\beta \end{bmatrix}^{2j+1} = \begin{bmatrix} \frac{\beta}{2}\Lambda^j & \frac{\lambda}{2}\Lambda^j a \\ \frac{\lambda}{2} a^\dagger \Lambda^j & -\frac{\beta}{2}\varphi^j \end{bmatrix}. \tag{19.146}$$

Now substituting (19.143) and (19.146) in (19.142) we get $U_2(t)$ in the form of a 2×2 matrix

$$\begin{bmatrix} \cos\left(\sqrt{\Lambda}\, t\right) - \frac{i\beta}{2}\frac{\sin\left(\sqrt{\Lambda}\, t\right)}{\sqrt{\Lambda}} & -i\frac{\lambda}{2}\frac{\sin\left(\sqrt{\Lambda}\, t\right)}{\sqrt{\Lambda}} a \\ -i\frac{\lambda}{2} a^\dagger \frac{\sin\left(\sqrt{\Lambda}\, t\right)}{\sqrt{\Lambda}} & \cos\left(\sqrt{\varphi}\, t\right) + \frac{i\beta}{2}\frac{\sin\left(\sqrt{\varphi}\, t\right)}{\sqrt{\varphi}} \end{bmatrix}. \tag{19.147}$$

The unitary matrix $U(t)$ is found from the product $U(t) = U_1(t)U_2(t)$ and is given by

$$\begin{bmatrix} e^{it(a^\dagger a+\frac{1}{2})}\left\{\cos\left(\sqrt{\Lambda}\, t\right) - \frac{i\beta}{2}\frac{\sin\left(\sqrt{\Lambda}\, t\right)}{\sqrt{\Lambda}}\right\} & -i\frac{\lambda}{2} e^{it(a^\dagger a+\frac{1}{2})}\frac{\sin\left(\sqrt{\Lambda}\, t\right)}{\sqrt{\Lambda}} a \\ -i\frac{\lambda}{2} e^{it(a^\dagger a-\frac{1}{2})} a^\dagger \frac{\sin\left(\sqrt{\Lambda}\, t\right)}{\sqrt{\Lambda}} & e^{it(a^\dagger a-\frac{1}{2})}\left\{\cos\left(\sqrt{\varphi}\, t\right) + \frac{i\beta}{2}\frac{\sin\left(\sqrt{\varphi}\, t\right)}{\sqrt{\varphi}}\right\} \end{bmatrix}. \tag{19.148}$$

Let us note that φ and Λ defined by operators, Eqs. (19.144) and (19.145), and these operators have the following properties which we can easily verify [2]:

$$\frac{\sin\left(\sqrt{\Lambda}\, t\right)}{\sqrt{\Lambda}} a = a\frac{\sin\left(\sqrt{\varphi}\, t\right)}{\sqrt{\varphi}}, \tag{19.149}$$

and

$$\cos\left(\sqrt{\Lambda}\, t\right) a = a\cos\left(\sqrt{\varphi}\, t\right). \tag{19.150}$$

We also find similar results for their Hermitian conjugates. Using these relations we can directly prove the unitariy of $U(t)$. From the time evolution operator, (19.148), we can determine any specific property of the system that we want. For instance we can find

$$\sigma_z(t) = U(t)\sigma_z(0)U^\dagger(t), \tag{19.151}$$

and then calculate $\langle \alpha, \downarrow |\sigma_z(t)|\alpha, \downarrow\rangle$ as before but now with the detuning parameter $\beta \neq 0$.

There are a number of other exactly solvable problems for the interaction of an atom with strong electromagnetic field where any number of photons can be emitted or absorbed. These have solvable Heisenberg equations for the state of the atom [19],[20].

Bibliography

[1] E.G. Harris, *A Pedestrian Approach to Quantum Field Theory*, (Wiley-Interscience, New York, 1972), Chapter 3.

[2] S. Stenholm, Quantum theory of electromagnetic fields interacting with atoms and molecules, Phys. Rep. 6, 1 (1973).

[3] H.A. Bethe and E.E. Salpeter, *Quantum Mechanics of One- and Two-Electron Atoms*, (Springer-Verlag, Berlin, 1957) p. 280.

[4] J.D. Jackson, *Classical Electrodynamics*, (John Wiley & Sons, New York, 1998).

[5] V. Weisskopf and E. Wigner, Berechnung der natürlichen Linenbreite auf Grund der Diracschen Lichttheorie, Z. Physik, 63, 54 (1930).

[6] M. Razavy, *Classical and Quantum Dissipative Systems*, (Imperial College Press, 2006).

[7] M.L. Goldberger and K.M. Watson, *Collision Theory*, (John Wiley & Sons, New York, 1964), Chapter 8.

[8] M. Razavy and E.A. Henley, Model for γ decay of atomic or nuclear systems, Can J. Phys. 48, 2399 (1970).

[9] M. Levy, On the validity of the exponential law for the decay of an unstable particle, Nuovo Cimento, 13, 115 (1959).

[10] W. Greiner *Quantum Mechanics, Special Chapters*, (Springer, Berlin, 1998).

[11] W.E. Lamb and R.C. Retherford, Fine structure of the hydrogen atom by a microwave method, Phys. Rev. 72, 241 (1947).

[12] H.A. Bethe, The electromagnetic shift of energy levels, Phys. Rev. 72, 339 (1947).

[13] This inconsistency has been mentioned in a number of references, e.g. in W. Greiner *Quantum Mechanics, Special Chapters*, (Springer, Berlin, 1998) p.156.

[14] B. Buck and C.V. Sukmar, Solution of the Heisenberg equations for an atom interacting with radiation, J. Phys. A 17, 877 (1984).

[15] B. Buck and C.V. Sukmar, Exactly soluable model of atom-phonon coupling showing periodic decay and revival, Phys. Lett., 81 A, 132 (1981).

[16] N.B. Narozhny, J.J. Sanchez-Mondragon and J.H. Eberly, Coherence versus incoherence: Collapse and reviaval in a simple quantum model, Phys. Rev. A 23, 236 (1981).

[17] E.T. Jaynes and F.W. Cummings, Comparison of quantum and semiclassical radiation theories with application to beam maser, Proc. IEEE, 51, 89 (1963).

[18] V. Vlatko, *Modern Foundations of Quantum Optics*, (Imperial College Press, London, 2005), Chapter 10.

[19] M. Sebawe Abdalla, M.M.A. Ahmed and A-S.F. Obada, Dynamics of a non-linear Jayens-Cummings model, Physica, A 162, 215 (1990).

[20] H. Iwasawa and K. Matsuo, Exact solutions of Heisenberg equations for multiphoton Jayens-Cummings model, J. Math. Phys. 36, 2720 (1995).

Chapter 20

Bell's Inequality

In 1935 in a celebrated paper, Einstein, Podolsky and Rosen, (EPR), discussed the subtle and paradoxical nature of quantum measurement and the conflict between the quantum theory and a realistic local theory of the physical world [1]–[22]. The publication of this paper generated a much needed debate about the meaning of completeness and objectivity in quantum theory. It also raised questions regarding the physical and philosophical interpretation of the theory, i.e. whether or not observations come about from some underlying deterministic and objective physical process, knowing well that the final predictions are probabilistic. Let us consider the reasoning of EPR regarding the incompleteness of quantum theory. The gedanken experiment advanced by EPR is about the correlation of coordinates and momenta of two particles that have interacted in the past and then separated to such an extent that they can considered as independent systems at the time that a measurement is performed on them. They suggested that a measurement of one of these variables for one of the particles of the pair can be done without disturbing the state of the second particle, then one can deduce the corresponding variable with 100% certainty.

20.1 EPR Experiment with Particles

Consider the following one-dimensional model which illustrate the gedanken EPR experiment. A system is composed of two particles and that these particles are allowed to interact between $0 \le t \le T$, but after $t > T$ there is no interaction between them. We assume that the states of the these particles were known for $t < 0$, and from this initial information we calculate the wave function of the system;

$$\Psi(x_1, x_2) = \sum_{n=1} \psi_n(x_2) u_n(x_1), \tag{20.1}$$

where $u_n(x_1)$ is an eigenstate of a Hermitian operator A

$$A u_n(x_1) = a_n u_n(x_1). \tag{20.2}$$

If the spectra of A is continuous, then (20.1) is replaced by

$$\Psi(x_1, x_2) = \int_{-\infty}^{\infty} \psi_\alpha(x_2) u_\alpha(x_1) d\alpha, \tag{20.3}$$

where α denotes the continuous eigenvalues of A. In particular for free unbounded motion of two particles we have the wave function

$$\Psi(x_1, x_2) = \int_{-\infty}^{\infty} \exp\left[\frac{i(x_1 - x_2 + x_0)p}{\hbar}\right] dp. \tag{20.4}$$

We can express $\Psi(x_1, x_2)$ in two different forms:

$$\Psi(x_1, x_2) = \int_{-\infty}^{\infty} \exp\left[\frac{-i(x_2 - x_0)p}{\hbar}\right] \exp\left[\frac{i x_1 p}{\hbar}\right] dp, \tag{20.5}$$

or

$$\begin{aligned} \Psi(x_1, x_2) &= \int_{-\infty}^{\infty} \left\{ \int_{-\infty}^{\infty} \exp\left[\frac{i(x - x_2 + x_0)p}{\hbar}\right] dp \right\} \delta(x_1 - x) dx \\ &= 2\pi\hbar \int_{-\infty}^{\infty} \delta(x - x_2 + x_0)\delta(x_1 - x) dx. \end{aligned} \tag{20.6}$$

By comparing (20.5) with (20.3) we find that

$$u_p(x_1) = \exp\left(\frac{i x_1 p}{\hbar}\right), \tag{20.7}$$

that is $u_p(x_1)$ is an eigenfunction of the linear momentum operator

$$p_1 = -i\hbar \frac{\partial}{\partial x_1}, \tag{20.8}$$

with the eigenvalue p. The same relation, (20.5), shows that

$$p_2 = -i\hbar \frac{\partial}{\partial x_2}, \tag{20.9}$$

is the momentum operator for particle 2 with the eigenfunction $\exp\left[-\frac{i(x_2 - x_1)p}{\hbar}\right]$ and the eigenvalue $-p$. Hence if we measure p_1 and we find p, then we can infer without disturbing the system that the momentum of particle 2 is $-p$. Note that after such a measurement $\Psi(x_1, x_2)$ is reduced to $\psi_p(x_2)u_p(x_1)$. On the other hand by writing (20.3) as

$$\Psi(x_1, x_2) = \int_{-\infty}^{\infty} \psi_x(x_2)u_x(x_1)dx, \tag{20.10}$$

we observe that

$$u_x(x_1) = \delta(x_1 - x), \tag{20.11}$$

is the eigenfunction of the position operator x_1 for the particle 1 with the corresponding eigenvalue $x_1 = x$. In the same way we find the function $\psi_x(x_2)$,

$$\psi_x(x_2) = 2\pi\hbar\delta(x - x_2 + x_0), \tag{20.12}$$

is the eigenfunction of the position operator x_2 (multiplied by $2\pi\hbar$) and has the eigenvalue $x_2 = x + x_0$. If we examine the wave function just before the measurement on 1 is performed, we find that there is no certainty regarding momentum of particle 2 because the wave function is a superposition of multiple momentum eigenstates of 1 and 2 as is evident from Eq. (20.4). Therefore the particle 2 must be in a definite state before the measurement of 1 takes place. However from the wave function description of 2 we cannot determine what its momentum is. We conclude that 2 has a definite momentum even though we cannot find this momentum from quantum mechanics. Using this simple two particle model, EPR made the following statements:

"If, without in any way disturbing the system (i.e. with a probability equal to unity) we can predict the value of a physical quantity, then there exists an element of physical reality corresponding to this physical quantity".
They also put forward the criterion for completeness of a physical theory.

"A theory is complete if every element of the physical reality has a counter part in the physical theory".

EPR's gedanken experiment has been criticized for a number of reasons (see Jammer's book [3]), among them one which is mathematical rather than physical or philosophical is the argument by Cooper about the question of the self-adjointness of the operators [2]. Let us examine whether the operator A in (20.2) is self-adjoint or not, a condition which is necessary for the validity of the expansions (20.2) and (20.3). In EPR's model the assumption of the separation between the particles after the interaction implies that the domain of the operator p_1 in (20.8) is not $-\infty \leq x \leq +\infty$, rather it is from some point, say $x = 0$ to $+\infty$. As we have seen in Sec. 3.9 in this domain the momentum operator is not self-adjoint. Thus the argument of separated system breaks down, either because in quantum-mechanical description these two systems cannot be completely separated, or if we assume that they are separated their momenta cannot be represented by self-adjoint operators. This objection can be bypassed if instead of measuring positions and momenta of the two particles we use some other dynamical variable such as spin degrees of freedom.

EPR Experiment with Spinors — A variation of the EPR gedanken experiment was proposed by Bohm and Aharonov in which the components of spin of a pair of particles, instead of momentum or position, were measured [4]. By studying this system we can make the argument much simpler, and may also extend it to the measurement of the polarization of a pair of photons, and in this way provide the simplest realization of the EPR experiment. In our discussion of Bell's inequality we will use Bohm's model to illustrate the difference between the results found from the quantum theory and those obtained from hidden variable theories.

In Chapter 4 we reviewed the concept of "state" of a system in classical mechanics. There, we considered measurable quantities of interest like $p_i(t)$ and $q_i(t)$, specify the state as they evolve in time. Thus in our classical description we have a one-to-one correspondence between dynamical properties of the objects as they are observed in the physical world and their mathematical representation in the theory. Furthermore, in classical dynamics we can enlarge the number of physical variables of the system, e.g. $p_i,\ q_i,\ L_i \cdots$ by performing an arbitrary number of measurements of these additional quantities without perturbing the state of the system. However in quantum theory a state of the system will be simultaneous eigenstates of a small set of observables associated with Hermitian commuting operators. Therefore even under ideal conditions we can determine only a few definite properties of the system. Additional measurements will disturb the system unless the observable happens to coincide with the eigenstate of the system before measurement.

Entanglement — If the state of two quantum systems cannot be expressed as the product of two separate states then we call that system "entangled" [5]. The concept of entanglement is central to the understanding of the theory of measurement in quantum mechanics. Consider a system with two degrees of freedom A and B. If the state vector can be written as

$$|\psi\rangle = |\alpha\rangle\,|\beta\rangle, \tag{20.13}$$

then each part or subsystem, $|\alpha\rangle$ for A and $|\beta\rangle$ for B, corresponds to two well-defined states. However once the two systems A and B interact, then the general state of a two part system cannot be written in this factorized form. The entangled form of the state vector, as we have seen for two spinors, is of the form

$$|\psi\rangle = \frac{1}{\sqrt{2}}\left(|\alpha_1\rangle\,|\beta_1\rangle \pm |\alpha_2\rangle\,|\beta_2\rangle\right), \tag{20.14}$$

In this entangled state there are strong correlations between A and B and this entangled state often embodies entirely new physical properties of the composite system that are not present in any of the two individual states A and B. Hence we can say that the states A and B have lost their individuality, or in other words the quantum-mechanical whole is different from the sum of its parts. If we measure each of these degrees of freedom separately we find that:

(a) - The probability that A is in $|\alpha_1\rangle$ and B is in $|\beta_1\rangle$ is $\frac{1}{2}$, and this is the same probability that A would be in $|\alpha_2\rangle$ and B in $|\beta_2\rangle$.

(b) - The probability that B is in $|\beta_1\rangle$ and A is in $|\alpha_2\rangle$ is zero, the same as the probability of A to be in $|\alpha_1\rangle$ and B to be in $|\beta_2\rangle$. From these observations we conclude that a measurement of A destroys the entangled state, and the aforementioned correlation is 100% regardless of the distance between the two systems A and B.

Decoherence — An important step in understanding the collapse of the wave function was taken when it was realized that the time-dependent linear wave equation is valid only for closed systems (see the discussion of the nonlinear Schrödinger–Langevin equation Sce. 17.12). Now in the process of measurement we are not dealing with an isolated closed quantum system, since the system under observation is coupled to the "classical" measuring apparatus. By assuming that the measuring apparatus plus the system are parts of a larger system, and that quantum mechanics is a valid theory for all systems, then the openness of the system that we want to measure is crucial to the understanding of quantum-to-classical transition [5]–[7].

For an analysis of the measurement process let us consider the simplest case of an entangled system which is composed of two spin $\frac{1}{2}$ particles one with spin up and the other down. We perform a Stern–Gerlach experiment on this system. We assume that the particle enters the magnetic field with the initial spin $|\alpha_x\rangle$;

$$|\alpha_x\rangle = \frac{1}{\sqrt{2}}\begin{bmatrix}1\\1\end{bmatrix} = \frac{1}{\sqrt{2}}(|\uparrow\rangle + |\downarrow\rangle) = \frac{1}{\sqrt{2}}(|\alpha_z\rangle + |\beta_z\rangle). \tag{20.15}$$

Now if we measure the spin of this particle we would find that either

$$|\alpha_x\rangle = \frac{1}{\sqrt{2}}(|\uparrow\rangle + |\downarrow\rangle)_{\text{after measurement}} \rightarrow |\uparrow\rangle = |\alpha_z\rangle, \tag{20.16}$$

or

$$|\alpha_x\rangle = \frac{1}{\sqrt{2}}(|\uparrow\rangle + |\downarrow\rangle)_{\text{after measurement}} \rightarrow |\downarrow\rangle = |\beta_z\rangle. \tag{20.17}$$

As we mentioned earlier, we want to treat the whole experimental set up, i.e. the system to be measured and the detector as a large quantum system, which can be described by the Schrödinger equation. When the particle with spin up, $|\uparrow\rangle$, passes through the positive channel of the Stern–Gerlach magnet then its passage is recorded by a detector which we want to describe by a wave function. Again for the sake of simplicity we assume that the wave function for the detector has only two parts, before the particle is detected the wave function is represented by $|\gamma_0\rangle$ and after, if the spin of the particle is $+\frac{1}{2}$ then it is $|\gamma_+\rangle$. Denoting the spin state of the particle by $|\alpha_z\rangle$, the total wave function of the whole set up before the passage is

$$|\psi_0\rangle = |\alpha_z\rangle|\gamma_0\rangle. \tag{20.18}$$

After the passage of the particle, the wave function becomes

$$|\psi_+\rangle = |\alpha_z\rangle|\gamma_+\rangle. \tag{20.19}$$

If the initial state of the particle is $|\beta_z\rangle$, then nothing is registered and the wave function before and after the measurement would be the same

$$|\psi_0\rangle = |\beta_z\rangle|\gamma_0\rangle. \tag{20.20}$$

So far we have assumed that the spin of the particle is an eigenstate of S_z. Now we ask that what happens if the particle has a spin state $|\alpha_x\rangle$. If this is the case then the initial state of the whole system, particle plus measuring instrument is

$$|\psi_0\rangle = |\alpha_x\rangle|\gamma_0\rangle = \frac{1}{\sqrt{2}}\left(|\alpha_z\rangle + |\beta_z\rangle\right)|\gamma_0\rangle. \tag{20.21}$$

After a time t this initial state evolves and because of the interaction, becomes entangled. From (20.19)–(20.21) we find that the wave function after measurement is

$$|\psi\rangle = \frac{1}{\sqrt{2}}\left(|\alpha_z\rangle|\gamma_+\rangle + |\beta_z\rangle|\gamma_0\rangle\right). \tag{20.22}$$

This result contradicts what we found earlier Eqs. (20.16),(20.17), since (20.18) implies that the wave function should collapse into either $|\alpha_z\rangle|\gamma_+\rangle$ or $|\beta_z\rangle|\gamma_0\rangle$, and the pointer does not show one of these two possibilities. The explanation lies in the complexity of the many-body wave function representing the measuring apparatus. Rather than what is given in Eq. (20.22) we should have a wave function which collapses either into one or the other of these states.

A Solvable Model Exhibiting Decoherence — Let us now consider in some detail the way that decoherence works. For this we start with a time-dependent formulation using a simple and solvable model very similar to the problem of the damped harmonic oscillator, Sec. 17.12. As we have seen in Sec. 18.1 the best approximation for a "classical" state is the coherent state. Therefore we will formulate the problem using the coherent state formalism. Keeping $|\alpha\rangle$ and $|\gamma\rangle$, notation but replacing $|\alpha\rangle|\gamma\rangle$ by a more descriptive notation of $|\alpha;\ \{\gamma_k\}\rangle$ representing the many degrees of freedom of the measuring device we write the many-body wave function $|\Psi_0\rangle$ as

$$|\Psi_0\rangle = C|\ \alpha_1(0); \{\gamma_k\} = 0\rangle + D\ |\alpha_2; \{\gamma_k\} = 0\rangle, \tag{20.23}$$

where

$$|\{\gamma_k\}\rangle \equiv |\gamma_1, \gamma_2 \cdots \gamma_j \cdots\rangle, \tag{20.24}$$

represent the many-body coherent state which becomes coupled to the entangled state $(|\alpha_1\rangle, |\alpha_2\rangle)$ when the measurement is taking place. Since (20.23) which represents the state vector of the entangled particles at $t = 0$ is not coupled to the many-body system, i.e. the measuring apparatus, we assume that this latter system is in its ground state, and we set all its excitations $\{\gamma_k\}$ equal to zero. Now by imposing the normalization condition on $|\psi_0\rangle$ we find

$$\langle\psi_0|\psi_0\rangle = |C|^2 + |D|^2 = 1. \tag{20.25}$$

Next we inquire about the selection of a suitable many-body Hamiltonian for the measuring apparatus. Since a general many-body problem cannot be solved exactly, we make two assumptions: First for the system that we want to measure, we choose a simple harmonic oscillator with the angular frequency Ω. Second we assume that this oscillator is linearly coupled to a large number of noninteracting oscillators. Setting $\hbar = 1$ we write the total Hamiltonian as

$$H = \Omega a^\dagger a + \sum_k \omega_k c_k^\dagger c_k + \sum_k \left(\lambda_k a^\dagger c_k + \lambda^* a c_k^\dagger\right), \tag{20.26}$$

where $a^\dagger$ and a are creation and annihilation operators for the coherent states of the particle and $c_k^\dagger$ and c_k are the corresponding operators for the particles forming the apparatus. The advantage of this Hamiltonian which is quadratic in these operators is that it can be diagonalized by a principal axis transformation, and the equations of motion for these operators are linear and thus exactly solvable [8]. Hamiltonians similar to (20.26) have been used where the many-body system is a general heat bath, or a one-dimensional scalar field [8],[9]. We have seen the latter model in our study of the damped harmonic oscillator. In order to determine the time evolution of the initial wave function $|\Psi_0\rangle$, Eq. (20.23), we write both parts of $|\Psi_0\rangle$ in terms of coherent states:

$$|\alpha; \{\gamma_k\}\rangle = \exp\left(-\frac{1}{2}|\alpha|^2\right) e^{\alpha a^\dagger} \prod_{k=1}^{\infty} \exp\left(-\frac{1}{2}|\gamma_k|^2\right) e^{\gamma_k c_k^\dagger} |0, \{0\}\rangle. \tag{20.27}$$

Since $|\alpha\rangle$ as well as $|\{\gamma_k\rangle\}$ are coherent states, their eigenvalues α and γ_k are complex numbers, and for each of these we can write the eigenvalue in terms of a phase and an amplitude. If the coupling between the two parts is weak, then the amplitude changes very slowly with time and we can assume that $|\alpha|^2$ and $|\gamma_k|^2$ are constants. In this approximation the time-dependent Schrödinger equation is

$$\begin{aligned} i\frac{\partial}{\partial t}|\alpha; \{\gamma_k\}\rangle &= i\left(a^\dagger \frac{\partial \alpha}{\partial t} + \sum_k c_k^\dagger \frac{\partial \gamma_k}{\partial t}\right)|\alpha; \{\gamma_k\}\rangle = H|\alpha; \{\gamma_k\}\rangle \\ &= \left\{\Omega a^\dagger a + \sum_k \omega_k c_k^\dagger c_k + \sum_k \left(\lambda_k a^\dagger c_k + \lambda^* a c_k^\dagger\right)\right\}|\alpha; \{\gamma_k\}\rangle \\ &= \left[\Omega \alpha a^\dagger + \sum_k \left\{\omega_k \gamma_k c_k^\dagger + \lambda \gamma_k a^\dagger + \lambda_k^* \alpha c_k^\dagger\right\}\right]|\alpha; \{\gamma_k\}\rangle. \end{aligned} \tag{20.28}$$

By comparing the coefficients of $a^\dagger|\alpha; \{\gamma_k\}\rangle$ and $c_k^\dagger|\alpha; \{\gamma_k\}\rangle$ on the two sides of (20.28) we find two linear coupled equations for α and γ_k;

$$i\frac{d\alpha}{dt} = \Omega\alpha + \sum_k \lambda_k \gamma_k, \tag{20.29}$$

and

$$i\frac{d\gamma_k}{dt} = \omega_k \gamma_k + \lambda_k^* \alpha. \tag{20.30}$$

There is a conservation law associated with these coupled equations. Thus if we write the complex conjugate of Eq. (20.29), multiply the result by another solution, say $\alpha'(t)$, then write (20.29) for $\alpha'(t)$ and multiply it by $\alpha^*(t)$ and add the resulting equations we obtain

$$\frac{d}{dt}\left[\alpha'(t)\alpha^*(t)\right] = \sum_k \left[\lambda_k \alpha^*(t)\gamma_k'(t) - \lambda_k^* \alpha'(t)\gamma_k^*(t)\right]. \tag{20.31}$$

Similarly from (20.30) we obtain

$$\frac{d}{dt}\left[\gamma_k'(t)\gamma_k^*(t)\right] = \left[\lambda_k^* \alpha'(t)\gamma_k^*(t) - \lambda_k \alpha^*(t)\gamma_k'(t)\right]. \tag{20.32}$$

By summing (20.32) over all k s and then combining the result with (20.31) we obtain the conservation law

$$\frac{d}{dt}\left[\alpha'(t)\alpha^*(t) + \sum_k \gamma_k^*(t)\gamma_k'(t)\right] = 0. \tag{20.33}$$

Equations (20.29) and (20.30) are similar to Eqs. (19.37) and (19.38) of the theory of natural line width, Sec. 19.1, and we can solve them exactly the same way. By writing $\alpha(t)$ and $\gamma_k(t)$ as Fourier transforms like Eqs. (19.41) and (19.42) we find

$$\omega\alpha(\omega) = \sum_k \lambda_k \gamma_k(\omega), \tag{20.34}$$

and

$$(\omega + \Omega)\gamma_k(\omega) = \omega_k \gamma_k(\omega) + \lambda_k^* \alpha(\omega). \tag{20.35}$$

Next we replace the summation over k by integration and for this purpose we introduce a density function $n(\omega)$ which is the average number of oscillators per unit frequency and thus this replacement takes the form of

$$\sum_k \to \int n(\omega) d\omega. \tag{20.36}$$

Following the method of contour integration of Sec. 19.1, we find $\alpha(t)$ and $\gamma_k(t)$ to be

$$\alpha(t) = \alpha(0) \exp\left[-i\left(\Omega + \Delta\omega\right)t - \frac{1}{2}\Gamma t\right], \tag{20.37}$$

and

$$\gamma_k(t) = -\lambda_k^* \left\{ \frac{\exp\left[-i\left(\Omega + \Delta\omega\right)t - \frac{1}{2}\Gamma t\right] - \exp(-i\omega_k t)}{\Omega + \Delta\omega - \omega_k - \frac{i}{2}\Gamma} \right\}, \tag{20.38}$$

where Γ and $\Delta\omega$ are given by

$$\Gamma = 2\pi |\lambda(\Omega)|^2 n(\Omega), \tag{20.39}$$

and

$$\Delta\omega = \mathcal{P}\int_0^\infty \frac{|\lambda(\omega)|^2 n(\omega)}{\Omega-\omega}d\omega. \qquad (20.40)$$

For a detailed discussion of the range of validity of the approximations that we have used here the reader is referred to the book of Omnès [10].

The Density Operator — The state of the whole system at time t can be described by the ket

$$|\Psi(t)\rangle = |\alpha(t), \{\gamma_k(t)\}\rangle. \qquad (20.41)$$

However a much more interesting case is the one where the initial state of the oscillator is a linear superposition of two macroscopically different states. We assume that these two states have different position with zero average velocity. Thus at $t = 0$ we have the total wave function

$$|\Psi(0)\rangle = C|\alpha_1(0), \{\gamma_k(0) = 0\}\rangle + D|\alpha_2(0), \{\gamma_k = 0\}\rangle. \qquad (20.42)$$

The reduced density operator for this initial state with no coupling to the group of oscillators $\{\gamma_k\}$ is

$$\begin{aligned}\rho(0) &= |C|^2|\alpha_1(0)\rangle\langle\alpha_1(0)| + |D|^2|\alpha_2(0)\rangle\langle\alpha_2(0)| \\ &+ CD^*|\alpha_1(0)\rangle\langle\alpha_2(0)| + DC^*|\alpha_2(0)\rangle\langle\alpha_1(0)|,\end{aligned} \qquad (20.43)$$

and this relation clearly shows the presence of quantum superposition. At a later time t, the density matrix can be written as a sum of two parts

$$\rho(t) = \rho_d(t) + \rho_{nd}(t), \qquad (20.44)$$

where $\rho_d(t)$ represents the diagonal part and $\rho_{nd}(t)$ the nondiagonal part of $\rho(t)$. Once we average $\rho(t)$ over the states of the oscillators $\{\gamma_k(t)\}$, we find that

$$\rho_d(t) = |C|^2\ |\alpha_1(t)\rangle\langle\alpha_1(t)| + |D|^2\ |\alpha_2(t)\rangle\langle\alpha_2(t)|, \qquad (20.45)$$

where we have used the result (see Eq. (18.63))

$$\langle\{\gamma_k(t)\}\,|\,\{\gamma_k(t)\}\rangle = 1. \qquad (20.46)$$

For the nondiagonal part, when we average over $\{\gamma_k(t)\}$ s we get the scalar products of different coherent states. Consider the scalar product

$$\langle\{\gamma_{k1}(t)\}\,|\,\{\gamma_{k2}(t)\}\rangle = \prod_{k=1}\langle\gamma_{k1}(t)|\gamma_{k2}(t)\rangle\,, \qquad (20.47)$$

which can be calculated with the help of Eq. (18.66)

$$\prod_{k=1}\langle\gamma_{k1}(t)|\gamma_{k2}(t)\rangle = \exp\left(-\sum_k \frac{1}{2}|\gamma_{k1}(t) - \gamma_2(t)|^2\right). \qquad (20.48)$$

We can simplify (20.48) using the conservation law (20.33), noting $\gamma_k(0) = 0$, and after simplification we have

$$\prod_{k=1} \langle \gamma_{k1}(t)|\gamma_{k2}(t)\rangle = \exp\left(-\frac{1}{2}|\alpha_1(0) - \alpha_2(0)|^2 - |\alpha_1(t) - \alpha_2(t)|^2\right). \quad (20.49)$$

With the help of this equation we find the time-dependence of the nondiagonal elements of the density matrix $\rho_{nd}(t)$;

$$\begin{aligned}\rho_{nd}(t) &= \{CD^* \, |\alpha_1(0)\rangle\langle\alpha_2(0)| + DC^*|\alpha_2\rangle\langle\alpha_1(0)|\} \\ &\times \exp\left[-\frac{1}{2}|\alpha_1(0) - \alpha_2(0)|^2 \left(1 - e^{-\Gamma t}\right)\right]. \end{aligned} \quad (20.50)$$

In order to relate α_1 and α_2 to the positions of oscillator we recall the definition of the coherent state operator in terms of x and p;

$$\alpha = \frac{1}{\sqrt{2}}\left(\Omega x + ip\right), \quad (20.51)$$

Assuming that the average momentum is zero, for the nondiagonal density matrix we have

$$\begin{aligned}\rho_{nd} &\sim \exp\left[-\frac{1}{2}|\alpha_1(0) - \alpha_2(0)|^2 \left(1 - e^{-\Gamma t}\right)\right] \\ &= \exp\left[-\frac{m\Omega^2}{4\hbar}\left(x_1(0) - x_2(0)\right)^2 \left(1 - e^{-\Gamma t}\right)\right], \end{aligned} \quad (20.52)$$

where in this relation m is the mass of the oscillator with the angular frequency Ω, and where we have written $\hbar$ explicitly. In order to have an idea about the order of magnitude of the quantum-to-classical transition time, which is the time that ρ_{nd} becomes vanishingly small, we consider the following extreme example. Let us take $m = 10^{-3}$ Kg, $t = \frac{2\pi}{\Omega} = 1$ s, $\Gamma^{-1} = 1800$ s, and $x_2(0) - x_1(0) = 10^{-6}$ m, then the decoherence time which is

$$\tau = \frac{4\hbar}{m\Omega\Gamma(x_1(0) - x_2(0))^2}, \quad (20.53)$$

is of the order of 10^{-8} seconds [10].

20.2 Classical and Quantum Mechanical Operational Concepts of Measurement

In his original paper on the new mechanics Heisenberg advanced the view that the atomic theory should deal only with the experimentally observable quantities. Later in his well-known book on quantum theory. Heisenberg considered the question of how the position and momentum can be measured. He suggested a gedanken experiment using a gamma ray microscope. Two years later in his lecture at the University of Chicago he expanded this ideas and also he discussed the measurement of the momentum of a particle [12]. He proposed that for the measurement of the distribution of momentum values of an electron in an atom, one may turn off the Coulomb force between the electron and the nucleus suddenly, and by a time-of-flight study made on the free electrons one can determine its momentum. By repeating the measurements one can determine the momentum distribution for an ensemble of atoms. In this section we examine this idea of Heisenberg about the measurement of momentum by turning on and off the interaction and expand it so that it can be used to get the desired initial wave function [13].

State Preparation in Classical Mechanics — In classical dynamics the initial state of a system is completely specified by the values of coordinates and momenta of all of the particles in the system given at $t = 0$, $q_i(0)$ and $p_i(0), i = 1, 2\cdots, N$, Eq. (1.47). Using these initial values we can integrate the Hamilton equations of motion (1.45) and (1.46) and thus obtain the future of the system $q_i(t)$ and $p_i(t)$. The preparation of the initial state in classical mechanics is trivial. For sake of simplicity we consider a one-dimensional motion, where we place a particle at $q(0)$ and give it a momentum $p(0)$. To bring the particle to the initial position $q(0)$ we set up a potential well $U_1(q)$ which has a minimum at $q = q(0)$, catch the particle in this potential and allow it to reach this minimum with the help of a small friction. Once the particle is in this position we subject it to an impulsive force derived from the potential $U_2(q)\delta(t)$. This force changes the momentum of the particle to $p(0)$ without affecting its position. After preparing the initial conditions in this way we turn on the potential $V(q)$ corresponding to the actual potential of the problem, turn off $U_1(q)$ and observe the evolution of the motion in time.

Preparing the State in a Quantum Mechanical System — Having considered the classical preparation, we now follow the same procedure for the quantum mechanical case. Let us assume that we want to prepare the initial state so that the initial wave function is of the form

$$\psi(q,0) = R(q)\exp\left(i\frac{S(q)}{\hbar}\right), \tag{20.54}$$

where $R(q)$ and $S(q)$ are both real functions of q. First, for the real wave function $R(q)$ consider the wave equation

$$ER(q) = \left(-\frac{\hbar^2}{2m}\frac{d^2}{dq^2} + U_1(q)\right)R(q), \tag{20.55}$$

and since $R(q)$ is known we find $U_1(q)$;

$$U_1(q) = E + \frac{1}{R(q)}\left(\frac{\hbar^2}{2m}\frac{d^2R(q)}{dq^2}\right). \tag{20.56}$$

We observe that $R(q)$ must be the wave function of a bound particle, and that it should have no nodes, i.e. it should be the ground state wave function of the potential $U_1(q)$. Therefore to prepare a real wave function we need to trap the particle in the potential $U_1(q)$ and let it reach the ground state by a small frictional force. Then apply an impulse of the form

$$U_2(q) = -S(q)\delta(t), \tag{20.57}$$

to the particle to get the wave function

$$\psi(q) = R(q)\exp\left(\frac{iS(q)}{\hbar}\right), \tag{20.58}$$

for $t > 0$.

Once this initial state is prepared, we turn off the potential $U_1(q)$, turn the actual potential $V(q)$, and follow how the wave function evolves in time [13].

A Gaussian Initial State — To illustrate this way of preparing the initial state let us consider a free particle which, at $t = 0$, can be represented by a Gaussian wave packet

$$\psi(q,0) = \frac{1}{(2\pi)^{\frac{1}{4}}}\frac{1}{\sqrt{\Delta q_0}}\exp\left[-\frac{1}{4}\frac{q^2}{(\Delta q_0)^2}\right]. \tag{20.59}$$

By solving the time-dependent Schrödinger equation

$$i\hbar\frac{\partial\psi(q,t)}{\partial t} = -\frac{\hbar^2}{2m}\frac{\partial^2\psi(q,t)}{\partial q^2}, \tag{20.60}$$

with the initial condition (20.59) we obtain the following wave function after a time T

$$\psi(q,T) = \frac{1}{(2\pi)^{\frac{1}{4}}}\frac{1}{\sqrt{\Delta q_0 + \frac{i\hbar T}{2m\Delta q_0}}}\exp\left[-\frac{1}{4}\left(\frac{q^2}{(\Delta q_0)^2 + \frac{i\hbar T}{2m}}\right)\right]. \tag{20.61}$$

Thus while the wave packet has retained its Gaussian shape, its width, Δq_0 has increased according to the relation

$$(\Delta q)^2 = (\Delta q_0)^2 + \frac{\hbar^2 T^2}{4m^2(\Delta q_0)^2}. \tag{20.62}$$

The probability density at $t = T$ is also given by a Gaussian function

$$\begin{aligned}\rho(q,T) &= |\psi(q,T)|^2 = \frac{1}{(2\pi)^{\frac{1}{2}}}\frac{1}{\sqrt{(\Delta q_0)^2 + \frac{\hbar^2T^2}{4m^2(\Delta q_0)^2}}}\\ &\times \exp\left[-\frac{1}{2}\left(\frac{q^2}{(\Delta q_0)^2 + \frac{\hbar^2T^2}{4m^2(\Delta q_0)^2}}\right)\right].\end{aligned} \tag{20.63}$$

Now we ask whether it is possible to start with a broad wave packet of the form (20.63) and prepare a Gaussian wave packet with a much smaller width Δq_0 by applying an impulsive force to it. In order to avoid the problem of time-reversal, we consider the broad wave packet having the initial form

$$\psi(q,-T) = \psi^*(q,T). \tag{20.64}$$

Following the method that we outlined in this section we write $\psi(q,-T)$ in terms of two real functions

$$\psi(q,-T) = R(q)\exp\left[\frac{iS(q)}{\hbar}\right]. \tag{20.65}$$

where $R(q)$ and $S(q)$ are given by

$$R(q) = \frac{1}{(2\pi)^{\frac{1}{4}}}\frac{1}{\sqrt{\Delta q_0}}\exp\left[-\frac{1}{4}\frac{q^2}{(\Delta q)^2}\right]. \tag{20.66}$$

and

$$S(q) = -\frac{\hbar T}{8m}\frac{q^2}{(\Delta q_0^2)(\Delta q)^2}. \tag{20.67}$$

We note that $R(q)$ is the ground state wave function for a simple harmonic oscillator, with the potential $U_1(q)$ in Eq. (20.56) given by

$$U_1 = \frac{\hbar^2 q^2}{2m(\Delta q_0)^2(\Delta q)^2}. \tag{20.68}$$

For a broad wave packet the spring constant

$$K = m\frac{\hbar^2}{(\Delta q_0)^2(\Delta q)^2}, \tag{20.69}$$

will be very small. From (20.67) we can determine the impulsive potential which in this case is a quadratic function of q;

$$U_1(q) = -S(q)\delta(t) = \frac{1}{8}Kq^2\delta(t), \tag{20.70}$$

and thus both $U_1(q)$ and $U_2(q)$ are harmonic oscillator type potentials, each with a small spring constant K.

20.3 Collapse of the Wave Function

In our discussion of the postulates of quantum mechanics (Chapter 4) and the comparison of these with the corresponding postulates of classical dynamics we noted that:

(a) - In classical mechanics each event can be determined from the laws of motion and the initial conditions.

(b) - In contrast, in quantum mechanics, one can make predictions about the relative probabilities of the occurrence of different events.
For instance suppose that in the Stern–Gerlach experiment we know that a spin $\frac{1}{2}$ particle which enters the inhomogeneous magnetic field is known to be an eigenstate S_x and when it leaves the field we measure S_z. Then the particle is deflected up or down, but we cannot predict which deflection will occur. If we have detectors in each of the channels of the Stern–Gerlach experiment, it is uncertain which of the two detectors will register the passage of an atom. This uncertainty can be traced back to the fourth postulate, or the measurement postulate, where we defined relative probabilities of different outcomes of measurements. According to postulate number 6 (Chapter 4) the evolution of operators in time (or evolution of the time-dependent wave function) between the measurements is completely deterministic, but it is the act of measurement that introduces the indeterminancy. In the Stern–Gerlach experiment, the wave function of the spin $\frac{1}{2}$ particle which was initially an eigenstate of S_x, after passage through the magnetic field splits into two parts, one corresponding to the spin up and the other to the spin down particles. Up to this point the motion is deterministic. However as soon as we proceed with the measurement of the arrival of a particle at the position of the counter, the particle interacts with the counter and then indeterminancy occurs. Here only one of the counters register the arrival of the particle, and the wave function becomes the eigenfunction of the spin up (or down). This process is known as the reduction or collapse of the wave function [10],[11].

Local Deterministic Description of Events — In quantum mechanics predictions are given in terms of probabilities as we have seen in postulate 4 of Chapter 5. One can inquire whether it is possible that some yet unknown and more fundamental theory, called "hidden variable" theory might be able to predict all dynamical quantities precisely, as ideally we can in classical dynamics. Our classical theories are founded on two basic assumptions:

(1) - Definite state of an object determines all measurable quantities such as position and momentum.

(2) - If two observers A and B are sufficiently far apart, a measurement made by A cannot influence the measurement made by B since local action cannot travel faster than the velocity of light.

20.4 Quantum versus Classical Correlations

If we examine correlations between the result of observations made by two observers we find that the prediction of classical mechanics can be different from that of quantum theory. To show this we discuss the simpler version of the EPR thought experiment which is due to Bohm and where one considers the measurement of the spin degree of freedom. Here the system that we observe consists of two spin $\frac{1}{2}$ particles which we label by i and j, and we assume that these two particles are in the singlet S state, i.e.

$$|\psi\rangle = \frac{1}{\sqrt{2}}\left(|i,\uparrow\rangle|j,\downarrow\rangle - |i,\downarrow\rangle|j,\uparrow\rangle\right). \tag{20.71}$$

Now suppose that this system breaks up and one of the particles moves one way and the other moves in the opposite direction. The two observers A and B measure the spin of these particles along any one of the directions specified by the unit vector $\hat{\mathbf{n}}_j$ ($j = 1,\ 2,\ 3$), and let us denote the angle between $\hat{\mathbf{n}}_i$ and $\hat{\mathbf{n}}_j$ by θ;

$$\hat{\mathbf{n}}_j \cdot \hat{\mathbf{n}}_i = \cos\theta. \tag{20.72}$$

The spin operators for these three directions are:

$$\mathbf{S}\cdot\hat{\mathbf{n}}_j = \frac{\hbar}{2}\boldsymbol{\sigma}\cdot\hat{\mathbf{n}}_j, \tag{20.73}$$

where the components of $\boldsymbol{\sigma}$ are the Pauli matrices. We can measure the eigenstates of the spin up and spin down of the particles $|j,\uparrow\rangle$ and $|j\downarrow\rangle$ along any three axes of detectors. In this way we obtain the probabilities $P_{j\uparrow\, i\uparrow}$, $P_{j\uparrow\, i\downarrow}$, $P_{j\downarrow\, i\uparrow}$, and $P_{j\downarrow\, i\downarrow}$ for the particles having spins up and down. To calculate these probabilities it will be convenient to choose $\hat{\mathbf{n}}_j$ to be along the z axis, $\hat{\mathbf{n}}_j = (0,\ 0,\ 1)$.
Since

$$\boldsymbol{\sigma}\cdot\hat{\mathbf{n}}_j = \sigma_z, \tag{20.74}$$

the up and down spin eigenstates are given by

$$|j,\uparrow\rangle = \begin{bmatrix}1\\0\end{bmatrix}, \tag{20.75}$$

$$|j,\downarrow\rangle = \begin{bmatrix}0\\1\end{bmatrix}. \tag{20.76}$$

For any other direction, say $\hat{\mathbf{n}}_i$, the unit vector $\hat{\mathbf{n}}_i$ is given by

$$\hat{\mathbf{n}}_i = (\sin\theta\cos\phi,\ \sin\theta\sin\phi,\ \cos\theta). \tag{20.77}$$

With this unit vector $\hat{\mathbf{n}}_i$ having the components shown in (20.77) we calculate $\boldsymbol{\sigma}\cdot\hat{\mathbf{n}}_i$;

$$\boldsymbol{\sigma}\cdot\hat{\mathbf{n}}_i = \begin{bmatrix}\cos\theta & e^{-i\phi}\sin\theta\\ e^{i\phi}\sin\theta & -\cos\theta\end{bmatrix}. \tag{20.78}$$

This matrix has two eigenvalues ± 1 with the eigenstates

$$|i,\uparrow\rangle = \begin{bmatrix} e^{-i\frac{\phi}{2}}\cos\frac{\theta}{2} \\ e^{i\frac{\phi}{2}}\sin\frac{\theta}{2} \end{bmatrix}, \tag{20.79}$$

and

$$|i,\downarrow\rangle = \begin{bmatrix} e^{i\frac{\phi}{2}}\sin\frac{\theta}{2} \\ e^{-i\frac{\phi}{2}}\cos\frac{\theta}{2} \end{bmatrix}. \tag{20.80}$$

By a judicious choice of the coordinates we can make the azimuthal angle for $\hat{\mathbf{n}}_i$, i.e. ϕ equal to zero. Now if the measurement of the spin of the particle j yields the value $\frac{\hbar}{2}$, then the particle i has to have a spin of $-\frac{\hbar}{2}$, i.e. it has to be in the state $\begin{bmatrix} 0 \\ 1 \end{bmatrix}$. By expanding $\begin{bmatrix} 0 \\ 1 \end{bmatrix}$ in terms of the eigenstates (20.79) and (20.80) with $\phi = 0$ we obtain

$$\begin{bmatrix} 0 \\ 1 \end{bmatrix} = \sin\left(\frac{\theta}{2}\right)\begin{bmatrix} \cos\frac{\theta}{2} \\ \sin\frac{\theta}{2} \end{bmatrix} + \cos\left(\frac{\theta}{2}\right)\begin{bmatrix} -\sin\frac{\theta}{2} \\ \cos\frac{\theta}{2} \end{bmatrix}. \tag{20.81}$$

The probability that the second measurement will give us a positive value for the spin is therefore

$$P_{\uparrow\uparrow}(\theta) = \sin^2\left(\frac{\theta}{2}\right). \tag{20.82}$$

Similarly we find the other probabilities to be

$$P_{\uparrow\downarrow}(\theta) = \cos^2\left(\frac{\theta}{2}\right), \quad P_{\downarrow\uparrow}(\theta) = \cos^2\left(\frac{\theta}{2}\right), \tag{20.83}$$

and

$$P_{\downarrow\downarrow}(\theta) = \sin^2\left(\frac{\theta}{2}\right). \tag{20.84}$$

These relations give us the predictions of quantum mechanics for measuring the spin component $S_{\theta j}$ of the particle j at an angle θ relative to the z-axis, having previously measured the spin components S_{zi} of the particle i.

Let us consider a source which emits pairs of correlated particles, for example a pair with zero total spin. One particle is sent to the observer A and the other to the observer B. Each observer independently chooses between various settings of the detector and then preforms an independent measurement of the particle's spin. Suppose that observers A and B can measure the components of spin along the three axes $\mathbf{n}_1$, $\mathbf{n}_2$ and $\mathbf{n}_3$ (which may not be mutually orthogonal). If the system obeys the rules of the hidden-variable theory, then these measurements cannot affect each other. As a result of such measurements the observer A finds out N_1 events where the components of spin are all up $(1\uparrow, 2\uparrow, 3\uparrow)$, where we have denoted the directions of $\mathbf{n}_1$, $\mathbf{n}_2$, $\mathbf{n}_3$ by $(1, 2, 3)$ respectively. But since the total spin of the pair is zero, B finds N_1 particles with spin down $(1\downarrow, 2\downarrow, 3\downarrow)$. Similarly A measures N_2 particles with components of spin along the axes 1 and 2 up and along 3 down or $(1\uparrow, 2\uparrow, 3\downarrow)$ and so on. In TABLE XVIII we have listed the complete set of results for eight nonoverlapping groups defined by the three components.

TABLE XVIII: Components of spin along the three axes (1, 2, 3) when the total spin is zero. Here it is assumed that three hidden variable are associated with each particle pair when these particles are emitted from the source, and these hidden variables do not change afterwards.

A	B	Frequency
$1\uparrow\ 2\uparrow\ 3\uparrow$	$1\downarrow\ 2\downarrow\ 3\downarrow$	N_1
$1\uparrow\ 2\uparrow\ 3\downarrow$	$1\downarrow\ 2\downarrow\ 3\uparrow$	N_2
$1\uparrow\ 2\downarrow\ 3\uparrow$	$1\downarrow\ 2\uparrow\ 3\downarrow$	N_3
$1\uparrow\ 2\downarrow\ 3\downarrow$	$1\downarrow\ 2\uparrow\ 3\uparrow$	N_4
$1\downarrow\ 2\uparrow\ 3\uparrow$	$1\uparrow\ 2\downarrow\ 3\downarrow$	N_5
$1\downarrow\ 2\uparrow\ 3\downarrow$	$1\uparrow\ 2\downarrow\ 3\uparrow$	N_6
$1\downarrow\ 2\downarrow\ 3\uparrow$	$1\uparrow\ 2\uparrow\ 3\downarrow$	N_7
$1\downarrow\ 2\downarrow\ 3\downarrow$	$1\uparrow\ 2\uparrow\ 3\uparrow$	N_8

Because the total spin is zero, for a particular value observed by B would mean the opposite result for the particle observed by A for that component. Thus we have found a way of measuring two components of spin for the particle A while we have disturbed it once. Let us use this fact to find the total number of particles observed by A and B for the case when A measures spin up along 1 and B measures spin up along axis 2, i.e. $(A\ 1\uparrow, B\ 2\uparrow)$. Then as was stated above the observer A will have information about the two components of the spin of the particle that he has detected, viz, $(A\ 1\uparrow, A\ 2\downarrow)$. Denoting the number of particles with these components by $\mathcal{N}(A\ 1\uparrow, A\ 2\downarrow)$, from TABLE XVIII we find

$$\mathcal{N}(A\,1\uparrow, A\,2\downarrow) = N_3 + N_4. \tag{20.85}$$

Similarly for $\mathcal{N}(1\uparrow, 3\downarrow)$ and $\mathcal{N}(2\uparrow, 3\downarrow)$ we have

$$\mathcal{N}(1\uparrow, 3\downarrow) = N_2 + N_4. \tag{20.86}$$

and

$$\mathcal{N}(2\uparrow, 3\downarrow) = N_2 + N_6, \tag{20.87}$$

where we have suppressed references to the observer A in (20.86) and (20.87). From these three relations we find

$$\mathcal{N}(2\uparrow, 3\downarrow) + \mathcal{N}(1\uparrow, 2\downarrow) - \mathcal{N}(1\uparrow, 3\downarrow) = N_3 + N_6 \geq 0. \tag{20.88}$$

If we divide (20.88) by N we find the average of these quantities

$$\mathcal{C}(2\uparrow, 3\downarrow) + \mathcal{C}(1\uparrow, 2\downarrow) - \mathcal{C}(1\uparrow, 3\downarrow) \geq 0. \tag{20.89}$$

This inequality is a simple version of Bell's inequality.

Let us emphasize that this relation is found on the assumption that the measurement by the observer B does not affect the result found by A, and that the value of the component of spin found for the particle observed by A must have existed prior to the measurement carried out by B.

A different and a more general method of deriving this inequality can be given in the following way [18]:

For this formulation we introduce the correlation coefficient $C(\theta)$ defined as the value of the product $S_{zi}S_{\theta j}$ averaged over a great number of measurements of such a pair of particles. From the definition of $C(\theta)$ it follows that

$$\begin{aligned} C(\theta) &= \frac{\hbar^2}{8}\left[P_{\uparrow\uparrow}(\theta) - P_{\uparrow\downarrow}(\theta) - P_{\downarrow\uparrow}(\theta) + P_{\downarrow\downarrow}(\theta)\right] \\ &= \frac{\hbar^2}{4}\left[\sin^2\left(\frac{\theta}{2}\right) - \cos^2\left(\frac{\theta}{2}\right)\right] = -\frac{\hbar^2}{4}\cos\theta, \end{aligned} \tag{20.90}$$

where we have used Eqs. (20.82) and (20.84) to write $C(\theta)$. Now let us formulate these probabilities in a way which is similar to the description given by classical statistical mechanics. Here the assumption is that $|\psi\rangle$ does not give a complete description of the system but there are some hidden variables collectively denoted by λ such that a complete description of the system by $|\psi, \lambda\rangle$ is possible. This classical theory will be deterministic and local. By being deterministic we we mean that the particles have a definite state which determines all of the properties of the motion in the course of time. The theory is also assumed to be local, i.e. the result of an experiment by the observer B does not depend on what observer the A measures.

The hidden variables are distributed with a classical probability density $\rho(\lambda)$, where $\rho(\lambda)d\lambda$ is the fraction of pairs of spin $\frac{1}{2}$ particles with λ lying between λ and $\lambda + d\lambda$. Thus

$$\int \rho(\lambda)d\lambda = 1, \quad \rho(\lambda) > 0. \tag{20.91}$$

If the total spin of the system is not zero we can derive a similar inequality for the correlation coefficient [18]. Let a be the component of spin of the particle i along the direction $\mathbf{n}_1$ and b be the component of the particle j along the direction $\mathbf{n}_2$. Consider the mean value of a obtained from a large number of individual measurements. This mean value which we denoted by $\mathcal{C}(\mathbf{n}_1, \mathbf{n}_1)$, Eq. (20.90), can now be written as an integral over λ

$$\mathcal{C}(\mathbf{n}_1, \mathbf{n}_2) = \int \langle a(\lambda, \mathbf{n}_1)\rangle \langle b(\lambda, \mathbf{n}_2)\rangle \rho(\lambda) d\lambda. \tag{20.92}$$

Now we make the following assumptions :

(a) - That the hidden variables appearing in the integral are independent of the directions $\mathbf{n}_1$ and $\mathbf{n}_2$ and

(b) - That the average of the product is equivalent to the product of the averages, or that the mean values of a and b over hidden variables are uncorrelated. The existence of a correlation implies that an individual measurement giving a could depend upon the hidden variables related to $\mathbf{n}_2$. Now the result of measurements of a and b can be either $\frac{\hbar}{2}$ or $-\frac{\hbar}{2}$, and thus we have the inequalities

$$|\langle a(\lambda, \mathbf{n}_1)\rangle| \le \frac{\hbar}{2}, \quad |\langle b(\lambda, \mathbf{n}_2)\rangle| \le \frac{\hbar}{2}. \tag{20.93}$$

Let us introduce the third unit vector $\mathbf{n}_3$ and use (20.92) to write an expression for $\mathcal{C}(\mathbf{n}_1, \mathbf{n}_3)$. Subtracting $\mathcal{C}(\mathbf{n}_1, \mathbf{n}_3)$ from $\mathcal{C}(\mathbf{n}_1, \mathbf{n}_2)$ we arrive at a result which is similar to (20.92);

$$\begin{aligned} &\mathcal{C}(\mathbf{n}_1, \mathbf{n}_2) - \mathcal{C}(\mathbf{n}_1, \mathbf{n}_3) \\ &= \int \left[\langle a(\lambda, \mathbf{n}_1)\rangle \langle b(\lambda, \mathbf{n}_2)\rangle - \langle a(\lambda, \mathbf{n}_1)\rangle \langle b(\lambda, \mathbf{n}_3)\rangle\right] \rho(\lambda) d\lambda. \end{aligned} \tag{20.94}$$

By introducing a new unit vector $\mathbf{n}$ and adding and subtracting the same quantity we can write (20.94) as

$$\begin{aligned} &\mathcal{C}(\mathbf{n}_1, \mathbf{n}_2) - \mathcal{C}(\mathbf{n}_1, \mathbf{n}_3) \\ &= \int \langle a(\lambda, \mathbf{n}_1)\rangle \langle b(\lambda, \mathbf{n}_2)\rangle \left[1 \pm \frac{4}{\hbar^2} \langle a(\lambda, \mathbf{n})\rangle \langle b(\lambda, \mathbf{n}_3)\rangle\right] \rho(\lambda) d\lambda \\ &- \int \langle a(\lambda, \mathbf{n}_1)\rangle \langle b(\lambda, \mathbf{n}_3)\rangle \left[1 \pm \frac{4}{\hbar^2} \langle a(\lambda, \mathbf{n})\rangle \langle b(\lambda, \mathbf{n}_2)\rangle\right] \rho(\lambda) d\lambda. \end{aligned} \tag{20.95}$$

From the inequalities (20.93) and Eq. (20.95) we obtain

$$\begin{aligned} &|\mathcal{C}(\mathbf{n}_1, \mathbf{n}_2) - \mathcal{C}(\mathbf{n}_1, \mathbf{n}_3)| \\ &\le \int \left[\frac{\hbar^2}{4} \pm \langle a(\lambda, \mathbf{n})\rangle \langle b(\lambda, \mathbf{n}_3)\rangle\right] \rho(\lambda) d\lambda \\ &+ \int \left[\frac{\hbar^2}{4} \pm \langle a(\lambda, \mathbf{n})\rangle \langle b(\lambda, \mathbf{n}_2)\rangle\right] \rho(\lambda) d\lambda. \end{aligned} \tag{20.96}$$

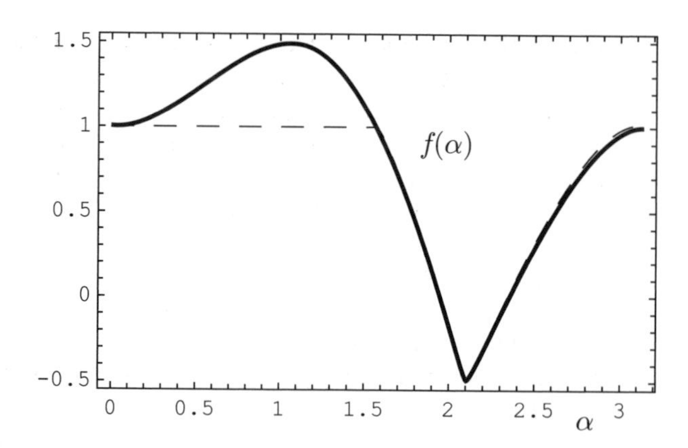

Figure 20.1: The function $f(\alpha)$ is plotted as a function of α. The solid line shows $f(\alpha)$ and the dashed line is a curve $f(\alpha) \le 1$, the latter is predicted by the hidden-variable theory.

This inequality and the fact that $\int \rho(\lambda) d\lambda = 1$ gives us

$$|\mathcal{C}(\mathbf{n}_1, \mathbf{n}_2) - \mathcal{C}(\mathbf{n}_1, \mathbf{n}_3)| \le \frac{\hbar^2}{2} \pm [\mathcal{C}(\mathbf{n}, \mathbf{n}_3) + \mathcal{C}(\mathbf{n}, \mathbf{n}_2)], \tag{20.97}$$

In obtaining this inequality which was first derived by Clauser, Horn, Shimony and Holt, (CHSH inequality) we have not assumed that the total spin of the system is zero. For the special case of zero spin, the two components of the spins of the two particles along the same direction are exactly opposite, so that

$$\mathcal{C}(\mathbf{n}, \mathbf{n}) = -\frac{\hbar^2}{4}. \tag{20.98}$$

Setting $\mathbf{n} = \mathbf{n}_3$ in (20.97) and using (20.98) we find Bell's inequality

$$|\mathcal{C}(\mathbf{n}_1, \mathbf{n}_2) - \mathcal{C}(\mathbf{n}_1, \mathbf{n}_3)| \le \frac{\hbar^2}{4} + \mathcal{C}(\mathbf{n}_2, \mathbf{n}_3). \tag{20.99}$$

Now if we choose the three vectors $\mathbf{n}_1$, $\mathbf{n}_2$ and $\mathbf{n}_3$ to be in the same plane, and denote the angle between $\mathbf{n}_1$ and $\mathbf{n}_2$ by α, and those between $\mathbf{n}_1$ and $\mathbf{n}_3$ by θ and $\theta - \alpha$ respectively, we can write (20.99) as

$$|\mathcal{C}(\theta) - \mathcal{C}(\alpha)| - \mathcal{C}(\theta - \alpha) \le \frac{\hbar^2}{4}. \tag{20.100}$$

This inequality is a consequence of any local deterministic hidden variable theory. To see whether (20.100) is compatible with the prediction of quantum mechanics, let us study the special case where $\theta = 2\alpha$. The quantal correlation functions are given by (20.90)

$$\mathcal{C}(\alpha) = -\frac{\hbar^2}{4} \cos\alpha, \quad \mathcal{C}(\theta) = -\frac{\hbar^2}{4} \cos\theta. \tag{20.101}$$

Now the hidden-variable theory is consistent with quantum mechanics if

$$f(\alpha) = (|\cos(\alpha) - \cos(2\alpha)| + \cos\alpha) \le 1. \quad (20.102)$$

In Fig. 20.1 $f(\alpha)$ is plotted as a function of α and shows that the Bell inequality is satisfied for $\frac{\pi}{2} < \alpha < \pi$, but is violated for $0 < \alpha < \frac{\pi}{2}$.

The correlation function has been measured in two-photon correlation experiment, e.g. those by Aspect and collaborators [19]-[20] (see also [21]). These are difficult experiments to perform, but they all to seem to indicate that Bell's inequality is indeed violated. These violations do not rule out the possibility of some sort of hidden variable theory, but then the theory must include nonlocal effects. Except for the standard quantum theory, so far, no other theory has been able to explain so accurately and so beautifully the structure and the behavior of atoms, molecules and radiation.

For a detailed discussion of the conceptual and philosophical problems of quantum mechanics the reader is referred to [10],[22].

Bibliography

[1] A. Einstein, B. Podolski and N. Rosen, Can quantum-mechanical description of physical reality be considered complete? Phys. Rev. 47, 777 (1935).

[2] J.L.B. Cooper, The paradox of separated systems in quantum theory, Proc. Cambridge Phil. Soc. 46, 620 (1950).

[3] M. Jammer, *The Philosophy of Quantum Mechanics*, (John Wiley & Sons, New York, 1974), Chapter 6.

[4] D. Bohm and Aharonov, Discussion of the experimental proof for the paradox of Einstein, Rosen and Podolski, Phys. Rev. 108, 1070 (1951).

[5] J.A. Wheeler and W. Zurek *Quantum Theory of Measurement*, (Princeton University Press, Princeton, 1987).

[6] W.H. Zurek, Decoherence, einselection, and the quantum origin of the classical, Rev. Mod. Phys. 75, 715 (2003).

[7] M.A. Schlosshauer, *Decoherence and the Quantum-to-Classical Transition*, (Springer, Berlin, 2007).

[8] M. Razavy, *Classical and Quantum Dissipative Systems*, (Imperial College Press, 2006).

[9] W.G. Unruh and W.H. Zurek, Reduction of a wave packet in quantum Brownian motion, Phys. Rev. D 40, 1071 (1989).

[10] R. Omnès, *The Interpretation of Quantum Mechanics*, (Princeton University Press, Princeton, 1994).

[11] A.I.M. Rae, *Quantum Mechanics*, Fifth Edition, (Taylor & Fransis, New York, 2008), Chapter 13.

[12] W. Heisenberg, *Physikalische Prinzipien der Quantentheorie*, (Hirzel, Leipzig, 1930)), English translation by , *The Physical Principles of Quantum Theory*, (University of Chicago Press, Chicago, 1930).

[13] W.E. Lamb, An operational interpretation of nonrelativistic quantum mechanics, Phys. Today, 22, 23, April (1969).

[14] W.E. Lamb, Quantum theory of measurement, Ann. New York Academy of Sciences, 480, 407 (1990).

[15] W. Greiner, *Quantum Mechanics*, (Springer-Verlag, New York, 1989), Chap. 17.

[16] J.S. Bell, On the Einstein-Podolski-Rosen paradox, Physics 1, 195 (1964), reprinted in J.S. Bell, *Speakable and Unspeakable in Quantum Mechanics*, (Cambridge University Press, Cambridge, 1987).

[17] D.M. Greenberger, M.A. Horne, A. Shimony and R.A. Holt, Proposed experiment to test local hidden-variabletheories, Phys. Rev. Lett. 23, 880 (1969).

[18] J.F. Clauser, M.A. Horne, A. Shimony and A. Zeilinger, Bell's theorem without inequality, Am. J. Phys. 58, 1131 (1990).

[19] A. Aspect, P. Grangier and G. Roger, Experimental test of test of realistic local theories via Bell's theorem, Phys. Rev. Lett. 47, 460 (1981).

[20] A. Aspect, J. Dalibard and G. Roger, Experimental test of Bell's inequalities using time-varying analyzers, Phys. Rev. Lett. 49, 1804 (1982).

[21] For a detailed review of the experimental evidence for violation of the Bell inequality see J.F. Clauser and A. Shimony, Rep. Prog. Phys. 41, 1881 (1978).

[22] J.S. Bell, *Speakable and Unspeakable in Quantum Mechanics*, (Cambridge University Press, London, 1987).

Index